*Volumes already published*

Volume 1: Trees I (1986)
Volume 2: Crops I (1986)
Volume 3: Potato (1987)
Volume 4: Medicinal and Aromatic Plants I (1988)
Volume 5: Trees II (1989)
Volume 6: Crops II (1988)
Volume 7: Medicinal and Aromatic Plants II (1989)
Volume 8: Plant Protoplasts and Genetic Engineering I (1989)
Volume 9: Plant Protoplasts and Genetic Engineering II (1989)

*Volumes in preparation*

Volume 10: Legumes and Oilseed Crops I
Volume 11: Somaclonal Variation in Crop Improvement I
Volume 12: Haploids in Crop Improvement I
Volume 13: Wheat
Volume 14: Rice

# Biotechnology in Agriculture and Forestry 9

# *Plant Protoplasts and Genetic Engineering II*

Edited by Y. P. S. Bajaj

With 141 Figures



Springer-Verlag Berlin Heidelberg New York
London Paris Tokyo Hong Kong

Professor Dr. Y. P. S. BAJAJ
A-137 New Friends Colony
New Delhi 110065, India



ISBN 3-540-50789-2 Springer-Verlag Berlin Heidelberg New York
ISBN 0-387-50789-2 Springer-Verlag New York Berlin Heidelberg

Library of Congress Cataloging-in-Publication Data. (Revised for vol. 2) Plant protoplasts and genetic engineering. Biotechnology in agriculture and forestry; 8– Includes bibliographies and index. 1. Plant protoplasts. 2. Plant genetic engineering. I. Bajaj, Y. P. S., 1936– . QK725.P578 1989 581.87'3 88-31197


Printed in Germany



Typesetting: K+V Fotosatz GmbH, Beerfelden.
Printing and binding: Konrad Triltsch, Graphischer Betrieb, Würzburg.
2131/3145-543210 – Printed on acid-free paper

*Dedicated to my wife,*
*Professor Dr. Satinder Bajaj*

# Preface

Genetic engineering through DNA recombinants and the in vitro manipulation of isolated protoplasts has recently attracted much attention in agricultural biotechnology, and has greatly advanced during the last 5 years. In an earlier book, *Plant Protoplasts and Genetic Engineering I*, methods for the isolation, fusion and culture of protoplasts were reviewed and the regeneration of complete plants from isolated protoplasts of rice, potato, soybean, linseed, cabbage, chicory, lettuce, butterbur, orchids, citrus and some other tree species, and interspecific and intergeneric somatic hybrids in *Lycopersicon, Petunia, Nicotiana, Solanum, Glycine, Citrus, Brassica, Medicago* and *Trifolium* spp. were discussed.

The present volume, *Plant Protoplasts and Genetic Engineering II*, deals with some of the newer techniques such as microinjections, electrofusion, flow cytometry, uptake and integration of DNA, nuclei, isolated chromosomes by plant protoplasts and the subsequent regeneration of transgenic plants. The literature on the DNA recombinants and genetic transformation, both *Agrobacterium*-mediated and direct gene transfer in agricultural crops and trees, such as poplars, is reviewed, and the uses of cytoplasts and miniprotoplasts in genetic manipulation are highlighted.

The recent studies on cereals and trees are of special significance. In the past genetic manipulation with these species was rather discouraging because of their inability to regenerate complete plants from isolated protoplasts. Now, due to technical advances, this hurdle has been removed and completely transformed plants of rice, maize and poplar have been obtained. This should pave the way, especially for wheat and forest trees. Genetic manipulation through the use of isolated protoplasts has far-reaching implications in agricultural and forest biotechnology, thus this book will be of particular interest to research scientists, teachers and advanced students of plant biotechnology, tissue culture, plant breeding and genetic engineering.

New Delhi, June 1989

Y. P. S. Bajaj
Series Editor

# Contents

XIII Direct Gene Transfer in Protoplasts of *Nicotiana plumbaginifolia*
F. SALA, M.L. MARCHESI, S. CASTIGLIONE, J. PASZKOWSKI, M. SAUL, I. POTRYKUS, and I. NEGRUTIU (With 2 Figures)



XIV Microinjection of Cells and Protoplasts: Integration of Foreign DNA
A. CROSSWAY (With 3 Figures)



XV Electrofusion: The Technique and Its Application to Somatic Hybridization
G.W. BATES (With 7 Figures)



XVI Versatile Apparatuses for Electrogene Mapping, Electrophoresis and Electrofusion
K. YOSHIDA and T. KONDO (With 12 Figures)

XXI Transfer of Mitochondria Through Protoplast Fusion
H. Ichikawa, L. Tanno-Suenaga, and J. Imamura
(With 6 Figures)



XXII Incorporation of the Firefly Luciferase Gene into Plant Cells
D.W. Ow and S.H. Howell (With 8 Figures)



XXIII Uptake of Viruses by Plant Protoplasts and Their Use as Transforming Agents
A.C. Cassells (With 7 Figures)



XXIV Uptake of Fungal Protoplasts by Plant Protoplasts
P.T. Lynch, S. Isaac, and H.A. Collin (With 13 Figures)

## List of Contributors

BABA, A., Department of Pure and Applied Sciences, College of Arts and Sciences, University of Tokyo, Komaba, Meguro-Ku, Tokyo, 153 Japan

BAJAJ, Y. P. S., A-137 New Friends Colony, New Delhi 110065, India

BATES, G. W., Institute of Molecular Biophysics, Florida State University, Tallahassee, FL 32306-3015, USA

CASSELLS, A. C., Department of Plant Science, University College, Cork, Ireland

CASTIGLIONE, S., Department of Genetics and Microbiology, University of Pavia, Via S. Epifanio, 27100 Pavia, Italy

COLLIN, H. A., Department of Botany, University of Liverpool, P.O. Box 147, Liverpool L69 3BX, United Kingdom

CROSSWAY, A., Calgene Incorporation, 1920 Fifth St., Davis, CA 95616, USA

DETMER, J. J., BioGrowth, Richmond, CA 94806, USA

ERIKSSON, T., Department of Physiological Botany, University of Uppsala, Box 540, 751 21 Uppsala, Sweden

FIROOZABADY, E., Plant Genetic Engineering Laboratory, New Mexico State University, Box 36L, Las Cruces, NM 88003, USA

GALBRAITH, D. W., Department of Plant Sciences, University of Arizona, Tucson, Arizona 85721, USA

GILES, K. L., Vipont Research Laboratories (Canada) Inc., c/o Plant Biotechnology Institute, National Research Council of Canada, 110 Gymnasium Road, Saskatoon, SK, Canada S7N OW 9

HANOLD, D., Department of Plant Pathology, Waite Agricultural Research Institute, Glen Osmond, South Australia 5064

HASEZAWA, S., Department of Pure and Applied Sciences, College of Arts and Sciences, University of Tokyo, Komaba, Meguro-ku, Tokyo 153, Japan

HERRERA-ESTRELLA, L., Centro de Investigación y de Estudios, Avanzados del I. P. N. Unidad Irapuato, Apartado Postal No. 629, 36500 Irapuato, Gto, México

HOWELL, S.H., Boyce Thompson Institute for Plant Research, Cornell University, Tower Road, Ithaca, NY 14853, USA

ICHIKAWA, H., Plantech Research Institute, Kamoshida 1000, Midori-ku, Yokohama 227, Japan

IMAMURA, J., Plantech Research Institute, Kamoshida 1000, Midori-ku, Yokohama 227, Japan

ISAAC, S., Department of Botany, University of Liverpool, P.O. Box 147, Liverpool L69 3BX, United Kingdom

KING, J., Department of Biology, University of Saskatchewan, Saskatoon, SK, Canada S7N OWO

KONDO, T., National Institute for Basic Biology, Okazaki 444, Japan

KUNG, S.D., Center for Agricultural Biotechnology, Maryland Institute of Biotechnology, The University of Maryland, College Park, MD 20742, USA

DE LAAT, A.M.M., Research Institute Ital, P.O. Box 48, 6700 AA Wageningen, The Netherlands

LANDGREN, C.R., Department of Biology, Middlebury College, Middlebury, VT 05753, USA

LURQUIN, P.F., Program in Genetics and Cell Biology, Washington State University, Pullman, WA 99164–4350, USA

LYNCH, P.T., Plant Genetic Manipulation Group, Department of Botany, University of Nottingham, Nottingham, NG7 2RD, United Kingdom

MARCHESI, M.L., Department of Genetics and Microbiology, University of Pavia, Via S. Epifanio, 27100 Pavia, Italy

MARETZKI, A., Hawaiian Sugar Planters' Association, P.O. Box 1057, Aiea, HI 96701, USA

MASCARENHAS, D., BioGrowth, Richmond, CA 94806, USA

MCCOWN, B.H., Department of Horticulture, University of Wisconsin-Madison, Madison, WI 53706, USA

METTLER, I.J., Sandoz Crop Protection Corporation, Palo Alto, 975 California Avenue, CA 94304-1104, USA

NEGRUTIU, I., Instituut voor Moleculaire Biologie, 65 Paardenstraat, 1640 Saint Genesius-Rode, Belgium

OW, D.W., Plant Gene Expression Center, U.S. Department of Agriculture, and University of California at Berkeley, 800 Buchanan Street, Albany, CA 94710, USA

PASZKOWSKI, J., Institut für Pflanzenwissenschaften, ETH-Zentrum, 8092 Zürich, Switzerland

PIERCE, D.A., EniChem, Monmouth Junction, NJ 08852, USA

POTRYKUS, I., Institut für Pflanzenwissenschaften, ETH-Zentrum, 8092 Zürich, Switzerland

RHODES, C.A., Sandoz Crop Protection Corporation, Palo Alto, 975 California Avenue, CA 94304-1104, USA

RICHARDS, C., Zoecon Research Institute, Sandoz Crop Protection Corporation, Palo Alto, CA 94304, USA

SALA, F., Department of Botany, University of Parma, Via Farini, Parma, Italy

SAUL, M., Institut für Pflanzenwissenschaften, ETH-Zentrum, 8092 Zürich, Switzerland

SAXENA, P.K., Department of Horticultural Science, University of Guelph, Guelph, Ontario, Canada N1G 2W1

SELLMER, J.C., Department of Horticulture, University of Wisconsin-Madison, Madison, WI 53706, USA

SIMPSON, J., Centro de Investigación y de Estudios, Avanzados del I.P.N. Unidad Irapuato, Apartado Postal No. 629, 36500 Irapuato, Gto, Mexico

SREE RAMULU, K., Research Institute Ital., P.O. Box 48, 6700 AA Wageningen, The Netherlands

SYBENGA, J., Department of Genetics, Agricultural University, Dreijenlaan 2, 6703 HA Wageningen, The Netherlands

SYŌNO, K., Department of Pure and Applied Sciences, College of Arts and Sciences, University of Tokyo, Komaba, Meguro-Ku, Tokyo 153, Japan

TANNO-SUENAGA, L., Plantech Research Institute, Kamoshida 1000, Midori-ku, Yokohama, 227 Japan

THOM, M., Hawaiian Sugar Planters' Association, P.O. Box 1057, Aiea, HI 96701, USA

TRAIL, F., Department of Plant Pathology, Cornell University, Geneva, NY 14456-0462, USA

VERHOEVEN, H.A., Research Institute Ital, P.O. Box 48, 6700 AA Wageningen, The Netherlands

WAARA, S., Department of Physiological Botany, University of Uppsala, Box 540, 751 21 Uppsala, Sweden

WALLIN A., Department of Physiological Botany, University of Uppsala, Box 540, 751 21 Uppsala, Sweden

WU, F.-S., Zoecon Research Institute, Sandoz Crop Protection Corporation, Palo Alto, CA 94304, USA

YAMAMOTO, M., Laboratory of Plant Pathology, College of Agriculture, Kyoto University, Kyoto 606, Japan

YOSHIDA, K., Botanical Institute, Faculty of Science, Hiroshima University, Hiroshima 730, Japan

# I Genetic Engineering and In Vitro Manipulation of Plant Cells – Technical Advances

Y. P. S. Bajaj [1]

## 1 Introduction

It is envisaged that the next agricultural revolution will be through in vitro technology. Tremendous progress made during the last decade has amply demonstrated that refinement of routine in vitro culture techniques and development of new biotechnologies such as somatic cell hybridization, transformation through *Agrobacterium*, direct gene transfer, electrofusion, electroporation, flow cytometry, uptake of nuclei, chromosomes, DNA, chloroplasts, bacteria, viruses, fungi, and the transfer of mitochondria by protoplasts and other recombinant DNA techniques have enabled the opening of new vistas for plant improvement (Potrykus et al. 1985a; Bajaj 1986, 1989; Crossway et al. 1986a; Perani et al. 1986; Cocking and Davey 1987; Schell 1987). Protoplasts have been shown to be excellent tools for genetic manipulation, since entire plants can be regenerated from them under suitable culture conditions – a prerequisite for genetic manipulation studies. The subject of isolation and culture of protoplasts has already been dealt with in detail and now there are a large number of species in which entire plants can be routinely regenerated (see Bajaj 1989).

In this chapter recent technical advances made in genetic engineering and in vitro manipulation of plant cells are reviewed and their impact on agricultural and forest biotechnology in the development of new genotypes and the improvement of plants highlighted.

## 2 Somatic Hybrids and Cybrids (Table 1)

Somatic hybridization has opened up several possibilities for the parasexual manipulation of plants; these aspects have been discussed by several workers (see Bajaj 1989). Since the first report on protoplast fusion-derived somatic hybrids between *Nicotiana tabacum* × *N. langsdorfii* (Carlson et al. 1972), numerous examples have become available (Table 1). Although the original aspirations and goals to combine the characters of any two species to obtain a novel plant with the desired agronomic traits has been rather discouraging, yet these studies during

[1] Former Professor of Tissue Culture, Punjab Agricultural University, Ludhiana, India
Present address: A-137 New Friends Colony, New Delhi 110065, India

Biotechnology in Agriculture and Forestry, Vol. 9
Plant Protoplasts and Genetic Engineering II (ed. by Y. P. S. Bajaj)

**Table 1.** Some recent examples of the interspecific and intergeneric somatic hybrids and cybrid plants obtained through protoplast fusion (see also Bajaj et al. 1987; Bajaj 1988)

| Somatic hybrid/cybrid | Reference |
|---|---|
| *Brassica napus* × *B. campestris* | Barsby et al. (1987) |
| *B. oleracea* × *B. campestris* | Robertson et al. (1987) |
| *B. napus* × *B. napus* | Chuong et al. (1988) |
| *Brassica oleracea* × *Moricandia arvensis* | Toriyama et al. (1987a) |
| *Citrus sinensis* × *Poncirus trifoliata* | Ohgawara et al. (1985) |
| Various *Citrus* species | Vardi et al. (1987) |
| *C. sinensis* × *Severinia disticha* | Grosser et al. (1988a) |
| *C. sinensis* × *P. trifoliata* | Grosser et al. (1988b) |
| *Daucus carota* × *D. capillifolius* | Matthews and Widholm (1986) |
| *D. capillifolius* × *D. carota* | Ichikawa et al. (1987) |
| *Lotus corniculatus* × *L. coimbrensis* | Wright et al. (1986) |
| *Lycopersicon esculentum* × *Solanum rickii* | O'Connell and Hanson (1986) |
| *L. esculentum* × *Solanum lycopersicoides* | Handley et al. (1986) |
| *Medicago sativa* × *M. falcata* | D'Hont et al. (1987) |
| *Nicotiana undulata* × *N. bigelovii* | Aviv et al. (1984b) |
| *N. rustica* × *N. sylvestris* | Aviv et al. (1984a) |
| *N. gossei* × *N. plumbaginifolia* | Medgyesy et al. (1985) |
| *N. debneyi* × *N. tabacum* | Asahi et al. (1988) |
| *Oryza sativa* × *Echinochloa oryzicola* | Terada et al. (1987) |
| *Petunia hybrida* × *P. parodii* | Boeshore et al. (1985) |
| *P. hybrida* × *P. inflata* | Clark et al. (1986) |
| *Populus* (hybrid) × *Hibiscus sabdariffa* | Ito et al. (1986) |
| *Sinapis turgida* × *Brassica oleracea* | Toriyama et al. (1987b) |
| *S. turgida* × *B. nigra* | |
| *Solanum melongena* × *Solanum sisymbrifolium* | Gleddie et al. (1986) |
| *S. tuberosum* × *S. brevidens* | Austin et al. (1986), Kemble et al. (1986), Vries de and Tempelaar (1987) |
| *S. tuberosum* × *S. phureja* | Puite et al. (1986) |

the last decade have certainly provided a useful tool for the induction of genetic variability and combinations which do not exist in nature (see Kumar and Cocking 1987). Somatic hybridization combines the whole genomes of two parents, and the resulting hybrids could be used as new amphidiploid plant varieties. Intergeneric and interspecific somatic hybrids may, however, result in completely sterile plants, and may not be of much use to a plant breeder as such. In a number of instances, characteristics of one parent are retained due to the elimination of chromosomes as well as to the sorting out of the plastomes of one parent. Even in more compatible combinations, due to genomic disharmony, instability, and unfavorable combinations, hybrids that possess complete genomes of two species may not be successful immediately as commercial crop varieties. Such somatic hybrids could be improved through conventional methods of breeding so as to produce elite hybrids or alien addition and substitution lines.

Some of the earlier-produced somatic hybrids such as potato × tomato (Melchers et al. 1978) and *Arabidobrassica* (Gleba and Hoffmann 1980) have not been of practical utility; nevertheless, these studies have highlighted the possibilities and the prospects of somatic hybridization in obtaining novel combinations of germplasm, and encouraged interest in this field.

For fusion studies, haploid protoplasts have certain definite advantages (Bajaj 1983). Fusion of haploid/diploid protoplasts could also be of great value in analytical methods of breeding polyploid crops. Different heterozygous diploid strains of potato can be fused to form completely heterozygous tetraploids and, being propagated vegetatively, they will become fixed immediately without any meiotic segregation (Wenzel et al. 1979). Recently, Pirrie and Power (1986) obtained a triploid gemetosomatic hybrid by fusing pollen tetrad protoplasts (haploid) of *N. glutinosa* with leaf protoplasts (diploid) of nitrate reductase-deficient *N. tabacum* and obtained triploid hybrids which were fertile and set normal seed.

The transfer of disease resistance to susceptible plants through protoplast fusion is a reality in *Nicotiana tabacum*×*N. glauca* (Evans et al. 1981) and *Solanum tuberosum*×*S. chacoense* (Butenko et al. 1982). Such experiments have also been successful with herbicide resistance (Binding et al. 1982).

The hybrid callus is expected to show greater variation because it generates the environment for enhancing chromosome breakage and reunion events (Larkin and Scowcroft 1981). The hybrid callus and somatic hybrids may provide the means for obtaining the genetic exchange needed between two genomes in wide hybridization.

The newer techniques, such as electrofusion, have considerably increased the frequency of fusion events (Zimmermann and Scheurich 1981); flow cytometry has also facilitated rapid selection of fused products and heterokaryons (Galbraith 1984) and viability determination.

Recently, much interesting information has accumulated on the cytoplasmic hybrids or "cybrid" plants which offer the opportunity for obtaining heterozygosity of extrachromosomal material (see Kumar and Cocking 1987, and Chap. XXI, this Vol.). A number of nucleo-cytoplasmic combinations can arise as a result of fusion. Cybrids, in contrast to conventional hybrids, possess nuclear genes from only one and cytoplasm from both parents. Cytoplasmic organelles, like chloroplasts and mitochondria, possess cp-DNA and mt-DNA respectively, which control the expression of male sterility (Levings et al. 1980), influencing photosynthetic apparatus and chlorophyll deficiency (Gleba et al. 1984), as well as carbon assimilation enzymes and streptomycin resistance. The transfer of male sterility from one species to another, through cytoplasmic hybrids, especially in otherwise incompatible combinations, is considered of great value for plant breeders for strengthening hybrid seed production programs. Therefore various methods have been evolved for the production of cybrids, such as

1. Inactivation or removal of the nucleus of one parent by X-irradiation (Zelcer et al. 1978) or by centrifugation through Percoll gradient (Maliga et al. 1982) and subsequent fusion with normal protoplasts.
2. Fusion of normal and subprotoplasts or microprotoplasts (Bilkey et al. 1982).
3. Elimination of chromosomes of one parent during proliferation of hybrid cells (Gleba and Hoffmann 1980).
4. Use of mutants and antimetabolites (Bourgin et al. 1986; Aviv and Galun 1986).

Following these methods, cybrids have been obtained in a number of plant species (Table 1, see also Bajaj et al. 1987). Tobacco and Brassica are two examples where extensive work has been done on cybridization which has far-reaching implications in plant breeding. In tobacco, for instance, cytoplasmic male sterile lines conventionally produced by backcrossing take 3–4 years, whereas protoplast fusion has enabled their production in about 1 year (Kubo 1985). In *Brassica*, Pelletier and his coworkers have done extensive work (see Pelletier 1989). By fusing protoplasts of a cytoplasmic male sterile line of *B. napus* with protoplasts of an atrazine-resistant line of *B. napus* they obtained a plant possessing the cDNA of *B. campestris* (Pelletier et al. 1983). Such hybrids cannot be produced by conventional means, and are now being incorporated into *Brassica* breeding programs in France. Recently, Chuong et al. (1988) employed fusion between protoplasts from haploid cytoplasmic atrazine-resistant (CATR) and haploid cytoplasmic male sterile (CMS) *Brassica napus* plants and obtained a diploid CMS/CATR cybrid. The hybrid nature of the cytoplasm was confirmed by comparing the E*co*RI restriction fragment patterns of chloroplast and mitochondrial DNA from the cybrid with the parental patterns. Protoplast-derived cybrids of other crops, such as potato (Binding et al. 1982) and carrot (Ichikawa et al. 1987), have also been produced.

To conclude, protoplast fusions-derived hybrids and cybrids have proved useful for transfer of disease resistance, male sterility, and for herbicide resistance in tobacco, *Brassica* and potato, and this work, when extended to other crops, will facilitate the induction of genetic variability not available in nature.

## 3 Recombinant DNA Technology and Transformation

Transfer of genes through protoplast fusion is more or less "juggling"; there is not much specificity, the outcome is unpredictable, and much selection is required. One the other hand, in direct gene transfer or recombinant DNA technology, a specific desirable gene-determining single trait can be incorporated into the system without concomitant transfer of undesirable traits (Wilke-Douglas et al. 1986).

During the last few years, tremendous progress has been made in the transformation of plants (for recent literature see Gheysen et al. 1985, Fraley et al. 1986, Shillito et al. 1986, Klee 1987, Schell 1987, Cocking and Davey 1987, and Chap. IV, this Vol.). The work has been summarized in Table 2.

Three methods are generally followed to bring about transformation in plants: (1) *Agrobacterium* as a vector, (2) virus vectors, and (3) uptake of DNA/direct gene transfer, which are discussed below.

### 3.1 *Agrobacterium* as a Vector for Transformation

*Agrobacterium tumefaciens*, a common soil bacterium, infects mostly the dicotyledonous plants and forms a crown gall tumor; this tumor when excised is

**Table 2.** Some examples of transformation through protoplasts, cells, and tissue cultures of agricultural crops and forest trees

| Plant | Vector/method | Reference |
|---|---|---|
| Tobacco | *Agrobacterium tumefaciens* | Krens et al. (1982), Herrera-Estrella et al. (1984), Horsch et al. (1985) |
| | Direct | Paszkowski et al. (1984), Negrutiu et al. (1987) |
| | Microinjection | Crossway et al. (1986b) |
| | Electroporation | Riggs and Bates (1986) |
| Tomato | *A. rhizogenes* | Shahin et al. (1986a) |
| Cucumber | *A. rhizogenes* | Trulson et al. (1986) |
| Potato | *A. tumefaciens* | Ooms et al. (1986) Also see Yamamoto, Chapter VII, this Volume |
| Cotton | *A. tumefaciens* | Firoozabady, Chapter VIII, this Volume |
| Carrot | Electroporation | Langridge et al. (1985), Ow et al. (1986) |
| Flax | *A. tumefaciens* | Basiran et al. (1987), Jordan and McHughen (1988) |
| Brassica | *A. rhizogenes* | Guerche et al. (1987) |
| | Microinjection | Miki et al. (1986), Neuhaus et al. (1987) |
| Alfalfa | *A. tumefaciens* | Shahin et al. (1986b) |
| Clover | *A. tumefaciens* | Webb (1986), White and Greenwood (1987) |
| Cowbean | Cowpea mosaic virus | Garcia et al. (1987) |
| *Lotus* | *A. tumefaciens* | Armstead and Webb (1987) |
| Moth bean | Direct | Kohler et al. (1987) |
| *Sesbania* | *A. tumefaciens* | Vlachova et al. (1987) |
| *Stylosanthes* | *A. tumefaciens* | Manners (1987) |
| Soybean | *A. tumefaciens* | Baldes et al. (1987) |
| | Electroporation | Christou et al. (1987) |
| Rye grass | Direct | Potrykus et al. (1985b) |
| Rice | Direct | Uchimiya et al. (1986) |
| Wheat | Direct | Lörz et al. (1985) |
| Maize | Electroporation | Fromm et al. (1986), Rhodes et al. (1988) |
| Sorghum | Electroporation | Ou-Lee et al. (1986) |
| Douglas-fir | *A. tumefaciens* | Dandekar et al. (1987) |
| Loblolly pine | *A. tumefaciens* | Sederoff et al. (1986) |
| Populus | *A. tumefaciens* | Fillatti et al. (1987) |

able to grow in vitro on a hormone-free medium (Braun 1943) and produces opines (six types are known). *Agrobacterium* contains Ti plasmid which is responsible for tumor induction (Zaenen et al. 1974) and involves transfer and integration of small DNA segments from this plasmid (Chilton et al. 1977; for details see Chap. IV, this Vol.). Because of this property, *Agrobacterium* has been used for transformation of plants in vitro, and is applicable mostly to dicot species.

*Agrobacterium rhizogenes*, used as an alternative to *A. tumefaciens*, causes hairly roots. It has the advantages that the roots are of clonal origin, cytologically stable, and in many cases can be regenerated to complete fertile plants. It has been employed for the transformation of *Brassica* (Guerche et al. 1987), cucumber (Trulson et al. 1986), tomato (Shahin et al. 1986a), etc. *Agrobacterium rhizogenes* has also been shown to increase the production of secondary metabolites in transformed hairy root cultures of *Atropa* (Kamada et al. 1986), *Datura* (Payne et al. 1987), and is likely to be the tool of choice for such studies in future (see Rhodes et al. 1987).

The infection with *Agrobacterium* is brought about by the use of

1. In vivo infection, and culture of wound callus (Zambryski et al. 1984);
2. Isolated protoplasts (Wullems et al. 1981);
3. Leaf disc infection (Horsch et al. 1985).

### 3.1.1 In Vivo Infection with *Agrobacterium*

The top leaves and apical meristem of the plants are removed and inoculation is effected with *Agrobacterium* at the wound site. No tumor tissue is formed with nononcogenic vectors, but wound callus proliferates in 3–4 weeks. This tissue can be excised and cultured in media supplemented with growth regulators (and antibiotic), and induced to regenerate transformed plants (Zambryski et al. 1984).

### 3.1.2 Co-Cultivation with Protoplasts

In this technique *Agrobacterium* is cultivated with isolated protoplasts for a few days, washed, and then grown in an antibiotic-containing medium to eliminate bacteria (Wullems et al. 1981). In regeneration medium, the callus differentiates to transformed shoots/complete plants. Some successful examples are tobacco (Krens et al. 1982; Hain et al. 1985), and potato (Ooms et al. 1986).

### 3.1.3 Leaf Disc Infection Method

Horsch et al. (1985) developed a simple and general method for transferring genes in plants. The surface-sterilized leaf discs of petunia, tomato, and tobacco, when inoculated with *A. tumefaciens* strain containing a modified tumor-inducing plasmid and cultured for 2 days and transferred to selective medium containing kanamycin, regenerated shoots within 2–4 weeks. The transformation was confirmed by the ability to form roots in a medium containing kanamycin. These authors emphasize that the method for producing transformed plants combines gene transfer, plant regeneration, and effective selection for transformants into a single process, and should be applicable to plant species that can be infected by *Agrobacterium* and regenerate from leaf explants.

This method has now been repeated in tomato (McCormick et al. 1986) and extended to other plant species effectively. Deak et al. (1986) infected shoot segments of alfalfa with *Agrobacterium*, and somatic embryos were induced to develop from transformed callus.

### 3.2 Viruses as Vectors

Brisson et al. (1984) achieved transformation in which part of the cauliflower mosaic virus genome was replaced by a selectable marker gene in such a way as to allow the virus to replicate normally. Shillito et al. (1986) emphasize that "such vectors may be useful in that they offer a way of placing multiple copies of the functional gene in each cell, but they have a narrow host range, and are not in general transmitted to the next generation".

Gemini viruses are another possibility, and the combination of a viral and a bacterial vector (agroinfection), in which viral functions are introduced via *A. tumefaciens* has also been studied (Grimsley et al. 1986).

### 3.3 Direct Gene Transfer/Uptake of DNA

Direct transfer of genes without involving any vector has recently attracted much attention. It avoids the need for the introduction of cloned DNA into the T-DNA of the *A. tumefaciens* Ti plasmid or its equivalent before insertion into the plant, which is intrinsic in *A. tumefaciens*-mediated methods. The direct gene transfer is thus time-saving and eliminates the possibility of mutation or rearrangement (Shillito et al. 1986; Negrutiu et al. 1987).

In earlier experiments, uptake of isolated Ti plasmid of *A. tumefaciens* into protoplasts was demonstrated (Davey et al. 1980; Krans et al. 1982). This involved first incubating protoplasts with DNA in the presence of PEG and $CaCl_2$) then plating the protoplasts in a selective medium.

These studies have led to the present state of the art regarding direct incorporation and expression of genes in plants (Paszkowski et al. 1984; Hain et al. 1985). It has been extended to monocots such as *Lolium multiflorum* (Potrykus et al. 1985b), *Triticum monococcum* (Lörz et al. 1985) and maize (Fromm et al. 1986). The efficiency of transformation has been increased by about 1000 times through a combination of heat shock, PEG treatment, and the use of electroporation (Shillito et al. 1985), or by the use of aphidocolin (Meyer et al. 1985).

Recently, Rhodes et al. (1988) regenerated a transformed maize plant from protoplasts isolated from embryogenic cell suspensions. The protoplasts were combined with plasmid DNA containing a gene coding for neomycin phosphotransferase (NPT II) next to the 35S promoter region of cauliflower mosaic virus. A high voltage electrical pulse was applied to the protoplasts, which were then grown on filters placed over feeder layers of maize suspension cells and selected for growth in the presence of kanamycin. Selected cell lines showed NPT II activity, and complete plants were regenerated. This work has attracted much attention because it is the first time that a cereal has been successfully transformed; this work will soon pave way for other cereals.

The literature on various other techniques for the uptake and incorporation of exogenous DNA has been reviewed (see Lurquin Chap. III, this Vol.), among which microinjection (Crossway et al. 1986a) assumes great significance for future studies. Microinjection has been successfully applied for the transformation of *Medicago* (Reich et al. 1986c), *Brassica* (Miki et al. 1986; Neuhaus et al. 1987), tobacco (Crossway et al. 1986b) etc.

Schell and coworkers (see Schell 1987) have shown that transformation can also be brought about by very simple techniques without involving refined mechanisms involved in the use of *Agrobacterium*. The direct injection with a hypodermic needle, of DNA coding for a selectable marker gene (NPTII) into the tillers of rye just underneath a developing inflorescence, apparently led to the uptake of the DNA in the genome of developing germ cells. From among 3000 seeds obtained from a few hundred injected plants, three independent transgenic plants that contained and expressed the transferred kanamycin-resistant gene were obtained (De la Pena et al. 1987). An even more direct way to introduce genes into cereals is suggested by the observation that mechanically isolated mature wheat embryos derived from dry seeds are able to take up DNA by inhibition of a DNA solution and express a chimeric NPTII gene transiently. These DNA-treated embryos can readily be cultured into full plants (Topfer et al. unpubl.).

## 3.4 Some Examples of Transgenic Plants

### 3.4.1 Engineering Herbicide Tolerance/Resistance in Transgenic Plants (Shah et al. 1986)

Recently, Shah et al. (1986) employed a different approach by following the observation that herbicide glyphosate is a potent inhibitor of enzyme 5-enolpyruvyl-shikimate-3-phosphate (EPSP) synthase in higher plants. They isolated a complementary DNA (cDNA) clone encoding EPSP synthase from a complementary DNA library of a glyphosate-tolerant *Petunia hybrida* cell line (MP4-G) that overproduces the enzyme. This line overproduces EPSP synthase messenger RNA as a result of a 20-fold amplification of the gene. A chimeric EPSP synthase gene was constructed with the use of the cauliflower mosaic virus 35S promoter to attain high level expression of EPSP synthase and introduced into *Petunia* cells. They observed that the transformed plants were tolerant to glyphosate (Shah et al. 1986). These results represent a major step toward establishing selective herbicide tolerance in crop plants.

### 3.4.2 Delay of Disease Development in Transgenic Plants (Abel et al. 1986)

In agriculture it is a common practice (cross-protection) to inoculate plants with mild strains of viruses to prevent more virulent strains from infecting the plant and causing severe disease symptoms. Cross-protection has been used to reduce yield losses in crops; however, there are a number of potential disadvantages to the use of cross-protection. Abel et al. (1986) have enumerated four disadvantages: "(1) the mild strain (protecting strain) might undergo mutation to a highly virulent form, leading to extensive crop losses; (2) the protecting strain might act in synergism with a nonrelated virus; (3) a protecting virus in one plant might be a severe pathogen in another; (4) the protecting strain may itself cause a small but significant loss in yield". According to these authors most of these objections could be overcome if cross-protection was engendered in plants "as the result of expression of a single viral gene, rather than as a result of infection with an intact virus". Their experiment showed that plants can be genetically engineered to suppress symptoms due to viral infection.

Abel et al. (1986) introduced a chimeric gene containing a cloned cDNA of the coat protein (CP) gene of TMV into tobacco cells on a Ti plasmid of *A. tumefaciens* from which tumor-inducing genes had been removed. Plants regenerated from transformed cells expressed TMV mRNA and CP as a nuclear trait. Seedlings from self-fertilized transgenic plants were inoculated with TMV and observed for development of disease symptoms. The seedlings that expressed the CP gene were delayed in symptom development and 10%–60% of the transgenic plants failed to develop symptoms. Increasing the concentration of TMV in the inoculum shortened the delay in the appearance of symptoms. These experiments show that plants can be genetically transformed for resistance to virus disease development, and provide a way for producing virus-resistant plants that should complement those used in classical plant breeding (Abel et al. 1986).

#### 3.4.3 Expression of Luciferase Gene in Plant Cells (Ow et al. 1986)

The luciferase gene from firefly (*Photinus pyralis*) is a novel tool for studying gene expression. This gene encodes an enzyme that catalyzes the light-producing, ATP-dependent oxidation of luciferin (DeLuca and McElroy 1978). Recently, in an elegant experiment, Ow et al. (1986) incorporated luciferase gene into carrot protoplasts by electroporation and into tobacco plants by using *Agrobacterium tumefaciens*. The extracts from electroporated carrot cells and from transgenic tobacco plants produced light when mixed with the substrates luciferin and ATP. Light produced by the action of luciferase was also detected in undisrupted leaves or cells in culture from transgenic plants incubated in luciferin and in whole transgenic plants "watered" with luciferin.

Moreover, a cell line (B-15-1) derived from transgenic plant and its extract from such cells grown in suspension culture showed moderate levels of luciferase activity. Ow et al. further emphasized that "for plant genetic engineering purposes, the luciferase gene can be particularly valuable as a reporter of gene expression, as a marker in genetic crosses, and as a probe for a variety of plant cell functions. The speed and ease by which the luciferase assay can be performed permit more rapid screening of large numbers of transgenic plants or transformed cell lines. Moreover luciferase can be used as a cell marker to identify cells and as a tag to follow the targeting of proteins to various destinations within and outside the cells".

## 4 Uptake and Incorporation of Organelles, Viruses, Bacteria, and Fungi by Protoplasts

### 4.1 Uptake of Isolated Nuclei, Micronuclei, and Chromosomes

Both intra- and interspecific nuclear transplantations were observed in *Petunia hybrida, Nicotiana tabacum*, and *Zea mays* protoplasts as early as 1973 (Potrykus and Hoffmann 1973). Later, Lörz and Potrykus (1978) improved the procedure by using PEG and Ca and obtained higher uptake frequency, but proof of the transfer and biological expression of the transferred genome were not obtained. These methods for the mass isolation and uptake of nuclei have now been refined during the last 5 years (Saxena et al. 1985a, b, 1986, 1987), and evidence for the transfer and function of a foreign genome following nuclear uptake has been provided. Saxena et al. (1986) induced uptake of isolated nuclei from *Vicia hajastana* cells into protoplasts of an auxotrophic cell line of *Datura innoxia* under the influence of PEG and Ca. The frequency of nuclear uptake varied. The prototrophic nuclear fusion products following nuclear uptake could be rescued by initial culture of the protoplasts in nonselective conditions and subsequent use of feeder cell layers to support the growth of surviving colonies. The presence of *Vicia* genomic DNA in some prototrophic clones was confirmed by dot-blot hybridization using *Datura* and *Vicia* DNA probes. Well-developed shoots typical of wild-type *Datura* could be regenerated employing an appropriate regeneration medium.

The uptake of isolated chromosomes is yet another area. Szabodas et al. (1981) provided cytological evidence for the uptake of wheat, parsley, and maize chromosomes into isolated protoplasts. Later, Dudits and Praznovsky (1985) remarked that the transplanted chromosomes were soon lost, but the genetic marker localized on that chromosome could be detected long afterwards.

With the availability of flow cytometry the uptake of chromosomes has been greatly facilitated. The chromosome isolation and flow sorting has been carried out in *Haplopappus* (De Laat and Blaas 1984), *Nicotiana* (Verhoeven et al. 1987), and *Petunia* (Conia et al. 1987). Only metaphase cells were suitable for chromosome isolation. De Laat and Blaas (1987) used an improved microinjecting system and metaphase chromosomes isolated from a partially synchronized kanamycin-resistant suspension culture of *Nicotiana plumbaginifolia* were microinjected into recipient wild-type protoplasts; however, their fate has not been reported.

Recently, attempts have been made to obtain micronuclei with a view to transferring single or specific chromosomes (Sree Ramulu et al. 1987). Amiprophos-methyl (APM) treatment of cells resulted in the accumulation of a large number of metaphases. The scattered single chromosomes or small groups of chromosomes decondensed, developed nuclear membranes and formed micronuclei in potato and carrot. Such micronuclei can be sorted out by flow cytometry. This opens prospects for application of micronuclei for transfer of specific intact chromosomes and gene mapping in plants (Sree Ramulu et al. 1987).

### 4.2 Uptake and Transfer of Chloroplasts and Mitochondria

The incorporation of spinach chloroplasts into animal cell cultures was reported as early as 1969 (Nass 1969). These cells divided normally, and the chloroplasts retained their structural integrity. Following this, Giles and Sarafis (1971, 1972) implanted *Nitella* chloroplasts into a hen's egg and demonstrated that these chloroplasts survive, remain metabolically active, maintain their morphological integrity, and divide repeatedly. This report on the succesful in-vitro culture of an organelle reaffirms the autonomy of chloroplasts. Potrykus (1973), as well as Bonnett and Eriksson (1974), reported the uptake of chloroplasts by albino *Petunia* and carrot protoplasts respectively. Carlson (1973) claimed incorporation of nonmutant tobacco chloroplasts into the cytoplasm of an albino protoplast. Since the albino trait is transmitted cytoplasmically, it is a mutant of the chloroplast-DNA and not of a nuclear gene; these externally supplied foreign chloroplasts escaped destruction and entered the cytoplasm. Whole plants were later regenerated from albino protoplasts containing these implanted chloroplasts. Potrykus (1973) transplanted *Petunia hybrida* chloroplasts in the albino protoplasts from an extrachromosomally inherited variegated plant, and expressed a valid criticism of Carlson's work.

In later studies there has been a shift in emphasis from direct uptake to transfer through protoplast fusion, and also by transformation through *Agrobacterium*, wherein much meaningful work has been done. However, chloroplast uptake and exchange continues to play a useful role in studies involving nuclear-chloroplast interaction, and for the physiology of isolated plastids as well as fot investigations of the transfer of herbicide resistance (see Chap. XVII, this Vol.).

Giles (1973) demonstrated for the first time a complementation involving the use of mutant lines of *Zea mays.* Fusion of protoplasts containing green chloroplasts from the striped mutant iojap, with protoplasts of the white-deficient mutant containing only white underdeveloped plastids, caused the greening of the white-deficient chloroplasts over a period of 72 h. Fusion of white and green protoplasts from the iojap plants caused similar greening of the mutant iojap plastids. Similarly, Shvidkaya and Gleba (1974) showed that it was possible to complement chlorophyll deficiency by protoplast fusion.

Glimelius et al. (1981) showed that albino protoplasts could be restored to photoautotrophy by protoplast fusions, while Medgyesy et al. (1985) used such a chloroplast mutant, together with gamma irradiation, to select for interspecific chloroplast transfer by production of cybrids of *Nicotiana* resulting from protoplast fusion. Menczel et al. (1982) used gamma irradiation for efficient transfer of chloroplasts during fusion.

Recently, Cseplo et al. (1986) used protoplast cultures of *Nicotiana plumbaginifolia* to isolate mutants resistant to antibiotics and to photosynthesis-inhibiting herbicides. They demonstrated that protoplast fusion was suitable to rescue the mutant chloroplasts from undesirable nuclear backgrounds, or to transfer them into different species.

The mitochondria can likewise be transferred during protoplast fusion, and some few traits coded on mitochondrial genomes have been reported to be transferred, such as toxin sensitivity of Texas male sterile cytoplasm in maize (Lonsdale 1987). Although the direct uptake and incorporation of isolated mitochondria has not been reported, the mDNA becomes incorporated during fusion of protoplasts. Mitochondrial genomes were first analyzed in *Nicotiana tabacum* cybrids by Belliard et al. in 1979. Since then, rearrangements of mt-DNA's have been reported in a number of somatic cybrids, including a recent report on carrot (Ichikawa et al. 1987).

### 4.3 Uptake of Bacteria and Viruses

The isolated protoplasts were demonstrated to uptake bacteria (*Rhizobium*) with a view to utilization for nitrogen fixation (Davey 1977); however, with the current information available on nif-gene functioning, such work on the uptake of whole bacteria by protoplasts has not advanced much further. However, the uptake of virus by protoplasts offers a number of possibilities for studying: (1) mechanism of infection, (2) nuclear-protein replication, (3) interaction among different viruses in the same host, and (4) virus as a vector for transformation. Tobacco mosaic virus was observed to infect the tomato fruit protoplasts (Cocking and Pojnar 1969), since then numerous reports have appeared on the uptake/infection of protoplasts (see Takebe 1983; Chap. XXIII, this Vol.), and various methods have been employed, i.e., poly-L-ornithine (Otsuki et al. 1972), polyethylene glycol (Cassells and Barlass 1978), and electroporation (Okada et al. 1986; Watts et al. 1987).

Recently, there has been increased interest in the use of viruses as vectors for genetic engineering studies, liposome-mediated delivery, which offers protection against nucleases, has been shown to be an important method for the introduction of nucleic acids into protoplasts (Lurquin 1984; Nagata 1984).

DNA virus offers alternative vectors to *A. tumefaciens*. Transformation achieved with a part of the cauliflower mosaic virus genome replaced by a selectable gene in such a way that the virus replicated normally (Brisson et al. 1984); similarly, combinations of virus and bacterial vectors (agroinfection) are being studied (Grimsley et al. 1986).

### 4.4 Uptake of Fungi

Fungal protoplasts through uptake, like the microbes, or through fusion with higher plant protoplasts, would enable the transfer of genes. Fungal protoplasts could also act as eukaryotic gene vectors for transformation. Constabel et al. (1982) attempted to transfer the ability for secondary metabolite synthesis in culture by fusing fungal protoplasts with those of *Catharanthus roseus*. Recent studies on PEG-mediated uptake/fusion of celery protoplasts with those of *Aspergillus nidulans* and *Fusarium oxysporum* have been successfully conducted (see Chap. XXIV, this Vol.). The celery protoplasts are large (10–40 μm), whereas fungal protoplasts are very small (0.5–7 μm), their uptake could be demonstrated by diamidino-2-phenylindole staining, and confirmed by ultrastructural studies (Lynch et al. 1986). After uptake of *A. nidulans*, the celery protoplasts showed improved level of viability, but there was a decrease in viability with *F. oxysporum*.

It has been known that fungal infections increase secondary metabolite synthesis of host plant cells (Mansfield et al. 1974); however, it remains to be seen if such a situation can be exploited in cell/protoplast system.

## 5 Newer Techniques for Genetic Engineering

### 5.1 Electrofusion

Electrofusion has greatly facilitated fusion methodology, and electrofusion chambers where the events can be better controlled are now commercially available. This method has certain definite advantages over chemical fusion in that it is (1) nontoxic, (2) encourages higher fusion frequency, and (3) is reproducible. Electric fusion is based on the reversible electrical breakdown of cell membrane (Zimmermann and Scheurich 1981; Scheurich et al. 1981). This breakdown results in a tremendous increase of the membrane permeability and conductivity. As a consequence, fusion is initiated where electric field pulse of 50 µs duration and of sufficiently high intensity induces reversible electrical breakdown of the membranes. Various modifications in technology have been employed (Watts and King 1984; Tempelaar and Jones 1985; Zachrisson and Bornman 1986) and currently electrofusion is being utilized routinely in a number of plant species, notable among which are tobacco (Bates 1985) and potato (De Vries and Tempelaar 1987).

Although chains of protoplasts are formed in electrofusion and in many cases this fusion product involves many protoplasts, which is not desirable, yet it is possible to control the dielectrophoretic adhesion of cells in such a way so as to produce aggregates consisting of two desired protoplasts (Fig. 3 of Zimmermann and Scheurich 1981). Recent work of Spangenberg and Schweiger (1986) has demonstrated that it is possible to select two protoplasts from two desired parents and to fuse them. The application of this method to fuse individually paired *Brassica* protoplasts in microdrop cultures has been found convenient. Nearly 50% of the fused pairs survived with 10% to 20% of these forming microcalli. Pairing and fusion of individual protoplasts provides a means of controlling the number and type of cells being fused, which also eliminates the need for selection systems.

### 5.2 Electroporation

Electroporation is an efficient method for the incorporation of foreign DNA into protoplasts, and thus for direct gene transfer into plants. The principle underlying electroporation is to subject the protoplasts briefly to electric pulse (Neumann et al. 1982) in order to momentarily induce gentle rupture of the membrane at places to form small pores sufficient to take in DNA molecules. The apparatus involved is rather simple (see Fig. 3, Chap. III, this Vol.), the voltage and the time range for application varies. Some workers (Fromm et al. 1986) employed a voltage peak value of 250–350 V with an RC constant in the ms range, while others (Langridge et al. 1985; Riggs and Bates 1986) used a voltage peak value of up to 2000 V with an RC constant in the µs range.

Electroporation has been successfully used for direct gene transfer in protoplasts of both monocots and dicots, i.e., tobacco, carrot, maize, sorghum, etc. (see Table 2), and for transient expression of genes (Fromm et al. 1985).

Laser micropuncture is another promising technique for gene transfer. Tao et al. (1987) have successfully transferred a plasmid gene for phosphotransferase into human cultured cells by producing small holes in the cell membrane with a laser beam. Transformation was 100 times more efficient and the gene was stably expressed in the human cells. Such a technique can be extended to plant cells.

## 5.3 Microinjection

Microinjection of foreign DNA into plant cells is a novel approach for direct gene transfer. Although microinjections have been used with animal cell cultures for over a decade (Jaenisch and Mintz 1974), the technique has recently been extended successfully to plant protoplasts. Through this technology it may also be possible to transfer subcellular organelles such as nuclei and chloroplasts (Crossway et al. 1986a). Moreover, extension of this technique to microcloning from microdissected plant chromosomes will aid gene isolation from plants as is done with mouse cells (Rohme et al. 1984).

Although the technique of microinjection was initiated as early as 1911 by Barber, who introduced chemicals and bacteria into *Nitella* and monitored their effect, it is only recently that Crossway and her colleagues have refined and extended it to plant cells/protoplasts by injecting foreign DNA (see Chap. XIV, this Vol.).

Various methods have been used for microinjection, i.e., surface attachment of protoplasts on a slide, by embedding in agarose and by using holding pipettes. The surface-attached technique generally involves attaching the cells/protoplasts to the surface of slide coated with polylysine (Reich et al. 1986b) and then injecting with microneedles. Such microinjected protoplasts are subsequently cultured (as attached) in liquid (Steinbiss and Stabel 1983), in agar or on filter paper feeder layers (Reich et al. 1986b) or in hanging drops (Crossway et al. 1986b). Some examples of successful cell transformation by intranuclear microinjections are *Nicotiana tabacum* (Crossway et al. 1986b), *N. debneyi* (Pasternak et al. 1986), *Medicago sativa* (Reich et al. 1986c), *Brassica napus* (Miki et al. 1986). Microinjection resulted in higher transformation frequency as compared to other available methods.

In addition to protoplasts, other systems for microinjection are single cells (Nomoura and Komamine 1986), pollen, pollen tubes (Hepher et al. 1985), immature embryos (Flavell and Mathias 1984) and chromosomes. Recently, De Laat and Blaas (1987) injected metaphase chromosomes isolated from kanamycin-resistant cell suspension of *Nicotiana plumbaginifolia* into recipient wild-type protoplasts and obtained visual evidence. Microinjection has also been used to bring about transformation in pollen embryos of *Brassica* (Neuhaus et al. 1987) and young tillers of rye (De la Pena et al. 1987).

## 5.4 Flow Cytometry

Flow cytometry, a relatively new technique extended to plant biotechnology during the last few years, has enabled the realization of a number of potential applications. This subject has been fairly well reviewed (Brown et al. 1986; Glimelius

1987). The technique has facilitated the isolation of somatic hybrid protoplasts (Galbraith 1984), sorting and uptake of chromosomes (De Laat and Blaas 1984), and viability determination, and has also been put to a number of other uses in plant research, which include determination of the DNA content of cells and thus the ploidy level (Galbraith and Shields 1982; Carlsberg et al. 1984; Sundberg and Glimelius 1986), analysis of the cell cycle (Meadows 1982; Galbraith et al. 1983; Anderson et al. 1985; Bergounioux et al. 1986), studies on organelles (Ashcroft et al. 1986; Petit 1986), detection of the transformed protoplasts (Tagu et al. 1987), score pollen mutants (Pinkel 1981), and detection of cells containing desired secondary metabolites (Brown 1984).

The principle underlying flow cytometry (Herzenberg and Sweet 1976; Melamed et al. 1979) is that a liquid jet containing protoplasts is excited by a laser beam, and there is scattering of light and emission of fluorescence which enables selection of the desired protoplasts.

Protoplasts, being isodiametrically shaped, are perhaps the most suitable material for flow cytometry, and recently extensive work has been done. Galbraith (see Chap. XVIII, this Vol.) applied flow cytometry for sorting and enrichment of heterokaryons. The overall concept is that the two populations of parental protoplasts are identified through the presence of both the fluorescent labels. Flow cytometry is then applied to sort out and enrich heterokaryon population. Hybrid callus is then obtained from materials from which hybrid plants are subsequently regenerated (Afonso et al. 1985; Glimelius 1987). Flow cytometry has also been applied for sorting of somatic hybrid protoplasts in various species of *Nicotiana* (Afonso et al. 1985; Ayers 1987; *Hordeum vulgare*×*Datura innoxia* (Alexander et al. 1985), and *Brassica campestris*×*Brassica oleracea* (Glimelius et al. 1986).

Generally two methods are employed for the isolation of protoplasts that contain fluorescent labels, either the fluorescence label can be added in the form of extrinsic probes, or use can be made of the naturally fluorescent compounds found in plant tissues, as in the case of chlorophyll. A number of fluorescent agents have been used, such as fluorescein diacetate, carboxyfluorescein diacetate, scopoletin, and fluorescein isothiocyanate (FITC).

Although several combinations of fluorescence labelling have been employed, the most popular has been the use of the natural chlorophyll autofluorescence. Combinations based on the fluorescence emission of fluorescein and rhodamine have been successfully employed for the selection of heterokaryons produced between labeled leaf protoplasts (Afonso et al. 1985).

The most attractive aspect of flow cytometry is the rapidity with which sorting/selection can be done — up to 5000 protoplasts can be screened per second.

### 5.5 Use of Miniprotoplasts and Cytoplasts

At high speed centrifugation of protoplasts through a density gradient, two main types of subprotoplasts are formed. The heavy subprotoplast (*miniprotoplast*) is characterized by a nucleus surrounded by some cytoplasm and a plasma membrane. The light subprotoplast (*cytoplast*) contains the major vacuole and the rest of the cytoplasm enclosed in a plasma membrane. The method for its isolation has been reviewed (Lörz et al. 1981; Archer et al. 1982; Bilkey et al. 1982; Lörz 1984; Hurlbut et al. 1985; also see Chap. XXVI, this Vol.).

The term miniprotoplast was coined by Wallin et al. (1978), and is equivalent to the other terminology, i.e., karyoplast (Lörz and Potrykus 1980), evacuolated protoplast (Griesbach and Sink 1983), or nucleoplast (Gleba and Sytnik 1984). Miniprotoplasts have the ability to regenerate the cell wall and divide, as in corn (Lörz et al. 1981), and regenerate plants, as in *Solanum niger* (Lesney et al. 1986).

Similarly, cytoplasts which are referred to as nucleus-free protoplasts (Wallin et al. 1978, 1979) have been observed to be formed in a number of plant species (see Chap. XXVIII, this Vol.) from various types of tissues/cells, for example pollen (Bajaj 1974), leaves (Wallin et al. 1978), epidermis (Bradley 1978), mesophyll cells (De and Swain 1983), hypocotyl (Spangenberg and Schweiger 1986), and cotton fibers (Gould et al. 1986); such subprotoplasts do not divide when cultured.

Subprotoplasts of various types have significant potential in cell manipulation and genetic engineering of plants as outlined by Wallin et al. in Chapter XXVIII, this Volume are summarized here:

Although cybrids can be obtained by fusion of normal protoplasts, it is desirable to have an efficient method for direct organelle transfer into plant cells. Thus an alternative method for transfer of plastomes and chondriomes is the fusion of cytoplasts with protoplasts. It has been shown that streptomycin resistance coded by chloroplasts could be transferred via cytoplasts (Maliga et al. 1982). Recently, Tan (1987) claimed to have obtained cybrids by fusion of cytoplasts from *Petunia hybrida* and protoplasts of *Lycopersicon peruvianum.*

Miniprotoplasts and cytoplasts may be unique materials for certain types of cell manipulation. For instance, in somatic hybridization, miniprotoplasts (which have negligible cytoplasm) might serve as recipients of cytoplasmic organelles. If a miniprotoplast is fused with a complete protoplast, a hybrid may be formed which has nuclei from both the parents, but presumably plastids and mitochondria from the complete protoplast only.

Miniprotoplasts can also facilitate the selection of fused products. By gradient centrifugation, bands of protoplasts, miniprotoplasts, and fusion products (protoplasts × miniprotoplasts) can be separated on the basis of this density. Thus Griesbach and Sink (1983), by fusing a miniprotoplast of *Petunia parodii* and a complete protoplast of *P. inflata*, created a system by which they obtained a band containing more than 50% fused products. Microprotoplasts have also been suggested to be used for microinjection studies (Burgess 1984).

## 6 Conclusions

The conventional methods of plant breeding have enabled gene transfer for crop improvement for centuries (Goodman et al. 1987), recent advances in genetic engineering and in vitro manipulation of cells (Bajaj 1987) provide additional tools of unprecedented dimensions to complement the efforts of breeders and to create gene pools hitherto not available in nature. Through recombinant DNA technology – transformation – plant regeneration (Wilke-Douglas et al. 1986), a specific gene can be incorporated, and a desired plant made to order.

During the past decade, tremendous progress has been made in the regeneration of complete plants from isolated protoplasts, production of somatic hybrids and cybrids, and the transformation of plant cells through *Agrobacterium* or direct gene transfer. The newer techniques, such as electrofusion, electroporation, microinjection, and flow cytometry, have facilitated the uptake and incorporation of DNA, specific isolated chromosomes, and desired genes into plant cells. The cloning of genes, production of transgenic plants/mutants showing resistance/tolerance to viruses, insects, herbicides, the successful genetic manipulation, and transformation of cereals such as rice and maize, cybrids for the induction of male sterility in crops, and increased production of pharmaceutical products by transformed hairy roots are some of the landmarks. Such studies have now been extended to forest trees. These achievements have far-reaching implications in agricultural and forest biotechnology.

## References

Abel PP, Nelson RS, De B, Hoffmann N, Rogers SG, Fraley RT, Bachy RN (1986) Delay of disease development in transgenic plants that express the tobacco mosaic virus coat protein gene. Science 232:738–743

Afonso CL, Harkins KR, Thomas-Compton MA, Krejci AE, Galbraith DW (1985) Selection of somatic hybrid plants in *Nicotiana* through fluorescence-activated sorting of protoplasts. Biotechnology 3:811–816

Alexander RG, Cocking EC, Jackson PJ, Jett JH (1985) The characterization and isolation of plant heterokaryons by flow cytometry. Protoplasma 128:52–58

Anderson LK, Stack SM, Fox MH, Chuanshan Z (1985) The relationship between genome size and synaptonemal complex length in higher plants. Exp Cell Res 156:367–378

Archer EK, Landgren CR, Bonnett HT (1982) Cytoplast formation and enrichment from mesophyll tissues of *Nicotiana* spp. Plant Sci Lett 25:175–185

Armstead IP, Webb KJ (1987) Effect of age and genotype of tissue on genetic transformation of *Lotus corniculatus* by *Agrobacterium tumefaciens*. Plant Cell Tissue Organ Cult 9:95–101

Asahi T, Kumashiro T, Kubo T (1988) Constitution of mitochondrial and chloroplast genomes in male sterile tobacco obtained by protoplast fusion of *Nicotiana tabacum* and *N. debneyi*. Plant Cell Physiol 29:43–49

Ashcroft RG, Preston C, Cleland RE, Critchley C (1986) Flow cytometry of isolated chloroplasts and thylakoids. Photobiochem Photobiophys 13:1–14

Austin S, Ehlenfeld MK, Baer MA, Helgeson JP (1986) Somatic hybrids produced by protoplast fusion between *Solanum tuberosum* and *S. brevidens:* phenotypic variation under field conditions. Theor Appl Genet 71:682–690

Aviv D, Galun E (1986) Does pretreatment by rhodamine 6-G affect the mitochondrial composition of fusion-derived *Nicotiana* cybrids? Plant Cell Rep 3:227–230

Aviv D, Bleichman S, Arzee-Gonen P, Galun S (1984a) Interspecific cytoplasmic hybrids in *Nicotiana*. Identification of plastomes and chondriomes in *N. sylvestris* + *N. rustica* cybrids having *N. sylvestris* nuclear genomes. Theor Appl Genet 67:499–504

Aviv D, Arzee-Gonen P, Bleichman S, Galun E (1984b) Novel alloplasmic *Nicotiana* plants by "donor-recipient" protoplast fusion: cybrids having *N. tabacum* or *N. sylvestris* nuclear genomes and either or both plastomes and chondriomes from alien species. Mol Gen Genet 196:244–253

Ayers NM (1987) Inheritance of organelles in somatic cell hybrids and cybrids of *Nicotiana*. Ph D Thesis, Univ Nebraska, Lincoln

Bajaj YPS (1974) Isolation and culture studies on pollen tetrad and pollen mother cell protoplasts. Plant Sci Lett 3:93–99
Bajaj YPS (1983) Haploid protoplasts. In: Giles KL (ed) Plant protoplasts. Academic Press, New York London, pp 113–141
Bajaj YPS (ed) (1986) Biotechnology in agriculture and forestry, vol 2. Crops I. Springer, Berlin Heidelberg New York Tokyo
Bajaj YPS (1987) In vitro biotechnologies for crop improvement. Int Symp Worksh Gene manipulation for plant improvement in developing countries, Kuala Lumpur
Bajaj YPS (ed) (1989) Biotechnology in agriculture and forestry, vol 8. Plant protoplasts and genetic engineering I. Springer, Berlin Heidelberg New York Tokyo
Bajaj YPS, Gill MS, Gosal SS (1987) Somatic hybridization – a biotechnological approach to plant breeding. In: Ganguli PK, Mandal AK (eds) Advances in plant breeding methods. Oxford and IBH Press, New Delhi
Baldes R, Moos M, Geider K (1987) Transformation of soybean protoplasts from permanent suspension cultures by cultivation with cells of *Agrobacterium tumefaciens*. Plant Mol Biol 9:135–146
Barsby TL, Yarrow SA, Kemble RJ, Grant I (1987) The transfer of cytoplasmic male sterility to winter-type oilseed rape (*Brassica napus* L.) by protoplast fusion. Plant Sci 53:243–248
Basiran N, Armitage P, Scott R, Draper J (1987) Genetic transformation of flax (*Linum usitatissimum*) by *Agrobacterium tumefaciens*: Regeneration of transformed shoots via a callus phase. Plant Cell Rep 6:396–399
Bates GW (1985) Electric fusion for optimum formation of protoplast heterokaryons in *Nicotiana*. Planta 165:217–224
Belliard G, Vedel F, Pelletier G (1979) Mitochondrial recombination in cytoplasmic hybrids of *Nicotiana tabacum* by protoplast fusion. Nature (London) 281:401–403
Bergounioux C, Perennes C, Miege C, Gadal P (1986) The effect of male sterility on protoplast division in *Petunia hybrida*. Cell cycle comparison by flow cytometry. Protoplasma 130:138–144
Bilkey PC, Davey MR, Cocking EC (1982) Isolation, origin and properties of enucleated plant microplasts. Protoplasma 110:147–151
Binding H, Jain SM, Fuiger J, Mordhorst G, Nehls R, Gressel J (1982) Somatic hybridization of an atrazine resistant biotype of *Solanum nigrum* with *Solanum tuberosum*. Theor Appl Genet 63:273–277
Boeshore ML, Hanson MR, Izhar S (1985) A variant mitochondrial DNA arrangement specific to *Petunia* stable sterile somatic hybrids. Plant Mol Biol 4:125–132
Bonnett HT, Eriksson T (1974) Transfer of algal chloroplasts into protoplasts of higher plants. Planta 120:71–79
Bourgin JP, Missonier C, Goujaud J (1986) Direct selection of cybrids by streptomycin and valine resistance in tobacco. Theor Appl Genet 72:11–14
Bradley PM (1978) Production of enucleated plant protoplasts of *Allium cepa*. Plant Sci Lett 13:287–290
Braun AC (1943) Studies on tumor inception in crown gall disease. Am J Bot 30:674–677
Brisson N, Paszkowski J, Penswick JR, Gronenborn B, Potrykus I, Hohn T (1984) Expression of a bacterial gene in plants using a viral vector. Nature (London) 310:511–514
Brown S (1984) Analysis and sorting of plant material by flow cytometry. Physiol Veg 22:341–349
Brown S, Renaudin JP, Prevot C, Guern J (1984) Flow cytometry and sorting of plant protoplasts: technical problems and physiological results from a study of pH and alkaloids in *Catharanthus roseus*. Physiol Veg 22:541–554
Brown S, Jullien M, Coutos-Thevenat P, Muller Ph, Renaudin JP (1986) Present developments of flow cytometry in plant biology. Biol Cell 58:173–178
Burgess J (1984) The revolution that failed. New Sci 1429:26–29
Butenko R, Kuchko A, Komarnitsky I (1982) Some features of somatic hybrids between *Solanum tuberosum* and *S. chacoense* and its $F_1$ sexual progeny. In: Fujiwara A (ed) Plant tissue culture 1982. Maruzen, Tokyo, pp 643–644
Carlsberg I, Glimelius K, Eriksson T (1984) Nuclear DNA content during the initiation of callus formation from isolated protoplasts of *Solanum tuberosum* L. Plant Sci Lett 35:225–230
Carlson PS (1973) The use of protoplasts for genetic research. Proc Natl Acad Sci USA 70:598–602
Carlson PS, Smith HH, Dearing RD (1972) Parasexual interspecific plant hybridization. Proc Natl Acad Sci USA 69:2292–2294

Cassells AC, Barlass M (1978) The initiation of TMV infection in isolated protoplasts by polyethylene glycol. Virology 87:459–462

Chilton MD, Drummond MH, Merlo DJ, Sciaky D, Montoya AL, Gordon MP, Nester EW (1977) Stable incorporation of plasmid-DNA into higher plant cells: the molecular basis of crown gall tumorigenesis. Cell 11:263–271

Christou P, Murphy JE, Swain WF (1987) Stable transformation of soybean by electroporation and root formation from transformed callus. Proc Natl Acad Sci USA 84:3962–3966

Chuong PV, Beversdorf WD, Powell AD, Pauls KP (1988) Somatic transfer of cytoplasmic traits in *Brassica napus* L. by haploid protoplast fusion. Mol Gen Genet 211:197–201

Clark E, Schnabelrauch L, Hanson MR, Sink KC (1986) Differential fate of plastid and mitochondrial genomes in *Petunia* somatic hybrids. Theor Appl Genet 72:748–755

Cocking EC, Davey MR (1987) Gene transfer in cereals. Science 236:1259–1262

Cocking EC, Pojnar E (1969) An electron microscopy study of the infection of isolated tomato fruit protoplasts by tobacco mosaic virus. J Gen Virol 4:305–312

Conia J, Bergounioux C, Perennes C, Brown S, Muller Ph, Gadal P (1987) Chromosomes from protoplasts of *Petunia hybrida*, flow analysis and sorting. Biol Cell 58 (Suppl): 5a

Constabel F, Kurz WGW, Kutney P (1982) Variation in cell cultures of periwinkle *Catharanthus roseus*. In: Fujiwara A (ed) Plant tissue culture 1982. Maruzen, Tokyo, pp 301–304

Crossway A, Hauptli H, Houck CM, Irvine JM, Oakes JV, Perani LA (1986a) Micromanipulation techniques in plant biotechnology. Biotechniques 4:320–334

Crossway A, Oakes JV, Irvine JM, Ward B, Knauf VC, Showmaker CK (1986b) Integration of foreign DNA following microinjection of tobacco mesophyll protoplasts. Mol Gen Genet 202:179–185

Cseplo A, Medgyesy P, Marton L (1986) In vitro induction, isolation and transfer of chloroplast mutations in *Nicotiana*. Proc Nuclear techniques and in vitro culture for plant improvement. IAEA, Vienna, pp 137–146

Dandekar AK, Gupta PK, Durzan DJ, Knauf V (1987) Transformation and foreign gene expression in micropropagated Douglas-fir (*Pseudosuga menziesii*). Bio/Technology 5:587–590

Davey MR (1977) Bacterial uptake and nitrogen fixation. In: Reinert J, Bajaj YPS (eds) Applied and fundamental aspects of plant cell, tissue, and organ culture. Springer, Berlin Heidelberg New York, pp 551–562

Davey MR, Cocking EC, Freeman J, Pearce N, Tudor I (1980) Transformation of *Petunia* protoplasts by isolated *Agrobacterium* plasmids. Plant Sci Lett 18:307–313

De DN, Swain D (1983) Protoplasts, cytoplasts and subprotoplasts from ripening tomato fruits: their nature and fusion properties. In: Sen SK, Giles KL (eds) Plant cell culture in corp improvement. Plenum, New York, pp 201–208

De la Pena A, Lörz H, Schell J (1987) Transgenic rye plants obtained by injecting DNA into young floral tillers. Nature (London) 325:274–276

De Laat AMM, Blaas J (1984) Flow cytometric characterization and sorting of plant chromosomes. Theor Appl Genet 67:463–467

De Laat AMM, Blaas J (1987) An improved method for protoplast microinjection suitable for transfer of entire plant chromosomes. Plant Sci 50:161–169

De Vries SE, Tempelaar MJ (1987) Electrofusion and analysis of potato somatic hybrids. In: Bajaj YPS (ed) Biotechnology in agriculture and forestry, vol 3. Potato. Springer, Berlin Heidelberg New York Tokyo, pp 211–222

Deak M, Kiss GB, Koncz C, Dudits D (1986) Transformation of *Medicago* by *Agrobacterium* mediated gene transfer. Plant Cell Rep 5:97–100

DeLuca M, McElroy WD (1978) Meth Enzymol 57:3

D'Hont A, Quetier F, Teoule E, Dattee Y (1987) Mitochondrial and chloroplast DNA analysis of interspecific somatic hybrids of a Leguminoseae: *Medicago* (alfalfa). Plant Sci 53:237–242

Dudits D, Praznonzky T (1985) Intergeneric gene transfer by protoplast fusion and uptake of isolated chromosomes. In: Zaitlin M, Day P, Hollaender A (ed) Biotechnology in plant science: Relevance to agriculture in the eighties. Acad Press, Orlando, pp 115–128

Evans DA, Flick CE, Jensen RA (1981) Incorporation of disease resistance into sexually incompatible somatic hybrids of the genus *Nicotiana*. Science 213:907–909

Fillatti JJ, Sellmer J, McCown B, Haissig B, Comai L (1987) *Agrobacterium* mediated transformation and regeneration of *Populus*. Mol Gen Genet 206:192–199

Flavell R, Mathias R (1984) Prospects for transforming monocot crop plants. Nature (London) 307:108–109
Fraley RT, Rogers SG, Horsch RB (1986) Genetic transformation in higher plants. CRC Crit Rev Plant Sci 4:1–46
Fromm ME, Taylor LP, Walbot V (1985) Expression of genes electroporated into monocot and dicot plant cells. Proc Natl Acad Sci USA 82:5824–5828
Fromm ME, Taylor LP, Walbot V (1986) Stable transformation of maize after gene transfer by electroporation. Nature (London) 319:791–793
Galbraith DW, Shields BA (1982) The effect of inhibitors of cell wall synthesis on tobacco protoplast development. Physiol Plant 55:25–30
Galbraith DW, Harkins KR, Maddox JM, Ayres NM, Sharma DP, Firoozabady E (1983) Rapid flow cytometric analysis of the cell cycle in intact plant tissues. Science 220:1049–1051
Galbraith DW (1984) Flow cytometric analysis of the cell cycle. In: Vasil IK (ed) Cell culture and somatic cell genetics. Acad Press, Orlando, pp 765–777
Garcia JA, Hille J, Vos P, Goldbach R (1987) Transformation of cowpea *Vigna unguiculata* with a full-length DNA copy of cowpea mosaic virus m-RNA. Plant Sci 48:89–98
Gheysen G, Dhaese P, Van Montague M, Schell J (1985) DNA flux across genetic barriers: The crown gall phenomenon. In: Hohn B, Dennis ES (eds) Genetic flux in plants. Springer, Berlin Heidelberg New York Tokyo, pp 12–47
Giles KL (1973) Complementation by protoplast fusion using mutant strain of maize. Plant Cell Physiol 13:207–210
Giles KL, Sarafis V (1971) On the survival and reproduction of chloroplasts outside the cell. Cytobios 4:61–74
Giles KL, Sarafis V (1972) Chloroplast survival and division in vitro. Nature New Biol 236:56–58
Gleba YY, Hoffmann F (1980) "Arabidobrassica": A novel plant obtained by protoplast fusion. Planta 149:112–117
Gleba YY, Sytnik KM (1984) Protoplast fusion, genetic engineering in higher plants. Springer, Berlin Heidelberg New York, pp 27–28
Gleba YY, Kolesnik NN, Meshkene IV, Cherep NN, Parokonny AS (1984) Transmission genetics of the somatic hybridization process in *Nicotiana* I. Hybrid and cybrids among the regenerates from cloned protoplast fusion products. Theor Appl Genet 69:121–128
Gleddie S, Keller WA, Setterfield G (1986) Production and characterization of somatic hybrids between *Solanum melongena* L. and *Solanum sisymbrifolium* Lam. Theor Appl Genet 71:613–621
Glimelius K (1987) Enrichment of somatic hybrids with flow cytometry. IAPTC Newslett 52:2–9
Glimelius K, Chen K, Bonnett HT (1981) Somatic hybridization in *Nicotiana*: Segregation of organellar traits among hybrid and cybrid progeny. Planta 153:504–510
Glimelius K, Djupsjobacka M, Fellner-Feldegg H (1986) Selection and enrichment of plant protoplasts heterokaryons of Brassicaceae by flow sorting. Plant Sci 45:133–141
Goodman RM, Hauptli H, Crossway A, Knauf CV (1987) Gene transfer in crop improvement. Science 236:48–54
Gould JH, Palmer RL, Dugger WM (1986) Isolation and culture of anucleate protoplasts from cotton (*Gossypium hirsutum*) fiber: assessment of viability and analysis of regenerated wall polymers. Plant Cell Tissue Organ Cult 6:61–72
Griesbach RJ, Sink KC (1983) Evacuolation of mesophyll protoplasts. Plant Sci Lett 30:297–301
Grimsley N, Hohn B, Hohn T, Walden R (1986) Agroinfection, a novel route for plant viral infection. Proc Natl Acad Sci USA
Grosser JW, Gmitter FG, Chandler JL (1988a) Intergeneric somatic hybrid plants from sexually incompatible woody species: *Citrus sinensis* and *Severinia disticha*. Theor Appl Genet 75: 397–401
Grosser JW, Gmitter FG, Chandler JL (1988a) Intergeneric somatic hybrid plants from sexually incompatible woody species: *Citrus sinensis* and *Severinia disticha*. Theor Appl Genet 75: 397–401
Grosser JW, Gmitter FG, Chandler JL (1988b) Intergeneric somatic hybrid plants of *Citrus sinensis* cv. *Hamlin* and *Poncirus trifoliata* cv. Plant Cell Rep 7:5–8
Guerche P, Jouanin L, Tepfer D, Pelletier G (1987) Genetic transformation of oilseed rape (*Brassica napus*) by the Ri T-DNA of *Agrobacterium rhizogenes* and analysis of inheritance of the transformed phenotype. Mol Gen Genet 206:382–386

Hain R, Stabel P, Czernilofski AP, Steinbiss HH, Herrera-Estrella L, Schell J (1985) Uptake, integration and genetic transmission of a selectable chimeric gene by plant protoplasts. Mol Gen Genet 199:161–168

Handley LW, Nickels RS, Cameron MW, Moore PP, Sink KC (1986) Somatic hybrid plants between *Lycopersicon esculentum* and *Solanum lycopersicoides*. Theor Appl Genet 71:691–697

Hepher A, Sherman A, Gates P, Boutler D (1985) Microinjection of gene vectors into pollen and ovaries as a potential means of transforming whole plants. In: Chapman GP, Mantell SH, Daniels RW (eds) The experimental manipulation of ovule tissues. Longman, New York, pp 52–63

Herrera-Estrella L, Broeck G van den, Maenhaust R, Van Montagu M, Schell J, Timko M, Cashmore A (1984) Light-inducible and chloroplast associated expression of a chimaeric gene introduced into *Nicotiana tabacum* using a Ti plasmid vector. Nature (London) 310:115–120

Herzenberg LA, Sweet RG (1976) Fluorescence-activated cell sorting. Sci Am 234:108–117

Horsch RB, Fry JB, Hoffmann NL, Wallroth M, Eichholtz D, Rogers SG, Fragley RT (1985) A simple and general method for transferring genes into plants. Science 227:1229–1231

Hurlbut P, Fellner-Feldegg H, Glimelius K (1985) Sorting of miniprotoplasts by flow cytometry. Hereditas Suppl 3:142

Ichikawa H, Tanno-Suenaga L, Imamura J (1987) Selection of *Daucus* cybrids based on metabolic complementation between X-irradiated *D. capillifolius* and iodoacetamide-treated *D. carota* by somatic cell fusion. Theor Appl Genet 74:746–752

Ito K, Tatemichi Y, Shibata M (1986) Regeneration of poplar-like plants from fusion between *Populus* (hybrid) and *Hibiscus*. Proc VI Int Congr Plant tissue and cell culture, Univ Minnesota, MN, Abstr, p 389

Jaenisch R, Mintz B (1974) Simian virus 40 DNA sequences in DNA of healthy adult mice derived from preimplantation blastocysts injected with viral DNA. Proc Natl Acad Sci USA 71:1250–1254

Jordan MC, McHughen A (1988) Glyphosate-tolerant flax plants from *Agrobacterium* mediated gene transfer. Plant Cell Rep 7:281–284

Kamada H, Okamura N, Satake M, Harada H, Shimomura K (1986) Alkaloid production by hairy root cultures in *Atropa belladonna*. Plant Cell Rep 5:239–242

Kemble RJ, Barsby TL, Wong RSC, Shepard JF (1986) Mitochondrial DNA rearrangements in somatic hybrids of *Solanum tuberosum* and *Solanum brevidens*. Theor Appl Genet 72:787–793

Klee H (1987) Agrobacterium-mediated plant transformation and its further applications to plant biology. Annu Rev Plant Physiol 38:467–486

Kohler F, Golz C, Eapen S, Kohn H, Schieder O (1987) Stable transformation of moth bean *Vigna acontifolium* via direct gene transfer. Plant Cell Rep 6:313–317

Krens FA, Molendijk L, Wullems GJ, Schilperoort RA (1982) In vitro transformation of plant protoplasts with Ti-plasmid DNA. Nature (London) 296:72–74

Kubo T (1985) Bull IWATA Tobacco. IWATA Exp Stn 17:130–133

Kumar A, Cocking EC (1987) Protoplast fusion: a novel approach to organelle genetics in higher plants. Am J Bot 74:1289–1303

Langridge WHR, Li BJ, Szalay AA (1985) Electric field mediated stable transformation of carrot protoplasts with naked DNA. Plant Cell Rep 4:355–359

Larkin PJ, Scowcroft WR (1981) Somaclonal variation – a novel source of variability from cell cultures for plant improvement. Theor Appl Genet 60:197–214

Lesney MS, Callow PW, Sink KC (1986) A technique for bulk production of cytoplasts and miniprotoplasts from suspension culture-derived protoplasts. Plant Cell Rep 5:115–118

Levings CS, Kim BD, Pring DR, Conde MF, Mans RJ (1980) Cytoplasmic reversion of cms-S, in maize: association with a transpositional event. Science 209:345–347

Lonsdale DM (1987) Cytoplasmic male sterility: a molecular perspective. Plant Physiol Biochem 25:265–271

Lörz H (1984) Enucleation of protoplasts: Preparation of cytoplasts and miniprotoplasts. In: Vasil IK (ed) Cell culture and somatic cell genetics of plants, vol 1. Academic Press, New York London, pp 448–453

Lörz H, Potrykus I (1978) Investigations on the transfer of isolated nuclei into plant protoplasts. Theor Appl Genet 53:251–256

Lörz H, Potrykus I (1980) Isolation of subprotoplasts for genetic manipulation studies. In: Ferenczy L, Farkas GL, Lazar G (eds) Advances in protoplast research. Pergamon, Oxford, pp 377–382

Lörz H, Paszkowski J, Dierks-Ventling C, Potrykus I (1981) Isolation and characterization of cytoplasts and miniprotoplasts derived from protoplasts of cultured cells. Physiol Plant 53:385–391

Lörz H, Baker B, Schell J (1985) Gene transfer to cereal cells mediated by protoplast transformation. Mol Gen Genet 199:178–182

Lurquin PF (1984) Liposomes as carriers for the transfer of nucleic acids into higher plant protoplasts. In: Beers RF, Bassett EG, Jr (eds) Cell fusion, gene transfer and transformation. Raven, New York, pp 227–236

Lynch PT, Isaac S, Collin HA (1986) Uptake of fungal protoplasts by plant protoplasts. Abstr VIth Int Congr Plant tissue and cell culture, Univ Minnesota, p 390

Maliga P, Lörz H, Lazar G, Nagy F (1982) Cytoplast-protoplast fusion for interspecific chloroplast transfer in *Nicotiana*. Mol Gen Genet 185:211–215

Manners JM (1987) Transformation of *Stylosanthes* spp. using *Agrobacterium tumefaciens*. Plant Cell Rep 6:204–207

Mansfield JW, Hargreaves JA, Boyle FC (1974) Phytoalexin production by live cells in broad bean leaves infected with *Botrytis cinerea*. Nature (London) 252:316–317

Matthews BF, Widholm JM (1986) Organelle DNA compositions and isoenzyme expression in an interspecific somatic hybrid of *Daucus*. Mol Gen Genet 198:371–376

McCormick S, Niedermeyer J, Fry J, Banason A, Horsch R, Fraley R (1986) Leaf disc transformation of cultivated tomato (*Lycopersicon esculentum*) using *Agrobacterium tumefaciens*. Plant Cell Rep 5:81–86

Meadows MR (1982) Characterization of cells and protoplasts of the B73 maize cell line. Plant Sci Lett 28:337–348

Medgyesy P, Golling R, Nagy F (1985) A light sensitive recipient for the effective transfer of chloroplast and mitochondrial traits by protoplast fusion in *Nicotiana*. Theor Appl Genet 70:590–594

Melamed MR, Mullaney PF, Mendelsohn ML (1979) Flow cytometry and sorting. J. Wiley & Sons, New York

Melchers G, Sacristan MD, Holder AA (1978) Somatic hybrid plants of potato and tomato regenerated from fused protoplasts. Carlsberg Res Commun 43:203–218

Menczel L, Galiba G, Nagy F, Maliga P (1982) Effect of radiation dosage on efficiency of chloroplast transfer by protoplast fusion in *Nicotiana*. Genetics 100:43–52

Miki BL, Reich TJ, Iyer VN (1986) Microinjection technology for cultured plant cells. Proc VI Int Congr Plant tissue and cell culture, Univ Minnesota, p 313

Nagata T (1984) Interaction of liposomes and protoplasts as a model system of protoplasts fusion. In: Beers RF Jr, Bassett EG (eds) Cell fusion, gene transfer and transformation. Raven, New York, pp 217–226

Nass MMK (1969) Uptake of isolated chloroplasts by mammalian cells. Science 165:1128–1131

Negrutiu I, Shillito R, Potrykus I, Biasini G, Sala F (1987) Hybrid genes in the analysis of transformation conditions. I. Setting up a simple method for direct gene transfer in plant protoplasts. Plant Mol Biol 8:363–373

Neuhaus G, Spangenberg G, Millelsten Sceid O, Schweiger HG (1987) Transgenic rapeseed plants obtained by the microinjection of DNA into microspore derived embryoids. Theor Appl Genet 75:30–36

Neumann E, Schaefer-Ridder M, Wang Y, Hofschneider PH (1982) Gene transfer into mouse lyoma cells by electroporation in high electric fields. EMBO J 1:841–845

Nomura K, Komamine A (1986) Embryogenesis from microinjected single cells in a carrot cell suspension culture. Plant Sci 44:53–58

O'Connell MA, Hanson MR (1986) Regeneration of somatic hybrid plants formed between *Lycopersicon esculentum* and *Solanum rickii*. Theor Appl Genet 72:243–249

Ohgawara D, Kobayashi S, Oghawara E, Uchimiya H, Ishii S (1985) Somatic hybrid plants obtained by protoplast fusion between *Citrus sinensis* and *Poncirus trifoliata*. Theor Appl Genet 71:1–4

Okada K, Nagata T, Takebe I (1986) Introduction of functional RNA into plant protoplasts by electroporation. Plant Cell Physiol 27:619–626

Ooms G, Bossen ME, Burrell M, Karp A (1986) Genetic manipulation of potato with *Agrobacterium rhizogenes*. Potato Res 29:367–379

Otsuki Y, Takebe I, Honda Y, Matsui C (1972) Ultrastructure of infection of tobacco mesophyll protoplasts by tobacco mosaic virus. Virology 49:188–194

Ou-Lee TM, Turgeon R, Wu R (1986) Expression of a foreign gene linked to either a plant-virus or a *Drosophilla* promoter, after electroporation of protoplasts of rice, wheat and sorghum. Proc Natl Acad Sci USA 83:6815–6819

Ow DW, Wood KV, DeLuca M, de Wet JR, Helinski DR, Howell SH (1986) Transient and stable expression of the firefly luciferase gene in plant cells and transgenic plants. Science 234:856–859

Pasternak TP, Melnikov PV, Gleba YY, Sytnik KM, Argentova VV, Kaliev AB, Andrianov VM, Piruzyan ES (1986) Genetic transformation of higher plant cells by DNA microinjections. Proc USSR Acad Sci Ser Biol 2:316

Paszkowski J, Shillito RD, Mandak SM, Hohn T, Hohn B, Potrykus I (1984) Direct gene transfer to plants. EMBO J 3:2717–2722

Payne J, Hamill JD, Robins RJ, Rhodes MJC (1987) Production of hyoscyamine by 'hairy root' cultures in *Datura stramonium*. Planta Med 53:474–478

Pelletier G (1989) Cybrids in oilseed Brassica crops. In: Bajaj YPS (ed) Biotechnology in agriculture and forestry, vol 10. Legumes and oilseed crops I. Springer, Berlin Heidelberg New York Tokyo (in press)

Pelletier C, Primand C, Vedel F, Chetrie P, Remy R, Rousselle P, Renard M (1983) Intergeneric cytoplasmic hybridization in *Cruciferae* by protoplast fusion. Mol Gen Genet 191:244–250

Perani L, Radke S, Wilke-Douglas M, Bossert M (1986) Gene transfer methods for crop improvement: Introduction of foreign DNA into plants. Physiol Plant 68:566–570

Petit P (1986) Flow cytofluorometry analysis of isolated plant and animal mitochondria. Biol Cell 58 (Suppl): 15a

Pinkel D (1981) On the possibility of automated scoring of pollen mutants. Environ Health Perspect 37:133–136

Pirrie A, Power JB (1986) The production of fertile, haploid somatic hybrid plant (*Nicotiana glutinosa* (N)+*N. tabacum* (2N) via gametic: somatic protoplast fusion. Theor Appl Genet 72:48–52

Potrykus I (1973) Transplantation of chloroplasts into protoplasts of *Petunia*. Z Pflanzenphysiol 70:364–366

Potrykus I, Hoffmann F (1973) Transplantations of nuclei into protoplasts of higher plants. Z Pflanzenphysiol 69:287–289

Potrykus I, Shillito RD, Saul MW, Paszkowski J (1985a) Direct gene transfer; State of the art and future potential. Plant Mol Biol Rep 3:117–128

Potrykus I, Saul MW, Petruska J, Paszkowski J, Shillito RD (1985b) Direct gene transfer to cells of a graminaceous monocot. Mol Gen Genet 199:183–188

Puite KJ, Roest S, Pijnacker LP (1986) Somatic hybrid plants after electrofusion of diploid *Solanum tuberosum* and *S. phureja*. Plant Cell Rep 5:262–265

Reich TJ, Iyer VN, Haffner M, Holbrook LA, Miki BL (1986a) The use of fluorescent dyes in the microinjection of alfalfa protoplasts. Can J Bot 64:1259–1267

Reich TJ, Iyer VN, Scobie B, Miki BL (1986b) A detailed procedure for the intranuclear microinjection of plant protoplasts. Can J Bot 64:1255–1258

Reich TJ, Iyer VN, Miki BL (1986c) Efficient transformation of alfalfa protoplasts by the intranuclear microinjection of Ti plasmids. Bio/Technology 4:1001–1003

Rhodes CA, Pierce DA, Mettler IJ, Mascarenhas D, Detmer JL (1988) Genetically transformed maize plants from protoplasts. Science 240:204–207

Rhodes MJC, Robins RJ, Hamill JD, Parr AJ, Walton NJ (1987) Secondary product formation using *Agrobacterium rhizogenes*-transformed "hairy root" cultures. IAPTC Newslett 53:2–15

Riggs CD, Bates GW (1986) Stable transformation of tobacco by electroporation: Evidence for plasmid concentration. Proc Natl Acad Sci USA 83:5602–5606

Robertson D, Palmer JD, Earle ED, Mutschler MA (1987) Analysis of organelle genomes in a somatic hybrid derived from cytoplasmic male-sterile *Brassica oleracea* and atrazine-resistant *B. campestris*. Theor Appl Genet 74:303–309

Rohme D, Fox H, Herrmann B, Frischauf AM, Edstrom JE, Mains P, Silver LM, Lehrach H (1984) Molecular clones of the mouse complex derived from microdissected metaphase chromosomes. Cell 35:783–788

Saxena PK, Fowke LC, King J (1985a) An efficient procedure for isolation of nuclei from plant protoplasts. Protoplasma 128:184–189

Saxena PK, Liu Y, Mii M, Fowke LC, King J (1985b) High nuclear yield from protoplasts of several plants. J Plant Physiol 121:193–197

Saxena PK, Mii M, Crosby WL, Fowke LC, King J (1986) Transplantation of isolated nuclei into plant protoplasts. Planta 168:29–35

Saxena PK, Liu Y, King J (1987) Nuclear transplantation into protoplasts: Optimal conditions for induction and determination of nuclear uptake. J Plant Physiol 128:451–460

Schell J St (1987) Transgenic plants as tools to study the molecular organization of plant genes. Science 237:1176–1183

Scheurich P, Zimmermann U, Schnabl H (1981) Electrically stimulated fusion of different plant cell protoplasts. Plant Physiol 67:849–853

Sederoff R, Stomp AM, Chilton WS, Moore LW (1986) Gene transfer into Loblolly pine by *Agrobacterium tumefaciens.* Bio/Technology 4:647–649

Shah DM, Horsch RB, Klee HJ, Kishore GM, Winter JA, Tumer NE, Hironaka CM, Sanders PR, Gasser CS, Aykent S, Siegel NR, Rogers SG, Fraley RT (1986) Engineering herbicide tolerance in transgenic plants. Science 233:478–481

Shahin EA, Sukhapinda K, Simpson RB, Spivey R (1986a) Transformation of cultivated tomato by a binary vector in *Agrobacterium rhizogenes*: transgenic plants with normal phenotypes harbor binary vector T-DNA, but no Ri-plasmid T-DNA. Theor Appl Genet 72:770–777

Shahin EA, Spielmann A, Sukhapinda K, Simpson RB, Yashar M (1986b) Transformation of cultivated alfalfa using disarmed *Agrobacterium tumefaciens.* Crop Sci 26:1235–1239

Shillito RD, Saul MW, Paszkowski J, Muller M, Potrykus I (1985) High efficiency direct gene transfer to plants. Bio/Technology 3:1099–1103

Shillito RD, Saul MW, Paszkowski J, Potrykus I (1986) Direct gene transfer to plants. IAPTC Newslett 48:5–15

Shvidkaya LG, Gleba YY (1974) Plating mesophyll protoplasts, isolated from different photosynthetic mutants of *Nicotiana tabacum* L. 3rd Int Congr Plant cell tissue culture, Abstr 195, Leicester Univ

Spangenberg G, Schweiger H-G (1986) Controlled electrofusion of different types of protoplasts and subprotoplasts including cell reconstitution in *Brassica napus* L. Eur J Cell Biol 41:51–56

Sree Ramulu K, Verhoeven HA, Gilissen LJW, Dijkhuis P, van der Valk HCPM (1987) Induction of micronuclei and their application for transfer of single, specific chromosomes in plant cells. Abstr 7th Int Protoplast Symp, Wageningen, p 33

Steinbiss HH, Stabel P (1983) Protoplast derived tobacco cells can survive capillary microinjection of the fluorescent dye Lucifer yellow. Protoplasma 116:223–227

Sundberg E, Glimelius K (1986) A method for production of interspecific hybrids within Brassiceae via somatic hybridization, using resynthesis of *Brassica napus* as a model. Plant Sci 43:155–162

Szabados L, Hadlaczky GY, Dudits D (1981) Uptake of isolated plant chromosomes by plant protoplasts. Planta 151:141–145

Tagu D, Bergounioux C, Perennes C, Brown S, Muller Ph, Gadal P (1987) Analysis by flow cytometry of the primary events of *Petunia hybrida* protoplast transformation. Plant Sci 51:215–223

Takebe I (1983) Protoplasts in plant virus research. Int Rev Cytol Suppl 16:89–111

Tan M-L MC (1987) Somatic hybridization and cybridization in Solanaceous species. Academisch Proefschrift. Vrije Univ te Amsterdam

Tao W, Wilkinson J, Stanbridge EJ, Berns MW (1987) Direct gene transfer into human cultured cells facilitated by laser micropuncture of the cell membrane. Proc Natl Acad Sci USA 84:4180–4184

Tempelaar MJ, Jones MGK (1985) Fusion characteristics of plant protoplasts in electric fields. Planta 165:205–216

Terada R, Kyozuka J, Nishibayashi S, Shimamoto K (1987) Plantlet regeneration from somatic hybrids of rice (*Oryza sativa* L.) and barnyard grass (*Echinochloa oryzicola* Vasing). Mol Gen Genet 210:39–43

Toriyama K, Hinata K, Kameya T (1987a) Production of somatic hybrid plants 'Brassicamoricandia', through protoplast fusion between *Moricandia arvensis* and *Brassica oleracea.* Plant Sci 48:123–128

Toriyama K, Kameya T, Hinata K (1987b) Selection of a universal hybridizer in *Sinapis turgida* Del. and regeneration of plantlets from somatic hybrids with *Brassica* species. Planta 170:308–313

Trulson AJ, Simpson RB, Shahin EA (1986) Transformation of cucumber (*Cucumis sativus* L.) plants with *Agrobacterium rhizogenes.* Theor Appl Genet 73:11–15

Uchimiya H, Fushimi T, Hashimoto H, Harada H, Syono K, Sugawara Y (1986) Expression of a foreign gene in callus derived from DNA-treated protoplasts of rice (*Oryza sativa* L.). Mol Gen Genet 204:204–207

Vardi A, Breiman A, Galun E (1987) *Citrus* cybrids: Production by donor recipient protoplast-fusion and verification by mitochondrial-DNA restriction profiles. Theor Appl Genet 75:51–58

Vlachova M, Metz BA, Schell J, de Bruijn FJ (1987) The tropical legume *Sesbania rostrata*: tissue culture, plant regeneration and infection with *Agrobacterium tumefaciens* and *rhizogenes* strains. Plant Sci 50:213–223

Wallin A, Glimelius K, Eriksson T (1978) Enucleation of plant protoplasts by cytochalasin B. Z Pflanzenphysiol 87: 330–340

Wallin A, Glimelius K, Eriksson T (1979) Formation of hybrid cells by transfer of nuclei via fusion of miniprotoplasts from cell lines of nitrate reductase deficient tobacco. Z Pflanzenphysiol 91:89–94

Watts JW, King JM (1984) A simple method for large scale electrofusion and culture of plant protoplasts. Biosci Rep 4:335–342

Watts JW, King JM, Stacey NJ (1987) Inoculation of protoplasts with viruses by electroporation. Virology 157:40–46

Webb KJ (1986) Transformation of forage legumes using *Agrobacterium tumefaciens*. Theor Appl Genet 72:53–58

Wenzel G, Schieder O, Przewozny T, Sopory SK, Melchers G (1979) Comparison of single cell culture-derived *Solanum tuberosum* L. plants and model for their application in breeding programs. Theor Appl Genet 55:49–55

White DWR, Greenwood D (1987) Transformation of the forage legume *Trifolium repens* L. using binary *Agrobacterium* vectors. Plant Mol Biol 8:461–469

Wilke-Douglas M, Perani L, Radke S, Bossert M (1986) The application of recombinant DNA technology toward crop improvement. Physiol Plant 68:560–565

Wright RL, Somers DA, McGraw RL (1986) Somatic hybridization between birdsfoot trefoil (*Lotus corniculatus*) and *L. coimbrensis*. Proc VIth Int Congr Plant tissue and cell culture, Univ Minnesota, p 86

Wullems GJ, Molendijk L, Ooms G, Schilperoort RA (1981) Differential expression of crown gall tumor markers in transformants obtained after in vitro *Agrobacterium tumefaciens*-induced transformation of cell wall regenerating protoplasts derived from *Nicotiana tabacum*. Proc Natl Acad Sci USA 78:4344–4348

Zachrisson A, Bornman CH (1986) Electromanipulation of plant protoplasts. Physiol Plant 67:507–516

Zaenen I, van Lerebeke N, Teuchy H, Montague M van, Schell J (1974) Supercoiled circular DNA in crown gall-inducing *Agrobacterium* strains. J Mol Biol 86:109–127

Zambryski P, Herrera-Estrella L, De Block M, Montague M van, Schell J (1984) The use of the Ti plasmid of *Agrobacterium* to study the transfer and expression of foreign DNA in plant cells: new vectors and methods. In: Setlow J, Hollaender A (eds) Genetic engineering, principles and methods. Plenum, New York, pp 253–278

Zelcer A, Aviv D, Galun E (1978) Interspecific transfer to cytoplasmic male sterility by fusion between protoplast of normal *Nicotiana sylvestris* and X-ray irradiated protoplasts of male sterile *N. tabacum*. Z Pflanzenphysiol 90:397–407

Zimmermann U, Scheurich P (1981) High frequency fusion of plant protoplasts by electric fields. Planta 151:26–32

# II Genetic Manipulation: Generative Versus Somatic

J. SYBENGA[1]

## 1 Introduction

The objective of plant breeding is the production of a genotype or an adapted combination of genotypes that meet specific performance requirements. The means to realize this objective are the introduction and manipulation of genetic variation. By far the large majority of all plant breeding still depends on the generative cycle, although in some instances effective use has been made of the somatic phase at the plant level for the induction of variation by mutagenesis and polyploidization. The rapid and large-scale development of molecular and in-vitro cell techniques (see Bajaj 1986) is based primarily on the somatic phase, and application to the generative cycle lags far behind. It appears as if this will not change for a considerable period to come. It is not reasonable to expect that the new somatic approaches will ever be able to replace generative approaches in all or even most phases of plant breeding (Borlaug 1983), but in several instances it will soon be, or is already, necessary to make a choice. The increasing ease with which molecular and cell manipulations techniques can be applied will reduce the barrier against their application. Whereas a few years ago, shortage of capital and progressiveness were thought to be the main obstacles to the introduction of modern somatic techniques in practical breeding institutions, this is clearly changing. Some of these obstacles will remain, but the emphasis in the choice between molecular and cell biological somatic versus generative approaches will be placed more and more on the requirements of the breeding program and the real merits of the techniques. This choice is not simple, because the development of the techniques, as well as the relation between costs and assets, is difficult to predict. Also, certain developments of generative (meiotic) manipulation promise results that cannot yet fully be appreciated (Sybenga 1983). Neither approach offers simple and cheap solutions for complex problems. Sophisticated generative techniques are conceptually more difficult, but technically less demanding. In total amount of work required for a positive end result, somatic and generative approaches may be comparable. It pays to make a careful analysis before the choice is made in any particular case. The following is an attempt to provide a preliminary basis for such a choice. To some extent it is a revision of an earlier report (Sybenga 1983).

[1] Department of Genetics, Agricultural University, Dreijenlaan 2, 6703 HA Wageningen, The Netherlands

Biotechnology in Agriculture and Forestry, Vol. 9
Plant Protoplasts and Genetic Engineering II (ed. by Y.P.S. Bajaj)

## 2 The Structure of a Plant Breeding Program

In a plant breeding program different steps or phases can be distinguished that are carried out in sequence (Table 1). The outcome of some of the phases may require a previous step to be repeated, either without change or after adjustment. Although the plant breeder is not always aware of it, an established breeding program tends to have several such "feedback lines". A trivial example of feedback is the adjustment of the objectives when the material appears to have properties not anticipated at the start. Another is the simple necessity to collect new material when the necessary genotypes cannot be constructed from the available stocks. There are, of course, much more subtle examples.

The different phases of the program do not have the same character. In some, only decisions are taken, in others systems are designed, in others again operations carried out. They also have different requirements. For design and decision primarily information is required, either of a general or of a more specific character, for instance on the genetic system of the species at hand, or selection techniques for such systems, etc. Testing of the starting material, or during selection, or of the end-result requires specific techniques. During several stages of a plant breeding program, the availability of information is a necessity, and the collection of this information, especially specific information on the material, is an important aspect of plant breeding.

In this chapter, the emphasis will be on the first section of Table 1 and more specifically on the second item: choice of approach in the construction of the desired genotype and reproductive system. This implies a discussion of the further steps in the breeding program, but not in detail. Selection techniques or the collec-

**Table 1.** The phases of a plant breeding program (cf. Sybenga 1983)

| I | II | III | IV |
|---|---|---|---|
| a) Formulation of objectives | a) Collection of material | a) Selection | a) Propagation/ maintenance |
| b) Choice of approach in construction of genotype and reproductive system | b) Testing and preselection of material | b) Testing | |
| c) Program design | c) Induction of additional genetic variation (mutation, transformation) | | |
| | d) Construction of special reproductive system | | |
| | e) Combination (hybridization, etc.) and recombination | | |

tion of specific information on the material, for instance, will not be discussed at any length.

A distinction is made between manipulation of the genotype and manipulation of the reproductive system, although it should not be forgotten that in most instances the reproductive system is directly controlled by the genotype. Production performance and reproductive performance are sufficiently different goals to justify a separate treatment. In relatively simple programs this is not necessary: the reproductive system at hand is left unchanged and the selection methods are adjusted to this system. In more complicated programs a change in the reproductive system may be considered. Hybrid varieties are an example, but autotetraploidy also automatically implies a genetically different reproductive system. It is important to recognize such implications.

## 3 Construction and Isolation of a Desired Genotype

### 3.1 Selection After (Large-Scale) Segregation: Generative

In the simplest programs, the base material can consist of cultivar collections with at least some heterogeneity, but in most programs heterozygous material is created by hybridization. From this material a segregating progeny is obtained on which selection can be practiced. Generative segregation may, in principle, be preceded by somatic hybridization, but in practice this is not yet a real alternative: the haploid cultures necessary to make a diploid hybrid are usually difficult to maintain and to work with, and protoplast fusion and regeneration or even in-vitro culture in general are not yet standard techniques for many important crop plants. In addition, in-vitro culture, especially when including a callus phase, is often genetically unstable. At times the resulting somaclonal variation may be exactly the purpose of including a callus phase, but this is a different subject, to be discussed later. For most purposes the hybrid is easily made by classical methods, if necessary supplemented by in-vitro culture of embryos (embryo rescue).

Meiotic recombination resulting in segregation, and selection in the segregating populations may be expected to remain the basis of most plant breeding programs (Borlaug 1983). This is mainly between chromosome recombination not involving crossing-over and crossing-over between genes wide apart in the same chromosomes. Within-chromosome recombination resulting from crossing-over between more closely linked genes does not receive much attention from plant breeders, mainly because of lack of time (Stam 1977). Effective recombination requires heterozygosity over several generations and this would cause an unacceptable delay in the release of marketable varieties. Extensive use of recombination can probably best be made when composing basic stocks as parental material for future breeding programs.

The possibilities of manipulating meiotic recombination have not been fully exploited or even studied. Within a species, recombination is regulated by three systems: (1) The number of chromosomes (the more chromosomes, the more recombination), (2) overall level of within-chromosome exchange, (3) distribution of within chromosome exchange. Grouping of genes in linkage groups, but also the localization of crossing-over and interference within chromosomes determine the distribution of recombination.

The number of chromosomes can be altered without increase in genetic material by centric fusion and centric split, and this results in changes in the level of recombination. Raising the overall level of crossing-over within chromosomes by external agents (Ihrke and Kronstad 1975) or by using special genotypes (Rose and Baillie 1979) is possible only to a very limited extent. Such approaches are not yet really effective, and selection from a larger population often has the same result.

Lifting a crossing-over localization system, resulting in random instead of localized crossing over, can have more effect. Several instances are known in which chiasma localization patterns have been thoroughly altered, usually by genetic means: hybrids between different but related species, inbreeding, mutations (Jones 1967, 1974). These possibilities have not even been explored satisfactorily. One disadvantage usually is the randomness of the effect, resulting in undesired, in addition to the wanted recombination. Further, the genes which enabled the recombination may not be desired in the final population, and may have to be removed again. More promising but even less explored are external agents disturbing normal restriction patterns. Temperature extremes may be a candidate, but these usually merely reduce recombination.

In special cases it may be possible to break linkages by splitting a chromosome into two or by the induction of translocations, by irradiation or by inserting an in-vitro callus phase. The process is so haphazard, that it is difficult to apply. The opposite, keeping specific genes together, is possible by combining two acrocentric chromosomes into one metacentric or by reciprocal translocations, but this again is only feasible in very special cases. Of some interest is the restriction of recombination between genes by placing a translocation break in between them. The application is limited to special cases, for instance in the use of certain systems applying male sterility in making hybrid varieties (this will be considered later in more detail). It is not really an aspect of using recombination and segregation in the production of a desired genotype.

Altering recombination in specific chromosomal segments is known to be possible in *Drosophila* by the introduction of certain inversions (Green and Green 1949) and in lower eukaryotes by selection of specific recombination genes (Simchen and Stamberg 1969). These possibilities again have been explored in plants to only a very limited extent, and it may be questioned if they are simpler to apply than methods of replacing one gene or allele by another by techniques to be described below.

There is one interesting application of an in-vitro somatic technique for exploiting generative segregation: raising plants from cells in the haplophase, usually young microspores. Although quite successful for isolating homozygous diploids (for which the chromosome number has to be doubled) from a heterozygous parent, where segregation plays a certain role, the technique has not been developed sufficiently for the recovery of scare recombinants in large-scale segregating populations, especially not in monocotyledons. For making homozygous lines directly from a segregating population, spontaneous haploids have been used for a long time (Chase 1952) and effective methods of selecting haploids have been developed (Chase 1952; Hermsen and Verdenius 1973). Their frequency is genetically determined (Kasha and Séguin-Swartz 1983), occasionally rather simple (haploid-inducing genes), and can be greatly increased by artificial means, especially by pollination with pollen of special donor species. Haploids produced by the "Bulbosum" techniques are based on a special form of somatic segregation and will be discussed below.

At the plant level, *somatic segregation* is usually not considered of much practical importance. Of course, spontaneous and induced mutations in a differentiated tissue of a plant will appear as somatic segregants (chimerism) but this is relevant primarily in another context. Similarly, somatic segregation of spontaneous genetic variants in callus cultures (somaclonal variation) will not be considered now. Somatic segregation, possibly resulting from somatic crossing-over, was reported by Jones in 1939, and continues to be observed, but its use is limited. For uncovering hidden recessive mutations, induction of mutation of the dominant allele is far better (Broertjes and van Harten 1988). Potentially more interesting is somatic segregation as first described in detail by Franzke and Ross (Ross 1965) for one special genotype of sorghum. The application of colchicine to seedlings of the variety "Experimental III" led to the segregation of homozygous diploid sectors on heterozygotes, by chromosome doubling after somatic reduction. Colchicine and other microtubule inhibitors like Benomyl can induce "reductional groupings" in most plants, but only in very few genotypes does this lead to balanced genomes (Sybenga 1955). Chloramphenicol (Yoshida and Yamaguchi 1973) is also able to effect chromosome reduction to some degree, but parafluorophenylanaline, known to induce haploidization in diploid strains of *Aspergillus nidulans*, does not appear to be effective in plants, including in-vitro callus and cell cultures, its main effect being the occasional induction of aneuploidy (Nitzsche 1980), like CIPC (Roth and Lark 1984). With an increased interest in spindle behavior and genome compartmentalization in the nucleus (Bennett 1982; Avivi et al. 1982), even though not all claims of systematical arrangements can be made hard, induced somatic segregation may become closer to application. Application to in-vitro cultures, where large-scale treatment and selection is in principle possible, may be interesting, but the great genetic instability may be a drawback when specific (homozygous) segregants are the aim.

An extreme type of somatic segregation is induced by amiprophos-methyl (Ramulu et al. 1988). The chromosomes are scattered throughout the cell and telophase condensation results in numerous micronuclei, many containing only a single chromosome. These can be isolated, sorted by flow cytometry and potentially used for transfer to protoplasts after irradiation or as such.

A special and practically applicable form of somatic segregation at the plant level occurs in certain hybrids of which the combination of *Hordeum vulgare* (barley) with *Hordeum bulbosum* is the most successful. At a critical stage of embryonic development, the genome of the latter is eliminated, resulting in haploid barley. This can be doubled to produce homozygous diploids in one generation (Lange 1971; Kasha and Reinbergs 1979). It is an effective way of recovering meiotic segregants and may occasionally lead directly to a variety. Application to wheat is being developed, but as yet somewhat less successful (Barclay 1975).

Quite a different form of somatic segregation is the segregation of cell organelles (plastids, mitochondria) after somatic hybridization (and mutagenesis, or transformation of the organelles). Usually, either by drift or natural selection, perhaps even actively, all except one genotype of the organelle are eliminated. In the meantime, especially with mitochondria but perhaps also with plastids, some form of fusion or other contact may lead to recombination between the circular DNA main chains, or to the exchange of plasmid-like DNA particles. As yet, it does not seem to be possible to direct this form of segregation, but the importance is obvious, in view of the critical role of mitochondria in metabolism. Emphasis, however, has till now been on mitochondrial plasmids that regulate cytoplasmic male sterility, important for hybrid seed production.

## 3.2 Introduction of Specific New Genes

### 3.2.1 Mutation

Almost without exception mutations are induced in the somatic phase, in the plant or in vitro. This results in chimerism and somatic segregation, sometimes followed by (diplontic) selection for or against mutated sectors.

(Co)dominant mutations or mutations of the dominant alleles in heterozygotes can be selected directly in the plant or the culture, but they are an exception. In vegetatively reproduced, ornamental plant breeding, however, they have had considerable application (Broertjes and van Harten 1988). Recessive mutations require either segregation (in the generative phase), or special techniques for recovery. Recovery of recessive mutations by induced loss of the homologous chromosome by CICP has been shown to be possible, but practical application is still far removed (Roth and Lark 1984). Although mutations are not induced fully at random and each type of agent has its own spectrum, mutagenesis is not a directed process and usually desired mutations are accompanied by undesired ones. The inability to strike a satisfactory balance between the necessity of extensive selection in large populations when the dose applied is low, and the negative effects of higher doses, is a major reason for the limited practical use of mutation induction. Good solutions for these problems are available (Dellaert 1979) for plant level mutagenesis, but seem to be neglected by most breeders. Mutation breeding of seed plants deserves wider application than it has now.

In-vitro mutagenesis can be quite promising provided good selection criteria are available. Low doses of mutagen, or none at all when somaclonal variation is sufficient, can be applied because low mutation frequency is compensated by effective selection in large cell populations. Good selection procedures, however, are still rare and other problems make the method less simple than sometimes expected. One bottle-neck is the protection of unmutated cells against the selective agent by surrounding cells in cell aggregates. Such cells may then make the impression of being resistant, while in reality they are not. Another problem is the often excessively high level of spontaneous mutation in the callus phase, which introduces too large a mutational load and for which somehow a solution must be found in all in-vitro breeding programs. Short callus phases and immediate formation of embryoids from protoplasts or isolated cells may be a good solution, but is not yet possible in the majority of plant species and genotypes. Then, however, external mutagenizing agents are necessary. Practical use of in-vitro instability as a mutagenizing agent (somaclonal variation, Larkin and Scowcroft 1981) has been applied on numerous occasions (Shepard et al. 1980; Shepard 1982; and many more reports). Gene mutations, chromosome structural, and numerical aberrations have all been observed. Although variants have been found, especially in the potato, that surpass the parent variety in performance in limited ecological situations, there are reasons not to be very optimistic. Especially in polyploids like the potato, wheat, and several other crop plants, gross chromosomal and gene aberrations will be tolerated by the buffering capacity of the polyploid, but have a (limited) semi-dominant expression. When this is the selected type, it may be expected to be accompanied by unfavorable side effects on productivity characteristics. In diploids such aberrations would be readily eliminated by natural selection. There is a parallel in the comparative mutation experiments on diploid and tetraploid vegetatively reproduced ornamentals, where, contrary to expectation, the observed mutation frequency was much higher in the tetraploid, but most mutations had serious drawbacks (Broertjes pers. commun.). Here too, the explanation could be greater tolerance of the polyploid against gross aberrations with dominant expression but undesired properties. Therefore, although in some instances positive properties are acquired, in highly bred cultivars the risk of exclusive negative results is great.

The application of somaclonal variation continues to receive attention. Almost any type of aberration can be encountered, and sometimes with high frequency. When endopolyploid cells present in the original explants are stimulated to proliferate, polyploid cultures result (see, for instance, Pijnacker et al. 1986). The same occurs when the spindle fails during mitosis in the callus. Is this failure only partial, then aneuploidy may result. Irregularities during DNA synthesis, possibly accompanied by ineffectiveness of the repair system, may result in chromosomal rearrangements, including translocations, inversions, deficiencies, duplications, and unstable chromosomal constructions which interfere with mitosis. Similarly, gene mutations may occur. Their frequency can be considerable, especially in some genotypes, and they may be based on single base substitutions or larger changes. Dennis et al. (1987) report on a somaclonal mutation of the maize alcohol dehydrogenase-1 gene, where a single base substitution resulted in an amino acid replacement changing the electrophoretic properties. Seed storage proteins were mutated in wheat in several instances (Larkin et al., 1984; Cooper et al. 1986). Selection in vitro for specific spontaneous or induced dominant mutations for resistance against biologically active substances and conditions has given positive and very important results: toxins produced by pathogens (Gengenbach et al. 1977; Bajaj 1981 and several later reports); stress conditions (heat, cold, salt, heavy metals: Chaleff and Parsons 1978); fungicides, herbicides (Weller et al. 1987), etc. Disease resistance obtained by this method has not yet led to varieties with field resistance, but herbicide (imidazole)-resistant maize lines will soon become available.

The recovery of exclusively dominant mutations when diploid or polyploid cultures are used is in itself an interesting aspect of these methods (Chaleff 1983), and possible because of the very effective selection procedures (provided these are available). After very high mutation rates, as are sometimes found during prolonged callus phases, codominant mutations (such as isoenzymes and seed protein variants) may appear with sufficient frequency to be isolated even without special selection methods. The main problem with their application is the high frequency of less desired, not readily recognized, but slightly or even seriously negative mutations induced simultaneously. The frequency of spontaneous (somaclonal) mutations increases considerably with the period of in-vitro culture. Often a drastic increase is observed after 3 months of culture. Culture conditions and composition of the medium may occasionally have a strong effect on genetic stability of the culture. In our experience with Haplopappus cell aggregate cultures an overdose of EDTA resulted in strongly elevated frequencies of chromosome rearangements.

A review of in-vitro mutagenesis with emphasis on animal cell cultures, but also very relevant for plant material, has been given by Smith (1985).

Direct in-vitro selection of recessive mutants is not possible in (eu)diploid cultures when the starting line is homozygous diploid. Then haploid cultures are in principle far superior, but these are difficult to maintain haploid or even to isolate and to grow. An interesting exception is the recovery of pure amylopectin mutants in haploid potato cultures. After irradiation of leaves of a stable monohaploid clone, adventitious plantlets were produced in vitro, which, even when doubled during the process, had mutated during the haploid phase. After regeneration, small tubers were formed which could be tested individually on starch composition. Thousands of tests were necessary to recover the mutants (Hovenkamp-Hermelink et al. 1987). This is an example where somaclonal variation may not be expected to lead to the desired mutant, because no callus phase was included.

It is clear that for mutation induction both plant level mutagenesis and somaclonal variation have their potential. The advantage of the high mutation frequencies of the latter is usually only apparent because direct application is problematic as a result of deleterious side effects. The experience in this respect of mutation breeders at the plant level should be a warning to in-vitro workers. Recovery of very special, including dominant mutants by large-scale in-vitro selection is the major advantage of in-vitro mutagenesis, provided the selection methods are available.

### 3.2.2 Introduction into the Genome from Outside

Entirely new genes or alleles that cannot be obtained by mutation of an existing gene can be introduced into the genome from outside. Depending on dominance and epistasis of genes already present, which are coding for functions incompatible with that of the new gene or allele, such original genes may have to be removed by deletion, inactivated by mutation or replaced by homologous recombination. Since the generative cycle permits subsequent rounds of recombination, it has been widely used to replace undesired genes by repeated backcrosses. When the allele to be introduced is recessive, intermittent generations of selfing have to be inserted to ensure that the allele is carried over. Closely linked neutral molecular markers such as restriction fragment length polymorphisms (RFLP) are very useful for monitoring the fate of recessive alleles in backcrossing programs without the necessity of intermittent selfed generations, but the development of a generally applicable system requires considerable input. In maize (Evola et al. 1986), tomato, and tobacco, RFLP banks have been constructed, and for other species they are being developed.

Even after extensive recombination it is practically impossible to transfer a gene without carrying along other, less readily detected other genes (hitch-hiking). When the transfer is between related forms of the same species, this may be acceptable. More problems arise with the transfer of genes between species. It may then be necessary to depart from the original objective of introducing only a single gene, and accept the necessity to adjust the genotype to the effects of the incorporated alien genes. One may then even go so far as to decide to change the original objective drastically and seek an alternative in the combination of the two species into an allopolyploid. This may be a legitimate choice when the original objectives were not very restricted. Namai et al. (1980) demonstrated that for the development of a fodder crop from *Brassica campestris* and *B. oleracea* both approaches would work.

When straight recombination cannot be directly realized, use may be made of one of the several intricate programs for transferring alien genes into a crop species by a special adaptation of the generative cycle. These have an old tradition, and have been used especially for transfer of dominant disease resistance genes. Because steps involving monosomics, additions, and substitutions of entire chromosomes are often necessary, allopolyploids have been used almost exclusively, primarily wheat (Feldman 1988).

As early as 1956, Sears published a report on the transfer of dominant leaf rust resistance from diploid *Aegilops umbellulata* to allohexaploid bread wheat via the addition of the *Aegilops* chromosome carrying the gene. The segment with the gene was transferred by translocation to an arbitrary wheat chromosome, using X-irradation. Part of this wheat chromosome is lost unless a very improbable interstitial translocation is induced, and in any case a segment of an alien chromosome is added. It appears that the negative effects of this cotransfer can be compensated by selection, and the method has been successfully applied on several occasions. The low frequency of the proper translocation requires thousands of preselected (resistant) plants to be checked for the transfer. More elegant, although in some respects more laborious, is the method developed by Riley et al. (1968) for the transfer by homeologous recombination of stripe rust resistance from *A. comosa* to wheat. It requires the introduction of genes (in this case from *A. speltoides*) for lifting the natural block against homeologous recombination. This results in a rather serious break up of the wheat genome, which has to be reconstructed later by backcrossing. The approach has been applied with considerable success, especially in wheat breeding (see, e.g., Koebner and Shepherd 1986, 1987). A third method has been less widely used, but deserves wider application. It involves the spontaneous interaction between the centromeres of breaking univalent chromosomes at meiosis (Lukaszewski and Gustafson 1983). When a monosomic of the recipient has a homologous monosomic addition of the alien chromosome carrying the desired gene, the two univalents will interchange (homologous) arms with a relatively high frequency. The process has occurred several times without being recognized in the progeny of wheat × triticale crosses, where the resulting wheat lines contained rye chromosomes replacing one of the wheat chromosomes, and in several instances a centromere translocation instead of a full substitution. Sometimes the negative effects of the large alien segment cannot be compensated by selection, for instance, when an undesired multicopy seed protein gene is present in the same segment as a desired resistance gene, like in 1RS translocated into wheat. Then crossing-over after induced allosyndesis in heterozygotes can remove the undesired gene (Koebner and Shepherd 1986). Induced mutations can be used to remove undesired single copy genes. With repeated allosyndetic recombination leading to a gradual decrease of the size of the segment transferred, one important disadvantage of transferring large segments is overcome. When entire arms or even large segments are carried over from an alien species, it is practically impossible to do so more than a few times without entirely upsetting the recipient's genetic balance. When the alien segment can be largely replaced by the original chromosome segment, the unbalance is greatly reduced.

For the construction of the starting point, an addition line, traditionally a hybrid, is made between the donor and the recipient species, followed by chromosome doubling and a series of backcrosses in which the chromosome to be added is maintained. When technically possible, part of the procedure can be replaced by fusion of protoplasts of somatic (diploid) tissue of the recipient and (haploid) tetrad cells of the donor (Pirrie and Power 1986). The resulting triploid is at least partly fertile and upon selfing or backcrossing yields addition lines. With the gradual increase in success with protoplast regeneration, also of graminaceous species, this technique may become of some importance. A somewhat related approach that appears to be promising is *asymmetric fusion* between one normal (recipient) protoplast and one (donor) protoplast that has received a high dose of radiation of hundreds of Grays. This destroys the nucleus from which small chromosome segments and DNA are released that can be integrated into the recipient nucleus by molecular transformation processes (Gupta et al. 1984; Dudits et al. 1987). With lower doses the normal nuclear functions are impaired and entire chromosomes or at least large segments with a centromere are transferred: chromosome additions result. When it is possible to recognize the added chromosomes either by the presence of a specific gene product or by chromosome morphology, a better controlled transfer of the desired gene can be realized. It could be possible with generative genetic means or perhaps by somatic crossing-over. Mass transfer of bulk DNA after complete nuclear disintegration although apparently possible (Dudits et al. 1987) has its drawbacks. With the increasing success of protoplast fusion and regeneration, including developments such as electrofusion (see, e.g., Bates et al. 1987a, b), large numbers of asymmetrical cell hybrids can be isolated, multiplied and tested.

The interest in the transfer of chromosome segments between species has decreased somewhat because primarily dominant genes for disease resistance were transferred, and this resistance appeared to be broken rapidly by mutant pathogen types. Centromere translocations are more specific than the other methods, and with the use of C-banding techniques (also applicable in the other cases) the chromosome segments can be recognized even when no gene expression is possible in the heterozygous intermediate. Then recessive genes can be transferred also, and, which is important for disease resistance, different independent resistance genes can be transferred in the same cycle. This considerably reduces the probability of resistance breakdown. The use of neutral selection markers, such as RFLP (Evola et al. 1986), will in principle also make it possible to carry out combined transfers (by any method) of different genes which cannot be distinguished or even recognized. Especially in species where transformation by isolated DNA (see below), with or without vector, is difficult, either because in-vitro culture is a bottleneck or the isolation or transfer of a gene, the use of RFLP may become very useful. The development of RFLP banks, however, requires a large investment and may be problematic in species with large amounts of evenly distributed repetitive DNA.

*Molecular transformation* when making use of in-vitro methods suffers from serious drawbacks: frequent technical failure at some important step in the process and genetic instability of the callus phase. The primary aim is to introduce alien DNA into the recipient genome (nuclear, mitochondrial, plastid) by some form of transformation. Several ways have been explored to carry the DNA to the target site: diffusion of naked DNA (modified to enhance incorporation, or unchanged), liposome encapsulation, bacterial plasmids (*Agrobacterium*) and phages with their derivatives, viral vectors. Of the latter only the cauliflower mosaic virus is a DNA virus and a potential candidate, but in spite of many efforts, not much success has been attained. Isolated bulk DNA applied to seeds was reported to be effective for transformation long ago, but more recently these positive results could not be repeated. According to Ohta (1986), De Wet et al. (1986), and others, bulk DNA mixed with pollen is highly effective on maize, especially for endosperm characters, but other reports are less positive. The claim that irradiation of pollen with high doses of ionizing radiation liberates DNA which can transform the egg subsequently (Pandey 1978) is apparently not correct, and the results must be explained by other mechanisms. Yet a fully satisfactory explanation of all observations on the effects of pollen irradiation has not been given. Transformation with cloned specific DNA without vector, but provided with special end segments, as is effective in animals (Palmiter et al. 1982), is also possible in plants. The DNA has to be introduced into protoplasts by fusion using polyethylene glycol with calcium ions, or by electroporation (Shillito et al. 1985) or by microinjection (Crossway et al. 1986). These methods can in principle be used for species which are nonhosts to *Agrobacterium*, but even microinjection is still possible only with protoplasts and many species can still not be sufficiently manipulated in vitro. For host species to *Agrobacterium*, the modified Ti plasmids from which the oncogenic genes have been removed but the sequences necessary for integration are still present are the most effective vector. The plasmid can be introduced in one piece or as binary vector, one component with the selected gene and the other with sequences necessary for the plasmid to function (Hoekema et al. 1983). Complete genes, reconstituted or native as desired, provided with the proper replication initiation sequences, promotors, etc. have been transferred. Protoplasts are not required when the plasmid is administered in the bacterium, and this is very convenient for many crop species. Successful transformants are now grown in the field. Especially herbicide resistance genes, that are relatively easily isolated and regulated, readily selected in vitro and commercially interesting, are already available in several crop plants after transformation. For monocotyledons, where only few genotypes permit plant regeneration from protoplasts, and where transformation of intact cells perhaps has to await effective microinjection methods, less success has been reported. Sooner or later for these, too, effective methods of transformation may be developed. Already, De la Peña et al. (1987) have shown that naked DNA injected into rye stem bases at the proper stage can occasionally be inserted into the host DNA in the sporogenous tissues.

With sufficiently high frequencies of transformation monitoring with the use of molecular techniques will be able to replace selection, either by analysis of bulk DNA or even by in-situ hybridization even when only few copies of the gene have been incorporated tandemly (Mouras et al. 1987). Recessive genes, or genes not expressed in vitro can then also be transferred effectively.

In addition to the direct use in a plant breeding program, transformation has a more theoretical use in the study of the effects of specific DNA sequences in the plant genome. These may be artificial, or extracted from specific sites in chromosomes with the use of transposable elements and possibly involved in quantitative characters.

More detailed reviews of the present state of the art of transformation of plant genomes (Gheysen et al. 1985; Bhatia et al. 1986), in-vitro mutagenesis (Maliga 1984; Smith 1985), somaclonal variation (Karp and Bright 1985), cybrids (Galun and Aviv 1983), asymmetric fusion (Gupta et al. 1984), etc. have been given elsewhere, and the subject will be considered in other chapters of this Volume.

*In conclusion* of this section: although the great success of transformation with isolated and cloned DNA may sometimes give the impression that there are no limits to its application, in several instances it is still necessary or even preferable to apply other techniques. This is most obvious for those cases where simple hybridization and backcrossing are available for transfer of genes from alien but sufficiently related species. In other instances too, however, transformation may not appear to be the preferred approach. The specification of the desired trait in terms of genes and even more so of DNA, including the proper promotors, enhancers, etc. is usually still impossible. In many species, especially most monocotyledons, the in-vitro techniques required cannot be successfully applied, or only in special genotypes, especially because of insufficient regeneration of plants from protoplasts. Also, the insertion process is often one of transposition rather than homologous recombination, and dominant and epistatic genes may have to be removed. In addition, multiple copies tend to be inserted when a concatenation process precedes integration; this may be desired in some cases but often it is not. Stability of the integrated segment, finally, is not guaranteed (Peerbolte 1986), and may sometimes be disappointing. Then, the generative cytogenetic approaches may be a better way to the goal, although they are neither technically nor conceptually simple. Especially in wheat, important practical results have been obtained (Feldman 1988). The danger exists that when the interest in these techniques decreases to below a critical level, insufficient expertise will be left over to carry out programs at the scale sufficient for success.

## 3.3 Gene Dose Effects

### 3.3.1 Multiple Copies

In eukaryotes several genes or gene clusters occur in multiple copies, usually in tandem (histone genes, rDNA, seed protein genes), or spread over the genome (some 5S-rRNA- and tRNA-genes). This is necessary for the production of large amounts of the primary gene product, which in these cases is also the end product. The numerous nontranscribed multiple copies known as repetitive DNA are not of much interest in the present context. Multiplication of normally single genes has been observed in cell cultures which have acquired resistance to an applied toxin. The best-analyzed case of spontaneous large-scale gene amplification in the somatic phase is that of methotrexate resistance in animal cell lines (Cowell 1982). The gene for dehydrofoliase, an alternative target for methotrexate, amplifies continuously under a methotrexate regime, as long as the selective pressure continues. It can do so in tandem, resulting in relatively stable megachromosomes, or in the form of small free fragments (double minutes) which disappear again gradually. The region amplified involves the large gene with introns and large flanking sequences, including a regulator and an autonomous origin of replication. A comparable example for plants is glyphosate (Roundup) tolerance in *Petunia* cell lines where elevated levels of EPSP synthase activity resulted from a 20-fold increase in gene copy number, established after stepwise selection on survival in increasing concentrations of glyphosate. The amplified gene could be transferred by transformation using Ti plasmids into leaf disk cells, from which resistant plants were regenerated via a callus phase (Shah et al. 1986). Toxin resistance in plant cell cultures and their regenerants may more often be based on gene amplification. Amplification leading to megachromosomes has earlier been observed without selective agent in *Nicotiana* hybrids (Gerstel and Burns 1966). In many plant species amplification of heterochromatin is quite common during evolution, and occasionally in the experiment large blocks may be seen to disappear (Gustafson et al. 1983). There is an obvious need for detailed studies of the amplification process and its potential for manipulation in plants.

Multiple integration in concatenated gene blocks or spread over the genome by nonhomologous transformation is also a form of gene amplification. It is not readily manipulated, as the requirements of regulation are complex, and multiple transformation is usually considered a problem rather than an asset, but applications in plant breeding may be found for herbicide, disease, pest or stress resistance, or for the production of specific substances. Chance success will probably precede a thoroughly planned result. Amplification through the generative cycle is in principle possible by unequal crossing-over, and this has been suggested as one of the causes of heterochromatin amplification, but manipulation seems hardly possible at present.

### 3.3.2 Duplication of Small Chromosome Segments

This has been realized in the experiment by different methods. Usually the inevitable excess of nondesired extra material is a disadvantage, but selection of the genetic background compensating for their negative effects is well possible (Hagberg pers. commun. 1985). Duplications have been important in evolution in most eukaryotes and many instances are known in which duplicated genes have acquired new properties (Ohno 1970). Even without such changes they may have a positive dosage effect. In barley, the duplication of the $\alpha$-amylase locus has been realized with the purpose of improving the malting properties (Hagberg 1965) with limited but gradually improving success. The simplest way of producing a small duplication is by symmetric interchange between homologous chromosomes in the somatic phase. The process, however, is almost entirely random and the chance of success small. Radiation, chemical mutagens, or destabilization in a callus phase may be used. Aberrant meiotic segregation of special translocations or the combination of two translocations gives the opportunity of more directed induction of duplications, provided enough of such rearrangements are available. This is so for maize, barley, and rye, for instance, where several directed duplications have been produced (Patterson 1973; Hagberg 1965; Sybenga and Verhaar 1980).

Duplication of specific chromosome segments, provided possible negative effects can be compensated, can also be used to construct special stable heterozygous allelic combinations, of potential interest in self-pollinating species. Their role for plant breeding has not been explored, but there are indications that in nature such duplications exist (Ramanna pers. commun.).

For the production of segment duplications molecular techniques do not appear to play a role and cellular approaches are not necessarily better than whole plant techniques.

### 3.3.3 Addition of Entire Chromosomes (Polysomy)

Extra chromosomes can have any form, shape, and composition as long as they are tolerated and not mechanically inefficient. Besides functioning as large duplications, they may serve as an intermediate for the transfer of specific genes (see above) or as the permanent vehicle of specific single or groups of genes, when they are otherwise inert or empty. There are several ways to produce trisomics (from triploids, translocations, etc.) but these are not stable in normal meiosis unless provided with a balancing system which works at the expense of reproductive efficiency. Generally, when a separate chromosome is large enough to function at meiosis, it is too large as a duplication, but there are a few exceptions. (1) Compensating trisomics in which one normal homolog is replaced by two partly homologous chromosomes, each carrying a small additional segment. These are readily produced by combining the proper translocations (Khush 1973; De Vries 1985). (2) A-B-chromosome translocations in which any small segment of a normal chromosome can be duplicated by transfer to a B-chromosome (Roman 1947; Beckett 1982; Birchler 1983). The desired effect can again be a dose effect or an interallelic interaction. Both of these two types of extra chromosomes have been proposed and applied for special forms of chromosome manipulation, but the application as a duplication has not had much attention. Especially the A-B-chromosome translocations carry more promise for chromosome manipulation than is usually realized.

Somatic transfer of chromosomes, either for use as a duplication or as an intermediate in gene transfers is in principle possible, but has not been as successful in plants as it has been in animals. Isolation is possible as condensed mitotic chromosomes from synchronized tissues or cultures, but not easy. Chromosomes can be sorted in a flowcytometer (De Laat and Blaas 1984), and other sorting techniques may become available. Incorporation in the cells is not yet possible by simple $Ca^{2+}$ coprecipitation on cell monolayers, as is routinely done in animal material. Fusion with polyethylene glycol (Szabados et al. 1981; Griesbach et al. 1982) is possible, but electroporesis is becoming more promising. The host cell must be at mitosis or nearly so because differences in condensation pattern can be destructive. Selective systems for isolating chromosome-transformed cells are almost a prerequisite, unless visually selected cells can be isolated and grown in a nurse culture which can later be removed (Hein et al. 1983). Microinjection of chromosomes is also a possibility.

### 3.3.4 Genome Multiplication

*1. Autopolyploidy.* Although many important crop species are allopolyploids and several are autopolyploids (Gottschalk 1976; Lewis 1980), it appears difficult to produce good artificial polyploids that can compete with their diploid progenitors. The main distinctive features of autopolyploids are: (1) gene dosage effects, some specific (including effects on contents of secondary metabolites, Lavania 1986), others general (gigas characteristics, lateness); (2) specific interallelic interactions, and (3) general interallelic interactions (heterosis), which are different for polyploids than for diploids because a larger number of different alleles can be present simultaneously. The genetic system is tetrasomic, which implies retarded release of genetic variation, an advantage in some cases, but problematic for selection and stabilization. Gene dose effects may often be negative in new polyploids and fertility especially is affected. Meiotic irregularities may further reduce fertility. Polyploids in general have a greater need for heterosis than diploids (Bingham 1980), and consequently for a higher level of heterozygosity. Combined with the need of a broad genetic background for overcoming the fertility problems by selection, as well as adjusting the genotype to the requirements of a polyploid crop in general, this places special demands on the degree of heterozygosity. The commonly applied method of somatic doubling, usually by colchicine but occasionally also spontaneously through a callus phase (Hermsen et al. 1981), results in low rather than high degrees of heterozygosity. Meiotic doubling using unreduced gametes, therefore, is a much better method, although possible only for crops or genotypes with a relatively high frequency of unreduced gametes (Skiebe et al. 1963; Hermsen and Ramanna 1981; McCoy 1982). Although not yet widely used for this purpose, protoplast fusion between genetically different lines may become a very attractive alternative.

*2. Allopolyploidy.* Gene dose effects, complex interallelic interactions, and heterosis are among the reasons to construct allopolyploids, but the main reason usually is the intact and stable combination of the characteristics of different, specific genotypes. The genetic system is disomic, and unlike in autotetraploids, the combined genomes are maintained intact during generative reproduction. The main reasons for the difficulty in producing successful new allopolyploids are genetic interactions between the genomes, resulting in disturbance of the subtle processes of generative reproduction: meiosis, fertilization, seed development. Gene dose effects may, as in autopolyploids, play a role also. Meiotic doubling again is superior, but a starting population produced by somatic doubling must normally precede meiotic doubling. Somatic hybridization is a promising method for the same reason: avoiding homozygosity is almost as important here as in autopolyploids. The expectation to produce successful allopolyploids between widely different species by somatic hybridization when they cannot be produced by generative hybridization is idle: the combination will be too unbalanced to be functional. Such combinations are of interest only as intermediates for transfer of genes or chromosome segments.

*In conclusion* for this section: the potential usefulness of highly amplified specific (short) DNA sequences for plant breeding may be considerable. Molecular and cellular approaches are the most effective for studying as well as for producing them.

Duplication of specific small chromosome segments and special chromosomes, perhaps especially A-B translocations, carries a certain promise and their production is typically a technique of generative manipulation.

Polyploidy, both auto- and allo- remains important, even when more difficult to exploit than sometimes understood. Their development is traditionally a technique of generative manipulation after somatic, plant level induction, but more emphasis should be placed on generative induction as well as on somatic hybridization.

## 4 Reproductive Systems

The reproductive system must fulfill two functions, which may sometimes be conflicting: the introduction, recombination, and removal of specific genes or gene combinations on one hand, and subsequent faithful reproduction of the new gene combinations on the other.

### 4.1 Natural Systems

*Disomy* is the most commonly occurring system in self- and cross-fertilizing diploids and allopolyploids. In most cases the plant breeder simply accepts it and attempts to apply it both for introduction and recombination and for maintenance of his selected genotypes. Recessive alleles are readily expressed in self-fertilizers, but, when deleterious, their removal is less simple in crossbreeders. In allopolyploids, undesired recessives will not be expressed when homologous genes are dominant and homozygous, and thus will not present a problem even in crossbreeders. For the same reason, expression of desired recessives does present a problem, except in highly diploidized species. Mutation of the dominant allele to a null allele may be a solution. In allopolyploids, direct introduction of genes from one of the original parents is an important possibility. Reconstruction of the allopolyploid (Gottschalk 1976) and subsequent hybridization with the cultivated form has been carried out on several occasions, but the difficulties should not be underestimated.

The change from one breeding system to another is not readily accomplished. For exploiting heterosis in a self-fertilizer, it is usually not well possible to turn it into a crossbreeder. (Artificial) hybrid varieties are, then, more promising. The opposite, turning a crossbreeder into a self-fertilizer, is seemingly easier. Acceptable inbred lines can often be constructed, but only with considerable effort. The advantage of homogeneity and rapid and uniform expression of introduced genes may compensate for reduced productivity. For field crops, where bulk production is a major goal, this may often not be acceptable, and then hybrid varieties may be the best solution. Especially in (allo)polyploids, which profit even more from heterosis than diploids, in spite of having their own stable heterozygosity, a shift from outbreeding to inbreeding is not desirable. In general, when a drastic change in the breeding system in considered, it will be preferably towards an artificial system, and towards outbreeding rather than inbreeding.

*Polysomy* is found normally in autopolyploids, usually autotetraploid species. The special characteristics of the genetic system are not the reason to introduce autopolyploidy, but the consequences are considerable (Sybenga 1972). For instance, with low levels of outbreeding, considerable heterozygosity is maintained. In contrast with allopolyploids, this heterozygosity is not stable. Selection is slow in polysomic systems (Wricke and Weber 1986), and recessive alleles may remain hidden for many generations of selection. In addition to gene imbalance causing reduced fertility, meiosis may be disturbed and lead to reduced fertility and, more important, be the origin of aneuploids in the progeny. These are often deviant and less productive. Apparently fertility in polyploids has many components (Gottschalk 1976), and simple selection is very indirect. It would be useful, if possible, to select for the different fertility components separately.

For reasons of both fertility and genetic stabilization, it has been attempted to transform the polysomic system of inheritance of autopolyploids into a disomic system (allopolyploidization). It is not sufficient to search for a simple gene or gene complex (like the Ph gene on chromosome 5B of wheat) without a basic difference in chromosome pairing affinity between the component genomes. Chromosomal rearrangements have been suggested as a basic differentiating mechanism for barley polyploids (Bender and Gaul 1966). These were not very successful in barley and rye, even when combined with slight pairing differences possibly existing within species (Sybenga 1973). The reason is mainly the necessity of introducing very large numbers of rearrangements in addition to the proper genotype. In spite of the negative results, attempts continue in barley (Meister and Bretschneider 1977; Scholz and Künzel 1981) and in maize, where some success appears to be realized (Doyle 1986). The possibility of differentiating only a limited number of chromosomes, resulting in segmental auto-allopolyploidy, has been insufficiently studied. The opposite, replacing two homeologous pairs by four homologous chromosomes, has been carried out by Sears (1966) in wheat, but application is not obvious. There are several more variants that need not be discussed.

This type of manipulating the genetic system offers definite possibilities, but is extremely laborious. It is typically a generative manipulation approach. Perhaps the introduction of specific meiosis genes by molecular transformation would some day be feasible, and in that case of considerable interest – at present, this is only fantasy.

## 4.2 Controlled or Limited Sexual Reproduction

### 4.2.1 Hybrid Varieties

Hybrid varieties are the most successful way of fully exploiting heterosis and at the same time producing uniform varieties into which new genes can be effectively introduced and expressed. Most are diploid, some allopolyploid, and a few (triploid sugar beet) are interploidy hybrids (diploid x tetraploid). Autotetraploid hybrid varieties are not yet available, but the seed-reproduced potato seems to be a good candidate.

There are several ways of making the production hybrid. The oldest method, hand emasculation of one of the parents and open (or hand) pollination by the other parent grown in alternate rows, is still practiced for maize, and for several horticultural crops where the plant structure is less favorable, but the price of the seed is high enough to justify the considerable expense of hand emasculation and pollination.

Very promising are gametocides. They are available for wheat, but very species-specific, and even within wheat there is a clear difference in response between varieties. In the most favorable cases application is possible during a long period directly after the initiation of the ear primordium. Treatment must be carried out carefully on the maternal rows, avoiding the pollinator rows. The development of resistant pollinator lines is the next step; in-vitro selection of dominant mutation would be a logical attempt, but the chance that the selected pollinator would be suitable for in-vitro culture is limited. Some effective gametocides have not been accepted for practical use because of potential hazards to field workers.

As long as effective gametocides are not generally available for other crops, other ways of automatic emasculation must be used where hand emasculation is not feasible. The most common is cytoplasmic male sterility, mostly based on plasmid-like DNA particles in the mitochondria, under the control of the nuclear genotype (Palmer et al. 1983). In addition to the male sterile cytoplasm in the maternal line, nuclear fertility restorer genes are required in the paternal line to prevent sterility in the final production variety. Good sterile cytoplasms are scarce and sometimes have very undesired side effects, which may be detected only after a considerable period of time and then appear to be disastrous. A notorious example is the Texas cytoplasm in maize that carried sensitivity to a *Helminthosporium* disease, leading to an almost complete loss of the maize crop in important maize-growing regions of the USA where this cytoplasm was in general use. New cytoplasms have been isolated, with the intention not to base the majority of a crop on a single cytotype again. The possibility of transfering the critical plasmid structures to the mitochondria in any selected genotype would be a very attractive solution. This is being attempted in several institutions, especially with crop species which are more favourable for somatic cell gentics than the Gramineae.

Male sterility conditioned by specific alleles of nuclear genes can be used in hybrid varieties only when the segregating male sterile type can be selected, or when self-regulating systems are applied. The first is again possible for some horticultural crops and has a definite application where the price of individual seeds is high. Close linkage with marker genes increases the applicability, especially when readily selectable but otherwise neutral markers are available (Rao et al. 1989). Such markers, however, are rare. Self-regulatory systems have been made technically operating in barley, wheat and maize, but without much practical success.

The construction of self-regulating systems has been attempted for barley, maize and rye, but not yet with sufficient success. The systems proposed are certainly of considerable potential interest, but the required input in labour for really satisfactory systems has nowhere been made available. The advantages are that the systems, once developed for one line, can be transferred simply by backcrossing to any other line, and special restorer genes are not necessary. The first proposal came from Ramage (1965) and was based on balanced tertiary trisomics. To a line that is homozygous for a recessive male sterility gene and a recessive (semi-)lethal selective marker, a translocation chromosome is added that has the dominant alleles of both these genes. This line is phenotypically normal, but the extra chromosome is not transmitted to the progeny through the pollen. Upon selfing, this chromosome is carried over through the egg, and a certain proportion of the progeny is therefore normal (fertile) again. The functional pollen has the recessive alleles only and when used to pollinate a male sterile line, entirely male sterile progeny is produced. The selective marker is necessary to distinguish between the male fertile maintainer line and segregating sterile plants, which have the (semi-)lethal marker gene. This is also carried over through the pollen, and this results in 50% lethality in the male sterile line. This line is used to produce the production variety by crossing with any good fertile male parent. Fertility is restored automatically. The translocation in the extra chromosome serves two purposes: it prevents transfer by recombination of the male sterile and marker genes from the extra chromosome to the normal chromosomes, and it prevents the substitution of one of the normal chromosomes by the extra chromosome.

For wheat Driscoll (1981) designed a somewhat different system based on the addition of an alien chromosome; Patterson (1973) used a duplication for the same purpose in maize, and Sybenga (1982) proposed a compensating trisomic for rye. None of these systems works satisfactorily, for a variety of reasons. One usually is a high level of male transmission (Janse 1987). The balance between sufficient reduction of male transmission and sufficient fertility and vigour of the maintainer line is difficult to strike. Selection of the proper chromosome and a genotype which favors its function is the main bottleneck. Recombination between the extra chromosome and one normal chromosome, transferring the markers to the normal complement is even more deleterious than male transmission, but it is usually more easily controlled. It is our experience with rye that in diploid outbreeders few inbred genotypes will sufficiently tolerate an extra chromosome or segment that is large enough not to be transmitted through the pollen.

Special gametocidal alien chromosomes as have been found in the progeny of some interspecific wheat/*Aegilops* hybrids (Maan 1975; Endo 1982; Tsujimoto and Tsunewaki 1988) are interesting in this respect. Single genes with reduced pollen transmission are also known. With the proper investment, the system may well be made to work. Except perhaps for transferring specific genes (pollen lethals, male sterility, lethal markers, for instance) and later perhaps the transfer of chromosomes, the role of molecular and cell biology seems to be very limited.

### 4.2.2 Self-Incompatibility

The most common natural way of optimizing heterozygosity outside sex differentiation for plants is self-incompatibility. It has not been possible yet to introduce self-incompatibility into species that do not have it by nature. Even where it exists, it cannot readily be manipulated. Wricke (1986) has based a potentially successful system on tempararily inactivating self-incompatibility by high temperature treatment in rye for producing (partial) inbred lines. In the field, self-incompatibility was restored and could be exploited in a special breeding system. It is still typically a generative approach also in its development, but in principle molecular transfer of incompatibility genes would be an interesting possibility.

### 4.2.3 Dioecy

This is not as rare an exception in plants as is sometimes assumed, and several economically important plants, e.g., hemp, spinach, asparagus are dioecious. Although for production and breeding purposes usually a switch to monoecy is preferred, sexual dimorphism still has some advantages. In asparagus, the cross between female XX and artificial male YY, produced by doubling haploids, uniformly produces the most productive male XY type, which at the same time is a hybrid (Randall and Rick 1945). It will probably pay to look for a link between sexual dimorphism and a system selectively eliminating the less productive sex in the production fields. Some may be based on sex-linked conditional lethals, others on chemical treatments, and others again on cytogenetic systems.

### 4.2.4 Special Systems of Sexual Reproduction

With the purpose of maintaining heterosis, these exist, for instance, in *Oenothera* and *Rhoeo*, based on a balanced, permanent system of translocation heterozygosity. Single inversions and translocations have frequently been observed to "float" in populations as stable polymorphisms, being maintained because of the favorable effect of the conservation of recombination-free heterotic gene blocks. Translocations are rather common in the Onagraceae, to which the genus

*Oenothera* belongs. In *Oenothera* several species have one or two translocations as polymorphisms, but in *O. lamarckiana* 12 chromosomes are involved in one large translocation complex and one pair only is still a bivalent (Cleland 1972). In other species of *Oenothera* and in *Rhoeo spathacea*, all chromosomes are included in the complex. Due to regular orientation, the complex segregates in two balanced chromosome groups without chromosome recombination and only limited, distal, crossing-over. There are special systems to ensure that one genome functions only through the male and the other only through the female. In effect it is closely related to large sex-associated translocation complexes such as occur in several insects, termites, copepods, etc. and a few plant species. The result is a reconstitution of the heterozygote in each new generation, with only very limited recombination. It has been possible to combine a series of translocations in barley (2n = 14), including all chromosomes (Sisodia and Shebeski 1965), but irregular segregration led to complete sterility. In *Tradescantia paludosa* (2n = 12), a species related to *Rhoeo*, Watanabe (1962) induced a set of translocations involving all chromosomes in one clone, applying a series of irradiations. In contrast to *Rhoeo*, orientation was quite irregular, perhaps as a result of a tendency of the centromeres to become amphitelic earlier than in *Rhoeo* (Sybenga and Rickards 1987). Individual translocations in *Tradescantia* do have considerable tendency to preferential balanced segregation, but this was insufficient in the complex. In *Pennisetum* also, multiple translocations were not functional (Brar and Minocha 1982). Later attempts by Scholz and Künzel (1981) with selected barley translocations in special genotypes were more promising. A combination of only a limited number of translocations may be a realistic goal. However, in order to construct a balanced permanent heterozygote, a selective system must be combined with the translocations. Two recessive lethals, one in each genome or set of chromosomes, is the simplest solution, but at the expense of fertility. The complex systems operating in nature are not readily copied. At the moment, it is typically a problem for the cytogeneticist and significant contributions from molecular and cell biology may not be expected.

### 4.2.5 Asexual Reproduction

For practical purposes three forms will be distinguished: natural vegetative reproduction by stolons, bulbs, etc.; in-vitro propagation; apomixis.

*1. Vegetative reproduction* by stolons, bulbs, runners, etc. is common in numerous economic plant species, especially in horticulture. There are several technical advantages and disadvantages, and from a genetic point of view the main characteristic is unaltered reproduction. When sexual reproduction is not suppressed, normal plant breeding remains possible. There is an increasing interest in improving the technical possibilities of vegetative reproduction in horticulture and forestry. As an alternative, much attention is paid to in-vitro propagation. Its use ranges from large-scale propagation of potted flower plants to forest tree reproduction, and as such is becoming of rapidly increasing, great economic interest. For field crops, its use is limited to breeding programs (e.g., sugar beets). There are several technical advantages to the method, including phytosanitary aspects and rapid reproduction, and genetically, the advantage is the same as that of the previous method.

2. *Apomictic seed production* would probably be the best method of reproduction when juvenile stages are not of importance, and provided it can be controlled at least to some extent. It permits the identical reproduction also when the seed is the main product or propagule. Apomixis is rare among cultivated plants. In the predominantly apomictic grass *Poa pratensis*, some sexual reproduction occurs, and although this implies the availability of genetic variation and consequently the possibility of selection, the obvious disadvantage is limited uniformity and segregation of numerous irregular chromosomal types, as the species is an aneuploid polyploid.

There are several ways in which apomixis is realized in plants, and some may be imitated experimentally. In a number of species of *Solanum* (Mok and Peloquin 1975; Hermsen and Ramanna 1981) and in some Brassicas and other Cruciferae, suppression of the first or second meiotic divisions, or fusion of division spindles, leads to the development of unreduced eggs, or even embryos, in special genotypes and after pollination with special pollinators. In unbalanced genotypes (hybrids, amphidiploids, polyploids) tendencies towards apomixis after meiotic breakdown are sometimes observed (Ellerström and Zagorcheva 1977; Mujeeb-Kazi 1981). As long as partial meiosis remains associated with the system, some segregation and gradual inbreeding are inevitable, but this approach may still be considered a first step towards manipulated apomixis. An interesting variant is the combination of asynapsis with incomplete meiosis (restitution after the first division), which results in complete absence of segregation. The genes are available in the potato (Jongedijk 1986), but it is not simple to introduce them into an acceptable cultivar.

There are other, relatively simple modifications of meiosis which result in fully apomictic reproduction and where pollination is necessary only for stimulation of embryo development (pseudogamy). One is found in a number of tetraploid *Allium* species (Gohil and Kaul 1981) and consists of an extra cycle of DNA synthesis just prior to meiosis, resulting in endo-reduplication. The sister chromatids form bivalents and meiosis follows a normal course, however, without reduction and segregation. It is well possible that only a very limited number of genes are involved and that polyploidy plays a role by affecting the relative durations of the phases of the premeiotic cycle.

Apospory (apomixis without meiosis) is common in numerous plant species and occurs in several different forms. Some of these have been genetically analyzed (Rutishauser 1967). It is of great importance that factors affecting the shift between sexual reproduction and apomixis are thoroughly studied in order to make it possible to introduce apomixis in a cultivar after introduction of genetic variation and after selection. Molecular transformation and perhaps also other cell manipulation techniques may potentially contribute in two ways: (1) the introduction of genes in already established apomicts; (2) the transfer of genes conditioning apomixis into a selected genotype.

*In conclusion* for this section: manipulation of the genetic system is a very important subject for plant breeding, but much of its potential has been seriously neglected. Only few contributions from molecular and cell biological techniques can be expected, but in some cases these may be important. Much can be expected from generative manipulation techniques, provided these are sufficiently implemented and carried out with the appropriate expertise.

## 5 Conclusions and Prospects

Classical methods will remain the basis of plant breeding, but both molecular/cell biological and special generative/cytogenetic approaches offer very promising new possibilities. The first are technically more demanding, the latter conceptually, but both are laborious. Each approach has its own specific applications and in a few cases they are competitive, especially for the transfer of genes from one species to another. For each specific case the choice has to be made separately on the basis of plant species, stage of techniques, genetic diversity, and capacity of the institution. Molecular and cell biological techniques offer promise also for the analysis of the breeding material, for providing genetic markers, and in a combination with generative approaches. In several areas of plant breeding sophisticated generative (cytogenetic) methods are becoming available which also deserve (renewed) attention because they offer possibilities of manipulation not available otherwise.

## References

Avivi L, Feldman M, Brown M (1982) An ordered arrangement of chromosomes in the somatic nucleus of common wheat, *Triticum aestivum* L. I and II. Chromosoma 86:1–26

Bajaj YPS (1981) Production of disease-resistant plants through cell culture – a novel approach. J Nucl Agric Biol 10:1–5

Bajaj YPS (ed) (1986) Biotechnology in agriculture and forestry 2. Crops I. Springer, Berlin Heidelberg New York Tokyo

Barclay IR (1975) High frequencies of haploid production in wheat (*Triticum aestivum*) by chromosome elimination. Nature 256:410–411

Bates GW, Nea LJ, Hasenkampf CA (1987) Electrofusion and plant somatic hybridization. In: Sowers AE (ed) Cell fusion. Plenum, New York, pp 479–496

Beckett JB (1982) An additional mechanism by which B chromosomes are maintained in maize. J Hered 73:29–34

Bender K, Gaul H (1966) Zur Frage der Diploidisierung autotetraploider Gerste. Z Pflanzenzücht 56:164–183

Bennett MD (1982) Nucleotypic basis of the spatial ordering of chromosomes in eukaryotes and the implications of the order for genome evolution and phenotypic variation. In: Dover GA, Flavell RB (eds) Genome evolution. Academic Press, London, pp 239–261

Bhatia CR, Viegas P, Bhagwat A, Mathews H, Notani NK (1986) Genetic transformation of plants. Proc Indian Acad Sci (Plant Sci) 96:79–112

Bingham ET (1980) Maximum heterozygosity in autopolyploids. In: Lewis WH (ed) Polyploidy. Biological Relevance. Plenum, New York, pp 471–489

Birchler JA (1983) Chromosomal manipulation in maize. In: Swaminathan MS, Gupta K, Sinha U (eds) Cytogenetics of crop plants. MacMillan, India, pp 379–403

Borlaug NE (1983) Contributions of conventional plant breeding to food production. Science 219:689–693

Brar DS, Minocha JL (1982) Multiple chromosomal interchanges in pearl millet. Theor Appl Genet 61:105–108

Broertjes C, van Harten AM (1988) Applied mutation breeding for vegetatively propagated crops. Dev Crop Sci 12. Elsevier, Amsterdam

Chaleff RS (1983) Isolation of agronomically useful mutants from plant cell cultures. Science 219:676–682

Chaleff RS, Parsons MF (1978) Direct selection in vitro for herbicide-resistant mutants of *Nicotiana tabacum*. Proc Natl Acad Sci USA 75:5104–5107

Chase SS (1952) Monoploids in maize. In: Gowen JW (ed) Heterosis. Iowa State College Press, Ames, Iowa, pp 389–399

Cleland RE (1972) Oenothera – cytogenetics and evolution. Academic Press, London

Cocking EC, Davey MR, Pental D, Power JB (1981) Aspects of plant genetic manipulation. Nature 293:265–270

Cooper DB, Sears RG, Lookhart GL, Jones BL (1986) Heritable somaclonal variation in gliadin proteins of wheat plants derived from immature embryo callus culture. Theor Appl Genet 71: 784–790

Cowell JK (1982) Double minutes and homogeneously staining regions: gene amplification in mammalian cells. Annu Rev Genet 16:21–59

Crossway A, Hauptli H, Houck CM, Irvine JM, Oakes JV, Perani LA (1986) Micromanipulation techniques in plant biotechnology. BioTechniques 4:320–334

De la Peña A, Lörz H, Schell J (1987) Transgenic rye plants obtained by injecting DNA into young floral tillers. Nature 325:274–276

Dellaert LMW (1979) Comparison of selection methods for specific mutants in self-fertilizing crops: theoretical approach. In: IAEA Symp on Seed Protein Improvement in Cereals and Grain Legumes. IAEA, Vienna, pp 57–74

Dennis ES, Bretell RIS, Peacock WJ (1987) A tissue culture-induced Adh 1 null mutant of maize results from a single base change. Mol Gen Genet 210:181–183

De Vries JN (1985) Isolation of telotertiary compensating trisomics from telocentric translocation trisomics and telo-substituted translocation heterozygotes of rye (*Secale cereale* L.). Genetica 68:47–58

De Wet JMJ, de Wet AE, Brink DE, Hepburn AG, Woods JA (1986) Gametophyte transformation in maize (*Zea mays*, Gramineae). In: Mulcahy DL, Mulcahy GB, Ottaviano E (eds) Biotechnology and Ecology of Pollen. Springer, Berlin Heidelberg New York Tokyo, pp 59–64

Doyle GG (1986) The allotetraploidization of maize. 4. Cytological and genetic evidence indicative of substantial progress. Theor Appl Genet 71:585–594

Driscoll CJ (1981) Perspectives in chromosome manipulation. Philos Trans R Soc Lond, Ser B 292:535–546

Dudits D, Maroy E, Praznovszky T, Olah Z, Gyorgyey J, Cella R (1987) Transfer of resistance traits from carrot into tobacco by asymmetric somatic hybridization. Proc Natl Acad Sci USA 84:8434–8438

Ellerström S, Zagorcheva L (1977) Sterity and apomictic embryosac formation in *Raphanobrassica*. Hereditas 87:107–120

Endo TR (1982) Gametocidal chromosomes of three *Aegilops* species in common wheat. Can J Genet Cytol 24:201–206

Evola SV, Burr FA, Burr B (1986) The suitability of restriction fragment length polymorphisms as genetic markers in maize. Theor Appl Genet 71:765–771

Feldman M (1988) Cytogenetic and molecular approaches to alien gene transfer in wheat. Proc 7th Int Wheat Genetics Symp Cambridge

Galun E, Aviv D (1983) Cytoplasmic hybridization – genetic and breeding applications. In: Evans DA, Sharp WR, Ammirato PV, Yamada Y (eds) Handbook of plant cell culture, Vol 1. Macmillan, New York, pp 358–392

Gengenbach BG, Green CE, Donovan CM (1977) Inheritance of selected pathotoxin resistance in maize plants regenerated from cell cultures. Proc Natl Acad Sci USA 74:5113–5117

Gerstel DU, Burns JA (1966) Chromosomes of unusual length in hybrids between two secies of *Nicotiana*. Chromosomes Today 1:41–56

Gheysen G, Dhaese P, van Montague M, Schell J (1985) DNA flux across genetic barriers: the crown gall phenomenon. In: Hohn B, Dennis ES (eds) Genetic Flux in Plants. Springer, Berlin Heidelberg, New York Tokyo, pp 11–47

Gohil RN, Kaul R (1981) Studies on male and female meiosis in Indian Allium. 2. Autotetraploid *Allium tuberosum*. Chromosoma 82:735–739

Gottschalk W (1976) Die Bedeutung der Polyploidie für die Evolution der Pflanzen. Gustav Fischer, Stuttgart

Green MM, Green KC (1949) Crossing-over between alleles at the lozenge locus in *Drosophila melanogaster*. Proc Natl Acad Sci USA 35:586–591

Griesbach RJ, Malmberg RL, Carlson PS (1982) Uptake of isolated lily chromosomes by tobacco protoplasts. J Hered 73:151–152

Gupta PP, Schieder O, Gupta M (1984) Intergeneric nuclear gene transfer between somatically and sexually incompatible plants through asymmetric protoplast fusion. Mol Gen Genet 197:30–35

Gustafson JP, Lukaszewski AJ, Bennett MD (1983) Somatic deletion and/or redistribution of telomeric heterochromatin in the genus *Secale* and *Triticale*. Chromosoma 88:293–298

Hagberg A (1965) Use of induced translocations in directed production of duplications. In: The use of induced mutations in plant breeding. FAO, IAEA, Rome Suppl, Rad Bot 5 Pergamon, pp 741–752

Hein T, Przewozny T, Schieder O (1983) Culture and selection of somatic hybrids using an auxotrophic cell line. Theor Appl Genet 64:119–122

Hermsen JGT, Ramanna MS (1981) Haploidy and plant breeding. Philos Trans R Soc London, Ser B 292:111–112

Hermsen JGT, Ramanna MS, Roest S, Bokelman GS (1981) Chromosome doubling through adventitious shoot formation on in vitro cultivated leaf explants from diploid interspecific potato hybrids. Euphytica 30:239–246

Hermsen JGT, Verdenius J (1973) Selection from *Solanum tuberosum* group Phureja of genotypes combining high-frequency haploid induction with homozygosity for embryo-spot. Euphytica 22:244–259

Hoekema A, Hirsch PP, Hooykaas PJJ, Schilperoord RA (1983) A binary plant vector strategy based on separation of vir- and T-region of the *Agrobacterium tumefaciens* Ti-plasmid. Nature 303:179–180

Hovenkamp-Hermelink JHM, Jacobsen E, Ponstein AS, Visser RGF, Vos-Scheperkeuter GH, Bijmolt EW, de Vries JN, Witholt B, Feenstra WJ (1987) Isolation of an amylose-free starch mutant in the potato (*Solanum tuberosum* L.). Theor Appl Genet 75:217–221

Ihrke CA, Kronstad WE (1975) Genetic recombination in maize as affected by ethylenediamidetetraacetic acid and dimethyl sulfoxide. Crop Sci 15:429–431

Janse J (1987) Certation between euploid and aneuploid pollen grains from a tertiary trisomic of rye, *Secale cereale* L. Genome 29:353–356

Jones DF (1939) Segregation of color and growth regulating genes in somatic tissues of maize. Proc Natl Acad Sci USA 22:163–166

Jones GH (1967) The control of chiasma distribution in rye. Chromosoma 22:69–90

Jones GH (1974) Correlated components of chiasma variation and the control of chiasma distribution in rye. Heredity 32:375–387

Jongedijk E (1986) The present state of research into the induction of apomixis in potato. In: Beekman ABG et al. (eds) Potato research of tomorrow. Proc Internat. Seminar, Wageningen, The Netherlands, Oct. 1985. Pudoc, Wageningen

Karp A, Bright SWJ (1985) On the causes and origins of somaclonal variation. Oxford Surv Plant Mol Cell Biol 2: 199–234

Kasha KJ, Reinbergs E (1979) Achievements with haploids in barley research and breeding. Proc John Innes Symp 4:215–230

Kasha KJ, Séguin-Swartz G (1983) Haploidy in crop improvement. In: Swaminathan MS, Gupta PK, Sinha U (eds) Cytogenetics of crop plants. MacMillan, India, pp 19–68

Khush GS (1973) Cytogenetics of aneuploids. Academic Press, London

Koebner RMD, Shepherd KW (1986) Controlled introgression to wheat of genes from rye chromosome arm 1RS by induction of allosyndesis. 1. Isolation of recombinants. Theor Appl Genet 73:197–208

Koebner RMD, Shepherd KW (1987) Allosyndetic recombination between a chromosome of *Aegilops umbellulata* and wheat chromosomes. Heredity 59:33–45

Laat AMM de, Blaas J (1984) Flow-cytometric characterization and sorting of plant chromosomes. Theor Appl Genet 67:463–467

Lange W (1971) Crosses between *Hordeum vulgare* L. and *H. bulbosum* L. II. Elimination of chromosomes in hybrid tissues. Euphytica 20:181–194

Larkin PJ, Ryan SA, Bretell RIS, Scowcroft WR (1984) Heritable somaclonal variation in wheat. Theor Appl Genet 67:443–455

Larkin PJ, Scowcroft WR (1981) Somaclonal variation – a novel source of variability from cell cultures. Theor Appl Genet 60:197–214

Lavania UC (1986) Genetic improvement of Egyptian henbane, *Hyoscyamus muticus* L. through induced tetraploidy. Theor Appl Genet 73:292–298

Lewis WH (ed) (1980) Polyploidy. Biological Relevance. Plenum, New York

Lukaszewski AJ, Gustafson JP (1983) Translocations and modifications of chromosomes in triticale×wheat hybrids. Theor Appl Genet 64:299–348

Maan SS (1975) Exclusive preferential transmission of an alien chromosome in common wheat. Crop Sci 15:278–292

Maliga P (1984) Isolation and characterization of mutants in plant cell culture. Annu Rev Plant Physiol 35:519–542

McCoy TJ (1982) The inheritance of 2 n pollen formation in diploid alfalfa *Medicago sativa*. Can J Genet Cytol 22:315–323

Meister A, Bretschneider H (1977) Präferentielle Paarung der Chromosomen autotetraploider Gerste – mathematisches Modell. Kulturpflanze 25:297–311

Mok DWS, Peloquin SJ (1975) The inheritance of three mechanisms of diplandroid (2n-pollen) formation in diploid potatoes. Heredity 35:295–302

Mouras A, Saul MW, Essad S, Potrykus I (1987) Localization by in situ hybridization of a low copy chimaeric resistance gene introduced into plants by direct gene transfer. Mol Gen Genet 207:204–209

Mujeeb-Kazi A (1981) Apomictic progeny derived from intergeneric *Hordeum-Triticum* hybrids. J Hered 72:284–285

Namai H, Sarashima N, Hosoda T (1980) Interspecific and intergeneric hybridization breeding in Japan. In: Tsumoda S, Hinata H, Gómez-Campo G (eds) *Brassica* crops and wild allies. Biology and breeding. Jpn Sci Soc Press, Tokyo, pp 191–203

Nitzsche W (1980) Chromosome reduction by halogenized amino acids in *Festuca-Lolium* hybrids. Z Pflanzenzücht 84:78–81

Ohno S (1970) Evolution by gene duplication, vol XVI. Springer, Berlin Heidelberg New York, pp 1–160

Ohta Y (1986) High efficiency genetic transformation of maize by a mixture of pollen and exogenous DNA. Proc Natl Acad Sci USA 83:715–719

Palmer JD, Shields CR, Cohen DB, Orton TJ (1983) An unusual mitochondrial DNA plasmid in the genus *Brassica*. Nature 301:725–727

Palmiter RD, Brinster RL, Hammer RE, Turmbauer ME, Rosenfeld MG, Birnberg NC, Evans RM (1982) Dramatic growth of mice that develop from eggs microinjected with metallothionein-growth hormone fusion genes. Nature 300:611–615

Pandey KK (1978) Gametic gene transfer in *Nicotiana* by means of irradiated pollen. Genetica 49:53–69

Patterson EB (1973) Genic male sterility and hybrid maize production. Proc 7th Meeting Maize and Sorghum Section, Eucarpia, Zagreb

Peerbolte R (1986) The fate of T-DNA during vegetative and generative propagation. Diss Leiden, 218 pp

Pirrie A, Power JB (1986) The production of fertile, triploid somatic hybrid plants (*Nicotiana glutinosa* (n)+*N. tabacum* (2n) via gametic somatic protoplast fusion. Theor Appl Genet 72:48–52

Pijnacker LP, Walch K, Ferwerda MA (1986) Behaviour of chromosomes in potato leaf tissue cultured in vitro as studied by BrdC-Giemsa labelling. Theor Appl Genet 72:833–839

Ramage RI (1965) Balanced tertiary trisomics for use in hybrid seed production. Crop Sci 5:177–178

Ramulu SK, Verhoeven HA, Dijkhuis P (1988) Mitotic dynamics of micronuclei induced by amiprophos-methyl and prospects for chromosome mediated gene transfer in plants. Theor Appl Genet 75:575–584

Randall TE, Rick CM (1945) A cytogenetic study of polyembryony in *Asparagus officinalis.* Am J Bot 32:560–569

Rao MK, Devi KU, Arundhati A (1989) Utilization of genetic male sterility in plant breeding. Plant Breeding (in press)

Riley R, Chapman V, Johnson R (1968) The incorporation of alien disease resistance in wheat by genetic interference with the regulation of meiotic chromosome synapsis. Genet Res 12:199–219

Roman H (1947) Mitotic non-disjunction in the case of interchanges involving the B-type chromosome in maize. Genetics 32:391–409

Rose AM, Baillie DL (1979) A mutation in *Caenorhabditis elegans* that increases recombination frequency more than three fold. Nature 281:599–600

Ross JG (1965) Somatic chromosome reduction and spectrum mutational effects after colchicine treatment of sorghum. In: The use of mutations in plant breeding. FAO, IAEA, Rome, pp 193–203, Suppl Rad Bot 5, Pergamon

Roth EJ, Lark KG (1984) Isopropyl-N-(3-chlorophenyl) carbamate (CIPC) induced chromosomal loss in soybean: a new tool for plant somatic cell genetics. Theor Appl Genet 86:421–431

Rutishauser A (1967) Fortpflanzungsmodus und Meiose apomiktischer Blütenpflanzen. Protoplasmatologia V1, $F_3$. Springer, Wien New York, pp 1–245

Scholz F, Künzel G (1981) Induced chromosome and gene mutations for specific goals in barley genetics and breeding. In: Induced mutations: a tool in plant research. IAEA, Vienna, pp 23–35

Sears ER (1956) The transfer of leaf-rust resistance from *Aegilops umbellulata* to wheat. In: Genetics in plant breeding. Brookhaven Symp Biol 9:1–22

Sears ER (1966) Nullisomic-tetrasomic combinations in hexaploid wheat. In: Riley R, Lewis KR (eds) Chromosome manipulations and plant genetics. Oliver and Boyd, Edinburgh, pp 29–45. Heredity Suppl 20 (1965)

Shah DM et al. (1986) Engineering herbicide tolerance in transgenic plants. Science 233:478–481

Shepard JF (1982) The regeneration of potato plants from leaf-cell protoplasts. Sci Am, May 1982, pp 112–121

Shepard JF, Bidney D, Shahin E (1980) Potato protoplasts in crop improvement. Science 208:17–24

Shillito RD, Saul MW, Paszkowski J, Müller M, Potrykus I (1985) High efficiency direct gene transfer to plants. BioTechnology 3:1099–1100

Simchen G, Stamberg J (1969) Genetic control of recombination in *Schizophyllum commune*: specific and independent regulation of adjacent and non-adjacent chromosomal regions. Heredity 24:369–381

Sisodia NS, Shebeski LH (1965) Synthesis of complete interchange stocks in barley (*H. vulgare* L.) Can J Genet Cytol 7:164–170

Skiebe K, Jahr W, Stein M (1963) Bedeutungen von Valenzkreuzungen für die Polyploidiezüchtung. Z Pflanzenzücht 50:26–33

Smith M (1985) In vitro mutagenesis. Annu Rev Genet 19:423–462

Stam P (1977) Selection response under random mating and under selfing in the progeny of a cross of homozygous parents. Euphytica 26:169–184

Sybenga J (1955) Variation in *Hordeum vulgare* L. and *Trifolium repens* L. after colchicine treatment. Diss Abstr 12:294

Sybenga J (1972) General cytogenetics. North Holland/Elsevier Amsterdam-London-New York

Sybenga J (1973) Allopolyploidization of autopolyploids. 2. Manipulation of the chromosome pairing system. Euphytica 22:433–444

Sybenga J (1982) Cytogenetic systems and chromosomal male sterility genes in hybrid rye breeding. Tagungsber Akad Landwirtschaftswiss DDR 198:161–163

Sybenga J (1983) Genetic manipulation in plant breeding: somatic versus generative. Theor Appl Genet 66:179–201

Sybenga J, Rickards GK (1987) The orientation of multivalents at meiotic metaphase I. A Workshop Report. Genome 29:612–620

Sybenga J, Verhaar HM (1980) Chiasma patterns in a translocation derived duplication heterozygote of rye. Chromosoma 80:147–162

Szabados L, Hadlaczky G, Dudits D (1981) Uptake of isolated plant chromosomes by plant protoplasts. Planta 151:141–145

Tsujimoto H, Tsunewaki K (1988) Gametocidal genes in wheat and its relatives. III. Chromosome location and effects of two *Aegilops speltoides*-derived gametocidal genes in common wheat. Genome 30:239–244

Watanabe H (1962) An X-ray induced strain of ring of 12 chromosomes in *Tradescantia paludosa*. Nature 193:603

Weller SC, Masiunas JB, Gressel J (1987) Biotechnologies of obtaining herbicide tolerance in potato. In: Bajaj YPS (ed) Biotechnology in agriculture and forestry 3. Potato. Springer, Berlin Heidelberg New York Tokyo, pp 281–297

Wricke G (1986) Hybrid breeding in rye by means of self-incompatibility. Persson G (ed) Proc Eucarpia Meeting on Rye, Svalöv 1985, pp 267–278

Wricke G, Weber WE (1986) Quantitative genetics and selection in plant breeding. W de Gruyter, Berlin

Yoshida H, Yamaguchi H (1973) Arrangement and association of somatic chromosomes induced by chloramphenicol in barley. Chromosoma 43:399–407

# III Uptake and Integration of Exogenous DNA in Plants

P. F. Lurquin[1]

## 1 Introduction

Representative organisms of the Monera, fungi, animals, and plants can be modified genetically via uptake of nonviral foreign DNA molecules. This phenomenon is now considered rather commonplace and the articles published on the subject are so numerous that it would be unfair to many authors to cite only a few. In the case of the Eukaryotae, this added genetic information has been shown to be expressed phenotypically in progeny cells and whole organisms and was thus stably maintained following mitosis, meiosis, fertilization, and morphogenesis.

As far as we know, such transgenic organisms cannot spontaneously appear in Nature since they would have to be the result of an exchange of genetic information between sexually incompatible species.

Quite exceptionally, it seems, this barrier to gene transfer does not exist between many dicotyledonous plant species and the bacterial genus *Agrobacterium*, the effects of which being either the disease known as crown gall or root proliferation at the inoculation site, depending on the *Agrobacterium* species involved (for review see Gheysen et al. 1985; see Chap. IV, this Vol.). As a matter of fact, we shall see that the understanding of this phenomenon at the molecular level has had far-reaching consequences on the DNA-mediated genetic modification of plants. Nevertheless, and without precluding the role that DNA transfer mechanisms may have played in the evolution of early life, it is probably safe to say that natural DNA exchange between modern and widely divergent organisms is an exception rather than a rule.

Interestingly, plants were the last shown to be transformable genetically by exogenous DNA, despite the fact that attempts to reach this goal were undertaken over 20 years ago. Clearly, in this case, a bold and direct approach did not pay off. Rather, it is via an extremely indirect route, the molecular study of crown gall, that the field eventually crystallized and became limited only by the imagination of the students of plant genetic transformation.

The aim of this chapter is to review briefly the historical aspects of this venture, to describe the current status of plant transformation research and finally to give the reader a precise idea of the fate of foreign DNA in higher plant protoplasts and transgenic plants regenerated from them. This portion of the article

[1] Program in Genetics and Cell Biology, Washington State University, Pullman, Washington 99164-4350, USA

Biotechnology in Agriculture and Forestry, Vol. 9
Plant Protoplasts and Genetic Engineering II (ed. by Y. P. S. Bajaj)

will specifically focus on direct gene transfer into protoplasts. *Agrobacterium*-mediated gene transfer is discussed in Chapter IV, this Volume. A detailed account of foreign gene expression in plant cells has been published (Lurquin 1987).

## 2 A Brief History of Plant Genetic Transformation

Early publications dealing with DNA uptake by germinating seeds, whole plants, and plant cells have been reviewed previously (Kleinhofs and Behki 1977; Lurquin 1977). The view held by the authors of these two papers was that no conclusive evidence existed at the time in favor of either plant transformation or even integration of foreign DNA within plant cell genomes. Ten years later, it turns out that the numerous studies which had claimed integration of massive amounts of foreign DNA (up to several hundreds of bacterial genome equivalents per host genome) in plants such as barley as well as actual genetic transformation of *Arabidopsis* by bacterial genes (Ledoux and Huart 1969; Ledoux et al. 1974) have never been reproduced by independent laboratories. It has been reasonably well demonstrated that artifacts due to bacterial contamination, analysis of crude DNA preparations and/or use of inadequate techniques were probably responsible for early claims of DNA integration in plants. In fact, it has been shown recently that even under the best of circumstances, the total amount of foreign DNA which can really become integrated within, for example, plant protoplasts, is several orders of magnitude less than the values reported in the early studies.

The case of the transformation of *Arabidopsis* thiamine-less mutants is also rather clear: contamination of mutant seed stocks used in these experiments by wild-type and/or other mutant seeds was invoked to explain the transformation data (Redei et al. 1976).

During the decade 1965–75, a few other reports (also reviewed in Kleinhofs and Behki 1977; Lurquin 1977), published essentially by three groups, similarly purported that bacterial genes were functional in a variety of plants and plant cells. Again, these observations could not be reproduced by others. Thus, the enthusiasm (often a reluctant one) generated by these results soon vanished, leaving a trail of frustration in many. Time had come to reconsider these questions and start anew. Fortunately, the lag period proved to be short. As a matter of fact, knowing what we now know about the differences between eukaryotic and prokaryotic promoter regions, the high transformation frequencies achieved by uptake of bacterial DNA (Ledoux et al. 1974) would not simply have required skilled experimenters, they would have had to be the result of a combination between repeated sheer luck and incredibly chance events such as systematic integration of a bacterial coding sequence downstream from a constitutive endogenous promoter. Nevertheless, these early investigators deserve recognition. After all, plants *are* transformable by foreign genes, and foreign DNA *is* integrated in their genome.

Fresh concepts were provided by the advent of the recombinant DNA technology, the isolation and cultivation of plant protoplasts and, of course, by molecular studies on crown gall and the Ti plasmid of *Agrobacterium*.

Once it was realized that a portion of pTi was present in the genome of crown gall cells (Chilton et al. 1977), that tobacco protoplasts could be transformed in vitro by *A. tumefaciens* cells (Marton et al. 1979a), and that protoplasts could be rendered permeable to plasmid DNA (Lurquin and Kado 1977), experiments were designed to test whether pTi DNA alone could effect crown gall transformation of protoplasts.

The Fifth International Protoplast Symposium held in Szeged, Hungary, in July 1979 was of particular interest in this respect. Two reports (Davey et al. 1979; Marton et al. 1979b) presented at this meeting provided preliminary evidence that *Petunia* and *Nicotiana* protoplasts incubated with pTi DNA under conditions known to enhance plasmid uptake gave rise to cell lines expressing genetic markers now known to be carried by the T-DNA of *A. tumefaciens* Ti plasmid.

It would be exaggerating to say that these results were instantly acclaimed. Nevertheless, pTi DNA transformation of protoplasts was later reproduced, perfected, and became widely accepted following a detailed molecular analysis (Krens et al. 1982).

In the meantime, the molecular biology of crown gall was progressing with giant steps (for review see Gheysen et al. 1985), and it became clear that, if properly engineered, the T-DNA portion of pTi should be an extremely useful vector for the transfer and expression of foreign genes in plant cells. This hypothesis was simultaneously demonstrated by three groups in 1983, thanks to the use of chimeric genes consisting of bacterial coding sequences determining antibiotic resistance under the control of the T-DNA nopaline synthase gene transcriptional control signals (Bevan et al. 1983; Fraley et al. 1983; Herrera-Estrella et al. 1983).

The "domestication", i.e., the deletion of crown gall-causing oncogenes from the T-DNA, subsequently allowed the regeneration of fertile transformed plants (reviewed in Lurquin 1987). The genetic transformation of dicotyledonous plants was thus demonstrated.

Another breakthrough was announced less than a year later. It was unequivocally demonstrated that a plant selectable marker carried by an *E. coli* replicon was integrated and expressed in plants derived from DNA-treated protoplasts (Paszkowski et al. 1984). Thus, it was demonstrated for the first time that genes expressable in plant cells need not be present between T-DNA borders to effect plant cell stable transformation. This report was followed by many others, basically presenting variations on the same theme and showing once and for all that transformation of protoplasts from dicots and monocots could be achieved in the presence of an efficient DNA uptake system and a selectable marker gene. A review of these studies constitutes the core of this chapter.

## 3 Uptake of Purified DNA by Seeds, Seedlings, and Plant Cells

There is presently no convincing evidence that seeds, seedlings, and plant cells in culture can take up, integrate, and express isolated DNA of foreign origin.

The use of such plant materials in DNA uptake experiments – rather than plant protoplasts – would only be justified in cases where plant regeneration from protoplasts is extremely difficult or presently impossible. However, there are several drawbacks to this approach. Firstly, even though translocation of foreign DNA could be demonstrated in seedlings, very intense degradation occurred (Lurquin 1977). Secondly, even if a few cells ended up transformed under those circumstances, it is extremely doubtful that the whole organism harboring them would survive the selection pressure necessary for the identification of transgenic material. Indeed, these plants would be mosaics containing a vast majority of untransformed cells. Furthermore, it is unlikely that germ cells would be arise from these few transformed cells, meaning that transmission of the foreign trait to the progeny would occur at an exceedingly low frequency.

An alternative approach would consist in treating germinating seeds or dissected embryos with DNA and subsequently plating them on nonselective callus induction medium. After a given growth period, totipotent (in some species) callus cells could be selected for the transformed trait and transgenic plants conceivably regenerated. This strategy would only work if germinating seeds or embryos were indeed able to take up and integrate DNA, a phenomenon which is as yet very poorly demonstrated, at best. As stated above, this approach has led to the publication of extremely controversial data and has all but been abandoned by the vast majority of the investigators. The last article to be published on the topic of DNA uptake by seedlings (at least to the knowledge of the author of this review) is that by Leber and Hemleben (1979). In this article it was shown that aseptically grown seedlings of *Matthiola incana* can incorporate bromodeoxyuridine-labeled homologous DNA into their nuclei, where it becomes covalently joined with chromosomal DNA. If such a phenomenon does indeed occur, it should be possible to directly transform seedlings with presently available vectors. This information is still lacking. As a matter of fact, such an attempt has yielded a negative result in the author's laboratory. In one experiment, several hundred seeds from *N. plumbaginifolia* were soaked in a 0.2 mg/ml solution of pMON 213 DNA (carrying a *nos-neo* chimeric gene) until germination occurred. After transfer to kanamycin medium, two anthocyanin-rich seedlings survived for a few weeks, while all other seedlings rapidly died. The two abnormal seedlings also died eventually (E. J. Perkins and P. F. Lurquin unpubl.).

Similarly, intact plant cells growing in tissue or suspension cultures have not been shown to be transformable by uptake of isolated DNA although they can be transformed by cocultivation with *Agrobacterium* (Muller et al. 1984; Pollock et al. 1985; An 1985). Recently, it has been demonstrated (Morikawa et al. 1986) that tobacco cells treated with pectinase did express TMV RNA after intracellular transfer mediated by an electric pulse. Evidently, such experiments could be attempted with DNA instead of RNA.

## 4 Uptake of Purified DNA by Gametes

It was suggested over a decade ago that pollen grains might act as vehicles for the transfer of foreign genes into plants (Hess et al. 1974). Later on, the same group reported experiments in which *Nicotiana glauca* pollen incubated with DNA iso-

**Table 1.** Studies describing the fate and biological effects of exogenous DNA in seedlings and pollen grains

| Plant species | Donor DNA | Effect of donor DNA | Reference |
|---|---|---|---|
| Seedlings | | | |
| *Matthiola incana* | [$^3$H]-adenine/5-BrdU-labeled homologous DNA | Uptake into nuclei; integration claimed after detection of hybrid CsCl buoyant density peak | Leber and Hemleben (1979) |
| Pollen | | | |
| *Petunia hybrida* | | | |
| *Nicotiana glauca* | R1 drd 16 plasmid [$^3$H]-DNA and chromosomal [$^3$H]-DNA from *Rhizobium leguminosarum* | Uptake suggested by autoradiography and CsCl gradient analysis | Hess et al. (1974) |
| *Nicotiana glauca* | *Nicotiana langsdorfii* | Tumor formation at wound side | Hess et al. (1976) |
| *Zea mays* | *Zea mays* | Expression of *I*, *Bz* and *Sh* dominant alleles in the recessive host | Ohta (1986) |
| *Secale cereale* | pLGVneo1103 (*nos-neo*) | Production of kanamycin-resistant seedlings | de la Pena et al. (1987)[a] |

[a] It could not be ascertained by de la Pena et al. which of the female or male pregametic cells took up the donor DNA

lated from *Nicotiana langsdorfii* and then used to pollinate emasculated *N. glauca* flowers led to the production of plants which responded to stem wounding by forming tumors at the wound site, a property of *N. glauca*×*N. langsdorfii* sexual hybrids. However, the authors avoided drawing strong conclusions regarding the transfer of genes from one species to the other, especially given the fact that unrelated DNA had a somewhat similar effect.

Ohta (1986) reported on the high transformation efficiency of *Zea mays* homozygous recessive for aleurone color and various endosperm morphological and color markers by a mixture of pollen and DNA extracted from a strain homozygous dominant for all genes. Unfortunately, the interpretation of these results is made difficult by the absence of a control involving DNA isolated from the recessive host.

Very convincing evidence that transgenic plants can be obtained through treatment of gametes with foreign DNA was subsequently provided by De la Peña et al. (1987). This approach also offers a solution (although a rather tedious one) to the problem of recalcitrant protoplast regeneration. In these experiments, the inflorescences of rye plants were manually injected with a solution of plasmid DNA carrying a chimeric gene consisting of the T-DNA nopaline synthase gene 5′- and 3′-control regions flanking the coding sequence of the Tn5 *neo* gene which codes for aminoglycoside antibiotic resistance. Pollination of the injected flowers was allowed and progeny seeds were plated on kanamycin medium. Two kanamycin-resistant seedlings isolated from 3000 planted seeds were shown to contain the *nos-neo* gene as well as the corresponding enzyme activity.

Although painstaking, this method – if widely applicable – does indeed bypass all problems of plant cell or protoplast culture and enables the direct recovery of transgenic plants from gametes treated in situ. Since male germ cells are not yet surrounded by a callose wall at the time of the injection, the authors conclude that male gametes are good candidates for DNA uptake in their system. Whether female gametes can also take up DNA remains to be seen. Table 1 summarizes these findings.

## 5 Uptake of Purified DNA by Protoplasts

Isolated DNA has been shown to be taken up by plant protoplast under essentially three sets of conditions including: (1) the use of chemical "helpers" known to act on both the DNA structure and the surface of plant protoplast, (2) the use of liposomes in which DNA can be encapsulated and (3) the transfer of DNA through the use of an electric pulse of short duration (electroporation).

### 5.1 DNA Uptake Promoted by Chemical Helpers

#### 5.1.1 Poly-L-Ornithine (PLO)

Plasmid DNA uptake stimulated by PLO was first demonstrated in cowpea protoplasts (Lurquin and Kado 1977) and later in carrot, periwinkle, and turnip protoplasts (Fernandez et al. 1978), as well as in barley protoplasts (Hughes et al. 1979).

Subsequently, this technique was used to transform for the first time *Petunia* protoplasts with pTi DNA (Davey et al. 1980). Similarly, PLO was shown to promote transformation of a cell-wall-deficient mutant of *Chlamydomonas reinhardii* (Rochaix and van Dillewijn 1982) and to allow the transfection of turnip protoplasts with cauliflower mosaic virus DNA (Yamaoka et al. 1982).

The exact mode of action of PLO is not understood. This polycation is known to slow down plasmid DNA degradation by deoxyribonucleases (Lurquin and Marton 1980), probably through complex formation, and is also known to strongly stimulate uptake and expression of viral RNA in plant protoplasts (Zaitlin and Beachy 1974), possibly through surface charge neutralization (Nagata and Melchers 1978). A detailed methodology describing the use of PLO in the DNA-mediated transformation of protoplasts has been published (Power et al. 1986).

#### 5.1.2 Polyethylene Glycol (PEG)

PEG is well known to cause protoplast fusion in the presence of calcium ions (Kao and Michayluk 1974). Although poorly understood, this phenomenon is most probably due to effects on the protoplast surface. PEG is also known to enhance DNA (Lurquin and Marton 1980) and RNA (Maule et al. 1980) uptake by plant protoplasts. Barker and Harrison (1982) demonstrated that the infectivity

TRANSIENT EXPRESSION OF THE CAT GENE IN PROTOPLASTS

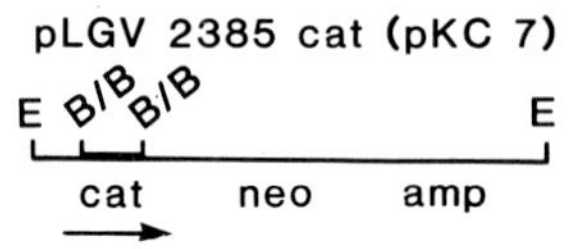


cat: E.coli cat/Bam cassette flanked by pNOS and tNOS

neo: Tn5 neo gene

B/B: Bgl II/Bam HI junction

| PROTOPLAST SOURCE | CPM [$^3$H]-CHLORAMPHENICOL ACETYLATED/30MIN. |
|---|---|
| N. plumbaginifolia | 4567 |
| Corn (cv. Golden Jubilee) | 1470 |
| Wheat (cv. Stephens) | 684 |

**Fig. 1.** Transient expression of the *cat* gene in protoplasts. The transformation vector consists of the pKC7-derived pLGV2385 (courtesy of J. Schell) into which we cloned the coding sequence of the Tn9 *cat* gene flanked by BamHI linkers were cloned. This gene is thus under the control of the *nos* promoter and terminator. Protoplasts were isolated from the mesophyll of *N. plumbaginifolia*, maize and wheat and incubated with pLGV2385-cat in the presence of calf thymus DNA and PEG/$CaCl_2$. *Cat* activity was measured according to the standard procedure involving radioactive chlorophenicol and thin layer chromatography. Spots corresponding to acetylated chlorophenicol (detected by autoradiography) were scraped and counted by liquid scintillation. Aspecific *cat* activity present in the protoplasts was measured after transformation of the the latter with pLGV2385 containing the *cat* gene in the reverse orientation. Radioactivity values given are equal to the difference between *cat* activities measured in the two types of experiments and thus represent net activity coded for by the vector (Paszty and Lurquin unpubl.)

of several viral RNA species was strongly increased if the protoplast incubation medium contained a mixture of PEG and $CaCl_2$.

Krens et al. (1982) and Draper et al. (1982) adapted this methodology to further demonstrate uptake, integration, and expression of purified pTi in tobacco protoplasts. These experiments clearly confirmed the existence of stable DNA-mediated transformation of plant protoplasts through covalent linkage between the donor DNA and plant nuclear DNA, often in a fashion quite different from that observed when the pTi T-DNA is transferred to plant nuclei via *Agrobacterium* infection (to be discussed later).

Later, Paszkowski et al. (1984) used the PEG technique to demonstrate for the first time that a bacterial vector harboring a dominant marker selectable in plant cells could become integrated within the nuclear genome of cells originated from DNA-treated protoplasts, and this in the absence of homology to either T-DNA borders or the host DNA. The PEG/$CaCl_2$-stimulated DNA uptake technique has recently been used to demonstrate expression of chimeric genes in three monocotyledonous species (Lörz et al. 1985; Potrykus et al. 1985a; Uchimiya et al. 1986a) and has been described in detail (Paszkowski and Saul 1986).

We have also used this technique to demonstrate transient expression of a chimeric *nos-cat* gene in mesophyll protoplasts isolated from tobacco, wheat, and maize (Fig. 1). Results showed that the *nos* promoter was much less active in monocot protoplasts than in tobacco protoplasts.

#### 5.1.3 Use of Calcium Phosphate Microcrystals

DNA can be coprecipitated upon formation of calcium phosphate microcrystals. This insoluble complex can be internalized by animal cells through endocytosis, after which it is dissociated, allowing DNA to reach the nucleus of the cell. This technique (Graham et al. 1974) is the one most commonly used to transfect and transform animal cells. It has been modified by Hain et al. (1985) for plant protoplast transformation, and in this version includes treatments of the DNA coprecipitate/protoplasts mixture with a fusogen (polyvinyl alcohol) and a high pH buffer, presumably in order to promote endocytosis of the calcium phosphate microcrystals.

### 5.2 Liposomes

Liposomes are microscopic lipid vesicles produced when phospholipids are dispersed in an aqueous phase. Depending on the conditions used, liposomes are multilamellar, unilamellar, neutral, positively or negatively charged. Liposomes large enough to allow entrapment of DNA can easily be generated and can interact with a variety of cells in such a way that plasmids will eventually be transferred to their nucleus (for review see Lurquin 1984). In the particular case of plant protoplasts, liposomes have been used to transfer both RNA and DNA, and have proven to be especially useful in the expression of viral RNA molecules (Caboche and Lurquin 1987).

Unilamellar liposomes composed of phosphatidylserine and cholesterol have been used to transfer and express in plant protoplasts a cloned cDNA copy of the potato spindle tuber viroid (Faustmann et al. 1986) and a *nos-neo* chimeric gene present in an *E. coli* plasmid (Deshayes et al. 1985). Efficient transfer of DNA from liposomes to intracellular compartments requires the action of PEG (Lurquin 1979). It is generally agreed upon that negatively charged unilamellar liposomes such as the ones mentioned above are best suited to achieve transfer of RNA and DNA in a biologically active form. Figure 2 shows that the initial liposome binding step is rather rapid but that transfer of the encapsulated DNA to the nuclear compartment occurs much more slowly. Werr and Lörz (1986) demonstrated that plasmid DNA transferred to protoplast nuclei can remain biologically active for several days in an unintegrated state and in a linear or open circular configuration.

Details of the liposome methodology have been published (Caboche and Lurquin 1987). Table 2 summarizes the various studies in which liposomes were used to transfer biologically active nucleic acids into protoplasts.

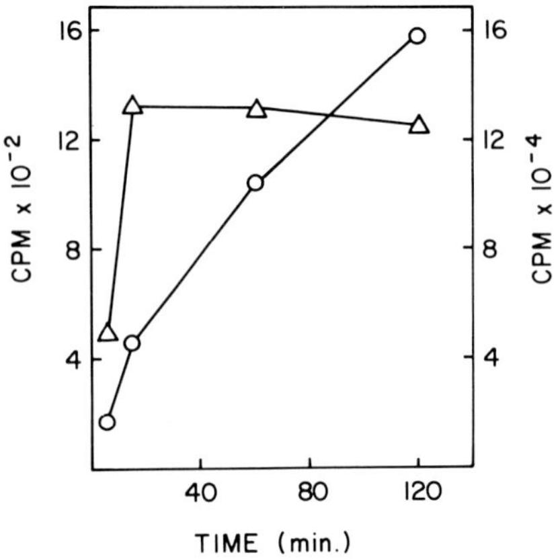


**Fig. 2.** Binding of lipsomes to protoplasts and transfer of encapsulated DNA to their nuclei. Phosphatidylserine/cholesterol reverse evaporation vesicles were loaded with nick-translated Adenovirus 2 [$^3$H]-DNA and incubated with protoplasts isolated from *N. tabacum* cv. Xanthi suspension cells. △ – △: total radioactivity associated with protoplasts; ○ – ○: radioactivity associated with purified nuclei isolated from treated protoplasts. A suboptimal concentration of PEG 6000 of 10% w/v was used to avoid heavy clumping of the protoplasts

**Table 2.** Use of unilamellar liposomes in the transfer and expression of genetic material into protoplasts

| RNA or DNA source | Protoplast source | Liposome type and composition | Reference |
|---|---|---|---|
| Tobacco mosaic virus (TMV) RNA | *N. tabacum* suspensions | Phosphatidylserine/cholesterol REV | Nagata et al. (1981) |
| TMV RNA | *Vinca rosea* suspension | Phosphatidylserine LUV | Fukunaga et al. (1981) |
| Turnip rosette virus RNA | *Brassica rapa* mesophyll | β-sitosterol/dicetylphosphate/ phosphatidylcholine REV | Rollo and Hull (1982) |
| TMV RNA | *N. tabacum* suspensions | Phosphatidylserine/cholesterol REV | Fraley et al. (1982) |
| Cowpea chlorotic mottle virus RNA | *Vigna sinensis* mesophyll | Phosphatidylserine/cholesterol REV | Christen and Lurquin (1983) |
| TMV RNA | *N. tabacum* mesophyll | Phosphatidylserine/cholesterol REV | Rouze et al. (1983) |
| Tabacco vein mottling virus RNA | *N. tabacum* mesophyll | Phosphatidylserine/cholesterol REV | Zhenghaixu et al. (1984) |
| pLGV23 neo DNA | *N. tabacum* mesophyll | Phosphatidylserine/cholesterol REV | Deshayes et al. (1985) |
| cDNA copy of potato spindle tuber viroid | *Solanum tuberosum* suspensions | Phosphatidylserine/cholesterol REV | Faustmann et al. (1986) |

LUV: large unilamellar vesicles
REV: reverse phase evaporation vesicles

Although there seems to be a general consensus regarding the high efficiency with which liposomes can deliver viral RNA molecules into protoplasts, the DNA-mediated transformation frequency observed after liposome transfer is not much higher than that reported in the case of the PEG method (Deshayes et al. 1985). Whether this is due to the total amount of DNA which can be transferred to individual protoplasts, or to the fact that only a fraction of the protoplast population is able to interact with liposomes, or to a combination of both factors is presently not clear. At any rate, attempts are being made at maximizing liposome-protoplast interactions (Cutler et al. 1984), notably through the use of pH-sensitive liposomes (Wang et al. 1986). This approach, in conjunction with liposome targeting with lectins (Sheehy and Lurquin 1983) could lead to higher transformation frequencies.

## 5.3 Electroporation

The term electroporation refers to the process by which macromolecules present in the extracellular medium are internalized by living cells following a brief electric pulse (Neumann et al. 1982). This DNA transfer procedure was found to be extremely efficient with protoplasts isolated from both dicots and monocots. The technique allows transfer of biologically active chimeric genes under essentially two sets of parameters consisting of: (1) a voltage peak value of 250–350 V with an RC constant in the millisecond range (Fromm et al. 1985, 1986; Ecker and Davis 1986) or (2) a voltage peak value of up to 2000 V with an RC constant

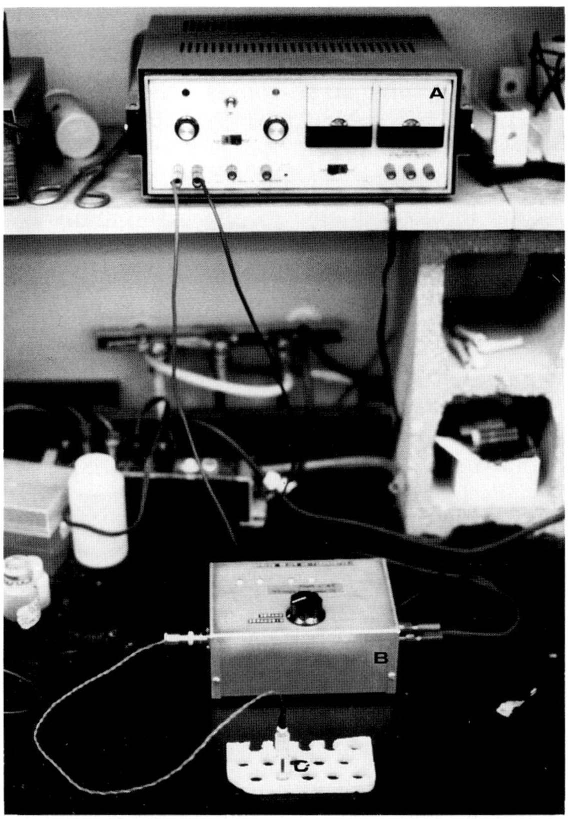


**Fig. 3.** Electroporation apparatus built at Washington State University. *A* Heathkit power supply (400 V max.); *B* box housing the capacitor and the charge/discharge circuits; *C* stainless steel electrodes built to fit a disposable spectrophotometer microcuvette (electrode gap = 0.9 cm). The leads and electrodes, their contacts being housed in a glass tubing which fits the opening of the cuvette, are autoclaved before use. The cuvettes are sterilized with 70% ethanol. This unit can be built for $130. It lacks flexibility in the sense that only pulse height can be adjusted. Pulse time can be changed by installing a different capacitor. Exponential pulses only can be delivered with this system. More sophisticated units are commercially available and cost a minimum of $1000

in the microsecond range (Langridge et al. 1985; Shillito et al. 1985; Riggs and Bates 1986; Ou-Lee et al. 1986).

The technique is compatible with high survival rates, requires minimum manipulation of the protoplasts, and necessitates equipment which can be built easily in any moderately well-installed workshop. Electroporation seems to be especially well adapted to the rapid assay of promotor strength in vivo because of the large amount of DNA which can be reproducibly transferred intracellular. For example, it has been shown that the promoter regions isolated from cauliflower mosaic virus DNA (the so-called 19S and 35S promoters) as well as the *copia* long terminal repeat promoter of *Drosophilla* are consistently more active than the T-DNA *nos* promoter (Fromm et al. 1985; Ou-Lee et al. 1986). Electroporation was also used to study the transient expression of a firefly luciferase gene in carrot protoplasts. It was demonstrated that this gene was active when cloned downstream from the 35S promoter, as indicated by luminescence of the protoplasts in the presence of added luciferin (Ow et al. 1986).

Figure 3 shows a simple electroporation unit built at Washington State University; it contains a 250 microfarad capacitor which we use to deliver 250 V pulses to protoplasts with a survival of at least 70% and easily measurable transient *CAT* activity.

### 5.4 Microinjection

Mammalian cells can be transformed at high efficiency by mechanical microinjection of DNA (Capecchi 1980). Similar results were reported in the case of tobacco mesophyll protoplasts (Crossway et al. 1986). In fact, the protoplast transformation frequency was so high (14%) that no selection was required in order to identify transformed cell lines by Southern blot hybridization. The major drawback of this technique is that it requires very sophisticated equipment and skilled, well-trained personnel.

Of all the DNA uptake techniques described above, the electroporation procedure may well be the easiest to use with the highest chance of success. Indeed, and except for the difficult microinjection technique, all the others produce transformation frequencies in the $10^{-5}$ range. The values reported for stable transformation by electroporation are much higher, in the $10^{-2}$ range. It seems quite likely that electroporation will become the DNA transfer technique par excellence, especially in "shotgun"-type experiments involving the use of uncloned nonselectable markers.

## 6 Nature of Expression Vectors Used in Direct Gene Transfer

Vectors used in gene expression experiments in plant protoplasts must fulfill two or three requirements, depending on the purpose of the investigation. If stable transformation is sought, the marker gene must be selectable in tissue culture and should be dominant in order to allow for the recovery of hemizygous transformants. This, of course, implies that the gene will have to be under the control of 5′- and 3′-regulatory sequences recognized as such in plant cells. In addition, the product of this gene should be assayable, in order to differentiate between true transformants and spontaneous mutants or revertants. To date, the most widely used dominant selectable marker is the *neo* gene isolated from bacterial transposon Tn5 and cloned downstream from either the T-DNA *nos* promoter or the cauliflower mosaic virus (CaMV) 19S or 35S promoter. The 3′-end terminator region is from either the *nos* gene or the common 19-S/35-S terminator from CaMV. These chimeric genes confer kanamycin resistance to transformed plant cells.

In cases where selection of transformants is not the goal of the experiment, any easily assayable gene can be used, provided it is, as above, active in plant cells. One such gene is constituted by the *CAT* coding sequence from bacterial transposon Tn9 under the control of either one of the above regulatory sequences.

**Table 3.** Representative examples of stable transformation of protoplasts

| DNA uptake conditions | Protoplast source | Donor plasmid | Selection | Reference |
|---|---|---|---|---|
| PLO | Petunia | pTi | Hormone autotrophy | Davey et al. (1980) |
| PEG/$CaCl_2$ | Tobacco | pTi | Hormone autography | Krens et al. (1982) |
| | Tobacco | pABDI (PVI-neo) | Kanamycin resistance | Paszkowski et al. (1984) |
| | *Triticum monococcum* | pBL-1103-4 (Pnos-neo) | Kanamycin resistance | Lörz et al. (1985) |
| | *Lolium multiflorum* | pABDI | Kanamycin resistance | Potrykus et al. (1985a) |
| | Rice | $pCT_2T_3$ (Pnos-neo) | Kanamycin resistance | Uchimiya et al. (1986a) |
| $Ca_3(PO_4)_2$ Microcrystals | Tobacco | pLGVneo 2103 (Pnos-neo) | Kanamycin resistance | Hain et al. (1985) |
| Liposomes | Tobacco | pLGV23neo (Pnos-neo) | Kanamycin resistance | Deshayes et al. (1985) |
| Electroporation | Maize | pCaMVNEO (P35-S-neo) | Kanamycin resistance | Fromm et al. (1986) |
| | Tobacco | pMON200 (Pnos-neo) | Kanamycin resistance | Riggs and Bates (1986) |
| | Carrot | pTi | Hormone autography | Langridge et al. (1985) |
| Microinjection | Tobacco | pCGN561 (Engineered T-DNA) | None | Crossway et al. (1986) |

PVI-neo, *neo* gene cloned downstream from the CaMV gene VI (or 19-S) promoter; Pnos-neo, *neo* gene cloned downstream from the *nos* T-DNA promoter; P35-S-neo, *neo* gene cloned downstream from the 35-S CaMV promoter responsible for the transcription of the full-length 35-S CaMV RNA used as intermediate in the viral genome replication

**Table 4.** Representative examples of transient gene expression studied in protoplasts

| Protoplast source | Gene transiently expressed | Reference |
|---|---|---|
| Carrot<br>Maize<br>Tobacco | Chloramphenicol acetyltransferase (*cat*) under control of either Pnos or PCaMV35-S | Fromm et al. (1985) |
| Carrot | *Cat* gene under control of Pnos or PCaMV35-S, in both orientations for the study of the inhibitory effect of antisense mRNA | Ecker and Davis (1986) |
| Rice<br>Wheat<br>Sorghum | *Cat* gene under control of PCaMV35-S or *copia* long terminal repeat promoter of *Drosophila* | Ou-Lee et al. (1986) |
| Carrot | Firefly liciferase gene under control of PCaMV35-S promoter | Ow et al. (1986) |

Electroporation was used to stimulate DNA uptake in all the above experiments

The advantage of this particular marker is that its product, chloramphenicol acetyltransferase, is rapidly and easily assayable (Sleigh 1986) and therefore, *CAT* can be considered a good "reporter" gene. Its disadvantage is that selection of transformants on chloramphenicol is difficult to reproduce (Pietrzak et al. 1986).

Since homology between the host genome and the expression vector is not necessary (see Sect. 5.1.2) for stable integration, the chimeric gene can be cloned in any convenient *E. coli* replicon such as pBR322, pUC19, or one of their derivatives.

Table 3 summarizes the studies in which stable transformation after direct gene transfer was demonstrated while Table 4 lists the reports in which transient expression of foreign genes in plant protoplasts (i.e., without selection of stable transformants) was studied.

## 7 Integration of Foreign DNA Taken up by Protoplasts

This section reviews the reports providing evidence that heterologous DNA present intracellularly after direct gene transfer can become integrated within the plant cell genome and is then stably maintained by covalent linkage to flanking plant DNA sequences.

### 7.1 Foreign DNA Is Integrated in the Nucleus

Foreign DNA present in cells derived from protoplasts transformed by direct gene transfer is integrated within plant genomic DNA, as indicated by Southern blot hybridization. This phenomenon occurs no matter what direct gene transfer technique is used, showing that as long as DNA is made to cross the plasma membrane, at least some of it will be stably maintained in the genome. So far, there is no evidence that foreign DNA can become integrated within the organelle genomes. In fact, given the eukaryotic nature of the promoter regions present in the chimeric genes used in direct gene transfer, it would have been unlikely that the transformants obtained in these studies could have been of the cytoplasmic type. Rather, the genetic analysis of transgenic plants regenerated from protoplasts transformed with a *nos-neo* chimeric gene clearly shows that the foreign marker is inherited as a single dominant Mendelian trait segregating in a 3:1 fashion in the $F_1$ generation of selfed kanamycin-resistant plants (Potrykus et al. 1985b; Hain et al. 1985; Deshayes et al. 1985; Riggs and Bates 1986). These results indicate the existence of a single functional copy of the foreign gene integrated in the nuclear genome of the transgenic plants which are thus hemizygous for the added trait. Interestingly, all the above studies also report the presence of several nonfunctional copies of the foreign gene, integrated as a block together with the functional gene. This tandem duplication of the integrated expression vector is indicative of other recombinational events occurring during transformation (see Sect. 7.2).

Uchimiya et al. (1986b) have also demonstrated that tobacco protoplasts transformed with a vector carrying an intact *nos* gene plus a *nos-neo* chimeric gene could be regenerated into plants expressing both traits (nopaline production and kanamycin resistance). In addition, these two traits remained linked in progeny plants, showing that the two marker genes originally contiguous on the vector were linked on the same chromosome. These experiments demonstrate that it will be possible to isolate plants transformed with a nonselectable marker, provided the latter is accompanied by a selectable gene on the same transformation vector.

## 7.2 Foreign DNA Undergoes Recombination upon Integration

Large DNA molecules such as pTi undergo cleavage when introduced into protoplasts by direct gene transfer, the result of which being that integration of DNA fragments of variable length occurs (Krens et al. 1982). In addition, scrambling and fusion of these DNA fragments is also observed, yielding complex integration patterns (Krens et al. 1985).

As indicated above, similar but less complicated integrative events are also observed when the expression vectors are much smaller than pTi. This is consistent with a much smaller target size for extracellular and intracellular nucleases. Of particular interest is the existence of tandem duplication of the vector. Deshayes et al. (1985) have hypothesized that vector concatenation occurred by homologous recombination prior to integration in the nuclear genome. Riggs and Bates (1986) also demonstrated that linearized plasmid introduced into protoplasts by electroporation could recircularize prior to integration.

In only one case was it shown that a plant expression vector could integrate as a single unscrambled copy. This phenomenon may be attributed to the low nuclease content of the *Brassica* protoplast used in these experiments (Paszkowski et al. 1986). These results show that the direct gene transfer techniques will prove invaluable in the study of recombinational events in plant cells.

### 7.3 Cotransformation

The term cotransformation characterizes the independent integration in the same transformed cell of two markers originally present in diffeent DNA molecules added together to the protoplast transformation medium. The PEG-mediated transformation technique (Krens et al. 1982) involves the use of calf thymus DNA as substrate for cellular nucleases. Peerbolte et al. (1985) discovered that tobacco protoplasts transformed to hormone autotrophy by pTi DNA in the presence of carrier calf thymus DNA possessed not only integrated copies of the T-DNA; in addition, up to 50% of these transformants also contained integrated calf thymus DNA sequences. This finding shows that two markers need not be present in the same DNA molecules in order to be transferred together at high frequency. The same authors also demonstrated that these unrelated DNA sequences were integrated in different chromosomes.

Pietrzak et al. (1986) basically confirmed the existence of cotransformation by using two markers selectable in plants (*neo* and *CAT*) present in two different plasmids.

Given the fact that markers introduced into plant protoplasts by cotransformation display meiotic segregation in regenerated plants (Peerbolte et al. 1985), whereas segregation of markers does not occur if they are originally present in the same molecules (Uchimiya et al. 1986b), it is clear that the two approaches do not have the same type of applicability.

## 8 Present Status and Conclusions

The genetic manipulation of plant protoplasts by direct gene transfer has clearly come of age after a long gestation period. The major limiting factor to the practical application of this technology is that protoplasts isolated from many crop plants are still refractory to regeneration. However, the great successes achieved in the past 2 or 3 years in the field of DNA transfer should renew or enhance interest in this difficult problem. In the meantime, the new technologies will allow to tackle basic problems of plant molecular genetics in an entirely different way, because of our present ability to transfer and follow the expression of reporter genes during such events as differentiation and morphogenesis. Also, transformation techniques will be invaluable in the rapid evaluation of promoters and other regulatory elements. As already mentioned above, recombination mechanisms in plants are now amenable to study, as are questions of gene targeting to specific locations in the genome and gene replacement. For example, we have constructed vectors carrying selectable markers which will enable us to study homologous recombination (and hence gene targeting) in tobacco and in maize due to the presence of a piece of host chromosomal DNA carried by the recombinant plasmids (Figs. 4, 5).

It has not been the purpose of this chapter to compare and contrast the merits of direct gene transfer vs. *Agrobacterium*-mediated gene transfer. Nevertheless, the coexistence of these two technologies warrants a few comments. The main

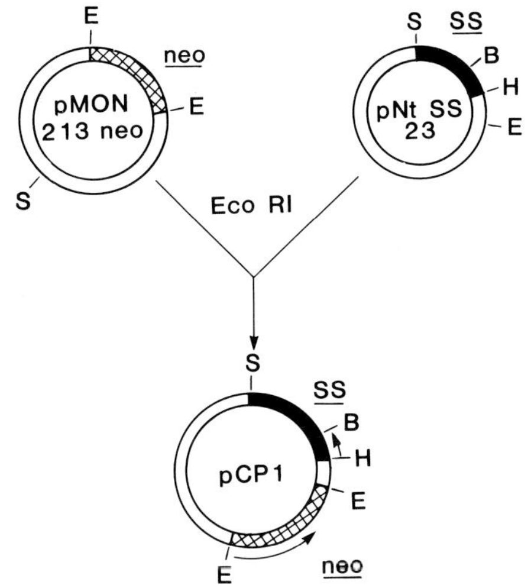


**Fig. 4.** Homologous recombination probe pCPl for use in direct gene transfer into *N. tabacum* protoplasts. The dominant selectable vector pMON213 (courtesy of S. Rogers) containing an EcoR I cassette carrying the *nos-neo* chimeric gene was cloned into pNtSS23 (courtesy of Dupont Co.), a recombinant plasmid containing a 1.2-kb 5′-end fragment of the tobacco gene coding for the small subunit (SS) of ribulose bisphosphate carboxylase (Paszty and Lurquin unpubl.). *Solid area* portion of the Rubisco SS gene; *hatched area* chimeric *neo* gene; *arrows* indicate direction of transcription. This vector contains a single Bgl II site (designated B) convenient for linearizing the plasmid inside its region of homology with the host chromosome

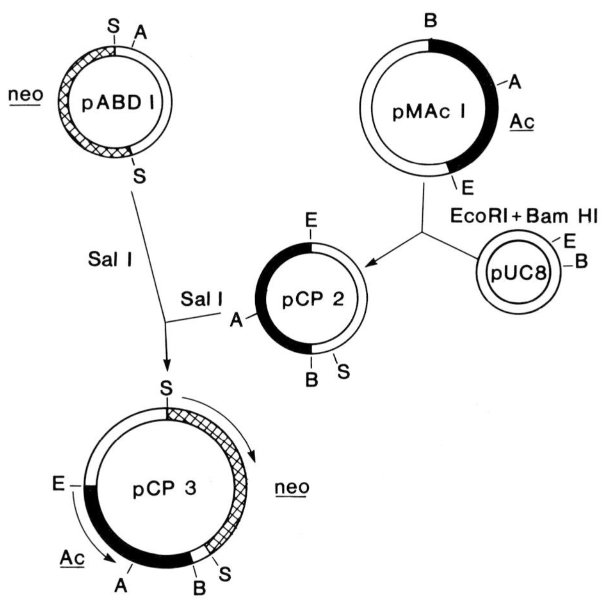


**Fig. 5.** Homologous recombination probe pCP3 for use in direct gene transfer into *Z. mays* protoplasts. This plasmid contains the *neo* gene flanked by the promoter and terminator regions of CaMV gene VI (courtesy of J. Paszkowski) as well as a 3-kb fragment from the 5′-end of a maize actin gene (courtesy of R. Meagher). The various construction steps are shown. We have also constructed a similar plasmid containing the *aphIV* gene for selection on hygromycin B (Paszty and Lurquin unpubl.) *Solid area* portion of the actin gene; *hatched area* chimeric *neo* gene; *arrows* indicate directions of transcription. This vector contains a single Ava I site (designated A) useful for linearizing the plasmid in its region of homology with the host chromosome

problem encountered with the first technique is the existence of recombination and tandem duplication of expression vectors before integration. This phenomenon is usually not observed when using the second technique. However, it remains that from a purely genetic point of view, as far as we know, plants regenerated from directly transformed protoplasts are indistinguishable from those regenerated from protoplasts transformed by cocultivation with *Agrobacterium*. In addition, the direct transformation technique requires less recombinant DNA manipulations and allows foreign gene assay without stable integration. Finally, this approach may eventually prove to be more flexible in the search for organelle transformation events.

*Acknowledgments.* This work was supported in part by a grant from the State of Washington High Technology Center.

## References


An G (1985) High efficiency transformation of cultured tobacco cells. Plant Physiol 79, 568–570

Barker H, Harrison D (1982) Infection of potato mesophyll protoplasts with five plant viruses. Plant Cell Rep 1:247–249

Bevan MW, Flavell RB, Chilton M-D (1983) A chimaeric antibiotic resistance gene as a selectable marker for plant cell transformation. Nature 304:184–187

Caboche M, Lurquin PF (1987) Liposomes as carriers for the transfer and expression of nucleic acids into higher plant protoplasts. In: Packer L, Douce R (eds) Methods in Enzymology. Academic Press, New York, Vol 148, pp 39–45

Capecchi MR (1980) High efficiency transformation by direct microinjection of DNA into cultured mammalian cells. Cell 22:479–488

Chilton M-D, Drummond MM, Merlo DJ, Sciaky D, Gordon MP, Nester EW (1977) Stable incorporation of plasmid DNA into higher plant cells: The molecular basis of crown gall tumorigenesis. Cell 11:263–271

Christen AA, Lurquin PF (1983) Infection of cowpea mesophyll protoplasts with cowpea chlorotic mottle virus (CCMV) RNA encapsulated in large liposomes. Plant Cell Rep 2:43–46

Crossway A, Oakes JV, Irvine JM, Ward B, Knauf VC, Shewmaker CK (1986) Integration of foreign DNA following microinjection of tobacco mesophyll protoplasts. Mol Gen Genet 202:179–185

Cutler AJ, Shargool PD, Kurz WGW, Constabel F (1984) Liposomes as vehicles for transferring low molecular weight compounds into periwinkle protoplasts. J Plant Physiol 117:29–40

Davey MR, Freeman J, Draper J (1979) Transformation of higher plant protoplasts by isolated *Agrobacterium* plasmids. Abstr Fifth Int Protoplast Symp, July 9–14 Szeged, Hungary, p 137

Davey MR, Cocking EC, Freeman J, Pearce N, Tudor I (1980) Transformation of *Petunia* protoplasts by isolated *Agrobacterium* plasmids. Plant Sci Lett 18:307–313

de la Pena A, Lörz H, Schell J (1987) Transgenic rye plants obtained by injecting DNA into young floral tillers. Nature 325:274–276

Deshayes A, Herrera-Estrella L, Caboche M (1985) Liposome mediated transformation of tobacco mesophyll protoplasts by an *Escherichia coli* plasmid. EMBO J 4:2731–2737

Draper J, Davey MR, Freeman JP, Cocking EC, Cox BJ (1982) Ti plasmid homologous sequences present in tissues from *Agrobacterium* plasmid-transformed *Petunia* protoplasts. Plant Cell Physiol 23:451–458

Ecker JR, Davis RW (1986) Inhibition of gene expression in plant cells by expression of antisense RNA. Proc Natl Acad Sci USA 83:5372–5376

Faustmann O, Kern R, Sanger HL, Muhlbach H-P (1986) Potato spindle tuber viroid (PSTV) RNA oligomers of (+) and (−) polarity are synthesized in potato protoplasts after liposome-mediated infection with PSTV. Virus Res 4:213–227

Fernandez SM, Lurquin PF, Kado CI (1978) Incorporation and maintenance of recombinant-DNA plasmid vehicles pBR313 and pCR1 in plant protoplasts. FEBS Lett 87:277–282

Fraley RT, Dellaporta SL, Papahadjopoulos D (1982) Liposome-mediated delivery of tobacco mosaic virus RNA into tobacco protoplasts: A sensitive assay for monitoring liposome-protoplast interactions. Proc Natl Acad Sci USA 79:1859–1863

Fraley RT, Rogers SG, Horsch RB, Sanders PR, Flick JS, Adams SP, Bittner ML, Brand LA, Fink DL, Fry JS, Galluppi GR, Goldberg SB, Hoffmann NL, Woo SC (1983) Expression of bacterial genes in plant cells. Proc Natl Acad Sci USA 80:4803–4807

Fromm ME, Taylor LP, Walbot V (1985) Expresssion of genes electroporated into monocot and dicot plant cells. Proc Natl Acad Sci USA 82:5804–5828

Fromm ME, Taylor LP, Walbot V (1986) Stable transformation of maize after gene transfer by electroporation. Nature 319:791–793

Fukunaga Y, Nagata T, Takebe I (1981) Liposome-mediated infection of plant protoplasts with tobacco mosaic virus RNA. Virology 113:752–760

Gheysen G, Dhaese P, Van Montagu M, Schell J (1985) DNA flux across genetic barriers: The crown gall phenomenon. In: Hohn B, Dennis ES (eds) Genetic flux in plants. Springer, Berlin Heidelberg New York Tokyo, pp 11–47

Graham FL, Van der Eb AJ, Heijneker HL (1974) Size and location of the transforming regions in human adenovirus type 5 DNA. Nature 251:687–691

Hain R, Stabel P, Czernilofsky AP, Steinbiss HH, Herrera-Estrella L, Schell J (1985) Uptake, integration, expression and genetic transmission of a selectable chimeric gene by plant protoplasts. Mol Gen Genet 199:161–168

Herrera-Estrella L, Depicker A, Van Montagu M, Schell J (1983) Expression of chimeric genes transferred into plant cells using a Ti-plasmid-derived vector. Nature 303:209–213

Hess D, Lorz H, Weissert E-V (1974) Uptake of bacterial DNA into swelling and germinating pollen grains of *Petunia hybridia* and *Nicotiana glauca*. Z Pflanzenphysiol 74:52–63

Hess D, Schneider G, Lorz H, Blaich G (1976) Investigations on the tumor induction in *Nicotiana glauca* by pollen transfer of DNA isolated from *Nicotiana langsdorfii*. Z Pflanzenphysiol 77:247–254

Hughes BG, White FG, Smith MA (1979) Fate of bacterial plasmid DNA during uptake by barley and tobacco protoplasts: II. Protection by poly-L-ornithine. Plant Sci Lett 14:303–310

Kao KN, Michayluk MR (1974) A method for high-frequency intergeneric fusion of plant protoplasts. Planta 115:355–367

Kleinhofs A, Behki R (1977) Prospects for plant genome modification by nonconventional methods. Annu Rev Genet 11:79–101

Krens FA, Molendijk L, Wullems GJ, Schilperoort RA (1982) In vitro transformation of plant protoplasts with Ti-plasmid DNA. Nature 296:72–74

Krens FA, Mans RMW, van Slogteren TMS, Hoge JHC, Wullems GJ, Schilperoort RA (1985) Structure and expression of DNA transferred to tobacco via transformation of protoplasts with Ti-plasmid DNA: Co-transfer of T-DNA and non T-DNA sequences. Plant Mol Biol 5:223–234

Langridge WHR, Li BJ, Szalay AA (1985) Electric field mediated stable transformation of carrot protoplasts with naked DNA. Plant Cell Rep 4:355–359

Leber B, Hemleben V (1979) Uptake of homologous DNA into nuclei of seedlings and by isolated nuclei of a higher plant. Z Pflanzenphysiol 91:305–316

Ledoux L, Huart R (1969) Fate of exogenous bacterial deoxyribonucleic acids in barley seedlings. J Mol Biol 43:243–262

Ledoux L, Huart R, Jacobs M (1974) DNA-mediated genetic correction of thiamineless *Arabidopsis thaliana*. Nature 249:17–21

Lörz H, Baker B, Schell J (1985) Gene transfer to cereal cells mediated by protoplasts transformation. Mol Gen Genet 199:178–182

Lurquin PF (1977) Integration versus degradation of exogenous DNA in plants: An open question. In: Cohn WE (ed) Progress in nucleic acid research and molecular biology, vol. 20. Academic Press, New York pp 161–207

Lurquin PF (1979) Entrapment of plasmid DNA by liposomes and their interactions with plant protoplasts. Nucleic Acids Res 6:3773–3784

Lurquin PF (1984) Entrapment of genetic material into liposomes and delivery to cells. In: Gregoriadis G (ed) Liposome technology. CRC Press, Boca Raton, pp 187–193

Lurquin PF (1987) Foreign gene expression in plant cells. In: Cohn WE, Moldave K (eds) Progress in nucleic acid research and molecular biology. Academic Press, New York, Vol 34, pp 143–188

Lurquin PF, Kado CI (1977) *Escherichia coli* plasmid pBR313 insertion into plant protoplasts and into their nuclei. Mol Gen Genet 154:113–121

Lurquin PF, Marton L (1980) DNA transfer experiments with plant protoplasts and bacterial plasmids. In: Farkas G, Dudits D, Ferenczy L (eds) Advances in protoplast research. Pergamon New York, pp 389–405

Marton L, Wullems GJ, Molendijk L, Schilperoort RA (1979a) *In vitro* transformation of cultured cells from *Nicotiana tabacum* by *Agrobacterium tumefaciens*. Nature 277:129–131

Marton L, Wullems GJ, Lurquin PF, Molendijk L, Schilperoort RA (1979b) Crown gall transformation of tobacco protoplasts by Ti plasmid DNA from *Agrobacterium tumefaciens*. Abstr Fifth Int Protoplast Symp July 9–14 Szeged, Hungary, p 136

Maule AJ, Boulton MI, Edmunds C, Wood KR (1980) Polyethylene glycol-mediated infection of cucumber protoplasts by cucumber mosaic virus and virus RNA. J Gen Virol 47:199–203

Morikawa H, Iida A, Matsui C, Ikegami M, Yamada Y (1986) Gene transfer into intact plant cells by electroinjection through cell walls and membranes. Gene 41:121–124

Muller A, Manzara T, Lurquin PF (1984) Crown gall transformation of tobacco callus cells by cocultivation with *Agrobacterium tumefaciens*. Biochem Biophys Res Commun 123:458–462

Nagata T, Melchers G (1978) Surface charge of protoplasts and their significance in cell-cell interaction. Planta 142:235–238

Nagata T, Okada K, Takebe I, Matsui C (1981) Delivery of tobacco mosaic virus RNA into plant protoplasts mediated by reverse-phase evaporation vesicles (liposomes). Mol Gen Genet 184:161–165

Neumann E, Schaefer-Ridder M, Wang Y, Hofschneider PH (1982) Gene transfer into mouse lyoma cells by electroporation in high electric fields. EMBO J 1:841–845

Ohta Y (1986) High-efficiency genetic transformation of maize by a mixture of pollen and exogenous DNA. Proc Natl Acad Sci USA 83:715–719

Ou-Lee T-M, Turgeon R, Wu R (1986) Expression of a foreign gene linked to either a plant-virus or a *Drosophila* promoter, after electroporation of protoplasts of rice, wheat and sorghum. Proc Natl Acad Sci USA 83:6815–6819

Ow DW, Wood KV, DeLuca M, De Wet JR, Helinski DR, Howell SH (1986) Transient and stable expression of the firefly luciferase gene in plant cells and transgenic plants. Science 234:856–859

Paszkowski J, Saul MW (1986) Direct gene transfer to plants. In: Weissbach H, Weissbach A (eds) Methods in enzymology, vol. 118. Academic Press, pp 668–675

Paszkowski J, Shillito RD, Saul M, Mandak V, Hohn T, Hohn B, Potrykus I (1984) Direct gene transfer to plants. EMBO J 3:2717–2722

Paszkowski J, Pisan B, Shillito RD, Hohn T, Hohn B, Potrykus I (1986) Genetic transformation of *Brassica campestris* var. *rapa* protoplasts with an engineered cauliflower mosaic virus genome. Plant Mol Biol 6:303–312

Peerbolte R, Krens FA, Mans RMW, Floor M, Hoge GHC, Wullems GJ, Schilperoort RA (1985) Transformation of plant protoplasts with DNA: Cotransformation of non-selected calf thymus carrier DNA and meiotic segregation of transforming DNA sequences. Plant Mol Biol 5:235–246

Pietrzak M, Shillito RD, Hohn T, Potrykus I (1986) Expression in plants of two bacterial antibiotic resistance genes after protoplast transformation with a new plant expression vector. Nucleic Acids Res 14:5857–5866

Pollock K, Barfield DG, Robinson SJ, Shields R (1985) Transformation of protoplast-derived cell colonies and suspension cultures by *Agrobacterium tumefaciens*. Plant Cell Rep 4:202–205

Potrykus I, Saul MW, Petruska J, Paszkowski J, Shillito RD (1985a) Direct gene transfer to cells of a graminaceous monocot. Mol Gen Genet 199:183–188

Potrykus I, Paszkowski J, Saul MW, Petruska J, Shillito RD (1985b) Molecular and general genetics of a hybrid foreign gene introduced into tobacco by direct gene transfer. Mol Gen Genet 199:169–177

Power JB, Davey MR, Freeman JP, Mulligan BJ, Cocking EC (1986) Fusion and transformation of plant protoplasts. In: Weissbach H, Weissbach A (eds) Methods in Enzymology, vol. 118. Academic Press, New York, pp 578–588

Redei GP, Acedo G, Weingarten H, Kier LD (1976) Has DNA corrected genetically thiamineless mutants of *Arabidopsis*? In: Dudits D, Farkas GL, Maliga P (eds) Cell Genetics in Higher plants. Akademic Kiado, Budapest, pp 91–94

Riggs CD, Bates GW (1986) Stable transformation of tobacco by electroporation: Evidence for plasmid concatenation. Proc Natl Acad Sci USA 83:5602–5606

Rochaix J-D, van Dillewijn J (1982) Transformation of the green alga *Chlamydomonas reinhardii* with yeast DNA. Nature 296:70–72

Rollo F, Hull R (1982) Liposome-mediated infection of turnip protoplasts with turnip rosette virus and RNA. J Gen Virol 60:359–363

Rouze P, Deshayes A, Caboche M (1983) Use of liposomes for the transfer of nucleic acids: Optimization of the method for tobacco mesophyll protoplasts with tobacco mosaic virus RNA. Plant Sci Lett 31:55–64

Sheehy RE, Lurquin PF (1983) Targeting of large liposomes with lectins increases their binding to plant protoplasts. Plant Physiol 72:386–390

Shillito RD, Saul MW, Paszkowski J, Muller M, Potrykus I (1985) High efficiency direct gene transfer to plants. Biotechnology 3:1099–1104

Sleight MJ (1986) A nonchromatographic assay for expresssion of the chloramphenicol acetyltransferase gene in eukaryotic cells. Anal Biochem 156:251–256

Uchimiya H, Fushimi T, Hashimoto H, Harada H, Syono K, Sugarawa Y (1986a) Expression of a foreign gene in callus derived from DNA-treated protoplasts of rice (*Oryza sativa* L.). Mol Gen Genet 204:204–207

Uchimiya H, Hirochika H, Hashimoto H, Hara A, Masuda T, Kasumimoto T, Harada H, Ikeda J-E, Yoshioka M (1986b) Coexpression and inheritance of foreign genes in transformants obtained by direct DNA transformation of tobacco protoplasts. Mol Gen Genet 205:1–8

Wang C-Y, Hughes KW, Huang L (1986) Improved cytoplasmic delivery to plant protoplasts via pH-sensitive liposomes. Plant Physiol 82:179–184

Werr W, Lörz H (1986) Transient gene expression in a Gramineae cell line. Mol Gen Genet 202:471–475

Yamaoka N, Furusawa I, Yamamoto M (1982) Infection of turnip protoplasts with cauliflower mosaic virus DNA. Virology 122:503–505

Zaitlin M, Beachy RN (1974) The use of protoplasts and separated cells in plant virus research. In: Lauffer MA, Bang FB, Maramorosch K, Smith KM (eds) Advances in virus research, vol 19. Academic Press, New York, pp 1–35

Zhengkaixu, Luciano CS, Ballard ST, Rhoads RE, Shaw JG (1984) Infection of tobacco protoplasts with liposome-encapsulated potyviral RNA. Plant Sci Lett 36:137–142

## Note Added in Proof

Since the submission of this review (early in 1987), several important contributions to the area of direct gene transfer into plants have been published. Some of these new developments are briefly described below.

A new reporter gene, *GUS* (beta-glucuronidase), was described by Jefferson (1987) and has since been used in many laboratories. Basically, *GUS* activity can be detected microscopically in individual cells from transgenic plants because of a simple histochemical staining process. This procedure does not require the use of radioisotopes and has the great advantage of allowing detection of chimeric gene expression at the cellular level.

An entirely new method of DNA delivery into plant cells, consisting in bombarding cells or plant parts with DNA-coated tungsten or gold particles, was also described (Klein et al. 1987, 1988; McCabe et al. 1988). This procedure is technically sophisticated as it requires the use of a high-velocity gun to accelerate the metal particles. Its advantages are that naked DNA can be introduced into cells (as opposed to protoplasts) or even meristems. Thus, this procedure bypasses one or several tissue culture steps and will be invaluable in the transformation of plants not amenable to protoplast isolation and regeneration. It has also been demonstrated that the latter steps can be avoided in the transformation of rice through the use of yet another DNA transfer system. Luo and Wu (1988) clearly showed that foreign DNA could be transferred through the pollen tube of decapitated rice florets. Up to 20% of the seeds recovered from such treated florets gave plants which expressed the foreign gene (*nptII*) and contained up to 300 copies of the donor DNA per genome.

Finally, geminivirus genomes were also shown to be usable as vectors for plant transformation. Ward et al. (1988) and Hayes et al. (1988) engineered DNA 1 from cassava latent virus (CLV) with a bacterial *CAT* gene and DNA A from tomato golden mosaic virus (TGMV) with a bacterial *neo* gene, respectively. In both cases, the systemic expression of the reporter gene could be followed in infected plants.

## References

Jefferson RA (1987) Assaying chimeric genes in plants. The GUS gene fusion system. Plant Molec Biol Rep 5:387–405

Klein TM, Wolf ED, Wu R, Sanford JC (1987) High velocity microprojectiles for delivering nucleic acids into living cells. Nature 327:70–73

Klein TM, Fromm M, Weissinger A, Tomes D, Schaaf S, Sletten M, Sanford JC (1988) Transfer of foreign genes into intact maize cells with high-velocity microprojectiles. Proc Natl Acad Sci USA 85:4305–4309

Luo ZX, Wu R (1988) A simple method for the transformation of rice via the pollen-tube pathway. Plant Molec Biol Rep 6:165–174

McCabe DE, Swain WF, Martinell BJ, Christou P (1988) Stable transformation of soybean (*Glycine max*) by particle acceleration. Bio/Technol 6:923–926

Hayes RJ, Petty ITD, Coutts RHA, Buck KW (1988) Gene amplification and expression in plants by a replicating geminivirus vector. Nature 334:179–182

Ward A, Etessami P, Stanley J (1988) Expression of a bacterial gene in plants mediated by infectious geminivirus DNA. EMBO J 6:1583–1587

# IV DNA Recombinants and Transformation of Agricultural Crops

J. SIMPSON and L. HERRERA-ESTRELLA[1]

## 1 Introduction

Biotechnology offers several potential benefits in agriculture. Two main areas of interest are: (1) the production of improved food crops and (2) the utilization of plants for the production of commercially important products, as is routine in yeasts and other microorganisms. In order to realize the potential of plant genetic engineering, systems which allow the transfer and expression of foreign genes into plant cells must be developed. To date, three main avenues of investigation have been followed to develop such systems, these are: the use of viral vectors, naked DNA transformation, and the natural gene transfer system of *Agrobacterium tumefaciens.*

Research on viral vector systems has focused mainly on two groups of viruses, the Gemini viruses and the caulimoviruses represented by the cauliflower mosaic virus (CaMv). CaMv has previously been used successfully to express foreign genes in plant cells (Brisson et al. 1984). However, due to its, pathogenic nature, the strong restrictions on genome size and the difficulties in transferring stably transformed characteristics from one generation to the next, it has so far proved difficult to produce vectors based on the genome of CaMv capable of efficient transformation and expression of foreign genes in plant cells. Gemini viruses are still under study and although they have a promising future, to date no results have been published concerning their usefulness as vectors. Therefore viruses will not be discussed in this chapter.

This chapter is mainly concerned with the *Agrobacterium* system, that has proved most useful and which could in theory be utilized to transform all dicotyledonous plants, and the naked DNA transformation techniques, which are proving useful in the transformation of several monocotyledonous crop plants. The development of selectable markers and expression systems have also been discussed in detail.

## 2 The *Agrobacterium* System

The soil bacterium, *Agrobacterium tumefaciens,* has the ability to infect most dicotyledonous plants usually at a site of wounding of the plant (Braun 1978,

[1] Centro de Investigación y de Estudios, Avanzados del I.P.N. Unidad Irapuato, Apartado Postal No. 629, 36500 Irapuato, Gto, México

Biotechnology in Agriculture and Forestry, Vol. 9
Plant Protoplasts and Genetic Engineering II (ed. by Y. P. S. Bajaj)

1982). The tissue around the wound then proliferates to form a neoplastic growth known as a crown gall tumor. This tumor tissue has the property of growing independently of the bacteria and, when excised from the plant, is able to grow axenically on hormone-free media (Braun 1943). Furthermore, the tumor tissue also produces a set of novel metabolites, termed opines. These opines are amino acid or sugar derivatives which cannot be metabolized by the plant cells themselves, but can, however, be utilized as a carbon and nitrogen source by the *A. tumefaciens* strain responsible for induction of the tumor (Tempé and Petit 1982). Since it was first demonstrated that *Agrobacterium* harbors a large plasmid (the Ti plasmid), which is directly responsible for tumor induction on plants (Zaenen et al. 1974; Van Larebeke et al. 1974; Watson et al. 1975), and that the mechanism of tumor formation involves transfer and stable integration of a small DNA segment from this plasmid (the T-DNA) (Chilton et al. 1977, 1978; Willmitzer et al. 1980) into the plant genome, this organism became the main target to develop plant transformation techniques.

The Ti or tumor-inducing plasmids are classified by the types of opine which the plant is induced to produce following infection by *Agrobacterium*; to date, six opine types have been determined: octopine, nopaline, mannopine, agropine, agrocinopine, and leucopine. Most studies have concentrated on octopine- and nopaline-type Ti plasmids. Detailed functional analysis of the octopine and nopaline Ti plasmids has been carried out by transposon insertion mutagenesis and from deletion mutants. These analyses revealed that two regions of the *Agrobacterium* Ti plasmid are essential for *Agrobacterium*-mediated transformation of plants: the T-DNA region and the *Vir* region. The T-DNA region is a defined segment bordered by specific sequences, which is transferred into the plant cell nucleus (or in some cases into the organellar genome) (Chilton et al. 1977, 1980; Thomashow et al. 1980; Lemmers et al. 1980; De Block et al. 1985). In nopaline-type strains a single DNA segment of 23 kb is transferred (Lemmers et al. 1980; Hepburn et al. 1983; Ursic et al. 1983), whereas in certain octopine-type strains, two DNA segments can be transferred either separately or together, one of 13.6 kb and one of approximately 7 kb (Thomashow et al. 1980; De Beuckeleer et al. 1981). Mutations in the T-DNA produce tumors of altered morphology or are attenuated in oncogenicity. The genes encoded on the T-DNA segments are not required for the transfer and integration of the T-DNA but are responsible for tumor formation after integration into plant DNA sequences.

Transposon insertion/deletion mutagenesis has also led to a correlation between the tumor phenotype and expression of some of the T-DNA-encoded genes. Normal crown gall tumors grow as unorganized undifferentiated tissue. Mutations in genes *1* and *2* induce tumors which produce an abundance of shoots, suggesting that shoot formation is normally suppressed by the proteins encoded by these genes in wild-type tumors. Mutations in gene *4* induce tumors producing an abundance of roots, suggesting that root formation is normally suppressed by the protein products of this gene (Garfinkel et al. 1981; Leemans et al. 1982; Joos et al. 1983). By analogy to plant growth regulators, gene *4* can be said to have a cytokinin-like effect, and genes *1* and *2* an auxin-like effect. The combination of these effects, i.e., suppressing root and shoot formation in the wild-type T-DNA, leads to the formation of a characteristic crown gall tumor. Recently, it has

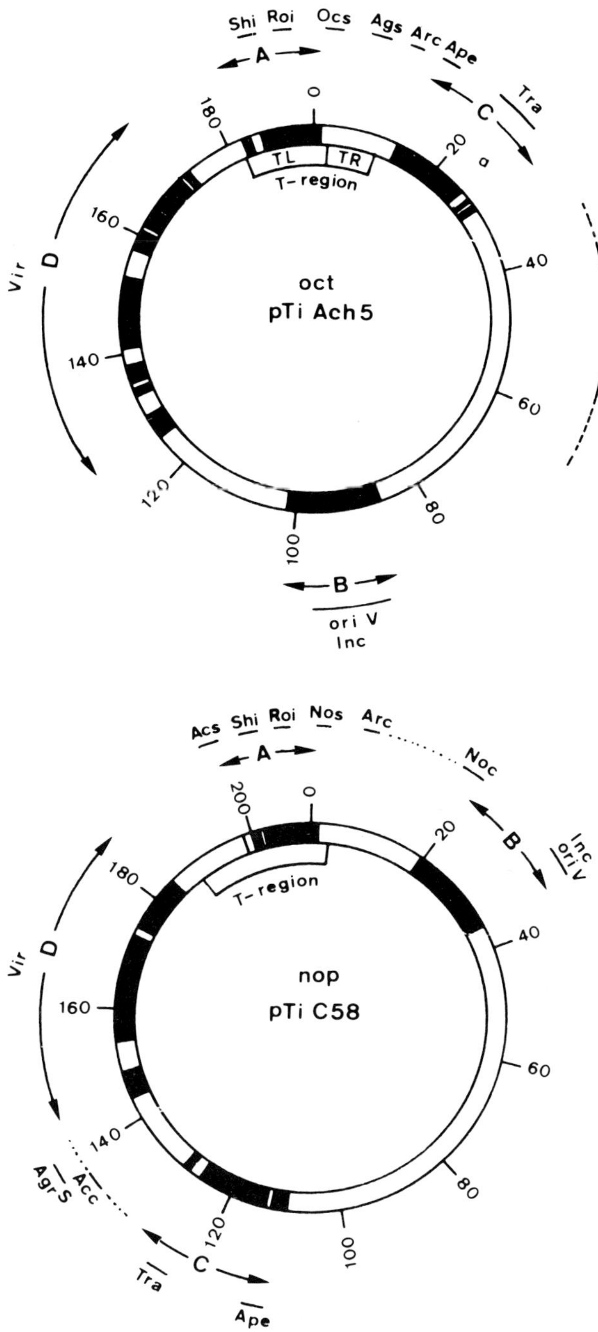


**Fig. 1.** Organization of the octopine-type Ti plasmid pTiAch5 and the nopaline-type Ti plasmid pTiC58. Regions of homology between the two plasmids are indicated in *black* (Engler et al. 1981). *A* T-DNA; *B* replication origin; *C* conjugative functions; *D Vir* region. The opine synthase genes are designated: *ocs* octopine; *agroc* agrocinopine; *nos* nopaline; *agr* agropine. (Holsters et al. 1980; De Greve et al. 1981)

been shown that these genes do encode enzymes that synthesize auxins and cytokinins (Schröder et al. 1983, 1984; Inzé et al. 1984).

The second essential Ti plasmid region is the *vir* region. Mutants, in this region are no longer capable of transforming plant tissue and producing tumors. This region of the Ti plasmid encodes the functions necessary for transfer and integration of the T-DNA. Although both *vir* and T-DNA regions are linked within the Ti plasmid, it has been shown that the *vir* region can also mediate T-DNA transfer in a *transacting* manner, since the T-DNA is transferred and integrated efficiently when the *vir* and T-DNA regions are on separate replicons (De Framond et al. 1983; Hoekema et al. 1983). Similarly, the octopine *vir* region can complement the nopaline *vir* region and vice versa.

The *vir* region is not thought to be transferred to plant cells since it has never been found present in the genome of transformed cells, (Lemmers et al. 1980; Zambryski et al. 1982; Thomashow et al. 1980; De Beuckeleer et al. 1981).

Figure 1 shows the organization of the Ti plasmid encoded genes, of an octopine- and a nopaline-type Ti plasmid deduced from these analyses (Holsters et al. 1980, 1983; De Greve et al. 1981).

### 2.1 Transfer and Integration of the T-DNA

Although Ti plasmid-derived systems are now widely used to transfer and express foreign genes in plants, the precise mechanism of T-DNA transfer is still unknown; however, some insight into the mode of transfer has been obtained by studying DNA sequences near the ends of the T-DNA region of the Ti plasmid and the junction of T-DNA and plant sequences generated upon integration of the T-DNA into plant DNA sequences. Comparison of the ends of the T-DNA from several tumor lines showed that the T-DNA boundaries are well conserved. These boundaries lie within or close to a 25-bp direct repeat which flanks the T-DNA in the Ti plasmid (Simpson et al. 1982; Yadav et al. 1982; Zambryski et al. 1982; Holsters et al. 1983).

Deletion analysis of the ends of the T-region showed that removal of the right 25-bp repeat eliminates the capacity of the Ti plasmid to transfer DNA to plant cells, whereas deletion of the left 25-bp repeat has little or no effect (Zambryski et al. 1982; Joos et al. 1983). These data indicate that the right 25-bp repeat plays a crucial role in the mechanism of T-DNA transfer. Moreover, it was shown that the right 25-bp sequence is a cis-acting element, and that DNA transfer is directional, depending upon the orientation of this sequence (Wang et al. 1984). This directional mechanism of T-DNA transfer resembles the mechanism used during conjugative DNA transfer common in bacteria, where the right 25-bp repeat would play a similar role to the origin of transfer (ori T) present in conjugative plasmids. Recent reports suggest that during T-DNA transfer single-stranded intermediates are formed which could represent the molecular form mobilized into the plant cell (Statchel et al. 1987).

## 2.2 *Agrobacterium rhizogenes*

A second *Agrobacterium* species which is being investigated in relation to developing plant transformation systems is *A. rhizogenes*. This species causes hairy root disease in plants, producing tumors and a proliferation of roots on susceptible plants. Tumors produced by *A. rhizogenes* regenerate at a high frequency to mature plants, unlike *A. tumefaciens* tumors. The tumor-inducing properties of *rhizogenes* strains are also carried on a large Ri plasmid, comparable to the Ti plasmid of *A. tumefaciens*. Two segments of Ri plasmid DNA, the TL-DNA and TR-DNA, are transferred to plant cells during infection. Homology and complementation studies involving *A. tumefaciens* strains have identified two auxin biosynthesis genes in the Ri plasmid TR-DNA (White et al. 1985). Four loci affecting virulence and tumor morphology have also been identified on the TL-DNA.

Several groups have been investigating whether changes in T-DNA structure were responsible for the high regeneration capacity of *A. rhizogenes*-transformed tumor cells. Tepfer (1984) has shown that at least the TL-DNA is present in plants regenerated from hairy root tumors. They also found that many such regenerated plants exhibit an abnormal phenotype, characterized by decreased apical dominance, adventitious rooting, and wrinkled leaves.

Taylor et al. (1985) compared T-DNA genes in regenerated plants of *N. tabacum* and *N. glauca* and found that the TR-DNA auxin biosynthetic genes were present in multiple copies in *N. tabacum* plants, but were undetectable in *N. glauca* plants. They also noted that expression of these genes in tobacco is at a low level or is suppressed. They suggested therefore that the auxin imbalance produced by the TR-DNA in tobacco is not sufficient to prevent differentiation and regeneration, although cells with low auxin production are more likely to undergo regeneration. However, in *N. glauca* the auxin biosynthetic genes block differentiation and regeneration, and this may only occur when these genes are lost from the TR-DNA.

Continuing investigation in this field should lead to a better understanding of the Ri T-DNA and to the development of efficient plant-transformation systems utilizing this system.

# 3 Vector Systems

The vector systems available for *Agrobacterium*-mediated plant transformation at present can be classified into two types: oncogenic and nononcogenic; the nononcogenic vectors can further be divided into cointegrative and binary type vectors.

## 3.1 Oncogenic Vectors

Oncogenic vectors are of limited use, since these vectors still produce crown gall tissue and mature, phenotypically normal plants cannot be regenerated from such tissue. The simplest oncogenic vectors are those which involve simply the insertion of foreign DNA sequences into the T-DNA by using intermediate vectors, in which the sequence of interest is flanked by T-DNA sequences. By conjugating the intermediate vector plasmid to *Agrobacterium* and allowing homologous recombination to occur between T-DNA sequences within the *Agrobacterium*, the sequence of interest becomes inserted between T-DNA borders. More useful and easier to use oncogenic vectors are those such as pGV3851, which is partially deleted for the oncogenic T-DNA functions (Zambryski et al. 1984). To construct this vector, the left internal region of the nopaline T-DNA was deleted and replaced by pBR322 sequences; however, the right T-DNA region is retained, including transcript 4 which is sufficient to induce tumors with abnormal shoot development.

Vectors based on pBR322 can be transferred by conjugation from *E. coli* to *Agrobacterium* and cointegrated into the T-DNA of pGV3851 by a single recombination event between the homologous pBR322 sequences present in both plasmids. pBR322 lacks the mobility (MOB) and transfer (TRA) functions necessary to undergo conjugative transfer to another bacterium; however, when a helper plasmid carrying the MOB and TRA functions is introduced in the same bacterium as pBR322, then pBR322 can undergo conjugation.

Since pBR322 cannot replicate in *Agrobacterium*, direct selection can be made for the specific T-DNA/pBR322 recombination events (Van Haute et al. 1983). Strains harboring foreign sequences can then be used to inoculate plants in vivo, and after a few weeks transformed crown gall tissue, often with shoot teratomas, can be obtained using this vector.

This system has been utilized to study the light-regulated expression of foreign genes in plant cells (Herrera-Estrella et al. 1984). However, the system has two major disadvantages: (1) not all cloned sequences may be expressed in such tumor tissue, or alternatively, they may be overexpressed giving a biased result; and (2) normal mature plants cannot be regenerated from this tissue.

### 3.2 Nononcogenic Vectors

The most important and widely used vectors are the nononcogenic vectors which allow the regeneration of transformed phenotypically normal plants. These vectors are available in two types, the cointegrative type and the binary type.

Cointegrative vectors, such as pGV3850 (Zambryski et al. 1983) are deleted for all the oncogenic functions encoded by the T-DNA only retaining T-DNA border sequences and often the opine synthase genes. The deleted sequences, as in the case of pGV3851, are replaced by pBR322 sequences. Utilizing the pBR322 mobilization system to allow conjugation from *E. coli* to *Agrobacteria*, and following a single recombination event between the pBR322 sequences of the cloning vehicle and pGV3850, the desired sequences become integrated between the T-DNA border regions (Fig. 2). The nononcogenic T-DNA is then transferred as normal to plant cells. However, in this case no crown gall tissue is formed, but the wound callus tissue may be excised and grown in culture medium. By stimulation with hormones, shoots or callus may be induced from this wound tissue. Transformed

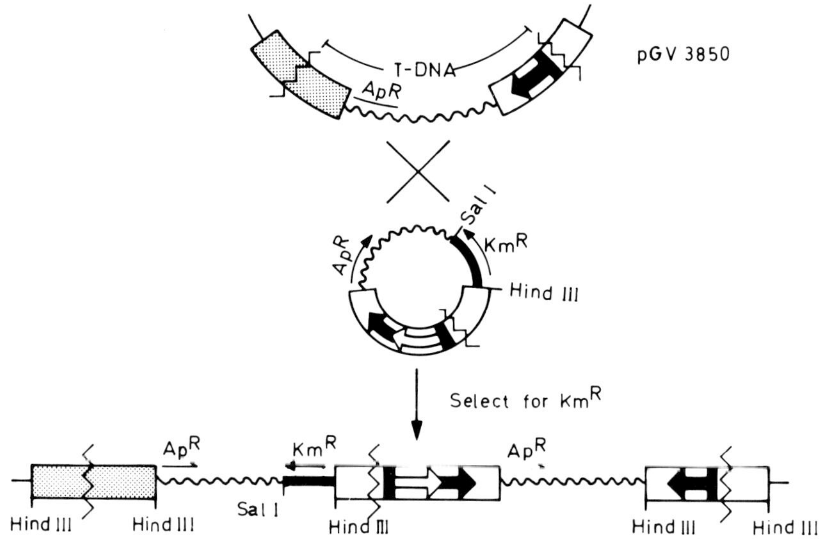


**Fig. 2.** Schematic representation of the homologous recombination event between pBR322 sequences of pGV3850 and the intermediate cloning vector. *Jagged lines* represent the T-DNA border sequences; *black arrow* nopaline synthase sequences; *white arrow* the foreign coding sequence to be expressed in plant cells. In this representation an extra T-DNA border is included in the intermediate vector; normally, however, this extra border is unnecessary

shoots may either be selected for by antibiotic resistance or screened for enzymatic activity. These shoots can eventually be isolated and regenerated to form plants.

A variation on the cointegrate vector system is the split-end vector system (Fraley et al. 1985); in this case, an octopine-type Ti plasmid derivative was constructed which contains only the TL-DNA left border, an antibiotic resistance gene from Tn*903*, the *vir* genes, and a sequence homologous to the intermediate vectors which are used as cloning vehicles. The main difference in this system is that the *E. coli* cloning vehicle contains a functional right T-DNA border. With such "split-end" vectors, after the recombination event between the homologous intermediate vector sequences, the two T-DNA border sequences are brought together on the same Ti plasmid, in the correct orientation reconstructing a functional T-DNA structure that allows transfer of the desired sequences to plant cells.

These cointegrate vectors retain the T-DNA and *vir* functions in a cis configuration within the same Ti plasmid; however, the final class of vectors, the binary vectors, have the *vir* and T-DNA functions on separate replicons. In this case the *vir* functions are supplied in trans for the transfer of the T-DNA.

### 3.3 Binary Vectors

This class of vectors is represented by the "Bin" series of plasmids (Bevan 1984). This system involves a host *Agrobacterium* strain containing a Ti plasmid deleted of its T-DNA and a cloning vehicle containing the T-DNA repeats, a selectable marker active in plant cells, and a sequence containing multiple restriction sites to facilitate cloning of desired sequences in between the T-DNA borders from *E. coli*. This cloning vehicle also carries an origin of replication which allows the plasmid to replicate and be stably maintained in *Agrobacterium* as well as in *E. coli*. Sequences of interest can be cloned into the vector and the plasmid mobilized by conjugation to *Agrobacterium*. In contrast to cointegrative vectors, no recombination event is required, the *vir* region of the T-DNA-deleted Ti plasmid acts in trans and the T-DNA from the cloning vehicle can be effectively transferred to plant cells. Several other binary vector systems applying the same principles have now been developed and shown to function efficiently (Hoekema et al. 1985; Van den Elzen 1985a; An 1986; Simoens et al. 1986).

## 4 Transfer and Expression of Foreign DNA in Plant Cells

Although it was demonstrated that the T-DNA of the Ti plasmid was transferred and stably integrated into the plant genome, it was at first uncertain as to whether foreign sequences inserted into the T-DNA would be cotransferred. Transposon insertion mutagenesis of the T-DNA made it possible to examine this. Calli transformed by *Agrobacteria* harboring a Ti plasmid containing a Tn7 insertion within the T-DNA were examined by Southern hybridization. It was shown that the inserted Tn7 sequence was indeed cotransferred with the T-DNA. After several such transformed lines had been analyzed, it could be concluded that almost any DNA sequence inserted within the T-DNA would be cotransferred. So far no size limit has been determined and sequences up to 50 kb have been successfully cotransferred (Hernalsteens et al. 1980).

The next question was, of course, whether these foreign sequences could be functionally expressed in plant cells. Initial attempts to express foreign genes in plants involved prokaryotic genes, such as the antibiotic resistance genes encoded by transposons, eukaryotic genes, such as the yeast alcohol dehydrogenase gene (Barton et al. 1983), or mammalian $\beta$-globin (Shaw et al. 1983) and interferon genes, all of which proved unsuccessful. In all these experiments, the promotor and 3′ regulatory sequences of the respective genes had been retained. These results showed that regulatory sequences of genes from bacteria or other eukaryotes are not normally functional in plant cells.

The logical answer to this lack of expression was to construct a vector containing the transcription initiation and termination sequences of a gene known to be functional in plant cells, with cloning sites in between these signals to insert the coding sequences of the foreign gene of interest. The first reported expression of an engineered gene in plant cells involved the promoter and 5′-flanking sequences of the nopaline synthase gene fused to the coding sequence and 3′ end of the octopine synthase (Herrera-Estrella et al. 1983a). The presence of octopine in plant tissue transformed using this construction confirmed that the chimeric *nos-ocs* gene was expressed.

These experiments opened the way for the successful expression of bacterial genes in plant cells. A similar construction was employed, but this time involving the fusion of the coding sequence of the chloramphenicol acetyl transferase or CAT gene from the bacterial plasmid pBR325 to the nopaline synthase promoter and 3′ sequences. An assay for CAT enzymatic activity showed that this construct was also functionally expressed in plants (Herrera-Estrella et al. 1983a). The expression of chimeric genes in plants then allowed the development of selectable markers to identify transformed plant tissue and greatly facilitated experiments in plant cell transformation in general.

## 5 Use of Selectable Reporter Genes in Plant Cells

Selectable markers are essential to all DNA-mediated transformation systems, yet developed and plant transformation systems are no exception. A selectable marker should: (1) be expressed at a particular cellular stage which allows selection of transformed cells and preferentially throughout the transformed organism, (2) produce a selectable change in the phenotype of the transformed organism.

Several dominant selectable marker systems have now been developed for plants (Bevan et al. 1983b; Fraley et al. 1983; Herrera-Estrella et al. 1983a, b). Vectors for expression in plant cells are constructed by putting the selectable marker gene under control of plant transcription regulation signals. The most widely used selectable marker gene in plant systems is that of the aminoglycoside phosphotransferase (3′) II [APH(3′) II], carried by transposon Tn5; this enzyme inactivates the aminoglycoside antibiotics of the neomycin family, i.e., G418, neomycin, and kanamycin. Many plant species are sensitive to these aminoglycosides, therefore the expression of the APH (3′) II gene in plant cells confers a resistant phenotype. Figure 3 shows a widely used vector, pLGV1103 neo containing the APH (3′) II (NPT II) gene fused to the nos promoter. Since all plant species may not be sensitive to similar compounds, it is necessary to develop several dominant selectable markers. Others which have proved useful are the dihydrofoliate reductase (DHFR), conferring resistance to methotrexate and carried by transposon Tn7 (Herrera-Estrella et al. 1983b) and the hygromycin phosphotransferase, giving resistance to hygromycin B and isolated from *E. coli* (Van den Elzen et al. 1985b; Waldron et al. 1985).

Reporter genes are those which encode easily detectable enzymatic activity and can supply information concerning the regulation or action of sequences from a different gene. This can be done by the use of fusions between regulatory sequences of a gene of interest and the coding sequence of a reporter gene. The expression of APH(3′) II and DHFR gene fusions can be measured not only by resistance of the plant cells to kanamycin and methotrexate but can, also be easily assayed for by their enzymatic activity and therefore they may also be used as reporter genes. The original marker genes used to identify *Agrobacterium*-transformed cells, the opine synthases, together with the chloramphenicol acetyl transferase, are the most widely used reporter genes to date (Otten and Schilperoort 1978; Herrera-Estrella et al. 1983a). New reporter genes are continuously being developed, such as those based on the Luciferase system (Ow et al. 1986). The use of such reporter genes has greatly facilitated the analysis of the regulatory

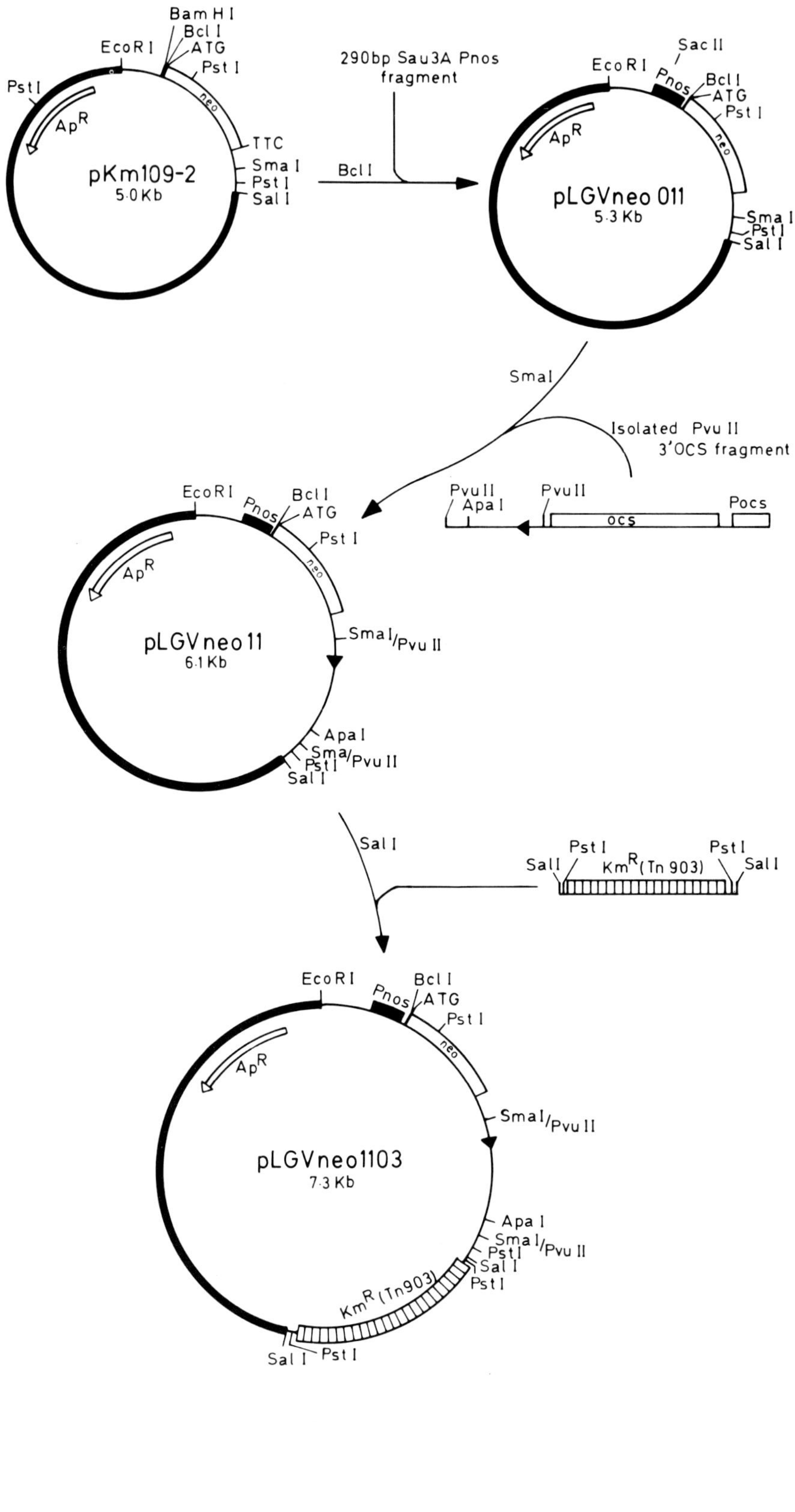
pKm109-2
5.0 Kb
Bam H I
Bcl I
ATG
Pst I
EcoR I
Pst I
neo
ApR
TTC
Sma I
Pst I
Sal I
290bp Sau3A Pnos fragment
Bcl I
pLGVneo 011
5.3 Kb
Sac II
EcoR I
Pnos
Bcl I
ATG
Pst I
Sma I
Pst I
Sal I
Sma I
Isolated Pvu II 3'OCS fragment
Pvu II
Apa I
Pvu II
ocs
Pocs
pLGVneo 11
6.1 Kb
EcoR I
Pnos
Bcl I
ATG
Pst I
Sma I/Pvu II
Apa I
Sma/Pvu II
Pst I
Sal I
Sal I
Pst I
Pst I
Sal I
KmR (Tn 903)
Sal I
pLGVneo1103
7.3 Kb
EcoR I
Pnos
Bcl I
ATG
Pst I
Sma I/Pvu II
Apa I
Sma I/Pvu II
Pst I
Sal I
Pst I
KmR (Tn903)
Sal I
Pst I

sequences which determine the qualitative and quantitative behavior of tissue-specific or inducible genes (see below Sect. 8).

## 6 Plant Transformation Systems

With the development of efficient dominant selectable marker genes, several *Agrobacterium*-mediated plant transformation systems have also been developed. For nononcogenic vectors, several methods may be employed and in most cases these are also applicable to oncogenic-type vectors.

Perhaps the longest-used technique is that of protoplast cocultivation. In this technique *Agrobacteria* are cultured with isolated mesophyll protoplasts for several days, after which time the protoplasts are washed and allowed to grow in media containing antibiotics to eliminate the *Agrobacteria* (Márton et al. 1979; Wullems et al. 1981). Eventually, the single cells regenerate to form calli from which transformed shoots can eventually in turn be regenerated.

A second useful technique is to make in-vivo infections. The top leaves and apical meristem of the plant are removed and inoculation is effected at the wound site. No tumor tissue forms with nononcogenic vectors but wound callus is regenerated in 3–4 weeks. This tissue can be excised and grown in media supplemented by hormones and selected on antibiotics, if a selectable marker is included in the vector. Again, shoots can be stimulated by the necessary hormone combinations and mature plants regenerated (Zambryski et al. 1984).

A third technique, which is rapidly becoming the method of choice for both speed and efficiency, is the leaf disc infection technique (Horsch et al. 1985). Leaf discs or pieces of any other explant are cocultivated with *Agrobacterium* for 2–3 days, and then placed directly on medium containing antibiotics that eliminate the bacteria and allow selection of the transformed cells.

Within 4–5 weeks small shoots can be observed, which upon transfer to the appropiate media can be regenerated to mature plants. An example of this is shown in Fig. 4. R1 plants often show Mendelian segregation patterns and by selection on kanamycin-containing medium, sensitive nontransformed seedlings can easily be distinguished from resistant, transformed seedlings, as shown in Fig. 5.

These techniques are all generally applicable to the model systems of tobacco and petunia; however, it is now becoming feasible to regenerate mature transformed plants from several important crop species, which functionally express the inserted foreign genes.

**Fig. 3.** Construction of a selectable marker system for use in plant cells. *Small black box* represents the nopaline synthase promoter sequence. *Open box* represents the NPT (II) coding sequence from T n 5. *Black triangle* represents the octopine synthase polyadenylation signal sequence and *hatched box* indicates the kanamycin resistance gene functional in bacteria and useful in selecting for Ti plasmid cointegrates a

b

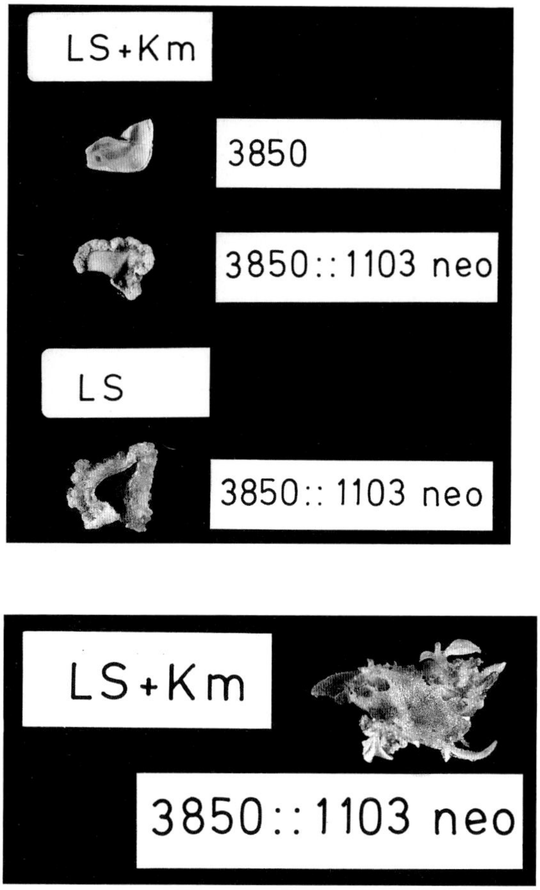


**Fig. 4a, b.** Selection of transformed callus following infection of tobacco leaf discs by *Agrobacterium*. **a** *Top sample* shows control tissue transformed with the pGV3850 vector alone and selected on kanamycin, no callus is formed. *Middle sample* shows tissue transformed with the *Agrobacterium* strain harboring the pGV3850 :: pLGVneo1103 cointegrate, grown under selection on kanamycin. Callus tissue forms around the edge of the disc. *Bottom sample* shows tissue transformed with the pGV3850 :: pLGV1103neo strain grown in the absence of kanamycin; here both transformed and untransformed callus tissue proliferates. **b** Regeneration of shoots from transformed callus tissue. Callus from leaf disc tissue is grown under selection on kanamycin and on media containing hormones to induce shoot formation

### 6.1 Transformation/Expression of Genes in Crop Plants

Almost all dicotyledonous plants are susceptible to infection by *Agrobacterium* and should therefore be amenable to *Agrobacterium*-mediated transformation. The main problem is to determine the optimum tissue culture conditions for the regeneration of mature transformed plants. Tomato and alfalfa are two dicotyledonous crop plants which have met with particular success in this area. Recent reports have demonstrated that several tomato varieties can be easily transformed and regenerated using a modified leaf disc infection technique (McCormick et al. 1986). This report describes efficient transformation of 16 different tomato lines utilizing indole acetic acid (IAA) and benzylamino purine (BAP) as hormones in the culture medium. In addition, they found that the cis form of *Agrobacterium* vector gave a fivefold higher transformation efficiency in comparison to binary type vectors.

Alfalfa is important as a leguminous food and fodder crop. Therefore much interest has centered on its transformation and regeneration. At least one species

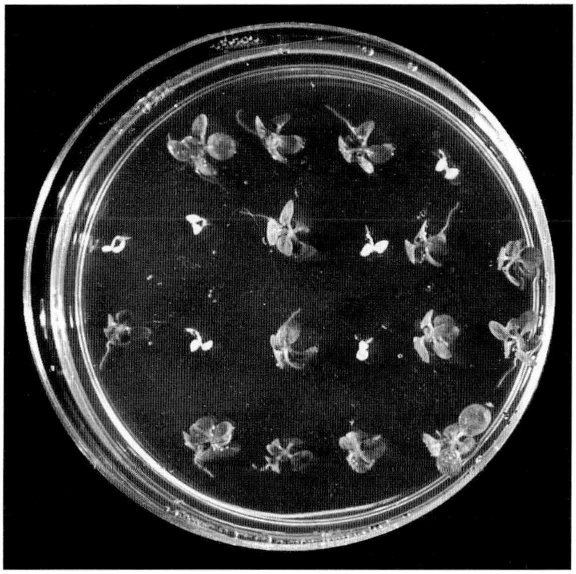

**Fig. 5.** In many cases new characteristics introduced into plant cells are transferred in a 3:1 Mendelian fashion in the Ri generation. Figure shows the results of germinating the Ri seeds of a tobacco plant transformed with the pLGV1103 plasmid and selected for kanamycin resistance. Kanamycin-resistant transformed seedlings are easily distinguished from nontransformed kanamycin-sensitive seedlings, which reach only the two-leaf stage before senescing and dying when selected on 100 μg/ml kanamycin

of alfalfa, *Medicago varia*, can be transformed by the *Agrobacterium* system and regenerated to produce mature transformed plants (Deak et al. 1986). The technique used involved cocultivation of alfalfa shoot segments with *Agrobacterium*. Approximately 4 weeks after infection fresh green tissue was observed. Somatic embryos were induced to develop from this callus tissue, from which mature alfalfa plants were regenerated.

Several other dicotyledonous plants can now be regenerated from protoplast, explants, or embryo cultures. These include crop plants such as soybean, potato, etc. (Ranch et al. 1985), and some woody species such as poplar. Poplar shows potential as a model system for woody species, and callus tissue of poplar has also recently been transformed using *Agrobacterium* (Parson et al. 1986). The transformation of soybean and potato are also under investigation, and it should not be long before mature transformed plants of all of these species and many more can be obtained.

## 6.2 Direct DNA Transformation Methods

Although several dicotyledonous plants are important as food crops, the main world food crops are the monocotyledons wheat, rice, and maize. Expression of foreign genes in these plants shares the problem of finding optimum tissue culture conditions, but in this case, efficient transformation techniques must also be developed, since in general monocotyledonous plants seem not to be susceptible to infection by *Agrobacteria*. Krens (1982) first described a technique for direct DNA transformation of plant cell protoplasts. This technique involves incubating protoplasts with naked DNA in the presence of polyethylene glycol (PEG) and

$CaCl_2$, the mixture is then diluted stepwise, after which the protoplasts are collected, resuspended in culture medium and plated in selective medium. Various modifications have been applied for different plant species, such as adding the PEG after the DNA, and heat shocking the protoplasts at 46 °C followed by chilling on ice. The average transformation frequency using this technique is between $10^{-2}$ and $10^{-5}$. Although first demonstrated for tobacco, cereal species can also be transformed using this technique (Lörz et al. 1985).

Fromm et al. (1985) demonstrated transient expression of foreign DNA in maize protoplasts by direct transfer of DNA using a different method, the electroporation technique. This technique involves passing a short pulse of electricity of high (200 V) voltage through the protoplasts, producing a reversible permeability on the outer membrane, and facilitating the entry of DNA molecules. The efficiency of this technique is relatively high ($10^{-1}-10^{-3}$) and can be used not only for transient expression studies but also to obtain stable transformation of maize and carrot cells. Probably many other monocot species (Fromm et al. 1986; Rhodes et al. 1988) will prove amenable to transformation by this technique.

Transient expression of foreign DNA sequences in other cereal crops, e.g., wheat and rice (Lörz et al. 1985; Potrykus et al. 1985), has also been reported. It has also been shown that mature rice plants can be regenerated from protoplasts (Yamada et al. 1986) and the transformation and regeneration of mature rice plants is now achieved (Toriyama et al. 1988).

## 7 Use of Reverse Genetics Techniques to Study Plant Gene Regulation

Before foreign genes could be transferred and expressed in plant cells, the only method to study plant gene expression was to study the levels of mRNA and the corresponding proteins. The advent of reverse genetics techniques for plants, however, has made it possible to isolate plant genes, to modify them in vitro, and to return them into the plant genome where their expression can be studied. Tobacco and petunia can be used as model systems to study expression of genes isolated from any plant background, including monocotyledonous plants. It was shown that plant genes contain CAAT and TATA box regulatory sequences similar to those previously identified in other eukaryotic systems (Breathnach and Chambon 1981; Messing et al. 1983). Studies of the *Agrobacterium*-encoded octopine and nopaline synthase genes provide some insight into the regulatory sequences necessary for constitutively expressed genes (Koncz et al. 1983; Shaw et al. 1984). However, many of the most important and interesting genes in plants are expressed in either a light-inducible, a tissue-specific, or a developmental fashion. The new recombinant DNA techniques meant that these regulated genes could also be studied in more detail.

One set of genes which has been extensively studied at the mRNA and protein levels is that of the ribulose-1, 5-bis-phosphate carboxylase small subunit (rbcS) genes. rbcS gene products are the most abundant in photosynthetic plant tissues, constituting up to 50% of total organelle protein (Bennett et al. 1984).

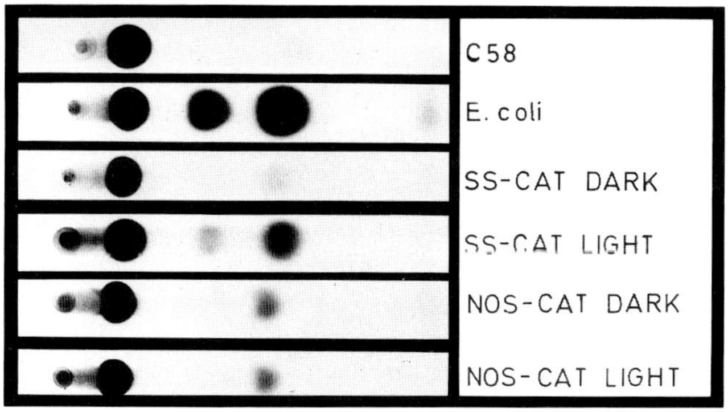


**Fig. 6.** Demonstration of light regulated gene expression in plant cells. Chloramphenicol acetyl transferase (CAT) activity is compared for plant tissue transformed by Nos-CAT and SS-CAT chimeric constructs. C58 (wild-type *Agrobacterium*) are included as negative and positive controls respectively

The expression of these genes is known to be inducible by light, and this regulation was also shown to be mainly at the transcriptional level (Gallagher and Ellis 1982; Silverthorne and Tobin 1984). Most of the important transcription regulation signals in both prokaryotic and eukaryotic systems studied previously were shown to reside in the 5′-flanking sequences of a gene. It was therefore a logical continuation to study these sequences in plant genes. To this end, Herrera-Estrella et al. (1984) fused a 900-bp sequence of the 5′-flanking region of a pea *rbcS* gene to the coding sequence of the *CAT* gene; they could then follow expression of the *CAT* sequence under control of the *rbcS* promoter in transformed plant tissue.

Results from these experiments demonstrated that the 900-bp *rbcS* flanking sequence was sufficient to confer light-inducible regulation on the *CAT* gene (Fig. 5). In addition, they noticed that the chimeric rbcS CAT construct was active only in green photosynthetically active tissue but not in white tissue. Similar studies utilizing another pea *rbcS* gene showed that deletions leaving only 40 bp of the 5′-flanking region including the TATA box were sufficient to confer light-dark regulation (Morelli et al. 1985). More detailed studies of the promoters of the pea *rbcS ss3.6* and chlorophyll a/b binding protein gene *AB80* promoters have shown that enhancer-like elements, already well documented in virus and animal cell systems (Serfling et al. 1985), also exist in plants (Timko et al. 1985; Fluhr et al. 1986; Simpson et al. 1986).

Other studies involving the chlorophyll a/b binding protein gene AB80 from pea have shown that 400 bp of the 5′-flanking sequence of this gene are sufficient to confer light-inducible and tissue-specific expression on a heterologous coding sequence (Simpson et al. 1985). More detailed studies of regulatory sequences of both the AB80 and ss-3.6 genes show that these genes are expressed mainly in chloroplast-containing tissues (Simpson et al. 1986). In addition, chimeric constructs involving AB80 sequences were shown to be under phytochrome control, whereas ss-3.6 constructs are regulated at least in part by some other photoreceptor system (Simpson et al. 1986).

**Table 1.** Summary of work done on some crop plant transformation (for more examples see Chap. III, this Vol.)

| Plant species | Plant tissue utilized | *Agro-bacterium-* mediated | Direct DNA transfer | Vector used | Comments | Reference |
|---|---|---|---|---|---|---|
| *Asparagus officinalis* (Asparagus) | Stem segments | – | | C58pTiC58 wild type | Transformed callus formed | Hernalsteens et al. (1984) |
| *Daucus carota* (Carrot) | Protoplasts | | – | pNosCAT pGMVCAT | Transformed callus tissue formed | Fromm et al. (1985) |
| *Glycine max* (Soybean) | Hypocotyls Embryonic axes | – | – | pNon9749 pTV1100 | Transformed plants | Hinchee et al. (1988) Mac Cabe et al. (1988) |
| *Gossypium hirsutum* (Cotton) | Hypocotyl segments | – | – | pCMC204 | Transformed plants | Umbeck et al. (1987) |
| *Junglans regia* (Walnut) | Somatic embryos | – | – | pCEN594 pCEN200 | Transformed plants | McGranahan et al. (1988) |
| *Lycopersicon esculentum* (Tomato) | Leaf discs | – | | pMON200SE or derivatives | Formation of transformed callus and complete transformed plants | McCormick et al. (1986), Fraley et al. (1985) (for vector) |
| *Medicago varia* (Alfalfa) | Stem segments | – | | pTiBO542/ pGA471 | Formation of transformed callus and transformed plants | Deak et al. (1986) |
| *Nicotiana tabacum* (Tobacco) | Leaf discs | – | | pTiBO542/ pGA472 | Formation of transformed callus and complete transformed plants | An et al. (1986) |
| | Protoplasts | | – | pLGV1103 derivative (see Fig. 3) | | Hain et al. (1985) |
| *Oryza sativa* (Rice) | Protoplast | | – | pCN | Transformed plants | Toriyama et al. (1988) |
| *Secale cereale* (Rye) | Floral tillers | | – | pEV1103 | Transformed plants | De la Peña et al. (1988) |
| *Solanum tuberosum* (Potato) | Leaf or stem segments | – | – | pTiBO542/ pGA472 | Formation of transformed callus and complete transformed plants | An et al. (1986) |
| *Triticum monococcum* (Wheat) | Protoplasts | | – | pLGV1103 | Stable transformed callus obtained | Lörz et al. (1985) |

## 7.1 Gene Regulation Studies in Crop Plants

Other genes which have been studied at the molecular level and which have great importance for the improvement of food crops are the storage proteins such as phaseolins, zeins, conglycinins, and lectins. These genes are expressed mainly in the seeds and at the earliest developmental stages of the plant. They are normally encoded in multigene families in the cell nucleus. Much work has been carried out on phaseolin, the major storage protein of the French bean *Phaseolus vulgaris.*

Certain of the phaseolin genes have been isolated and introduced into tobacco plants, where expression studies were carried out. Transformed tobacco plants were shown to produce no phaseolin in mature leaves, whereas seeds contained high levels of phaseolin. Furthermore, the phaseolin found in tobacco seeds was properly glycosylated and expressed in endosperm and embryo tissues, as occurs in bean seeds. Therefore, this gene is properly expressed and the protein undergoes the correct modifications in a different plant species from which it originated. Proteolytic cleavage of phaseolin occurs during tobacco seed germination, mirroring the processes occurring in beans, and suggesting that translation and processing functions are conserved between the Leguminosae and Solanaceae (Sengupta-Gopolan et al. 1985).

Similar experiments have been carried out utilizing a gene encoding the $\alpha$ subunit of the $\beta$-conglycinin protein isolated from soybean. This gene was transferred to *Petunia* plants by *Agrobacterium*-mediated transformation, and the transformed plants analyzed for the presence of the $\beta$-conglycinin $\alpha$ subunit. The transformed petunia plants were found to produce high levels $\beta$-conglycinin only in petunia seeds. This protein when isolated from petunia was found to be assembled into a multimeric form similar to the assembly of $\beta$-conglycinin in soybean (Beachy et al. 1985).

## 7.2 Engineering Plants for Agricultural Benefit

Many losses in agriculture occur due to infection of crop species by fungi, bacteria, viruses, or to attack by insects. The use of herbicides either to clear fields or to destroy weeds is also an uneconomical practice. The benefits of producing disease-resistant plants or plants tolerant to a specific herbicide are therefore obvious. Much research has been concentrated in these areas, and it is now possible to produce virus-resistant and herbicide-tolerant plants in the model systems of tobacco and petunia.

A widely used broad spectrum herbicide called glyphosate has been investigated by Shah et al. (1986). These workers selected *Petunia hybrida* cells for tolerance to increasing levels of glyphosate. They related this tolerance to the overproduction of enolpyruvyl shikimate-3-phosphate (EPSP), which is inhibited by glyphosate. The EPSP gene from the herbicide-tolerant cells was isolated, cloned into a plant expression vector utilizing the constitutively expressing cauliflower mosaic virus 35s promoter, and transferred into petunia cells. Transformed petunia callus could be selected directly in the presence of glyphosate, whereas nontransformed controls could not grow in the presence of glyphosate. When transgenic plants were regenerated from this callus, they tolerated spraying of commercial glyphosate herbicide at two to four times the dose required to kill wild-type controls. In theory, this gene could be used to produce herbicide-tolerant crop plants and allow selective killing of weeds.

A second line of research has produced virus-resistant plants. In this case the researchers utilized the knowledge that for many years agriculturalists have inoculated plants with mild viral strains to prevent infection by much more virulent forms. By isolating the tobacco mosaic virus, TMV, coat protein gene and expressing it in transgenic tobacco plants (Abel et al. 1986), plants resistant to infection by virulent TMV were produced, in other words a similar occurrence to the viral cross-protection took place. The use of transgenic plants to prevent virus infection has several advantages over inoculation by mild viral strains, i.e., the mild strain may still cause low level infection, may mutate and become virulent, may be virulent on nonrelated plant species, or may interact with a second virus to produce an infection. It could therefore prove economically very important to exploit the potential to produce virus-resistant crop plants with none of the disadvantages of using live virus inoculants.

## 8 Variability of Transformed Plants

When examining gene expression of transformed plants, there are several factors which may influence the expression of foreign genes and must be taken into account. These include problems of copy number, chromosomal localization, and plant-to-plant variability. Since the mechanism of T-DNA transfer is not completely understood, it is unknown how many T-DNA copies will be transferred and where they will be integrated within the chromosome. Results to date suggest that T-DNA integration is completely random (Zambryski et al. 1982), and that in general one to five T-DNA copies per cell are transferred (Jones et al. 1985). Some cases, however, show many integrated copies, even up to 10–15. When transformed plants are allowed to self-fertilize and the $F_1$ generation is studied, in general progeny are found to contain the foreign sequences in a 3:1 ratio of those containing the sequence to those which do not (De Block et al. 1984). This is the classic result for the transfer of a single dominant Mendelian trait. This suggests that either only a single copy of the sequence is present, or that, if more than one copy is present, they are closely linked on the same chromosome, or if they are on different chromosomes, only one copy is active.

Recent reports have shown that transforming plants using chimeric gene constructions with no promoter signals can result at a low frequency in the expression of the foreign inserted sequences presumably due to the effects of integration of the foreign sequences close to regulatory elements which have allowed the genes to be expressed (Teeri et al. 1986). This demonstrates that sequences surrounding the inserted sequences may influence their expression. Recent reports have shown that in transformed petunia plants, the chromosomal position of integration affected the quantitative levels of expression, but not the qualitative levels (Jones et al. 1985). Quantitative differences can vary widely from one plant to the other and this seems to be independent of copy number. In general, transcriptional fusions are expressed lower than translational fusions, and foreign sequences are expressed much lower than endogenous genes of the same type. There may also be differences due to the developmental stage of the plant and the environment in which it is grown.

## 9 Conclusions

Within the span of only 5 years it has not only become possible but routine practise to transform plant tissue. This has made feasible the manipulation of plant genes and the study of plant gene regulation at the molecular level.

Initially only the model systems of petunia and tobacco were amenable to efficient manipulation. However, extensive research into tissue culture techniques and new methods of plant cell transformation have now allowed foreign genes to be expressed in certain agriculturally important crop species. In addition, several of these can be stably transformed and mature regenerated plants obtained.

The field of genetic engineering in plants is rapidly expanding and within the next 5–10 years it is likely that engineered plants will undergo field tests and become commercially available. This means that not only shall we move toward the goal of better food resources, but will also gain valuable information about how plant genes are regulated and plant development controlled.

*Acknowledgments.* Thanks are due to Albert Verstraet and Karel Spruyt for figures and photographs, and to Gabriela Padilla N. and Leticia Anguiano J. for typing the manuscript.

## References

Abel PP, Powell P, Nelson RS, De B, Hoffman N, Rogers SG, Fraley RT, Beachy RN (1986) Delay of disease development in transgenic plants that express the tobacco mosaic virus coat protein gene. Science 232:738–743

An G (1986) Development of plant promoter expression vectors and their use for analysis of differential activity of nopaline synthase promoter in transformed tobacco cells. Plant Physiol 81:86–91

Barker RF, Idler KB, Thompson DV, Kemp JD (1983) Nucleotide sequence of the T-DNA region from the *Agrobacterium tumefaciens* octopine Ti plasmid pTi15995. Plant Mol Biol 2:335–350

Barton KA, Binns AN, Matzke AJM, Chilton D-D (1983) Regeneration of intact tobacco plants containing full length copies of genetically engineered T-DNA, and transmission of T-DNA to R1 progeny. Cell 32:1033–1043

Beachy RN, Chen Z-L, Horsch RB, Rogers SG, Hoffman JJ, Fraley RT (1985) Accumulation and assembly of soybean $\beta$-conglycinin in seeds of transformed petunia plants. EMBO 4 (12): 3047–3053

Bennett J, Jenkings GI, Hartley MR (1984) Differential regulation of the accumulation of the light-harvesting chlorophyll a/b complex and ribulose bisphosphate carboxylase/oxygenase in greening pea leaves. J Cell Biochem 25:1–13

Bevan M (1984) Binary *Agrobacterium* vectors for plant transformation. Nucl Acids Res 12:8711–8721

Bevan M, Barnes WM, Chilton M-D (1983a) Structure and transcription of the nopaline synthase gene region of T-DNA. Nucl Acids Res 11:369–385

Bevan MW, Flavell RB, Chilton M-D (1983b) A chimaeric antibiotic-resistance gene as a selectable marker for plant cell transformation. Nature (Lond) 304:184–187

Braun AC (1943) Studies on tumor inception in crown gall disease. Am J Bot 30:674–677

Braun AC (1978) Plant tumours Biochim Biophys Acta 516:167–191

Braun AC (1982) A history of the crown gall problem. In: Kahl G, Schell J (eds) Molecular biology of plant tumours. Academic Press, New York, pp 155–210

Breathnach R, Chambon P (1981) Organization and expression of eukaryotic split genes coding for proteins. Annu Rev Biochem 50:349–383

Brisson N, Paszkowski J, Penswick JR, Gronenborn B, Potrykus I, Hohn T (1984) Expression of a bacterial gene in plants using a viral vector. Nature (Lond) 310:511–514

Chilton M-D, Drummond MH, Merlo DJ, Sciaky D, Montoya AL, Gordon MP, Nester EW (1977) Stable incorporation of plasmid-DNA into higher plant cells: the molecular basis of crown gall tumorigenesis. Cell 11:263–271

Chilton M-D, Drummond MH, Merlo DJ, Sciaky D (1978) Highly conserved DNA of Ti-plasmids overlaps T-DNA, maintained in plant tumors. Nature (Lond) 275:147–149

Deak M, Kiss GB, Koncz C, Dudits D (1986) Transformation of *Medicago* by *Agrobacterium*-mediated gene transfer. Plant Cell Rep 5:97–100

De Beuckeleer M, Lemmers M, De Vos G, Willmitzer L, van Montagu M, Schell J (1981) Further insight on the transferred-DNA of octopine crown gall. Mol Gen Genet 183:283–288

De Blaere R, Bytebier B, De Greve H, Deboek F, Schell J, van Montagu M, Leemans J (1985) Efficient octopine Ti plasmid-derived vectors for *Agrobacterium*-mediated gene transfer to plants. NAR 13 (13):4777–4788

De Block M, Herrera-Estrella L, van Montagu M, Schell J, Zambryski P (1984) Expression of foreign genes in regenerated plants and their progeny. EMBO J 3:1681–1689

De Block M, Schell J, van Montagu M (1985) Chloroplast transformation by *Agrobacterium tumefaciens.* EMBO J 4:1367–1372

De Block M, Botterman J, Vandewiele M, Dockx J, Thoen C, Gossele V, Rao Movva N, Thompson C, Van Montagu M, Leemans J (1987) Engineering herbicide resistance in plants by expression of a detoxifying enzyme. EMBO J 6(9):2513–2518

De Frammond AJ, Barton KA, Chilton M-D (1983) Mini Ti: a new vector strategy for plant genetic engineering. Biotechnology 1:262, 269

De Greve H, Decraemer H, Seurinck J, Van Montagu M, Schell J (1981) The functional organization of the octopine *Agrobacterium tumefaciens* plasmid pTiB6S3. Plasmid 6:235–248

De Greve H, Dhaese P, Seurinck J, Lemmers M, van Montagu M, Schell J (1982) Nucleotide sequence and transcript map of the *Agrobacterium tumefaciens.* Ti plasmid-encoded octopine synthase gene. J Mol Appl Genet 1:499–512

De la Peña A, Lörz H, Schell J (1987) Transgenic rye plants obtained by injecting DNA into young floral tillers. Nature 325:274–276

Depicker A, Stachel S, Dhaese P, Zambryski P, Goodman HM (1982) Nopaline synthase: transcript mapping and DNA sequence. J Mol Appl Genet 1:561–574

Engler G, Depicker A, Meanhout R, Villarroel-Mandiola R, van Montagu M, Schell J (1981) Physical mapping of DNA base sequence-homologies between an octopine and nopaline Ti-plasmid of *Agrobacterium tumefaciens.* J Mol Biol 152:183–208

Fluhr R, Kuhlemeier C, Nagy F, Chua N-H (1986) Organ specific and light-induced expression of plant genes. Science 232:1106–1112

Fraley RT, Rogers SG, Horsch RB, Sanders P, Flick J, Adams S, Bittner M, Brand L, Fink C, Fry J, Gallupi G, Goldberg S, Hoffmann N, Woo S (1983) Expression of bacterial genes in plant cells. Proc Natl Acad Sci USA 80:4803–4807

Fraley RT, Rogers SG, Horsch RB, Eichholtz DA, Flick JS, Fink CL, Hoffmann NL, Sanders PR (1985) The SEV system: a new disarmed Ti plasmid vector system for plant transformation. Biotechnology 3:629–635

Fromm ME, Taylor LP, Walbot V (1985) Expression of genes transferred into monocot and dicot plant cells by electroporation. Proc Natl Acad Sci USA 82:5824–5828

Fromm ME, Taylor LP, Walbot V (1986) Stable transformation of maize after gene transfer by electrophoresis. Nature (Lond) 319:791–793

Gallagher TF, Ellis RJ (1982) Light-stimulated transcription of genes for two chloroplast polypeptides in isolated pea leaf nuclei. EMBO J 1:1493–1498

Garfinkel DJ, Simpson RB, Ream LW, White F, Gordon MP, Nester EW (1981) Genetic analysis of crown gall: Fine structure map of the T-DNA by site directed mutagenesis. Cell 27:143–153

Gielen J, De Beuckeleer M, Seurinck J, Dechboeck F, De Greve H, Lemmers M, van Montagu M, Schell J (1984) The complete nucleotide sequence of the TL-DNA of *Agrobacterium tumefaciens* plasmid pTiAch5. EMBO J 3:825–846

Gorman CM, Moffat LF, Howard BH (1982) Recombinant genomes which express chloramphenicol acetyltransferase in mammalian cells. Mol Cell Biol 2:1044–1051

Hain R, Stabel P, Czernilofsy AP, Steinkiss HH, Herrera-Estrella L, Schell J (1985) Uptake, integration, expression and genetic transmission of a selectable chimaeric gene by plant protoplasts. Mol Gen Genet 199 (2):161–168

Hepburn AG, Clarke LE, Blundy KS, White J (1983) Nopaline Ti plasmid, pTiT37, T-DNA insertions into a flax genome. J Mol Appl Genet 2:211–224

Hernalsteens JP, van Vliet F, De Beukeleer M, Depicker A, Engler G, Lemmers M, Holsters M, van Montagu M, Schell J (1980) The *Agrobacterium tumefaciens* Ti plasmid as a host vector system for introducing foreign DNA in plant cells. Nature (Lond) 287:654–656

Hernalsteens JP, Thia-Toong L, Schell J, van Montagu M (1984) An *Agrobacterium*-transformed cell culture from the monocot *Asparagus officinalis*. EMBO J 3:3039–3041

Herrera-Estrella L, Depicker A, van Montagu M, Schell J (1983a) Expression of chimaeric genes transferred into plant cells using a Ti plasmid-derived vector. Nature (Lond) 303:209–213

Herrera-Estrella L, De Block M, Nessens E, Hernalsteens J-P, van Montagu M, Schell J (1983b) Chimaeric genes as dominant selectable markers in plant cells. EMBO J 2:987–995

Herrera-Estrella L, van den Broeck G, Maenhaust R, van Montagu M, Schell J, Timko M, Cashmore A (1984) Light-inducible and chloroplast associated expression of a chimaeric gene introduced into *Nicotiana tabacum* using a Ti plasmid vector. Nature (Lond) 310:115–120

Hinchee MAW, Connor-Ward DV, Newell CA, McDonnell RE, Sato SJ, Grasser CS, Fischhoff DA, Re DB, Fraley RT, Horsch RB (1988) Production of transgenic soybean plants using *Agrobacterium*-mediated DNA transfer. Biotechnology 6:915–922

Hoekema A, Hirsch PR, Hooykas PJ, Schilperoort RA (1983) A binary plant vector strategy based on separation of *vir* and T-region of the *Agrobacterium tumefaciens* Ti plasmid. Nature (Lond) 303:179–181

Hoekema A, van Haaren MJ, Fellinger AJ, Hooykas PJJ, Schilperoort RA (1985) Non-oncogenic plant vectors for use in the *Agrobacterium* binary system. Plant Mol Biol 5:85–95

Hoffman LM, Donaldson DD, Bookland R, Rashka K, Herman EM (1987) Synthesis and protein body deposition of maize 15-kd zein in transgenic tobacco seeds. EMBO J 6:3213–3221

Holsters M, Silva B, van Vliet F, Genetello C, De Block M, Dhaese P, Depicker A, Inzé D, Engler G, Villaroel R, van Montagu M, Schell J (1980) The functional organization of the nopaline *A. tumefaciens* plasmid pTiC58. Plasmid 3:212–230

Holsters M, Villaroel R, Gielen J, Seurinck J, De Greve H, van Montagu M, Schell J (1983) An analysis of the boundaries of the octopine TL-DNA in tumours induced by *Agrobacterium tumefaciens*. Mol Gen Genet 190:35–41

Horsch RB, Fry JE, Hoffmann NL, Eichholtz D, Rogers SG, Fraley RT (1985) A simple and general method for transferring genes into plants. Science 227:1229–1231

Inzé D, Follin A, van Lijsebettens M, Simoens C, Genetello C, van Montagu M, Schell J (1984) Genetic analysis of the individual T-DNA genes of *Agrobacterium tumefaciens*; further evidence that two genes are involved in indole-3-acetic acid synthesis. Mol Gen Genet 194:265–274

Jefferson RA (1987) Assaying chimeric genes in plants: the GUS gene fusion systems. Plant Mol Biol Rep 5:387–405

Jones JDG, Dunsmuir P, Bedbrook J (1985) High level expression of introduced chimeric genes in regenerated transformed plants. EMBO J 4:2411–2418

Joos H, Inzé D, Caplan A, Sormann M, van Montagu M, Schell J (1983) Genetic analysis of T-DNA transcripts in nopaline crown galls. Cell 32:1057–1067

Klein TM, Wolf ED, Wu R, Sanford JC (1987) High-velocity projectiles for delivering nucleic acids into living cells. Nature 327:70–74

Klein TM, Gradziel T, Fromm ME, Sanford JC (1988) Factors influencing gene delivery into *Zea mays* cells by high-velocity projectiles. Biotechnology 6:559–563

Koncz C, De Greve H, André D, Deboeck F, van Montagu M, Schell J (1983) The octopine synthase genes carried by Ti plasmids contain all signals necessary for expression in plants. EMBO 2:1597–1603

Koukolíková-Nicola A, Shillito RD, Hohn B, Wang K, van Montagu M, Zambryski P (1985) Involvement of circular intermediates in the transfer of T-DNA from *Agrobacterium tumefaciens* to plant cells. Nature (Lond) 313:191–196

Krens FA, Molenclijk L, Wullems GJ, Schilperoort RA (1982) In vitro transformation of plant protoplasts with Ti plasmid DNA. Nature (Lond) 296:72

Leemans J, Shaw C, Deblaere R, De Greve H, Hernalsteens J-P, Maes M, van Montagu M, Schell J (1981) Site-specific mutagenesis of *Agrobacterium* Ti plasmids and transfer of genes to plant cells. J Mol Appl Genet 1:149–164

Leemans J, Deblaere R, Willmitzer L, De Greve H, Hernalsteens J-P, van Montagu M, Schell J (1982) Genetic identification of functions of TL-DNA transcripts in octopine crown galls. EMBO J 1:147–152

Lemmers M, De Beuckeleer M, Holsters M, Zambryski P, Depicker A, Hernalsteens JP, van Montagu M, Schell J (1980) Internal organization, boundaries and integration of Ti-plasmid DNA in nopaline crown gall tumours. J Mol Biol 144:353–376

Li BJ, Langridge WHR, Szalay AA (1985) Somatic embryogenesis and plantlet regeneration in soybean *Glycine max*. Plant Cell Rept 4:344–367

Lörz H, Baker B, Schell J (1985) Gene transfer to cereal cells mediated by protoplast transformation Mol Gen Genet 199:178

Márton L, Wullems GJ, Molendijk L, Schilperoort RA (1979) In vitro transformation of cultured cells from *Nicotiana tabacum* by *Agrobacterium tumefaciens*. Nature (Lond) 277:129–130

McCormick S, Niedermeyer J, Fry J, Banason A, Horsch R, Fraley R (1986) Leaf disc transformation of cultivated tomato (*L. esculentum*) using *Agrobacterium tumefaciens*. Plant Cell Rep 5:81–86

McCabe DE, Swain WF, Martinell BJ, Christou P (1988) Stable transformation of soybean (*Glycine max*) by particle acceleration. Biotechnology 6:923–926

McGranahan GH, Leslie CA, Uratsu SL, Martin LA, Dandeker AM (1988) *Agrobacterium*-mediated transformation of walnut somatic embryos and regeneration of transgenic plants. Biotechnology 6:800–804

Messing J, Geraghty D, Heidecker G, Hu N-T, Kridl J, Rubenstein I (1983) Plant gene structure. In: Kosuge T, Meredith CP, Hollaender A (eds) Genetic engineering of plants, and agricultural perspective. Plenum, New York, pp 211–227

Morelli G, Nagy F, Fraley RT, Rogers SG, Chua N-H (1985) A short conserved sequence is involved in the light-inducibility of a gene encoding ribulose 1,5-bisphosphate carboxylase small subunit of pea. Nature (Lond) 315:200–204

Nagy F, Morelli G, Fraley RT, Rogers SG, Chua N-H (1985) Photo regulated expression of a pea rbcS gene in leaves of transgenic plants. EMBO J 4:3063–3068

Otten LA, Schilperoort RA (1978) A rapid microscale method for the detection of lysopine and nopaline dehydrogenase activities. Biochim Biophys Acta 527:497–500

Ow DW, Wood KV, De Luca M, De Wet JR, Helinski DR, Howell SH (1986) Transient and stable expression of the firefly luciferase gene in plant cells and transgenic plants. Science 234:856–859

Parsons TJ, Sinkar VP, Stetler RF, Nester EW, Gordon MP (1986) Transformation of poplar by *Agrobacterium tumefaciens*. Biotechnology 4(6):533–537

Potrykus I, Saulk M, Petruska J, Paszkowski J, Shillito R (1985) Direct gene transfer to cells of a graminaceous monocot. Mol Gen Genet 199:183–188

Ranch JP, Oglesby L, Zielinski AC (1985) Plant regeneration from embryo-derived tissue cultures of soybeans. In Vitro Cell Dev Biol II:653–658

Reiss B, Sprengel R, Will H, Schaller H (1984) A new sensitive method for qualitative and quantitative assay of neomycin phospho-transferase in crude cell extracts. Gene 30:211–218

Rhodes CA, Pierce DA, Mettler IJ, Mascarenhas D, Detmer JL (1988) Genetically transformed maize plants from protoplasts. Science 240:204–207

Schröder G, Klipp W, Hillebrand A, Whring R, Koncz C, Schröder J (1983) The conserved part of the T-region in Ti-Plasmids expresses four proteins in bacteria. EMBO J 2:403–409

Schröder G, Waffenschmidt S, Wiler EW, Schröder J (1984) The T-region of Ti plasmids codes for an enzyme synthesizing indole-3-acetic acid. Eur J Biochem 138:387–391

Sengupta-Gopalan C, Reichert NA, Barker RF, Hall TC, Kemp JD (1985) Proc Natl Acad Sci USA 82:3320–3324

Serfling E, Jasin M, Schaffner W (1985) Enhancer and eukaryotic gene transcription. Trends Genet 1:224–230

Shah DM, Horsch RB, Klee HJ, Kishore GM, Winter JA, Tumer NE, Hironata CM, Sanders PR, Gasser CS, Ag Kent S, Siegel NR, Rogers SG, Fraley RT (1986) Engineering herbicide tolerance in transgenic plants. Science 233:478–481

Shaw CH, Leemans J, Shaw CH, van Montagu M, Schell J (1983) A general method for the transfer of cloned genes to plant cells. Gene 23:315–330

Shaw CH, Watson MD, Carter GH, Shaw CH (1984) The right hand copy of the nopaline Ti-plasmid 25 bp repeat is required for tumour formation. Nucl Acids Res 12:6031–6041

Silverthorne J, Tobin EM (1984) Demonstration of transcriptional regulation of specific genes by phytochrome action. Proc Natl Acad Sci USA 81:1112–1116

Simoens C, Alliotte Th, Mendel R, Müller A, Schiemann J, van Lijsebettens M, Schell J, van Montagu M, Inzé D (1986) A binary vector for transferring genomic libraries to plants. Nucl Acids Res 14:8073–8090

Simpson RB, O'Hara PJ, Kwok W, Montoya AL, Lichtenstein C, Gordon MP, Nester EW (1982) DNA from the A6S/2 crown gall tumor contains scrambled Ti-plasmid sequences near its junctions with the plant DNA. Cell 29:1005–1014

Simpson J, Timko MP, Cashmore AR, Schell J, van Montagu M, Herrera-Estrella L (1985) Light-inducible and tissue-specific expression of a chimaeric gene under control of the 5′-flanking sequence of a pea chlorophyl a/b-binding protein gene. EMBO J 4:2723

Simpson J, van Montagu M; Herrera-Estrella L (1986) Photosynthesis associated gene families: Differences in response to tissue-specific and environmental factors. Science 233:36–38

Stachel SE, Timmerman B, Zambryski P (1987) Activation of *Agrobacterium tumefaciens* Vir gene expression generates multiple single-stranded T-strand molecules from the pTi A6 T-region: requirement for 5′ vir D gene products. EMBO J 4:857–863

Taylor BH, Amasino RM, White FF, Nester EW, Gordon MP (1985) T-DNA analysis of plants regenerated from hairy root tumours. Mol Gen Genet 201:554–557

Teeri T, Herrera-Estrella L, Depicker A, van Montagu M, Palva ET (1986) Identification of plant promoters in situ by T-DNA mediated transcriptional fusions to the *npt*-II gene. EMBO J 5:1755–1760

Tempé J, Petit A (1982) Opine utilization by *Agrobacterium*. In: Kahl G, Schell J (eds) Molecular biology of plant tumours. Academic Press, New York, pp 451–459

Tepfer D (1984) Transformation of several species of higher plants by *Agrobacterium rhizogenes*: sexual transmission of the transformed genotype and phenotype. Cell 37:959–967

Thomashow MF, Nutter R, Montoya AL, Gordon MP, Nester EW (1980) Integration and organization of Ti-plasmid sequences in crown gall tumours. Cell 19:729–739

Thomashow L, Reeves S, Thomashow MF (1984) Crown gall oncogenesis: evidence that a T-DNA gene from *Agrobacterium* Ti plasmid pTiAG encodes an enzyme that catalyses synthesis of indoleacetic acid. Proc Natl Acad Sci USA, p 5071

Timko MP, Kausch AP, Castresana C, Fassler J, Herrera-Estrella L, van den Broeck G, van Montagu M, Schell J, Cashmore AR (1985) Light regulation of plant gene expression by an upstream enhancer-like element. Nature (Lond) 318:579–582

Toriyama K, Arimoto Y, Uchimaya H, Hinata K (1988) Transgenic rice plants after direct gene transfer into protoplasts. Biotechnology 6:1072–1074

Umbeck P, Johnson G, Barton K, Swain W (1988) Genetically transformed cotton (*Gossypium hirsutum*). Biotechnology 5:263–266

Ursic D, Slightom JL, Kemp JD (1983) *Agrobacterium tumefaciens* T-DNA integrates into multiple sites of the sunflower crown gall genome. Mol Gen Genet 190:494–503

van den Elzen P, Lee KY, Townsend J, Bedbrook JR (1985a) Simple binary vectors for DNA transfer to plant cells. Plant Mol Biol 5:149–154

van den Elzen PJM, Townsend J, Lee KY, Bedbrook JR (1985b) A chimaeric hygromycin resistance gene as a selectable marker in plant cells. Plant Mol Biol 5:299–302

Van Haute E, Joos H, Maes M, Warren G, van Montagu M, Schell J (1983) Intergeneric transfer and exchange recombination of restriction fragments cloned in pBR322: a novel strategy for the reversed genetics of Ti plasmids of *Agrobacterium tumefaciens*. EMBO J 2:411–418 van Larebeke N, Engler G, Holsters M, van den Elsacker S, Zaenen I, Schilperoort RA, Schell J (1974) Large plasmid in *Agrobacterium tumefaciens* essential for crown gall-inducing ability. Nature (Lond) 252:169–170

Waldron C, Murphy EB, Roberts JL, Gustafson GD, Armour SL, Malcom SK (1985) Resistance to hygromycin B: a new marker for plant transformation studies. Plant Mol Biol 5:103–108

Wang K, Herrera-Estrella L, van Montagu M, Zambryski P (1984) Right 25-bp terminus sequences of the nopaline T-DNA is essential for and determines direction of DNA transfer from *Agrobacterium* to the plant genome. Cell 38:455–462

Watson B, Currier TC, Gordon MP, Chilton M-D, Nester EW (1975) Plasmid required for virulence of *Agrobacterium tumefaciens*. J Bacteriol 123:255–264

White FF, Taylor BH, Huffman GA, Gordon MP, Nester EW (1985) Molecular and genetic analysis of the transferred DNA regions of the hairy root-inducing plasmid of *Agrobacterium rhizogenes*. J Bact Biol 164(1):33–44

Willmitzer L, De Beuckeleer M, Lemmers M, van Montagu M, Schell J (1980) DNA from Ti-plasmid is present in the nucleus and absent from plastids of plant crown-gall cells. Nature (Lond) 287:359–361

Wullems GJ, Molendijk L, Ooms G, Schilperoort RA (1981) Differential expression of crown gall tumor markers in transformants obtained after in vitro *Agrobacterium tumefaciens*-induced transformation of cell wall regenerating protoplasts derived from *Nicotiana tabacum*. Proc Natl Acad Sci USA 78:4344–4348

Yadav NS, Vanderleyden J, Bennett DR, Barnes WM, Chilton M-D (1982) Short direct repeats flank the T-DNA on a nopaline Ti plasmid. Proc Natl Acad Sci USA 79:6322–6325

Yamada Y, Yang Z-Q, Tang D-T (1986) Plant regeneration from protoplast derived callus of rice (*Oryza sativa* L.). Plant Cell Rep 5:85–88

Zaenen I, van Lerebeke N, Teuchy H, van Montagu M, Schell J (1974) Supercoiled circular DNA in crown gall-inducing *Agrobacterium* strains. J Mol Biol 86:109–127

Zambryski P, Holsters M, Kriger K, Depicker A, Schell J, van Montagu M, Doodman HM (1980) Tumor DNA structure in plant cells transformed by *A. tumefaciens*. Science 209:1385–1391

Zambryski P, Depicker A, Kriger K, Goodman H (1982) Tumor induction by *Agrobacterium tumefaciens*: analysis of the boundaries of T-DNA, J Mol Appl Genet 1:361—370

Zambryski P, Joos H, Genetello C, Leemans J, van Montagu M, Schell J (1983) Ti plasmid vector for the introduction of DNA into plant cells without alteration of their normal regeneration capacity. EMBO J 2:2143–2150

Zambryski P, Herrera-Estrella L, De Block M, van Montagu M, Schell J (1984) The use of the Ti plasmid of *Agrobacterium* to study the transfer and expression of foreign DNA in plant cells: new vectors and methods. In: Setlow J, Hollaender A (eds) Genetic engineering, principles and methods (Vol. 6). Plenum, New York, pp 253–278

# V Genetically Transformed Maize (*Zea mays* L.) Plants from Protoplasts

C.A. RHODES[1], D.A. PIERCE[2], I.J. METTLER[1], D. MASCARENHAS[3], and J.J. DETMER[3]

## 1 Introduction

Agriculturally important cereal crops, including maize, have been difficult to engineer genetically using current techniques for gene insertion. With few exceptions (Graves and Goldman 1986; Grimsley et al. 1987), most of the graminaceous crops are not readily susceptible to infection by *Agrobacterium tumefaciens*, which is a vector for gene transfer commonly used with many dicot species. Genes can be transferred directly into protoplasts, without an *Agrobacterium* vector, by methods that permit DNA to cross the plasmalemma (Krens et al. 1982; Shillito et al. 1985; Riggs and Bates 1986; Fromm et al. 1986). Stable transformation of maize cells has been achieved through direct uptake of DNA into protoplasts that had been permeabilized by electroporation (Fromm et al. 1986; Pierce et al. 1987), but until recently (Rhodes et al. 1988a), no plants had been recovered from maize protoplasts. We now describe regeneration of maize plants derived from protoplasts into which a gene encoding neomycin phosphotransferase II (NPT II) was introduced via electroporation. NPT II permits plant cells to grow on inhibitory levels of the antibiotic kanamycin (Fraley et al. 1983; Fromm et al. 1986) and can be used as a dominant marker to select for transformed cells. Following is the account of our published work on the regeneration of transformed plants from maize protoplasts (Rhodes et al. 1988b).

## 2 Methodology and Experimental Results

Protoplasts were isolated from an embryogenic cell suspension culture of maize inbred A188 (Rhodes et al. 1988a). The cell culture was initiated as callus from immature embryos 18 months before these experiments were begun, and had been grown as a suspension culture in liquid N6ap medium (Rhodes et al. 1988a) for approximately 1 year. At this time, plants were easily regenerated by transferring callus to N6ap or MS (Murashige and Skoog 1962) medium without auxin.

[1] Sandoz Crop Protection Corporation, Palo Alto, 975 California Avenue, CA 94304-1104, USA
[2] EniChem, Monmouth Junction, NJ 08852, USA
[3] BioGrowth, Richmond, CA 94806, USA

Biotechnology in Agriculture and Forestry, Vol. 9
Plant Protoplasts and Genetic Engineering II (ed. by Y.P.S. Bajaj)

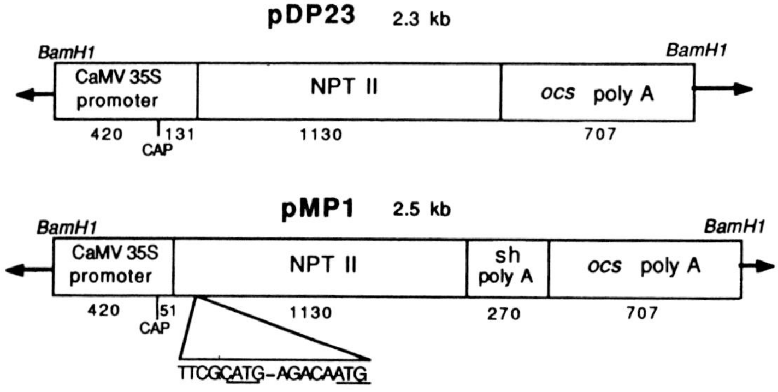


**Fig. 1.** Diagram of pDP23 and pMP1. The 35S promoter fragment in pDP23 (Pierce et al. 1987) extends from the Bam HI site at −420 bp to the Dde I site at +131 bp with respect to the start of RNA transcription (Guilley et al. 1982). The fragment carrying the NPT II gene includes 15 bp of 5′-noncoding region and 343 bp of 3′-noncoding region extending to the Sal I site in Tn5. The false start ATG 16 bp upstream from the original Tn5 version has been deleted. The 3′ region of the *ocs* gene was recovered as a 707-bp Pvu II fragment (Barker et al. 1983). The cassette in pMP1 contains a version of the 35S promoter which includes 51 bp of CaMV sequence downstream from the start of RNA transcription (Pierce et al. 1987). For pMP1, the sequence in the 5′-noncoding region of NPT II in pDP23 was changed from 5′-TTCGC*ATG*-3′ to 5′-AGACA*ATG*-3′. The 3′ region of pMP1 includes a 270 bp Rsa I fragment from the 3′ region of the sucrose synthase (sh) gene (Werr et al. 1985) fused at its 3′ end to the Pvu II fragment of the *ocs* terminator. The NPT II cassettes in pDP23 and pMP1 are bordered by Bam HI sites and were inserted into the Bam HI site of pUC19 (Rhodes et al. 1988b).

Freshly prepared protoplasts were suspended at densities of $3\times10^6$ to $6\times10^6$ per milliliter of N6ap medium (450 mosM, pH 8). In some experiments, protoplasts were incubated for 2 to 4 min at 43 °C and then cooled by ice water before proceeding. This heat treatment did not affect protoplast viability. Protoplasts were portioned into microfuge tubes (0.5 ml per tube) and the following solutions were added to each tube: 75 µl of 2 M KCl; 20 µl of sonicated calf thymus DNA (2.5 mg/ml); 0.5 ml of N6ap medium, pH 8.0; and 20 to 40 µg of supercoiled plasmid DNA.

Two plasmid constructions, pDP23 and pMP1 (Fig. 1) were used, each of which included a bacterial NPT II gene from Tn5 as a selectable marker for transformation. Plasmid pDP23 (Pierce et al. 1987) contained the 35S promoter region of cauliflower mosaic virus (CaMV) (Guilley et al. 1982), including 131 base pairs of untranslated leader sequence and the 3′-polyadenylation region of the octopine synthase gene (*ocs*) from the Ti plasmid of *Agrobacterium tumefaciens* (Barker et al. 1983). The pMP1 plasmid contained a similar version of the CaMV 35S promoter, but with only 51 base pairs of leader sequence (Pierce et al. 1987) and selected changes in the sequence upstream of the ATG start signal in the NPT II coding region, as well as a modified 3′-terminal region (Fig. 1).

Each mixture of protoplasts and DNA was immediately transferred to sterile 0.4-cm plastic cuvettes fitted with aluminum foil electrodes (Fromm et al. 1985). An electrical field of 200 to 300 V (500 to 750 V/cm) was applied at room temperature via discharge of a 1200 µF capacitor that had been previously charged with an electrophoresis power supply. Five minutes later, electroporated protoplasts were diluted with 2 vol. of N6ap medium (pH 5.8). Percentages of viable protoplasts decreased with increasing field strength. Application of 250 (625 V/cm) caused a 30% to 50% decrease in viability.

Protoplasts were plated on Millipore filters (0.8 µm pore size) over N6ap medium (adjusted to 450 mosM with mannitol) with Black Mexican Sweet (BMS) suspension cells as a nurse culture. Each feeder layer plate consisted of 0.5 g of BMS cells suspended in 2 ml of liquid N6ap medium and spread over 20 ml of agar-solidified N6ap (Rhodes et al. 1988a). Millipore support filters with protoplasts were transferred after 7 days to N6ap medium of reduced osmolarity (300 mosM) with fresh BMS feeder layers. Kanamycin sulfate (50 to 150 mg/l) (Sigma) was included in the growth medium after the initial 7 days to inhibit growth of cells not expressing the NPT II gene. Filters were transferred again at day 14 to N6ap medium (180 mosM) with kanamycin and without any BMS feeder layers (Fig. 2A). Three weeks after electroporation, individual calli that were resistant to growth inhibition by kanamycin were selected from the filter and transferred directly to N6ap medium containing kanamycin. After an additional 2 weeks of growth, there was sufficient callus tissue (approximately 800 mg) of each kanamycin-resistant cell line to assay for NPT activity, as well as to regenerate plants (Fig. 2B). This protocol of 4 weeks of selection for resistance to kanamycin resulted in virtually no nontransformed "escapes".

Experiments were done to identify the optimum concentration of kanamycin to use in this selection system. Growth of a nontransformed suspension was nearly stopped after three weeks in kanamycin at 50 mg/l, whereas the growth rate of one transformed cell suspension in kanamycin at 500 mg/l was 77% of its growth rate without kanamycin. Low levels of NPT II expression were correlated with slow growth of some transformed cell lines on kanamycin at 150 mg/l. There was some inhibition of BMS feeder cell growth on medium containing kanamycin at 100 mg/l, but not enough in 7 days to visibly affect protoplast growth. We recommend using kanamycin at 100 mg/l in this selection system, as this concentration effectively inhibited growth of control calli without requiring extremely high levels of NPT II expression in transformants.

This selection system provided immediate immobilization of protoplasts, reducing the probability of individual transformed clones splitting into two or more cell lines or of protoplast fusions that could create chimeric cell lines. Only one presumptive chimeric clone was obtained among approximately 200 selected cell lines.

Transformation frequencies were calculated by comparing the number of kanamycin-resistant, NPT II-positive calli to the number of calli grown without selection from replicate plates receiving the same treatment. Transformation frequencies were as high as 5% of callus-forming protoplasts. More typical frequencies were lower, approximately 0.5% to 1%. Heat treatment of protoplasts before electroporation was sometimes effective in increasing transformation frequency as much as threefold (Fig. 2A). This result was not consistent; less vigorous protoplasts showed no effect with heat treatment. Electroporation with 250 V (625 V/cm) produced higher transformation frequencies than did lower voltages, despite greater reductions in percentage of viable protoplasts (30% to 50% decrease) than were seen with lower voltages.

**Fig. 2 A, B.** Selection of kanamycin-resistant A188 callus and regeneration of plants. A188 protoplasts were electroporated in a solution containing the plasmid pMP1 (20 to 40 μg/ml) and selected for growth in the presence of kanamycin, as described in the text. **A** Callus grown from electroporated protoplasts after 19 days on medium containing kanamycin. Filter *1* shows minimal growth from control protoplasts, which were electroporated without pMP1. Filter *2* shows callus grown from protoplasts which were electroporated in the presence of pMP1. Protoplasts on filter *3* were subjected to a heat treatment (2 min at 43 °C) before electroporation under the same conditions used with protoplasts on filter *2*. **B** Mature plant regenerated from protoplast-derived callus selected for kanamycin resistance. As is typical for regenerated maize plants, this plant was about half as tall as seed-grown plants (Rhodes et al. 1988 b)

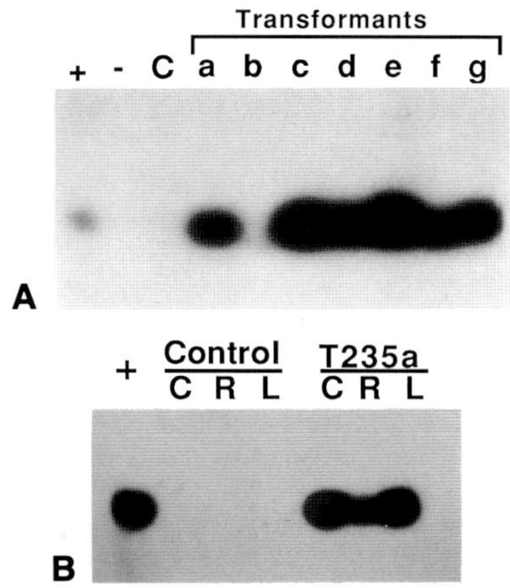


**Fig. 3A, B.** NPT II expression in transformed A188 callus and plants. Extracts (20 μl) of callus and plant tissues were electrophoresed through a polyacrylamide gel and assayed for NPT II activity (Reiss et al. 1984). **A** Extract of a BMS cell line transformed with the NPT II gene was the positive control in the first lane (+). The third lane (*C*) contained extract from A188 callus grown from protoplasts electroporated without pMP1 or pDP23 present. Lanes *a* through *g* contain extracts from kanamycin-resistant callus selected in three separate transformation experiments. **B** NPT II expression in transformed plants. Extracts from callus (*C*) and from roots (*R*) and leaves (*L*) of a transformed plant regenerated from this cell line (T235a) contained NPT II activity, whereas no NPT II activity was present in extracts of equivalent nontransformed control tissues (Control lanes *C, R, L*). (Rhodes et al. 1988b)

Cell extracts of kanamycin-resistant calli and regenerated plants were tested for NPT II activity by means of a standard in vitro assay (Reiss et al. 1984). Callus from protoplasts porated without DNA encoding NPT II showed no NPT II activity, whereas callus from protoplasts porated in the presence of either pDP23 or pMP1 had phosphorylation activity that migrated with the activity in transformed BMS callus (Fig. 3A). NPT II activity was stable in callus grown without kanamycin for at least 10 months. Relative specific activity of NPT II was calculated from liquid scintillation counting of the amount of radioisotope-labeled kanamycin phosphate bound to the P81 phosphocellulose paper, adjusted for the amount of protein in the original extract loaded per lane. The mean value for relative specific activity of NPT II from 58 cell lines transformed with pMP1 was 155, compared to a mean of 34 for 34 cell lines transformed with pDP23. This corresponds well with relative values obtained with transient expression assays of these two plasmids in BMS protoplasts (Mettler et al. unpubl.). Specific activities of NPT II in leaf and root tissues of plants were 1.8 and 2.0 times higher, respectively, than levels in transformed callus (Fig. 3B).

Southern blot analysis of DNA from callus and plants with NPT II activity was performed to confirm the presence of the NPT II gene. The 0.9-kb Pst I fragment which carries the NPT II coding sequence was isolated and labeled with $^{32}P$ by nick translation, then hybridized to Bam HI-restricted genomic DNA from several maize cell lines and plant tissue (Fig. 4). The NPT II probe did not hybridize to DNA from nontransformed A188 callus, but did hybridize to the ex-

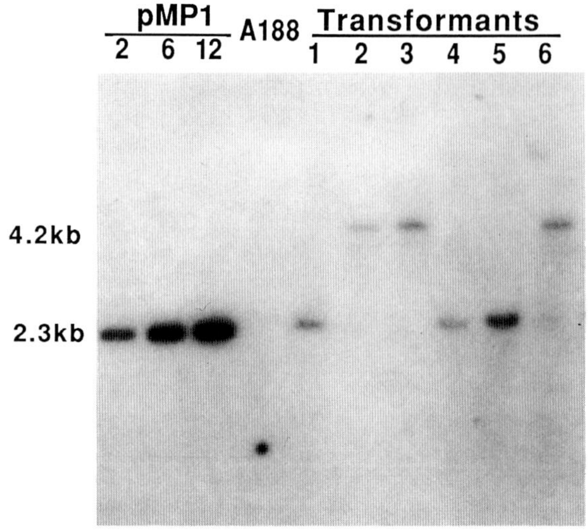


**Fig. 4.** Southern blot analysis of DNA from A188 callus and regenerated plants. DNA was isolated from a plant (lane *1*) and callus lines (lanes *2–6*) with NPT II activity and from control callus (A188) by modification of a previous method (Dellaporta et al. 1983). Tissues were frozen in liquid nitrogen, ground to a powder, and homogenized in buffer containing 50 mM Tris, pH 8.0, 50 mM EDTA, 250 mM NaCl, and 10 mM $\beta$-mercaptoethanol. After extraction with phenol and chloroform, polysaccharides were removed by treatment of the aqueous phase with 0.1 volume ethanol for 10 min at 0 °C, then sedimenting the polysaccharides at 10000 g for 5 min at 4 °C (Schröder and Schröder 1982). Nucleic acids were then precipitated with ethanol from the aqueous phase. The DNA was cut with the restriction enzyme Bam HI, electrophoresed through an agarose gel, and transferred to a nitrocellulose membrane (Southern 1975). The blot was hybridized to a $^{32}$P-labeled 0.9 kb fragment that contained the NPT II gene. The first three lanes were loaded with Bam HI-restricted pMP1 at the equivalent of 2, 6, and 12 copies per haploid maize genome ($4.5 \times 10^9$ bp) (Rhodes et al. 1988b)

pected 2.35-kb DNA fragment of Bam HI-digested pMP1, which carries the NPT II gene cassette (lanes designated pMP1 and A188 in Fig. 4). DNA from a regenerated plant and five kanamycin-resistant cell lines, derived from three separate experiments, hybridized to the NPT II DNA probe (Fig. 4, lanes T1 through T6). The expected fragment size of 2.35 kb was not always observed. Transformants 3, 4, and 6 apparently contained the NPT II gene cassette rearranged in a way that altered the position of the Bam HI restriction sites. These three cell lines were recovered from protoplasts which received different treatments, thus eliminating the possibility that their similar hybridization patterns were the result of sampling the same transformant more than once. None of these rearrangements correlated with relative NPT II activity in these calli. Southern blot analysis of unrestricted DNA of these transformants revealed no hybridization to the NPT II gene probe at the expected size of single-copy plasmids, an indication that the transforming DNA was integrated into the plant genomes. The estimated number of NPT II gene copies was less than five per transformant genome, as judged by a comparison with copy number standards included on the same hybridized filter (Fig. 4). Relative specific activities of NPT II, which ranged from 34 to 594 for these transformants, were not directly correlated with the number of NPT II gene copies per genome.

Regeneration of shoots occurred readily in approximately 18% of the transformed calli when transferred to medium without 2,4-dichlorophenoxyacetic acid. At this time, 29 cell lines from five separate experiments have produced plants which were transferred to soil. Fifty-two plants derived from 12 different cell lines have reached maturity. None have yet produced viable pollen; only one plant produced silks, but did not produce seed when outcrossed. Insufficient numbers of control plants have grown to maturity to permit conclusions as to whether the processes of protoplasting, transformation, or the cell line itself, might be responsible for lack of fertility. Efforts continue to regenerate fertile plants from these transformed cultures.

## 3 Summary and Conclusions

We have demonstrated that maize protoplasts can be genetically transformed and grown to mature plants that express the inserted gene. Electroporation is a convenient method of transformation, both because of its simplicity and the reasonable rates of transformation that can be achieved. Although the transformed plants from these electroporation experiments did not produce seed, it is likely that the infertility problems are associated with the cell line itself. These cultures are now over 2 years old, and it is not unusual for maize cell cultures of this age to carry genetic changes serious enough to cause sterility (Edallo et al. 1981; Rhodes et al. 1986). Other work (Rhodes et al. 1988a) suggests that this technique should be applicable to other maize cell cultures and genotypes.

As it exists now, this system permits testing of tissue-specific gene promoters and of gene expression in nonchimeric maize plants. Our data on levels of NPT II activity in callus and plants transformed with 35S-NPT II-*ocs* gene constructs indicate that expression in plants can be as high or higher than in callus.

*Acknowledgments.* We thank our co-workers at Stauffer Chemical Co., especially R. Lachmansingh, K. Lowe, P. Olson, and L. Pomeroy for their many contributions to this research. We also appreciate the assistance of D. Gray with the illustrations.

## References

Barker RF, Idler KB, Thompson DV, Kemp JD (1983) Nucleotide sequence of the T-DNA region from the *Agrobacterium tumefaciens* octopine Ti plasmid pTi5955. Plant Mol. Biol 2:335

Dellaporta S, Wood J, Hicks J (1983) A plant DNA minipreparation: version II. Plant Mol Biol Rep 1:19

Edallo S, Zucchinali C, Perenzin M, Salamini F (1981) Chromosomal variation and frequency of spontaneous mutation associated with in vitro culture and plant regeneration in maize. Maydica 26:39

Fraley R, Rogers S, Horsch R, Sanders P, Flick J, Adams S, Bittner M, Brand L, Fink C, Fry J, Galluppi G, Goldberg S, Hoffman N, Woo S (1983) Expression of bacterial genes in plant cells. Proc Natl Acad Sci USA 80:4803

Fromm M, Taylor LP, Walbot V (1985) Expression of genes electroporated into monocot and dicot plant cells. Proc Natl Acad Sci USA 82:5824

Fromm MF, Taylor LP, Walbot V (1986) Stable transformation of maize after gene transfer by electroporation. Nature (London) 319:791

Graves ACF, Goldman S (1986) The transformation of *Zea mays* seedlings with *Agrobacterium tumefaciens*: Detection of T-DNA specific enzyme activities. Plant Mol Biol 7:43

Grimsley N, Hohn T, Davies JW, Hohn B (1987) *Agrobacterium*-mediated delivery of infectious maize streak virus into maize plants. Nature (Lond) 325:177

Guilley H, Dudley RK, Jonard G, Balazs E, Richards K (1982) Transcription of cauliflower mosaic virus DNA: Detection of promoter sequences and characterization of transcripts. Cell 30:763

Krens FH, Molendijk L, Wullems G, Schilperoort RA (1982) In vitro transformation of plant protoplasts with Ti-plasmid DNA. Nature (Lond) 296:72

Mettler IJ, Lachmansingh AR, Detmer JJ, Pierce DA, Mascarenhas D (unpublished data)

Murashige T, Skoog F (1962) A revised medium for rapid growth and bioassays with tobacco tissue cultures. Physiol Plant 15:473

Pierce DA, Mettler IJ, Lachmansingh AR, Pomeroy LM, Weck EA, Mascarenhas D (1987) Effect of 35S leader modifications on promoter activity. In: UCLA Symposia on Molecular and Cellular Biology, New Series (Liss, New York), vol. 62, pp 301–310

Reiss B, Sprengel R, Will H, Schaller H (1984) A new and sensitive method for qualitative and quantitative assay of neomycin phosphotransferase in crude cell extracts. Gene 30:211

Rhodes CA, Phillips RL, Green CE (1986) Cytogenetic stability of aneuploid maize tissue cultures. Can J Genet Cytol 28:374

Rhodes C, Lowe K, Ruby K (1988a) Plant regeneration from protoplasts isolated from embryogenic maize cell cultures. Biotechnology 6:56

Rhodes C, Pierce I, Mettler D, Mascarenhas D, Detmer J (1988b) Genetically transformed maize plants from protoplasts. Science 240:204

Riggs CD, Bates GW (1986) Stable transformation of tobacco by electroporation: Evidence for plasmid concatenation. Proc Natl Acad Sci USA 83:5602

Schröder G, Schröder J (1982) Hybridization selection and translation of T-DNA encoded mRNAs from octopine tumors. Mol Gen Genet 185:51

Shillito RD, Saul MW, Paszkowski J, Muller M, Potrykus I (1985) High efficiency direct gene transfer to plants. Biotechnology 3:1099

Southern E (1975) Detection of specific sequences among DNA fragments separated by gel electrophoresis. J Mol Biol 98:503

Werr W, Frommer WB, Maas C, Starlinger P (1985) Structure of the sucrose synthase gene on chromosome 9 of *Zea mays* L. EMBO J 4:1373

# VI Protoplast Culture and Transformation Studies on Rice

S. HASEZAWA, A. BABA, and K. SYŌNO[1]

## 1 Introduction

Rice (*Oryza sativa* L.), which is one of the representative crops of Gramineae plants, has been cultivated for the past 10000 years. Rice grain contains a high amount of protein and amino acids and is highly nutritious. The plant has the highest yield per unit area of all the small grains and has the advantage in that even a small area can feed a large population. At present, it is the most important world crop, together with wheat, and supplies about half of the world population with the principal food in the temperate and the tropical zones. Most other staple cereals – barley, corn, rye, oat – belong to the Gramineae. Many monocotyledonous grasses such as forage plants also belong to the same family. Therefore, the studies of tissue culture and genetic manipulation in Gramineae plants are of great importance for the agricultural industry. The cultivation and regeneration of protoplasts is important as a prerequisite for the use of various technologies for breeding such as gene transfer and somatic cell fusion.

It is well known that virulent strains of *Agrobacterium tumefaciens* attach to the plant cell wall and introduce a piece of its Ti plasmid DNA into the plant cell. This piece of Ti plasmid DNA (T-DNA) is integrated in the nuclear genome of the plant cell. Ti plasmid, a natural vector for plants, is currently the most efficient system for transferring foreign genetic materials into plant cells. However, as Gramineae plants are not susceptible to tumor formation by *A. tumefaciens*, it is generally believed that the Ti plasmid can be used as a vector for dicotyledonous plants only. For Gramineae plants, alternative transformation procedures have to be developed. At present, several transformation systems independent of *Agrobacterium* infection, such as electroporation and direct gene transfer, have been devised. Protoplasts are usually used in these systems. Therefore, the cultivation and regeneration of Gramineae plants are especially important.

However, the cultivation techniques of Gramineae protoplasts have not yet been developed sufficiently as compared with those of Solanaceae. Although some examples of cultivation of Gramineae protoplasts had been reported, the results in reports before 1984, especially those on rice protoplasts, were insufficient for practical use due to low division frequencies resulting from inadequate plant material and technical difficulties. Recently, several research groups in-

[1] Department of Pure and Applied Sciences, College of Arts and Sciences, University of Tokyo, Komaba, Meguro-Ku, Tokyo, 153 Japan

Biotechnology in Agriculture and Forestry, Vol. 9
Plant Protoplasts and Genetic Engineering II (ed. by Y.P.S. Bajaj)

cluding ours have been concerned in the development of cultivation of rice protoplasts which are suitable for use in experiments in genetic modification of plant cells.

In this chapter, our work on the improvement of the cultivation and transformation of rice protoplasts is summarized. Recent progress, prospects, and the significance of the culture and genetic modification of Gramineae plants are also discussed.

## 2 Isolation and Culture of Rice Protoplasts

Although several papers concerning isolation and culture of rice protoplasts have been published before 1984 (Table 1), division frequencies were low or not described, and/or clear photographic evidence of protoplast division was not presented. In such cases the possibility cannot be excluded that the colony was derived from undigested cells. However, the recent advances in the cultivation of rice protoplasts are remarkable. Many reports on high frequency of protoplast division and whole plant regeneration from the protoplasts, published since 1985, are shown in Table 1. We also obtained rice protoplasts which were capable of dividing vigorously enough for successful experiments in genetic modification. The methods of isolation and cultivation of rice protoplasts are shown here.

### 2.1 Preparation of Rice Protoplasts

Rice protoplasts are isolated from suspension cultures of *Oryza sativa* L. maintained in 80 ml Murashige and Skoog's (MS) medium (1962) supplemented with 1 mg/l 2,4-dichlorophenoxyacetic acid (2,4-D) in 300-ml Erlenmeyer flasks on a reciprocal shaker (80 excursion/min) at 26 °C. Five ml of the culture are transferred to fresh medium at weekly intervals, and the 5-day culture is used as the source of protoplasts. This line of rice culture has been maintained for many years. It was first obtained as follows. Rice (*Oryza sativa* L. C5924) seeds were sterilized, washed, and germinated on MS basal medium. After a week of germination, root explants were excised from the seedlings and cultured on MS agar medium containing 1 mg/l 2,4-D and 0.1 mg/l kinetin. The callus cultures, named Oc, initiated from them were successively subcultured every month on the same medium. After 4 years, the callus was transferred to the liquid medium as described above and subcultured every week. The callus grew rapidly as small cell aggregates uniformly suspended in the liquid medium.

The conditions of protoplast isolation are as described below (Scheme 1). The cells are treated with the enzyme solution containing 1% Cellulase Onozuka RS (Yakult Honsha Co., Ltd., Tokyo, Japan) and Pectolyase Y23 (Seishin Pharmaceutical Co., Ltd., Tokyo, Japan) in 0.56 M mannitol (pH 5.3) at 30 °C. After 1 h enzyme treatment and filtration with 40 – 60 μm pore-size nylon mesh, freshly isolated protoplasts are washed with 0.56 M mannitol solution and cultured.

**Table 1.** Isolation and culture of rice protoplasts

| Plant material *Oryza sativa* L. cultivar | Enzyme mixture | Culture medium | Division frequency % | Growth response | Reference |
|---|---|---|---|---|---|
| Lian Kiang Mi Tsao (Pollen callus) | Pectinase (1.5%) + Cellulase Onozuka R-10 (5%) + potassium dextran sulfate (0.2%) in 0.8 M mannitol | Original medium + 2,4-D (0.5 mg/l) + BA (0.3 mg/l) | 0.1 | Cluster | Somatic Hybridization Research Group and Cytobiochemistry Research Group (1975) |
| (Seedling) | Macerozyme (1%) + Cellulase (2%) in 0.7 M mannitol | Modified Nagata and Takebe (1971) and Maeda (1965) | – | – | Lai and Liu (1976) |
| (Leaf sheaths and callus) | Pectinase (2%) in 0.45 M mannitol and (or Pectinase (2%) + Cellulase Onozuka SS (3%) in 0.45 M mannitol | Original combination | – | Callus and roots | Deka and Sen (1976) |
| A-58 etc. (callus) | Macerozyme R-10 (0.4%) + Cellulase Onozuka R-10 (2%) + Driselase (1%) in 0.7 M mannitol | Murashige and Skoog (MS) medium (1962) + 2,4-D (4 mg/l) + BA (2 mg/l) | – | Callus | Niizeki and Kita (1981) |
| B370 (callus) | Macerozyme R-10 (0.5%) + Cellulase Onozuka R-10 (1%) in 0.5 M mannitol (+ repeated cellulase treatment) | Gamborg et al. (1975) | – | Somatic hybrid cells between rice × pea | Bajaj (1983) |
| Norin 8 (nitrate reductase-defficient callus) | Macerozyme R-10 (1%) + Cellulase Onozuka RS (4%) in 0.3 M mannitol | AA medium (Müller and Grafe 1978) + 2,4-D (2 mg/l) + zeatin (0.1 or 2 mg/l) | 0.1 – 1 | Callus | Wakasa et al. (1984) |
| A-58 (callus) | Macerozyme R-10 (1%) + Cellulase Onozuka RS (4%) in 0.7 M mannitol | Modified Kao and Michayluk | – | Somatic hybrid callus from rice and soybean | Niizeki et al. (1985) |
| Chinsurah Boro II (cytoplasm) and Taichang 65 (nucleus) (suspension culture) | Macerozyme R-10 (0.2%) + Cellulase Onozuka R-10 (4%) + $CaCl_2 \cdot 2H_2O$ (0.1%) in 6.5% mannitol | Modified AA, $NO_3$ and B5 | 10 – 20 | Callus and roots | Toriyama and Hinata (1985) |

**Table 1.** (Continued)

| Plant material *Oryza sativa* L. cultivar | Enzyme mixture | Culture medium | Division frequency % | Growth response | Reference |
|---|---|---|---|---|---|
| Nihonbare and Sasanishiki (suspension culture) | Macerozyme R-10 (1%) + Cellulase Onozuka RS (4%) + $CaCl_2 \cdot 2H_2O$ (0.5%) + potassium dextransulfate (0.5%) in 0.4 M mannitol | Original combination + 2,4-D (1 mg/l) | 10 | Plantlets | Fujimura et al. (1985) |
| Moroberekan (callus) | Macerozyme (1%) + Pectolyase (0.5%) + Cellulase Onozuka R-10 (1%) + Driselase (0.6%) in 5.5% mannitol and 5.5% sorbitol | Kao and Michayluk | – | Plantlets | Coulibaly and Demarly (1986) |
| C 5924 (suspension culture) | Pectolyase Y23 (0.1%) + Cellulase Onozuka RS (1%) in 0.55 M mannitol | MS + 2,4-D (0.5 mg/l) + kinetin (0.3 mg/l) | 5 – 10 | Transformed callus | Baba et al. (1986) |
| Taipei 309 and Fujisaka 5 (suspension culture) | Pectolyase Y23 (0.1%) + Cellulase Onozuka RS (1%) + 5 mM MES + CPW salts in 13% mannitol | Modified Kao (1977) with agarose (+ heat shock) | 20 – 40 | Plants | Abdullah et al. (1986) |
| Yamahoushi (suspension culture) | Same as Toriyama and Hinata (1985) | Same as Toriyama and Hinata (1985) | – | Plants | Toriyama et al. (1986) |
| A-58, Fujiminori, Toyotama etc. (suspension culture) | Macerozyme R-10 (2%) + Pectolyase Y23 (0.1%) + Cellulase Onozuka RS (2%) + Driselase (1%) + calf serum (0.8%) + $CaCl_2 \cdot 2H_2O$ (80 mM) + $MgCl_2$ (0.125 mM) + MES (0.5 mM) in LS medium containing 0.3 M glucose | Original combination + 2,4-D (2 mg/l) | 10 | Plants | Yamada et al. (1986) |
| Taipei 309 (suspension culture) | Same as Abdullah et al. (1986) | Same as Abdullah et al. (1986) with agarose (+ heat shock) | 45 | Callus | Thompson et al. (1987) |
| Nipponbare, Iwaimochi, Norin 14, and Fujisaka 5 (suspension culture or callus) | Macerozyme R-10 (1%) + Cellulase Onozuka RS (4%) in 0.4 M mannitol | Nurse culture | 10 | Plants | Kyozuka et al. (1987) |
| Same as Kyozuka et al. (1987) | Same as Hayashi et al. (1986) | Same as Hayashi et al. (1986) | – | Plants (in field) | Ogura et al. (1987) |

**Scheme 1.** Methods for isolation of rice protoplasts (Baba et al. 1986)

*Oryza sativa* cells cultured in MS medium
containing 1 mg/l 2,4-D

|

Add enzyme solution
1 h incubation at 30 °C

|

Filtrate through nylon mesh
centrifugation

|

Wash with mannitol solution twice

|

Spheroplast treatment

| Enzyme solution | |
|---|---|
| Cellulase Onozuka RS | 1% |
| Pectolyase Y-23 | 0.1% |
| Mannitol | 0.56 M |
| pH | 5.3 |

| Culture medium for protoplasts | |
|---|---|
| MS medium containing 0.5 mg/l 2,4-D and 0.3 mg/l kinetin | |
| Mannitol | 0.56 M |
| pH | 5.5 |

## 2.2 Protoplast Culture

The protoplasts (Fig. 1) are suspended in MS medium supplemented with 1% sucrose, 0.56 M mannitol and 0.5 mg/l 2,4-D and 0.3 mg/l kinetin (pH 5.5) at a density of $2.5 \times 10^5$ protoplasts/ml. The protoplasts isolated from single cells are small-sized (ca. 20 µm in diameter) and densely cytoplasmic. However, they are easily fused spontaneously, so the sizes of protoplasts vary. Extensive filtration through nylon mesh is needed to remove large fusion bodies before the cultivation of protoplasts.

In our experiments, about 80% of the protoplasts were viable after a week and about 10% of them showed the typical figures of protoplast division (Fig. 1 B–D). Anomalous figures of protoplast division were also observed (Fig. 1 E, F). After 3 weeks, suspensions of 10- to 50-cell colonies derived from the protoplasts are diluted every 3 days with the medium used for protoplast culture except that mannitol is omitted and the sucrose concentration is 3%. After two dilutions, the colonies are embedded in 0.6% agar medium of the same composition. At 40 to 50 days after the isolation of the protoplasts, many small calli are observed on agar plates.

# 3 Recent Progress in Culture and Regeneration of Rice Protoplasts (1984–1987)

## 3.1 Division Frequency

Recent progress in the cultivation of rice protoplasts has been remarkable (Table 1). Many reports on this subject, including ours, made it possible to use rice pro-

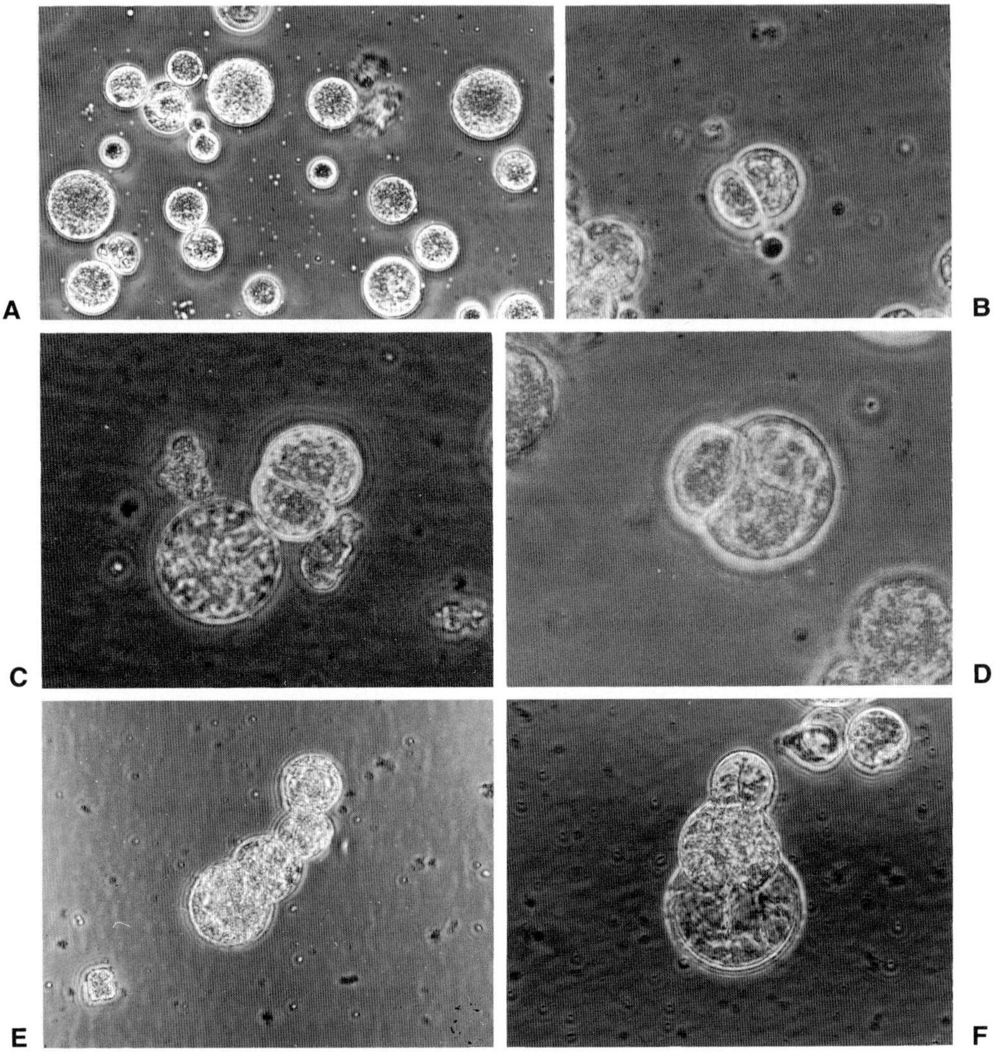


**Fig. 1A–F.** Various stages in the isolation and culture of rice protoplasts. **A** Freshly isolated protoplasts. **B–F** Various figures observed in the division of rice protoplasts. **B, C** Typical figures of protoplast division (×400). **D** A second division of protoplast (×400). **E, F** Anomalous division figures of protoplasts (**E** ×192, **F** ×256). These figures were observed at 5 days after cultivation (Baba et al. 1986)

toplasts for experiments such as gene transfer. High frequencies of protoplast division (more than 10%) were reported in them. Clear photographic evidence of protoplast division was also reported (Toriyama and Hinata 1985; Baba et al. 1986; Abdullah et al. 1986), so colony formation from rice protoplasts was confirmed. Recently, division frequency of rice protoplasts was as high as 45% (Thompson et al. 1987).

In these reports, protoplast isolation was achieved by enzyme solutions consisting of highly concentrated cellulase and maceration enzymes, using cultured cells as the source of protoplasts. In many cases rice protoplasts, treated with powerful enzyme solutions over relatively short times, were suspended at a density of $10^5 - 10^6$/ml and cultured in liquid medium containing 2,4-D. The first division, involving typical and anomalous division, was observed within a week.

### 3.2 Regeneration from Rice Protoplasts

Regeneration of plantlets and plants from rice protoplasts was also reported in several papers of Table 1 (Fujimura et al. 1985; Coulibaly and Demarly 1986; Abdullah et al. 1986; Toriyama et al. 1986; Yamada et al. 1986; Kyozuka et al. 1987). In these cases embryogenic cell suspensions were used as the source of protoplasts. Rice protoplasts, isolated by the enzyme solutions with mineral or other supplements, were cultured in the liquid medium. After 40–50 days, colonies derived from protoplasts were transferred to the medium without phytohormones, leading to differentiation. The line of rice cultivar used for material, the subculture length of suspension culture as the source of protoplasts, as well as the composition of culture media, especially $NH_4^+$, influenced the differentiation ability of callus derived from protoplasts (Table 1).

## 4 Transformation of Rice Protoplasts by Spheroplast Treatment

Tumor-inducing (Ti) plasmid, present in virulent strains of *Agrobacterium tumefaciens*, is frequently used as the vector to dicotyledonous plants. The transfer of Ti plasmid from *Agrobacterium* to plant cells, occurring through the infection of plants with this bacterium, results in the formation of crown galls. However, most of monocotyledonous plants involving Gramineae are generally reported to be resistant to *Agrobacterium* infection; therefore such plants are unable to be transformed using Ti plasmid as the vector. The spheroplast method as described in our Chapter X, this Volume, entitled Transformation in *Vinca*, made it possible to bypass the dependence on *Agrobacterium* infection process for the transfer of Ti plasmid to plant cells. We have actually transformed rice protoplasts by the spheroplast treatment as follows.

### 4.1 Spheroplast Treatment

Transformation of rice cells using the protoplasts was accomplished as described in Section 2. The procedures of preparation of *Agrobacterium* spheroplasts and the spheroplast treatment are almost the same as described in Chapter X, this Volume. *Agrobacterium* spheroplasts are also introduced into rice protoplasts by polyethylene glycol (PEG) 4000 (Wako Pure Chemical Industries Ltd., Osaka, Japan) (Scheme 2). After the spheroplast treatment, the protoplasts are washed

**Scheme 2.** Introduction of *Agrobacterium* spheroplasts into rice protoplasts

*Oryza sativa* protoplasts | *Agrobacterium tumefaciens* spheroplasts

Mix

Add PEG solution

Dilute gradually with 0.56 M mannitol

Wash with 0.56 M mannitol 5–10 times

Wash and culture in a liquid medium

PEG solution

PEG 4000 20% (w/w) in MS basal medium

five to ten times with 0.56 M mannitol solution by 2 min centrifugation at 900 rpm. (The treatment of high pH-high calcium solution is omitted in this case, because this treatment does not have a good effect on rice protoplasts). They are then washed with the culture medium for rice protoplasts containing 2,4-D, as described above, and resuspended in the same medium.

### 4.2 Culture of Protoplasts Treated with Spheroplasts

The protoplasts are cultured in the medium supplemented with 200 mg/l vancomycin to eliminate the rest of the spheroplasts. Three ml of protoplast suspension is cultured in each Corning plastic culture flask. After 3 weeks, an equivalent amount of MS basal medium is added to the flask and the diluted protoplast suspension is divided equally into two flasks. The dilution with MS basal medium is repeated three times every 3 days. Some of the protoplasts continue to divide in the hormone-free medium (Fig. 2). Finally, they are washed with MS basal medium and embedded in the same medium solidified with 0.6% agar. Some of the colonies, treated with spheroplasts of virulent strains A208 or A277, continue to grow, forming larger colonies in MS basal plates. After 1 month, visible colonies are observed on the agar plate (Fig. 3). They are picked up and cultured individually for analysis of transformants. (A few colonies are also obtained from nontreated protoplasts or protoplasts treated with nonvirulent *Agrobacterium* spheroplasts. However, they rarely appear and grow very slowly).

### 4.3 Analysis of Transformants

In our experiments, the obtained clones were used for opine detection by the same methods as described in Chapter X, this Volume. The results are summarized in Table 2. Nopaline synthase activities were also detected by the same method as described in the same chapter. After incubation of enzyme preparation with

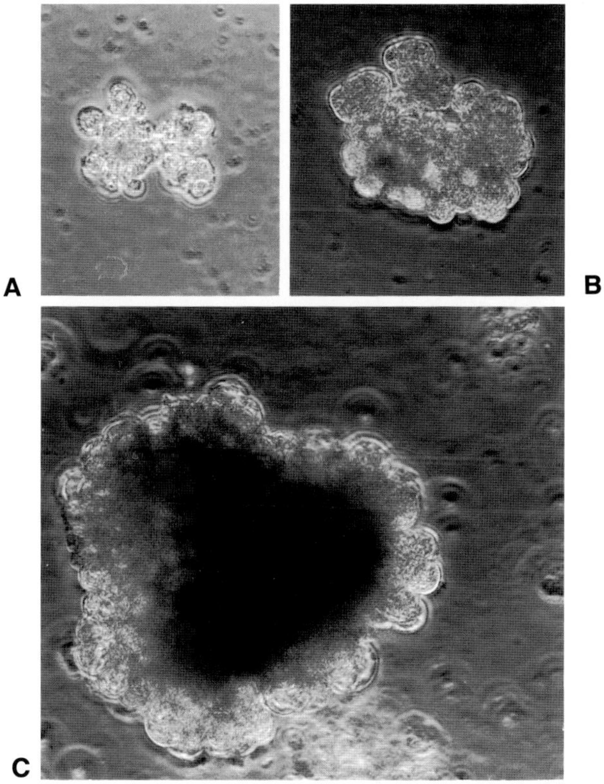

**Fig. 2 A–C.** Compact colonies growing in MS basal medium after the spheroplast treatment (×200). These figures were observed at 3-, 4-, and 6-week culture after the protoplast isolation, respectively (Baba et al. 1986)

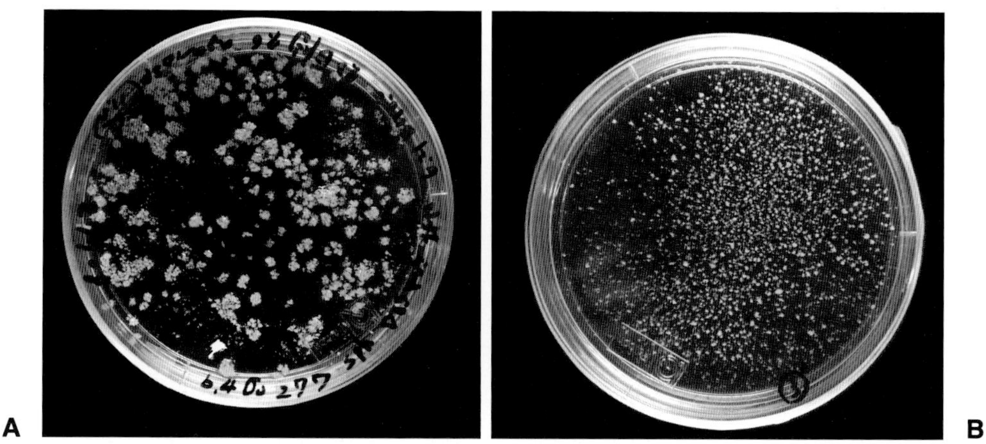

**Fig. 3 A, B.** Colonies formed on hormone-free agar plates. **A** Rapidly growing colonies observed on a plate treated with *Agrobacterium* spheroplast. **B** Slowly or nongrowing colonies observed on a nontreated plate (Baba et al. 1986)

**Table 2.** Frequency of transformation of *Oryza sativa* protoplasts by *Agrobacterium tumefaciens* (Baba et al. 1986)

| Experiments | Transformation systems | Strains of *Agrobacterium* | Number of colonies on hormone-free medium / Number of treated protoplasts | Frequency of putative transformants | Number of opine-producing clones / Number of clones examined for opine | Transformation frequency |
|---|---|---|---|---|---|---|
| 1. | Spheroplast treatment | A 208 | $2607/3.6 \times 10^6$ | $7.24 \times 10^{-4}$ | 4/20 | $1.45 \times 10^{-4}$ |
| 2. | Spheroplast treatment | A 277 | $1798/3.0 \times 10^6$ | $5.99 \times 10^{-4}$ | 3/19 | $0.95 \times 10^{-4}$ |
| 3. | Spheroplast treatment | A 479 | $18/7.5 \times 10^5$ | $0.24 \times 10^{-4}$ | 0/2 | 0 |
| 4. | Non-treatment | – | $34/3.75 \times 10^5$ | $0.90 \times 10^{-4}$ | 0/5 | 0 |

**Table 3.** Detection of radioactivities and its ratios incorporated into nopaline by NDH of rice transformants (Baba et al. 1986)

| Strains | CPM | | | Fluorescences by phenanthraquinone |
|---|---|---|---|---|
| | $^3H$ | $^{14}C$ | $^3H/^{14}C$ | |
| Tobacco teratoma (3n-3) – positive control | 57746 | 7719 | 7.48 | + + |
| Rice-habituated cells derived from protoplasts – negative control | 738 | 23 | 32.1 | – ~ ± |
| Oc 208- 1-1 | 2333 | 120 | 19.4 | ± |
| - 2–9 | 1720 | 54 | 31.9 | ± |
| - 7–2 | 5844 | 919 | 6.36 | + |
| -12–1 | 4098 | 540 | 7.59 | + |

radioactive substrates of nopaline synthase ($^3$H)-arginine and ($^{14}$C)-$\alpha$-ketoglutaric acid, the reaction mixture was purified by the ion-exchange columns and separated by thin-layer chromatography. The radioactivities from ($^3$H)-arginine in the extracts of positive control (tobacco teratoma clone 3n-3) and several putative transformant clones (Oc 208-7-2 and -12-1) derived from spheroplast-treated protoplasts co-migrated with authentic nopaline. The ratios of ($^3$H)/($^{14}$C) of these clones were almost the same as that of the positive control (Table 3). These results indicate that both ($^3$H)-arginine and ($^{14}$C)-$\alpha$-ketoglutaric acid were incorporated into nopaline in these clones, so that the rice cells were transformed by the spheroplast treatment in this case.

## 5 Introduction and Expression of Foreign Genes in Gramineae Plants

Recently, transformations have been accomplished in Gramineae plants using several techniques besides the spheroplast method. These transformations have been successful using direct gene transfer by the chemical or electrical method. The former is polyethylene glycol (PEG)- or polyvinyl alcohol (PVA)-induced uptake of DNA into monocot protoplasts and the latter is electric pulse-induced uptake, designated electroporation.

The reports of introduction of foreign information into Gramineae cells are summarized in Table 4. The genes of amino-glycoside phosphotransferase [APH(3′)II] or chloramphenicol acetyltransferase (CAT), which were introduced into wheat (Lörz et al. 1985), Italian ryegrass (Potrykus et al. 1985), maize (Fromm et al. 1986; Howard et al. 1987), and rice (Uchimiya et al. 1986; Ou-Lee et al. 1986), were expressed in these host cells. In these cases, in addition to nopaline synthase promoter (pNOS) the promoters originating in cauliflower mosaic virus (CaMV) or *Drosophila* were also employed. These results suggest that expression of foreign genes linked to various promoters is possible in Gramineae plants. The introduction of foreign DNA into host genome DNA seems to be achieved by the mechanism of random insertion.

**Table 4.** Genetic modification of Gramineae protoplasts

| Introduced foreign gene (origin) | Origin of promoter | Host plant | Gene expression | Reference |
|---|---|---|---|---|
| Aminoglycoside phosphotransferase type II gene – APH (3′) II (Tn 5) | Nopaline synthase promoter (pNOS) | *Triticum monococcum* (wheat) | Kanamycin resistance | Lörz et al. (1985) |
| APH (3′) II (Tn 5) | Cauliflower mosaic virus (CaMV) gene VI | *Lolium multiflorum* (Italian ryegrass) | G 418 resistance | Potrykus et al. (1985b) |
| APH (3′) II (Tn 5) | pNOS | *Oryza sativa* L. | Kanamycin resistance | Uchimiya et al. (1986) |
| Chloramphenicol acetyltransferase – CAT gene (Tn 9) | CaMV 35S promoter and *Drosophila copia* long-terminal repeat promoter | *Oryza sativa* L., *Triticum monococum* and *Sorghum bicolor* | CAT activity | Ou-Lee et al. (1986) |
| APH (3′) II (Tn 5) | CaMV 35S promoter | *Zea mays* L. | Kanamycin resistance | Fromm et al. (1986) |
| CAT gene (Tn 9) | Maize alcohol dehydrogenase promoter | *Zea mays* L. | CAT activity | Howard et al. (1987) |
| APH (3′) II (Tn 5) | CaMV 35S promoter | *Oryza sativa* L. | – | Luo and Wu (1988) |
| APH (3′) II (Tn 5) | CaMV 35S promoter | *Oryza sativa* L. | G418 resistance | Toriyama et al. (1988) |
| APH (3′) II (Tn 5) | CaMV 35S promoter | *Oryza sativa* L. | Kanamycin resistance | Zhang et al. (1988) |

## 6 Summary and Conclusion

During the last 5 years, we have been developing spheroplast methods for transformation of plant protoplasts independent of *Agrobacterium* infection (see Chap. X, this Vol.), using as experimental material, however, *Vinca rosea*, which is susceptible to *Agrobacterium* infection. To prove one of the merits of the spheroplast method for transformation of plant protoplasts independent of *Agrobacterium* infection, we tried to transform rice cells, which this bacterium can not infect. We have improved on the cultivation of rice protoplasts to make the system work for screening large numbers of cultured protoplasts for rare transformation events and demonstrated the occurrence of nopaline synthase activity in cells derived from the protoplasts treated with *Agrobacterium* spheroplasts. This shows that transformation of rice cells could be obtained by this method (Baba et al. 1986).

Now, besides ours, several other systems of rice protoplasts have been developed and the regeneration of whole plant from protoplasts has also been demonstrated. As for transformation, direct gene transfer has also succeeded in rice protoplasts as described above (Uchiminya et al. 1986; Ou-Lee et al. 1986). In these cases, the activities of marker genes such as APH(3′)II and CAT with pNOS were detected. As pNOS actually function in rice cells, the process of *Agrobacterium* infection is suggested to be one of the causes for the absence of crown gall formation on monocot plants.

In any case, the useful promoters for the Gramineae (Table 4), separated from *Agrobacterium* Ti plasmid or cauliflower mosaic virus, made the genetic modification of Gramineae cells easy. In the near future, it seems that the introduction of practically valuable genes instead of marker genes will be tried using Gramineae protoplasts and these promoters. As for the next step, the transmission of introduced genes through sexual cross to offsprings is the significant problem in the application of plant breeding. It was reported that the functional gene, introduced by direct gene transfer, is actually inherited in tobacco plants (Potrykus et al. 1985a). The transmission of introduced foreign genes also seems to be promising in Gramineae plants such as rice.

Another direction of genetic modification of rice protoplasts is somatic hybridization. Cell division or callus formation were successful in somatic hybrids of combinations such as rice×pea (Bajaj 1983), rice and soybean (Niizeki et al. 1985), and rice and barnyardgrass. These trials seem to be promising in cases of introduction of genetic information concerning plural genes.

In conclusion, basic techniques for genetic manipulation in the Gramineae have been greatly studied, and the appearance of genetic-engineered Gramineae plants seems to be only a question of time.

## References

Abdullah R, Cocking EC, Thompson JA (1986) Efficient plant regeneration from rice protoplasts through somatic embryogenesis. Biotechnology 4:1087–1090

Baba A, Hasezawa S, Syōno K (1986) Cultivation of rice protoplasts and their transformation mediated by *Agrobacterium* spheroplasts. Plant Cell Physiol 27:463–471

Bajaj YPS (1983) Survival of somatic hybrid protoplasts of wheat×pea and rice×pea subjected to −196 °C. Indian J Exp Biol 21:120–122

Coulibaly MY, Demarly Y (1986) Regeneration of plantlets from protoplasts of rice, *Oryza sativa* L. Z Pflanzenzücht 96:79–81

Deka PC, Sen SK (1976) Differentiation in calli originated from isolated protoplasts of rice (*Oryza sativa* L.) through plating technique. Mol Gen Genet 145:239–243

Fromm ME, Taylor LP, Walbot V (1986) Stable transformation of maize after gene transfer by electroporation. Nature 319:791–793

Fujimura T, Sakurai M, Akagi H, Negishi T, Hirose A (1985) Regeneration of rice plants from protoplasts. Plant Tissue Cult Lett 2:74–75

Gamborg OL, Shyluk J, Kartha KK (1975) Factors affecting the isolation and callus formation in protoplasts from the shoot apices of *Pisum sativum* L. Plant Sci Lett 4:285–292

Hayashi Y, Kyozuka J, Shimamoto K (1986) Plant regeneration from protoplasts of rice (*Oryza sativa*). Jpn J Breed 36:46–47

Howard EA, Walker JC, Dennis ES, Peacock WJ (1987) Regulated expression of an alcohol dehydrogenase 1 chimeric gene introduced into maize protoplasts. Planta 170:535–540

Kao, KN (1977) Chromosomal behaviour in somatic hybrids of soybean – *Nicotiana glauca*. Mol Gen Genet 150:225–230

Kao KN, Michayluk MR (1975) Nutritional requirements for growth of *Vicia hajastana* cells and protoplasts at a very low population density in liquid media. Planta 126:105–110

Kyozuka J, Hayashi Y, Shimamoto K (1987) High frequency plant regeneration from rice protoplasts by novel nurse culture methods. Mol Gen Genet 206:408–413

Kyozuka J, Shimamoto K, Ogura H (1989) Regeneration of plants from rice protoplasts. In: Bajaj YPS (ed) Biotechnology in agriculture and forestry 8. Protoplasts and genetic engineering I. Springer, Berlin Heidelberg New York Tokyo, pp 108–123

Lai KL, Liu LF (1976) On the isolation and fusion of rice protoplasts. J Agric Assoc China 93:10–29

Luo Z, Wu R (1988) A simple method for the transformation of rice via the pollen-tube pathway. Plant Mol Biol Rep 6:165–174

Lörz H, Baker B, Schell J (1985) Gene transfer to cereal cells mediated by protoplast transformation. Mol Gen Genet 199:178–182

Maeda E (1965) Callus formation and isolation of single cells from rice seedlings. Proc Crop Sci Soc Japan 34:139–147

Müller AJ, Grafe R (1978) Isolation and characterization of cell lines of *Nicotiana tabacum* lacking nitrate reductase. Mol Gen Genet 161:67–76

Murashige T, Skoog F (1962) A revised medium for rapid growth and bioassays with tobacco tissue cultures. Physiol Plant 15:473–497

Niizeki M, Kita F (1981) Cell division of rice and soybean and their fused protoplasts. Jpn J Breed 31:161–167

Niizeki M, Tanaka M, Akada S, Hirai A, Saito K (1985) Callus formation of somatic hybrid of rice and soybean and characteristics of the hybrid callus. Jpn J Genet 60:81–92

Ogura H, Kyozuka J, Hayashi Y, Koba T, Shimamoto K (1987) Field performance and cytology of protoplast-derived rice (*Oryza sativa*): high yield and low degree of variation of four *Japonica* cultivars. Theor Appl Genet 74:670–676

Ou-Lee TM, Turgeon R, Wu R (1986) Expression of a foreign gene linked to either a plant-virus or a *Drosophila* promoter, after electroporation of protoplasts of rice, wheat, and sorghum. Proc Natl Acad Sci USA 83:6815–6819

Potrykus I, Paszkowski J, Saul MW, Petruska J, Shillito RD (1985a) Molecular and general genetics of a hybrid foreign gene introduced into tobacco by direct gene transfer. Mol Gen Genet 199:169–177

Potrykus I, Saul MW, Petruska J, Paszkowski J, Shillito RD (1985b) Direct gene transfer to cells of a graminaceous monocot. Mol Gen Genet 199:183–188

Somatic Hybridization Research Group and Cytobiochemistry Research Group, Peking Institute of Botany, Academia Sinica (1975) Isolation and culture of rice protoplasts. Sci Sin 18:779–784

Thompson JA, Abdullah R, Chen W-H, Gartland KMA (1987) Enhanced protoplast division in rice (*Oryza sativa* L.) following heat shock treatment. J Plant Physiol 127:367–370

Toriyama K, Hinata K (1985) Cell suspension and protoplast culture in rice. Plant Sci 41:179–183

Toriyama K, Hinata K, Sasaki T (1986) Haploid and diploid plant regeneration from protoplasts of anther callus in rice. Theor Appl Genet 73:16–19

Toriyama K, Arimoto Y, Uchimiya H, Himata K (1988) Transgenic rice plants after direct gene transfer into protoplasts. BioTechnology 6:1072–1074

Uchimiya H, Fushimi T, Hashimoto H, Harada H, Syōno K, Sugawara Y (1986) Expression of a foreign gene in callus derived from DNA-treated protoplasts of rice (*Oryza sativa* L.). Mol Gen Genet 204:204–207

Wakasa K, Kobayashi M, Kamada H (1984) Colony formation from protoplasts of nitrate reductase-deficient rice cell lines. J Plant Physiol 117:223–231

Yamada Y, Zhi-Qi Y, Ding-Tai T (1986) Plant regeneration from protoplast-derived callus of rice (*Oryza sativa* L.). Plant Cell Rep 5:85–88

Zhang HM, Yang H, Rech EL, Golds TJ, Davis AS, Mulligan BJ, Cocking EC, Davey MR (1988) Transgenic rice plants produced by electroporation-mediated plasmid uptake into protoplasts. Plant Cell Rep 7:379–384

# VII Transformation in Potato

M. YAMAMOTO[1]

## 1 Introduction

Potato (*Solanum tuberosum* L.) is an important world food crop. It is produced in 130 countries where three quarters of the world's population live. It is grown as a major source of food in most countries with a temperate climate. World production of potatoes is 254861000 metric tons (United Nations 1982). Until recently, most world potato production was in Europe, but it has now increased most rapidly in Far Eastern countries, such as China, India, and Indonesia. China alone produces 56 million tons of potatoes, Indonesia 9 million tons, and Turkey 3 million tons. At present, potato consumption in developing countries is doubling every 10 to 15 years (International Potato Center, CIP 1984, United Nations 1982).

According to Hawkes (1978), potato is undoubtedly of ancient origin; it was domesticated in South America and was been dispersed over a considerable area by the Spaniards who arrived in the 16th century. Scot (1976) stated that the Inca Indians were the first to cultivate potato in 200 B.C. The potato appeared in Europe during the last quarter of the 16th century. It was first said to have been brought to Russia by Peter the Great at the end of the 17th century (Harris 1978), and was introduced from Ireland into the United States in 1718. At present, the geographic range of the potato is worldwide.

The most characteristic feature of potato is its high productivity and its wide usage. Potato yield in Japan was 3727000 tons in a cultivated area of 130150 ha in 1985 (Japan Agriculture Yearbook 1987).

Potato is susceptible to a great many diseases caused by bacteria, fungi, viruses, mycoplasma, viroids, and nematodes (Rich 1983). Late blight (*Phytophthora infestans*) is one of the oldest and the most serious diseases of potato. In 1845 it caused the Irish famine, one of the most traumatic tragedies stemming from plant diseases. About one million people died of starvation in Ireland, and another million and a half emigrated to the US, Canada, and other countries (Large 1940; Western 1971). There is a great need to produce late blight-resistant potato, and various biotechnologies are being applied for its improvement (Bajaj 1987a). In this chapter our work on transformation of potato with a view to induce late blight tolerance is detailed and the contemporary work in this field is summarized.

[1] Laboratory of Plant Pathology, College of Agriculture, Kyoto University, Kyoto 606, Japan
Present address: 25 Minami-Gosho-cho, Okazaki, Sakyo-ku, Kyoto 606, Japan

Biotechnology in Agriculture and Forestry, Vol. 9
Plant Protoplasts and Genetic Engineering II (ed. by Y.P.S. Bajaj)

## 2 Transformation Studies in Potato for Resistance to *Phytophthora infestans*

To know the nature of resistance to the invasion of a pathogen is the first step for the breeding of resistant plants. I have been studying the mechanism of resistance in potatoes to the invasion of *Phytophthora infestans* (Fig. 1) from physiological and biochemical aspects since 1951, and established the importance of DNA in the resistant cultivar for the nature of resistance in 1966, and published the results in 1968 (Yamamoto 1968).

The resistance of potatoes to the invasion of *P. infestans* is divided into two categories, i.e., hypersensitivity and field resistance. Bills (1977) and Kuć et al. (1979) have reviewed research on the potato-*P. infestans* system. The expression of single gene host resistance in potato to *P. infestans* involves a hypersensitive reaction that has been extensively studied (Tomiyama et al. 1976; Yamamoto 1982). The hypersensitivity is influenced by R genes derived from *Solanum demissum*, each showing a Mendelian inheritance pattern. The other genetic system is composed of multiple genes, which appear to be additive in effect and are inherited in polygenic fashion. This system is called field resistance. Field resistance is race-non-specific, and is a combination of different factors, the sum of which determine the actual level of resistance to the pathogen under field conditions. In this chapter, only hypersensitivity is described.

The initial event of the transition from compatibility to incompatibility involves mutual recognition by the potato and the *P. infestans* race, which begins to operate shortly after their encounter. There was no difference between the incompatible and compatible races at the time required for penetration.

Resistant potato cells were damaged immediately after penetration by an incompatible race of *P. infestans*. This is the usual hypersensitive effect of *P. infestans* on potato plants. Protoplasmic streaming of potato cells infected by an incompatible race of *P. infestans* creased within 30 min after penetration. In con-

**Fig. 1.** Scanning electron micrograph of the potato late blight fungus, *Phytophthora infestans*. (Yamamoto 1985)

trast, potato cells infected by a compatible race coexist with the pathogen for at least 3 days.

The sequence of events which lead to hypersensitive death of the cell appear to be governed by metabolic rate. Under phase-contrast microscopy, the resistant cultivar shows protoplasmic currents during the early stage of penetration of the pathogen (Yamamoto 1974, 1982). The acceleration of protoplasmic current in infected cells of highly resistant cultivars is accompanied by a stimulation of the energy cycle. Although respiration and enzymatic activities increase immediately after the invasion of the pathogen, these activities are nonspecific, hence occur irrespective of incompatibility or compatibility of the potato cultivar to the race of *P. infestans.* Phenylpropanoid (phenol compound) content, respiration, and peroxidase activity are enhanced in tissue near the hypersensitive flecks (Yamamoto 1970; Yamamoto et al. 1975, 1986a). Rishitin ($C_{14}H_{22}O_4$) and other phytoalexins are found in potato tissues infected with *P. infestans* (Tomiyama et al. 1968a, b). Doke and Tomiyama (1980) claim the importance of glucan on hypersensitivity. Garas and Kuć (1981) reported that lectins elicit terpenoid accumulation in potatoes infected with *P. infestans.* Flor (1971) developed the gene-for-gene theory from his studies of the relationship between flux and rust.

This theory implies that the genes in the plant that condition the reaction can be identified only by their interaction with a specific race of the pathogen, and genes in the pathogen that condition pathogenicity can be identified only by their counterpart with a specific cultivar of the plant.

Modification of cells by genetic engineering renders genes manipulable in ways, which were not possible before. In this connection, plant protoplasts have the remarkable property of being able to take up macromolecules and particles such as protein, DNA, viruses, bacteria, and even whole nuclei and isolated chloroplast.

The introduction of DNA molecule into protoplasts has been attempted and Yamaoka et al. (1982) demonstrated that CaMV DNA transfected turnip protoplast.

However, hitherto, the literature on the uptake of exogenous DNA into potato plants has been limited. The following fundamental data might be useful for considering this problem in connection with breeding of potatoes resistant to late blight fungus.

## 3 Techniques and Material Used for Transformation Studies

### 3.1 Plants and Fungi

The potato interspecific hybrid cv. 96-56 ($R_1$ gene), SH-469 ($R_4$ gene), 1506-b(9) ($R_1$ $R_4$ genes) and Greta ($R_1$ $R_3$ $R_4$ genes) were used as resistant potato for *P. infestans* and common cv. Irish Cobbler, Norin No. 1, and Nagasaki-aka (r gene) were used as susceptible ones. *P. infestans* used were race 0 and race 1. Race 0 is compatible to gene r and shows susceptible lesion to common cultivars and hypersensitive flecks to potatoes having R genes. Race 1 is compatible to r gene and $R_1$ gene potatoes, and incompatible to other R gene potatoes.

### 3.2 Tissue Culture, Isolation of Free Cells and Protoplasts from Potatoes

Tissues of potato tubers (5×5×2 mm) from cv. Irish Cobbler, Norin No. 1, hybrid 96-56 and SH-469 were immersed in Antiformin for 5 min, thoroughly rinsed with sterilized water, and cultured on Murashige-Skoog medium (MS) for 40 days or 2 months at 20 °C. In order to examine the effect of exogenous DNA, MS medium containing 1% DNA of Norin No. 1 was prepared, and 40 day-old calli of Norin No. 1, Irish Cobbler of hybrid 96-56 were cultured on this medium for 2 weeks, and inoculated with *P. infestans* (Yamamoto and Kitami 1971).

Free cells were prepared from calli cultured on MS solution (Irish Cobbler, hybrid 96-56, and SH-469) by shake culture for 2 weeks at 120 rpm. Potato protoplasts were obtained as follows. The leaf surface of potatoes (cv. Irish Cobbler and hybrid 96-56) was first sterilized with 70% ethanol for 15 s and then 0.2% mercuric chloride for 15 s. After rinsing with sterilized water, the lower epidermis was removed. The tissue was incubated in protoplast solution which contained 9.1% Mannitol, 1% Cellulase and 0.05% Macerozyme (Yakult Co.), and the pH adjusted at 5.8 with 0.2 N KOH prior to sterilization. Leaf materials were kept in the enzyme solution at 30 °C for 3 h without agitation until a mesophyl protoplast suspension was formed (modified Takebe and Otsuki 1969).

### 3.3 Preparation of DNA

Nuclear DNA was extracted by the method of Thomas and Abelson (1966), Shigematsu et al. (1966), or Hall and Spiegelman (1961) from leaves of the hybrids Chloroplast DNA was prepared by the method of Kolodner and Tewari (1975).

Sometimes, tomato (cv. Fukuju No. 2), jimsonweed (*Datura alba* Nees) and tobacco (*Nicotiana tabacum* L. cv. Xanthi) were also used as the source of DNA fraction. Electrophoresis of partially purified DNA (final concentration was 100 μg/ml) was run at 40 V for 16 h to separate a different size of DNA, 1750–2550 kDa. The extracts from these slab gels were rubbed on leaves of a susceptible potato cultivar with an adsorbant cotton swab. In some experiments, DNA was digested by restriction enzymes. Sometimes, potato leaves of a susceptible cultivar were infiltrated with hybrid DNA solution, kept in the moist chamber at 20 °C for 2 h, and washed with tap water in order to prevent DNA remaining outside tissues.

### 3.4 Assay for Fleck and Lesion Formation on Potato Leaves

Zoospore suspension of *P. infestans* (concentration varied in each experiment) was inoculated on the dorsal epidermis of each leaf (especially around midrib). The inoculated leaves were incubated in a moist chamber at 20 °C for 4–10 h in the dark after rubbing with exogenous DNA solution (0.5–1 mg/ml) with a brush or a piece of absorbent cotton swab, and the number of flecks and lesions were counted under optical microscope at 68–72h after inoculation.

### 3.5 Incorporation of DNA into Potato Cells

In order to examine the incorporation of DNA into potato cells, the resistant hybrid DNA solution (1 mg/ml) was added with propidium iodine (PI) or 4′-6-diamidino-2-phenyliodole (DAPI) and incubated at 20 °C for 30 min. The DNA-PI or DNA-DAPI complexes were then purified by ethylalcohol precipitation and finally dissolved in N-buffer (0.05 M sucrose, 0.1 mM $CaCl_2$, 0.6 mM spermidin, 0.8 mM phenylmethylsulfonylfluoride, 0.05% 2-mercapto-ethanol, 10 mM Piper pH 6) to give a concentration of 1 mg/ml. The DNA-fluorescent dye complexes were rubbed onto the dorsal epidermis of potato leaves (cv. Irish Cobbler) with the aid of absorbant cotton swab and then cut with a razor blade. The sections were examined under epifluorescent microscope equipped with a 100 W Hg lamp (Olympus BHK-BKF type) (Yamamoto and Hatta 1974).

### 3.6 Destruction of DNA by Physical and Chemical Treatments

*1. Thermal Treatments.* DNA from the hybrid 96-56 and cv. Norin No. 1 were treated at 50, 60, 70, 80, 90, and 100 °C for 10 min and cooled immediately.

*2. UV Irradiation.* DNA solution (100 µg/ml) was treated with RNase at 37 °C for 30 min and protease at 37 °C for 10 min, successively. After treatment with phenol, precipitated DNA was dissolved in 0.01 M phosphate buffer (pH 7.2) and passed through Sepharose 2B at 9–9.6 ml/h flow rate. One ml of the DNA solution (100 µg) so obtained was placed on a Petri dish with a diameter of 2.7 cm and exposed to UV irradiation at room temperature (Yamamoto et al. 1975, 1979).

*3. DNase Treatment.* DNase II (from bovine spleen, dissolved at 400 µg/ml in 0.1 M acetate buffer solution of pH 4.6, containing 1 mM $MgCl_2$ was used. Two ml (200 µg/ml) of DNA solution added to 2 ml of this DNase solution incubated for 25 min at 37 °C. After incubation the mixture was poured into ethanol. The resultant precipitate was dissolved in 0.01 M citrate buffer (pH 7.2) and fractionated through a Sephadex G-25 column at a flow rate of 240 ml/h and collected in 2-ml fractions (Yamamoto et al. 1979).

*4. Thermal Neutron Irradiation.* 5 mg of the DNA were placed in a polyethylene capsule containing a small golden leaf (7.71–7.96 mg for the measurement of thermo-luminescent flux) and 5mg of the thermo-luminescence dosimeter (for the measurement of $\gamma$ rays) and the capsule was placed in polyvinyl tube. The DNA in the capsule in an irradiation tube was irradiated with the thermal neutrons in a $D_2O$ facility at 5000 kW/h. The flux of thermoluminescence was $2.07-2.14 \times 10^9 n/cm^2$ (Yamamoto et al. 1979).

*5. Restriction Enzyme Digestion.* DNA (1750–2550 kDa) was digested by restriction enzymes, *Eco*RI, *Hae*III, and *Bam*III. Sometimes, the restriction enzymes were used twice or three times, i.e., *Eco*RI-*Sal*I or *Eco*RI-*Kpn*T, and *Eco*RI-

*Sal*I-*Pvu*II or *Eco*RI-*Sal*I-*Kpn*I, and subjected to biological assay. For digestion, 1 unit of enzyme was added to 1 μg of distilled water and treated for 3 h at 37 °C. Restriction enzymes were produced by Takara Shuzo Co. (Yamamoto 1982, 1983, 1986a).

### 3.7 Some Physiological Characters of Potato Plants

*1. Protein Content.* Soluble protein in potato leaves was prepared by the method of Stahmann and Staples (1964). Electrophoresis was conducted by Davis and Ornstein method (1964) with modifications. Protein content was measured by the method of Lowry et al. (1951) using Bovin serum albumin as standard.

*2. Enzymatic Activities.* Phenylalanine ammonia lyase (PAL) was measured by the method of Zucker (Zucker 1963). Peroxidase activity was assayed by the Wood and Barbara method (Wood and Barbara 1971b).

*3. Rishitin Content.* For extraction and estimation of rishitin, the method of Tomiyama et al. (1968a) and Wood and Barbara (1971a) was used.

*4. Fraction-I-Protein (FIP).* FIP in potatoes with different susceptibility to *P. infestans* was investigated by isoelectrofocusing (Uchimiya et al. 1979). Electrophoresis gel was 8 M urea, 9% polyacrylamide, 1% pharmalyte (Pharmacia Fine Chemicals, Sweden) (pH 5–8). Thickness of gel was 2 mm. After electrophoresis was run at 300 V, small subunits were examined.

### 3.8 Wounds on Potato Leaves Caused by Rubbing with Exogenous DNA

To examine the wounds on potato leaves due to rubbing exogenous DNA, specific fluorescence of ethidium bromide bound to DNA was examined under the epifluorescence microscope (Nikon epifluorescence microscope, XF-EF type). Wounds on potato leaves were examined by scanning electron microscope (JSMUS 3 type). Fixation was made by 3% glutaraldehyde.

## 4 Results

### 4.1 Appearance of Hypersensitive Flecks and the Increase of Metabolic Activities of Potatoes After Treatment with Hybrid DNA

#### 4.1.1 Fleck Formation by DNA Treatment

Potato leaves of susceptible cultivars treated with DNA fraction isolated from resistant hybrid and other plants resistant to *Phytophthora infestans* showed coexistence of hypersensitive flecks and susceptible lesions (Nozu et al. 1976, 1977;

Yamamoto 1968, 1974, 1982, 1983, 1985; Yamamoto and Matsuo 1976; Yamamoto and Otsuka 1971).

The time required for cell death in susceptible cultivars was shortened after treatment with the DNA fraction isolated from the resistant plant, and mycelial growth of *P. infestans* in resistant hybrid tissue occurred after treatment with the DNA fraction from the susceptible cultivar (Yamamoto and Matsuo 1976). The frequency of co-existence of flecks and lesions decreased when DNA had been kept after 10 min at temperatures higher than 90 °C (Yamamoto and Hatta 1974). DNA from herring sperm did not affect fleck formation (Yamamoto 1968). Hypersensitive flecks appeared even when DNA-treated susceptible leaves were washed within 10 min of application. The higher molecular weight DNA fraction causes more flecks than low molecular weight DNA. When potato tuber discs of hybrid ($R_1$ $R_4$, $R_1$ $R_3$ $R_4$ genes) and common cultivars (r gene) are treated with culture filtrate of *P. infestans* (race 0 and race 1), more spots are observed on the hybrid tuber disc than susceptible cultivars (24 h treatment). An increase in brown spots is observed on discs of cv. Benimaru after application of DNA from the hybrid 96-56, and before the application of a culture filtrate from *P. infestans* (Nozu et al. 1976, 1977).

### 4.1.2 Effect of Nuclear and Chloroplast DNA on the Formation of Fleck on Potato Leaves

DNA from potato nuclei or chloroplast was dissolved in distilled water (50 μg/ml) and rubbed on leaves of Irish Cobbler, and 4 h later, the leaves were inoculated with zoospores of *P. infestans* (race 0). As shown in Table 1, Irish Cobbler leaves which had been contacted with nuclear DNA from the hybrid showed 44.9% of flecks. This is almost the same as that obtained in previous experiments with exogenous DNA. On the other hand, leaves rubbed with chloroplast DNA of the resistant hybrid showed a percentage of flecks less than 3. Thus, the chloroplast

**Table 1.** Induction of flecks by a compatible race of *Phytophthora infestans*, race 0, on Irish Cobbler leaves which had been rubbed with nuclear and chloroplast DNA from resistant hybrid (cv. Greta) prior to inoculation with the fungus[a]. (Yamamoto and Kuroiwa 1986)

| Treatment | Cultivar inoculated | Hypersensitive fleck[b] | Susceptible lesion[b] |
|---|---|---|---|
| Untreated control | Irish Cobbler | 0 (0) | 135 (100) |
| | Greta | 121 (100) | 0 (0) |
| Rubbed with | | | |
| nuclear DNA | Irish Cobbler | 89 (44.9) | 104 (55.0) |
| chloroplast DNA | Irish Cobbler | 4 (3.0) | 128 (97.0) |

[a] Spore suspension of *P. infestans* was applied 4 h after rubbing with DNA (50 μg/ml). Nuclear DNA was extracted by Hall and Spiegelman method, and chloroplast DNA was extracted by Kolodner and Tewar method

[b] Scored 72 h after inoculation. Number in parenthesis represents percentage

DNA seemed to have almost negligible activity in the induction of resistant response. Experiments were further extended to determine whether nuclear and chloroplast DNA are translocated to nuclei of potato cells.

In the dorsal epidermal cells on which DNA-PI or DNA-DAPI complex of the hybrid Greta had been rubbed, fluorescence appeared in the nuclei and around the cell wall 120 min after rubbing. At 60 min after rubbing, the section of potato leaves treated with DNA-DAPI complex showed paleblue fluorescence in the nuclei and around the cell wall. In the case of the DNA-DAPI complex from the hybrid chloroplast, however, no fluorescence was observed in nuclei even 2 h after rubbing.

Since the large subunit of Fraction-I-Protein (FIP) is considered to be related to chloroplast DNA and the small subunit to nuclear DNA, FIP from potatoes with different susceptibility to *P. infestans* was observed by isoelectrofocusing. Irish Cobbler (r gene) lacks 1 band in the small subunit of FIP which shows the band specific to the hybrid. No change was observed in the bands of small subunit 7 days after the DNA of the hybrid 96-56 was rubbed on the leaves of Irish Cobbler.

### 4.1.3 Increase of Metabolic Activities in Potato Applied with Exogenous DNA

The quantity of RNA in potato tubers (cv. Benimaru), decreased 6–12 h after inoculation with *P. infestans* (race 0), but the quantity increased in the resistant hybrid 1506-b(9). After treatment with a DNA fraction from the hybrid, the amount of RNA in cultivars inoculated with *P. infestans* increased three times, by 12 h after inoculation, as compared to the cv. Benimaru without treatment (Nozu et al. 1976).

An increase in the quantity of soluble protein, phenylalanine ammonia lyase (PAL), and peroxidase is observed in extracts from resistant cultivars (Yamamoto et al. 1970). A new protein band was recognized among the proteins isolated from leaves of the susceptible cultivar which had been treated with a DNA fraction from a resistant hybrid prior to inoculation. The new protein band coincided with the Rf value of a protein isolated from the resistant hybrid 6 h after it had been inoculated with an incompatible race of *P. infestans* (Figs. 2, 3) (Yamamoto and Konno 1976). An increase of peroxidase activity in leaves inoculated with a com-

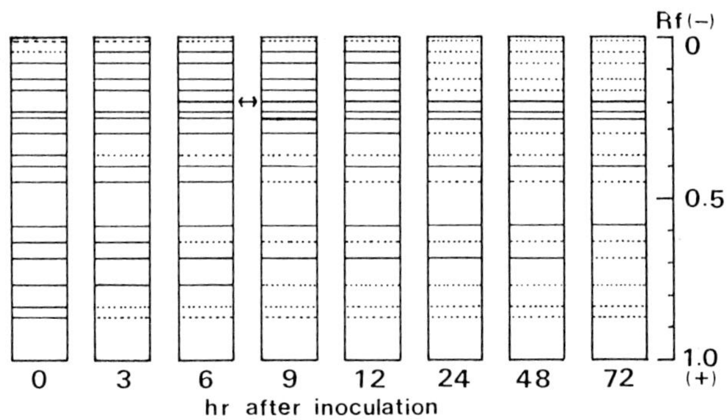


**Fig. 2.** Electrophoretic patterns of soluble proteins of potato interspecific hybrid 96-56 infected with *Phytophthora infestans*. *Arrows* show the new protein band. (Yamamoto and Konno 1976)

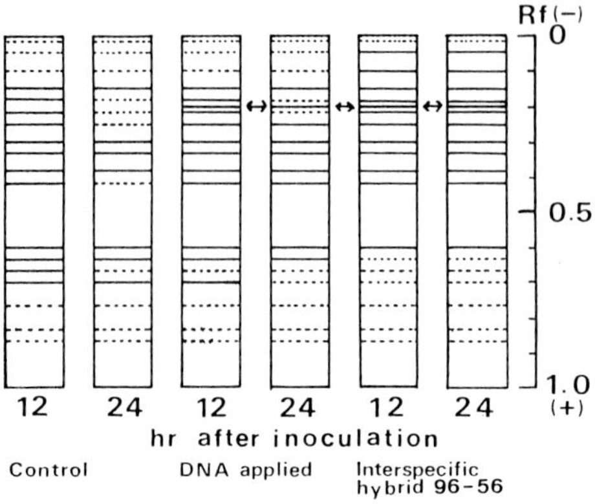


**Fig. 3.** Electrophoretic patterns of soluble proteins in potato cultivar Benimaru (r gene) infected with *Phytophthora infestans* (race 0) after applying the DNA fraction of interspecific hybrid 96-56 ($R_1$ gene). *Arrows* show the new protein band. Control: 0.15 M NaCl applied. The DNA fraction contains 500 μg DNA per ml. (Yamamoto and Konno 1976)

patible race of *P. infestans* is similar to that observed after treating plants with DNA extracted from resistant hybrids, and to that observed after inoculation with an incompatible race of the fungus, as is also the pattern of PAL activity observed after treating the plants with resistant hybrid DNA, and to that observed after inoculation with an incompatible race of *P. infestans*. Rishitin has been detected in the calli of the hybrid 1506-b(9), infected with *P. infestans*, but not in calli of the cv. Benimaru, whether infected with the fungus or not. A large amount of rishitin was recognized in tubers of the hybrids inoculated with the fungus. Rishitin was also observed in tubers of cv. Benimaru treated with a DNA fraction from the hybrid before its being inoculated with *P. infestans* (Yamamoto et al. 1975).

## 4.2 Incorporation of Exogenous DNA into Potato Cells

As preliminary experiments, young leaves from potato cuttings were soaked in $H^3$-thymidine solution (2 °C/ml) for 4 days to trace radioactive DNA. When leaves were incubated in 4% solution of radioactive DNA, microautoradiographs showed that tritiated DNA was incorporated into the epidermal and the apical tissues. The epidermis rubbed with DNA was washed with water 5, 10, 30 min and 1, 2.5, and 14 h after rubbing, and then *P. infestans* was inoculated. The number of flecks decreased when leaves were washed within 10 min after rubbing DNA (Yamamoto et al. 1969).

A few flecks appeared on inoculated leaves of Irish Cobbler if the hybrid Greta DNA solution had been placed without rubbing, but flecks increased when this DNA solution was rubbed with carborandum or cotton swab. By fluorescent microscope observation, DAPI-DNA complex was recognized in the nuclei of leaf cells of Irish Cobbler 2 h after rubbing DNA with carborandum and cotton swab, but not in those where the DAPI-DNA complex was just placed as a drop (Yamamoto 1986a, b).

When exogenous DNA was rubbed on potato leaf epidermis with a brush or piece of cotton swab, epidermal hairs and neighboring tissues were destroyed, as seen with a scanning electron microscope (Fig. 4) (Yamamoto 1983). A specific fluorescence of ethidium bromide combined with DNA is observed on these hairs

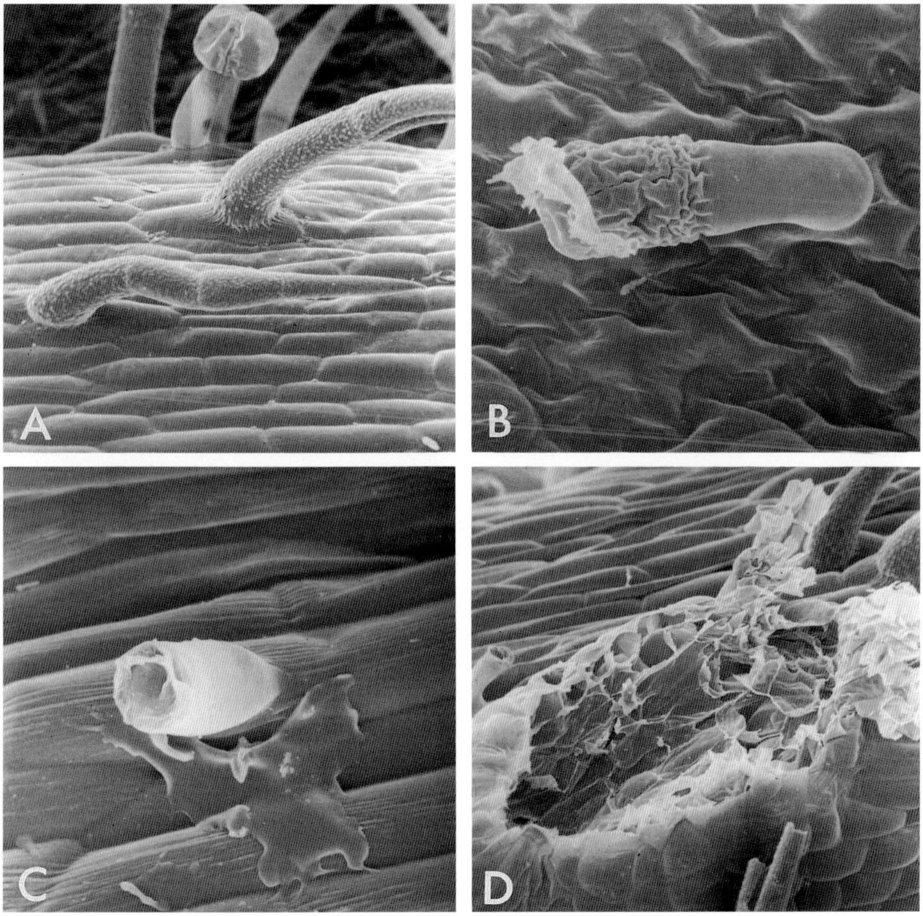


**Fig. 4A–D.** Observations by scanning electron microscope on potato epidermis after rubbing exogenous DNA. **A** Before rubbing DNA. **B–D** Destroyed hair and neighboring part after rubbing DNA. **A, D.** ×150. **B, C** ×450. Cultivar used: Irish Cobbler. Observed by scanning electron microscope (JSMUS type 3). (Yamamoto 1983)

and neighboring tissues by use of the epifluorescent microscope (Yamamoto 1982). Phase contrast and epifluorescent micrographs of cross-sections of epidermal and subepidermal cells from potato leaves on which PI-DNA complexes had either been rubbed or not are shown in Fig. 4. The sections were incubated in acetate buffer without enzyme, with RNase, or with DNase plus RNase. In the ventral and dorsal epidermal cells, to which PI-DNA complex (of potato origin) had not been applied, no fluorescence was present, but dorsal epidermal cells to which the PI-DNA complex had been applied showed fluorescence 120 min after application. The fluorescence was also present in nuclei and near-subepidermal cells. Treatment of RNase did not change the intensity of fluorescence, but when DNase was added to the sample after treatment with RNase, the intensity of the fluorescence decreased strikingly in the cytoplasm (Yamamoto et al. 1984;

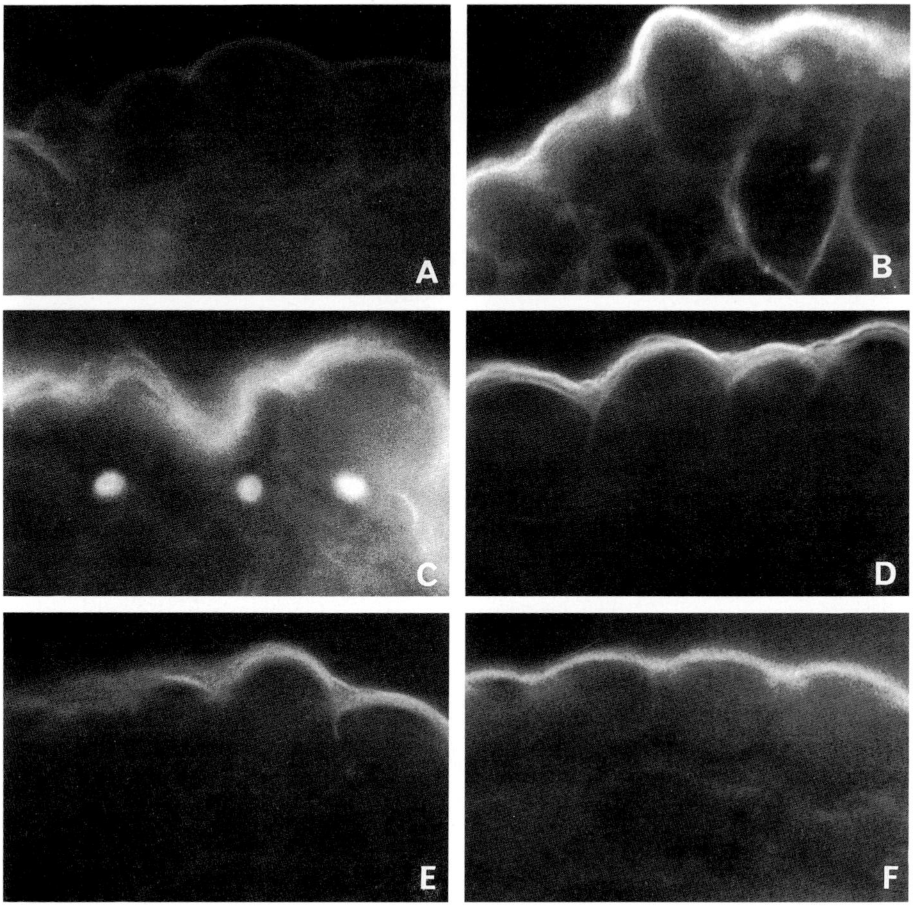


**Fig. 5A–F.** Fluorescent photomicrographs of cross-sections of the dorsal epidermis of a potato leaf immediately (**A**) and 120 min after rubbing on the DAPI-DNA complexes (of potato origin) (**B–D**), the DAPI (**E**) or DAPI-DNA complexes (of calf-thymus origin) (**F**). **C** and **D** are the respective fluorescent images after RNase and DNase treatment. (Yamamoto et al. 1984)

Yamamoto and Kuroiwa 1986). These results suggest that the exogenous PI-DNA complexes were incorporated into the cell nuclei via the damaged cell wall within 120 min of application. Fluorescence micrographs of sections of epidermis before and after the application of the DAPI-DNA (of potato origin) complexes are shown in Fig. 5. Fluorescence micrographs of leaf sections treated with DAPI and those treated the DAPI-DNA (calf-thymus origin) complexes are also shown. Fluorescence was detected in the cell nuclei (Yamamoto et al. 1984). When tissues that had incorporated the DAPI-DNA complexes were treated with RNase, fluorescent intensity did not change. DNase treatment of the same tissue, however, caused the complete disappearance of fluorescence in the leaf cells and the intercellular space. To determine whether the nuclei were stained with the DAPI released from DAPI-DNA complexes (of potato origin) and whether the incorpora-

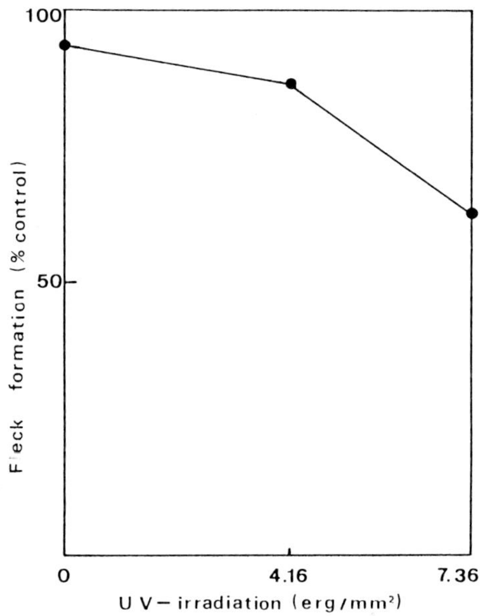


**Fig. 6.** Formation of hypersensitive flecks of late blight on epidermis of potato leaf petiole midribs (cultivar Norin No. 1) when treated with the DNA fraction from the hybrid 96-95 irradiated with UV light at various intensities. Number of flecks observed in the control (100%) was 202 and 239 in two experiments respectively. (Yamamoto et al. 1979)

tion of DNA in epidermal cell is species-specific, DAPI at a concentration of 1 mg/ml was rubbed on the surface of the potato leaf. The concentration of DAPI used was equivalent to that of the DAPI released from the DAPI-DNA complexes during purification. In both cases, no fluorescence was observed in epidermal cells up to 120 min after the chemical was rubbed on. This suggested that the DAPI-DNA complexes (potato origin) were only taken up by the cell nuclei through the intercellular space.

### 4.3 Destruction of DNA by Physical and Chemical Treatments

*1. Thermal Treatments.* The degree of coexistence of flecks and lesions decreased by heating the DNA at more than 90 °C for 10 min (Yamamoto and Hatta 1974).

*2. UV Irradiation.* When the DNA solution (100 μg DNA/ml) was exposed to $4.6 \times 10^3$ ergs/mm$^2$ of UV light, the number of flecks appeared on plants treated under conditions of darkness was 94.1% of the number appearing on the plants given nontreated DNA. When the exposure was $7.16 \times 10^3$ ergs/mm$^2$, the relative appearance of flecks was reduced to 61.6% (Yamamoto et al. 1979) (Fig. 6).

*3. DNase Treatment.* The number of flecks formed on the plants treated with fraction No. 19 and 38 of the DNA fraction which had been obtained from DNase II digestion, was 26% and 3.1% of the number formed on the control plants, respectively. DNA contained in fraction 19 and 38 seemed to be fragmented by DNase (Yamamoto et al. 1979).

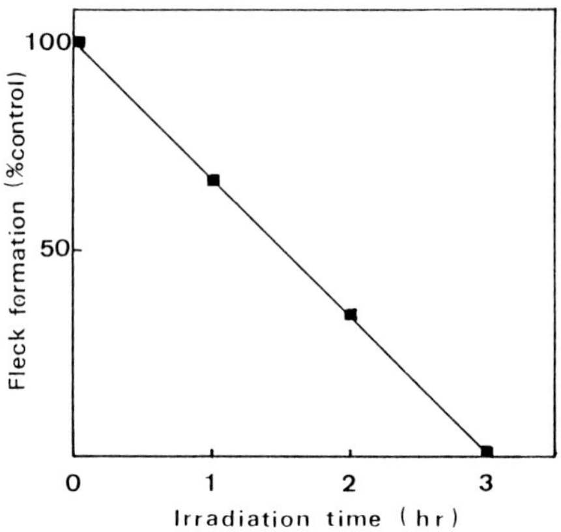


**Fig. 7.** Fleck formation of potato (cultivar Norin No. 1) treated with the DNA fraction from hybrid 96-56 irradiated with the malneutrons for varying periods of time. $\phi$ (thermo-luminescence flux) = $2.07-2.14\times10^2$ n/cm$^2$. Number of flecks observed in the control (100%) was 205 and 253 in two experiments respectively. (Yamamoto et al. 1979)

*4. Thermal Neutron Irradiation.* After irradiation of the DNA for 1 h with thermal neutron, the fleck formation of the plants treated with the solution was 67.1% of that appearing on plants given nonirradiation. After 2 h irradiation, the percentage decreased to 40% and after 3 h irradiation, no flecks were formed (Yamamoto et al. 1979) (Fig. 7). The DNA fraction from the hybrid 96-56 and irradiated with thermal neutrons for 3 h (completely ineffective in inducing fleck formation) was placed on Sephadex G-25. An absorption at 260 nm in the low molecular weight region, indicating that DNA was broken into small doses by $\gamma$-rays was detected; even if the thermal neutron was filtered with LIF cover, this dose did not affect the irradiated DNA in fleck formation.

**Table 2.** Fleck formation of potato DNA treated with restriction enzymes. (Yamamoto 1986)

| Plots | Types of lesions and flecks | Hypersensitive flecks | Susceptible lesions |
|---|---|---|---|
| Control | S-CV (nontreated with R-DNA) | 0/354 (0) | 354/354 (100) |
| | R-DNA (nontreated with restriction enzymes) ... S-CV | 145/313 (46.3) | 168/313 (53.7) |
| R-DNA (treated with restriction enzymes) ... S-CV | *Hae* III | 4/427 (0.9) | 423/427 (99.0) |
| | *Hha* I | 2/184 (1.1) | 182/184 (98.9) |
| | *Eco* RI | 73/161 (45.3) | 88/161 (54.7) |
| | *Bam* HI | 9/167 (5.4) | 158/167 (94.6) |

N.B. ( ) Percentage
Figures are given as data 72 h after inoculation
R-CV Resistant interspecific hybrid 96-56 ($R_1$ gene)
S-CV Susceptible cultivar Irish Cobbler

**Table 3.** Induction of flecks by a compatible race of *Phytophthora infestans*, race 0, on Irish Cobbler leaves which had been rubbed with resistant hybrid nuclear DNA digested with *Eco*RI and *Kpn*I or *Sal*I[a]. (Yamamoto and Kuroiwa 1986)

| Treatment | | | Cultivar inoculated | Fleck[b] | Lesion[b] |
|---|---|---|---|---|---|
| Nontreated control | | | Greta | 1140 (100) | 0 (0) |
| | | | Irish Cobbler | 0 (0) | 963 (100) |
| Rubbed with hybrid DNA | No treatment with restriction enzymes | | Irish Cobbler | 1055 (46.7) | 1205 (55.3) |
| | Treated with restriction enzymes | *Eco*RI + *Kpn*I | Irish Cobbler | 0 (0) | 1609 (100) |
| | | *Eco*RI + *Sal*I | Irish Cobbler | 1060 (75.2) | 350 (24.8) |

[a] DNA from the hybrid Greta ($R_1R_3R_4$ genes) was extracted by Hall and Spiegelman method, and digested with *Eco*RI first and then with *Kpn*I or *Sal*I. Leaves were rubbed with these digested and nondigested DNA and were inoculated with zoospore suspension 4 h later

[b] Scored 72 h after inoculation. Numbers in parenthesis are percentage of the control

From these results, DNA seemed to be less effective when applied after thermal treatment, treatment with DNase, or after exposure to UV irradiation, or thermal neutrons.

*5. Restriction Enzyme Digestion.* DNA from hybrid 96-56 ($R_1$ gene), treated with various restriction enzymes (*Eco*RI, *Bam*III, *Hha*I, or *Hae*III), and rubbed onto potato leaves of cv. Irish Cobbler leaves treated with digested DNA produced a few flecks upon inoculation with *P. infestans*, indicating that the fleck formation acitvity of DNA was destroyed by the *Hae*III treatment (Yamamoto 1982, 1983) (Table 2).

DNA from the hybrid Greta ($R_1$ $R_3$ $R_4$ genes) was digested with two different restriction enzymes. *Eco*RI-*Sal*I-digested DNA induced hypersensitive flecks and the *Eco*RI-*Kpn*I-digested one did not (Yamamoto and Kuroiwa 1986) (Table 3). DNA from the hybrid Greta was digested with the three different restriction enzymes. Leaves treated with *Eco*RI-*Sal*I-*Kpn*I-digested DNA showed no or few flecks, and leaves rubbed with *Eco*I-*Sal*-*Hae*III, *Eco*RI-*Sal*I-*Hind*III-digested DNA showed a few flecks (Yamamoto 1986a).

## 5 Other Studies on Transformation in Potato

Recently there has been much interest in the biotechnology of potato (see Bajaj 1987b). Various approaches have been followed for genetic engineering and transformation of plants (Fraley and Horsch 1986). With the refinement in protoplast technology, Ramulu et al. (1983) reported phenotypic variation and ploidy level of plants regenerated from protoplasts of tetraploid potato (*Solanum tuberosum* L. var. Bintje), and Austin et al. (1985) described transfer of resistance to potato roll virus from *S. brevidens* into *S. tuberosum* by protoplast fusion. Karp and her coworkers have done extensive work on potato culture and transformation (see Ooms et al. 1985, 1986; Burrell et al. 1985; Ooms and Lenton 1985). These studies deal with genetic transformation in potato cultivars with T-DNA from disarmed *Agrobacterium*, genetic manipulation with *A. rhizogenes* (Ooms et al. 1986), expression of shoot-inducing Ti TL-DNA in differentiated tissues (Burrell et al. 1985), genetic modification of potato development using RI-T-DNA (Ooms et al. 1985).

A cDNA library has been established for potato, allowing fishing of nuclear genes from one variety to insert into another (Eckes et al. 1985). Shahin and Simpson (1986a, b) introduced a foreign gene encoding neomycin phosphotransferase (NPT-II), specifying resistance to kanamycin into potato cells via a disarmed *Agrobacterium tumefaciens*. Phenotypically normal transgenic plants were recovered from these resistant calli and the foreign gene was expressed in leaves, stems, petioles, roots, and stolons of these plants. Likewise Jaynes et al. (1986) regenerated transformed potato plants either by induction of hairy roots using *A. rhizogenes* followed by isolated root culture, or by the use of disarmed *A. tumefaciens* vectors and cultured leaf discs. Recently Ball (1987) has reviewed studies on molecular genetics of potato and has emphasized the importance of transformation.

## 6 Significance of the Work and Its Prospects

As described at the beginning of this chapter, leaves of susceptible cultivar inoculated with *P. infestans* after treatment with DNA from the resistant potato hybrid cultivars developed hypersensitive flecks together with susceptible lesions. Leaves of susceptible cultivar with DNA from a resistant hybrid showed increases in peroxidase and phenylalanine ammonia lyase activities, RNA content, and formation of new protein and rishitin (Yamamoto 1970; Yamamoto and Nakao 1976; Yamamoto et al. 1975). These are similar to phenomena observed in resistant potato plant inoculated with an incompatible race of *P. infestans*.

We demonstrated that the exogenous PI-DNA or DAPI-DNA penetrated the cell wall and taken up by the nucleus within 120 min after rubbing on the leaf (Yamamoto and Kuroiwa 1986; Yamamoto et al. 1984). The taking up of DNA in the cell nucleus seems to be the initial step in these metabolic activities.

The most important contribution of our research is the establishment of incorporation of exogenous DNA and the demonstration of increased metabolic activities after application of exogenous DNA to plant tissues (Yamamoto 1974, 1982, 1983, 1986). As yet we have not obtained adult potato plants resistant to *P. infestans* by the application of exogenous DNA from resistant potato hybrids, but it will certainly be possible to obtain these because we have already obtained resistant tobacco plants from protoplast-derived calli selected for their resistance to *Pseudomonas* and *Alternaria* toxins (Thanutong et al. 1983).

Recently many examples of plant transformation using protoplasts and calli have been reported (see Chap. I, this Vol; Fluhr et al. 1986; Kato 1986; Melchers et al. 1987; Nelson et al. 1987; Nomura and Komamine 1986; Ooms and Lenton 1986; Potrykus and Hoffman 1973; Sukhepind et al. 1987; Trulson et al. 1986) and this is an area of considerable importance for potato improvement.

## References

Austin SM, Baur M, Ehlenfeld M, Kamierzak PJ, Helgesson H (1985) Intra-specific fusion in *Solanum tuberosum*. Theor Appl Genet 71:172–175

Bajaj YPS (ed) (1987a) Biotechnology and 21st century potato. In: Bajaj YPS (ed) Biotechnology in agriculture and forestry 3. Potato. Springer, Berlin Heidelberg New York Tokyo, pp 3–22

Bajaj YPS (1987b) Biotechnology in agriculture and forestry 3. Potato. Springer, Berlin Heidelberg New York Tokyo

Ball SG (1987) Molecular genetics of potato. In: Bajaj YPS (ed) Biotechnology in agriculture and forestry 3. Potato. Springer, Berlin Heidelberg New York Tokyo pp 155–173

Bills DD (1977) Interactions between *Phytophthora infestans* and potato host. Am Chem Soc Symp 26:47–60

Burrell MM, Twell D, Karp A, Ooms G (1985) Expression of shoot-inducing Ti-TL-DNA in differentiated tissues of potato (*Solanum tuberosum* cv. Maris Bard). Plant Mol Biol 5(4):213–223

Davis JW, Ornstein L (1964) Disc electrophoresis II. Method and application to human protein. Ann NY Acad Sci 121:404–427

Doke N, Tomiyama K (1980) Suppression of the hypersensitive response of potato tuber protoplasts to hyphal wall components by water-soluble glucans isolated from *Phytophthora infestans*. Physiol Plant Pathol 16:177–186

Eckes P, Schell J, Willmitzer L (1985) Organic specific expression of three leaf/stem specific cDNA's from potato is regulated by light and correlated with plastid development. Mol Gen Genet 199:216–224

Facciotti D, O'Neal JK, Lee S, Shoemaker CK (1985) Light inducible expression of a chimeric gene in soybean tissue transformed with *Agrobacterium*. Biotechnology 3:241–246

Flor HH (1971) Current status of the gene-for-gene concept. Annu Rev Phytopathol 9:275–296

Fluhr RC, Kluhenneier C, Nagy F, Chua Nam-Hai (1986) Organospecific and light induced expression of plant genes. Science 232:1106–1112

Fraley RT, Horsch RB (1986) Genetic transformation in higher plants. CRC Crit Rev Plant Sci 4:1–16

Garas N, Kuć J (1981) Potato lectin lyses zoopores of *Phytophthora infestans* and precipitates elicitors of terpenoid accumulation produced by the fungus. Physiol Plant Pathol 18:227–237

Hall BD, Spiegelman S (1961) Sequence complementarity of T2 specific RNA. Proc Natl Acad Sci USA 47:137–149

Harris PW (1978) The potato crop. The scientific bases for improvement. Chapman and Hall, London

Hawkes JG (1978) History of potato. In: Harris PW (ed) The potato crop. Chapman and Hall, London, pp 1–4

International Potato Center (CIP) (1984) Potatoes for the developing world. Lima, Peru, CIP, pp 1–148

Japan Agriculture Yearbook (1987) Japan Agriculture Yearbook Association, Tokyo, pp 288–291

Jaynes JM, Espinoza NO, Dodds JH (1986) Genetic engineering of potato, p 153. Abstr Int Congr Plant tissue and cell culture. Univ Minnesota, Minneapolis, Aug 3–8, 1986

Kato T (1986) Potato breeding by cell fusion. Potato Sci 6:80–81

Kolodner R, Tewari KK (1975) The molecular size and conformation of the chloroplast DNA from higher plants. Biochim Biophys Acta 402:372–390

Kuć J, Henfling J, Garas N, Doke N (1979) Control of terpenoid metabolism in the potato-*Phytophthora infestans* interaction. J Food Bot 42:508–511

Large EC (1940) The advance of the fungi. Holt, New York

Lowry OH, Rosenbrough NJ, Farr AL, Randall RJ (1951) Protein measurements with the Folin phenol reagents. J Biol Chem 193:265–275

Melchers G, Sacristan MD, Holder AA (1978) Somatic hybrid plants of tomato regenerated from fused protoplast. Carlsberg Res Commun 43:203–218

Nomura K, Komamine A (1986) A cell technological approach to somatic embryogenesis from single cells of carrot. VI Int Congr Plant tissue and cell culture. Univ Minnesota Minneapolis Abstr Aug. 3–8, 1986, p 344

Nelson RS, Cressen GP, Bright WJ (1983) Plant regeneration from protoplast of *Solanum brevidens*. Plant Sci Lett 30:355–362

Nozu M, Yamamoto M, Konno K (1976) Occurrence of resistant reaction by exogenous DNA fraction on potato leaves infected with *Phytophthora infestans*. Bull Fac Agric Shimane Univ 10:80–85

Nozu M, Yamamoto M, Ofune S, Nishimura A, Bando T (1977) Pathophysiological studies on potato late blight, with special reference to the induction of resistant reaction by the application of DNA fraction from the interspecific hybrid. Bull Fac Agric Shimane Univ 11:28-33

Ooms G, Lenton LR (1985) T-DNA genes to study plant development: precosious tuberisation and enhanced cytokinins in *A. tumefaciens*-transformed potato. Plant Mol Biol 5:205–212

Ooms G, Karp A, Burrell MM, Twell D, Roberts J (1985) Genetic modification of potato development using RI T-DNA. Theor Appl Genet 70(4):440–446

Ooms G, Bossen ME, Burrell MM, Karp A (1986) Genetic manipulation in potato with *Agrobacterium rhizogenes*. Potato Res 29(3):367–379

Potrykus I, Hoffmann P (1973) Transplantation of nuclei into protoplasts of higher plants. Z Pflanzenphysiol 70:364–366

Ramulu KS, Dijkhuis P, Roest S (1983) Phenotypic variation and ploidy level of plants regenerated from protoplasts of tetraploid potato (*Solanum tuberosum* L. cv. Bintje). Theor Appl Genet 66:329–338

Rich AV (1983) Potato diseases. Academic Press, New York, pp 1–238

Scot JD (1976) Praise the potato! Reader's Digest, Dec 1976, pp 205–212

Shahin EA, Simpson RB (1986a) Gene transfer system for potato. Hort Sci 21:1199–1201

Shahin EA, Simpson RB (1986b) Introduction and expression of foreign gene in potato (*Solanum tuberosum* L.), p 152. Abstr VI Int Congr Plant tissue and cell culture. Univ Minnesota, Minneapolis, Aug 3–8, 1986

Shigematsu A, Mizusawa Y, Hirai T (1966) Incorporation of two radioactive amino acids into the nucleic acid fractions in tobacco leaves infected with Tobacco Mosaic Virus. Virology 28:331–338

Stahmann MA, Staples RC (1964) Changes in protein and several enzymes susceptible bean leaves after infection by the bean rust fungus. Phytopathology 54:760–764

Sukhepind K, Spivey R, Simpson RB, Shahin SA (1987) Transgenic tomato (*Lycopersicon esculentum* L.) transformed with a binary vector in *Agrobacterium rhizogenes*: non-chimeric origin of callus clone and low copy numbers of integrated vector T-DNA. Mol Gen Genet 206(3):491–497

Tabeke I, Otsuki Y (1969) Infection of tobacco mesophyll protoplasts by tobacco mosaic virus. Proc Natl Acad Sci USA 64:843–849

Thanutong PI, Furusawa I, Yamamoto M (1983) Resistant tobacco plants from protoplast-derived calluses selected for their resistance to *Pseudomonas* and *Alternaria* toxins. Theor Appl Genet 66:209–215

Thomas CA Jr, Abelson J (1966) The isolation and characterization of DNA from bacteriophage. Nucleic Acid Res. Harper & Row, New York London, pp 553–561

Tomiyama K, Ishizaka N, Sato N, Masamune T, Katsui N (1968a) "Rishitin". A phytoalexin-like substance. Its role in the defense reaction of potato tubers to infection. Biochemical regulation in diseased plants or injury. Japan Phytopathological Society, Tokyo, pp 289–295

Tomiyama K, Sakuma T, Ishizaka N, Sato N, Katsui N, Takasugi N, Masamune R (1968b) A new antifungal substance isolated from resistant potato tuber tissue infected by pathogen. Phytopathology 58:115–116

Tomiyama K, Doke N, Lee HS (1976) Mechanism of hypersensitive cell death in host-parasite interaction. In: Tomiyama K, Daly JM, Uritani I, Oku H, Ouchi S (eds) Biochemistry and cytology of host-parasite interaction. Kodansha and Elsevier, Tokyo Amsterdam, pp 136–142

Trulson AJ, Simpson RB, Shahin EA (1986) Transformation of cucumber (*Cucumis sativus* L.) plants with *Agrobacterium rhizogenes*. Theor Appl Genet 73(1):11–15

Uchimiya H, Chen K, Wildman SG (1979) A microelectro-focussing method for determining the large and small composition of fraction I protein. Plant Sci Lett 14:377–394

United Nations (1982) Statistical Yearbook. United Nations, pp 516–517

Western JH (1971) Diseases of Crop Plants. Wiley, New York

Wood KR, Barbara DJ (1971a) Specificity of rishitin and phytuberin accumulation by potato. Phytopathology 61:969–971

Wood KR, Barbara DJ (1971b) Virus multiplication and peroxidase activity in leaves of cucumber mosaic virus. Physiol Plant Pathol 1:73–81

Yamamoto M (1968) Hypersensitivity of potatoes to the invasion of *Phytophthora infestans*. In: Hirai T, Hidaka Z, Uritani I (eds) Biochemical Regulation in Diseased Plants or Injury. Phytopath Soc Jpn, Tokyo, pp 335–342

Yamamoto M (1970) On the peroxidase activity of potato leaves infected with *Phytophthora infestans*. Bull Fac Agric Shimane Univ 4:14–18

Yamamoto M (1974) Potato late blight, with special reference of potatoes to the invasion of *Phytophthora infestans.* Rev Plant Prot Res 7:45–56

Yamamoto M (1982) Recent approach to studies of potato late blight, with special reference to the nature of resistance of potatoes to the invasion of *Phytophthora infestans.* Forsch a d Geb Pflkr 9(1):1–37

Yamamoto M (1983) Induction of resistance of potatoes to *Phytophthora infestans* by exogenous potato DNA, with special reference to the incorporation of rubbed DNA into potato tissues. Ann Phytopathol Soc Jpn 49:331–337

Yamamoto M (1985) Studies on Potato Late Blight. Report on the Fundamental Scientific Research. General Research (C) Ministry of Education and Culture, No. 59560044. Lab Plant Path Coll Agric Kyoto Univ (in Japanese), pp 1–111

Yamamoto M (1986a) Restriction enzyme treatment of DNA from potato hybrid resistant to *Phytophthora infestans* and appearance of flecks on susceptible potato cultivar (in Japanese). Kinki-Chugoku Agric Res 71:16–18

Yamamoto M (1986b) Incorporation of DNA from potato hybrid resistant to *Phytophthora infestans* on susceptible variety leaves placing with or without carborandum and adsorbent cotton swab (in Japanese). Kinki-Chugoku Agric Res 72:9–11

Yamamoto M, Hatta S (1974) Studies on the compatibility of potato varieties and races of *Phytophthora infestans* (Mont.) DeBary, with special reference to the DNA fraction of the suscept and pathogen. Bull Fac Agric Shimane Univ 8:28–40

Yamamoto M, Kitami K (1971) Observation on the invasion of *Phytophthora infestans* to the cultured tissues of potatoes. Ann Phytopathol Soc Jpn 37:58–62

Yamamoto M, Konno K (1976) Changes in soluble proteins in potato leaves infected with *Phytophthora infestans.* Plant Cell Physiol 17:843–846

Yamamoto M, Kuroiwa T (1986) Induction of hypersensitive response to a compatible race of *Phytophthora infestans* by pre-treatment of potato leaves with DNA extracted from a resistant cultivar. Ann Phytopathol Soc Jpn 52:8–14

Yamamoto M, Matsuo S (1976) Involvement of DNA in resistance of potatoes to invasion of *Phytophthora infestans.* Nature 259(5538):63–64

Yamamoto M, Nakao K (1976) Changes of protein, PAL and PO activities in potato leaves infected with *Phytophthora infestans*, with special reference to the DNA fraction of different varieties. In: Tomiyama K, Daly JM, Uritani I, Oku H, Ouchi S (eds) Biochemistry and Cytology of Plant-Parasite Interaction. Kodansha and Elsevier, Tokyo Amsterdam, pp 195–197

Yamamoto M, Otsuka N (1971) Investigation of the DNA of potato leaves in relation to the resistance against the invasion of *Phytophthora infestans.* Ann Phytopathol Soc Jpn 37:84–90

Yamamoto M, Nozu M, Shigematsu A (1969) Experiments on DNA containing fraction obtained from potatoes in relation to the hypersensitivity of resistant plant to the invasion of *Phytophthora infestans.* Bull Fac Agric Shimane Univ 3:1–5

Yamamoto M, Otsuka N, Imamura K, Kitami K (1971) Effect of DNA from the resistant potatoes on the *Phytophthora* infectivity to the susceptible cultivar. Symp Cell Biol 22:51–57

Yamamoto M, Matsuo K, Konno K, Bando T (1975) On the DNA fraction from plants and hypersensitivity of potatoes to the invasion of *Phytophthora infestans.* Proc Symp Current Topics in Plant Pathology. Akademia Kiado, Budapest, Hungary, pp 53–56

Yamamoto M, Matsuo K, Konno K (1979) Further evidence on activity of the DNA fraction from a resistant hybrid inducing localized flecks on *Phytophthora infestans*-infected potato plants. Bot Mag Tokyo 92:139–143

Yamamoto M, Kuroiwa T, Nishibayashi S (1984) Further evidence for the uptake of exogenous DNA into potato leaf cells. Plant Cell Physiol 25:665–670

Yamaoka N, Furusawa I, Yamamoto M (1982) Infection of turnip protoplasts with cauliflower mosaic virus. Virology 122:503–505

Zucker M (1963) Sequential induction of phenylalanine ammonia lyase and lyase inactivating system in potato disc. Plant Physiol 43:363–374

# VIII Transformation in Cotton

E. FIROOZABADY

## 1 General Account

Genetic transformation systems developed for a variety of plants (see the review by Fraley et al. 1986) have the potential to be used to improve the agronomic performance of crop species. Cotton breeders have selected for resistance to pests and stresses, earliness, fiber characteristics, and morphological traits. It is envisaged that during the next decade, biotechnology (Bajaj and Gill 1987) and, in particular, genetic transformation will open new avenues for the production of improved cultivars of cotton. A number of potential applications of genetic engineering for cotton improvement will be discussed in a later section.

### 1.1 Economical Importance of Cotton as a World Crop

There are four domesticated species of cotton: *Gossypium arboreum* L. and *G. herbaceum* L., – both diploids, are native to the Old World. *G. arboreum* ranges generally from India to Burma, Indonesia, and China. *G. herbaceum* occurs in Africa through the eastern Mediterranean region as far east as India (Fryxell 1980). *G. hirsutum* (Upland cotton) and *G. barbadense* (extra long-staple, Egyptian, and Pima cotton), both allotetraploids, evolved in the New World. Upland cotton supplies about 90% and *G. barbadense* 9% of the current world production of fiber.

Cotton is grown virtually around the world in tropical and subtropical latitudes (Table 1). Annually, about 16 million tons of cotton fiber and about 33 million tons of cotton seed are produced worldwide (Table 2). The United States is a major cotton-producing country, accounting for about 14% of the world's production (Table 1). Cotton is grown in the southern part of the country from the Carolinas to California, and is the most important industrial crop and the nation's fourth leading crop after corn, soybean, and wheat. Table 3 presents USDA estimates for 1986 US lint production. The value of US production for farmers in 1986 for Upland cotton was $ 2353109000 and for Pima $ 87411000; and in 1985 was $ 3577714000 and $ 67696000 for Upland and Pima cotton respectively (USDA 1987).

Plant Genetic Engineering Laboratory, New Mexico State University, Box 3GL, Las Cruces, NM 88003, USA

Biotechnology in Agriculture and Forestry, Vol. 9
Plant Protoplasts and Genetic Engineering II (ed. by Y. P. S. Bajaj)

**Table 1.** Distribution and production of cotton fiber in 1986 in major producing countries (Cotton: World Statistics 1987)

| World/country | Area harvested ($10^6$ ha) | Yield (kg/ha) | Production ($10^6$ metric tons) |
|---|---|---|---|
| World | 30.12 | 505 | 15.20 |
| Africa | 3.87 | 345 | 1.34 |
| Egypt | 0.44 | 918 | 0.41 |
| North and Central America | 3.72 | 631 | 2.35 |
| Mexico | 0.15 | 925 | 0.14 |
| USA | 3.44 | 620 | 2.13 |
| South America | 2.97 | 421 | 1.25 |
| Brazil | 1.92 | 389 | 0.75 |
| Asia | 19.28 | 518 | 9.98 |
| China, P.R. | 4.40 | 789 | 3.47 |
| India | 7.50 | 215 | 1.61 |
| Iran | 0.19 | 556 | 0.11 |
| Pakistan | 2.42 | 541 | 1.31 |
| Turkey | 0.57 | 817 | 0.47 |
| USSR | 3.33 | 749 | 2.49 |
| Europe | 0.29 | 992 | 0.29 |
| Greece | 0.21 | 931 | 0.20 |
| Australia | 0.15 | 1353 | 0.20 |

**Table 2.** World cotton seed and lint production, 1983 – 1986

| Year | Area harvested[a] (1000 ha) | Cotton seed[b] (1000 mt) | Cotton lint[b] (1000 mt) |
|---|---|---|---|
| 1983 | 32159 | 27594 | 14331 |
| 1984 | 34950 | 35093 | 18255 |
| 1985 | 35165 | 32273 | 17149 |
| 1986 | 31986 | 29449 | 15505 |

[a] FAO 1986; [b] FAO 1987

**Table 3.** USDA estimates of US cotton fiber production in 1986

| Crop/region | Area harvested (1000 acres) | Yield (lbs/acre) | Production (1000 bales) |
|---|---|---|---|
| Upland[a] | | | |
| Southeast | 771.9 | 455 | 731.2 |
| Delta | 2508.0 | 669 | 3495.0 |
| Southwest | 4251.0 | 357 | 3161.2 |
| West | 1311.0 | 1131 | 3090.0 |
| USA | 8841.9 | 569 | 10477.4 |
| Pima[b] | | | |
| USA | 111.1 | 890 | 205.9 |

[a] USDA 1986; [b] USDA 1987. Pima cotton is produced only in Arizona, New Mexico, and Texas

Cotton is a multipurpose crop that supplies five basic products: lint, oil, meal, seed hulls, and linters. The lint is the most important product of the cotton plants and provides much of the high-quality fiber for the textile industry. Oil, the most important by-product of the seed, is used for shortening, cooking oil, salad dressing, mayonnaise, and margarine. Cotton oil accounts for the second largest production of oil of all oilseed crops (FAO 1986). The meal from glandless seeds or from seeds after the removal of gossypol (a yellow polyphenolic pigment contained in all plant parts of the predominant cotton cultivars) is used as a protein supplement for livestock. Glandless cotton seeds have been used for human consumption, e.g., as a popped seed snack called TAMU nuts, a flour used for production of a high-protein bread, and as a meat extender. Seed hulls are used for cattle feed. Linters, the short fibers removed from seeds before crushing, are an important source of industrial cellulose. The National Cotton Council of America claims over 100000 uses for cotton, which makes it a truly remarkable resource.

## 1.2 Application of Genetic Engineering in Cotton

Improvements for yield, fiber quality, uniform maturity, short stature, and earliness are the traits on which cotton breeders have placed much emphasis. Also, breeding for resistance to pathogens, nematodes, insects, and spider mites is an important consideration in many breeding programs. However, sources of significant resistance still are not available for many pests. Genetic engineering is among the most important components of plant biotechnology which will have implications in cotton improvements. With this method, single gene-inherited traits such as resistance to herbicides, insects, fungi, nematodes, viruses, and bacteria can be transferred to commercial cultivars of cotton.

Over 100 insect species damage cotton in the cotton belt (Berger 1969). Controlling insect pests is critical, and chemical pesticides are extensively used. In 1985, nearly $ 1 billion was spent worldwide for insecticides applied to cotton (Farm Chemicals 1985). The estimated crop loss due to insects, after chemical treatments, in major cotton-producing countries was high in the 1960's (Table 4). However, the crop loss has been reduced due to improvements of the chemicals.

**Table 4.** Estimates of percent loss (after pesticide treatment) in cotton production due to insects worldwide (Cramer 1967)

| Country | Loss from Insects (%) |
|---|---|
| USA | 17.0 |
| Mexico | 16.0 |
| Brazil | 15.0 |
| Egypt | 13.0 |
| Turkey | 12.0 |
| Pakistan | 16.0 |
| India | 16.0 |
| World | 16.1 |

**Table 5.** Estimated reduction (after pesticide treatment) in 1986 United States cotton yields resulting from insect and mite damage (King et al. 1987)

| Common name | Scientific name | Bales lost | Percent loss |
|---|---|---|---|
| Boll weevil | *Anthonomous grandis* (Boheman) | 189339 | 1.93 |
| Bollworm-<br>tobacco budworm | *Heliothis zea* (Boddie)<br>*H. virescens* (F.) | 216738 | 2.20 |
| Cotton fleahopper | *Pseudatomoscelis seriatus* (Reuter) | 84811 | 0.86 |
| Lygus bugs | *Lygus* spp. | 79008 | 0.80 |
| Leaf perforator | *Bacculatrix thurberiella* (Busck) | 1909 | 0.01 |
| Pink bollworm | *Pectinophora gossypiella* (Saunders) | 20836 | 0.21 |
| Spider mites | *Tetranychus* spp. | 37036 | 0.37 |
| Thrips | *Frankliniella* spp. | 26644 | 0.27 |
| Armyworms | *Spodoptera* spp. | 2483 | 0.02 |
| Minor pests | | 101466 | 1.03 |
| New pests | | 1122 | 0.01 |
| Total | | 761392 | 7.76 |

**Table 6.** Average percentage losses (from 1956 through 1980) in cotton production caused by infectious diseases and nematodes in the USA (Ridgway et al. 1984)

| Disease | Av. Loss (%) |
|---|---|
| Seedling diseases | 2.93 |
| *Verticillium* wilt | 2.56 |
| *Fusarium* wilt | 0.81 |
| *Phymatotrichum* root rot | 1.06 |
| Bacterial blight | 1.09 |
| Boll rots | 2.51 |
| Nematodes | 1.92 |
| Total | 14.73 |

For example, in the US in 1985, the loss after chemical treatments was only 7% (Table 5). Insect-resistant cotton plants are extremely useful, as crop loss would be minimal due to efficient control, costs of chemical treatment are omitted, and insecticide contaminations of the environment are eliminated.

Weed control is another important aspect in cotton production. Worldwide, over 100 weed species have been reported in cotton. The estimated loss for 1976 US cotton crop was $ 317/ha, and the estimated cost of chemical weed control for 1980 was $77/ha in Texas and $156/ha in the Mississippi Delta area (Ridgway et al. 1984).

Herbicide-resistant cotton lines will have advantages. Herbicides can be applied directly on the crop, thus controlling weeds more efficiently and reducing crop loss to minimum. Farmers will be able to use a new generation of more environmentally safe broad spectrum herbicides.

Several bacterial, fungal, and nematode parasites infect cotton and contribute to reduced yield (Table 6). Also, at least 16 viral diseases have been reported on cotton (Watkins 1981), that are mostly important in Africa, Asia, Central and South America, and some parts of the USA.

With genetic transformation systems, genes for resistance to herbicides, insects and pathogens can be transferred to commercial cultivars of cotton. Certainly, this area of research, as well as other biotechnological aspects (Bajaj and Gill 1987) that bring genetic variabilities to cotton germplasm (i.e., Bajaj and Gill 1985), will complement breeding efforts in producing improved cultivars of cotton.

### 1.3 Systems of Plant Genetic Engineering

*A. tumefaciens* provides a natural gene-transfer mechanism that can be utilized to transfer a specific DNA sequence into the genomes of cells of many dicotyledonous plants (Fraley et al. 1986). The development of successful systems for producing transformed plants has primarily depended on: (1) the engineering of *Agrobacterium tumefaciens* strains containing specially modified Ti-plasmid vectors, and (2) development of cell culture techniques that permit efficient DNA delivery, selection of transformed cells, and regeneration of normal transgenic plants. Transformed plants have been regenerated in tobacco and petunia (Horsch et al. 1985), tomato (McCormick et al. 1986; Fillatti et al. 1987), potato (Shahin and Simpson 1986), alfalfa (Shahin et al. 1986), and cotton (Firoozabady et al. 1987b; Umbeck et al. 1987) using *Agrobacterium*-mediated transformation techniques.

There are three major techniques for transformation of plant cells by *A. tumefaciens*: (1) wounding and inoculation of intact or decapitated plants (Klee et al. 1982); (2) inoculation of explants in vitro, e.g., inverted stem (Barton et al. 1983), hypocotyl (Firoozabady et al. 1986), leaf-disc (Horsch et al. 1985), or cotyledon-section transformation methods (Firoozabady et al. 1987a, b, 1988); and (3) cocultivation of plant protoplasts and bacteria (Marton et al. 1979; Firoozabady and Galbraith 1984).

## 2 Transformation and Plant Regeneration in Cotton

A major obstacle to the application of *Agrobacterium*-mediated transformation to cotton has been the absence of a high efficiency plant regeneration method (Davidonis and Hamilton 1983; Bajaj and Gill 1986; Shoemaker et al. 1986; Gawel et al. 1986; Trolinde and Goodin 1987). A second barrier to cotton transformation has been the optimization of methods for using a selectable marker such as kanamycin resistance. In a system such as tobacco, transformed tissue at any stage of development can be selected on high levels of kanamycin, up to 1000 mg/l (Firoozabady et al. in preparation), while in cotton, such levels are toxic. Conversely, several factors can lead to growth of nontransformed cotton calli, or to chimeric calli consisting of predominantly nontransformed tissues. These factors include the use of low kanamycin levels, selection in later stages of callus proliferation, and use of explants such as hypocotyls that are only in partial contact with the selective media (unpubl.).

Cotton transformation has been reported by Zhou et al. (1983) through injection of DNA directly into embryos in immature cotton (*G. hirsutum* L.) bolls. Because the injected DNA was obtained from Sea Island cotton (*G. barbadense* L.) plants, the putative transformed plants were identified by phenotypes such as boll size and fiber length. The absence of any discrete foreign gene precluded analysis by DNA hybridization to genomic Southern blots, so these results must await molecular confirmation. Mutation or physiological changes might also account for the variations observed.

Firoozabady et al. (1986) reported transformation of cotton via *Agrobacterium* using hypocotyl as explants. However, the hypocotyl inoculation resulted in high levels of nontransformed regenerated plants. Recently, Umbeck et al. (1987) reported preliminary results on *Agrobacterium*-mediated cotton (*G. hirsutum* L.) transformation using similar hypocotyl inoculations with a higher success, although the technologies for transformation, selection, and regeneration were not thoroughly presented.

In the following sections (also reported by Firoozabady et al. 1987b), a simple, reproducible, and efficient transformation/regeneration system is described which overcomes the above difficulties. *A. tumefaciens* strains containing avirulent Ti plasmids were used to transform cotton cotyledon tissues. The method provides significant modifications to the general transformation and selection regime which allows efficient regeneration of transgenic cotton plants with normal morphologies.

## 2.1 Transformation of Cotton Cotyledon Tissues

*A. tumefaciens* strains LBA4404 (Hoekema et al. 1983) carrying the binary vector pH 575 (Fig. 1, Firoozabady et al. 1987b) and strain 1592, derived from *A. tumefaciens* strain 15955 by replacing $T_L$-DNA of 15955 with the T-DNA region of pH 575 to generate a disarmed *cis* vector (Firoozabady et al. 1987b), were cultured on Luria broth agar medium containing 5 g/l NaCl and 250 μg/ml streptomycin and 25 μg/ml kanamycin for selection. The bacteria were scraped off the agar medium and suspended to a concentration of $\approx 10^8$ cells/ml in a cotton callus initiation liquid medium ($G_2$) containing MS salts, 100 mg/l *myo*inositol, 0.4 mg/l thiamine HCl, 5 mg/l 2iP, 0.1 mg/l NAA, 3% glucose, pH 5.9.

To transform cotton tissues, cotyledon pieces ($\approx 0.5$ cm$^2$ surface area) from sterile 12-day-old Coker 201 seedlings (Firoozabady and DeBoer 1986) were dipped in the *A. tumefaciens* suspensions in Petri dishes and gently shaken for a few seconds to ensure contact of all cotyledon edges with the bacterial cultures. The cotyledon pieces were then blotted dry and placed on Whatman # 1 filter paper on callus initiation medium $G_2$ solidified with 0.2% Gel-rite. Inclusion of the filter paper was not necessary for transformation but greatly reduced bacterial overgrowth on plant tissues. Tissues were incubated at 25 °C with a 16-h photoperiod (90 μE/m$^2$/s).

After 3 days cocultivation, cotyledon pieces were transferred to Petri plates (without the filter paper) containing the same medium supplemented with 500 mg/l carbenicillin and 15–35 mg/l kanamycin sulfate. Tissues were incubated at 30 °C with a 16-h photoperiod (90 μE/m$^2$/s).

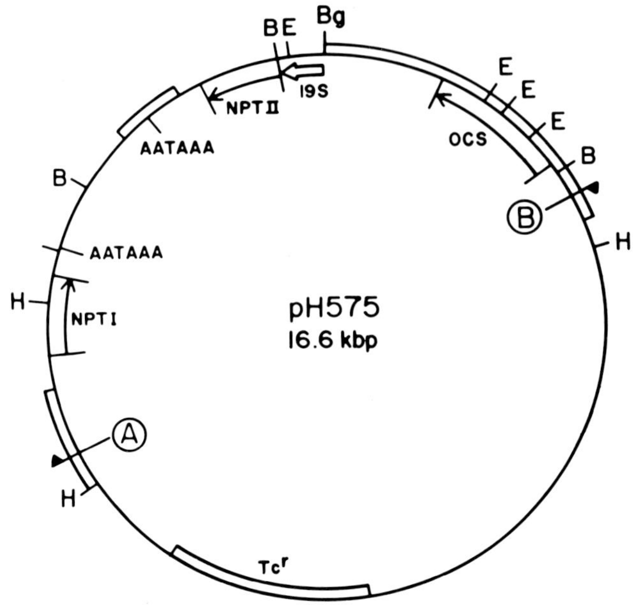


**Fig. 1.** Map of pH575. Resistance to kanamycin is conferred to transformed plant cells by the NPT II gene from bacterial transposon Tn5, under control of the cauliflower mosaic virus (CaMV) 19S promoter. A unique *Bgl*II recognition site (Bg), situated between the octopine synthase (ocs) and 19S/NPT II genes, is used for the introduction of foreign genes. Abbreviations *A* and *B* (*circled*), the 25 bp border repeats, which specify T-DNA ends; *AATAAA*, RNA polyadenylation signals; *NPT I* bacterial kanamycin resistance; *Tc$^r$* tetracyclin resistance; *B Bam*HI; *E Eco*RI; *H Hind*III. Additional details of the construction are in Firoozabady et al. (1987b)

After 7–10 days, cotyledon pieces inoculated with bacteria initiated transformed kanamycin-resistant microcallli (0.5 mm) at wound sites (Fig. 2A), while no callus grew on 25 mg/l kanamycin from control untreated tissues or from tissues treated with LBA4404 alone. Normally, 3 to 10 kanamycin-resistant calli were obtained per cotyledon section (Fig. 2B). Sections plated on medium with no kanamycin produced callus on all wound sites. After 3–4 weeks, 2–4 mm kanamycin-resistant calli were excised from the original explants, transferred to fresh medium, and incubated under lower light intensities (10 $\mu E/m^2/s$). Generally, more than 95% of the calli continued to be kanamycin-resistant. Similar results were obtained when the experiments were repeated. A total of 221 putative transformed calli from one experiment were tested for octopine, and 75%–98% were octopine-positive, depending on experimental treatments (Firoozabady et al. 1987b).

On lower levels of kanamycin (15 mg/l), some nontransformed calli proliferated very slowly but eventually turned brown and died. The percentage of kanamycin-resistant, octopine-positive calli was increased using higher levels of kanamycin (25 or 35 mg/l) for selection. There was no difference between 25 or 35 mg/l kanamycin selection regimes for transformation parameters measured. All the kanamycin-resistant calli tested were positive for neomycin phosphotransferase II (NPT II) using ELISA assays (Firoozabady et al. 1987b).

**Fig. 2. A–E** Regeneration of transgenic cotton plants. **A** A kanamycin-resistant microcallus (shown with *arrow*) developed at wound sites 10 days after coculture (bar = 0.7 mm). **B** A cotyledon section showing several calli on 25 mg/l kanamycin 3 weeks after coculture (bar = 1 mm). **C** Embryogenic calli developed 6–8 weeks after transfer to medium $G_0$ (bar = 2 mm). **D** Somatic embryos with different morphologies (bar = 2 mm). **E** Regenerated transgenic plants after transfer to soil

These results suggest that 25 mg/l kanamycin is sufficient to inhibit the growth of nontransformed cotton cells under the conditions applied. Also, cotyledons are desirable explants for transformation and selection, since they provide callus proliferation only on the wound sites, remain in contact with the selective media, and consequently, produce predominantly kanamycin-resistant calli due to expression of NPT II protein.

## 2.2 Regeneration of Transgenic Plants

Six to seven weeks after inoculation, calli were placed and maintained on embryogenic medium ($G_2$ with no hormones) under kanamycin selection (25 mg/l). On this medium, more than 80% of the calli became embryogenic, i.e., friable cream-colored granular calli that produced somatic embryos (Fig. 2C). Somatic embryos varied considerably in size and morphology (Fig. 2D), and some matured faster than others. Mature somatic embryos were transferred to GRMgn medium, which is a lower ionic strength medium (Stewart and Hsu 1977) modified by addition of 0.1 mg/l $GA_3$, 0.01 mg/l NAA and 0.5% glucose instead of sucrose (with no antibiotics), and incubated at 30 °C, 16-h photoperiod (90 µE/$m^2$/s) to germinate and form plantlets. Approximately, 30% of the embryos germinated on GRMgn medium and formed plantlets. Plantlets were transferred to $GA_7$ Magenta cubes containing GRMgn medium. To date, 252 transformed (kanamycin-resistant and/or octopine-positive) plantlets have been produced using different plasmids and 155 of these regenerated into normal plants.

Plants with three to five leaves were transferred to a commercial peat moss/perlite mixture, maintained in a growth chamber at 25 °C night, 30 °C day, 16-h photoperiod (150 µE/$m^2$/s), were gradually hardened off, repotted in soil, and moved to a greenhouse (Fig. 2E) for further analyses. The whole process from infection until transfer of transgenic plants to soil takes about 6–8 months for Coker 201. Most of regenerated transgenic plants have normal morphology, are fertile, and set seed.

## 2.3 Analyses of Transformed Plants

To determine transformation frequencies, embryos or plants regenerated from kanamycin-resistant calli were analyzed for octopine production and kanamycin resistance (Firoozabady et al. 1987b). The percentages of octopine-positive plants regenerated from octopine-positive calli varied from 75% to 85% and kanamycin-resistant plants from 85% to 100% (Table 7). Interestingly, 100% of plants regenerated from octopine-negative, kanamycin-resistant calli transformed with LBA4404 (pH 575) were octopine-positive and kanamycin-resistant. However, in the population of plants regenerated from octopine-negative calli transformed with 1592, only 30% were kanamycin-resistant and 10% were octopine-positive (Table 7). All the octopine-positive plants tested were kanamycin-resistant, where-

**Table 7.** Transformation frequencies of embryos or plantlets regenerated from kanamycin-resistant Coker 201 calli (partly from Firoozabady et al. 1987b)

| Bacteria | Octopine production by calli | No. embryos (plantlets) tested | Transformation frequencies (%)[a] | |
|---|---|---|---|---|
| | | | Kanamycin-resistance | Octopine-positive |
| LBA4404(pH575) | + | 13 | 100 | 85 |
| LBA4404(pH575) | – | 18 | 100 | 100 |
| 1592 | + | 53 | 85 | 75 |
| 1592 | – | 110 | 30 | 10 |

[a] Transformation frequencies were determined as the percentages of embryos or plantlets that were kanamycin-resistant or octopine-positive. In one experiment all calli (produced from leaf discs on 25 mg/l kanamycin) tested showed detectable levels of NPT II protein

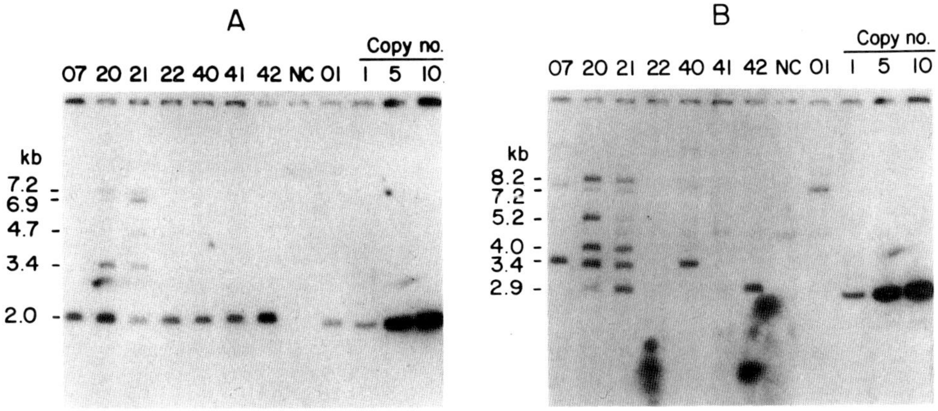


**Fig. 3. A, B.** Southern hybridization analyses of DNA from transgenic cotton plants. DNA blot hybridized to $^{32}$P-labeled *Bam*HI nick-translated NPT II fragment **(A)** and synthetic ocs transcript **(B)**. *NC* is nontransformed control plant. Plant *01* (positive control) had additional 4.3 kb DNA sequences from *Bacillus thuringiensis* and showed the expected 7.2 kb *Bam*HI fragment **(B)**. Several plants including NC showed an endogenous hybridizing faint band ($\approx$4.5 kb) with the ocs probe **(B)** (Firoozabady et al. 1987b)

as not all kanamycin-resistant plants assayed were octopine-positive (Table 7). These results have several possible interpretations: some calli are chimeric and some nontransformed cells have been carried through selection; one, or both, of the ocs and NPT II genes are not expressed in some plants; the expression of these genes is below detection levels in the assays used; the ocs gene is not present in the kanamycin-resistant plants; or a combination of these phenomena exists.

DNA was isolated (Firoozabady et al. 1987b) from 20 plants. All were kanamycin-resistant and all but three were octopine-positive. DNA blot hybridizations showed the expected *Bam*HI fragment containing the NPT II gene in all the plants tested (data for seven of these are presented in Fig. 3A). This was predictable since tissues were maintained under kanamycin selection. Plants 22 and 41 were octopine-negative, predictably they did not show the expected band when hybridized with the ocs probe (Fig. 3B), indicating the ocs gene is not present in these plants. Plants 20, 21, and 42 showed the expected 2.9 kb *Bam*HI fragment containing ocs gene. Plants 20 and 21 regenerated from the same original microcalli, showed identical patterns of additional bands with both probes, suggesting rearranged copies of NPT II and ocs genes. These results suggest that plants 20 and 21 are clones and derived from a single transformed cell. Plants 07 and 40, regenerated from the same original calli, showed a higher molecular weight band (3.4 kb) when DNA was probed with the ocs transcripts. One possible explanation is that the *Bam*HI recognition site next to T-DNA B border is methylated and the enzyme cuts at an external *Bam*HI site in the plant genome, $\approx$200 bp from the border. If this is the case, then plants 07 and 40 also appear to be derived from a single cell. Similar results were obtained by Sukhapinda et al. (1987), where transgenic tomato plants derived from the same callus obtained via cotyledon inoculation by *A. rhizogenes* showed identical banding patterns.

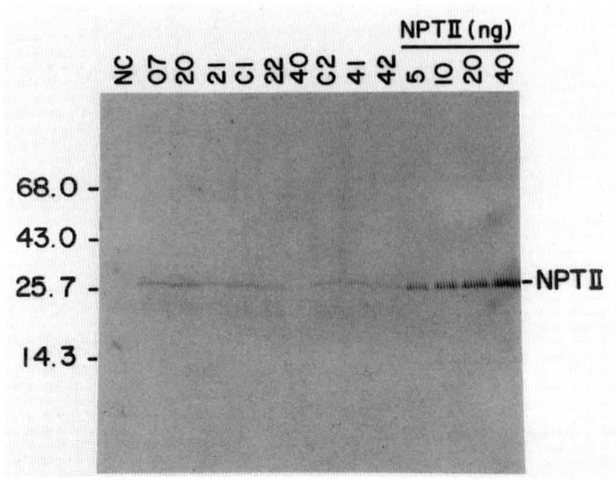


**Fig. 4.** Immunoblot analyses of NPT II protein. *NC* is the extract from leaves of a regenerated plant from uninoculated tissue (control). *C1* and *C2* are extracts from calli transformed with LBA4404(pH575) and 1592, respectively. 5–40 ng purified NPT II was used as standard. Extracts from leaves of all transgenic plants and calli showed the expected protein band. Extract from plant no. 40 showed a very faint band not visible in the figure (Firoozabady et al. 1987b)

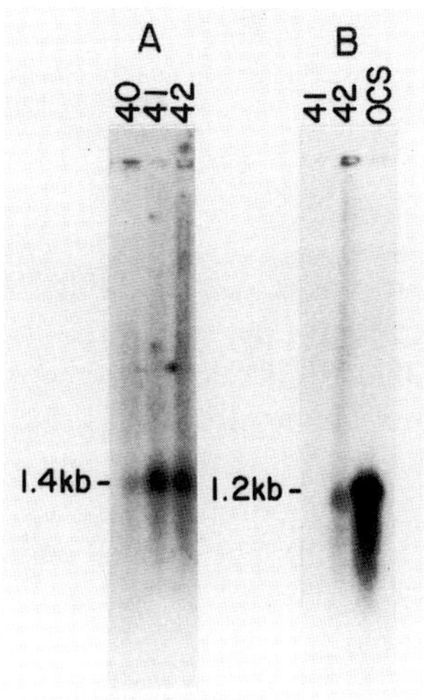


**Fig. 5. A, B** Northern blot analyses of total RNA from transgenic cotton plants. RNA blot hybridized with nick-translated NPT II (**A**) or ocs (**B**) DNA fragments. Plants *40, 41,* and *42* showed the expected 1.4 kb NPT II RNA band. Plant *41*, octopine-negative, predictably did not show the expected (1.2 kb) ocs RNA band (**B**) whereas plant *42*, octopine-positive, showed the expected band. Lane ocs is total RNA isolated from LBA4404 (pH575) used as positive control (Firoozabady et al. 1988)

When eight of the transgenic plants were tested for NPT II protein using ELISA assay (Firoozabady et al. 1987b), the amount of protein ranged from 16 to 50 ng per mg total soluble protein. The same plants also contained protein of the expected molecular weight in the immunoblot analysis (Firoozabady et al. 1987b, Fig. 4).

Northern blot analyses of total RNA isolated (Firoozabady et al. 1988) from transgenic cotton plants showed correlation between NPT II RNA and protein expression, and between ocs RNA expression and octopine production. Plants 22 and 41, which were ocs negative on Southern blots and in octopine assays, lacked ocs RNA on Northern blots (a part of results is shown in Fig. 5).

R1 progeny population from self-pollination of plant no. 21 showed a 4:1 positive: negative segregation for kanamycin resistance, octopine, and NPT II production, which is close to an expected Mendelian pattern of inheritance for single dominant genes for ocs and NPT II in this plant. There was a good correla-

**Table 8.** Kanamycin resistance, octopine, and NPT II production by R1 progeny population from self-pollination of transgenic plant no. 21 (Firoozabady et al. unpubl.)

| Seedling no. | Kanamycin resistance | Octopine production | NPT II protein, ng/mg soluble protein |
|---|---|---|---|
| 1 | + | + + | 43.0 |
| 2 | + | + + | 46.0 |
| 3 | − | − | 0.0 |
| 4 | + | + + | 41.2 |
| 5 | + | + + | 73.7 |
| 6 | + | + + | 52.8 |
| 7 | + | + + | 34.0 |
| 8 | + | + | 36.0 |
| 9 | − | − | 0.0 |
| 10 | + | + | 105.5 |
| 11 | − | − | 0.0 |
| 12 | + | + | 38.0 |
| 13 | + | + | 36.7 |
| 14 | + | + | 14.2 |
| 15 | + | + + + | 52.3 |
| 16 | + | + + | 41.7 |
| 17 | + | + + + | 24.2 |
| 18 | − | − | 0.0 |
| 19 | + | + + | 90.0 |
| 20 | + | + + | 77.0 |

tion between octopine, NPT II production, and kanamycin resistance in the progeny population (Table 8).

## 3 Conclusions and Prospects

The transformation/regeneration system presented here is simple, efficient, reproducible, and rapid. The method is a new research tool for both application in the cotton industry, and basic research studies on gene expression, developmental, and physiological processes. The application of genetic transformation to cotton improvement for insect and herbicide resistance is unique and extremely desirable at the present time. According to a survey (Deterling 1987), many biotechnology companies are interested in transferring genes for resistance to insects or herbicides to cotton. We have transferred insecticidal crystal protein genes from *B. thuringiensis* (Bt) strains to cotton (unpubl.). These are either Bt strains that attack larvae of the *Heliothis* species, which include cotton bollworms and tobacco budworms, or the strains that are effective against the boll weevil and other coleoptera.

Genes have been isolated that confer resistance to glyphosate (Fillatti et al. 1987), bromoxynil (Deterling 1987), and sulfonylurea herbicides. Production of transgenic cotton lines resistant to these herbicides or others would allow growers to control weed and grass problems with an over-the-top treatment and gain more effective control which means less crop loss or higher yield.

The high frequencies of transformation would permit investigators to look among a large number of transformants for high expressing individuals, e.g., for higher insecticidal activity or higher tolerance to herbicides. Superior individuals will be used directly as new cultivars or for breeding into other cultivars.

Genetic engineering of cotton is costly and time-consuming. It probably takes another 3 – 5 years to bring new technology from the lab into the hands of growers. In this contest, development of hybrid cotton lines is very important for purpose of transgenic variety protection. Fortunately, research in this area has been successful and commercial production of $F_1$ hybrid cotton lines will be possible in the near future.

Somaclonal variations generated via tissue culture and high frequency of plant regeneration from callus might be useful in cotton. This is a major key to the application of tissue culture in cotton breeding programs. For example, selection for resistance to drought, salinity, heat, cold, herbicide, and pathotoxins can be practiced. Also, among regenerants (transformants), one can look for better fiber quality, i.e., length, strength, and uniformity; earliness; and uniformly short stature determinate flowering plant suitable for mechanical harvesting. Somaclonal variants such as "glanded plant-glandless seeds" are desirable; plants are relatively insect-resistant and seeds are gossypol-free. Also, the removal of the cyclopropene ring from malvic acid could potentially be induced using mutagens on tissue culture material. The removal of cyclopropene ring is desirable since cyclopropenoid fatty acids are known to exert toxic or adverse biological effects when ingested by a variety of animals (Scarpelli 1974).

Work needs to be done in the following areas before cotton transformation/regeneration technologies can be used extensively. (1) Efforts are required to extend plant regeneration from callus to various cultivars or species of cotton grown commercially in all parts of the world. (2) Work needs to be aimed at shortening the regeneration period so that the regenerated plants would have reduced potential somaclonal variation and maximum similarity to the original cultivar with addition of the transferred trait(s). (3) Efforts need to be extended in the areas of gene isolation and tissue-specific promoters for cotton, since we will soon be looking for other gene(s) to transfer to cotton. How specific genes are expressed in cotton will also be important.

With the increasing efforts to refine the cotton regeneration system and the progress that has been made in the identification and isolation of useful genes, coupled with the transformation system described, the next several years should bring about new cotton germplasm resistant to insects and herbicides and a dramatic expansion in our understanding of plant gene structure and function at the molecular level. The wealth of interesting genetic, physiological, and agronomic problems available for study in cotton ensures that there will be a rapid transition from the molecular analysis of gene structure to using transformation methods in more biological and agricultural applications.

*Acknowledgments.* The author wishes to thank D. DeBoer, L. Amerson and K. Rashka for their excellent technical assistance, D. Merlo for providing *Agrobacterium* strains, and J. Rowe, J. Cramer, and J. Ingle for their critical review of the manuscript. This is Agrigenetics Advanced Science Company manuscript no. 81.

## References

Bajaj YPS, Gill MS (1985) In vitro induction of genetic variability in cotton (*Gossypium* spp.). Theor Appl Genet 70:363–368

Bajaj YPS, Gill MS (1986) Micropropagation and germplasm preservation of cotton (*Gossypium* spp.) through shoot tip and meristem culture. Indian J Exp Biol 24:581–583

Bajaj YPS, Gill MS (1987) Biotechnology of cotton improvement. In: Crocomo OJ, Sharp WR, Evans DA, Bravo JE, Tavares FCA, Paddock EF (eds) Biotechnology of Plants and Microorganisms. Ohio State Univ Press, Columbus, pp 118–151

Barton KA, Binns AN, Matzke AJ, Chilton M-D (1983) Regeneration of intact tobacco plant containing full length copies of genetically engineered T-DNA, and transmission of T-DNA to R1 progeny. Cell 32:1033–1043

Berger J (1969) The world's major fibre crops, their cultivation and manuring. Conzett & Huber, Zürich

Cotton: World Statistics (1987) Bulletin of the International Cotton Advisory Committee Vol 40:25–39

Cramer HH (1967) Plant protection and world crop production. Bayer Leverkusen, W-Germany

Davidonis GH, Hamilton RH (1983) Plant regeneration from callus tissue of *Gossypium histutum* L. Plant Sci Lett 32:89–93

Deterling D (1987) Genetic engineers promise pest-resistant varieties. Progr Farmer 102:Cotton 12–13

FAO (ed) (1986) 1985 FAO production yearbook 39:151–155

FAO (ed) (1987) Monthly Bull Stat 10: Tab 4, 14

Farm Chemicals (ed) (1985) A look at world pesticide markets. Farm Chem 148:26–34

Fillatti JJ, Kiser J, Rose R, Comai L (1987) Efficient transfer of a glyphosate tolerance gene into tomato using a binary *Agrobacterium tumefaciens* vector. Biotechnology 5:726–730

Firoozabady E, Galbraith DW (1984) Presence of a plant cell wall is not required for transformation of *Nicotiana* by *Agrobacterium tumefaciens*. Plant Cell Tissue Org Cult 3:175–188

Firoozabady E, DeBoer D, Maroney M (1986) Transformation and regeneration of cotton, *Gossypium hisrutum* L. Tailoring Genes for Crop Improvement: An Agricultural prospective. Univ Cal, Davis, Abstr p 24

Firoozabady E, DeBoer DL, Murray EE, Merlo DJ, Adang MJ, Halk EL (1987a) Transformation of cotton (*Gossypium hirsutum* L.) by *Agrobacterium tumefaciens* and regeneration of transgenic plants. In Vitro 23:67A

Firoozabady E, DeBoer DL, Merlo DJ, Halk EL, Amerson LN, Rashka KE, Murray EE (1987b) Transformation of cotton (*Gossypium hirsutum* L.) by *Agrobacterium tumefaciens* and regeneration of transgenic plants. Plant Mol Biol 10:105–116

Firoozabady E, DeBoer DL, Amerson LN, Halk EL, Murray EE, Rashka KE, Merlo DJ (1988) Regeneration of cotton (*Gossypium hirsutum* L.) plants transformed by *Agrobacterium tumefaciens*. In: Angarita AZ (ed) Int Congr Plant tissue culture-tropical species, Phytoclone, Bogota, Colombia (in press)

Fraley RT, Rogers SG, Horsch RB (1986) Genetic transformation in higher plants. CRC Crit Rev Plant Sci 4:1–46

Fryxell PA (1980) The natural history of the cotton tribe. Texas A & M Univ Press, College Stn, Texas

Gawel NJ, Rao AP, Robacker CD (1986) Somatic embryogenesis from leaf and petiole callus cultures of *Gossypium hirsutum*. Plant Cell Rep 5:457–459

Hoekema A, Hirsch PR, Hooykaas PJJ, Schilperoort RA (1983) A binary plant vector strategy based on separation of *vir*- and T-region of the *Agrobacterium tumefaciens* Ti-plasmid. Nature (Lond) 303:179–180

Horsch RB, Fry JE, Hoffmann NL, Wallroth M, Eichholtz D, Rogers SG, Fraley RT (1985) A simple and general method for transferring genes into plants. Science 227:1229–1231

King EG, Phillips JR, Hend RB (1987) 40th annual conference report on cotton insect research and control. 1987 Proc Beltwide Cotton Prod Res Conf, pp 170–192

Klee HJ, Gordon MD, Nester EW (1982) Complementation analysis of *Agrobacterium tumefaciens* Ti-plasmid mutations affecting oncogenicity. J Bacteriol 150:327–331

Marton L, Wullems GJ, Molendijk L, Schilperoort RA (1979) In vitro transformation of cultured cells from *Nicotiana tabacum* by *Agrobacterium tumefaciens*. Nature (Lond) 277:129–131

McCormick S, Niedermeyer J, Fry J, Barnason A, Horsch R, Fraley R (1986) Leaf disc transformation of cultivated tomato (*L. esculentum*) using *Agrobacterium tumefaciens*. Plant Cell Rep 5:81–84
Ridgway RL, Bell AA, Veech JA, Chandler JM (1984) Cotton protection practices in the USA and world. In: Kohel RJ, Lewis CF (eds) Cotton. Agronomy No. 24, American Society of Agronomy Madison, Wisconsin, pp 265–365
Scarpelli DG (1974) Mitogenic activity of sterculic acid, a cyclopropenoid fatty acid. Science 185:958–959
Shahin EA, Simpson R (1986) Gene transfer system for potato. HortScience 21:1199–1201
Shahin EA, Spielmann A, Sukhapinda K, Simpson R, Yashar M (1986) Transformation of cultivated alfalfa using disarmed *Agrobacterium tumefaciens*. Crop Sci 26:1235–1239
Shoemaker RC, Couche LJ, Galbraith DW (1986) Characterization of somatic embryogenesis and plant regeneration in cotton (*Gossypium hirsutum* L.). Plant Cell Rep 3:178–181
Sukhapinda K, Spivey R, Simpson RB, Shahin EA (1987) Transgenic tomato (*Lycopersicon esculentum* L.) transformed with a binary vector in *Agrobacterium rhizogenes*: Non-chimeric origin of callus clone and low copy numbers of integrated vector T-DNA. Mol Gen Genet 206:491–497
Trolinder NL, Goodin JR (1987) Somatic embryogenesis and plant-regeneration in cotton (*Gossypium hirsutum* L.). Plant Cell Rep 6:231–234
Umbeck P, Johnson G, Barton K, Swain W (1987) Genetically transformed cotton (*Gossypium hirsutum* L.) plants. BioTechnology 5:263–266
USDA-Economic Research Service (ed) (1986) Cotton and wool situation and outlook yearbook, vol 46, p 6
USDA-National Agricultural Statistical Service (ed) (1987) Crop production, vol CrPr 2-2 (5-87), pp A-16, A-17
Watkins GM (1981) Compendium of cotton diseases. Am Phytopathol Soc, St Paul, Minn
Zhou G-Y, Weng J, Zeng Y, Huang J, Qian S, Liu G (1983) Introduction of exogenous DNA into cotton embryos. Meth Enzymol 101:433–481

# IX Transformation in *Populus* spp.

J.C. SELLMER and B.H. MCCOWN[1]

## 1 Introduction

The application of gene transfer technologies to agricultural crops is expanding at a frenzied rate. Conversely, the application of such biotechnological approaches to woody, perennial crop plants (fruit trees, ornamental, forestry, and plantation crops) is occurring at a much slower pace, primarily due to a general lack of research effort on such crops. However, successful transformation of *Pinus taeda* (Loblolly pine; Sederoff et al. 1986), *Pseudotsuga menziesii* (Douglas-fir; Dandekar et al. 1987), and *Populus* species and hybrids (poplars and aspens; Parsons et al. 1986; Minocha et al. 1986; Fillatti et al. 1987; Pythoud et al. 1987) has been reported using *Agrobacterium*-mediated vector systems within the past few years.

The genus *Populus* was the first woody crop plant to be transformed and regenerated with a gene of direct economic importance (Fillatti et al. 1987). Presently, poplars remain in the forefront as a model for formalizing transformation parameters and techniques for application to other woody plants. This chapter will summarize these efforts and indicate some of the emphases being made in future research.

### 1.1 Importance of *Populus*

Poplars are of worldwide economic importance in both historical and contemporary terms as sources of wood fiber for construction, furniture, packaging, esthetics, energy, and pulp. The versatility of poplar is due to its great natural diversity and geographic distribution. The genus *Populus* of the family Salicaceae is organized into five sections (Turanga, Leucoides, Leuce, Aigeiros, and Tacamahaca) based on leaf, flower, and fruit morphology, and geographical distribution. Representatives of the five sections are distributed throughout the Northern and overlap into the Southern Hemisphere. The size and worldwide economic value of the sections vary from the single-species section of Turanga with only regional value in Central Africa and Western Asia to the Aigeiros section, which is distributed worldwide and makes up 90% of the cultivated poplar genotypes (FAO 1979). Three sections of *Populus* (Leuce, Tacamahaca, and Aigeiros) have had the

[1] Department of Horticulture, University of Wisconsin-Madison, Madison, WI 53706, USA

Biotechnology in Agriculture and Forestry, Vol. 9
Plant Protoplasts and Genetic Engineering II (ed. by Y.P.S. Bajaj)

greatest economic utilization. Selected poplar species from these sections are used for a myriad of industrial purposes; however, recent interest in poplars has focused on energy and pulp production.

Studies were undertaken in the early 1970's on short-rotation intensive culture (SRIC) of selected trees on marginal agricultural land for energy and pulp production. The SRIC goal is to maximize annual woody biomass production per acre under intensive management practices including fertilization, short harvest cycles (2–10 years), and continued production by coppice shoots from cut stumps. The criteria for suitable trees for SRIC plantations include: high annual growth, tolerance to competition, desirable wood quality, amenability to water and nutrient management, capability for clonal propagation, resistance to diseases and insects, wide climatic adaptability, and good crown architecture. No tree species is entirely suited for the SRIC plantations; however, *Populus* demonstrates the greatest overall potential. The disadvantage of many of the poplars centers on disease resistance and weed competition. Weed competition during the first 2 years under SRIC conditions, and numerous fungal diseases such as *Septoria musiva, Marssonina*, and *Melampsora* can drastically reduce the vigor and lifespan of numerous poplar clones (USDA Forest Service 1980).

### 1.2 Poplar Improvement

Genetic improvement is integral to the successful use of *Populus* in such programs as SRIC. *Populus* has been the focus of genetic improvement programs worldwide (i.e., selection, plant breeding) since the late 1920's (FAO 1979). Early poplar tree breeding focused on phenotypic selection of clones and hybrids exhibiting increased growth rate, biomass production, crown architecture characteristics, and disease resistance. Natural and artificial poplar hybrids have been very successful in fulfilling both local and widespread tree improvement objectives. Valuable interspecific and intraspecific hybrids have been developed for forestry, such as the *P. euramerican* hybrids (*P. deltoides*×*P. nigra* crosses) (FAO 1979).

Rapid progress in woody perennial crop improvement is inordinately more difficult than in herbaceous crops, and *Populus* is no exception. In addition to the generally large plant size, the breeder must face long juvenile periods (usually 5–10 years for *Populus*), strong seasonal dormancy cycles, and a dirth of background genetic information on inheritance of important characters. Thus, the genetic improvement of trees becomes long-term and utilizes progeny selection based on phenotype rather than on an identifiable genotype.

Modern genetic manipulation using biotechnology promises to help overcome some of these difficulties, especially for a genus like *Populus* (Haissig et al. 1987). Because poplars can be clonally propagated (Ahuja 1986), the necessity of waiting through long juvenile periods for seed production is eliminated and genetic transformation of juvenile stock becomes an appropriate approach. However, selection in the juvenile growth phase may be impossible for characters expressed only in the mature growth phase. This problem is lessened with such SRIC species as *Populus*, since the crop is harvested when the plants are physiologically young. Because of the ongoing selection programs for *Populus*, generally superior genotypes now exist and provide the genetic background ideal for insertion of one or two important genetic characters (e.g., pest resistance, herbicide tolerance) using biotechnology. Finally, *Populus* appears to be one of the more manipulable woody genera in in vitro culture systems. Thus the techniques which are well developed for herbaceous systems may be more readily utilizable with *Populus* than with other woody plants, where microculture itself (especially plant regeneration) becomes a major obstacle.

## 2 Genetic Transformation of *Populus*

### 2.1 General Strategies

Ideally, the most appropriate candidate for genetic manipulation at the allele level is a highly selected and proven individual genotype. Further, such selection must be performed with plants well beyond the seedling stage of growth. Unfortunately, this presents another major obstacle for woody species in that the most responsive tissues for in vitro manipulation are from seedlings. One way to circumvent this problem is to use stabilized shoot cultures as the source of tissues for in vitro-based techniques. Such shoot cultures can be derived from plants selected in the mature phase of growth. Furthermore, the establishment of shoot cultures appears at least in part to rejuvenate the tissues and thus restore this in vitro responsiveness (McCown 1985). Thus in a genetic transformation program using trees, the following strategy has proven useful:

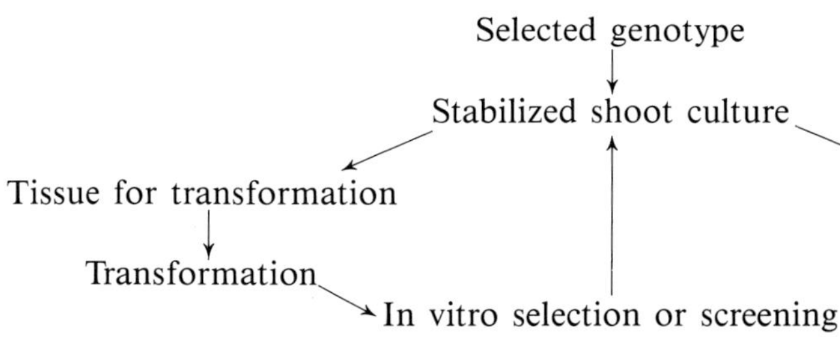


Although the application of gene transfer technologies in *Populus* began such a short time ago (1985), several research groups have reported success in transformation and selection of *Agrobacterium* cocultivated poplar tissues (Table 1). Furthermore, stable expression of transferred genes has been reported in actively growing poplar regenerates and callus tissue. This chapter will review the most documented example of successful genetic transformation of *Populus* and then discuss other approaches that may be equally or more successful in the future.

**Table 1.** Summary of reported transformation successes in *Populus* species

| *Populus* species/cvs | *Agrobacterium tumefaciens* Vector genes | Technique | Level of success | Reference |
|---|---|---|---|---|
| *P. trichocarpa* × *P. deltoides* (H11) | A281-agropine and A348-octopine | Green stem segments | Transformation, selection, and stable expression | Parsons et al. (1986) |
| *P. alba* × *P. grandidentata* 'Cranden' (NC-5339) | C58/587/85-NPT II′ and *aro* A | Leaf discs | Transformation, regeneration, selection, stable expression, and greenhouse screening | Fillatti et al. (1987)<br>Riemenschneider et al. (1988) |
| *P. tremuloides* | pGV3850-NPT II′ and *CAT* | Leaf discs and stem segments | Transformation, regeneration, selection, and stable expression | Minocha et al. (1986) |
| | *Agrobacterium rhizogenes* | | | |
| *P. trichocarpa* × *P. deltoides* (H11) | R1600 and R1601-NPT II′ | Green stem segments | Transformation, regeneration, selection, and stable expression | Pythoud et al. (1987) |

## 2.2 *Agrobacterium*-Mediated Incorporation of Kanamycin Resistance and Glyphosate Tolerance into *Populus* via Leaf Disc Cocultivation

Poplars are very sensitive to competition from weeds during the first 1 to 3 years of growth in newly planted plantations. In addition, poplars are very sensitive to most herbicides that might be used in weed control. A logical approach then is to develop poplars that are genetically tolerant to environmentally safe herbicides so that the option of aerial spraying for weed control of newly established plantations can be realized. Unfortunately, direct screening or selection for glyphosate (a proven, broad-spectrum, widely utilized herbicide) tolerant regenerates in vitro from cocultivated leaf tissue has not been generally employed. As a result, a selectable marker system for resistance to the antibiotic kanamycin has been employed. The neomycin phosphotransferase II′ gene (NPT II′) imparts kanamycin resistance by phosphorylating (inactivating) the kanamycin taken up by plants. Selection and screening occurs by regenerating and growing cocultivated tissues and regenerates on medium containing predetermined lethal doses of kanamycin (50–200 μg/ml). The kanamycin-resistant plants can then be analyzed for stable incorporation and expression of the glyphosate tolerance gene within the plant genome and screened for glyphosate tolerance in the greenhouse and field.

### 2.2.1 Experiments with Clone NC-5339 (*Populus alba*×*Populus grandidentata* Crandon)

A cooperative project was launched to transfer kanamycin resistance and glyphosate tolerance to selected genotypes of *Populus*. The gene construct was developed at Calgene, Inc. (Davis, California) and was a binary vector (Fig. 1) containing two chimeric kanamycin resistance genes (NPT II′) and a chimeric *aro* A gene [coding for a mutant form of the enzyme 5-enolpyruvylshikimate 3-phosphate synthase (EPSP) less sensitive to the action of the herbicide glyphosate] under the control of a mannopine synthase promoter (Comai et al. 1985). The genes were transferred to leaf pieces obtained from shoot cultures of *Populus* by using a wild-type oncogenic strain of *Agrobacterium tumefaciens* (C58). The protocol for previously reported transformation procedures (Fillatti et al. 1987) is presented below.

1. Three days prior to the transformation experiment, make tobacco feeder plate medium and dispense 15–20 ml per Petri plate. Tobacco feeder medium consists of MS medium (Murashige and Skoog 1962) with 0.2 mg/l 2,4-D (2,4-Dichlorophenoxyacetic acid) and 0.6% (w/v) Sigma agar. Dispense 5 ml aliquots of a log phase tobacco suspension culture (*Nicotiana tabacum* var. *xanthi* or *N. tabacum* Plant e) onto each feeder plate, and place a piece of sterile Whatman #1 filter paper soaked in liquid feeder plate medium over the tobacco 2 days before the transformation experiment. Seal and place these plates in a well-lighted growth room until the day of the transformation experiment.
2. The day before the transformation experiment, dispense 5 ml aliquots of mg/l broth (Parsons et al. 1986) into test tubes (the number of tubes should be deter-

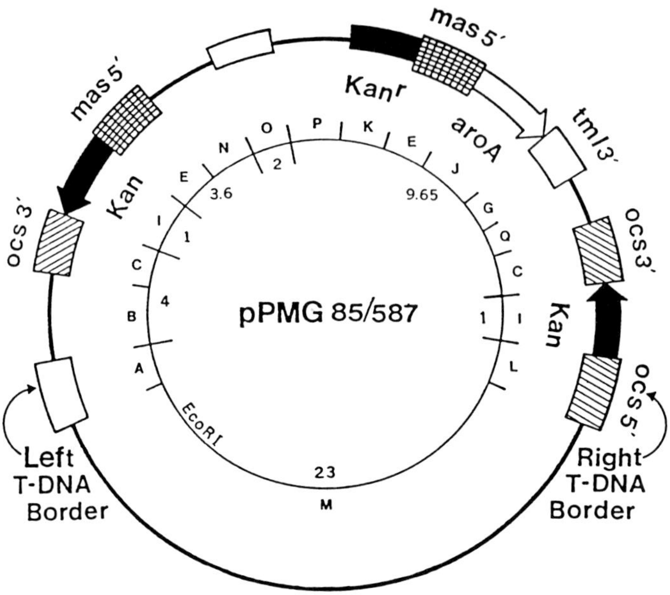


**Fig. 1.** Genetic map of pPMG85/587. (Fillatti et al. 1987)

mined based on bacterial strains used and amount of transformation events planned). Inoculate each tube with a single colony of *Agrobacterium*. Flame and seal each test tube, and place them at an angle in test tube racks. Place the racks in a shaker/incubator at 30 °C and 200 rpm overnight.

3. On the day of the experiment, harvest healthy, expanded leaves from the upper $\frac{1}{2}$ of 1–2-month-old shoot cultures of *Populus* NC-5339 grown on Woody Plant Medium (WPM) (Lloyd and McCown 1980) containing 0.01 μM benzyladenine (BA). Place the leaves into Petri plates containing 5–10 ml of sterile distilled water.

4. Excise leaf margins and midveins from each leaf using sterile surgical scissors or a scalpel. Place 20 to 30 leaf segments 10–30 mm long and 5–10 mm wide on each tobacco feeder plate.

5a. Collect and combine test tubes containing like bacterial strains into 100-ml sterile fleakers. Estimate the bacterial population by spectrophotometer readings at 550 nm. Measure the optical density (O.D.) using 3 or more 1/10 dilutions of the stock bacteria. Calculate the average O.D. using a standard O.D. × bacteria/ml conversion chart previously established for the *Agrobacterium* strain being used.

5b. A bacterial population of $8-12 \times 10^6$ cells/ml for the C58 background strain is suggested. Dilute the *Agrobacterium* stock based on the estimated population.

6a. Dispense 10 ml of *Agrobacterium* stock into sterile Petri plates. Remove leaf tissue from corresponding feeder plates and place into *Agrobacterium* plates. Spread tissue out in bacteria insuring all leaf surfaces are covered. Agitate plates frequently for 30 min.

6b. After 30 min remove tissue piece by piece, carefully and quickly blotting excess bacteria from the tissue on sterile paper towels.

6c. Lay tissue adaxial side up (upside down) on feeder plates, seal plates and place in culture room in the dark for 48–72 h.
7. Remove tissue at 48–72 h after inoculation and place on regeneration medium, adaxial side down. Regeneration medium: MS with 4 µM BA and 4 µM Zeatin (ZEA), 500 mg/l Carbenicillin and 250 mg/l Cefotaxime, 0.3% (w/v) agar and 0.1% (w/v) gelrite, and 0 mg/l (nonselective regeneration) or 60 mg/l kanamycin direct selection during regeneration. Fifteen pieces of tissue per regeneration plate are recommended to allow leaf expansion and prevent pockets of unchecked bacteria to form. Seal plates and place on culture shelf in dark for the first 2 weeks of regeneration because Cefotaxime is light-sensitive. Watch for condensation build-up in plates, as this obstructs viewing the tissues and may lead to contamination.
8. Watch the plates closely and score for regeneration after 4 weeks. Liberal transferring of tissue and removal of dead or highly contaminated tissue is suggested to prevent complete bacterial overgrowth of the plates.
9. Regenerated shoots are removed and placed onto 30 ml of growth medium/jar; MS or WPM 0.1–0.4 µM BA, 0.3% (w/v) agar and 0.1% (w/v) gelrite, and 500 mg/l Carbenicillin.
10. Post-regeneration screening of shoots regenerated on kanamycin-free medium are challenged with 60 mg/l kanamycin once the shoot apices are actively growing and the harvested shoots have rooted on MS or WPM 0.01 µM BA. These shoots are challenged by adding filter-sterilized aliquots of kanamycin to equal a final concentration of 60 mg/l within each culture vessel. Kanamycin-sensitive shoots will bleach from the apex down and then the apex will die.
11. Leaves from kanamycin-resistant shoots are excised, homogenated, and analyzed for the presence of the NPT II′ enzyme coded for by the NPT II′ gene (Reiss et al. 1984) (Fig. 2). Similar leaf extracts are also analyzed for the presence of the bacterial EPSP synthase protein encoded for by the inserted *aro* A gene (Fig. 3).
12. A final measure in assuring the stable incorporation of the NPT II′ and *aro* A genes involves DNA isolation from the plants exhibiting production of the two target proteins. Southern hybridization of the isolated DNA with radiolabelled DNA probes homologous to the two inserted genes is performed. Probing for the

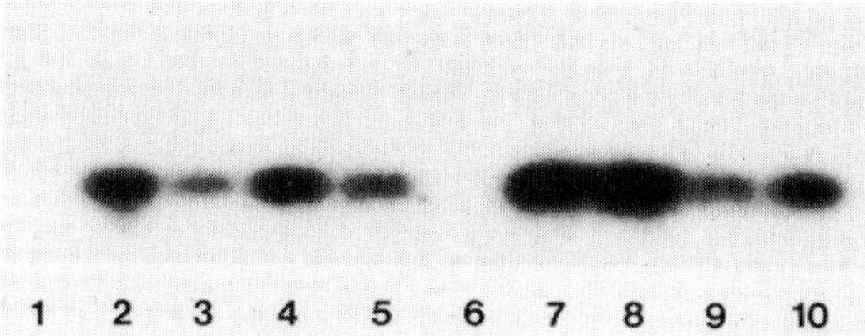


**Fig. 2.** Detection of NPT II′ enzyme activity in transformed *Populus* NC-5339 plants and shoots. Lane *1*, leaf extract from an untransformed *Populus* NC-5339 plant; lane *2*, leaf extract from a positive tomato control; lanes *3–10* leaf extracts from A-1 type shoots; Normal-kanamycin-resistant. (Fillatti et al. 1987)

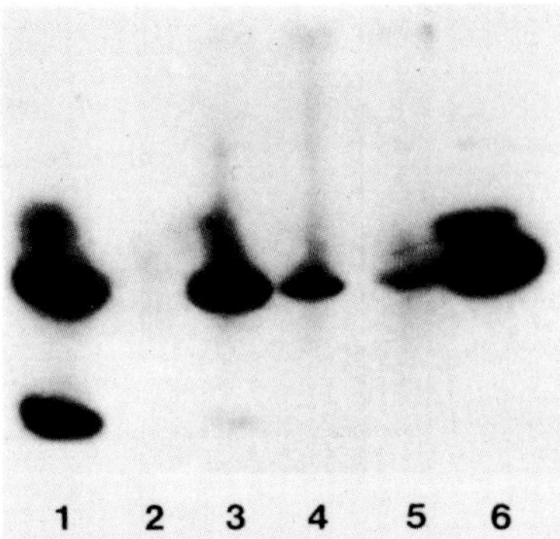


**Fig. 3.** Detection of the bacterial EPSP synthase protein in transformed *Populus* NC-5339 plants by Western blot analysis. Lane *1*, leaf extract from a tomato plant transformed using *Agrobacterium tumefaciens* strain LBA4404 pPMG 587/85; lane *2*, leaf extract from a control untransformed *Populus* NC-5339 plant; lanes *3* and *4*, leaf extracts from transformed *Populus* NC-5339 plants; lane *5*, 250 mg purified EPSP synthase protein added to untransformed *Populus* NC-5339 leaf extract; lane *6*, 250 mg purified EPSP synthase protein and no leaf extract. (Fillatti et al. 1987)

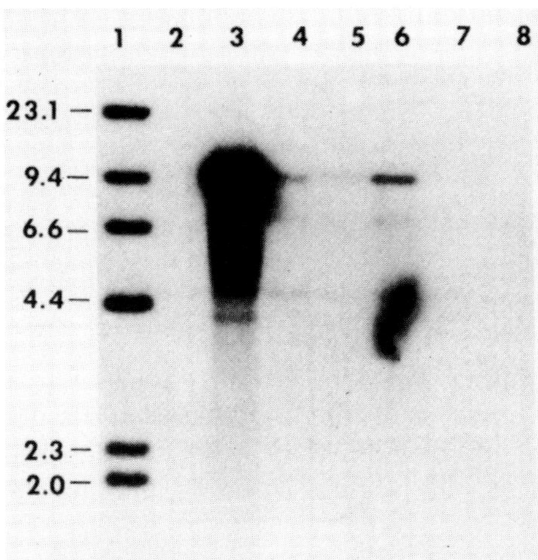


**Fig. 4.** Southern blot analysis of the *aro* A gene in transformed poplar. Lane *1*, coliphage lambda HindIII standard; *2*, control poplar; *3* and *4*, A-11 and A-33 are transformed poplar lines of teratoma type; *5* and *6*, 75-12 75-15 are normal *aro*A-positive; lanes *7* and *8*, 39-3 and 30-115 are normal, *aro*A-negative lines. (Fillatti et al. 1987)

presence of the genes in the plant genome provides evidence for successful stable integration of the 2 transferred genes (Fig. 4).

This experiment and follow-up experiments in our laboratory (see Section 2.2.1) resulted in the recovery of regenerates exhibiting two differing morphologies: Normal cultures, shoots visually indistinguishable from parent plants and teratomic cultures, shoots exhibiting a deformed, rootless growth pattern on hormone-free medium. Using nonselective regeneration coupled with screening at the shoot stage resulted in the recovery of two differing phenotypes within each of the above morphologies: Normal-kanamycin-sensitive and Normal-kanamycin-tolerant; Teratomic-kanamycin-sensitive and Teratomic-kanamycin-tolerant

**Fig. 5.** Examples of three of the four phenotypes of NC-5339 recovered from transformation experiments with a wild-type oncogenic *Agrobacterium tumefaciens* strain (C58) carrying a binary plasmid construct encoding resistance to kanamycin. *Left* Teratomic-kanamycin-sensitive. *Middle* Normal-kanamycin-sensitive, and *right* normal-kanamycin-resistant

(Fig. 5). The recovery of the two visually distinct morphologies and two distinct phenotypes of NC-5339 is in a large part due to the *Agrobacterium tumefaciens* construct (C58/587/85) used. This is an oncogenic wild-type *Agrobacterium* strain containing a binary or miniplasmid which carries the kanamycin resistance/glyphosate tolerance genes. A binary plasmid is an engineered plasmid, smaller than the native bacterial Ti plasmid. The binary plasmid remains in the bacterial cell as an autonomous replicon, separate from the Ti plasmid carrying the oncogenic bacterial T-DNA sequence. The T-DNA region of the Ti plasmid contains the opine, auxin (*tms*), and cytokinin (*tmr*) genes, which result in a shooty teratomic phenotype on plant tissue transformed by a C58 *Agrobacterium* strain (Amasino and Miller 1982). During the transformation process, the T-DNA border regions of the Ti plasmid and similar border regions on the binary plasmid interact during the natural excision and insertion process mediated by the excitation of the virulence (vir) region of the Ti plasmid (Nester et al. 1984; Fraley et al. 1986). Thus when the T-DNA region of the Ti plasmid is excised and inserted into the wounded plant cell, the binary plasmid may also be inserted. Single and multiple copies of genes have been found within the genome of transformed plants, including both fragments and whole genes together, suggesting that in some cases the insertion process was derailed or some form of processing occurred which resulted in nonexpression of certain gene segments (Fraley et al. 1986). Thus each of the NC-5339 phenotypes may represent either correct or incorrect insertion and/or expression of the oncogenic T-DNA segment and/or the binary plasmid genes encoding kanamycin resistance.

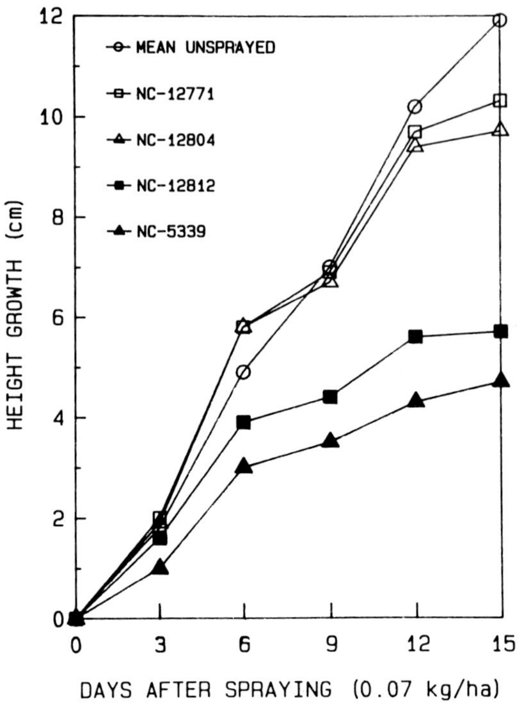


**Fig. 6.** Height growth of plants from three transformed subclones and the parent clone, after treatment with 0.07 kg/ha Round-up (active ingredient), compared to height growth of unsprayed plants. Height growth of all subclones and parent clone plants was reduced by day 3 as compared to untreated plants. However, plants of subclones NC-12771 and NC-12804 still grew significantly better than the parent clone after treatment with 0.28 kg/ha Round-up. Growth of plants from other subclones was intermediate between NC-12804 and NC-12812. Photograph courtesy of Don Riemenschneider and reprinted from Riemenschneider et al. 1988)

The normal, kanamycin-tolerant regenerates have been tested in the greenhouse for tolerance to glyphosate sprays by D. Riemenschneider at the USDA North Central Forestry Lab at Rhinelander, WI (Fig. 6). The highest tolerance noted was at 0.28 kg/ha, a tolerance significantly above that of the untransformed controls which demonstrated highly reduced growth 5 days after spraying with 0.07 kg/ha (Riemenschneider et al. 1988). These plants appeared normal in all other respects.

### 2.2.2 Experiments with Other Genotypes of *Populus*

After demonstrating that foreign genes can be transferred to a selected poplar clone and expressed in recovered plants, we became interested in how broadly applicable this technique was in *Populus*. Thus we conducted the same transformation experiments on five other poplar clones representing a total of three of the five taxonomic sections of *Populus* (Table 2). Expansion of transformation to a diverse group of poplar clones has resulted in variable responses to regeneration, selection, and transformation between the clones and their taxonomic sections. As would be expected, the regenerative potential of all six poplar clones decreased markedly under selection with 60 mg/l kanamycin during regeneration as compared to regeneration on nonselective medium (Table 3). Thus kanamycin appeared to be a potent inhibitor of shoot regeneration in *Populus*.

**Table 2.** *Populus* clones used in this research. These clones represent three of the five sections of the genus *Populus*

1 *Populus tremula* Erecta L. European aspen (PTX)
Taxonomic section: Leuce subsection; Trepidae
Commercial use: Match sticks, paper, plywood, ruminant feed, and ornamental street trees
Habit: pyramidal, highland species found in dense clonal stands
Advantages: rapid growth, root suckers heavily, and tolerant to environmental extremes
Disadvantages: difficult to propagate by cuttings
Miscellaneous: Taxonomic relatives include *P. tremuloides* (quaking aspen) and *P. grandidentata* (Michx.) (big tooth aspen) of North America

2 *Populus alba*×*P. grandidentata* Crandon (NC-5339)
Taxonomic section: Leuce subsection; Albidae×Trepidae hybrid
History: 1 of 4 natural hybrid varieties identified in Iowa. Sherrill, Shimek, and Hansen are the other hybrids
Distribution: Widely distributed in North America and Europe
Commercial use: Recommended for SRIC plantations
Habit: Adaptable to upland and low areas
Advantages: Fast growth rate, and disease- and pest-resistant
Disadvantages: Difficult to root as dormant hardwood cuttings and not as productive as the other three clones

3 *Populus deltoides*×*P. nigra* Eugenei (*P. euramericana*) (NC-5326)
Taxonomic section: Aigeiros hybrid
History: Male clone of a natural hybrid 100 years ago
Distribution: widely planted and heavily used commercial cultivar
Commercial use: Recommended for SRIC plantations
Habit: Self-pruning (natural cladoptosis) and broadly adapted to environmental extremes
Advantages: Roots easily, has a moderate growth rate, shows low susceptibility to *Septoria* and moderate susceptibility to *Marssonina*

4 *Populus nigra*×*P. laurifolia* Strathglass (NC-5272)
Taxonomic section: Aigeiros×Tacamahaca hybrid
History: Female clone of an artificial hybrid by Stout and Schreiner 1924–1926
Distribution: eastern North America and the great Lake states
Commercial use: SRIC production and tree improvement research
Habit: Branchy, fastigiate growth
Advantages: Roots easily, and is fast growing in the upper lake states
Disadvantages: High susceptibility to *Septoria musiva*

5 *Populus nigra* betulifolia×*P. trichocarpa* (NC-5331)
Taxonomic section: Aigeiros×Tacamahaca hybrid
History: Female clone of an artificial hybrid by Stout and Schreiner 1924–1926
Distribution: Field tested throughout the US
Commercial use: SRIC production and tree improvement research
Habit: Numerous moderately angled branches, moderate growth rate in the northern lake states
Advantages: Roots easily and has low susceptibility to *Melampsora* rust
Disadvantages: Medium to high susceptibility to *Septoria*

6 *Populus maximowiczii*×*P. trichocarpa* Androscoggin (NC-11390)
Taxonomic section: Tacamahaca hybrid
History: Male clone from an artificial hybrid by Stout and Schreiner 1924–1926
Distribution: Field tested throughout the US
Commercial use: One of the most commonly used cvs. for production in the eastern US
Habit: Fast growth rate
Advantages: Low susceptibility to *Melampsora* rust and roots easily
Disadvantages: Medium to high susceptibility to *Septoria* leaf spot and canker

**Table 3.** Summary of shoot regeneration, selection, and screening data for 5 diverse *Populus* clones used in genetic transformation experiments. Regeneration data is from leaf discs that were cocultivated with *Agrobacterium* containing a kanamycin resistance gene (Strain C58/587 or C58/587/85) or were not cocultivated with *Agrobacterium* (control). Regeneration was attempted on MS medium with 4 μM each of BA and Zeatin with selection at 60 mg/l kanamycin or 0 mg/l. Screening occurred on MS or WPM medium with 0.01 μM BA containing 60 mg/l kanamycin

| Clone | Selection/ screening | Total # of explants | % of explants regenerating | # of shoots recovered | # surviving screening |
|---|---|---|---|---|---|
| NC-5331 | Select | 24 | 0 | 0 | 0 |
| | control | 11 | 0 | 0 | 0 |
| | Screen | 23 | 87 | 503 | 21 |
| | control | 11 | 100 | 76 | 0 |
| NC-5272 | Select | 14 | 0 | 0 | 0 |
| | control | 14 | 0 | 0 | 0 |
| | Screen | 25 | 44 | 266 | 18 |
| | control | 12 | 0 | 0 | 0 |
| NC-11390 | Select | 24 | 0 | 0 | 0 |
| | control | 23 | 0 | 0 | 0 |
| | Screen | 29 | 26 | 62 | 9 |
| | control | 32 | 50 | 61 | 0 |
| *P. tremula* Erecta | Select | 64 | 0 | 0 | 0 |
| | control | 10 | 0 | 0 | 0 |
| | Screen | 63 | 27 | 76 | 14 |
| | control | 10 | 60 | 13 | 0 |
| NC-5326 | Select | – | – | – | – |
| | control | – | – | – | – |
| | Screen | 19 | 0 | 0 | 0 |
| | control | – | – | – | – |

When compared to the previous experiments with clone NC-5339 (see Section 2.2.1), the response of the other poplar clones differed in three major respects.

1. Except for clone NC-5272, cocultivation with the oncogenic *Agrobacterium* did not generally stimulate shoot regeneration on nonselective (kanamycin-free) medium. Fillatti et al. (1987) suggested that cocultivation with an oncogenic bacterial strain in synergy with exogenous media hormones led to enhanced regeneration of NC-5339. The overall hormonal concentration did not enhance the regeneration of NC-5326 and *P. tremula* Erecta discs and may have further stressed these already sensitive tissues. NC-11390 and NC-5331 tissues showed greater regeneration in the absence of cocultivation, a response in contrast to that of NC-5339.
2. Except for clone NC-5339, shoot regeneration was not observed on medium containing kanamycin. This result was probably due to the lack of incorporation and/or expression of the kanamycin resistance genes in the cocultivated leaf tissue and/or to the harshness of the selective regeneration environment. Likewise, only NC-5339 cocultivated tissues regenerated shoots exhibiting the teratomic phenotype. The strong inhibitory effect of kanamycin on shoot regeneration in poplars, coupled with the existing stresses of bacterial infection and high exogenous hor-

mone levels, creates a very adverse environment for single cells or cell colonies to develop, even if they contain kanamycin resistance. Sensitivity to kanamycin during regeneration has also been noted on root regeneration from callus formed on cocultivated small green stem sections of *P. trichocarpa*×*P. deltoides* (H11) (T. Parsons pers. commun.). Furthermore, this inhibition of regeneration decreases the potential for recovery of all possible transformation events and the corresponding morphologies produced.

Putatively transformed shoots were recovered from five of the six clones when post-regeneration screening methods were used. However, the response of the clones differed and could be roughly divided into two groups. Regenerates of NC-5339 and *P. tremula* Erecta developed along cut leaf edges with *P. tremula* Erecta regenerates, developing slowly from bleached callus tissue of wounded primary and secondary vein ends on nonselective media. Cocultivated *P. tremula* Erecta regenerates responded similarly to verified NC-5339 transformants. Putative transformants remained dark green, succulent, and actively growing, whereas kanamycin-sensitive shoots exhibited bleaching of the apical and the newly developing leaves. The bleaching response suggests an interruption in the synthesis of chlorophyll in chloroplasts. This corresponds to the mode of action of kanamycin which binds to the 70S ribosome, causing misreading of mRNA in bacteria (Maniatas et al. 1984). Sensitive shoots of NC-5339 and *P. tremula* Erecta died after 3 weeks on 100 mg/l kanamycin.

Leaf pieces of NC-11390 (Fig. 7) and NC-5331 demonstrated extensive shoot regeneration on the leaf surface but not at cut leaf edges (where transformation would be expected to occur), while NC-5272 showed limited regeneration on leaf surfaces, and NC-5326 did not regenerate at all. However, actively growing regenerates from these cocultivated clones exhibited no visible signs of bleaching at 100 mg/l kanamycin, but shoot growth and leaf formation ceased. At 200 mg/l kanamycin, partial signs of bleaching of shoots and leaves occurred. Bleaching was difficult to visualize in NC-5272, NC-5331, and NC-11390, due to the slower growth and yellow appearance of these clones in contrast to NC-5339 and

**Fig. 7.** Extensive regeneration of *Populus* NC-11390 on medium without kanamycin. Regeneration occurs over the entire leaf surface, but not at the cut edges where transformation is expected

*P. tremula* Erecta grown on 0.01 μM BA medium. These results suggest a variable response to kanamycin sensitivity among the poplar clones. This variable response may be related to a difference among the different genotypes of *Populus* in the uptake and translocation of kanamycin through the plant and/or a natural tolerance to kanamycin by specific genotypes. This variation in response to kanamycin further emphasizes that the use of kanamycin cannot be generalized and that inherent variation among genotypes can further aggravate the inaccuracies of screening with kanamycin.

3. Positively transformed shoots were obtained only with the NC-5339 and *P. tremula* Erecta clones. Neomycin phosphotransferase II′ (NPT II′) enzyme activity assays were carried out on crude leaf extracts of the surviving poplar clones (Reiss et al. 1984). Only NC-5339 and *P. tremula* Erecta clones were found to contain NPT II′ enzyme activity (Fig. 8A, B). None of the other poplar clones selected as positive at 200 mg/l kanamycin demonstrated NPT II′ activity. Similar results were noted for tissues regenerated from poplar H11 explants cocultivated with various *A. tumefaciens* vectors containing kanamycin resistance genes (T. Parsons pers. commun.).

Several representatives of the Leuce section of poplars have been successfully transformed via *A. tumefaciens* and these include: NC-5339 (Fillatti et al. 1987), *P. tremula* Erecta (Sellmer unpubl.), and *P. tremuloides* (Minocha et al. 1986). Parsons et al. (1986) has been the only reported success of transformation of a hybrid representing the Aigeiros and Tacamahaca sections of *Populus*. Natural infectivity of *Agrobacterium* may certainly play a role in expanding the success of gene transfer to other poplar hybrids and species. The successful transformation of the hybrid H11 (Parsons et al. 1986) was the result of using two *Agrobacterium* strains which showed natural infectivity on poplar H11. Our unsuccessful application by *Agrobacterium* strains which succeeded in transforming Leuce section clones (NC-5339 and *P. tremula* Erecta) to clones from other poplar sections may be the result of incompatibility between the plant and the bacteria. Neshe

A 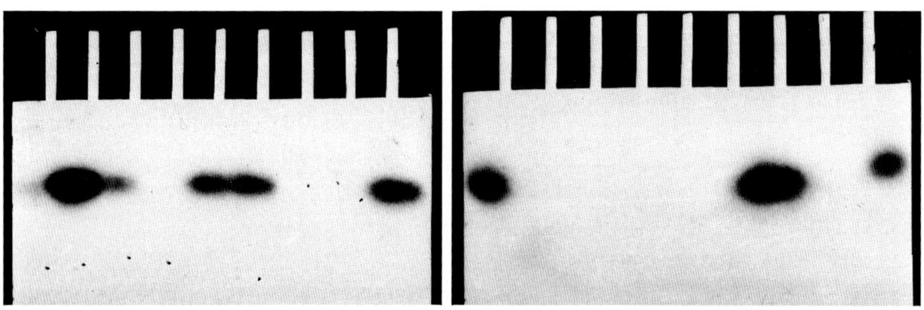 B

**Fig. 8. A** An NPT II′ enzyme activity assay autoradiograph of putatively transformed and selected NC-5339 leaf tissue from regenerates. Lanes *2*, *5*, *6*, and *10* contain positively transformed tissue; lanes *1*, *3*, and *7* contain noncocultivated and tissue from positively selected shoots not expressing NPT II′ activity respectively. **B** An NPT II′ autoradiograph of *P. tremula* Erecta leaf tissue. Lane *1* contains a verified NC-5339 transformant (positive control); lane *2* is a blank lane; lanes *4*, *5*, *6*, and *9* contain cocultivated but kanamycin-sensitive leaf tissue; and lanes *7*, *8*, and *10* contain putatively transformed and positively selected leaf tissue et al. (1987) describe a differential response of Leuce section poplar cultivars to natural infection of *Agrobacterium tumefaciens*. Furthermore, Nester et al. (1984) noted that bacterial strains are capable of inducing variable responses in different cultivars of a single species and may further induce different responses on different tissues of a single clone. Thus the genotype, phenotype, and growth habit of the poplars and the variable infectivity of the bacteria are both probable determining factors in successful transformation of poplars.

### 2.2.3 Bacterial Contamination in Regenerates

Several laboratories have noted problems with removing *Agrobacterium* and other bacterial contaminants from cocultivated tissues. We have spent countless hours trying to recover cocultivated tissues, regenerates and selected stable transformed shoots from this recurring bacterial build-up. Our attempts with antibiotics (e.g., Rifampicin, Carbenicillin, Cefotaxime) singularly or in combination and cultural practices (e.g., sucrose minus media, yeast extract with antibiotic treatments) have had variable results. Riemenschneider et al. (1988) suggest two methods for the removal of bacterial contaminants from cocultivated plants: (1) Repeated subculturing of the actively growing transformed shoot tip, and more effectively (2) rooting and returning the transformed shoots to the greenhouse under a rapid growing regime and then reestablishment of the active shoot tips in sterile culture.

### 2.2.4 Overall Poplar Transformation Conclusions

The conclusions from this work are as follows:

1. Foreign genes can be transferred to *Populus* via *A. tumefaciens* and normal plants can be recovered expressing the foreign gene traits. However, these approaches appear to be most applicable to the Leuce section only.
2. Genotype responses to kanamycin as a selection and/or screening agent are variable, so experimentation with other selective and screening agents are suggested to insure identification of transformants and to be able to differentiate between variable expression in regenerates.
3. Selection during regeneration inhibits the regeneration potential of cocultivated tissues, and further reduces the potential for recovery of all expression events which develop during transformation with an oncogenic binary plasmid vector system. Post-regeneration screening reduces the stress on transformed cells allowing for the recovery of all possible expression events, including the teratomic morphologies and variable levels of kanamycin tolerance. Both classes of expression will be useful in examining gene expression in trees at maturity.
4. Survey experiments to determine the natural infectivity and the compatibility between the plant genotypes and bacterial strains to be used is a prerequisite to overall success in transformation within a plant genus.
5. Tolerance in trees to herbicides can be gained by genetic transformation.

Research is underway to enhance tolerance to glyphosate by increasing foreign gene expression and by circumventing differential gene expression induced by

*Agrobacterium* transformation systems. A technique of interest for enhancing glyphosate tolerance includes targetting the modified EPSP synthase to the chloroplast where glyphosate activity is proposed to be the strongest. In order to target the modified EPSP protein to the chloroplast, a leader sequence or transit peptide coding DNA region must be inserted into the *aro* A gene reading frame.

### 2.3 Alternate Approaches to Gene Transfer in *Populus*

Pythoud et al. (1987) have successfully transformed and regenerated *P. trichocarpa*×*P. deltoides* (H11) using a "supervirulent" oncogenic *Agrobacterium rhizogenes* (A4) strain containing a nopaline synthase gene (NOS) and NPT II′ gene and the *vir* region of the pTiBo542 plasmid of *A. tumefaciens in trans.* Pythoud's procedure consisted of regenerating roots from callus of cocultivated stem segments (Parsons et al. 1986) from which transformed shoots eventually differentiated. Regeneration occurred on hormone-free medium and the regenerates were tested by enzyme assay.

With the recent reported success of plant regenerates from protoplasts of *Populus* (Russell and McCown 1986), gene transfer directly into poplar protoplasts becomes a realistic alternative. The most logical approach would be to employ electroporation technology (Fromm et al. 1985). Work in this aspect has just begun and, if successful, the approach would provide an alternative with considerably more flexibility and much lower complexity than the *Agrobacterium*-based systems.

### 2.4 Conclusions

The potential for improvement via gene transfer techniques in plantation tree production is exciting, especially in short rotation intensive tree crop production for energy and pulp where specific herbicide, insect, and disease resistance requirements have been identified. Initial experiments in the transformation and expression of glyphosate tolerance in *Populus* indicates the feasibility of tree improvement through biotechnology. However, two immediate biological questions result from this early work:

1. Is *Agrobacterium* oncogenicity required for successful transformation of trees? If true, then the physiological basis of this requirement, the degree of oncogenicity, and its relationship to the virulence of *Agrobacterium* are all intriguing questions.
2. Are the Leuce section poplars (*Populus alba, Populus tremula, Populus tremuloides*, and hybrids) uniquely more susceptible to infection and transformation by *Agrobacterium tumefaciens* than the other poplar sections (Tacamahaca, Aigeiros, and hybrids)? If this is true, then what physiological or biochemical factors within the Leuce section are controlling this activity? Phenolic activity (Bolton et al. 1986) may be a likely target for further study.

Answers to these two questions will undoubtedly advance both the science and technological aspects of woody plant transformation. However, for tree physiologists one important question remains to be answered: What is the stability of foreign gene expression in trees? How will expression be influenced by strong seasonal dormancy cycles and long-term physiological phase change under natural environmental conditions? In what tissue (i.e., cambium?, bark?) will these genes be expressed? Interestingly, answering such questions will not directly contribute to biotechnology, but will yield considerable insight into general gene expression in woody plants.

## References


Ahuja MR (1986) Aspen. In: Evans DA, Sharp WR, Ammirato PV (eds) Handbook of plant cell culture, vol 4. MacMillan, Riverside, NJ, pp 626–651

Amasino RJ, Miller CO (1982) Hormonal control of tobacco crown gall tumor morphology. Plant Physiol 69:389–392

Bolton GW, Nester EW, Gordon MP (1986) Plant phenolic compounds induce expression of the *Agrobacterium tumefaciens* loci needed for virulence. Science 232:983–985

Comai L, Facciotti D, Hiatt WR, Thompson G, Rose RE, Stalker DB (1985) Expression in plants of a mutant aroA gene from *Salmonella typhimurium* confers tolerance to glyphosate. Nature (Lond) 317:741–744

Dandekar AK, Gupta PK, Durzan DJ, Knauf V (1987) Transformation and foreign gene expression in micropropagated Douglas-fir (*Pseudotsuga menziesii*). Biotechnology 5:587–590

FAO (ed) (1979) Poplars and willows in wood production and land use. FAO, Rome, For Ser 10

Fillatti JJ, Sellmer J, McCown B, Haissig B, Comai L (1987) *Agrobacterium* mediated transformation and regeneration of *Populus*. Mol Gen Genet 206:192–199

Fraley RT, Rogers SG, Horsch RB (1986) Genetic transformation in higher plants. CRC Crit Rev Plant Sci 4(1):1–46

Fromm M, Taylor T, Walbot V (1985) Expression of genes transferred into monocot and dicot plant cells by electroporation. Proc Natl Acad Sci USA 82:5824–5828

Haissig BE, Nelson ND, Kidd GD (1987) Trends in the use of tissue culture in forest improvement. Biotechnology 5:52–59

Lloyd G, McCown B (1980) Commercially-feasible micropropagation of mountain laurel, *Kalmia latifolia* by use of shoot-tip culture. Proc Int Plant Prop Soc 30:421–427

Maniatis T, Fritsch DF, Sambrook J (1984) Molecular cloning: a laboratory manual. Cold Springs Harbor Press, New York, p 73

McCown BH (1985) From gene manipulation to forest establishment: shoot cultures of woody plants can be a central tool. TAPPI J 68:116–119

Minocha SC, Noh EW, Kausch AP (1986) Tissue culture and genetic transformation in *Betula papyrifera* and *Populus tremuloides*. TAPPI J 68:116–119

Murashige T, Skoog F (1962) A revised medium for rapid growth and bio-assays with tobacco tissue cultures. Physiol Plant 15:473–497

Neshe X, Michel MF, Digat B (1987) Population heterogeneity of *Agrobacterium tumefaciens* in galls on *Populus* L. from a single nursery. Appl Environ Microbiol 53:655–659

Nester EW, Gordon MP, Amasino RJ, Yanofsky MK (1984) Crown Gall: A molecular and physiological analysis. Annu Rev Plant Physiol 35:387–413

Parsons TJ, Sinkar VP, Stettler RF, Nester EW, Gordon MP (1986) Transformation of poplar by *Agrobacterium tumefaciens*. Biotechnology 4:533–536

Pythoud F, Sinkar VP, Nester EW, Gordon MP (1987) Increased virulence of *Agrobacterium rhizogenes* conferred by the vir region of pTiBo542: Application to genetic engineering of poplar. Biotechnology 5:1323–1327

Reiss B, Sprengel R, Will H, Schaller H (1984) A new sensitive method for qualitative and quantitative assay of neomycin phosphotransferase in crude cell extracts. Gene 30:211–218

Riemenschneider DR, Haisig BE, Sellmer J, Fillatti JJ (1988) Expression of a bacterial herbicide tolerance gene in hybrid poplar. In: Ahuja MR (ed) Somatic Cell Genetics of Woody Plants, Proc IUFRO Somatic Cell Genetics Working Party S2.04–07, August 10–13, 1987, Großhansdorf, Federal Republic of Germany. Kluwer Academic Publishing, Dordrecht Boston London, p73–80

Russel JA, McCown BH (1986) Culture and regeneration of *Populus* leaf protoplasts isolated from non-seedling tissue. Plant Sci 46:133–142

Sederoff R, Stomp AM, Chilton WS, Moore LW (1986) Gene transfer into loblolly pine by *Agrobacterium tumefaciens.* Biotechnology 4:647–649

Sellmer JC (1987) Microculture dynamics and *Agrobacterium*-mediated gene transfer in *Populus.* MS Thesis Univ Wisconsin-Madison

USDA (ed) Forest Service (1980) Energy and wood from intensively cultured plantations: research and development program. Gen Tech Rep NC-28

# X Transformation in *Vinca rosea* L.: An Example of Spheroplast Method for Gene Transfer

S. HASEZAWA and K. SYŌNO[1]

## 1 Introduction

Higher plant cells have firm cell walls that have been obstacles for cell manipulations. The method for removing cell walls by commercial enzymes was developed (Takebe et al. 1968). The spheres of cytoplasm isolated as a result were called protoplasts. Protoplasts are also obtained from Gram-positive or -negative bacteria by lysozyme, an enzyme for digesting bacterial cell walls. However, in many cases of Gram-negative bacteria, lysozyme treatments yield incomplete bacterial protoplasts accompanied by the remains of cell walls, designated spheroplasts (Birdsell and Cota-Robles 1967).

Recently, several techniques for transformation through the introduction of foreign genetic informations into plant cells have been designed (see Chaps. I, IV, this Vol.). Among these we have tried to use bacterial spheroplasts as a tool which transports genetic information into plant protoplasts. Bacterial spheroplasts are desirable vessels which transport genetic information into plant protoplasts. The genetic material within the spheroplasts is protected by cell membranes and safely transferred into plant cytoplasm without interference from DNase and other obstacles. Spheroplasts offer a simple and effective method of gene transfer to plant cells, because it needs no DNA isolation from bacteria and realizes direct gene transfer of cloned recombinant DNA from microorganisms to higher plant cells. Furthermore, no special instruments are needed for the introduction of genes by spheroplasts.

*Agrobacterium tumefaciens*, harboring tumor-inducing plasmid (Ti plasmid), is a causative agent to induce the tumor named crown gall on many dicotyledonous plants. The tumorous state of crown gall tissue has been shown to result from the transfer of a part of Ti plasmid (T-DNA) from the bacteria to the plant cells. This fact suggests the possibility of introducing some useful genes into plant cells, using the Ti plasmid as a vehicle. At present, various available vectors to plant cells have been constructed by the modification of Ti plasmid. When *A. tumefaciens* carrying the engineered Ti plasmid is used to infect plant cells, some genes inserted in the T-DNA region are transferred to plant chromosomes together with the T-DNA fragment.

[1] Department of Pure and Applied Sciences, College of Arts and Sciences, University of Tokyo, Komaba, Meguro-ku, Tokyo, 153 Japan

Biotechnology in Agriculture and Forestry, Vol. 9
Plant Protoplasts and Genetic Engineering II (ed. by Y.P.S. Bajaj)

The transformation system depends on the infection ability of *Agrobacterium* (Marton et al. 1979); however, it is impossible to attempt to induce Ti plasmid into Gramineae plants and many monocotyledonous plants which are not susceptible to *Agrobacterium* in nature. On the contrary, in the spheroplast method, the introduction is accomplished physically. Therefore, it is easy to attempt to transfer spheroplasts to any plant protoplasts without restriction of the host range of bacteria.

In our experimental system, *A. tumefaciens* was chosen as the source of spheroplasts. Transformants obtained by treatment with *A. tumefaciens* spheroplasts are expected to be selected on hormone-free medium, because crown gall cells, natural transformants by the bacteria, are known to grow on the medium without hormone. On the other hand, we selected *Vinca rosea* L. [*Catharanthus rosea* (L.) G. Don], a perennial herb belonging to the Apocynaceae, as the source of plant protoplasts. It has paired, untoothed leaves and five-lobed, bell-shaped flowers. The slightly asymmetric petals create a pinwheel effect. The plant produces medicinally important indole alkaloids such as vinblastin and vincristin. *V. rosea* has often been used for the study of crown gall. The plant is resistant to relatively high temperature. This character has been used for the elimination of *A. tumefaciens* bacteria, which are very sensitive to temperature, from the infected crown gall by heat treatment at 42 °C for 5 days, when the crown galls are cultured in vitro aseptically. Furthermore, the cultured cells of this plant are very friable and grow rapidly, and stable protoplasts can be easily obtained from them. These properties are appropriate for the experimental materials.

In this chapter we will describe the technical aspect of in vitro transformation of *Vinca* cells mediated by *Agrobacterium* spheroplasts, an example of successful transformation using spheroplasts (Hasezawa et al. 1981; Okada et al. 1985). The other possibilities of gene transfer by the spheroplast method will also be discussed.

## 2 Techniques

### 2.1 Preparation of *Agrobacterium* Spheroplasts

*Agrobacterium tumefaciens* (Fig. 1 A), harboring octopine or nopaline type Ti plasmid, is maintained at 26 °C on a 0.8% Nutrient Broth (Difco Lab., MI, USA) medium solidified with 1.5% Bacto-agar (Difco). Spheroplasts are prepared from the logarithmic growth stage of bacteria cultured overnight in AB minimal medium (Chilton et al. 1974) at 30 °C. At the late logarithmic stage, 500 mg/l of carbenicillin (Taito Pfizer, Tokyo, Japan) is added. Two hours later, bacteria are harvested by centrifugation (10000 rpm, 10 min, room temperature) and washed with Tris buffer (pH 8.2) containing 20 mM Tris-HCl, 10 mM 2Na-EDTA, and 0.4 M sorbitol. The cells are then treated at 30 °C with a lysozyme solution containing 400 mg/l lysozyme (Boehringer Mannheim GmbH, FRG) dissolved in the same buffer. After 2 h incubation almost all the bacteria are converted to spheroplasts, as monitored by a transmission electron microscope (Fig. 1 B).

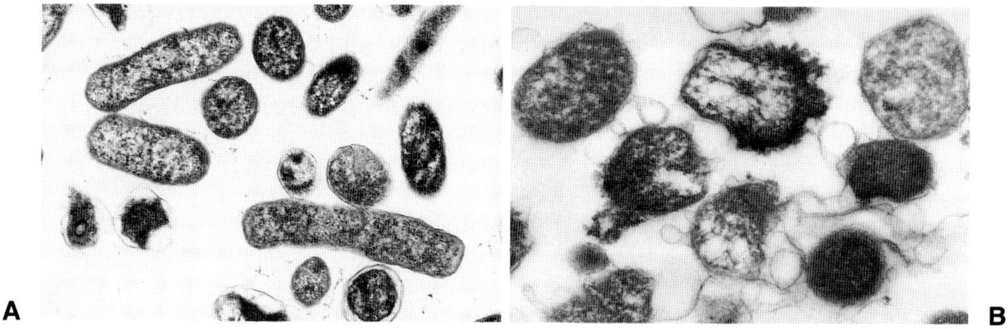

**Fig. 1. A** *Agrobacterium tumefaciens*, a Gram-negative soil bacterium (×7 500). **B** Spheroplasts of *Agrobacterium* after the lysozyme treatment (×10 500). (Courtesy of Dr. C. Matsui)

**Scheme 1.** Preparation of *Agrobacterium* spheroplasts

Suspend with AB medium

│ overnight-incubation at 30 °C

Add carbenicillin (final 500 mg/l) (1)

│ 2-h incubation at 30 °C

Harvest and wash with Tris-HCl buffer (pH 8.2) (2)

│ centrifugation (10000 rpm, 10 min)

Resuspend with Tris-HCl buffer (2)
and add lysozyme (final 400 mg/l) (3)

│ 2-h incubation at 30 °C

│ centrifugation (5000 rpm, 10 min)

Wash with 0.4 M sorbitol

│ centrifugation (5000 rpm, 10 min)

Spheroplast treatment

Stock solution

1. 5 g/l carbenicillin (×10)
2. 20 mM Tris-HCl (pH 8.2)
   10 mM 2Na-EDTA
   0.4 M sorbitol
3. 4 g/l lysozyme (×10)

AB medium

| | |
|---|---|
| $K_2HPO_4$ | 3 g |
| $NaH_2PO_4$ | 1 g |
| $NH_4Cl$ | 1 g |
| $MgSO_4 \cdot 7H_2O$ | 0.3 g |
| KCl | 0.15 g |
| $CaCl_2$ | 0.01 g |
| $FeSO_4 \cdot 7H_2O$ | 2.5 mg |
| glucose | 5 g/l |

Freshly prepared spheroplasts are harvested by centrifugation (5000 rpm, 10 min, room temperature) and washed with the Tris buffer and 0.4 M sorbitol solution. Finally, they are resuspended in 0.4 M sorbitol or mannitol solution with gentle pipetting at respective concentration and then used for spheroplast treatment.

## 2.2 Isolation of *Vinca* Protoplasts

*Vinca* protoplasts are easily isolated from suspension cultures of *Vinca rosea* L. maintained in 50 ml Murashige and Skoog's (MS) medium (1962) supplemented with 0.5 mg/l 2,4-dichlorophenoxyacetic acid in 300-ml Erlenmeyer flasks on a reciprocal shaker (80 excursions/min) at 26 °C. One-tenth of the culture is transferred to fresh medium at weekly intervals, and the 5-day culture is used as the source of protoplasts. The conditions of protoplast isolation are as described by Hasezawa et al. (1981) (Scheme 2), except that 1% Cellulase Onozuka RS (Yakult Honsha Co., Ltd., Tokyo, Japan) and 0.1% Pectolyase Y-23 (Seishin Pharmaceutical Co., Ltd., Tokyo, Japan) in place of 2% Cellulase Onozuka R-10 and 0.05% Pectolyase Y-23 are used for the enzyme mixture. After 45 min enzyme treatment, freshly isolated protoplasts are washed with 0.4 M mannitol solution and then used for spheroplast treatment.

**Scheme 2.** Isolation of protoplasts from cultured *Vinca* cells

*Vinca* suspension culture

| centrifugation (900 rpm, 2 min)

Suspend with enzyme solution

| 45 min ~ 1-h incubation at 30 °C

Wash with 0.4 M mannitol

|

Wash and resuspend with protoplast medium

|

Spheroplast treatment

| Enzyme solution | |
|---|---|
| Cellulase Onozuka RS | 1% (w/v) |
| Pectolyase Y 23 | 0.1% (w/v) |
| Mannitol | 0.4 M |
| pH | 5.5 |

| Protoplast medium | |
|---|---|
| Murashige and Skoog's medium supplemented | |
| IAA | 1 mg/l |
| (or 2,4-D | 0.5 mg/l) |
| Sucrose | 1% (w/v) |
| Mannitol | 0.4 M |
| pH | 5.5 |

### 2.3 Spheroplast Treatment

*Agrobacterium* spheroplasts can be introduced into *Vinca* protoplasts by polyethylene glycol (PEG) 4000 (Wako Pure Chemical Industries Ltd., Osaka, Japan) (Kao and Michayluk 1974) or polyvinyl alcohol (PVA) No. 203 (Kuraray Co., Ltd., Kurashiki, Japan) (Nagata 1978). A 1000-fold excess of *Agrobacterium* spheroplasts is mixed with *Vinca* protoplasts to make up a total volume of 0.5 ml in a 10-ml centrifuge tube. After 10 min incubation at room temperature, 0.5 ml of PEG or PVA solution is added to the mixture dropwise to 10% ~ 20% (w/w) at final concentration and mixed by gentle swirling. After 10 min incubation, these polymers are removed by dilution with 0.05 M glycine-NaOH buffer (pH 10.5) containing 0.05 M $CaCl_2$ and 0.4 M mannitol (high pH-high calcium solution) and incubated for 20 min at 30 °C. Subsequently the protoplasts are washed 5 – 10 times with 0.4 M mannitol solution by centrifugation (900 rpm, 2 min) and then resuspended with protoplast medium.

**Scheme 3.** Spheroplast treatment. Introduction of *Agrobacterium* spheroplasts into *Vinca* protoplasts with PEG (or PVA)

*Vinca* protoplasts — *Agrobacterium* spheroplasts

mix

Add PEG (or PVA) solution (final 20% PEG or 10% PVA)

10-min incubation at room temperature

Add high pH-high calcium solution

20-min incubation at 30 °C

Wash with 0.4 M mannitol 5 ~ 10 times

Wash and resuspend with protoplast medium

PEG (or PVA) solution

| | |
|---|---|
| Polyethylene glycol 4000 | 40% (w/w) |
| or | |
| Polyvinyl alcohol No. 203 | 20% (w/w) |
| in protoplast medium | |

### 2.4 Culture of Protoplasts and Selection of Transformants

After washing, the protoplasts are cultured in the protoplast medium at a density of $5 \times 10^4 - 1 \times 10^5$ protoplasts/ml. After antibiotic treatment or heat treatment at 40 °C for 4 days to eliminate the rest of the free spheroplasts, the protoplasts are kept at 26 °C. After 3 – 4 weeks, they are diluted by addition of hormone-free MS medium at weekly intervals. Finally, the colonies are embedded in the hor-

**Scheme 4.** Culture and selection methods of protoplasts

Protoplasts after spheroplast treatment
(in a liquid medium, $5 \times 10^4$/ml)
|
Add vancomycin (200 mg/l) and carbenicillin (250 mg/l)
|
Culture for 2 – 3 weeks at 26 °C
|
Dilute with MS basal medium (2 – 3 times, every week)
|
Wash and suspend with MS basal medium
|
Culture in agar (0.6%) MS medium for a month at 26 °C
|
Pick up grown-up colonies for further cultivation on MS agar medium and use for the detection of nopaline or NDH activities

mone-free MS medium solidified with 0.6% agar in plastic Petri dishes (Falcon, CA, USA). Surviving colonies are further transferred to and maintained on hormone-free MS agar medium, and used for the following analysis.

## 2.5 Analysis of Transformants

*1. Detection of Opine and Opine Synthase Activity.* Opines, octopine or nopaline, are extracted and purified with ion-exchange resin, they are identified with

**Scheme 5a.** Extraction of opine

Fresh tissue
|
Ground with 80% EtOH in a mortar
| centrifuge 8000 rpm, 10 min
Supernatant
|
Pass through a Amberlite IR-120B ($H^+$ form) column
| wash with 80% EtOH and water
| elute with 2 N $NH_4OH$
Pass through a Dowex 50W ($H^+$ form) column
| wash with water
| elute with 1 N pyridine
Opine fraction

**Scheme 5b.** Identification of opine

| | Ascending solvent system (v/v) | |
|---|---|---|
| Paper chromatography | 1 Ethanol : water | (1 : 1) |
| | 2 n-Butanol : acetic acid : water | (12 : 3 : 5) |
| | 3 n-Butanol : pyridine : water | (1 : 1 : 1) |
| Thin-layer chromatography | 1 Ethanol : water | (1 : 1) |
| | 2 Water saturated with n-butanol | |
| High voltage electrophoresis | 0.05 N formic acid adjusted to pH 3.6 with $NH_4OH$ | |
| | run at 1000 V | |

**Scheme 6.** Detection of radioactivity incorporated into nopaline. (Otten and Schilperoort 1978)

Fresh tissues

Homogenize with extraction buffer

centrifugation

Add incubation buffer to the supernatant (1 : 1)

Add 2 μCi of $^{14}C$-arginine to 1 ml of mixture
(or 2 μCi of $^{14}C$-$\alpha$-ketoglutaric acid and 10 μCi of $^{3}H$-arginine)

2-h incubation at 25 °C

Pass through ion-exchange columns

TLC analysis

HPLC analysis

Scintillation-counter analysis

| Extraction buffer | | Incubation buffer | |
|---|---|---|---|
| Tris-HCl | 0.1 M | L-arginine | 60 mM |
| Sucrose | 0.5 M | $\alpha$-Ketoglutaric acid | 60 mM |
| Ascorbic acid | 0.1% (w/v) | NADH | 16 mM |
| Cysteine-HCl | 0.1% (w/v) | Sodium phosphate | 0.2 M |
| pH | 8.0 | pH | 6.8 |

phenanthrenquinone reagent after separation by high voltage electrophoresis or thin-layer chromatography (Hasezawa et al. 1981).

Opine synthase activities are assayed by using radioisotope substrates. Preparation of crude opine synthase is prepared according to Otten and Schilperoort (1978). The purified reaction mixture is separated by thin-layer chromatography, and quantitative analysis of opine synthase activity can be carried out by counting radioactivities incorporated in opine by using a radiochromatogram scanner and a scintillation counter (Okada et al. 1985).

*2. Southern Hybridization.* Preparation of DNA is carried out using cetyltrimethylammonium bromide or CsCl-centrifugation. The prepared sample DNA's are digested with respective restriction endonuclease. After electrophoresis on agarose gel, they are transferred to nitrocellulose filter. Then they are hybridized with respective probes from Ti plasmid. Using this method, T-DNA structures of transformants were examined (Okada et al. 1985).

## 2.6 Preparation of *E. coli* Spheroplasts

Spheroplasts of *Escherichia coli* have the possibility of being a desirable tool of gene transfer to plant cells in different use from those of *Agrobacterium* (see Sect.

**Scheme 7.** Preparation of *E. coli* spheroplasts

Suspend with Fraser and Jerrel's medium
| 12-h incubation at 37 °C
Harvest and wash with 0.01 M Tris-HCl buffer (pH 8.0) (1)
|
Resuspend with Tris-HCl buffer supplemented with 0.5 M sucrose (2)
| 30-min incubation at room temperature
Add lysozyme (final 20 mg/l) (3)
| 5 – 10-min incubation at room temperature
Dilute with Tris-HCl buffer (1 : 1) (1)
|
Add 2Na-EDTA solution (final $10^{-3}$ M) (4)
| 10 ~ 15-min incubation at room temperature
| centrifugation (5000 rpm, 10 min)
Wash with mannitol or sorbitol solution
| centrifugation (5000 rpm, 10 min)
Spheroplast treatment

*Stock solution*

1. 0.01 M Tris buffer (pH 8.0)
2. 0.01 M Tris buffer (pH 8.0)
   0.5 M sucrose
3. 400 mg/l lysozyme (×20)
4. 0.01 M 2Na-EDTA (×10)

| Fraser and Jerrel's medium | |
|---|---|
| $Na_2HPO_4$ | 10.5 g |
| $KH_2PO_4$ | 4.5 g |
| $NH_4Cl$ | 1.0 g |
| $MgSO_4 \cdot 7H_2O$ | 0.6 g |
| Casamino acid | 15.0 g |
| Glycerol | 30.0 g |
| 1 M $CaCl_2$ solution | 0.3 ml/l |

2.6). *E. coli* spheroplasts are prepared from the logarithmic stage of bacteria cultured in Fraser and Jerrel's medium (1953) at 37 °C. The bacteria are harvested by centrifugation (10000 rpm, 10 min, room temperature) and washed with Tris-HCl buffer (pH 8.0). The collected bacteria are resuspended in Tris buffer supplemented with 0.5 M sucrose. After 30 min incubation at room temperature, 20 mg/l lysozyme at final concentration is added. The cells are incubated at room temperature for 10 min, and then the same volume of the Tris buffer without sucrose is added and mixed well. Subsequently, 2Na-EDTA solution is added to them. After 15 min incubation at room temperature, the spheroplasts (Fig. 2) are harvested by centrifugation (5000 rpm, 10 min) and washed with 0.4 M mannitol solution. Finally, they are resuspended in 0.4 M mannitol solution with gentle pipetting and then used for spheroplast treatment.

## 3 Technical Problems in the Spheroplast Method

### 3.1 Preparation of Spheroplasts

Special attention should be given in the preparation of bacterial spheroplasts as follows. (1) It is desirable for the stock solutions of chemical reagents used for the yields of spheroplasts to be prepared as described below in Schemes 1 and 2. It is convenient to divide them into small vessels for each experiment after sterilization. Before lysozyme treatment, the carbenicillin pretreatment is indispensable for the preparation of *Agrobacterium* spheroplasts, but not for that of *E. coli*. At the time of the lysozyme treatments, occasional gentle shaking by hand is effective for good yields of spheroplasts. (2) After the lysozyme treatment, gentle manipulation is indispensable for handling spheroplasts. Therefore, spheroplasts must be collected by centrifugation below 5000 rpm. Their pellets must be separated patiently by gentle pipetting until complete separation, when they are washed (spheroplasts are liable to break and aggregate on rough treatment). (3) Yields of bacterial spheroplasts are monitored by a phase contrast microscope. As an easy method, the state of spheroplast yield can be judged by the appearance when the spheroplast suspension is added dropwise to distilled water. Observation

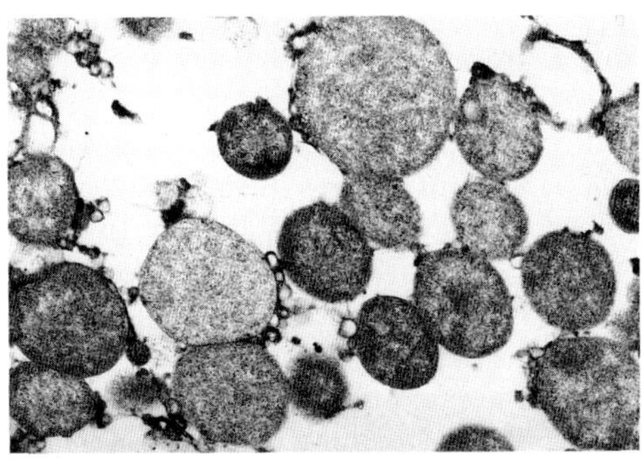

**Fig. 2.** Spheroplasts of *E. coli* after the lysozyme treatment (×9250). (Courtesy of Dr. C. Matsui)

by a transmission electron microscope is, of course, necessary for detailed research in spheroplasts (Hasezawa et al. 1983).

### 3.2 Elimination of Bacteria

After the spheroplast treatment and washing, it is necessary to eliminate the rest of the free spheroplasts. In the case of *Agrobacterium* spheroplasts, the heat treatment described above is effective for sterilization. However, in general, antibiotic treatments are available for elimination of bacteria. On the one hand, many antibiotics, such as tetracycline and gentamycin, cannot eliminate the bacteria completely when they are used at concentrations nontoxic to plant protoplasts. On the other hand, some antibiotics, vancomycin, cefotaxim, carbenicillin etc., and their combination are effective in sterilizing *Agrobacterium* as shown above. Fosfomycin is also effective in the case of *E. coli.*

### 3.3 Selection of Transformants

The transformants have the ability to grow without plant hormones as crown gall tissues. Therefore, this ability has been used for the selection of transformants. This method functions when habituation does not occur. In the case of *Vinca* cells, the frequency of colony formation on hormone-free MS medium was $10^{-3}-10^{-4}$. This frequency was not so different from the frequencies in the cases of tobacco by our results (unpublished data) and those of another group (Hain et al. 1984). They also used *Agrobacterium* spheroplasts harboring the chimeric plasmid consisting of Ti plasmid and a kanamycin-resistant gene for the transformation. Kanamycin resistance is also available as a selectable marker when the plants used for the source of protoplasts have no natural kanamycin-resistant ability, as several plant cells in the Gramineae (unpublished data).

The colonies selected as described above are analyzed. In the case of *Vinca*, opines, opine-synthase activities, and T-DNA could be detected (Hasezawa et al. 1981; Okada et al. 1985).

## 4 Mechanism of Introduction of Spheroplasts into Protoplasts

In the presence of PEG or PVA, *Agrobacterium* spheroplasts can be easily introduced into *Vinca* protoplasts. The process of introduction of spheroplasts into protoplasts was examined by transmission electron microscope (Hasezawa et al. 1983). Various figures in the uptake of spheroplasts were observed as follows: (1) Spheroplasts adhered tightly onto cell membrane of protoplasts (Fig. 3A). (2) Various steps of spheroplasts engulfed by endocytosis (Fig. 3B–D). (3) Spheroplasts lost their integrity and components seemed to collapse in the cytoplasm of protoplast (Fig. 3E). (4) Spheroplasts trapped between two fusing protoplasts (Fig. 3F) (cases of this type were rarely observed), and cases of fusion between spheroplasts and protoplasts could not be found.

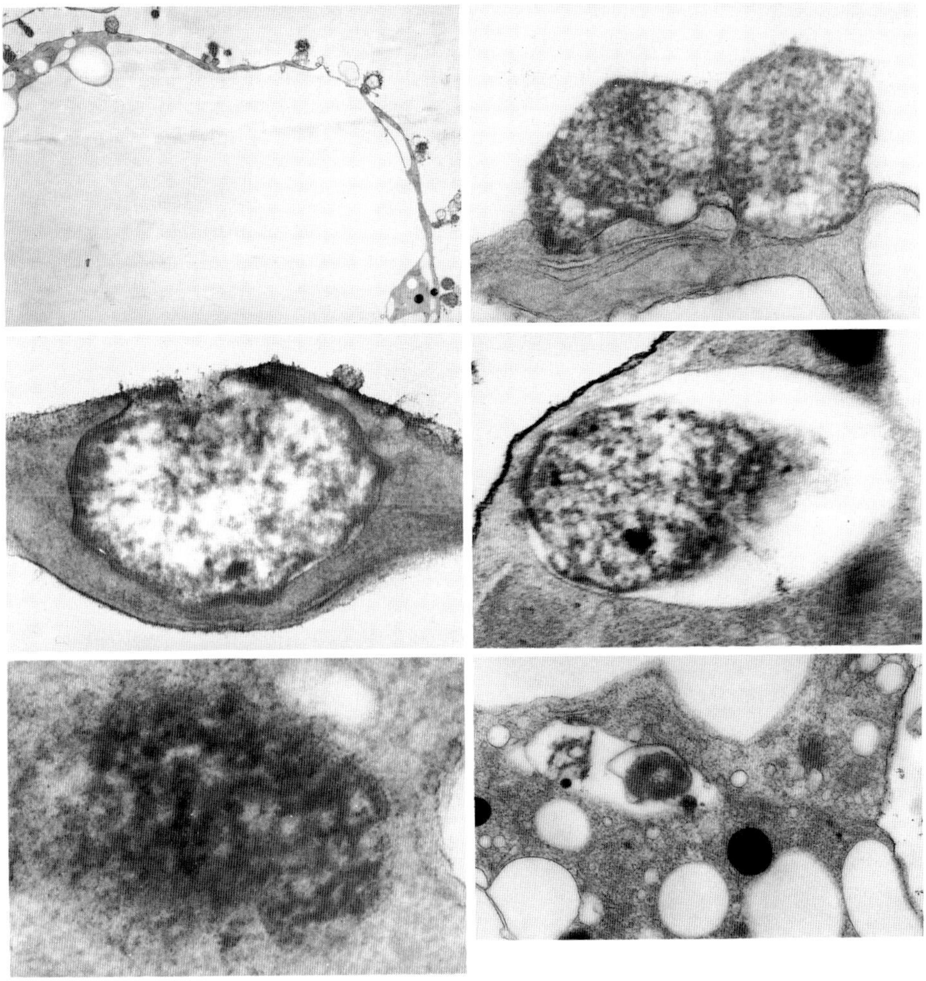


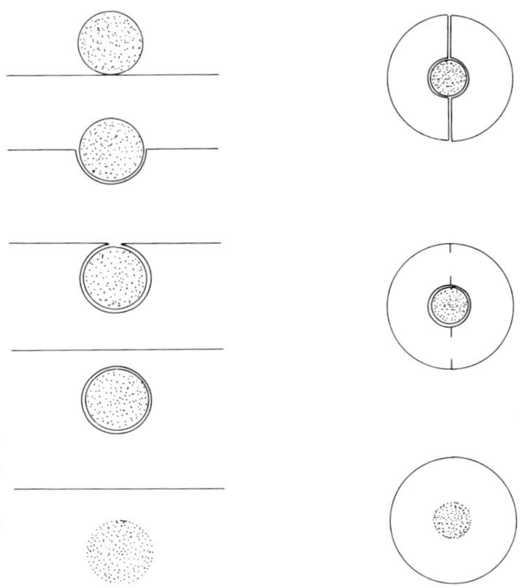


**Fig. 3. A–G** Process of uptake of *Agrobacterium* spheroplasts by PEG treatment. **A** A wide view of a thin-sectioned *Vinca* protoplast with *Agrobacterium* spheroplasts adhered onto the surface of the protoplast (×2125). **B** A magnification of two of these spheroplasts (×29750). **C** A spheroplast at a stage of invagination (×34000). **D** A spheroplast within the plant cytoplasm (×39950). **E** Collapse of spheroplast in the cytoplasm (×51000). **F** A spheroplast entrapped between the fusing protoplasts (×15300). (Courtesy of Dr. C. Matsui). **G** Two types of introduction of spheroplasts into protoplast. Introduction of bacterial spheroplasts into plant protoplasts are roughly classified into two types. The endocytotic process (*left*) and the process entrapped between the fusing protoplasts (*right*). The former is observed predominantly

These figures show two types of introduction of spheroplasts into protoplasts as illustrated in Fig. 3G. After 10 min incubation with fusogen, all the examples mentioned above could be observed. The introduction of spheroplasts induced by fusogen proceeds quite rapidly, and spheroplasts were taken up through the endocytotic process in many cases. The number of spheroplasts incorporated into protoplasts could be counted using 4,6-diamidino-2-phenylindole (DAPI) staining observation (Kuroiwa and Suzuki 1980). The best results were obtained using 20% PEG or 10% PVA at final concentrations for the uptake of spheroplasts. About 10–20 spheroplasts were usually observed in a protoplast (Hasezawa et al. 1983).

## 5 Introduction of *E. coli* Spheroplasts into Protoplasts

Spheroplasts of *E. coli* can be easily prepared by the short-time lysozyme treatment described above. The spheroplasts are also introduced into plant protoplasts by PVA treatment. *E. coli* is the most readily available bacterium for the hosts of recombinant plasmids. The idea illustrated in Fig. 4, that recombinant DNA's cloned and amplified in *E. coli* are transferred directly into plant protoplasts, resulting in transformation of plant cells, is thought to be possible. Actually, *E. coli* spheroplast-mediated transfer of cloned recombinant plasmids has already been carried out in yeast, animal, and plant cells (Table 1). As fusogen for the introduction of relatively large spheroplasts such as those of *E. coli*, PVA was more effec-

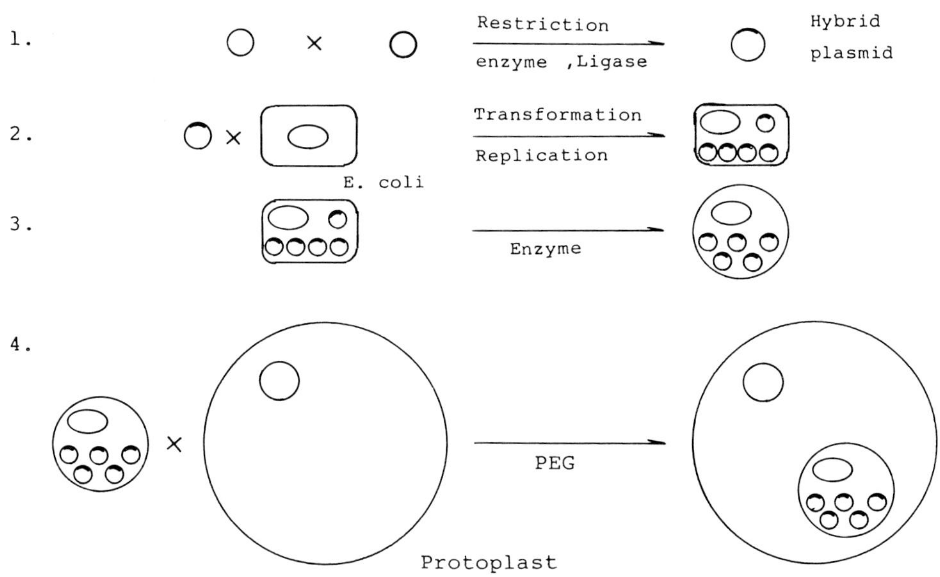


**Fig. 4.** A scheme for the introduction process of cloned recombinant DNA into eukaryotic cells with bacterial spheroplasts. *1* construction of a recombinant DNA in vitro; *2* transfer of the recombinant DNA into bacterial cells; *3* digestion of the cell wall by lysozyme treatment after the amplification of the recombinant DNA; *4* introduction of bacterial spheroplasts carrying the cloned genes into eukaryotic cells. (Syōno and Hasezawa 1985)

**Table 1.** Examples of transformation through the introduction of *E. coli* spheroplasts. (Syōno and Hasezawa 1985)

| Recipient cells | Transferred genes | Reference |
|---|---|---|
| Yeast ($trp^-$) | pBR 313 – TRP 1 | Kingsman et al. (1979) |
| Monkey cell CV-1 | pBR 322 – SV 40 | Schaffner (1980) |
| African green monkey cells ($TK^-$) | pBR 325 – HSV ($TK^+$) | Sandri-Goldin et al. (1981) |
| Monkey cell CV-1<br>Rat cell FR3T3 | pBR 322 – SV 40 (early gene) | Rassoulzadegan et al. (1982) |
| Plant protoplast (*Orychophragmus violaceus*) | pBR 322 – CaMV | Tanaka et al. (1984b) |

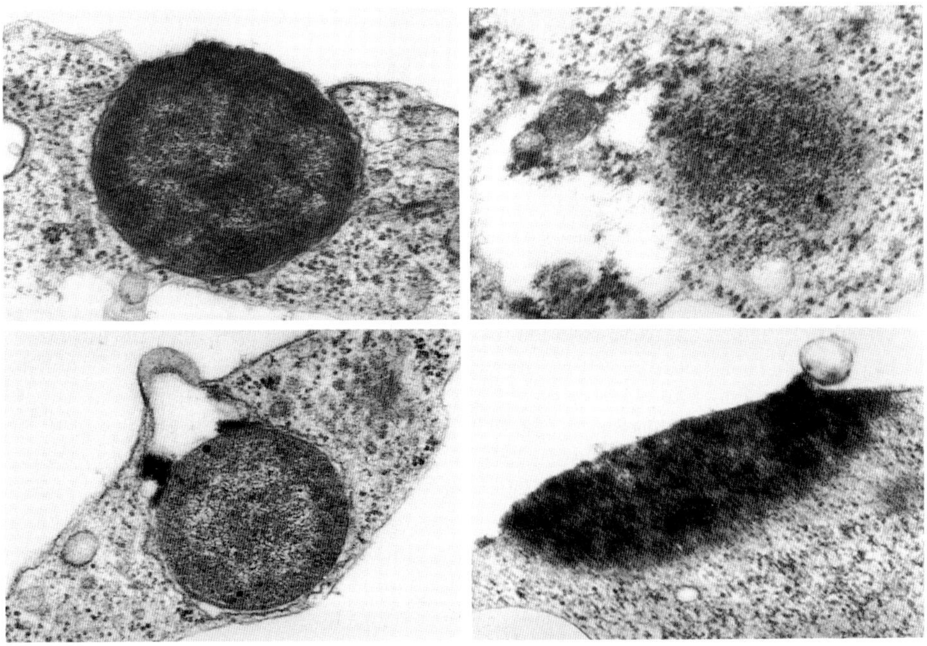

A C B D

**Fig. 5. A–D** Process of uptake of *E. coli* spheroplasts by PVA treatment. **A** A stage of invagination (×25500). **B** A spheroplast enclosed within vesicle in the plant cytoplasm (×26250). **C** Collapse of spheroplast in the plant cytoplasm (×25500). **D** Fusion between *E. coli* spheroplast and *Neurospora crassa* slime cells (×31500). (Courtesy of Dr. C. Matsui)

tive than PEG, which induced extreme dehydration of plant cytoplasm. In the presence of 10% PVA, *E. coli* spheroplasts were introduced by the same mechanism observed in the case of *Agrobacterium* spheroplasts (Matsui et al. 1983). Almost all of them were taken up into plant protoplasts by the endocytotic process (Fig. 5A–C). However, in the case of introduction into *Neurospora crassa* slime cells, a fusion between *E. coli* spheroplast and the plant cytoplasm was observed (Fig. 5D) (Tanaka et al. 1984a).

## 6 Summary and Conclusion

*V. rosea* protoplasts and *A. tumefaciens* spheroplasts harboring Ti plasmid were mixed and treated with PEG or PVA, which facilitated the introduction of spheroplasts into plant protoplasts endocytotically. This treatment can be applied to the introduction of other bacteria and their spheroplasts into plant protoplasts. After the treatment at 40 °C for 4 days or antibiotics such as carbenicillin, vancomycin, and cefotaxim, *A. tumefaciens* was found to be completely eliminated. Fosfomycin was effective in the case of *E. coli.* Among treated protoplasts, some colonies grew on hormone-free medium. The frequency of colony formation was $10^{-3}-10^{-4}$. In these putative transformants, opines, opine synthase activities, and T-DNA were detected. These results showed that *V. rosea* protoplasts treated with *A. tumefaciens* spheroplasts were truly transformed by the introduction of a Ti plasmid into protoplasts mediated by the spheroplasts.

The transformation system depending on the ability of infection of *A. tumefaciens* cannot be applied to Gramineae plants resistant to the infection of the bacteria. In spheroplast method, the introduction processes of genetic materials are accomplished physically and independent of host range of bacteria. Actually, we have tried and succeeded in the transformation of rice protoplasts by spheroplast method (Baba et al. 1986; see Chap. VI, this Vol.). Thus, bacterial spheroplasts can be utilized as desirable vehicles for the introduction of genetic information into plant cells. Besides our work, such trials by spheroplast method have been attempted by several groups in plant cells (Hain et al. 1984; Freeman et al. 1984; Millman and Lurquin 1985).

The spheroplast method enables us to design various kinds of experiments of gene transfer. In this connection, the gene transports have been accomplished by the fusion between bacterial spheroplasts and eukaryotic cells such as yeast and animal cells (see Table 1). Bacteria-to-bacteria fusion has also been attempted. In addition to uses in cell technological and genetic engineering studies, spheroplasts are thought to be useful in various biochemical studies, many successful results having already been obtained (see the review by Syōno and Hasezawa 1985). The spheroplast method has many advantages and is practically useful in various kinds of investigations. On the other hand, it has a few difficulties in use, such as the elimination step of bacteria. We are now grappling with such technical problems for the purpose of making this method a better tool for various studies.

## References

Baba A, Hasezawa S, Syōno K (1986) Cultivation of rice protoplasts and their transformation mediated by *Agrobacterium* spheroplasts. Plant Cell Physiol 27:463–471

Birdsell DC, Cota-Robles EH (1967) Production and ultrastructure of lysozyme and ethylenediaminetetraacetate-lysozyme spheroplasts of *Escherichia coli.* J Bacteriol 93:427–437

Chilton M-D, Currier TC, Ferrand SK, Bendich AJ, Gordon MP, Nester EW (1974) *Agrobacterium tumefaciens* DNA and PS8 bacteriophage DNA not detected in crown gall tumors. Proc Natl Acad Sci USA 71:3672–3676

Fraser D, Jerrel EA (1953) The amino acid composition of T3 bacteriophage. J Biol Chem 205:291–295

Freeman JP, Draper J, Davey MR, Cocking EC, Gartland KMA, Hardring K, Pental D (1984) A comparison of methods for plasmid delivery into plant protoplasts. Plant Cell Physiol 25:1353–1365

Hain R, Steinbiß H-H, Shell J (1984) Fusion of *Agrobacterium* and *E. coli* spheroplasts with *Nicotiana tabacum* protoplasts. Direct gene transfer from microorganism to higher plant. Plant Cell Rep 3:60–64

Hasezawa S, Nagata T, Syōno K (1981) Transformation of *Vinca* protoplasts mediated by *Agrobacterium* spheroplasts. Mol Gen Genet 182:206–210

Hasezawa S, Matsui C, Nagata T, Syōno K (1985) Cytological study of the introduction of *Agrobacterium tumefaciens* spheroplasts into *Vinca rosea* protoplasts. Can J Bot 61:1052–1057

Kao KN, Michayluk MR (1974) A method for high-frequency intergeneric fusion of plant protoplasts. Planta 115:355–367

Kingsman AJ, Clarke L, Mortimer RK, Carbon J (1979) Replication in *Saccharomyces cerevisiae* of plasmid pBR 313 carrying DNA from the yeast trp 1 region. Gene 12:139–146

Kuroiwa T, Suzuki T (1980) An improved method for the demonstration of in situ chloroplast nuclei in higher plants. Cell Struct Funct 5:195–197

Marton L, Wullems GJ, Molendijk L, Schilperoort RA (1979) In vitro transformation of cultured cells from *Nicotiana tabacum* by *Agrobacterium tumefaciens*. Nature 277:129–131

Matsui C, Hasezawa S, Tanaka N, Syōno K (1983) Introduction of *Escherichia coli* cells and spheroplasts into *Vinca* protoplasts. Plant Cell Rep 2:30–32

Millman RA, Lurquin PF (1983) Study of plant protoplast-bacterial spheroplast and cell interaction by flow cytometry. J Plant Physiol 117:431–440

Murashige T, Skoog F (1962) A revised medium for rapid growth and bioassays with tobacco tissue cultures. Physiol Plant 15:473–497

Nagata T (1978) A novel cell-fusion method of protoplasts by polyvinyl alcohol. Naturwissenschaften 65:263–264

Okada K, Hasezawa S, Syōno K, Nagata T (1985) Further evidence for the transformation of *Vinca rosea* protoplasts by *Agrobacterium tumefaciens* spheroplasts. Plant Cell Rep 4:133–136

Otten LABM, Schilperoort RA (1978) A rapid microscale method for the detection of lysopine dehydrogenase activities. Biochim Biophys Acta 525:497–500

Rassoulzadegan M, Binétruy B, Cuzin F (1982) High frequency of gene transfer after fusion between bacteria and eukaryotic cells. Nature 295:257–259

Sandri-Goldin RM, Goldin AL, Levine M, Glorioso JC (1981) High frequency transfer of cloned herpes simplex virus type 1 sequences to mammalian cells by protoplast fusion. Mol Cell Biol 8:743–752

Schaffner W (1980) Direct transfer of cloned genes from bacteria to mammalian cells. Proc Natl Acad Sci USA 77:2163–2167

Syōno K, Hasezawa S (1985) Spheroplast – As a tool of gene transfer. In: Cheremisinoff PN, Ouellette RP (eds) Biotechnology – Application and research. Technomic Pub, Lancaster: 463–476

Takebe I, Otsuki Y, Aoki S (1968) Isolation of tobacco mesophyll cells in intact and active state. Plant Cell Physiol 9:115–124

Tanaka N, Fukunaga Y, Hasezawa S, Syōno K, Matsui C (1984a) Endocytic uptake of *Escherichia coli* spheroplasts by *Neurospora crassa* slime cells. – An ultrastructural study. Appl Microbial Biotechnol 19:296–299

Tanaka N, Ikegami M, Hohn T, Matsui C, Watanabe I (1984b) *E. coli* spheroplast-mediated transfer of cloned cauliflower mosaic virus DNA into plant protoplasts. Mol Gen Genet 195:378–380

# XI Transformation in *Hyoscyamus*

D. HANOLD[1]

## 1 Introduction

### 1.1 The Need for Transformation Studies

Until recently, for the improvement of agricultural crops the traditional way of breeding and selection has been the only choice. However, with the development of new biotechnology methods (see Bajaj 1986), the alternative of directed genetic modification has become available. If used wisely, such manipulation techniques could become a powerful tool to solve agricultural problems. In vitro transformation is one of these techniques, and it has several advantages over traditional plant breeding. If genes for specific desirable properties of the plant can be isolated, they can be directly introduced into plant cells in one step, and whole plants possessing the new property can be regenerated, rather than having to go through the traditional tedious and time-consuming process of crossing and selection necessary so far (Cocking 1985). With the help of a suitable vector, desirable genes present in one species of plants – or even bacteria or animals – could be transferred to completely unrelated species with which natural exchange of genetic information cannot be achieved, making it possible to cross genetic borders imposed by nature (Wullems and Schilperoort 1985).

Therefore, transformation can provide a faster, more accurate, and much more versatile tool for the genetic improvement of crop plants than was at our disposal in the past. It is, however, necessary and important to explore this area thoroughly in order to assess the potential and limitations of this method so that it may be used safely for the benefit of mankind.

### 1.2 Distribution and Importance of *Hyoscyamus*

Plants of the genus *Hyoscyamus* – commonly known as henbanes – are distributed throughout Europe, the Mediterranean countries, North Africa, and into southwestern and central Asia (Morley 1970), and Australia (Black 1957). About 20 species are known. Like other Solanaceous plants, they produce a range of alkaloids, the major ones being hyoscyamine and scopolamine. During the Middle

[1] Department of Plant Pathology, Waite Agricultural Research Institute, Glen Osmond, South Australia 5064

Biotechnology in Agriculture and Forestry, Vol. 9
Plant Protoplasts and Genetic Engineering II (ed. by Y. P. S. Bajaj)

Ages *H. niger*—Black Henbane – was a hallucinogenic ingredient of sorcerers' potions and poisons. Because of its anesthetic and sedative properties, henbane was also collected and cultured for medicinal purposes in Europe, and during World War II even in the USA (Hill 1952), but now it has mostly been replaced by other pharmaceuticals. *H. muticus* – Egyptian Henbane – is still being smoked in North Africa and India for intoxication (Hill 1952). The names "henbane" and "*Hyoscyamus*" (from Greek "hog bean") allude to the belief that these plants poison animals (Black 1957). Current interest centers around *Hyoscyamus* as a model for the plant sciences (see also Strauss 1989).

### 1.3 A Model for Genetic Engineering

*H. muticus* provides a good system to investigate and establish techniques for the improvement of commercially important crop plants. Protoplasts can be isolated from glasshouse plants, sterile shoot cultures, and cell suspension cultures, and they can be cultured and regenerated into whole plants (Lörz et al. 1979). Protoplasts, especially from suspension cultures, provide large, uniform populations of synchronized cells. These single cells can be transformed, mutagenized, or otherwise manipulated, and the effects and efficiency of these treatments can be observed on resulting whole plants representing pure genetic clones of one such cell (Cocking 1985).

A further advantage of *Hyoscyamus* is that haploid plants can be obtained from anther culture (Wernicke et al. 1979). Haploidy can easily be verified by counting the number of chloroplasts in the stomatal guard cells of the lower leaf epidermis – five to six for haploid, eight to ten for diploid plants (Wernicke et al. 1979). Genetic manipulation of haploid protoplasts and subsequent diploidization of the tissue using colchicine or by spontaneous diploidization in tissue culture before regeneration yields plants that are homozygous for the altered gene(s). This is particularly useful if the genetic property under investigation is recessive.

The value of the *H. muticus* system can be exemplified by its use in some of the earliest experiments to systematically mutagenize protoplasts (Strauss et al. 1981; Gebhardt et al. 1981). Haploid *H. muticus* protoplasts were treated with the mutagen nitrosoguanidine, and it was possible to isolate nitrate reductase negative, amino acid auxotrophic, and temperature-sensitive mutants. Some of these mutants were used by Potrykus et al. (1982) and Jia et al. (1983) for complementation studies after protoplast fusions. *H. muticus* has also been used for establishing a system for protoplast transformation by *Agrobacterium tumefaciens* (Hanold 1983) and this is described here.

## 2 In Vitro Transformation of Protoplast-Derived *H. muticus* Cells by *A. tumefaciens*

### 2.1 General Outline

The Ti-plasmids from *A. tumefaciens* (Smith and Townsend) Conn can be used as vectors for genetic modification of plants (Nester and Kosuge 1981; Wullems and Schilperoort 1985). To obtain a sufficiently high number of transformed lines after gene transfer, an in vitro transformation system for single cells is desirable. The presence of opines and/or T-DNA in the plant cells can be used as proof of a successful transformation. In a method previously established for protoplast-derived cells of *Nicotiana tabacum* L., an octopine-inducing strain of *A. tumefaciens* was used, and transformation was confirmed by detection of octopine. A modification of this technique for use with *Hyoscyamus muticus* L. is described here (Hanold 1983).

## 2.2 Experimental

*1. Agrobacterium Strains and Culture.* Strain C58 was a gift of Professors J. Schell and M. van Montagu (University of Ghent, Belgium). Strains LBA4058 and LBA4011 were obtained from Professor R. Schilperoort (University of Leiden, The Netherlands). Strain LBA4058 contains pTiT37 *tra*$^c$ in a cured Ach5 bacterium, while LBA4011 is Ach5 cured of its Ti plasmid (Klapwijk et al. 1978). Bacteria were grown in 10 g/l Tryptone (Difco Chemicals), 5 gl/l yeast extract, 8 g/l NaCl, with or without agar at 25–29 °C. The cell density of bacterial cultures was estimated from measurement of absorption at 660 nm.

*2. Protoplast Isolation and Culture of Callus and Shoots.* The method for protoplast preparation was modified after Lörz et al. (1979). All solutions except agar containing media were filter-sterilized. *H. muticus* plants raised from seed in a glasshouse were kept in growth chambers at 95% relative humidity for 1 week prior to leaf harvest at 2500 lx for 15 h, 22 °C, in moist but not wet soil. Only young plants prior to floral initiation were used. Young leaves of main and side shoots were washed and sterilized for 10 min in 0.01% (w/v) $HgCl_2$ containing a few drops of the detergent Tween 80 (Serva Chemicals), then washed three times in sterile distilled water. They were cut as finely as possible with a very sharp blade in osmoticum consisting of 3 vol of 0.4 M mannitol and 1 vol of 0.16 M $CaCl_2$, 0.5% (w/v) [2(N-morpholino)] ethane sulfonic acid (MES, Sigma Chemicals), pH adjusted to 5.7 with KOH. The slices were vacuum infiltrated in fresh osmoticum and then incubated for 1–2 h at 30 °C in 1% (w/v) cellulase Onozuka R-10 (Yakult Chemicals), 1% (w/v) Pektinol (Roehm Pharma), 0.2% (w/v) hemicellulase (Sigma Chemicals) dissolved in the osmoticum. Undigested tissue was removed by filtration over a 100 µm steel net. During centrifugation at about 50 g for 10 min the protoplast population separated into three subpopulations according to their buoyant density: (a) floating protoplasts with a low buoyant density at the air-water interface, (b) protoplasts with the buoyant density of the osmoticum, which were still suspended in the liquid and (c) a sediment consisting of protoplasts with a higher buoyant density and cell debris. Fraction b was diluted 1:1 with 0.16 M $CaCl_2$ plus 0.5% (w/v) MES (pH 5.7) to reduce the buoyant density of the osmoticum, and centrifuged as above to sediment protoplasts. Fraction c was purified by underlayering it with 22% (w/v) sucrose, 0.5% (w/v) MES (pH 5.7) and centrifuged at 150 g. Treating the three fractions separately turned out to be least harmful for the protoplasts. The three fractions were combined, washed two times with the osmoticum, and suspended in DPD medium (Durand et al. 1973) containing 5 mg/l p-chlorophenoxy acetic acid (pCPA) and 2.5 mg/l kinetin, 2% (w/v) sucrose, 0.3 M mannitol (pH 5.7), to give a final cell density of approximately $5 \times 10^4$/ml. The cultures were kept in the dark at 12 °C overnight, then in the dark at 25 °C for 2 or 3 days, followed by light of 500 lx at 25 °C. On the average about 10% of the isolated protoplasts divided. Colonies were first plated on the DPD medium described above but without mannitol and solidified with 0.8% (w/v) agar. Later the medium was switched to LS (Linsmaier and Skoog 1965) with the same additions. Shoots regenerating were separated from the calli and cultured further as described by Wernicke and Thomas (1980). Non-rooting transformed shoots were grafted onto wild-type plants by inserting them into a vertical pocket cut along the side of the stem.

*3. Infection Procedure.* a) *A. tumefaciens strain C58.* A single colony of strain C58 was suspended in broth and grown for 3 days as described above. The protoplast-derived cells were inoculated with the stationary phase bacterial culture 7 – 10 days after isolation, when they had started to divide. The ratio of bacteria per protoplast originally was about 100. The bacteria were added as a suspension to the cells in culture dishes. After 2 days, a concentrated sterile solution of carbenicillin (Pyopen, Beecham Laboratories) was added to the cell cultures to give a final concentration of 1 mg/ml. When the cell clusters reached the 50 – 100 cell stage (after 3 – 4 weeks), they were washed once with medium and plated in a hormone-free soft-agar layer (about 0.4%, w/v) containing 1 mg/ml carbenicillin on top of carbenicillin-free, hormone-free medium. The initial population density was not diluted. The culture conditions during the infection procedure were identical to those described above for the untreated protoplasts. The calli were treated further as mentioned above omitting hormones from the medium. Control cultures received the same treatment, however, no bacteria were added.

b) *A. tumefaciens strain LBA4058.* In this series of experiments a minor modification of the method described by Marton et al. (1979) was used with exponentially growing cultures of *A. tumefaciens* strain LBA4058. A single colony of bacteria was suspended and grown for 48 h, diluted 1:10 and grown for another 24 h, then washed once in physiological salt solution, resuspended in physiological salt, and added to the protoplasts at a ratio of 400 bacteria per protoplast. The cells were diluted to 1/15 of the initial population density when they were plated on agar medium. Two controls were used: incubation without bacteria, and incubation with the avirulent *A. tumefaciens* strain LBA4011.

*4. Nopaline Assay.* After three to five passages of the calli on hormone-free medium, the presence of nopaline after an incubation allowing nopaline dehydrogenase activity in the tissue was assayed by the method of Otten and Schilperoort (1978) with two modifications. Mercaptoethanol (0.1%, v/v) was added to the incubation mixture to protect the nopaline dehydrogenase. Because several compounds in *Hyoscyamus* interfere with the detection of opines, electrophoresis was carried out in acetic acid/formic acid/water (12:15:63, pH 0.8), which gave a good separation of contaminating fluorescent compounds from the nopaline spots. Pure nopaline used as a standard was a gift from E. Messens (University of Ghent, Belgium).

*5. Hybridization Studies.* Hin dIII-digestions of total plant nucleic acid extracts were separated on 0.8% agarose gels, transferred to nitrocellulose filters (Southern 1975), and hybridized against nick-translated T-region probe of pTi C58 (Maniatis et al. 1982). A cosmid clone of pTi C58 T-region was kindly supplied by Drs. T. and B. Hohn, Friedrich Miescher Institute, Basel.

## 2.3 Results and Discussion

Three nopaline-positive calli out of 15 hormone-independent clones were recovered from the first set of two experiments using *A. tumefaciens* strain C58 (Fig. 1). One of these clones lost the ability to produce nopaline after 6 months in culture, but showed no conspicuous changes in its pattern of growth. From the second set of three experiments using strain LBA4058, out of 7104 calli growing without exogenously supplied hormones 84 calli were assayed; none was nopaline-positive.

The three cell lines which initially produced nopaline were not teratomatous in appearance; rather, they grew as unorganized calli. No roots and only very few shoots were formed. It was not possible to obtain shoots for grafting onto wild-type plants to continue plant regeneration. On the other hand, many of the hormone-independent, nopaline-negative lines were highly morphogenic, and shoots could be isolated for grafting. Recloning of the transformed cell lines was tried, but it proved impossible to isolate either protoplasts or single cells from the callus material.

In the transformation experiments with LBA4058, the number of hormone-independent, nopaline-negative calli was far above the normal frequency of habituation (i.e., spontaneous change to the hormone-independent state) in *H. muticus* (Tables 1 and 2) which was below $10^{-5}$.

Most of the calli obtained in the control cultures, both untreated and treated with LBA4011, died very soon after plating them without hormones. Another 10% of the calli ceased to proliferate after two or three passages on hormone-free medium. Only two cell lines treated with LBA4011 continued growing and were maintained and treated like the other hormone-independent calli derived from incubation with virulent bacteria.

Protoplast-derived cells ($2 \times 10^5$) were treated with *A. tumefaciens* strain C58 in two independent experiments. The experiments yielded two and one transformed clone(s), respectively, and a few hormone-independent, nopaline-negative calli. In both experiments no controls grew without hormones. Three independent

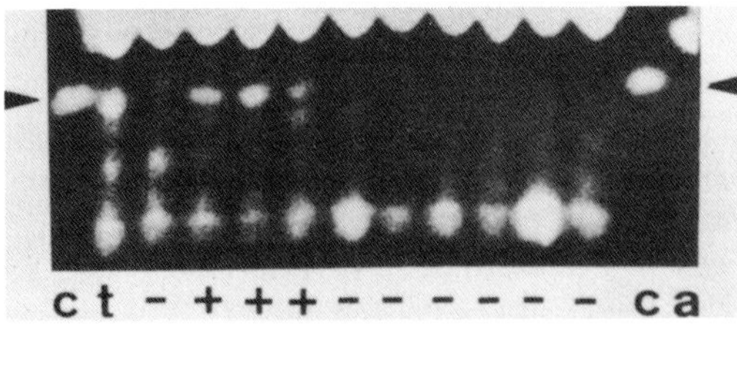


**Fig. 1.** Electrophoresis for nopaline test. (Hanold 1983). +, nopaline present (third from the left ceased nopaline production after 6 months); −, no nopaline detected; *c* control, pure nopaline; *t* tumor extract; *a* pure arginine

**Table 1.** Survey of the transformation experiments. (Hanold 1983)

| Experiment | Treatment | Cells treated | Hormone-independent clones | Frequency of hormone independence | Nopaline[a]-positive clones assayed |
|---|---|---|---|---|---|
| 1 | C58 | $10^5$ | 6 | $6\times10^{-5}$ | |
| 1 | untreated | $5\times10^3$ | 0 | – | 3/15 |
| 2 | C58 | $10^5$ | 9 | $9\times10^{-5}$ | |
| 2 | untreated | $5\times10^3$ | 0 | – | |
| 3 | LBA4058 | $8\times10^4$ | 1450 | $1.8\times10^{-2}$ | |
| 3 | LBA4011 | $3\times10^4$ | 0 | – | |
| 3 | untreated | $3\times10^4$ | 0 | – | |
| 4 | LBA4058 | $1.5\times10^5$ | 1337 | $9\times10^{-3}$ | |
| 4 | LBA4011 | $5\times10^4$ | 1 | $1.5\times10^{-5}$ | 0/84 |
| 4 | untreated | $5\times10^4$ | 0 | – | |
| 5 | LBA4058 | $1.9\times10^5$ | 4317 | $2.3\times10^{-2}$ | |
| 5 | LBA4011 | $5\times10^4$ | 1 | $1.5\times10^{-5}$ | |
| 5 | untreated | $5\times10^4$ | 0 | – | |

[a] Only lines able to grow without added hormones were tested

**Table 2.** Summarized results. (Hanold 1983)

| | Experiments 1 and 2 (C58) | Experiments 3 to 5 (LBA4058) |
|---|---|---|
| Frequency of hormone independence | $7.5\times10^{-5}$ | $1.7\times10^{-2}$ |
| Increase in frequency of hormone-independent lines approximately | 7-fold | 2000-fold |
| Nopaline-positive lines in calli checked | 20% | <1.2% |
| Frequency of nopaline-positive lines | $1.5\times10^{-5}$ | $<2\times10^{-6}$ |

experiments were carried out with strain LBA4058, in which a total of $4.2\times10^5$ cells were treated. Both the high frequency of hormone-independent colonies without detectable nopaline, and the behavior of controls in all three experiments was similar (Tables 1, 2).

Preliminary hybridization experiments between isolated callus DNA and T-region probe have confirmed the transformation of all three initially nopaline-positive and one of the nopaline-negative cell lines (Fig. 2, only two nopaline-positive lines shown). Three of the hormone-independent, nopaline-negative cell lines derived from incubation with strain LBA4058 have been tested so far, and did not show hybridization to T-region. These hybridization results have yet to be consolidated and extended.

The *A. tumefaciens* strain C58 gave three genetically transformed cell lines using the presence of nopaline and hybridization as criteria, and one which showed weak hybridization only (Fig. 2, lane 2). Strain LBA4058 showed no transforming activity in *H. muticus* despite the fact that more cells were treated than with C58. In contrast, C58 treatment of *N. tabacum* cells yields less transformants than

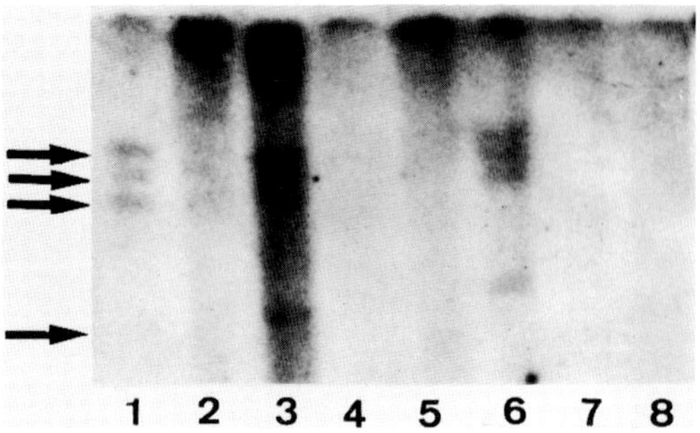


**Fig. 2.** Hybridization of plant DNA digests against T-region probe. *1* positive control pTi C58; *2* nop−(from incubation with strain C58); *3* nop+(from C58); *4* nop−(from LBA4058); *5* negative control (from LBA 4011); *6* nop+, later nop−(from C58); *7, 8* nop−(from LBA4058)

treatment with LBA4058 (G. Wullems pers. commun.). Statistical analysis of the data obtained from the second set of experiments using strain LBA4058 shows with a probability of 95% less than 4% of the hormone-independent calli contain nopaline. Amongst the hormone-independent, nopaline-negative lines one cannot exclude the existence of partial or abortive transformants which might lack, for example, information for nopaline production. However, there could also be habituated cell lines, because even if habituation occurred rarely under normal culture conditions it could be enhanced by the transformation procedure. If this were the case, then a high frequency of hormone independence should appear as well after the treatment with the avirulent strain LBA4011 – this was not observed. Another explanation could be that these cell lines are mosaic cultures with a very low percentage of transformed cells, so that it was not possible to detect the nopaline with the method applied. One such case seems to be the line assayed in lane 2, Fig. 2. Further clarification of this point can be expected by extended hybridization studies between isolated callus DNA and T-region probe.

It is also worth mentioning that the procedure leading to transformation in this work was different from the one successful with *N. tabacum* (Marton et al. 1979). In the experiments yielding transformants stationary phase bacterial cultures were used instead of exponentially growing ones. No dilution or change of culture medium was made until the time of plating, i.e., the hormone concentrations were reduced in one step. In the unsuccessful experiments the hormones were reduced gradually.

## 3 Summary and Conclusions

Transformation of plant cells is a new powerful tool for genetic improvement of plants. To gain further knowledge about its use, studies on in vitro transformation of protoplasts have been undertaken using *Hyoscyamus muticus* as a model species.

Protoplasts of *H. muticus* were treated with *A. tumefaciens* strain C58 in vitro and four transformed cell lines were established. Three of these cell lines initially produced nopaline, although one lost this ability after 6 months in culture. A highly increased frequency of hormone-independent growth was observed after incubation with *A. tumefaciens* strain LBA4058. No transformants, however, could be detected.

The results demonstrate that in vitro transformation with *A. tumefaciens* is successful for *H. muticus*. A method previously designed for transformation of *N. tabacum* protoplasts had to be adapted to suit the new species. It is therefore likely that this method with the respective changes required can also be used to transform other plant species.

## References

Bajaj YPS (1986) Biotechnology in agriculture and forestry 2. Crops I. Springer, Berlin Heidelberg New York Tokyo

Black J (1957) Flora of South Australia. Part IV. K.M. Stevenson, Government Printer, Adelaide

Cocking EC (1985) Applications of protoplast technology. In: Pilet P (ed) The Physiological Properties of Plant Protoplasts. Springer, Berlin Heidelberg New York Tokyo, pp 6–11

Durand J, Potrykus I, Donn G (1973) Plantes issues de protoplastes de pétunia. Z Pflanzenphysiol 69:26–34

Gebhardt C, Schnebli V, King P (1981) Isolation of biochemical mutants using haploid mesophyll protoplasts of *Hyoscyamus muticus*. II. Auxotrophic and temperature-sensitive clones. Planta 153:81–89

Hanold D (1983) In vitro transformation of protoplast-derived *Hyoscyamus muticus* cells by *Agrobacterium tumefaciens*. Plant Sci Lett 30:177–183

Hill A (1952) Economic botany. McGraw-Hill, London

Jia J, Potrykus I, Lázár G, Saul M (1983) *Hyoscyamus-Nicotiana* fusion hybrids selected via auxotroph complementation and verified by species-specific DNA hybridisation. In: Potrykus I, Harms C, Hinnen A, Hütter R, King P, Shillito RA (eds) Protoplasts 1983, Poster Proc 6th Int Protoplast Symp Basel, Birkhäuser, pp 110–111

Klapwijk P, Scheuldermann T, Schilperoort R (1978) Coordinated regulation of octopine degradation in *Agrobacterium tumefaciens*: evidence for a common regulatory gene and separate operons. J Bacteriol 136:775–785

Linsmaier E, Skoog F (1965) Organic growth factor requirement of tobacco tissue culture. Physiol Plant 18:100–127

Lörz H, Wernicke W, Potrykus I (1979) Culture and plant regeneration of *Hyoscyamus* protoplasts. Planta Med 36:21–29

Maniatis T, Fritsch E, Sambrook J (1982) Molecular cloning, a laboratory manual. Cold Spring Harbor, New York

Marton L, Wullems G, Molendijk L, Schilperoort RA (1979) In vitro transformation of cultured cells from *Nicotiana tabacum* by *Agrobacterium tumefaciens*. Nature 277:129–131

Morley B (1970) Wild flowers of the world. Rainbird Reference Books, p 38

Nester E, Kosuge T (1981) Plasmids specifying plant hyperplasias. Annu Rev Microbiol 35:531–565

Otten L, Schilperoort RA (1978) A rapid microscale method for the detection of lysopine and nopaline dehydrogenase activity. Biochem Biophys Acta 527:497–500

Potrykus I, Shillito R, Jia J, Lázár G (1982) Auxotroph complementation via protoplast fusion in *Hyoscyamus muticus* and *Nicotiana tabacum*. In: Lurquin P, Kleinhof A (eds) Genetic engineering in eucaryotes. Proc NATO Advanced Study Institute, Pullman 1982, Plenum, New York

Southern E (1975) Detection of specific sequences among DNA fragments separated by gel electrophoresis. J Mol Biol 98:503–518

Strauss A (1989) *Hyoscyamus* species: In vitro culture and the production of tropane alkaloids. In: Bajaj YPS (ed) Biotechnology in agriculture and forestry 7. Medicinal and aromatic plants II. Springer, Berlin Heidelberg New York Tokyo, pp 286–314

Strauss A, Bucher F, King P (1981) Isolation of biochemical mutants using haploid mesophyll protoplasts of *Hyoscyamus muticus*. I. A nitrate non-utilising clone. Planta 153:75–80

Wernicke W, Thomas E (1980) Studies on morphogenesis from isolated plant protoplasts: shoot formation from mesophyll protoplasts of *Hyoscyamus muticus* and *Nicotiana tabacum*. Plant Sci Lett 17:401–407

Wernicke W, Lörz H, Thomas E (1979) Plant regeneration from leaf protoplasts of haploid *H. muticus* L. produced via anther culture. Plant Sci Lett 15:239–249

Wullems G, Schilperoort RA (1985) Plant protoplast transformation by *Agrobacterium* in relation to plant biotechnology. In: Fowke L, Constable F (eds) Plant Protoplasts, CRC Press, Boca Raton, Florida, USA, pp 205–229

# XII Genetic Manipulation in *Brassica*

F. TRAIL[1], C. RICHARDS[2], and F.-S. WU[2]

## 1 Introduction

Members of the family Cruciferae are sources of condiments, fodder, spices, and vegetables. Their contributions to human food supply range from the edible roots of the radish plant (*Raphanus sativa*) to the seeds of *Brassica napus*, which provide 10% of the world's vegetable oil (Austin 1986). The six species of *Brassica* commonly cultivated worldwide are: *B. campestris* L. (turnip, turnip rape, chinese cabbage), *B. carinata* Braun (Abyssinian mustard), *B. juncea* Coss. (leaf mustard), *B. napus* (oil rape, canola, rutabaga), *B. nigra* Koch (black mustard), and *B. oleracea* L. (cabbage, cauliflower, broccoli, brussel sprouts). The family Cruciferae also includes the genera *Sinapis* (mustard), *Eruca* (salad greens) and *Crambe* (source for industrial oils). The less familiar *Moricandia* has been the object of recent studies as a physiological intermediate to C3 and C4 plants (Hunt et al. 1987). *Arabidopsis* is a common weed, but an important model plant for genetic and developmental studies because of its small genome size and rapid life-cycle (Meyerowitz and Pruitt 1985).

Traditional breeding of *Brassica* spp. has centered around the development of polyploid crops through interspecific and intergeneric crosses. The combined works of Morinaga (1934) and U (1935) revealed the genetic relationships of the six economically important species: *B. juncea* ($2n = 36$, AABB), *B. napus* ($2n = 38$, AACC), and *B. carinata* ($2n = 34$, BBCC) are the natural amphidiploid hybrids of combinations of *B. campestris* ($2n = 20$, AA), *B. nigra* ($2n = 16$, BB) and *B. oleracea* ($2n = 18$, CC). Knowledge of this relationship led to attempts to improve *B. napus* by exploiting a wider range of genetic material through the creation of artificial amphidiploids. This approach has now been used for intergeneric crosses in most of these species, resulting in the transfer of clubroot (Lammerick 1970; Chiang and Crete 1983) and blackleg resistance (Roy 1984), resistance to triazine herbicides (Beversdorf et al. 1980), cytoplasmic male sterility (Bannerot et al. 1974), and self-incompatibility (Mackay 1977). However, these wide crosses often result in a higher incidence of sterile offspring (Olsson 1960), chlorophyll deficiency (Bannerot et al. 1977), reduced hybrid seed production (Matsuzawa and Sarashima 1986), and aborted embryos (Bannerot et al. 1974).

In the last two decades, cytoplasmically inherited traits have become increasingly important to *Brassica* breeding programs due to the rise in commercial seed

[1] Dept. of Plant Pathology, Cornell University, Geneva, New York 14456-0462, USA
[2] Zoecon Research Institute, Sandoz Crop Protection Corporation, Palo Alto, California, USA

Biotechnology in Agriculture and Forestry, Vol. 9
Plant Protoplasts and Genetic Engineering II (ed. by Y.P.S. Bajaj)

production. Cytoplasmic male sterility (CMS) is useful for hybrid seed production in crops with unstable or limited numbers of self-incompatibility genes, as it eliminates the tedious and unreliable task of hand pollination (Ogura 1968). In *Brassica* spp., the CMS trait is associated with the mitochondrial genome (Pelletier et al. 1983). Several cytoplasms harboring male sterility are available, including the *ogu* cytoplasm, introduced from *Raphanus sativa* (Bannerot et al. 1974). Unfortunately, the chloroplasts of this cytoplasm do not function normally in the presence of a *Brassica* nucleus. Symptoms of this incompatibility include stunting and chlorosis when plants are exposed to low temperatures, a condition referred to as cold-chlorosis (Bannerot et al. 1977). Due to the maternal inheritance of cytoplasmically encoded traits, traditional breeding methods have not been able to introduce healthy chloroplasts into the *ogu* cytoplasm.

New methods of gene transfer and improved methods of plant regeneration are being developed which may overcome barriers that limit the applications of traditional breeding programs. This chapter will summarize recent work on the genetic manipulation of *Brassica* spp. that use cellular and subcellular manipulation to introduce new traits. Advances in protoplast fusion research of *Brassica* spp. will be described, followed by a review of studies in genetic transformation.

## 2 Protoplast Fusion

The formation of protoplasts has made fusion, electroporation, and nuclear injection (see Sect. 3.2) of plant cells possible. Individual plant cells are totipotent, thus they have the potential to regenerate new plants. Fusion of protoplasts of two different cell lines, followed by regeneration of hybrid plants, can be used to combine cytoplasms and nuclei. The technique has the potential to extend intercrop breeding across wide genetic gaps that limit traditional methods. However, in practice, the limitations of cell fusion have been similar to those found in traditional breeding systems. While sexual incompatibility can be bypassed by cell fusion, genetic incompatibilities are still manifested during regeneration. Furthermore, difficulties arising from linkages between deleterious and desirable genes cannot be overcome by fusion methods. However, cytoplasmically inherited traits from different cell lines have been successfully combined by fusion techniques. Cell lines formed by fusion of cytoplasms are called cybrids. Currently, the synthesis of cybrid lines is the only means of combining cytoplasmically inherited cellular components. Cell-fusion studies in *Brassica* crops have centered around efforts to combine such traits as mitochondrial-encoded CMS and chloroplasts resistant to cold-induced chlorosis or to triazine herbicides (Pelletier et al. 1983; Chetrit et al. 1985; Yarrow et al. 1986; Barsby et al. 1987; Menczel et al. 1987). The following sections will describe the progress in protoplast fusion of *Brassica* spp.

### 2.1 Fusion Methods

Two methods of protoplast fusion are in use at the present time: electrofusion and chemical fusion. High frequency electrofusion of plant protoplasts was first de-

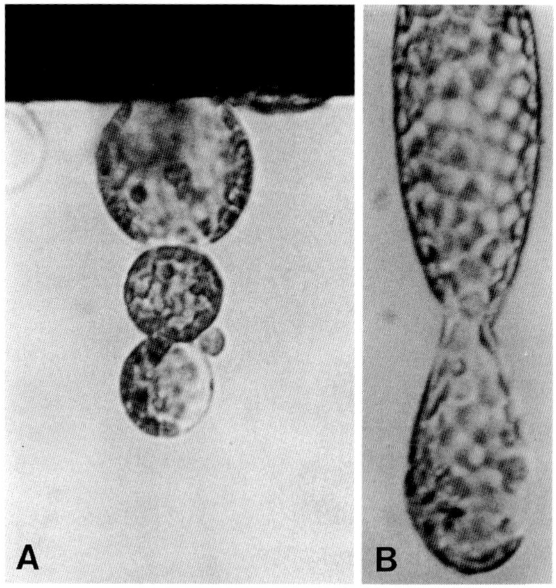


**Fig. 1 A, B.** Electrofusion of *Brassica oleracea* mesophyll protoplasts. **A** Pearl-chain formation induced by an alignment frequency of 500 kHz at 4 V. Note the flattened interface between the cells. **B** Fusion of two cells following a 50-μs pulse of 24 V (interelectrode distance = 200 μm). Temporary breakdown of the membranes allows the cells to coalesce at the interface

scribed by Zimmermann and Scheurich (1981). This method incorporates an alternating current to align the protoplasts in chains, reminiscent of a string of pearls, and establishes intermembrane contact (Fig. 1 A). Fusion is then induced by a short direct-current pulse which causes temporary membrane breakdown and allows the cells to coalesce (Fig. 1 B). Electrofusion has been used to fuse intraspecific (*B. napus*+*B. napus*), interspecific (*B. napus*+*B. campestris*) and intergeneric (*B. napus*+*Primula acaulis*) protoplasts (Zachrisson and Bornman 1984) with fusion rates up to 20%. Regeneration of fusion products was not attempted. Electrofusion has been applied to fuse individually paired *Brassica* protoplasts in microdrop cultures (Spangenberg and Schweiger 1986). Nearly 50% of the fused pairs survived with 10% to 20% of these forming microcalli. Pairing and fusion of individual protoplasts provides a means of controlling the number and type of cells being fused. In addition, the fusion of isolated cells eliminates the need for selection systems.

Chemical fusion is accomplished with several types of long-chain polymers, including dextran (Kameya 1983) and polyethyleneglycol (PEG) (Kartha et al. 1974; Yarrow et al. 1986). The PEG-induced fusion of a *B. oleracea* hypocotyl protoplast with an evacuolated, enucleated mesophyll protoplast is illustrated in Fig. 2. In a high calcium, high pH buffer, these high molecular weight molecules cause the agglutination of protoplasts. Figure 2 A shows the agglutination step, similar to pearl-chain formation in electrofusion. Fusion is completed upon dilution with low-salt culture medium (Fig. 2 B). The addition of dimethylsulfoxide has been shown to enhance the fusion frequency of *Brassica napus* cells (Menczel and Wolfe 1984). Chemical fusion rates are generally reported between 1% and 10%, although rates as high as 20% have been cited (Menczel et al. 1987). Due to the high toxicity of fusogens and to lower efficiency, chemical fusion ex-

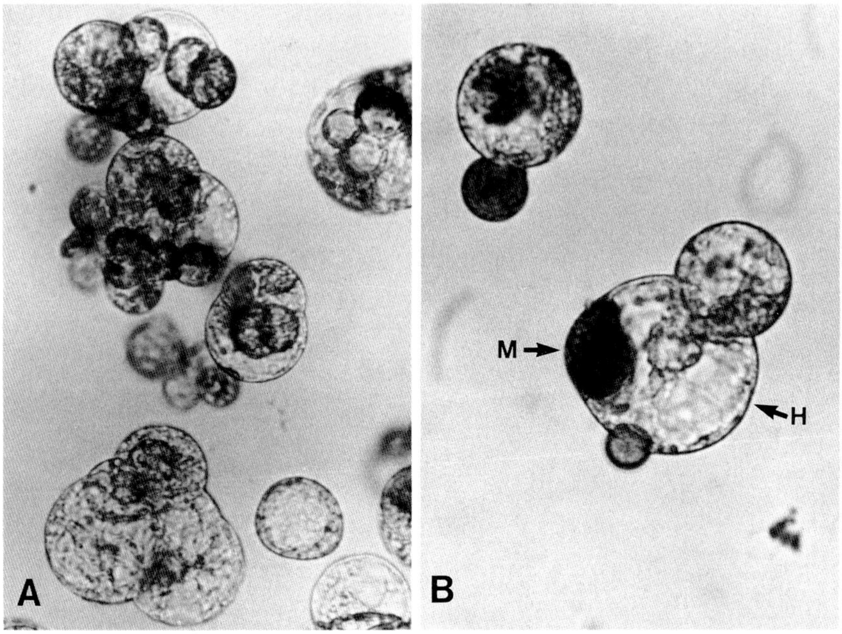


**Fig. 2A, B.** PEG-induced fusion of *Brassica oleracea* mesophyll (*M*) and hypocotyl (*H*) protoplasts. Mesophyll cells have been enucleated and evacuolated as shown in Fig. 3. **A** Aggregation of cells in solution of 44% PEG and 15% DMSO. **B** Fusion of cells following dilution with culture medium

periments usually have lower fusion rates and cell viability than electrofusion procedures. This increase in cell death can be reduced by minimizing exposure of cells to the fusion solutions. Despite higher fusions frequencies, greater cell viability, and other benefits of electrofusion over chemical methods (see Zachrisson and Bornman 1986), the former is still not as widely used as the latter. This is due to the high costs of the electrofusion equipment and to the low volumes of cells. Recently, however, Negrutiu et al. (1986) have developed a method for electrofusion of protoplasts in larger volumes.

The problem of cytoplasmic and nuclear incompatibilities may be overcome by fusing cytoplasts (enucleated protoplasts) with protoplasts. Cytoplasts have been isolated and electrofused with protoplasts, or karyoplasts (cells from which the cytoplasm has been removed; Spangenberg and Schweiger 1986). Fusion and survival rates were slightly lower than those of intact protoplasts fused under similar conditions. Cytoplasts can be easily formed by centrifugation in a self-forming Percoll gradient, as illustrated in Fig. 3. Cytoplasts form a dark band after centrifugation, while nuclei are pulled to the bottom (Fig. 3A, B). An additional advantage to this method is the elimination of vacuoles from the cytoplasts, since these intracellular structures have been cited as interfering with the fusion process (Zachrisson and Bornman 1986). As the nuclei are pulled down out of the protoplasts, vacuoles which are less dense than the cytoplasm are left in the upper phase (Fig. 3C). We have used chemical fusion to combine mesophyll cyto-

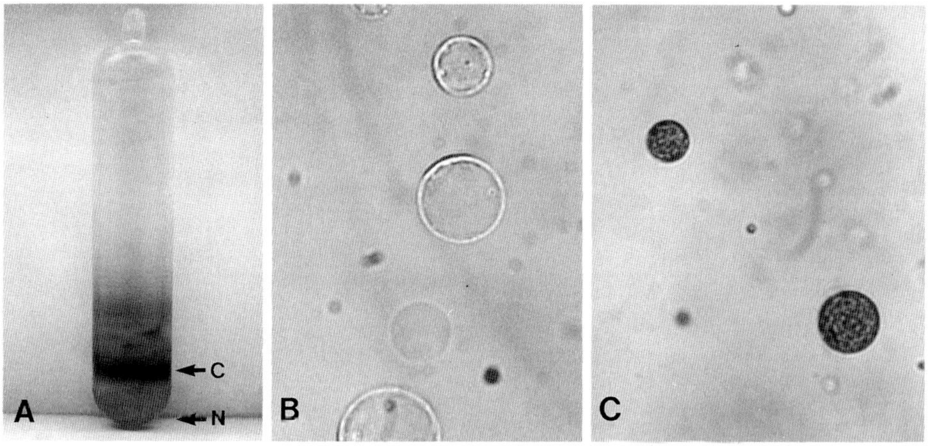


**Fig. 3 A–C.** Enucleation of mesophyll protoplasts by centrifugation in self-forming Percoll gradients. The initial Percoll concentration was 31%. **A** Gradients following centrifugation. Cytoplasts (*C*). Nuclei (*N*). **B** Vacuoles distributed in the upper layer. **C** Cytoplasts from band in **A**. (Procedure modified from Griesbach and Sink 1983)

plasts of *B. oleracea* var. *alboglabra* with hypocotyl protoplasts of rapid-cycling lines of *B. oleracea* (Wu et al. 1987). Although enucleation frequency of mesophyll cells in the Percoll gradients was estimated to be about 80%, it was not possible to distinguish products of hypocotyl-cytoplast fusion from hypocotyl fusion with intact mesophyll cells. The enucleation procedure must be optimized before it can be used confidently to eliminate all nuclei. Fusion of cellular components is a promising method for increasing the frequency of segregation of desirable combinations of organelles and nuclei.

## 2.2 Selection of Fusion Products

Several methods have been used to select and enrich for *Brassica* fusion products. Successful procedures include the use of various biochemical markers, cultural conditions which favor the fusion product, nuclear inactivation of one or more of the parental lines, and visual identification of hybrids through staining or distinct morphological characteristics.

The availability of biochemical mutants in higher plants is limited due to the effort required to isolate and characterize them. An approach towards use of biochemical markers for fusion experiments has been to provide two selectable markers in a single cell line, thus creating a "universal hybrid" which can be fused with a wild-type partner of choice. One marker must be dominant, to select against the unfused wild-type cells. The other marker is counter-selectable, for elimination of the single "universal hybrid" cells. A double-mutant line of *Sinapis turgida* has been prepared for fusion with wild-type *Brassica* species (Toriyama et al. 1987a). The dominant marker confers resistance to 5-methyltryptophan (5 MT); the counterselectable marker prohibits the cells from using nitrate as the sole nitrogen source. After fusion, cells are grown with nitrate as the only nitrogen source to eliminate all contaminating single *Sinapis* protoplasts. Thereafter, a medium containing 5 MT is used to eliminate the wild-type protoplasts. In two experiments, the *S. turgida* line was fused with *B. nigra* and *B. oleracea* protoplasts. Several somatic hybrid lines were selected and their hybrid nature was confirmed by isozyme analysis and chromosome cytology (see Sect. 2.4). Although plantlets were obtained from the cross with *B. oleracea*, they died before maturing.

Double inactivation methods offer an alternative to selection protocols that rely on preexisting biochemical mutations. Parental cell lines are impaired in different vital metabolic pathways which are complemented in the fusion products. Impairments commonly include treatment of one cell line with cytoplasm-directed biochemical inhibitors, such as iodoacetate (IOA), while the other cell line is disrupted in nuclear function. Gamma irradiation of protoplasts is often used to disrupt nuclear function as it causes chromosomal breaks which inhibit regeneration. Where nuclear recombination between fusion partners is desirable, limited exposure of cells to radiation may lead to increased recombination and reassortment of single chromosomes and pieces of chromosomes (Pontecorvo 1971; Negrutiu et al. 1984). Treatment of cells with iodoacetate and/or gamma-irradiation does not consistently inactivate either chloroplasts or mitochondria. The mitochondria from a CMS *B. napus* line were combined in a single cytoplasm with the atrazine-resistant chloroplasts from a fertile line of the same species using double inactivation (Barsby et al. 1987). Hypocotyl protoplasts from the CMS line were treated with IOA and mesophyll protoplasts from the fertile line were exposed to gamma-irradiation. After fusion, only mesophyll-hypocotyl hybrid cells harboring both an active cytoplasm and an active nucleus were capable of regeneration. Enrichment for the desired hybrid plants harboring CMS-mitochondria and atrazine-resistant chloroplasts was large enough that no additional direct selection was necessary.

Inactivation schemes are not always necessary for enrichment of fusion products. As mentioned in the previous section, manipulation and pairing of single cells for fusion can also eliminate the need for post-fusion selection. In some cases, one parent cell line may be left untreated, as it will not regenerate under the culture conditions provided. We have found that use of glucose as the sole osmoticum in regeneration medium will select against growth of unfused mesophyll protoplasts in *Brassica oleracea*, whereas hypocotyl cells will fluorish. In an experiment designed to reconstruct *B. napus* by fusing protoplasts of *B. oleracea* and *B. campestris*, selection was based on optimizing the conditions for regeneration of *B. napus* (Schenck and Röbbelen 1982). From 1236 different fusion combinations of 77 different cell lines, 12 plants were finally regenerated, three of which had the *B. napus* chromosome complement ($2n = 38$). Low regeneration frequency makes assessment of this type of selection difficult.

Mechanical isolation can involve micromanipulation by hand or by a mechanical cell-sorting apparatus (flow cytometer). The more labor-intensive means of mechanical sorting is micromanipulation. A description of the procedure for micromanipulation of fused cells is given by Patnaik et al. (1982). Prestained cells can be fused and fusion products easily distinguished by their stain combinations.

Several nontoxic, fluorescent stains have been introduced in recent years. Most of the stains listed here are organelle-specific and have been used successfully to label *Brassica* cells. Rhodamine 123 is a mitochondrial-specific stain, especially good for the rapidly growing tissue found in hypocotyl and root-tip cells (Wu 1987). Carboxyfluorescein, fluorescein isothiocyanate and fluorescein diacetate will stain the cytoplasm (Patnaik et al. 1982; Barsby et al. 1984). Nitro-blue tetrazolium can be used for staining photosynthetically active chloroplasts (Robertson and Earle 1987), although autofluorescence of chloroplast pigments is usually used when mesophyll protoplasts are fusion partners. Unstained cells can be visually identified for micromanipulation as products of mesophyll-hypocotyl fusion by the presence of both cytoplasmic strands visible in hypocotyl cells and the chloroplasts of mesophyll cells. There is a time constraint in distinguishing both stained cells and mesophyll-hypocotyl fusion products. Upon exposure to light, the stains begin to fade and plastids in hypocotyl cells begin to produce chlorophyll. In our experiments, cultures were kept in the dark except during manipulation and unstained mesophyll-hypocotyl fusion products could be distinguished for up to 2 days. A flow cytometer was used to isolate fused *Brassica* protoplasts (Pauls and Chuong 1986; Glimelius et al. 1986). Fused cells were detected by the fluorescence of chlorophyll in unstained mesophyll cells of one parent and the carboxyfluorescein fluorescence from stained hypocotyl protoplasts of the other parent. The flow cytometer is set to measure fluorescence intensity at two different wavelengths. It will separate fusion products harboring labels from both parent cells, each harboring one label. However, emission spectra of the two labels must not overlap (Pauls and Chuong 1986). Use of the automated flow cytometer can provide as much as 80% concentration of fusion products (Yarrow et al. 1986). Fusion products can first be concentrated by a flow cytometer and then micromanipulated into individual cell cultures. Concentration of products of chemical fusion is desirable because of low chemical fusion frequencies.

## 2.3 Regeneration

Successful protoplast regeneration procedures have been reported for *Brassica* hypocotyl (Glimelius 1984; Chuong et al. 1985; Barsby et al. 1986), mesophyll (Kartha et al. 1974; Kolenbach et al. 1982; Pelletier et al. 1983; Robertson and Earle 1986), root (Xu et al. 1982), and stem cortex tissue (Klimaszewska and Keller 1987). Genotype appears to play an important role in regeneration efficiency of unfused cells (Jourdan and Earle 1986). Williams and Hill (1986) have developed rapid-cycling populations of many of the agronomically important cruciferous crops. From seed to seed, these populations require 4 weeks under specified conditions. We have found protoplasts and stem explants of rapid-cycling *B. oleracea* and *B. napus* to regenerate in an accelerated and prolific way in comparison to other genotypes. Use of these lines in experiments may facilitate aspects of genetic manipulation. Figure 4 illustrates the regeneration sequence for rapid-cycling *B. oleracea* hypocotyl protoplasts. Despite the abundance of regeneration protocols for unfused cells, many fusion experiments have culminated in aborted plantlets and low frequency of regeneration (Schenck and Robbelen 1982; Sundberg and Glimelius 1986; Barsby et al. 1987; Robertson et al. 1987; Terada et al. 1987; Toriyama et al. 1987a). Reasons for this bottleneck are not yet clear due to the difficulty in distinguishing inappropriate cultural conditions from the expression of incompatibility among combined nuclei and cytoplasms. A detailed discussion of the role of somatic incompatibility in fusion experiments is presented by Harms (1983).

Several studies have involved intergeneric fusion with a *Brassica* species as one partner (Hoffmann and Adachi 1981; Toriyama et al. 1987a, b). Such studies are aimed at increasing the gene pool available for improvement of *Brassica* crops. Toriyama et al. (1987b) fused protoplasts of suspension culture cells of *Moricandia arvensis* with *B. oleracea* hypocotyl protoplasts. Regenerants exhibited the morphological characteristics of both parents and were shown to be hybrids by isozyme analysis. "*Arabidobrassica*" was formed by the fusion of protoplasts from *Arabidopsis thaliana* and from *B. campestris* (Gleba and Hoffmann 1978, 1980; Hoffmann and Adachi 1981). Over a 15-month period, 305 hybrid regenerants with intermediate morphological characteristics were recovered. Chromosome analysis of the regenerants revealed complements ranging from various levels of aneuploidy to true amphidiploids (having full chromosome sets from both parents). Hybrids containing chromosomes predominantly from one parent were the most common. These experiments illustrate the difficulties in using protoplast fusion for nuclear recombination. Much more work needs to be done to expand our understanding of chromosomal stabilization with a fused cell if this technique is to become useful for plant manipulation.

Presently, the only agriculturally useful product generated from protoplast fusion experiments involving *Brassica* spp. has been the combination of mitochondria and chloroplasts from two different cytoplasms. Segregation and recombina-

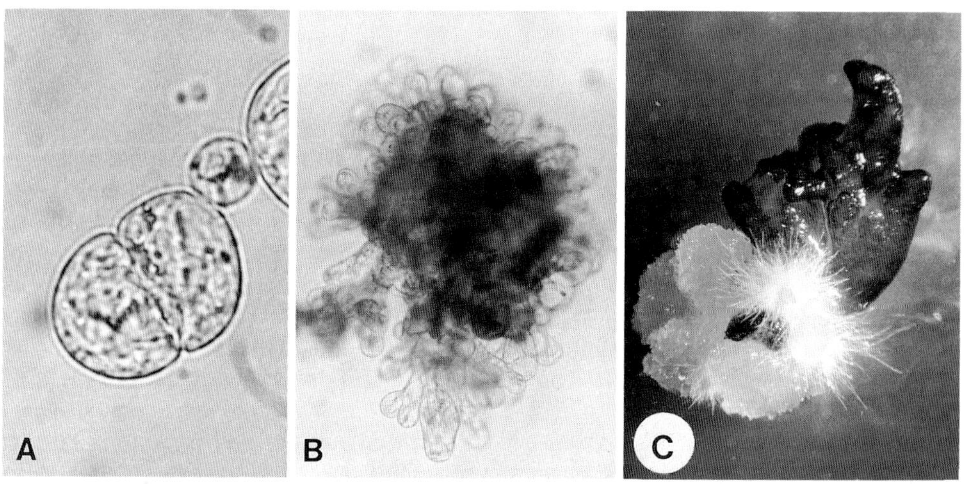


**Fig. 4. A–C.** Regeneration of rapid-cycling *Brassica oleracea* hypocotyl protoplasts. **A** First cellular division 2 days after isolation. **B** Microcalli in liquid culture 14 days after isolation. **C** Callus with a regenerating shoot 27 days post-isolation. Roots are present, but obscured by a mass of root hairs

tion of organelles can occur during regeneration of fusion products. At this time, there is no means of directing segregation of particular combinations of mitochondria and chloroplasts, although several workers have documented independent segregation of mitochondria and chloroplasts from fusion experiments (Pelletier et al. 1983; Chetrit et al. 1985; Barsby et al. 1987; Morgan and Maliga 1987). These studies have revealed extensive recombination of mitochondrial genomes while confirming the stability of the chloroplast genomes during organelle segregation. In one study involving the fusion of a CMS and a fertile line of *Brassica napus*, chloroplast segregation occurred within 20 cell generations following fusion, while mitochondrial segregation took somewhat longer and recombination was detected (Morgan and Maliga 1987). Furthermore, the combination of chloroplasts from the fertile line and mitochondria from the CMS line was obtained in 37 of 42 clones regenerated. The cultural and genetic factors that contribute to organelle selection and stabilization in this study remain unknown.

### 2.4 Analysis of Regenerants

The plants and calli regenerated from fusion products must be analyzed to reveal their nuclear and cytoplasmic components. Large numbers of the regenerants can show culture-induced morphological variation which is not related to new combinations of traits (Newell et al. 1984; Karp and Bright 1985; Evans and Sharp 1986). Methods of detecting these somaclonal variants and genetically analyzing the hybrids include peptide and isozyme characterization; chromosome counting and cytology; determination of nuclear DNA content by flow cytometry; restriction fragment length polymorphisms (RFLP's) of chloroplast, mitochondrial or nuclear DNA; cellular resistance to herbicides or other chemicals; and morphological characteristics of regenerated plants. All but the last of these methods can be used to detect culture-induced variation.

Detection of isozyme patterns is particularly suitable for fusion experiments because enzyme activity can be readily observed in a very small amount of tissue. The choice of enzyme is critical, with high activity and genotype-specific patterns being minimal requirements. Usually several enzymes are analyzed to detect and eliminate those that are highly variable. Among the isozymes used for detection of *Brassica* fusion products are acid phosphatase, aspartate aminotransferase, esterase, lactate dehydrogenase, leucine aminopeptidase, phosphoglucomutase, 6-phosphogluconate dehydrogenase, and phosphoglucose isomerase (Gleba and Hoffmann 1978; Sundberg and Glimelius 1986; Taguchi and Kameya 1986; Toriyama et al. 1987a, b). Detection of hybrid isozymes in analzyed tissue confirms a hybrid origin, whereas detection of both parental isozyme types may indicate chimeric tissue. Isozyme analysis can only confirm the presence of pieces of parental genomes. It cannot be used to detect full chromosome complements.

Chromosome cytology exploits the physical appearance of the chromosomes to determine ploidy and to detect multiple fusions. Physical appearance is useful in detecting hybrids only if the chromosomes of the fusion parents are distinct. Furthermore, *Brassica* spp. are noted for their small chromosomes and counting these can be a tedious task. Chromosome morphology was used to monitor chromosome loss from *Arabidopsis+Brassica* hybrid calli over a period of 15 months (Hoffmann and Adachi 1981). Two observations were made regarding the stability of chromosomes following fusion of these distantly related species. First, the fusion products underwent significant intrachromosomal recombination not observed in regenerating parent cell lines. In addition, chromosome loss and rearrangements in one hybrid line appeared to be stabilized by organogenesis. Studies on the progeny of these hybrids have not yet been reported.

Automated flow cytometry can be used to estimate the ploidy level of cells as a rapid alternative to chromosome counting. Sundberg and Glimelius (1986) used flow cytometry to determine the ploidy levels of a single hybrid shoot of *B. oleracea+B. campestris*. DNA content of the regenerated shoot and of the parental lines was measured after staining the nuclei with propidium iodide. The hybrid shoot contained approximately double the sum of the DNA of the parents. Large numbers of cells are needed for this type of DNA analysis, thus it is unsuitable for assessment of calli.

Hybridization of DNA probes to restriction digests of cellular DNA can be used to detect genotype-specific chloroplast, mitochondrial and nuclear DNA sequences. Hybrids or somaclonal variants are detected as changes in the size of defined DNA restriction fragments (RFLP's) which are separated by electrophoresis and visualized by hybridization to specific probes. In the analysis of fusion products, heterogeneity in RFLP's may be indicative of heterogeneity of other characters as well. This type of analysis has been used to distinguish *Raphanus* and *Brassica* cytoplasms (Pelletier et al. 1983; Barsby et al. 1987; Kemble 1987). The mitochondrial and chloroplast genomes of *Brassica* spp. have been well characterized. The CMS and fertile cytoplasms from different *Brassica* species can be distinguished by RFLP patterns from the mitochondrial and chloroplast DNA (Kemble 1987). Organelle complements from hybrid plants and recombination between mitochondrial genomes can be determined (Chetrit et al. 1985). Intermitochondrial recombination must then be distinguished from recombination within a single mitochondrial species. DNA from specific restriction fragments of *B. campestris* mitochondrial DNA were used as probes to determine the parental origins of somatic hybrids of *B. oleracea+B. campestris* (Robertson et al. 1987).

When biochemical mutants are available as fusion parents, drug or herbicide resistance can be used to select early in the regeneration process, during the first divisions, or later in the greenhouse and field (Barsby et al. 1987; Robertson et al. 1987; Toriyama et al. 1987a). Tests for atrazine resistance can be done on regenerated plants by applying a paste of the herbicide onto leaves. Resistant leaves will not be damaged by this treatment, whereas sensitive leaves will turn chlorotic and die (Barsby et al. 1987). Sensitivity to herbicides and drugs may vary with tissue type, and stage of plant development.

Morphological variations of trichomes, leaves, and flowers have been used to identify hybrids among the regenerated plants (Taguchi and Kameya 1986; Terada et al. 1987; Toriyama et al. 1987b), although the presence of somaclonal variants in regeneration experiments may limit the use of this method. In our laboratory, regenerated hybrid plants from fusion between wrinkled-leaved and smooth-leaved *B. oleracea* plants were found to have intermediate leaf morphologies (Fig. 5). The hybrid origin of these plants was confirmed by hybridization of

**Fig. 5.** Intermediate morphology of a somatic hybrid (*center*) as compared to parental genotypes (*flanking*). Rapid-cycling *Brassica oleracea* (CrGC-3; Williams and Hill 1986) grown from seed (*left*). *B. oleracea* spp. *gemmifera* cv. Prince Marvel Brussel Sprouts regenerated from unfused protoplasts (*right*)

restriction digests of total cellular DNA to organelle-specific probes (S. Garger and C. Richards, unpubl.).

## 2.5 Improvement of Cytoplasmic Male Sterile *Brassica* Cultivars

As previously mentioned, cell-fusion studies in the *Brassica* spp. have centered around efforts to combine mitochondrial-encoded CMS with chloroplasts resistant to cold-induced chlorosis or to the triazine herbicides. These efforts have been rewarded by the recent successes of several workers (Pelletier et al. 1983; Chetrit et al. 1985; Yarrow et al. 1986; Barsby et al. 1987; Menczel et al. 1987) using three different CMS systems. The *ogu* cytoplasm, introduced into the *Brassica* spp. from *Raphanus sativa*, is the most stable of these three systems. The *nap* CMS system was obtained from *B. napus* (Shiga and Baba 1971, 1973; Thompson 1972). The *pol* CMS system occurred naturally in *B. napus* (Fu 1981). The latter two cytoplasms become partially fertile at high temperatures (Fan and Stefansson 1986), and are thus less desirable to plant breeders. However, only the *ogu* cytoplasm is associated with the problem of cold-induced chlorosis.

The work of Pelletier et al. (1983) is the first report of successful recombination of organelles in the *Brassica* spp. In two experiments, these workers introduced CMS mitochondria of the *ogu* cytoplasm into fertile *B. campestris* and *B. napus* lines harboring atrazine resistant and cold-chlorosis resistant chloroplasts, respectively. Without pretreatment of the parent protoplasts, or post-fusion selection, they obtained six regenerated cybrid plants containing the desirable combinations of organelles. Five of these regenerants were described in a later publication in greater detail (Chetrit et al. 1985). By crossing these cybrid lines with restorer lines of *B. napus*, which harbor genes that prevent the expression of CMS, they were able to confirm that male sterility of these cybrids was neither culture-induced, nor a new character due to recombination between parental lines. Menczel et al. (1987) modified the Pelletier procedure by using gamma-irradiated and iodoacetate-inactivated cells to limit contaminating parental regenerants. Six lines of cybrids with the desirable chloroplasts/mitochondrial combination were regenerated and these traits were stably inherited to the F1 generation.

The *nap* (Yarrow et al. 1986) and *pol* (Barsby et al. 1987) mitochondria have been introduced into cytoplasms with atrazine-resistant chloroplasts. Since *nap* and *pol* CMS cytoplasms have a greater variety of restorer lines available in *Brassica* spp., they are still desirable for breeding, despite their environmental instability.

These experiments make available new and agriculturally important combinations of cytoplasmically inherited traits which would not have been attainable by traditional breeding methods. In addition, they have confirmed the association of CMS with the mitochondrial genome. The work of Pelletier et al. (1983), Chetrit et al. (1985) and Menczel et al. (1987) also confirms that the cold-sensitivity of the *ogu* cytoplasm is determined by poor functioning of *Raphanus sativa* chloroplasts with the *Brassica* nucleus. An understanding of these relationships will make efforts to combine these cytoplasmic traits with other desirable cytoplasmically inherited traits simpler and more efficient.

## 3 Transformation

Gene transfer methods provide several advantages over protoplast-fusion techniques. As cloned DNA sequences are introduced into cells, the risk of expression of incompatibility and deleterious linked genes is reduced. In addition, with the increasing availability of expression signals, genes can be introduced across much wider genetic gaps than is currently possible with fusion experiments. However, the techniques are limited to traits for genes which have been isolated and cloned and as yet, the technique has not been successfully applied to transfer genes into organelles.

The development of reliable protocols for regeneration of protoplasts and tissue explants of several *Brassica* spp. has made this group of crops ripe for genetic manipulations. Currently, gene transfer techniques in *Brassica* spp. are being optimized by the use of marker genes, primarily drug-resistance markers. Methods of selective gene transfer include *Agrobacterium*-mediated transformation, electroporation, nuclear injection and transformation using viral vectors. The application of these systems to *Brassica* spp. are described below.

### 3.1 *Agrobacterium*-Mediated Transformation

The *Brassica* species that have been transformed with wild-type *A. tumefaciens* include *B. campestris*, *B. hirta*, *B. juncea*, *B. napus*, *B. nigra*, and *B. oleracea* (Holbrook and Miki 1985; Mathews et al. 1985; 1986, Ooms et al. 1985). Successful transformation of *B. napus* by *A. rhizogenes* has also been reported (Ooms et al. 1985; Tanaka et al. 1985; Guerche et al. 1987). For induction of wild-type symptoms of *A. rhizogenes* or *A. tumefaciens*, hypocotyls from seedlings or stems from older plants are wounded and inoculated by coating the wound with a logarithmic phase culture of the *Agrobacterium* strain. This method has been successful for both tumorigenesis and for hairy-root production. *Brassica* plants 8 to 10 weeks old have been found to be optimally susceptible to gall formation at all internodes. Seedling tissue appears to be less responsive (Holbrook and Miki 1985). Tumors and adventitious roots can be removed and cultured axenically to regenerate transformed plants. Cultured tumors will frequently sprout shoots characterized by their lack of apical dominance, a "shooty" appearance, and the inability to root normally (Mathews et al. 1985). Cultured adventitious roots regenerate more easily than do normal root cultures. Plants regenerated from segments of these roots exhibit the "rooty" phenotype, a characteristic which shows Mendelian inheritance (Guerche et al. 1987). Transformed adventitious roots may be easily maintained in continuous organ culture. They differ from other root cultures by exhibiting more vigorous growth and reduced apical dominance. These features are desirable for large-scale secondary metabolite production (Guerche et al. 1987).

Strains of *A. tumefaciens* produce a variety of opines. *Brassica* spp. have been transformed by strains that produce mannopine, agropine, octopine, and nopaline. Two studies have indicated reduced susceptibility of *B. napus* to octopine-producing strains (Holbrook and Miki 1985; Pua et al. 1987). Plants of *B. nigra*, *B. campestris*, *B. oleracea*, and *B. hirta* inoculated with octopine-producing strains showed 27% to 100% tumor production, while *B. napus* plants inoculated under the same conditions produced tumors on less than 8% of the inoculated plants (Holbrook and Miki 1985). The basis for this resistance is not known.

Tumor induction on *Brassica* hosts establishes that these species are transformable by *Agrobacterium*. Tumor-inducing genes on the T-DNA can be replaced by marker genes, such as those conferring drug resistance. These marker genes can be used, in turn, to establish in vitro inoculation, selection and regeneration conditions. The final step is to introduce agriculturally desirable genes by these established protocols. In vitro inoculation protocols generally involve explants dipped in *Agrobacterium*, freed from the transforming bacteria after transformation has occurred by addition of antibiotics, and regenerated under appropriate selection conditions. *B. oleracea* and *B. napus* regenerate most reliably and prolifically from stem strips, as described by Klimaszewska and Keller (1985). Four drugs commonly used in plant transformation experiments as dominant selectable markers are: hygromycin, G418, kanamycin, and chloramphenicol. While these drugs have been used effectively in selection of transformants in many other plants (Fraley et al. 1986), success in *Brassica* spp. has been limited. In our laboratory, concentrations of G418 as low as 5 mg/l stimulated secretion of a black residue from stem explants (Trail and Wu 1986). This residue was thought to suppress the growth of transformed cells. Hygromycin and chloramphenicol allowed minimal shoot development and did not cause any visible secretion (Trail and Wu 1986). However, *Brassica* spp. have high endogenous levels of chloramphenicol-acetyltransferase which interferes with the assays used to monitor the expression of chloroamphenicol resistance (Balazs and Bonneville 1987). Several workers have reported difficulties in selecting kanamycin-resistant *Brassica* transformants from stem explants, although the reason for this phenomenon has not yet been elucidated (Holbrook and Miki 1985; Pua et al. 1987). It must be remembered that different types of tissues respond uniquely to various antibiotics. Kanamycin has been used successfully to select resistant *Brassica* protoplasts (Guerche et al. 1986), yet it appears to be unsuitable for stem explant tissue. We have found that *Brassica* explants are sensitive to the presence of *Agrobacterium* in tissue culture. Frequent rinsing of inoculated explants to keep bacterial populations to a minimum seems to prevent this damage. Hypocotyl explants from etiolated seedlings are more sensitive to the bacteria than are green tissues.

Pua et al. (1987) have recently reported the regeneration of transgenic plants of *B. napus* from stem strips. The transforming T-DNA contained the mouse gene for dihydrofolate reductase (dhfr). The dhfr gene was linked to the 35S promoter from the Cauliflower Mosaic Virus (CaMV) to produce a functional gene that confers resistance to methotrexate. Transformed plants were regenerated from 10% of the inoculated explants on 0.01 ml/l methotrexate. Expression of dhfr was confirmed by Northern analysis of the mRNA isolated from putative transformants. Progeny of two confirmed transformants showed stable Mendelian inheritance of the introduced gene at a single chromosomal site. This is the first report of stable *Agrobacterium*-mediated transformation of a *Brassica* species by an engineered Ti plasmid.

### 3.2 Electroporation and Nuclear Injection

Direct gene transfer can be accomplished by electroporation or by injection of DNA into plant protoplasts. Both methods have been used successfully to deliver DNA into *Brassica* cells, although their most promising applications are to monocots and other plants not sensitive to *Agrobacterium*-mediated transformation. At this time, neither method transforms plant cells with the efficiency of the *Agrobacterium*-mediated transformation techniques. However, electroporation is a fast and accurate test of gene expression when regeneration of stable transformants is not necessary.

Electroporation facilitates the uptake of DNA by large numbers of protoplasts. The same principles of pore formation apply to electroporation as apply to electrofusion, except that in electroporation pores are used to get macromolecules into cells, rather than to fuse them. A short, direct-current pulse is applied to newly isolated protoplasts in the presence of DNA (Fromm et al. 1985). Temporary pores are formed in the membrane, allowing DNA to penetrate the cell. Expression of marker genes introduced by electroporation into *B. napus* protoplasts has been reported by Guerche et al. (1986). *B. napus* mesophyll cells were electroporated in the presence of a plasmid containing a neomycin phosphotransferase (NPTII) gene conferring resistance to kanamycin. After 11 days in culture, selection was begun through the introduction of 100 mg/l kanamycin. Two resistant plants were regenerated, one of which contained the NPTII gene. This is the first published report of a successful transformation event in *Brassica*.

Spangenberg et al. (1986) describe the injection of plasmid DNA encoding the NPTII gene into isolated karyoplasts (protoplasts from which the cytoplasm has been removed). Some injected karyoplasts were then electrofused to cytoplasts (cells from which the nuclei have been removed). Gene expression was detected 5 to 10 days following microinjection by immunofluorescent staining of cells exposed to antibodies against NPTII. While 60% to 70% of the injected cells showed a positive reaction to the NPTII antibodies, 25% regenerated into microcalli. Since these cellular manipulations are far more labor intensive and less reliable than *Agrobacterium*-mediated transformation systems, their usefulness for genetic manipulation of these crops appears limited.

### 3.3 Viruses

There are a few reports of the use of viruses as vehicles for transporting foreign DNA into plant cells. Most of this work has been done with the Cauliflower Mosaic Virus (CaMV). CaMV has several features which make it a suitable model plant virus for this type of study. First, CaMV is one of the few plant viruses containing double-stranded DNA. The genome of CaMV has been cloned and cDNA derived from RNA transcripts has been shown to infect cells (Howell et al. 1980). Furthermore, the genes encoded by the RNA transcripts have been studied and functions of several open reading frames (ORFs) are known. CaMV ORF VI codes for the main protein of the viral inclusion bodies. ORFs II and VII can be deleted without disturbing viral infectivity. In one study, ORF II was replaced by the dhfr gene. *B. campestris* plants infected with the engineered virus developed systemic infections (Brisson et al. 1984). Dhfr was found to be present in infected plants as monitored by antibodies to dhfr. In a recent report, ORF VI was replaced with the bacterial gene for NPTII (Paszkowski et al. 1986), however, the resulting hybrid vector was noninfective. In the presence of an unaltered CaMV genome, as a helper plasmid, protoplasts were transformed with the NPTII hybrid vector and transformants were recovered under kanamycin selection. Two colonies were maintained stably in the presence of kanamycin for 12 months, although no plants were regenerated.

Grimsley et al. (1986) combined the use of viruses and T-DNA to introduce an altered CaMV genome into *B. napus* and *B. campestris* plants. They infected the crowns of plants with a disarmed strain of *Agrobacterium tumefaciens* harboring multiple copies of the CaMV genome. Systemic viral infection occurred only when the viral genome was between the T-DNA borders. Presumably, the viral DNA was able to escape from the T-DNA once inside the plant cells, and produce a systemic infection. The term "Agroinfection" has been applied to this type of transformation. Agroinfection of *B. campestris* has been used to study virulence mutants of *A. tumefaciens*. A copy of the CaMV genome was cloned between the T-DNA borders of various *Agrobacterium* virulence mutants. Viral symptoms were detected upon successful transfer of the T-DNA into the plant cell. This system is far more sensitive for detecting T-DNA transfer than monitoring tumor formation. At the present time, viral vectors appear to be more useful as tools to understanding host-pathogen interactions than as applications to crop improvement.

## 4 Conclusion

Gene transfer techniques can be applied to *Brassica* crops to introduce agronomically useful genes from a wide variety of sources. These genes are becoming increasingly available. Two genes for resistance to the herbicide glyphosate have been isolated (Comai et al. 1985; Shah et al. 1986), as well as a gene for an insecticidal protein (Vaeck et al. 1987), and a gene for *Fusarium* wilt resistance (Shahin and Spivey 1986). Recently, a gene for self-incompatibility has been isolated from *B. oleracea* (Nasrallah et al. 1985). Production of commercial F1 hybrid seed has been facilitated through the use of self-incompatible cultivars, and the exploitation of limited numbers of these genes from the *Brassica* spp. through genetic transfer should be profitable. Still, little is known about the transfer, integration, and expression of genes in different genetic backgrounds.

Fusion techniques provide a means of combining cytoplasmically inherited characteristics. Characterization of chloroplast, mitochondrial, and nuclear DNA has been a significant contribution to our ability to determine the genetic components of manipulated cells. An understanding of the segregation mechanisms of organelles may expand the possible combinations of chloroplast and mitochondrial-encoded traits available for crop improvement by this method. Fusion experiments also have the potential of allowing transfer of characteristics which have not been attributed to particular sets of genes across wide genetic gaps. However, cell fusion has not been proven useful in combining nuclear traits. Lack of an understanding of somatic incompatibilities and factors which affect chromosome stability have limited the usefulness of this technique to agricultural improvements.

Much progress has been made in our ability to genetically manipulate *Brassica* crops. The broad range of crops delineated by the six key *Brassica* species reflect their global importance. As new sources of breeding material disappear, improvement of these crops will rely increasingly on the transfer of agronomically important traits across wide genetic gaps, and thus on the methods that make these transfers possible.

*Acknowledgments.* The authors gratefully acknowledge M. L. Christianson, R. Rodriguez, and J. J. Witcosky for their help in the preparation of this manuscript.

## References


Austin RB (1986) Oilseed rape. In: Austin RB, Flavell RB, Henson IE, Lowe HJB (eds): Molecular biology and crop improvement: a case study of wheat, oilseed rape and faba beans. Univ Press, Cambridge, pp 70–87

Balazs E, Bonneville JM (1987) Chloramphenicol acetyl transferase activity in *Brassica* spp. Plant Sci 50:65–68

Bannerot H, Boulidard L, Cauderon Y, Tempé J (1974) Transfer of cytoplasmic male sterility from *Raphanus sativa* to *Brassica oleracea*. Proc Eucarpia Meet Cruciferae Scottish Hortic Res Inst, Invergowrie, GB, pp 52–54

Bannerot H, Boulidard L, Chupeau Y (1977) Unexpected difficulties met with the radish cytoplasm. Eucarpia Cruciferae Newslett 2:2–16

Barsby TL, Yarrow SA, Shepard JF (1984) Heterokaryon identification through simultaneous fluorescence of fluorescein isothiocyanate and tetramethylrhodamine isothiocyanate labelled protoplasts. Stain Technol 59:217–220

Barsby TL, Yarrow SA, Shepard JF (1986) A rapid and efficient alternative procedure for regeneration of plants from hypocotyl protoplasts of *Brassica napus*. Plant Cell Rep 5:101–103

Barsby TL, Chuong PV, Yarrow SA, Wu SC, Coumans M, Kemble RJ, Powell AD, Beversdorf WD, Pauls KP (1987) The combination of Polima CMS and cytoplasmic triazine resistance in *Brassica napus*. Theor Appl Genet 73:809–814

Beversdorf WD, Weiss-Lerman J, Erikson LR, Souza-Machado V (1980) Transfer of cytoplasmically inherited triazine resistance from birds-rape to cultivated oilseed rape (*Brassica campestris* and *B. napus*). Can J Genet Cytol 22:167–172

Brisson N, Paszkowski J, Penswick JR, Groennborn B, Potrykus I, Hohn T (1984) Expression of a bacterial gene in plants by using a viral vector. Nature (Lond) 310:511–514

Chetrit P, Mathieu C, Vedel F, Pelletier G, Primard C (1985) Mitochondrial DNA polymorphism induced by protoplast fusion in Cruciferae. Theor Appl Genet 69:361–366

Chiang MS, Crete R (1983) Transfer of resistance to race 2 of *Plasmodiophera Brassicae* from *Brassica napus* to cabbage (*B. oleracea* spp. *Capitata*). V. The inheritance of resistance. Euphytica 32:479–483

Chuong PV, Pauls KP, Beversdorf WD (1985) A simple culture method for *Brassica* hypocotyl protoplasts. Plant Cell Rep 4:4–6

Comai L, Facciotti D, Miatt MR, Thompson G, Rose RE, Stalker DM (1985) Expression in plants of a mutant aroA gene from *Salmonella typhimurium* confers tolerance to glyphosate. Nature (Lond) 317:741–744

Evans DA, Sharp WR (1986) Applications of somaclonal variation. Biotechnology 4:528–532

Fan Z, Stefansson BR (1986) Influence of temperature on sterility of two cytoplasmic male-sterile systems in rape (*Brassica napus* L.). Can J Plant Sci 66:221–227

Fraley RT, Rogers SG, Horsh RB (1986) Genetic transformation in higher plants. CRC Crit Rev Plant Sci 4:1–46

Fromm M, Taylor LP, Walbot V (1985) Expression of genes transferred into monocot and dicot plant cells by electroporation. Proc Natl Acad Sci USA 82:5824–5828

Fu TD (1981) Production and research of rapeseed in the People's Republic of China. Eucarpia Cruciferae Newslett 6:6–7

Gleba YY, Hoffmann F (1978) Hybrid cell lines *Arabidopsis thaliana+Brassica campestris*: no evidence for specific chromosome elimination. Mol Gen Genet 165:257–264

Gleba YY, Hoffmann F (1980) "*Arabidobrassica*": A novel plant obtained by protoplast fusion. Planta 149:112–117

Glimelius K (1984) High growth rate and regeneration capacity of hypocotyl protoplasts in some *Brassicaceae*. Physiol Plant 61:38–44

Glimelius K, Djupsjobacka M, Fellner-Feldegg H (1986) Selection and enrichment of plant protoplast heterokaryons of *Brassicaceae* by flow sorting. Plant Sci 45:133–141

Griesbach RJ, Sink KC (1983) Evacuolation of mesophyll protoplasts. Plant Sci Lett 30:297–301

Grimsley N, Hohn B, Hohn T, Walden R (1986) "*Agroinfection*", an alternative route for viral infection of plants by using the Ti plasmid. Proc Natl Acad Sci USA 83:3282–3286

Guerche P, Charbonnier M, Jouanin L, Pelletier G (1986) Direct gene transfer in *Brassica napus*. Eucarpia Cruciferae Newslett 11:98–99

Guerche P, Jouanin L, Tepfer D, Pelletier G (1987) Genetic transformation of oilseed rape (*Brassica napus*) by the Ri T-DNA of *Agrobacterium rhizogenes* and analysis of inheritance of the transformed phenotype. Mol Gen Genet 206:382–386

Harms CT (1983) Somatic incompatibility in the development of higher plant somatic hybrids. Q Rev Biol 58:325–353

Hoffmann F, Adachi T (1981) "*Arabidobrassica*": Chromosomal recombination and morphogenesis in asymmetric intergeneric hybrid cells. Planta 153:586–593

Holbrook LA, Miki BL (1985) *Brassica* grown gall tumourigenesis and in vitro of transformed tissue. Plant Cell Rep 4:329–332

Howell SH, Walker LL, Dudley RK (1980) Cloned Cauliflower Mosaic Virus DNA infects turnips (*Brassica rapa*). Science 208:1265–1267

Hunt S, Smith AM, Woolhouse HW (1987) Evidence for a light-dependent system for reassimilation of photo-respiratory $CO_2$, which does not include a C4 cycle, in the C3-C4 intermediate species *Moricandia arvensis*. Planta 171:227–234

Jourdan PS, Earle ED (1986) Influence of genotype on the regeneration of plants from seedling mesophyll protoplasts of three *Brassica* species (Abstr). Cruciferae genetics Worksh III, Univ Guelph, Ontario, Can

Kameya T (1983) Studies on plant cell fusion by dextran: Effect of pH, inorganic salts and electric stimulus. Cytologia 48:873–878

Karp A, Bright SWJ (1985) On the causes and origins of somaclonal variation. Oxford Surv Plant Mol Cell Biol 2:199–234

Kartha KK, Michayluk MR, Kao KN, Gamborg OL, Constabel F (1974) Callus formation and plant regeneration from mesophyll protoplasts of rape plants (*Brassica napus* L. cv. Zephyr). Plant Sci Lett 3:265–271

Kemble RJ (1987) A rapid, single leaf, nucleic acid assay for determining the cytoplasmic organelle complement of rapeseed and related *Brassica* species. Theor Appl Genet 73:364–370

Klimaszewska K, Keller WA (1985) High frequency plant regeneration from thin layer explants of *Brassica napus*. Plant Cell, Tissue Org Cult 4:184–197

Klimaszweska K, Keller WA (1987) Plant regeneration from stem cortex protoplasts of *Brassica napus*. Plant Cell, Tissue Org Cult 8:225–233

Kolenbach HW, Wenzel G, Hoffmann F (1982) Regeneration of *Brassica napus* plantlets in cultures from isolated protoplasts of haploid stem embryos as compared with lead protoplasts. Z Pflanzenphysiol 105:131–142

Lammerick J (1970) Inter-specific transfer of clubroot resistance from *Brassica campestris* L. to *B. napus* L. N Z J Agric Res 13:105–110

Mackay GR (1977) The introgression of S-alleles into forage rape, *Brassica napus* L. from turnip, *Brassica campestris* L. spp. *rapifera*. Euphytica 26:511–519

Mathews H, Bhatia CR, Mitra R, Krisha TG, Rao PS (1985) Regeneration of shoots from *Brassica juncea* (Linn) Czern and Coss cells transformed by *Agrobacterium tumefaciens* and expression of nopaline dehydrogenase genes. Plant Sci 39:49–54

Mathews H, Rao PS, Bhatia CR (1986) Transformation of *Brassica juncea* by *Agrobacterium tumefaciens* harbouring plasmid pTiT37 and its 'rooty' mutant pTiT37.14a/a. J Genet 65:37–44

Matsuzawa Y, Sarashima M (1986) Intergeneric hybridization between *Raphanus sativa* L. and *Brassica nigra* Koch and alloplasmic radish derivative. Jpn J Breed 36:122–130

Menczel L, Wolfe K (1984) High frequency of fusion induced in freely suspended protoplast mixtures by polyethylene glycol and dimethylsulfoxide at high pH. Plant Cell Rep 3:196–198

Menczel L, Morgan A, Brown S, Maliga P (1987) Fusion-mediated combination of Ogura-type cytoplasmic male sterility with *Brassica napus* plastids using X-irradiated CMS protoplasts. Plant Cell Rep 6:98–101

Meyerowitz EM, Pruitt RE (1985) *Arabidopsis thaliana* and plant molecular genetics. Science 229:1214–1218

Morgan A, Maliga P (1987) Rapid chloroplast segregation and recombination of mitochondrial DNA in Brassica cybrids. Mol Gen Genet 209:240–246

Morinaga T (1934) Interspecific hybridization in *Brassica*. VI. The cytology of F1 hybrids of *B. juncea* and *B. nigra*. Cytologia 6:62–67

Nasrallah JB, Kao T-H, Goldberg ML, Nasrallah ME (1985) A cDNA clone encoding an S-locus-specific glycoprotein from *Brassica oleracea*. Science 318:263–267

Negrutiu I, Jacobs M, Cattoir-Reynaerts A (1984) Progress in cellular engineering of plants: biochemical and genetic assessment of selectable markers from cultured cells. Plant Mol Biol 3:289–302

Negrutiu I, DeBrouwer D, Watts JW, Sidorov VI, Dirks R, Jacobs M (1986) Fusion of plant protoplasts: a study using auxotrophic mutants of *Nicotiana plumbaginifolia*, Viviani. Theor Appl Genet 72:279–286

Newell CA, Rhoads ML, Bidney DL (1984) Cytogenetic analysis of plants regenerated from tissue explants and mesophyll protoplasts of winter rape, *Brassica napus* L. Can J Genet Cytol 26:752–761

Ogura H (1968) Studies of the new male sterility in japanese radish with special reference to the utilization of this sterility towards the practical raising of hybrid seed. Mem Fac Agric Kagoshima Univ 6:39–78

Olsson G (1960) Species crosses within the genus *Brassica*. 1. Artificial *Brassica juncea* Coss. Hereditas 46:171–223

Ooms G, Bains A, Burrell M, Karp A, Tuell D, Wilcox E (1985) Genetic manipulation in cultivars of oilseed rape (*Brassica napus*) using *Agrobacterium*. Theor Appl Genet 71:325–329

Paszkowski J, Pisan B, Shillito RD, Hohn T, Hohn B, Potrykus I (1986) Genetic transformation of *Brassica campestris* var. *rapa* protoplasts with an engineered Cauliflower Mosaic Virus genome. Plant Mol Biol 6:303–312

Patnaik G, Cocking EC, Hamill J, Pental D (1982) A simple procedure for the manual isolation and identification of plant heterokaryons. Plant Sci Lett 24:105–110

Pauls KP, Chuong PV (1986) Flow cytometric identification of *Brassica napus* protoplast fusion products. Can J Bot 65:834–838

Pelletier C, Primand C, Vedel F, Chetrie P, Remy R, Rousselle P, Renard M (1983) Intergeneric cytoplasmic hybridization in Cruciferae by protoplast fusion. Mol Gen Genet 191:244–250

Pontecorvo G (1971) Induction of directional chromosome elimination in somatic cell hybrids. Nature (Lond) 230:367–369

Pua E-C, Mehra-Palta A, Nagy F, Chua N-H (1987) Transgenic plants of *Brassica napus* L. Biotechnology 5:815–817

Robertson D, Earle ED (1986) Plant regeneration from leaf protoplasts of *Brassica oleracea* var. *italica* cv. green comet broccoli. Plant Cell Rep 5:61–64

Robertson D, Earle ED (1987) Nitro-blue tetrazolium: A specific stain for photosynthetic activity in protoplasts. Plant Cell Rep 6:70–73

Robertson D, Palmer JD, Earle ED, Mutschler MA (1987) Analysis of organelle genomes in a somatic hybrid derived from cytoplasmic male-sterile *Brassica oleracea* and atrazine-resistant *B. campestris*. Theor Appl Genet 74:303–309

Roy NN (1984) Interspecific transfer of *Brassica juncea*-type high blackleg resistance to *Brassica napus*. Euphytica 33:295–304

Schenck HR, Röbbelen G (1982) Somatic hybrids by fusion of protoplasts from *Brassica oleracea* and *B. campestris*. Z Pflanzenz 89:278–288

Shah DM, Horsch RB, Klee HJ, Kishore GM, Winter JA, Tumer NE, Hironaka CM, Sanders PR, Gasser CS, Aykent S, Siegel NR, Rogers SG, Fraley RT (1986) Engineering herbicide tolerance in transgenic plants. Science 233:478–481

Shahin EA, Spivey R (1986) A dominant gene for *Fusarium* wilt resistance in protoplast-derived tomato plants. Theor Appl Genet 73:164–169

Shiga T, Baba S (1971) Cytoplasmic male sterility in rape plants (*Brassica napus* L.). Jpn J Breed 21:16–17

Shiga T, Baba S (1973) Cytoplasmic male sterility in oilseed rape, *Brassica napus* L., and its utilization to breeding. Jpn J Breed 23:187–197

Spangenberg G, Schweiger H-G (1986) Controlled electrofusion of different types of protoplasts and subprotoplasts including cell reconstitution in *Brassica napus* L. Eur J Cell Biol 41:51–56

Spangenberg G, Neuhaus G, Schweiger H-G (1986) Expression of foreign genes in a higher plant cell after electrofusion-mediated cell reconstitution of a microinjected karyoplast and a cytoplast. Eur J Cell Biol 42:236–238

Sundberg E, Glimelius K (1986) A method for production of interspecific hybrids within *Brassiceae* via somatic hybridization, using resynthesis of *Brassica napus* as a model. Plant Sci 43:155–162

Taguchi T, Kameya T (1986) Production of somatic hybrid plants between cabbage and chinese cabbage through protoplast fusion. Jpn J Breed 36:186–189

Tanaka N, Hayakawa M, Mano Y, Ohkawa H, Matsui C (1985) Infection of turnip and radish storage roots with *Agrobacterium rhizogenes*. Plant Cell Rep 4:74–77

Terada R, Yamashita Y, Nishibayashi S, Shimamoto K (1987) Somatic hybrids between *Brassica oleracea* and *B. campestris*: selection by the use of iodoacetamide inactivation and regeneration ability. Theor Appl Genet 73:379–384

Thompson KF (1972) Cytoplasmic male sterility in oilseed rape. Heredity 29:253–257

Toriyama K, Kameya T, Hinata K (1987a) Selection of a universal hybridizer in *Sinapis turgida* Del. and regeneration of plantlets from somatic hybrids with *Brassica* species. Planta 170:308–313

Toriyama K, Hinata K, Kameya T (1987b) Production of somatic hybrid plants, "Brassicomoricandia", through protoplast fusion between *Moricandia arvensis* and *Brassica oleracea*. Plant Sci 48:123–128

Trail F, Wu FS (1986) Transformation in *Brassica* spp.: Sensitivity to antibiotics in vitro and tumor formation in vivo (Abstract). VIth Int Congr Plant tissue and cell culture, Univ Minnesota, Minneapolis, p 203

U N (1935) Genome analysis in *Brassica* with special reference to the experimental formation of *B. napus* and peculiar mode of fertilization. Jpn J Bot 7:389–452

Vaeck M, Reynaerts A, Hofte H, Jansens S, De Beuckeleer M, Dean C, Zabeau M, Van Montagu M, Leemans J (1987) Transgenic plants protected from insect attack. Nature (Lond) 328:33–37

Williams P, Hill CB (1986) Rapid-cycling populations of *Brassica*. Science 232:1385–1389

Wu FS (1987) Localization of mitochondria in plant cells by vital staining with rhodamine 123. Planta 171:346–357

Wu FS, Richards C, Trail F (1987) Calli formation and regeneration of hypocotyl protoplasts fused with evacuolated and enucleated mesophyll protoplasts in *Brassica oleracea* (Abstract). In Vitro 23:45A

Xu ZH, Davey MR, Cocking EC (1982) Plant regeneration from root protoplasts of *Brassica*. Plant Sci Lett 24:117–121

Yarrow SA, Wu SC, Barsby TL, Kemble RJ, Shepard JF (1986) The introduction of CMS mitochondria to triazine tolerant *Brassica napus* L., var. "Regent", by micromanipulation of individual heterokaryons. Plant Cell Rep 5:415–418

Zachrisson A, Bornman CH (1984) Application of electric field fusion in plant tissue culture. Physiol Plant 61:314–320

Zachrisson A, Bornman CH (1986) Electromanipulation of plant protoplasts. Physiol Plant 67:507–516

Zimmermann U, Scheurich P (1981) High frequency fusion of plant protoplasts by electric fields. Planta 151:26–32

# XIII Direct Gene Transfer in Protoplasts of *Nicotiana plumbaginifolia*

F. SALA[1,2], M.L. MARCHESI[2], S. CASTIGLIONE[2], J. PASZKOWSKI[3], M. SAUL[3], I. POTRYKUS[3], and I. NEGRUTIU[4]

## 1 Introduction

Different methodologies to transfer foreign genes into higher plants have recently been developed. These include: (1) insertion of the gene into the T-DNA region of the Ti plasmid of *Agrobacterium tumefaciens* (or of the Ri plasmid of *A. rhizogenes*) followed by infection of tissues (Gheysen et al. 1985); (2) direct gene transfer from bacterial plasmids to protoplasts (Paszkowski et al. 1984; Shillito et al. 1985); (3) insertion of the gene into a plant virus used to infect plants (Brisson et al. 1984); (4) microinjection of plasmid or nuclear DNA into plant nuclei (Reich et al. 1986); (5) injection of DNA into young floral tillers (De La Pena et al. 1987); (6) delivery of DNA into plant cells using high-velocity microprojectiles (Klein et al. 1987).

Unfortunately, none of these is of universal application. In most cases, having been established for specific plant species, the methodologies have to be adapted to different plants or, sometimes, cannot be adapted at all. In fact, the most widely used system, the one based on the *Agrobacterium* plasmids (pTi or pRi), is not yet amenable for plants of the utmost agricultural importance, such as the cereal plants. The use of viral vectors is limited by different problems, among them the narrow host range of the plant DNA viruses. Microinjection is still laborious. Direct gene transfer appears of more universal use but finds its limitations in the necessity for regeneration systems from protoplasts, which, again, are critical for many plants, including cereals. The potentialities of DNA injection and of microprojectiles have yet to be defined.

Thus, further experimentation is demanded in order to extend the possibility of transferring genes to all plants of experimental or agricultural interest.

In our case, after setting up a highly effective methodology for direct gene transfer to protoplasts of *Nicotiana tabacum* (Paszkowski et al. 1984; Potrykus et al. 1985; Shillito et al. 1985) based on a sequence of treatments including heat shock, incubation with PEG, and electroporation, it was soon learnt that this was not equally effective with protoplasts of the closely related species *N. plumbaginifolia.*

[1] Department of Botany, Univ. of Parma, Via Farini, Parma, Italy
[2] Department of Genetics and Microbiology, Univ. of Pavia, Via S. Epifanio, 27100 Pavia, Italy
[3] Institut für Pflanzenwissenschaften, ETH-Zentrum, 8092 Zürich, Switzerland
[4] Instituut voor Moleculaire Biologie, 65 Paardenstraat, B-1640 Saint Genesius-Rode, Belgium

Biotechnology in Agriculture and Forestry, Vol. 9
Plant Protoplasts and Genetic Engineering II (ed. by Y.P.S. Bajaj)

In an attempt to optimize conditions for direct gene transfer to this plant, numerous transformation parameters were analyzed and simple techniques were established. The result is the definition of two different protocols, one based on a simple PEG treatment to protoplasts in the presence of $Mg^{2+}$ and the other on electroporation. In both cases the efficiency of transformation is increased with protoplasts derived from both leaves and cultured cells, although it remains more than ten fold lower than that obtained, with a similar protocol, with *N. tabacum*. Possibly this indicates a species-specific component for "competence" of transformation.

A procedure for transient expression of a foreign gene introduced by a PEG treatment into protoplasts was also established.

In this chapter we report the optimized methodologies and define the influence of the major factors in the transformation schemes.

## 2 A Simple Chemical Procedure for the Stable Transformation of *Nicotiana plumbaginifolia* Protoplasts

When protoplasts of *N. plumbaginifolia* were transformed according to the standard protocol of Shillito et al. (1985) involving heat shock, incubation with PEG, and electric field treatment, under conditions yielding the highest transformation rates with *N. tabacum*, the transformation frequencies were two orders of magnitude below those reported for tobacco (Negrutiu et al. 1987). No improvement was observed when the resistance and/or capacitor values were varied for electroporation. Interestingly, controls without heat treatment and without electroporation gave similar results (unpubl.).

It was also evident that the electroporation step considerably reduced (usually below 10% of controls without treatments) the survival rate of *N. plumbaginifolia* protoplasts (Negrutiu et al. 1987).

Thus, we performed a series of experiments aimed at defining essential parameters of the protocol. These led to a simple chemical procedure, based solely on the synergic interaction of $MgCl_2$ and PEG (Negrutiu et al. 1987).

**Table 1.** Protocol for stable transformation of *Nicotiana plumbaginifolia* through a PEG treatment of protoplasts

| Operation | Description |
|---|---|
| 1. Cell growth | Leaves (2–5 cm in length) from sterile shoot cultures or cell suspension cultures (exponential phase of growth)[a] |
| 2. Protoplast isolation | With 1% cellulase RS+1% macerozyme R10 dissolved in K3 medium containing 0.4 M sucrose (for leaf pieces) or in MS medium containing 0.4 M mannitol+1.0 mM MES (for cultured cells)<br>Incubate overnight at 26 °C in the dark<br>Filter with a 63-μ mesh and float (leaf protoplasts) or sediment (cultured cells)[b] |
| 3. Protoplast wash | Wash twice[b] in $W_5$ and store in the same solution |

**Table 1** (continued)

| Operation | Description |
|---|---|
| 4. Transformation | Replace $W_5$ with MaMg (25 mM $MgCl_2$, 0.1% MES and 0.4 M mannitol, pH 5.6)[c] at $1.5 \times 10^6$ protoplast per ml<br>Take 0.3 ml aliquots and add, in the order, calf thymus DNA and linear plasmid (final concentrations: 50 and 10 µg/ml, respectively[d]). Alternatively, mix the two DNA's before addition to the protoplast suspension<br>Add 72% PEG to a final concentration of 24%[e]<br>Mix gently and incubate at room temperature (22–25 °C) for 20 min |
| 5. Protoplast wash | Dilute the protoplast suspension with 5–10 volumes of $W_5$ or 0.2 M $CaCl_2 \cdot 2H_2O$ and pellet[b] |
| 6. Protoplast culture | Resuspend protoplasts in K3 medium containing 0.4 M glucose (final protoplast concentration: $3 \times 10^4$ per ml) and transfer into Petri dishes<br>Incubate at 26 °C in the dark |
| 7. Selection for kanamycin-resistant colonies | After 8 days count colonies composed of 8–16 cells, pellet, resuspend at $3 \times 10^3$ colonies per ml in K3 medium containing 0.4 M glucose + 0.8% agar and transfer 3 ml to the center of a Petri dish (9 cm in diameter)<br>After agar has solidified, add 12 ml of K3 medium containing 0.4 M glucose + 50 µg/ml kanamycin<br>Incubate at 26 °C<br>After 10 days, replace liquid medium with K3 medium containing 0.2 M glucose + 50 µg/ml kanamycin<br>After 10 days, count visible and healthy colonies |
| 8. Regeneration of kanamycin-resistant plants | Transfer colonies to Petri dishes containing RMOP medium containing 50 µg/ml kanamycin + 0.8% agar and incubate at 26 °C in the light (5000–10000 lx)<br>After about 1 month transfer calli to RP medium + 50 µg/ml kanamycin + 0.8% agar. Continue incubation in the same conditions<br>After 1 month (or more) transfer the differentiated shoots to RSA medium + 50 µg/ml kanamycin + 0.8% agar<br>Fully developed plantlets can be maintained and propagated in MS medium + 0.8% agar (without hormones) or transferred to soil |

[a] Sterile shoot cultures were initiated by sterilizing *N. plumbaginifolia* seeds in 70% ethanol for 30 s followed by 2 min in 0.5% NaOCl + 0.04% Teepol and by six washes in sterile water. Seeds were then treated with 0.5 mg/ml $GA_3$ for 2 h and germinated in vitro in MS medium + 0.8% agar (no hormones). Shoots were grown and propagated, by apical bud transplantation, on the same medium. Young in vivo-grown plants may also be used, after leaf sterilization with 1% NaOCl + 0.04% Teepol for 20 min. Cell suspension cultures were obtained from R. Shields, Unilever laboratory, Sharnbrook, Bedford, UK (Barfield et al. 1985) and serially subcultured in MS medium supplemented with 0.5 mg/l 2,4-D and 0.5 mg/l 6-BAP.

[b] Protoplasts are floated or washed by low speed centrifugation.

[c] Alternatively, use $W_5$ + 15–25 mM $MgCl_2$. This is recommended especially for cell suspension cultures.

[d] We use a sterile solution of calf thymus DNA at 1 mg/ml and pABD1 or pHP23, linearized with SmaI or NdeI, respectively, and stored at 0.5 mg/ml in water.

[e] PEG 4000 or 6000 are dissolved at 72% concentration in MaMg, pH corrected to 7–9, sterilized by autoclaving at 0.8 Atm (118 °C) for 20 min and stored at −20 °C.

Our materials are from the following sources: Cellulase "Onozuka" RS and Macerozyme R10 from Yakult Honsha Co, Tokio, Japan; Zeatin Riboside, 6-BAP (6-Benzyl-Aminopurine), 2,4-D (2,4-dichlorophenoxyacetic acid), NAA (naphthalene acetic acid) and MES (2-N-morpholino-ethane sulfonic acid) from Sigma; Calf thymus DNA from Sigma or Serva. Polyethyleneglycol CMS4 and CMS6 (PEG 4000 and 6000) from Merck. Teepol 610 from BDH. Agar, of commercial grade, from Difco.

**Table 2.** Composition of the culture media used in the transformation and culture of *N. plumbaginifolia* protoplasts

| Compound | K3 (mg/l) | RMOP (mg/l) | RP (mg/l) | R'SA (mg/l) | MS (mg/l) |
|---|---|---|---|---|---|
| Macronutrients | | | | | |
| $NH_4NO_3$ | 250 | 1650 | – | 1650 | 1650 |
| $(NH_4)_2SO_4$ | 134 | – | 233 | – | – |
| $KNO_3$ | 2500 | 1900 | 273 | 1900 | 1900 |
| $Ca(NO_3)_2 \cdot 4H_2O$ | – | – | 416 | – | – |
| $CaCl_2 \cdot 2H_2O$ | 716 | 440 | – | 440 | 440 |
| $KH_2PO_4$ | – | 170 | 271 | 170 | 170 |
| $NaH_2PO_4 \cdot H_2O$ | 150 | – | – | – | – |
| $MgSO_4 \cdot 7H_2O$ | 250 | 370 | 57 | 370 | 370 |
| $Mg(NO_3)_2 \cdot 6H_2O$ | – | – | 392 | – | – |
| Micronutrients | | | | | |
| $MnSO_4 \cdot 4H_2O$ | – | 22.3 | 4.46 | 22.3 | 22.3 |
| $MnSO_4 \cdot H_2O$ | 10 | – | – | – | – |
| $ZnSO_4 \cdot 7H_2O$ | 2 | 8.6 | 1.72 | 8.6 | 8.6 |
| $H_3BO_3$ | 3 | 6.2 | 1.24 | 6.2 | 6.2 |
| KI | 0.75 | 0.83 | 0.166 | 0.83 | 0.83 |
| $CuSO_4 \cdot 5H_2O$ | 0.025 | 0.025 | 0.005 | 0.025 | 0.025 |
| $Na_2MoO_4 \cdot 2H_2O$ | 0.25 | 0.25 | 0.05 | 0.25 | 0.25 |
| $CoCl_2 \cdot 6H_2O$ | 0.025 | 0.025 | 0.005 | 0.025 | 0.025 |
| Na-Fe-EDTA | 43 | – | 35 | – | – |
| $FeSO_4 \cdot 7H_2O$ | – | 27.85 | – | 27.85 | 27.85 |
| $Na_2$-EDTA | – | 37.25 | – | 37.25 | 37.25 |
| Vitamins | | | | | |
| Inositol | 100 | 100 | 100 | – | 100 |
| Nicotinic acid | 1 | – | 1 | 1 | 0.5 |
| Pyridoxine HCl | 1 | – | 1 | 1 | 0.5 |
| Thiamine HCl | 10 | 1 | 1 | 1 | 0.1 |
| Biotine | – | – | 0.01 | 0.01 | – |
| Ca-Pantothenate | – | – | 1 | 1 | – |
| L-amino acids | | | | | |
| Cysteine HCl | – | – | – | 1 | – |
| Glycine | – | – | – | – | 2 |
| Organic acids (adjusted to pH 5.5 with $NH_4OH$) | | | | | |
| Na-pyruvate | 20 | – | – | – | – |
| Citric acid | 40 | – | – | – | – |
| Malic acid | 40 | – | – | – | – |
| Fumaric acid | 40 | – | – | – | – |
| Hormones | | | | | |
| 6-BAP | 1 | 1 | – | – | 0.5 |
| NAA | 1 | 0.1 | – | – | – |
| 2,4-D | – | – | – | – | 0.5 |
| Zeatine riboside | – | – | 0.25 | – | – |
| Sugars | | | | | |
| Sucrose | [a] | – | 500 | 10000 | 30000[b] |
| Glucose | [a] | 36030 | – | – | – |
| Mannitol | – | – | 18000 | – | – |
| Xylose | 250 | – | – | – | – |

[a] Sugar is sucrose or glucose at the concentration given in Tables 1 and 5.
[b] Sucrose is substituted by 0.4 M mannitol for protoplast isolation.
pH is corrected to 5.6 (with NaOH) in all media. Agar (0.8%) is added for solid media. Sterilization is at 0.8 Atm (118 °C) for 20 min.

The main steps are summarized in Table 1, while Table 2 describes the necessary culture media. The procedure makes use of protoplasts and of a minimal DNA vector, an *E. coli* plasmid containing, in our case, the kanamycin-resistance gene from Tn5 and gene expression signals from the gene VI of CaMV.

The source of the protoplasts does not appear to be critical, since comparable transformation frequency is obtained with protoplasts derived from leaves of cultured cells.

Critical parameters and treatments, as well as those which turned out to be of no significant effect on the established protocol are given below.

## 2.1 Parameters of Critical Effect on Transformation Frequency

These are related to the following steps:

*1. Incubation Solution for Protoplast Transformation.* Different salt solutions were tested as resuspending media for protoplast transformation. These were:
MaMg: 0.4 M mannitol, 25 mM $MgCl_2$, 0.1% MES (morpholinoethanesulphonic acid), pH 5.6.
$W_5$: 154 mM NaCl, 125 mM $CaCl_2 \cdot 2H_2O$, 5 mM KCl; 5 mM glucose, pH 5.6–6 (Menczel et al. 1981).
MaCa: 0.4 M mannitol, 15 mM $CaCl_2 \cdot 2H_2O$, pH 5.6.
HeNa/F: 10 mM Hepes, pH 7.1, 5 mM $CaCl_2$, 150 mM NaCl, 0.2 M mannitol (Fromm et al. 1985).

MaMg consistently proved to be the most satisfactory medium, with a clearcut effect of $Mg^{2+}$, whose role appeared to be specific. There is an optimum $MgCl_2$ concentration at 20–30 mM (Negrutiu et al. 1987): thus 25 mM $MgCl_2$ was adopted in the transformation protocol of Table 1. These conditions are similar to those established for *N. tabacum* except that, in the latter case, the optimal $MgCl_2$ concentration in the MaMg solution is shifted to 15 mM (Negrutiu et al. 1987).

It should be noticed that the MaMg salt solution does not appear to be exclusive: it can effectively be replaced by $W_5$, provided $MgCl_2$ (at 15–25 mM) is added. Furthermore, recent results have shown that the Hepes-buffered saline solution (HeNa/F) utilized by Fromm et al. (1985) for their electroporation experiments could also substitute the MaMg salt mixture in the protocol of Table 1. This is surprising, since HeNa/F contains 5 mM $CaCl_2$ but no $MgCl_2$.

*2. DNA Concentration, Form and Type.* Plasmid concentrations of 5–10 μg/ml and carrier DNA at 50 μg/ml (usually calf-thymus DNA with an average size of 4–40 kbp) were established as most satisfactory. Linearized plasmids were the most efficient, while two different plasmids, pABD1 and pHP23, transformed at similar rates (Table 3). The use of circular plasmids and the delection of carrier DNA had drastic additive effects in reducing the transformation frequency.

*3. PEG Treatment of the Protoplast/DNA Mixture.* PEG was the essential factor in the established chemical procedure for direct gene transfer. No transformants were found in the absence of PEG, while, as Table 3 shows, the higher the PEG

**Table 3.** Effect of different experimental conditions on the frequency of transformation for kanamycin resistance in leaf protoplast cultures of *Nicotiana plumbaginifolia*

| Transformation conditions | | | Treated proto- plasts ($\times 10^5$) | Survival % | Transformants | | |
|---|---|---|---|---|---|---|---|
| | | | | | Total number | % of protoplast- derived colonies | % of treated protoplast |
| 1. DNA additions[a] | | | | | | | |
| Linear pABD1 | Linear pHP23 (µg/ml) | Calf thymus DNA | | | | | |
| – | – | 50 | 5 | 18.0 | 0 | 0 | 0 |
| 10 | – | 50 | 5 | 17.8 | 24 | 0.017 | 0.0048 |
| 10 | – | – | 5 | 16.1 | 2 | 0.001 | 0.0004 |
| 50 | – | – | 5 | 15.8 | 3 | 0.002 | 0.0006 |
| – | 10 | 50 | 3[b] | 16.3 | 19 | 0.023 | 0.0042 |
| 10 | – | 50 | 3[b] | 14.8 | 21 | 0.028 | 0.0042 |
| Circular pABD1 | (µg/ml) | Calf thymus DNA | | | | | |
| 10 | | 50 | 5 | 15.5 | 0 | 0 | 0 |
| 50 | | 10 | 5 | 17.8 | 0 | 0 | 0 |
| 2. PEG concentration[a] | | | | | | | |
| PEG CMS4 | | | | | | | |
| 8% | | | 7.2 | – | 2 | – | 0.0003 |
| 13% | | | 10.5 | – | 7 | – | 0.0007 |
| 20% | | | 9.0 | – | 22 | – | 0.0024 |
| 24% | | | 10.0 | – | 32 | – | 0.0032 |
| 27% | | | 10.0 | – | 46 | – | 0.0046 |
| 3. Electroporation[c] | | | | | | | |
| PEG | | Volts/0.3 cm | | | | | |
| – | | 100 | 8.0 | 9.8 | 29 | 0.037 | 0.0036 |
| – | | 170 | 8.0 | 4.7 | 39 | 0.103 | 0.0049 |
| 8% | | 100 | 8.0 | 8.9 | 3 | 0.004 | 0.0004 |
| 16% | | 100 | 8.0 | 6.9 | 20 | 0.036 | 0.0025 |
| 24% | | 100 | 8.0 | 3.0 | 17 | 0.075 | 0.0021 |

[a] Transformation conditions were as described in Table 1, except that DNA or PEG were varied as indicated.

[b] In these experiments protoplasts were from cell suspension cultures.

[c] Electroporation was performed as described in Table 4 with a 940-µF capacitor charged at the voltage indicated in the table. Electrode distance was 0.3 cm. This corresponds to an electric field intensity of 330 and 560 V/cm and RC time constant of 230 ms in the two tested conditions.

concentration the more transformants were produced. A 24% final PEG concentration is suggested as most satisfactory, since a 72% stock solution can be easily prepared and handled. Going above this value is technically difficult, in terms of handling, conservation, and dilution of the transformation mixture.

Molecular weight of PEG does not appear to be critical. However a PEG CMS 4000 or 6000 (at pH 7–9) turned out more satisfactory than lower molecular weight PEG.

## 2.2 Parameters with no Significant Effect

These included:

a) The protoplast source (leaves or cultured cells) and the ploidy level of *N. plumbaginifolia.*

b) The protoplast isolation procedure (enzyme composition and time of incubation), provided the protoplasts are healthy and thus able to withstand the PEG treatment.

c) The heat shock at 45 °C for 5 min which was effective in the formerly established protocol for *N. tabacum* (Shillito et al. 1985). Under the established conditions it had a neutral effect on the frequency of transformation of *N. plumbaginifolia* and was thus deleted.

d) The time interval between the preparation of protoplasts (in $W_5$) and the transfer of protoplasts into the MaMg salt solution. In fact, protoplasts could be transformed immediately or kept at room temperature in $W_5$ up to 6 h before proceeding with the transformation protocol.

e) The duration of incubation with PEG, between 20 min and 6 h.

## 2.3 Parameters with Negative Effect

DB (dichlorobenzonitrile) selectively blocks cell wall synthesis. Its addition to the protoplast preparation, in an attempt to delay cell wall appearance, had no positive effect, and also strongly affected the survival of cultures.

## 2.4 Molecular Evidence and Progeny Analysis

Transformed *N. plumbaginifolia* clones have acquired the Tn5 kanamycin gene at specific physical locations within the genome. This was shown by Southern blot analysis, following extraction of DNA from several transformed clones (Fig. 1).

Progeny analysis of a representative sample of regenerated transformed plants has also shown that the gene is stably inherited into the progeny and segregates as a Mendelian trait, usually monogenic.

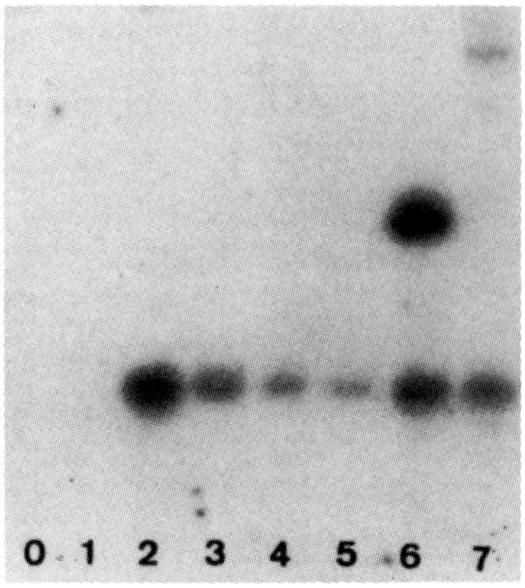


**Fig. 1.** Integration pattern of the Tn5 kanamycin resistance gene in plants of *N. plumbaginifolia* regenerated from leaf protoplasts. Plant DNA was digested with EcolR5, 5–10 μg loaded on a 0.8% agarose gel, and probed with the EcoR5 1.2 kb fragment of pABD1. Lane *0* wild type control; lane *1* a hygromycin B transformant; lanes *2–7* plants from different transformation events

## 3 Stable Transformation of *Nicotiana plumbaginifolia* by Electroporation

Fromm et al. (1986) recently described an electroporation protocol effective on maize protoplasts which did not require PEG treatment. This alternative protocol was tested on *N. plumbaginifolia* protoplasts and proved to be effective (Table 4).

In fact, as shown in Table 3, 0.1% of the protoplast-derived colonies (0.005% of treated protoplasts) were transformed by a 170-V pulse; transformation is di-

**Table 4.** Stable transformation of *Nicotiana plumbaginifolia* by electroporation

| Operation | Description |
|---|---|
| 1. Cell growth | As in Table 1 |
| 2. Protoplast isolation | As in Table 1 |
| 3. Protoplast wash | As in Table 1 |
| 4. Transformation | Replace $W_5$ with HeNa/F salt mixture (10 mM Hepes, pH 7.1, 5 mM $CaCl_2$, 150 mM NaCl, 0.2 M mannitol) at 1.2–1.5 $\times 10^6$ protoplasts per ml. Transfer 0.3 ml aliquotes into appropriate wells and add, in the order, calf thymus DNA and linear plasmid (final concentrations: 50 and 10 μg/ml, respectively[a]). Alternatively, mix the two DNAs before addition to the protoplast suspension. Mix gently and electroporate at 500–800 Volts/cm, 900–1000 μF |
| 5. Protoplast culture | As in Table 1[b] |
| 6. Selection for kanamycin-resistant colonies | As in Table 1 |
| 7. Regeneration of kanamycin-resistant plants | As in Table 1 |

[a] Solution of DNA are prepared as in Table 1.
[b] It is not necessary to wash protoplasts before culturing.

rectly dependent on the voltage used, although the higher voltages are detrimental to survival. PEG is not only unnecessary, but its addition considerably reduces both survival and total number of transformants.

The efficiency obtained with this simplified electroporation protocol parallels that obtained with the chemical procedure described above. Thus, this method may be proposed as an alternative simple way for gene transfer into protoplasts of *N. plumbaginifolia.*

## 4 Transient Expression of a Foreign Gene Introduced by a PEG Treatment into *Nicotiana plumbaginifolia* Protoplasts

Fromm et al. (1985) developed an electroporation-mediated gene transfer system that allows the monitoring of foreign gene activity within hours, possibly before the integration of the gene into the cellular genome.

With *N. plumbaginifolia* we have found that, for this purpose, electroporation can be effectively substituted by an appropriate PEG treatment. In fact, in the experimental conditions summarized in Table 5, the chloramphenicol acetyltransferase (CAT) gene has been introduced into protoplasts and shown to be active after 24 h (Fig. 2). The procedure is extremely easy and the activity is considerably high, as judged from the autoradiographic analysis of enzyme activity.

**Table 5.** Transient expression of a foreign gene introduced by a PEG treatment into protoplasts of *Nicotiana plumbaginifolia*

| Operation | Description |
|---|---|
| 1. Cell growth | As in Table 1 |
| 2. Protoplast isolation | As in Table 1 |
| 3. Protoplast wash | As in Table 1 |
| 4. Transformation | Replace $W_5$ with a modified MaCa solution (25 mM $CaCl_2$, 0.1% MES and 0.5 M mannitol, pH 5.6) at $2.0-2.5 \times 10^6$ protoplasts per ml<br>Take 0.3 ml aliquots and add circular plasmid[a] at a final concentration of 10 µg/ml<br>Add PEG to a final concentration of 20% – 24%[b]<br>Mix gently and incubate at room temperature for 5 – 20 min |
| 5. Protoplast culture | Add 4 ml of K3 medium containing 0.4 M glucose and transfer into a 5-cm Petri dish<br>Incubate at 26 °C |
| 6. Assay of CAT activity | After 24 h pellet the protoplasts, wash in $W_5$ and assay for CAT activity[c] |

[a] Plasmid constructs based on the CaMV 35S promoter and the coding region of the chloramphenicol acetyltransferase gene are very efficient.
[b] PEG 4000 or 6000 is dissolved in the transformation buffer at 3 times the final concentration, pH is corrected to 7 – 9. The solution is sterilized by autoclaving at 0.8 Atm (118 °C) for 20 min and stored at – 20 °C.
[c] CAT activity is assayed in protoplast extracts following the procedure described by Gorman et al. (1982) and by Fromm et al. (1985).

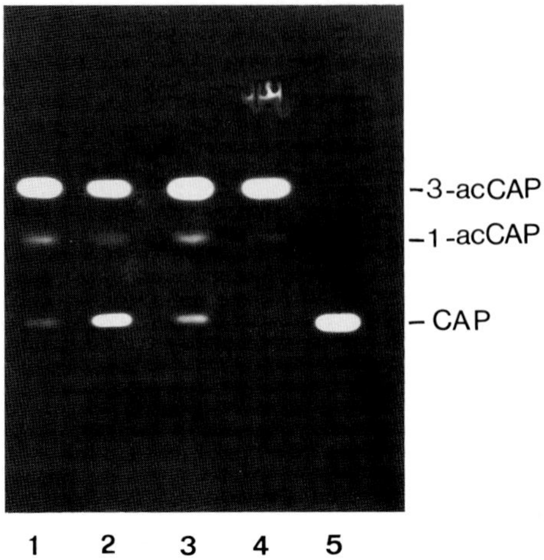


**Fig. 2.** Expression of the CAT gene introduced into *Nicotiana plumbaginifolia* protoplasts by the PEG treatment described is Table 4. CAT activity was measured 24 h after treating the protoplasts with the p19SCAT and p35SCAT constructs. The autoradiogram of the chromatographically separated products of the CAT activities are shown. The positions of untreated chloramphenicol (CAP) and of mono-acetylated chloramphenicol (1-acCAP and 3-acCAP) are indicated. *Lanes* CAT expression after transformation with p35SCAT (lanes *1, 3, 4*) or with p19SCAT (lane *2*) and in a control sample without PEG treatment (lane *5*)

## 5 Discussion

The high efficiency protocol for direct gene transfer to protoplasts of *N. tabacum* that was used as a starting point was based on three essential steps: heat shock to protoplasts, treatment with 8% PEG, and electroporation (Shillito et al. 1985). Our attempt to optimize conditions for *N. plumbaginifolia* have shown that these may differ for different, although closely related, plant species and have led to considerable simplified transformation procedures. Heat shock and electroporation have been deleted from the initial protocol, since they consistently proved to have no significant effect on the frequency of transformation of *N. plumbaginifolia* and now the procedure merely relies on a synergic interaction of PEG (at high dose) and $MgCl_2$.

Comparative experiments with protoplasts of *N. tabacum* have also brought to a similar simplification of the former protocol: here too electroporation was dispensable if PEG concentration was increased to 20%–24% (Negrutiu et al. 1987).

The comparison showed that the $Mg^{2+}$/PEG procedure is the method of choice for both *Nicotiana* species and that a species-specific susceptibility for transformation seems to operate, with $SR_1$ tobacco exhibiting a much higher efficiency than *N. plumbaginifolia*. In fact, in the newly established condition, transformation rates with *N. plumbaginifolia* remain at least an order of magnitude below those routinely obtained with *N. tabacum*.

The results have also shown that an appropriate PEG treatment can lead to transient expression of a foreign gene and that, under the specific conditions described by Fromm et al. (1985), *N. plumbaginifolia* can also be transformed by a simple electroporation treatment.

It should be clear that the reported protocols are just a summary of our present-day experience and should be intended as a starting point for further improvement, in respect to both protocol simplification and to efficiency of transformation.

Although it is conceivable that conditions will be further improved for *N. plumbaginifolia*, our impression is that the system now reacts as if our protoplast population was composed of "competent" (a limited number) and "noncompetent" individuals. A point in favour of this is represented by the fact that a comparable number of transformants is obtained when the protoplast resuspending salts are changed or when the chemical PEG treatment is substituted by the electroporation protocol.

Furthermore, the observation that transient expression is very efficient in *N. plumbaginifolia* strongly suggests that the limiting factor in the stable transformation within this species is not DNA uptake but rather the stable and functional integration of the gene into the plant genome.

## References


Barfield DG, Robinson SJ, Shields R (1985) Plant regeneration from protoplasts of long term haploid suspension cultures of *N. plumbaginifolia*. Plant Cell Reports 4:104–107

Brisson N, Paszkowski J, Penswick JR, Gronenborn B, Potrykus I, Hohn T (1984) Expression of a bacterial gene in plants by using a viral vector. Nature 310:511–514

De La Pena A, Lörz H, Schell J (1987) Transgenic rye plants obtained by injecting DNA into young floral tillers. Nature 325:274–276

Fromm ME, Taylor LP, Walbot V (1985) Expression of genes transferred into monocot and dicot plants cells by electroporation. Proc Natl Acad Sci USA 82:5824–5828

Fromm ME, Taylor LP, Walbot V (1986) Stable transformation of maize after gene transfer by electroporation. Nature 319:791–793

Gorman CM, Moffat LF, Howard BH (1982) Recombinant genomes which express chloramphenicol-acetyltransferase in mammalian cells. Molec Cell Biol 2:1044–1051

Gheysen G, Dhaese P, Van Montagu M, Schell J (1985) DNA flux across genetic barriers: the crown gall phenomenon. In: Hohn B, Dennis E (eds) Plant Gene Research. Genetic flux in plants. Springer, Vienna, pp 12–47

Klein TM, Wolf ED, Wu R, Sanford JC (1987). High velocity microprojectiles for delivering nucleic acids into living cells. Nature 327:70–73

Menczel L, Nagy F, Kiss Z, Maliga P (1985) Streptomycin-resistant and sensitive somatic hybrids of *N. tabacum*+*N. knitiana*: correlation of resistance to *N. tabacum* plastids. Theor Appl Genet 59:191–195

Negrutiu I, Shillito RD, Potrykus I, Biasini MG, Sala F (1987) Hybrid genes in the analysis of transformation conditions. I. Setting up a simple method for direct gene transfer in plant protoplasts. Plant Mol Biol 8:363–373

Paszkowsky J, Shillito RD, Saul MW, Mandak V, Hohn J, Hohn B, Potrykus I (1984) Direct gene transfer to plants. EMBO J 3:2717–2722

Potrykus I, Shillito RD, Saul MW, Paszkowski J (1985) Direct gene transfer. State of the art and future potential. Plant Mol Biol Rep 3:117–128

Reich TJ, Iyer VN, Miki BL (1986) Efficient transformation of alfalfa protoplasts by the intranuclear microinjection of Ti plasmids. Biotechnology 4:1001–1004

Shillito RD, Saul MW, Paszkowski J, Müller M, Potrykus I (1985) High efficiency direct gene transfer to plants. Biotechnology 3:1099–1103

# XIV Microinjection of Cells and Protoplasts: Integration of Foreign DNA

A. CROSSWAY[1]

## 1 Introduction

A number of plant transformation techniques are currently available, none of which has proven to be universally applicable. These methods include *Agrobacterium*-mediated gene transfer (Nester et al. 1984), direct DNA uptake via chemical mediation (Krens and Schilperoort 1984), liposome fusion (Deshayes et al. 1985) or electroporation (Fromm et al. 1986), and direct DNA microinjection (Crossway et al. 1986c; Reich et al. 1986c). These methods are currently limited by either the host-range restrictions of *Agrobacterium* or by the species restrictions on plant regeneration from protoplasts. These barriers are gradually being overcome; the most notable example is, perhaps, the regeneration of rice from protoplasts (Coulibaly and Demarly 1986; Yamada et al. 1986; Abdullah et al. 1986; Toriyama et al. 1986; also see Kyozuka et al. 1989). Transformation techniques are now being applied to a variety of regenerable plant cell types (especially those with intact cell walls) in order to broaden the range of plant species which can be transformed. As both the transformation and cell culture technologies advance (see Bajaj 1986), a repertoire of plant transformation technologies will be built that will allow selection of the method most suited to a particular cell type or plant species based on the efficiency of transformation.

The overall efficiency of any plant transformation method is the product of several factors: (1) the frequency of cell survival following the transformation treatment, (2) transformation frequency (which is the product of DNA integration frequency and expression frequency), and (3) regeneration frequency with a particular plant cell culture regime. Of primary consideration in the recovery of a plant expressing a foreign gene is the integration of an intact copy of the gene into the recipient plant genome. These aspects of plant cell transformation via microinjection are examined here.

## 2 Techniques

### 2.1 Historical Summary

Micromanipulations were used in biological research as early as 1859 (Schmidt 1859). Although micromanipulation of living plant cells began in 1911 (Barber

[1] Calgene Incorporation, 1920 Fifth St., Davis, CA 95616, USA

Biotechnology in Agriculture and Forestry, Vol. 9
Plant Protoplasts and Genetic Engineering II (ed. by Y.P.S. Bajaj)

1911a, b), its application to plant transformation was dependent on the development of cell culture and regeneration technology and on gene isolation and expression technologies using molecular biology. These developments, which led to the current efforts on intracellular microinjection in plants, have been reviewed previously (Crossway et al. 1986a) and are summarized below.

Successful application of microinjection to animal cell transformation preceded similar work with plants by approximately a decade. The first successful transformation of some cells of mature mice via microinjection of purified DNA into the blastocoel cavity of embryos was demonstrated in 1974 (Jaenisch and Mintz 1974). By the early 1980's, several groups had reported transformation of entire mice following intranuclear injection of eggs with cloned DNA (Gordon et al. 1980; T.E. Wagner et al. 1981; Harbers et al. 1981; E.F. Wagner et al. 1981; Brinster et al. 1981). Transgenic mice can now be routinely obtained at an average frequency of about 27% (Brinster et al. 1985) and work has quickly progressed to allow the transformation of more economically important animals such as rabbits, sheep, and pigs (Hammer et al. 1985).

## 2.2 Plant Cell Transformation

The techniques and strategies developed for use with animals have proven applicable to plants, with some modifications, and the problems associated with that technology transfer have been reviewed elsewhere (Crossway et al. 1986a). Intense work over the past 5 years has resulted in the development of suitable methods for immobilization of different plant cell types during microinjection and for their subsequent culture (see below and Crossway et al. 1986a). Positive pressure applied to microtools via volume changes of liquid or gas in attached tubing has generally been used to effect delivery of aqueous DNA solutions into cells. More recently, iontophoresis has been shown to deliver DNA molecules without additional aqueous volume by application of electrical pulses to the DNA solution in the microtools (Lo 1983; Hepher et al. 1985); this technique may prove advantageous in some plant applications.

Plant researchers patterned their immobilization strategies for microinjection of plant cells after techniques developed by animal microinjection groups. Surface-immobilization allowed for very fast rates of injection (up to thousands of cells per hour) with surface-adherent animal cells (Capecchi 1980; Graessmann et al. 1980; Kopchick et al. 1981). Because plant cells and protoplasts do not attach to surfaces as they grow, other methods have been developed for immobilizing plant cells in a two dimensional array for microinjection. Glass surfaces have been coated with polylysine (Steinbiss and Stabel 1983; Rasmussen et al. 1984; Gordon-Kamm and Bushnell 1985; Reich et al. 1986b) allowing protoplast attachment for microinjection (Fig. 1). Plant protoplasts immobilized in this way have been cultured as attached in liquid (Steinbiss and Stabel 1983) or embedded in agar (Reich et al. 1986a) or have been subsequently removed and cultured on filter paper feeder layers (Reich et al. 1986b). Alternatively, protoplasts have been embedded in agar or agarose as a means of immobilization during injection and for subsequent culture of the injected protoplasts (Griesbach 1983; Tilton and Russell

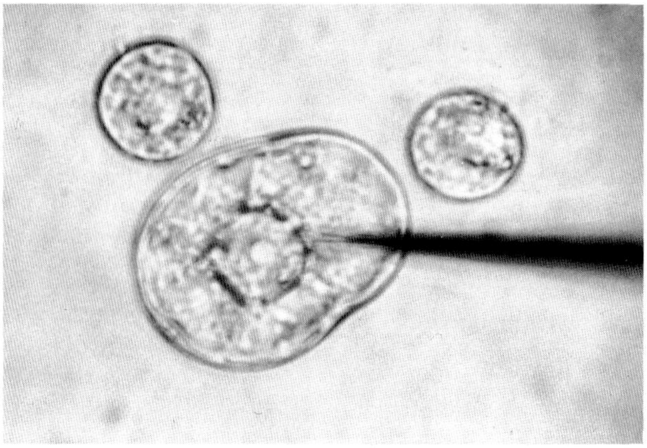

**Fig. 1.** Microinjection of *Nicotiana plumbaginifolia* suspension protoplast surface-immobilized using poly-L-lysine. (Courtesy of R. Rasmussen, Calgene Inc., Davis, CA)

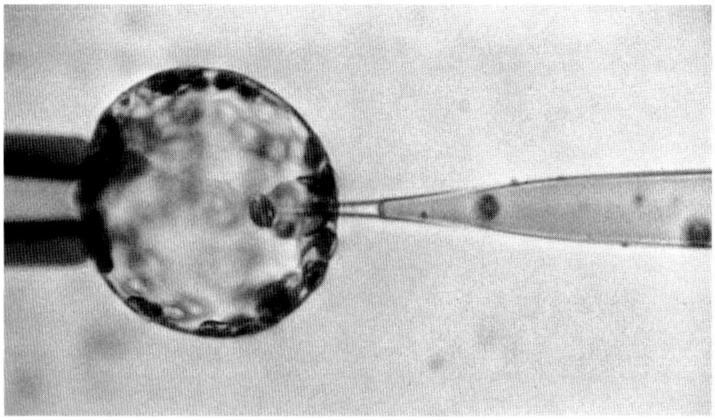

**Fig. 2.** Microinjection of *Nicotiana tabacum* mesophyll protoplast using holding pipette immobilization

1983; DeLaat and Blass 1985; Aly and Owens 1985; Lawrence and Davies 1985). Nonetheless, rates of injection attained with surface-immobilized plant protoplasts have only been about 100 per hour at best (Steinbiss and Stabel 1983; Griesbach 1985; Lawrence and Davies 1985; Reich et al. 1986b).

The holding pipette method devised by animal workers has also been successfully adapted as an immobilization method for plant protoplast microinjection (Fig. 2). Protoplasts injected by this method have then been cultured using methods which promote cell growth while maintaining the individual injected cell colonies as distinct entities. Injected protoplasts have been grown in a feeder culture system in which a nylon mesh ring embedded in agarose separates the injected protoplasts from the feeder culture (a protoplast suspension overlaying the agarose) (Morikawa and Yamada 1985). Alternatively, cells have been grown isolated in hanging drops of less than 0.25 microliter (Crossway et al. 1986b), or in 30-nanoliter lying drops under oil (Spangenberg et al. 1986). Protoplast injection rates using the holding pipette technology range from 10–100 per hour (Steinbiss and Stabel 1983; Morikawa and Yamada 1985; Lawrence and Davies 1985; Spangenberg et al. 1986); injection rates using holding pipette immobilization are much more dependent on the skill of the injector than with surface immobilization. However, each of these immobilization methods may have unique applications that are useful for plant transformation. Several variations on these two basic microinjection systems have been described (Crossway et al. 1986a) and the feasibility of plant cell transformation via protoplast microinjection has now been demonstrated with both (Crossway et al. 1986c; Reich et al. 1986c).

## 2.3 Technique of Microinjection

The holding pipette method of microinjection followed by hanging drop culture (Fig. 3) is used here to illustrate the techniques involved in plant protoplast

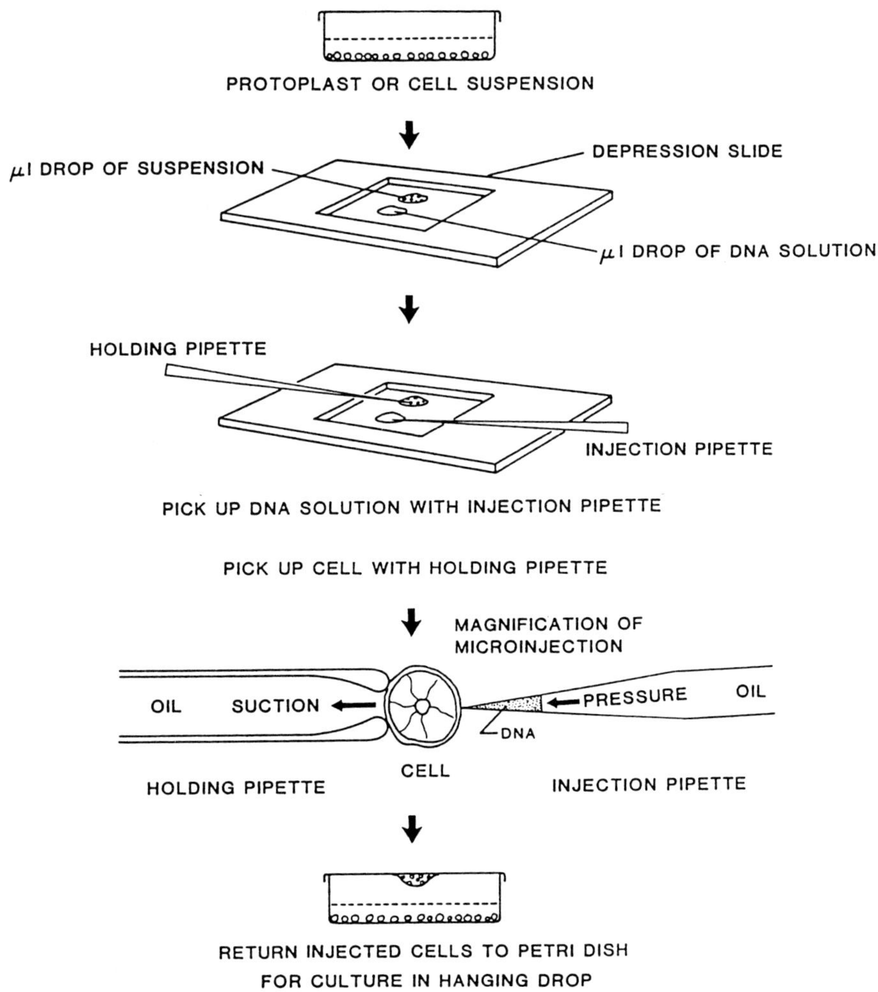


**Fig. 3.** Microinjection by the holding pipette method, followed by hanging drop culture. (Crossway et al. 1986b)

microinjection. Protoplasts are prepared from leaf explants or suspension cultures using protocols optimized for the particular plant species. Preculturing of the protoplasts prior to injection is performed in sealed Petri dishes in the dark for 1 – 7 days to allow partial cell wall regeneration which facilitates handling during injection. The cloned DNA to be injected is suspended in a buffer, such as 1XTE (10 mM Tris, pH 8; 1 mM EDTA), at the desired DNA concentration (typically in the range of 0.01 – 1 mg/ml) for injection. A drop of protoplasts in culture medium and a separate drop of the DNA solution are placed into depression slides (Crossway and Houck 1985). Injection pipettes (tip external diameter approximately 0.2 μm) and holding pipettes are pulled from glass capillary tubing on a pipette puller. Holding pipettes (5 – 10 μm internal diameter) are then manufactured on a microforge (Crossway and Houck 1985).

Microinjections are performed as described by T. E. Wagner et al. (1981). Protoplasts are picked up individually on the holding pipette by gentle suction and turned so that the nucleus is readily accessible to the injection pipette (Fig. 2). For intranuclear injections, the pipette is inserted into the nucleus and successful injection is assessed by observation of slight nuclear swelling. The quantity of DNA solution injected into each protoplast (typically 1 – 10 picoliters) is estimated by meniscus displacement. Following microinjection, the injection pipette is removed and the protoplast is placed at the bottom of the medium drop by releasing suction on the holding pipette. After all the protoplasts in the drop are injected, the slide is transferred to a laminar flow hood for culture preparation.

Hanging drop culture allows for the propagation of microinjected protoplasts separate from noninjected protoplasts by suspension of the injected protoplasts from the lid of a Petri dish in a drop of culture medium (Fig. 3). Normal protoplast culture densities ($10^4 - 10^6$ per ml) are obtained by using very small drops of medium for the culture of small numbers of injected protoplasts (Crossway et al. 1986b, c). The individual calli produced from single injected protoplasts can then be subcultured separately onto agar-solidified medium for plant regeneration and transformation assays, such as Southern hybridization analysis.

## 3 Uptake and Integration of DNA

### 3.1 Parameters Affecting Transformation

Plant cells contain large vacuoles which animal cells do not have; this feature has been postulated as one reason for the difficulty in developing microinjection techniques for plants (Steinbiss and Stabel 1983; Griesbach 1985; Morikawa and Yamada 1985). Evacuolation of the protoplasts prior to injection (Griesbach 1983; Lawrence and Davies 1985; Spangenberg et al. 1986) and monitoring of injection with fluorescent dye (Steinbiss and Stabel 1983) have been used to avoid this problem. In one case (Spangenberg et al. 1986), evacuolated protoplasts (karyoplasts) were then fused with cytoplasts (enucleated protoplasts) to reconstitute the cell prior to culture. However, efficient plant cell transformation has been achieved without resorting to these techniques (Crossway et al. 1986c; Pasternak et al. 1986). This suggests that the vacuole is not a major limiting factor for plant transformation.

Animal cell transformation literature has shown that many parameters can affect transformation efficiency following microinjection (Brinster et al. 1985). In particular, microinjection into the nucleus appears to be essential for achieving high transformation rates. DNA may be rapidly degraded in the cytoplasm but not in the nucleus (Wyllie et al. 1977; Yamaizumi et al. 1983). This is reflected in the transformation rates observed for transgenic mice produced following nuclear and cytoplasmic injections (about 25% and less than 1%, respectively; see Hammer et al. 1985; Brinster et al. 1985). Attempts to transform mammalian cell cultures via nuclear and cytoplasmic injections resulted in similar disparities (Kopchick et al. 1981).

Thus, nuclear targeting was of concern to plant microinjectionists as they developed their transformation systems. Steinbiss and Stabel (1983) monitored localization by coinjecting the fluorescent dye Lucifer Yellow and found it difficult to target the nucleus of their surface-attached protoplasts. Using holding pipette immobilization, Morikawa and Yamada (1985) likewise found it difficult to monitor targeting by co-injection of Lucifer Yellow or several other fluorescent dyes. However, they could target the nucleus using berberine hydrochloride, which has intense fluorescence only when in contact with DNA. Other workers have found fluorescent dyes more useful for nuclear targeting when the plant cells have been stained prior to microinjection (Hepher et al. 1985; Reich et al. 1986a). Nomarski optics also appear useful for improving nuclear resolution (Nomura and Komamine 1986). On the other hand, Crossway et al. (1986c) found bright field optics to be sufficient for successful nuclear targeting using holding pipette immobilization, which allows each cell to be rotated for optimal orientation to the injection pipette.

The importance of intranuclear targeting is perhaps best assessed via plant cell transformation following microinjection. Although the data have not been published, Steinbiss et al. (1985) reported obtaining only a single transformed callus from several hundred tobacco protoplasts injected; they attributed this partly to their difficulty in nuclear targeting. Crossway et al. (1986c) made a direct comparison between cytoplasmic and intranuclear injections (Tables 1 and 2). Transformation was assessed by Southern hybridization analysis of tobacco calli cul-

**Table 1.** Summary of pCGN561 intranuclear microinjection experiments with tobacco mesophyll protoplasts (Crossway et al. 1986b)

| Exp. # | # Injected protoplasts | # Alive day 3 | # Calli produced | # Containing 561 DNA |
|---|---|---|---|---|
| 1 | 14 | 11 | 2 | 1 |
| 2 | 21 | 19 | 1 | 0 |
| 3 | 25 | 24 | 24 | 3 |
| 4 | 20 | 15 | 15 | 2 |
| 5 | 33 | 28 | 14 | 2 |
| Total | 113 | 97 | 56 | 8 |

**Table 2.** Summary of pCGN561 cytoplasmic microinjection experiments with tobacco mesophyll protoplasts (Crossway et al. 1986c)

| Exp. # | # Injected protoplasts | # Alive day 3 | # Calli produced | # Containing 561 DNA |
|---|---|---|---|---|
| 1 | 26 | 10 | 1 | 0 |
| 2 | 31 | 22 | 13 | 1 |
| 3 | 43 | 20 | 7 | 1 |
| 4 | 33 | 21 | 5 | 1 |
| 5 | 38 | 14 | 5 | 0 |
| 6 | 32 | 18 | 3 | 0 |
| 7 | 31 | 15 | 2 | 0 |
| 8 | 40 | 28 | 13 | 0 |
| 9 | 30 | 16 | 4 | 0 |
| Total | 304 | 164 | 53 | 3 |

tured from individual protoplasts microinjected with bacterial DNA (see Crossway et al. 1986c for description of plasmid CGN561). DNA sequences homologous to the injected DNA were found in 14% of the calli produced from intranuclear injections (Table 1) but in only 6% of the calli produced from cytoplasmic injections (Table 2). Additional cytoplasmic injections by these same researchers (Crossway et al. unpubl.) have not resulted in any more transformants. Although the numbers are small, these plant data mimic the animal data (Brinster et al. 1985) in that occasional cytoplasmic injection experiments yield relatively high transformation rates, but the average rate of transformation over many experiments is low. Reproducibly high transformation frequencies apparently require precise targeting of the foreign DNA to the nucleus.

This interpretation is supported by the efficient transformation that Reich et al. (1986c) achieved using the fluorescent dye Hoechst 33258 to visualize the nucleus of surface-attached protoplasts. In this study, transformation was assessed by measurement of nopaline synthase activity in calli cultured from individual alfalfa protoplasts microinjected with a variety of Ti plasmids. Nopaline synthase activity was detected in 15%–26% of the calli produced from intranuclear injections (Table 3). Although results from cytoplasmic injections were not reported, the authors had purposefully designed their injection protocol to maximize intranuclear injection. The only other report to date of plant cell transformation by microinjection (Pasternak et al. 1986) likewise incorporates nuclear targeting of DNA injections. In this report, Pasternak's group assessed transformation by callus growth on selective media following microinjection of *Nicotiana debneyi* protoplasts with a variety of plasmids containing antibiotic resistance or phytohormone genes. In three experiments, transformation frequencies ranged from 9%–14% of injected cells (Table 3). So, although limited at present, the plant data support the requirement for intranuclear injection to achieve reproducibly high transformation frequencies.

The concentration of DNA injected is another parameter affecting transformation rates. For animal cells, the frequency of integration improves as the DNA

**Table 3.** Plant cell transformation by intranuclear microinjection of plasmid DNA into protoplasts

| Species | Plasmid[a] form | # of experiments | # Injected protoplasts | # Calli produced | # Calli transformed | Transformation frequency | | Reference |
|---|---|---|---|---|---|---|---|---|
| | | | | | | % of calli | % Injected protoplasts | |
| *Nicotiana tabacum*[b] (tobacco) | S | 5 | 113 | 56 | 8 | 14 | 7 | Crossway et al. (1986c) |
| | L-A | 1 | 51 | 9 | 6 | 67 | 12 | [c] |
| | S-A | 1 | 51 | 8 | 0 | 0 | 0 | [c] |
| | L-B | 1 | 131 | 38 | 6 | 16 | 5 | [c] |
| *Nicotiana debneyi*[d] | S | 1 | 96 | – | 11 | – | 11 | Pasternak et al. (1986) |
| | S | 1 | 111 | – | 10 | – | 9 | Pasternak et al. (1986) |
| | S | 1 | 93 | – | 13 | – | 14 | Pasternak et al. (1986) |
| *Medicago sativa*[e] (alfalfa) | S | 3 | 523 | 65 | 17 | 26 | 3 | Reich et al. (1986c) |
| | S | 3 | 694 | 116 | 17 | 15 | 2 | Reich et al. (1986c) |
| | S | 4 | 813 | 106 | 18 | 17 | 2 | Reich et al. (1986c) |
| | S | 3 | 492 | 90 | 18 | 20 | 4 | Reich et al. (1986c) |
| *Brassica napus*[e] (rapeseed) | L | 2 | 149 | 27 | 16 | 59 | 11 | Miki et al. (1986) |

[a] S = supercoiled, L = linearized; -A = DNA concentration 1 mg/ml, -B = DNA concentration 0.01 mg/ml.
[b] Transformation assessed by Southern hybridization analyses.
[c] Crossway et al. unpubl.; 6.9 kb plasmid.
[d] Transformation assessed by callus growth on antibiotic-containing or hormone-free medium.
[e] Transformation assessed by nopaline synthase activity.

concentration is increased, but injection of very large amounts of DNA (greater than 2 ng/μl) is toxic (Brinster et al. 1985). Little has been reported on this aspect for plant cell injections. However, DNA concentrations that have been used in plant cell microinjection experiments (Crossway et al. 1986c; Reich et al. 1986c) are one to two orders of magnitude higher than those used in animal cell injections, yet comparable transformation rates were achieved. In another plant study (Crossway et al. unpubl.; Table 3), a 100-fold increase in DNA concentration of a linearized 6.9 kilobase (kb) plasmid resulted in a three to four times higher transformation frequency. If there is an optimal DNA concentration for plant cells at which integration increases at the expense of cell viability, it has not yet been observed in plant microinjection studies.

The form (supercoiled or linear) of the injected DNA may also be important in plant cell transformation by microinjection. Animal studies have shown that integration of supercoiled DNA is fivefold less efficient than integration of linear molecules (Brinster et al. 1985). Although no comprehensive studies have been conducted on this aspect of plant cell microinjection, no transformants were obtained from injection of the supercoiled form of a 6.9 kb plasmid which yielded 67% transformation when injected in its linear form (Crossway et al. unpubl.; Table 3). Furthermore, the highest transformation rates achieved to date by plant microinjection groups have been with linear DNA injections (Table 3).

It is common to report plant transformation frequencies based on the number of calli produced from protoplasts of a particular species. However, these plating efficiencies can vary dramatically from species to species. Table 3 lists injection experiments from three different research groups using four different plant species with differing protoplast plating efficiencies. It is interesting that transformation frequencies based on the percentage of injected protoplasts (i.e., the overall efficiency of transformation) are not dramatically different from one research group to another or from one plant species to another. Perhaps when more plant data are accrued species differences will become obvious, as large differences have been found with different animal genotypes (Brinster et al. 1985). The plant data presently available may not represent fully optimized systems for any particular species and may thereby obscure differences. Nonetheless, parameters such as DNA form and DNA concentration appear to have the most effect on transformation frequencies, and therefore the most potential for improving transformation rates.

### 3.2 DNA Integration

Other insights into plant transformation might be gained by looking at how DNA is integrated into the genome following microinjection. This has been investigated to a very limited extent in plants. Several groups have injected Ti plasmids or engineered plasmids containing the T-DNA borders of *Agrobacterium tumefaciens*; in *Agrobacterium*-mediated gene transfer these T-DNA borders are essential to integration of foreign DNA (Nester et al. 1984). The data from microinjection are similar to those for chemically induced DNA uptake (Krens and Schilperoort 1984) in that the T-DNA borders are not preferentially used for DNA integration.

In a study by Crossway et al. (1986c), injected cells were cultured in isolation eliminating the need for antibiotic selection. The bacterial plasmid injected was a 29.3 kb circular DNA molecule containing T-DNA borders. After 6 months, each protoplast had produced enough callus tissue to assay for the presence of the foreign DNA. Southern hybridization analysis performed with DNA isolated from these calli showed that random portions of the injected plasmid were integrated into the tobacco genome without use of the T-DNA borders present on the plasmid.

Reich et al. (1986c) assessed Ti plasmid integration by Southern analysis of calli produced from microinjected alfalfa protoplasts. Their data suggest that integration of the supercoiled plasmid did not occur at the right T-DNA border because all lines contained an intact fragment which would have been destroyed by such an integration event. Furthermore, they found a number of fragments which hybridized to regions of bacterial plasmid DNA outside of the T-DNA. Transformation frequencies were equivalent regardless of whether the plasmid contained T-DNA borders, even when coinjected with a functional virulence region (genes responsible for induction of T-DNA transfer by *Agrobacterium*). Although Pasternak et al. (1986) also injected a series of T-DNA vectors into *Nicotiana debneyi* suspension protoplasts, no information on the integration patterns was presented.

Little information on the number of integrated copies of injected sequences has been reported for plants. In the study by Crossway et al. (1986c) using a 29.3 kb plasmid, some of the transformed calli had fewer than one copy of the bacterial DNA per genome equivalent while others had as many as two copies. Experiments with a different plasmid (6.9 kb) produced some calli which contained up to 20 copies of the injected sequences per genome (Crossway et al. unpubl.). Thus, it is possible to integrate multiple copies as has been shown for animal microinjection (Brinster et al. 1985). Although good expression of foreign genes has been observed in calli produced from microinjected protoplasts (Reich et al. 1986c), as yet there are no reports on the inheritance, stability, and expression of the integrated foreign gene in whole plants regenerated from microinjected protoplasts.

## 4 Future Directions

Plant transformation via microinjection has thus far been limited to protoplast systems, but the major crop species, the cereals, have been notably difficult to regenerate from protoplasts. To recover plants from species not responsive to protoplast culture, other plant materials with intact cell walls are being targeted for transformation via microinjection. Many plant species can now be regenerated from single cells and/or multicellular clumps of cells grown in suspension cultures. Nomura and Komamine (1986) used a combination of surface-attached and holding pipette immobilization to inject single plasmolyzed carrot suspension cells with about 50% survival. This suggests that it will be possible to microinject plant cells with intact walls and thus extend transformation techniques to a number of species not currently transformable, such as monocots.

Microinjection of pollen grains, for example, is appealing because plants could be recovered without an intervening cell culture step. This would require development of pollination and fertilization techniques for microinjected pollen. Because internal compartments (i.e., nuclei) are generally obscured by the thick pollen walls, attempts are being made to microinject into the pollen tubes as the grains germinate (Flavell and Mathias 1984; Hepher et al. 1985). Another potential microinjection target is microspores (immature pollen grains). Because of their thin undeveloped walls, it is easier to see their internal compartments for targeting of injection than with mature pollen grains. Because microspores must presently be cultured via embryogenesis to obtain plants, the primary advantage of microinjecting pollen would be lost unless methods could be developed to allow in vitro maturation followed by fertilization. Immature embryos or zygotes might also be useful for microinjection of species for which protoplast culture is difficult. Attempts are being made to inject immature embryos in situ on the plant, in cultured ovules, or in an isolated state prior to culturing them to plants (Flavell and Mathias 1984; Steinbiss et al. 1985).

Application of microinjection techniques to cellular systems such as these is a logical move as there are currently several methods capable of transforming plants via protoplasts. Thus, transformation of additional plant species requires advances in protoplast regeneration or development of methods for transforming intact cells of those species recalcitrant to current transformation methods. Microinjection will give a boost to our transformation capabilities if intact cells can be transformed; this would avoid the species restrictions of both *Agrobacterium*-mediated transformation and of protoplast regeneration required for direct DNA uptake.

## 5 Protocol

1. Culture of donor plant material (plants, suspensions, etc.).
2. Isolation of cells or protoplasts.
3. Attachment of desired gene to plant gene regulatory sequences.
4. Isolation of cloned DNA and suspension in injection buffer.
5. Manufacture of micropipettes.
6. Preparation of slides with cells or protoplasts and DNA solution.
7. DNA microinjection into individual cells or protoplasts.
8. Culture of microinjected cells or protoplasts.
9. Regeneration of transgenic plants.
10. Analysis of transformation:
    a) Southern hybridization for DNA integration.
    b) Northern hybridization for mRNA production.
    c) Western hybridization for protein production.
    d) Enzyme activity assays.
    e) Phenotypic analysis.

*Acknowledgments.* The author thanks Drs. Christine K. Shewmaker, Holly Hauptli and Robert M. Goodman for critical review of the manuscript.

## References


Abdullah R, Cocking EC, Thompson JA (1986) Efficient plant regeneration from rice protoplasts through somatic embryogenesis. Biotechnology 4:1087–1090

Aly MAM, Owens LD (1985) Protoplast isolation and culture for use with microinjection. In: Glalau GA (ed) Abstracts First International Congress of Plant Molecular Biology. 27 October–2 November 1985, Athens, Univ Georgia, USA, p 106

Bajaj YPS (ed) (1986) Biotechnology in agriculture and forestry 2. Crops I. Springer, Berlin Heidelberg New York Tokyo

Barber MA (1911a) A technic for the inoculation of bacteria and other substances into living cells. J Infect Dis 8:348–360

Barber MA (1911b) The effect on the protoplasm of *Nitella* of various chemical substances and of microorganisms introduced into the cavity of the living cell. J Infect Dis 9:117–129

Brinster RL, Chen HY, Trumbauer M, Senear AW, Warren R, Palmiter RD (1981) Somatic expression of *Herpes* thymidine kinase in mice following injection of a fusion gene into eggs. Cell 27:223–231

Brinster RL, Chen HY, Trumbauer ME, Yagle MK, Palmiter RD (1985) Factors affecting the efficiency of introducing foreign DNA into mice by microinjecting eggs. Proc Natl Acad Sci USA 82:4438–4442

Capecchi MR (1980) High efficiency transformation by direct microinjection of DNA into cultured mammalian cells. Cell 22:479–488

Coulibaly MY, Demarly Y (1986) Regeneration of plantlets from protoplasts of rice, *Oryza sativa* L. Z Pflanzenz 96:79–81

Crossway A, Houck CM (1985) A microassay for detection of DNA and RNA in small numbers of plant cells. Plant Mol Biol 5:183–190

Crossway A, Hauptli H, Houck CM, Irvine JM, Oakes JV, Perani LA (1986a) Micromanipulation techniques in plant biotechnology. Biotechniques 4:320–334

Crossway A, Houck CM, Facciotti D (1986b) The potential of micromanipulation techniques for plant improvement. In: Proc Int Symp on Nuclear Techniques and In Vitro Culture for Plant Improvement, Vienna, Austria, 19–23 August 1985, IAEA, Vienna, pp 471–479

Crossway A, Oakes JV, Irvine JM, Ward B, Knauf VC, Shewmaker CK (1986c) Integration of foreign DNA following microinjection of tobacco mesophyll protoplasts. Mol Gen Genet 202:179–185

DeLaat AMM, Blass J (1985) Potential applications for microinjection of plant protoplasts. In: Glalau GA (ed) Abstracts First International Congress of Plant Molecular Biology. Athens, Univ Georgia, USA, 27 October–2 November 1985, p 111

Deshayes A, Herrera-Estrella L, Caboche M (1985) Liposome-mediated transformation of tobacco mesophyll protoplasts by an *Escherichia coli* plasmid. EMBO J 4:2731–2737

Flavell R, Mathias R (1984) Prospects for transforming monocot crop plants. Nature 307:108–109

Fromm ME, Taylor LP, Walbot V (1986) Stable transformation of maize after gene transfer by electroporation. Nature 319:791–793

Gordon JW, Scangos GA, Plotkin DJ, Barbosa JA, Ruddle FH (1980) Genetic transformation of mouse embryos by microinjection of purified DNA. Proc Natl Acad Sci USA 77:7380–7384

Gordon-Kamm WJ, Bushnell W (1985) Intranuclear microinjection of isolated protoplasts using differential interference contrast microscopy. In: Glalau GA (ed) Abstracts First International Congress of Plant Molecular Biology. Athens, Univ Georgia, USA, 27 October–2 November 1985, p 109

Graessmann A, Graessmann M, Müller C (1980) Microinjection of early SV40 DNA fragments and T-antigen. Methods Enzymol 65:817–825

Griesbach RJ (1983) Protoplast microinjection. Plant Mol Biol Rep 1:32–37

Griesbach RJ (1985) Advances in the microinjection of higher plant cells. Biotechniques 3:348–351

Hammer RE, Pursel VG, Rexroad Jr CE, Wall RJ, Bolt DJ, Ebert KM, Palmiter RD, Brinster RL (1985) Production of transgenic rabbits, sheep and pigs by microinjection. Nature 315:680–683

Harbers K, Jahner D, Jaenisch R (1981) Microinjection of cloned retroviral genomes into mouse zygotes: integration and expression in the animal. Nature 293:540–542

Hepher A, Sherman A, Gates P, Boulter D (1985) Microinjection of gene vectors into pollen and ovaries as a potential means of transforming whole plants. In: Chapman GP, Mantell SH, Daniels RW (eds) The experimental manipulation of ovule tissues. Longman, New York, pp 52–63

Jaenisch R, Mintz B (1974) Simian virus 40 DNA sequences in DNA of healthy adult mice derived from preimplantation blastocysts injected with viral DNA. Proc Natl Acad Sci USA 71:1250–1254

Kopchick JJ, Ju G, Skalka AM, Stacey DW (1981) Biological activity of cloned retroviral DNA in microinjected cells. Proc Natl Acad Sci USA 78:4283–4387

Krens FA, Schilperoort RA (1984) Ti-plasmid DNA uptake and expression by protoplasts of *Nicotiana tabacum*. In: Vasil IK (ed) Cell culture and somatic cell genetics of plants, Vol 1: Laboratory procedures and their application. Academic Press, New York, pp 522–532

Kyozuka J, Shimamoto K, Ogura H (1989) Regeneration of plants from rice protoplasts. In: Bajaj YPS (ed) Biotechnology in agriculture and forestry 8. Protoplasts and genetic engineering I. Springer, Berlin Heidelberg New York Tokyo

Lawrence WA, Davies DR (1985) A method for the microinjection and culture of protoplasts at very low densities. Plant Cell Rep 4:33–35

Lo CW (1983) Transformation by iontophoretic microinjection of DNA: Multiple integrations without tandem insertions. Mol Cell Biol 3:1803–1814

Miki BL, Iyer VN, Reich TJ (1986) Transformation of higher plant cells by intranuclear microinjection. Proc VI Int Congr Plant Tissue and Cell Culture, August 3–8, 1986, Minneapolis, Minnesota

Morikawa H, Yamada Y (1985) Capillary microinjection into protoplasts and intranuclear localization of injected materials. Plant Cell Physiol 26:229–236

Nester EW, Gordon MP, Amasino RM, Yanofsky MF (1984) Crown gall: A molecular and physiological analysis. Annu Rev Plant Physiol 35:387–413

Nomura K, Komamine A (1986) Embryogenesis from microinjected single cells in a carrot cell suspension culture. Plant Sci 44:53–58

Pasternak TP, Melnikov PV, Gleba YY, Sytnik KM, Argentova VV, Kaliev AB, Andrianov VM, Piruzyan ES (1986) Genetic transformation of higher plant cells by DNA microinjections. Proc USSR Acad Sci Ser Biol 2:316

Rasmussen R, Ward B, Knauf V (1984) Microinjection of immobilized protoplasts. Plant Physiol Suppl 75:141

Reich TJ, Iyer VN, Haffner M, Holbrook LA, Miki B (1986a) The use of fluorescent dyes in the microinjection of alfalfa protoplasts. Can J Bot 64:1259–1267

Reich TJ, Iyer VN, Scobie B, Miki BL (1986b) A detailed procedure for the intranuclear microinjection of plant protoplasts. Can J Bot 64:1255–1258

Reich TJ, Iyer VN, Miki BL (1986c) Efficient transformation of alfalfa protoplasts by the intranuclear microinjection of Ti plasmids. Biotechnology 4:1001–1003

Schmidt HD (1859) On the minute structure of the hepatic lobules, particularly with reference to the relationship between the capillary blood vessels, the hepatic cells, and the canals which carry off the secretion of the latter. Am J Med Sci 37:2–40

Spangenberg G, Neuhaus G, Schweiger H-G (1986) Expression of foreign genes in a higher plant cell after electrofusion-mediated cell reconstitution of a microinjected karyoplast and a cytoplast. Eur J Cell Biol 42:236–238

Steinbiss H-H, Stabel P (1983) Protoplast derived tobacco cells can survive capillary microinjection of the fluorescent dye Lucifer Yellow. Protoplasma 116:223–227

Steinbiss H-H, Stabel P, Topfer R, Hirtz RD, Schell J (1985) Transformation of plant cells by microinjection of DNA. In: Chapman GP, Mantell SH, Daniels RW (eds) Experimental manipulation of ovule tissues. Longman, London, pp 64–75

Tilton VR, Russell SH (1983) Microinjection of plant cells. II. Cell isolation, preparation and initial post-injection culture. Plant Physiol Suppl 72:3

Toriyama K, Hinata K, Sasaki T (1986) Haploid and diploid plant regeneration from protoplasts of anther callus in rice. Theor Appl Genet 73:16–19

Wagner EF, Stewart TA, Mintz B (1981) The human $\beta$-globin gene and a functional viral thymidine kinase gene in developing mice. Proc Natl Acad Sci USA 78:5016–5020

Wagner TE, Hoppe PC, Jollick JD, Scholl DR, Hodinka RL, Gault JB (1981) Microinjection of a rabbit $\beta$-globin gene into zygotes and its subsequent expression in adult mice and their offspring. Proc Natl Acad Sci USA 78:6376–6380

Wyllie AH, Gurdon JB, Price J (1977) Nuclear localisation of an oocyte component required for the stability of injected DNA. Nature 268:150–152

Yamada Y, Zhi-Qi Y, Ding-Tai T (1986) Plant regeneration from protoplast-derived callus of rice (*Oryza sativa* L.). Plant Cell Rep 5:85–88

Yamaizumi M, Horwich AL, Ruddle FH (1983) Expression and stabilization of microinjected plasmids containing the *Herpes simplex* Virus thymidine kinase gene and Polyoma Virus DNA in mouse cells. Mol Cell Biol 3:511–522

# XV Electrofusion: The Technique and Its Application to Somatic Hybridization

G. W. BATES[1]

## 1 Introduction

The use of electric fields for inducing protoplast fusion (electrofusion) has its origins in biophysical studies of cell membranes. However, recent work showing electrofusion to be simpler, quicker, and often more efficient than chemically induced fusion is attracting the interest of geneticists and tissue-culture practitioners. Indeed, electrofusion may become the fusion technique of choice for species whose cells exhibit a severe cytotoxic response to polyethylene glycol (PEG). Interest in this area has been further heightened by recent demonstrations that electric pulses can be used to introduce foreign DNA into plant cells ("electroporation": Fromm et al. 1986; Riggs and Bates 1986). This chapter provides an overview of electrofusion and its application to plant somatic hybridization.

The first report of protoplast electrofusion is that of Senda et al. (1979). By positioning microelectrodes at the ends of pairs of adhering *Rauwolfia* protoplasts, they induced fusion with 5–12-μAmp DC pulses. Fusion yields were restricted to single protoplast pairs. However, Zimmermann and coworkers (Scheurich and Zimmermann 1981; Vienken et al. 1981; Zimmermann and Scheurich 1981) soon showed how batches of protoplasts could be fused by electric fields. Zimmermann's protocol is used by most groups working with electrical fusion. In this protocol fusion is a two-step process. First, the protoplasts are introduced into a small fusion chamber (Fig. 1) containing parallel wires or plates, which serve as electrodes; then a low-voltage, rapidly oscillating AC field is applied. This AC field causes the protoplasts to become aligned into chains of cells ("pearl chains") between the electrodes (Fig. 2). Cell alignment, which is complete in a few minutes, creates the cell-to-cell contacts that are a prerequisite of fusion. Once alignment is complete, fusion is induced by application of one to two brief, high-voltage DC pulses (Fig. 3). These DC pulses induce a reversible breakdown of the plasma membrane's structure. When membrane breakdown occurs at sites of cell contact, the ensuing membrane reorganization leads to cell fusion. This entire process, from the introduction of the protoplasts into the fusion chamber to their transfer to culture media, can be completed in 5 min or less.

The work of Zimmermann's group indicated that electrofusion was effective with protoplasts of many species and cell types, and the fusion efficiencies report-

[1] Department of Biological Science and Institute of Molecular Biophysics, Florida State University, Tallahassee, FL 32306-3015, USA

Biotechnology in Agriculture and Forestry, Vol. 9
Plant Protoplasts and Genetic Engineering II (ed. by Y. P. S. Bajaj)

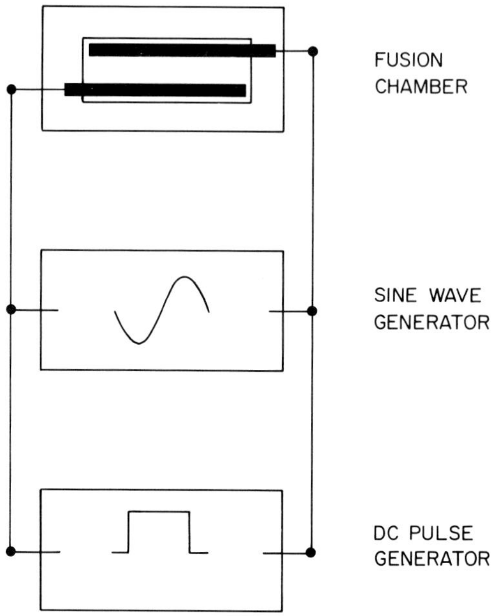


**Fig. 1.** Schematic representation of the organization of electrofusion equipment. A fusion chamber containing two or more parallel electrodes is connected to a high-frequency oscillator (sine-wave or AC-field generator) and a DC-pulse generator. An electrical gate (not shown) is also often included so that the AC field can be disconnected from the circuit during delivery of the DC pulse

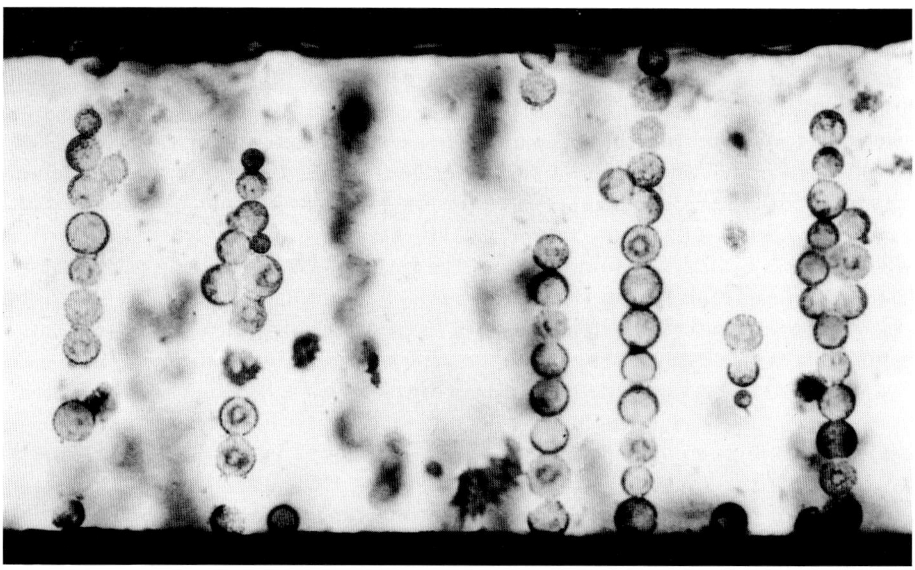

**Fig. 2.** Tobacco mesophyll protoplasts aligned into pearl chains under the influence of an AC field (100 V/cm, 0.6 MHz). The electrodes are located at the top and bottom of the field of view. Because of the depth of the fusion chamber, some protoplast chains are out of focus

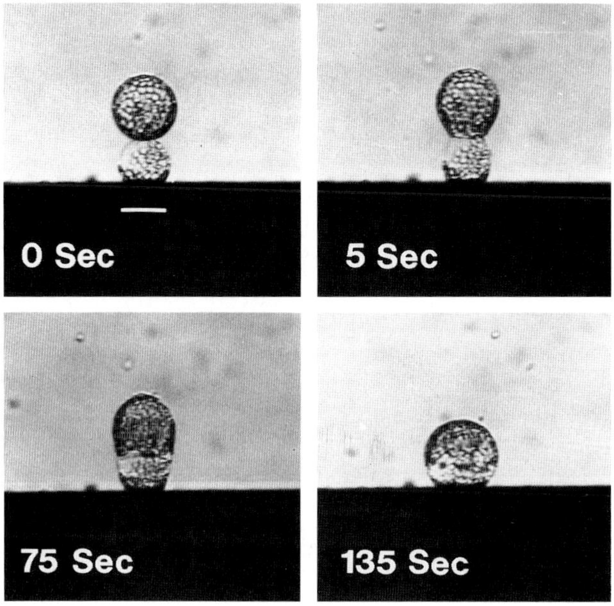


**Fig. 3.** Fusion of dielectrophoretically aligned oat mesophyll protoplasts following application of a single DC pulse (800 V/cm, 15 µs duration). The pulse was given just after the photograph labeled 0 s was taken. (Bates et al. 1983)

ed were remarkably high, up to 100% (Scheurich and Zimmermann 1981; Vienken et al. 1981; Zimmermann and Scheurich 1981). However, no attempts to culture electrically fused protoplasts were reported. This omission raised serious questions about their viability. A number of groups have now regenerated somatic hybrids from electrically fused protoplasts, thus settling this question. Successful culture depended partly on careful attention to hybrid identification and selection systems. Success also depended on modifying the original electrofusion equipment and protocols to make it compatible with protoplast culture. Because appreciation of these developments depends on an understanding of the biophysical basis of electrofusion, this chapter begins with a brief discussion of this subject (see Zimmermann 1982; Bates et al. 1987c; for detailed reviews) before turning to the application of electrofusion to protoplast fusion and somatic hybridization.

## 2 Biophysics of Electrofusion

### 2.1 Cell Alignment

The alignment of cells into pearl chains in response to AC fields is known as mutual dielectrophoresis (Pohl 1978) and is due to a field-induced separation of cell-surface charges. As a result, cells that are close together are attracted to each other and form pairs and chains of cells that radiate outward perpendicularly from the electrodes (Fig. 2). The formation of cell-to-cell contacts depends on several factors, including the frequency and voltage of the AC field, the shape of the electrodes, and the composition of the medium.

AC field frequencies in the range of 0.5 – 1 MHz (megahertz or million cycles per second) are used for the alignment of plant protoplasts. The voltage required for cell alignment depends on the shape of the electrodes and the distance separating them. With wire electrodes spaced 0.5 mm apart, a 5 – 10-V AC field produces good dielectrophoresis. If the electrodes are farther apart, the voltage must be correspondingly higher. (To account for the electrode spacing, voltages used for dielectrophoresis and cell fusion are usually reported as field strengths, i.e., V/cm. In the example above, the AC field strength would be 100 – 200 V/cm). Ordinarily, the electrodes used for dielectrophoresis are wires (or needles) because they result in a nonuniform electrical field. The more nonuniform the field, the lower the AC voltage required to produce cell alignment by dielectrophoresis. It should be pointed out, however, that plate electrodes can also be used (Watts and King 1984; Tempelaar and Jones 1985a, b).

The dielectrophoretic force is greatest in media of low conductivity. Therefore, electrofusion is carried out in media containing an inert osmoticum, such as mannitol or sucrose, with little or no added salts. A small amount of $CaCl_2$ (0.1 – 0.5 mM) is sometimes added (Watts and King 1984; Spangenberg and Schweiger 1986, Nea et al. 1987), as it often improves fusion and reduces cell lysis. However, $Ca^{2+}$ concentrations in excess of 1 mM greatly reduce dielectrophoresis. Increasing the AC voltage to overcome this effect leads to heating of the medium, and the resulting turbulence disrupts the cell contacts.

The role of dielectrophoresis is to create cell-to-cell contacts. Alternative approaches are available that achieve the same end. Cell contacts can be formed simply by use of very high cell densities (Morikawa et al. 1986), or cell aggregation can be induced by addition of small amounts of PEG to the medium (Chapel et al. 1986). In either case, electrofusion is achieved by application of only the high-voltage DC pulse. Both these approaches sacrifice some control over the formation of cell contacts and probably result in a somewhat lower efficiency of fusion. However, they have the advantage of requiring simpler electronic equipment and fusion chambers, and fusion can be carried out in media containing normal levels of salts.

## 2.2 Electric Field-Induced Membrane Breakdown and Fusion

Electric field-induced membrane breakdown has been studied by Zimmermann (1982) and by Kinosita and Tsong (1977, 1979). The essential observation is that cell membranes undergo a dramatic increase in permeability when exposed to high-voltage DC pulses. Experiments using radiolabeled tracer molecules suggest that discrete, nanometer-sized pores form in the membrane in response to these high-voltage pulses (Kinosita and Tsong 1977) – hence the term electroporation. The size and number of the pores increase with increasing voltage and pulse length (Zimmermann 1982). DC pulses of 1000 V/cm (10 – 50 μs duration) are usually effective for the electroporation and electrofusion of plant protoplasts. At room temperature the pores remain open only seconds; pore closing restores normal membrane properties.

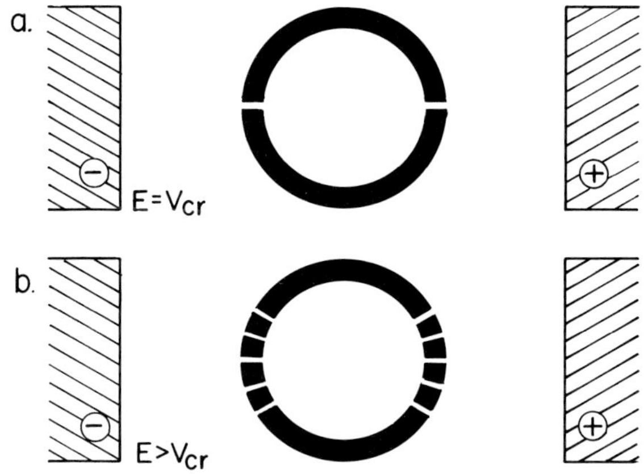


**Fig. 4a, b.** Relationship of the magnitude of an applied electric field (*E*) to the formation and distribution of pores in the plasma membrane. **a** When the voltage of the applied field is just sufficient to induce pore formation ($V_{cr}$ or critical voltage), the pores form only at the poles of the cell (area of the membrane orthogonal to the electrodes). In dielectrophoretically aligned protoplasts, such pulses lead to protoplast fusion because the poles are the sites of cell-to-cell contact. **b** Application of higher-voltage (or longer) pulses induces pore formation over more of the cell surface. In aligned protoplasts, pore formation in membrane patches outside of cell-to-cell contact sites increases the chance of protoplast lysis. (After Zimmermann et al. 1984)

Cell fusion results when electroporation occurs at intercellular contact sites. As indicated in Fig. 4, the pulse-induced pores form preferentially at the poles of the cell (Zimmermann 1982; Mehrle et al. 1985). These are precisely the points of intercellular contact when protoplasts are aligned under the influence of AC fields. Thus, the two-step electrofusion protocol (AC field-induced alignment followed by DC pulse-induced fusion) leads to highly efficient protoplast fusion. As long as the pores quickly reseal, cell viability is not affected. However, the reversibility of pore formation does depend on the length and voltage of the DC pulse. Large or excessively long pulses result in cell lysis. In practice, efficient electrofusion is always accompanied by some cell lysis (10%–20% is probably acceptable).

## 2.3 Electrical Equipment

A number of commercial electrofusion units are available (BTX Biotechnologies and Experimental Research Inc., San Diego, CA, USA; D.E.P. Systems Inc., Troy, MI, USA; GCA Precision Corp., Chicago, IL, USA). The electronic apparatus in all of them appears to give reasonably good fusion. (Not all fusion chamber designs work well, however; this point will be discussed more fully below). The commercial fusion devices are expensive, and several home-made fusion devices have been described in the literature (Zimmermann and Scheurich 1981; Bates et al. 1983; Watts and King 1984; Zachrisson and Bornman 1984) that can be built for a fraction of the cost of the commercial machines. A sine-wave generator (oscillator) is used to produce the AC field, and DC pulses of variable voltage and pulse length can be delivered by the stimulators used in most neurophysiology laboratories. One difficulty is that the power of these units must be fairly high (or they must be amplified). This is especially true when a large fusion chamber (one capable of holding 0.25 – 1 ml of cell suspension) is used. A simple and low-cost way to produce high-voltage DC pulses is by capacitive discharge (Watts and King 1984; Riggs and Bates 1986). The capacitor can be charged with an electrophoresis power supply and then discharged through the cell preparation by means of an electric switch (Riggs and Bates 1986). In this case, pulse length is controlled by the size of the capacitor.

## 3 Application of Electrofusion to Plant Protoplasts

### 3.1 Optimal Electrical Field Settings for Protoplast Fusion

Several groups have characterized electrofusion with respect to electrical parameters. Although Zimmermann's group (Scheurich and Zimmernann 1981; Vienken et al. 1981; Zimmermann and Scheurich 1981) indicated that electrofusion yields could approach 100%, subsequent studies have shown routine fusion yields to be lower, in the range of 10% – 50%. Still, these yields are considerably higher than those ordinarily obtained with PEG.

In interpreting data on electrofusion yields, two factors must be kept in mind. First, not all protoplasts in the fusion chamber become aligned into chains by the AC field, and in chambers with wire electrodes, fusion occurs preferentially in chains located where the wires are closest together. Thus, there is always a background of unfused protoplasts. Second, conditions favoring a high percentage of fusion also favor the formation of multicellular fusions over the fusion of pairs of cells. This is also the case for PEG-induced fusion (Constabel et al. 1975).

Electrofusion exhibits a DC-voltage threshold. Such a threshold is expected theoretically (Zimmermann 1982) and has been demonstrated experimentally (Chapel et al. 1984; Bates 1985; Tempelaar and Jones 1985 a). Working with *Nicotiana tabacum* mesophyll and *N. plumbaginifolia* suspension-culture protoplasts, Bates (1985) found no fusion in response to 500-V/cm DC pulses. Optimal fusion was obtained with 1000-V/cm pulses (Fig. 5). Increasing the voltage to 2000 V/cm shifted the balance of fusion products from binucleate to multinucleate cells without improving the percentage of fusion. Chapel et al. (1984) and Tempelaar and Jones (1985 a) obtained comparable results for the electrofusion of *N. tabacum* and *Solanum brevidens* mesophyll protoplasts. For a given DC voltage, increasing the duration of the pulse also improves fusion (Tempelaar and Jones 1985 a). Square-wave pulses 50 – 100 μs in duration (1000 – 2000 V/cm) generally yield maximum fusion. Longer pulses lyse the cells.

Although fewer studies have documented the effects on fusion of changing the AC voltage, the data of Watts and King (1984) and Bates (1985) suggest that AC fields of 150 V/cm give maximum fusion. Electrofusion yields also increase as the

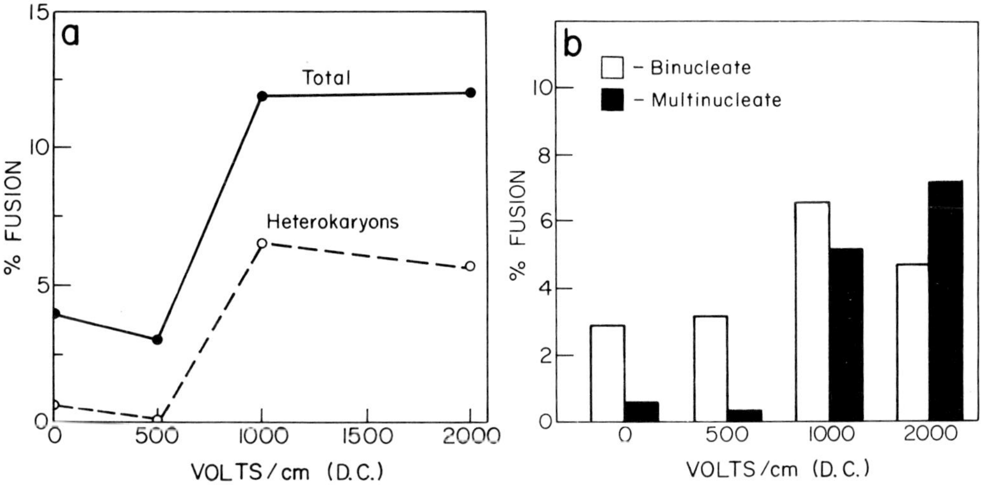


**Fig. 5a, b.** Characteristics of fusion as the DC voltage is varied. A mixed population of *Nicotiana tabacum* mesophyll and *N. plumbaginifolia* suspension-cell protoplasts was introduced into a D.E.P. Systems fusion chamber (shown in Fig. 7a) and aligned in an AC field of 150 V/cm, 0.6 MHz. Fusion was induced by application of two DC pulses (each 50 μs long) of the voltage indicated in the graph. Fusion was assessed by staining of the cells with modified carbol-fuchsin. Fusion of mesophyll and suspension-cell protoplasts also made it possible to determine the number of homokaryons and heterokaryons that were formed. **a** Overall fusion and heterokaryon formation; **b** proportions of bi- and multinucleate cells. (Bates 1985)

density of the protoplast suspension increases from $10^4$ to $10^6$ protoplasts/ml (Chapel et al. 1984; Watts and King 1984; Bates 1985; Tempelaar and Jones 1985a), because the formation of cell-to-cell contacts is inefficient at low cell densities. Most groups use cell densities ranging from $5 \times 10^5$ to $10^6$ protoplasts/ml.

Conditions that maximize the fusion percentage (high-strength pulses and high cell densities) can result in 30%–60% of the fusion products' being multinucleate (Chapel et al. 1984; Bates 1985; Tempelaar and Jones 1985a). However, these conditions also result in the maximum absolute yield of binucleate fusion products. Thus, in practice, optimal yields of fusion products suitable for culture are probably realized under conditions that maximize the overall fusion percentage (Bates 1985; Tempelaar and Jones 1985a). One must be prepared, however, to identify and discard polyploids when plants are finally regenerated.

## 3.2 Effects of Cell Type and Chemical Pretreatments on Electrofusion Efficiency

Protoplast source also affects electrofusion. Using a standard set of conditions, Tempelaar and Jones (1985a) compared the fusion of protoplasts isolated from leaves, roots, hypocotyls, and suspension cultures of nine different species. Considerable differences were observed. For example, 55% of the aligned *Datura* suspension-culture protoplasts were fused, whereas root tip protoplasts of *Solanum brevidens* gave only 15% fusion. These cell type differences may be partly attributable to differences in cell size. Theoretically, the threshold voltage for membrane breakdown and fusion is inversely proportional to cell size (Zimmermann 1982). However, Tempelaar and Jones (1985a) found that protoplasts from certain sources exhibited distinct differences in fusion yield despite being similar in size. In general, mesophyll protoplasts were found to fuse more readily than suspension-culture protoplasts. These findings suggest that factors other than size are also important.

Tempelaar and Jones (1985a) suggest that these differences in fusion are due to inherent differences in the plasma membranes of protoplasts from different sources. Nea and Bates (1987), working with *Daucus* suspension-culture protoplasts, found that using different enzymes for cell wall digestion (Driselase versus Cellulysin) led to dramatic differences in the subsequent efficiency of electrofusion (see Fig. 6a). Moreover, brief treatment of the protoplasts with pronase or proteinase K greatly improved the efficiency of fusion (Fig. 6a; see also Ruzin and McCarthy 1986; Spangenberg and Schweiger 1986; Nea and Bates 1987). These data indicate that the method of protoplast isolation can also affect electrofusion efficiency and suggest that membrane-surface proteins are one factor limiting the fusion of certain protoplast types.

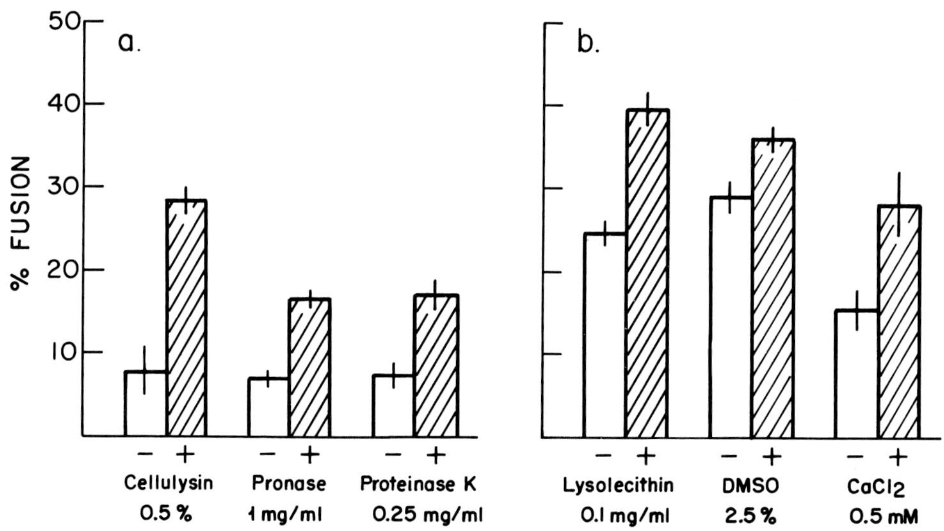


**Fig. 6a, b.** Effects of enzymes and small molecules on the electrofusion of suspension-cell protoplasts of *Daucus carota*. Protoplasts were isolated by a 5-h digestion with 2% driselase, given various treatments, and then subjected to electrofusion. (−) indicates results obtained for control protoplasts, (+) indicates treated protoplasts. The data represent the average ±SE of 4−7 determinations. **a** Protoplasts were treated for 5 min with pronase or proteinase K and then washed just prior to fusion. In the case of the cellulysin treatment, the treated protoplasts (+) were isolated by a 5-h digestion in 1.5% driselase +0.5% cellulysin. Control protoplasts (−) were isolated in 2% driselase alone. **b** Protoplasts were treated with lysolecithin, or DMSO, or $CaCl_2$. The lysolecithin-treated protoplasts were washed prior to fusion; however, in the cases of DMSO and $CaCl_2$, the agents were left in the medium during fusion

Other chemical treatments that have been found to improve the efficiency of electrofusion include lysolecithin, $Ca^{2+}$, and DMSO (Fig. 6b; see also Watts and King 1984; Spangenberg and Schweiger 1986; Nea and Bates 1987; Tempelaar et al. 1987). Both DMSO and lysolecithin increase electric field-induced cell lysis substantially; however, low concentrations of $Ca^{2+}$ (0.5 – 1 mM) do not have this effect and thus are a useful additive to the fusion medium (Nea and Bates 1987; Tempelaar et al. 1987).

In mixtures of protoplasts, Bates (1985) found fusion to be random. Thus, when two types of protoplasts were equally represented in the original mixture, fusion led to the recovery of equal numbers of homokaryons and heterokaryons. However, this may not be the case if the two types of protoplasts have very different inherent susceptibilities to fusion. Tempelaar and Jones (1985b) exploited inherent differences in fusion sensitivity of two types of protoplasts to direct fusion toward heterokaryon formation. A low-fusing cell type (*N. rustica* suspension-culture protoplasts) and a high-fusing cell type (*S. brevidens* mesophyll protoplasts) were mixed in various ratios and fused. Protoplast mixtures containing a three- to fivefold excess of *N. rustica* protoplasts resulted in the formation of twice as many hetero-fusions as were obtained for 1 : 1 mixtures of the protoplasts. Apparently the greater potential of the *S. brevidens* protoplasts to undergo fusion extends to a greater ability to fuse with other cell types.

### 3.3 Comparison with the Efficiency of the PEG-Induced Fusion

The uniformly high fusion percentages obtained and the ease and speed of electrofusion compare favorably with PEG-induced fusion. Recently, there have been some direct comparisons between these two fusion techniques. Bates (1985) found a twofold improvement of electrofusion over the use of PEG for *Nicotiana* protoplasts (Table 1). Also working with *Nicotiana* protoplasts, Negrutiu et al. (1986) found that electrofusion gave rates of hybrid formation two- to fourfold higher than those of most PEG fusion protocols and rates approximately equal to the

**Table 1.** Percentages and types of fusions obtained with different chambers and fusion techniques. Mesophyll protoplasts of *Nicotiana tabacum* were mixed in equal numbers with *N. plumbaginifolia* suspension-cell protoplasts. For electrofusion, the protoplasts were fused as described in Fig. 5. The PEG fusion procedure is that of Kao (1975). Fusion percentages and the fraction of heterokaryons were determined as described in Fig. 5. The "mass-fusion" chamber is that shown in Fig. 7c. The D.E.P. "micro-fusion" chamber is shown in Fig. 7a. Control indicates protoplasts that were mixed together and then fixed and stained without first being subjected to electric fields or PEG treatment. "n" is the number of trials; each percentage is followed by the S.E. (Bates 1985)

| | Fusions (%) | |
|---|---|---|
| | Total | Heterokaryons |
| Mass-fusion chamber | 36.8 ± 2.1 | 19.5 ± 1.0 |
| D.E.P. Micro-fusion chamber | 15.4 ± 2.3 | 6.2 ± 0.1 |
| PEG-treated | 17.0 ± 1.0 | 10.0 ± 0.5 |
| Control | 6.8 ± 1.3 | 0.2 ± 0.1 |

best ones. Moreover, the best PEG protocols reduced protoplast survival more than did electrofusion. However, the crucial test for electrofusion will be whether it is effective with species resistant to fusion by PEG. Such studies are just beginning. For example, Ghose et al. (1986) found that protoplasts from a variety of forage legumes that fuse poorly with PEG (0.1%–2%) gave electrofusion frequencies up to 10%.

### 3.4 Experimental Designs

Electrofusion differs from chemically induced fusion in requiring the use of a chamber containing a pair (or many pairs) of electrodes. Good fusion chambers are crucial for successful fusion and for subsequent culture and hybrid selection. Most groups make their own chambers, and a wide variety of designs have been described (Bates et al. 1983; Watts and King 1984; Zachrisson and Bornmann 1984; Bates 1985; Puite et al. 1985; Koop and Schweiger 1985; Morikawa et al. 1986; Naton et al. 1986). Fusion chambers are also available commercially.

The first chambers described were small (Fig. 7a, b), holding only a few microliters (or tens of microliters) of cell suspension (Bates et al. 1983; Bates 1985; Koop and Schweiger 1985; Zachrisson and Bornman 1984); probably because small chambers are easier to build. Small chambers also force the use of microculture techniques, and such techniques are not successful with all plant species. A number of chamber designs have now been described that are capable of handling 0.25–1 ml of protoplast suspension (Fig. 7c, d; see also Watts and King 1984; Bates 1985; Puite et al. 1985; Tempelaar and Jones 1985a; Chapel et al. 1986; Naton et al. 1986). Such volumes greatly facilitate protoplast culture.

Bates (1985) reported using a mass-fusion chamber consisting of a series of thin Au/Pd wires laid parallel to each other on a polycarbonate plastic slide (Fig. 7c). The wires, spaced 0.5 mm apart, were connected to posts at the ends of the slide such that each wire would be of electrical polarity opposite to that of its neighbors. A polycarbonate ring glued down over the wires formed a well to hold the protoplast suspension. Half-milliliter aliquots of protoplasts ($5 \times 10^5$ protoplasts/ml) could be fused in this chamber with efficiencies in the range of 20%–40% (Table 1). After fusion, culture was initiated simply – the protoplasts were pipetted out of the chamber and diluted with culture medium. No washing or reisolation of the protoplasts was required.

Watts and King (1984) reported using a series of flat, tinned copper plates as electrodes (Fig. 7d). The plates were held parallel and 5 mm apart by plastic spacers. Alternate plates were connected together electrically to create opposing electrodes of opposite polarity. A somewhat similar electrode system has been described by Puite et al. (1985). Chambers employing such plate electrodes are easy to build, will accommodate 0.3–0.5 ml of cell suspension, and can simply be lowered into a Petri dish containing the protoplasts. After fusion, the protoplasts are diluted with culture medium to initiate cultures. With their electrode system, Watts and King (1984) reported fusion efficiencies of up to 90% for tobacco protoplasts. However, Tempelaar and Jones (1985b) and Nea et al. (1987), working

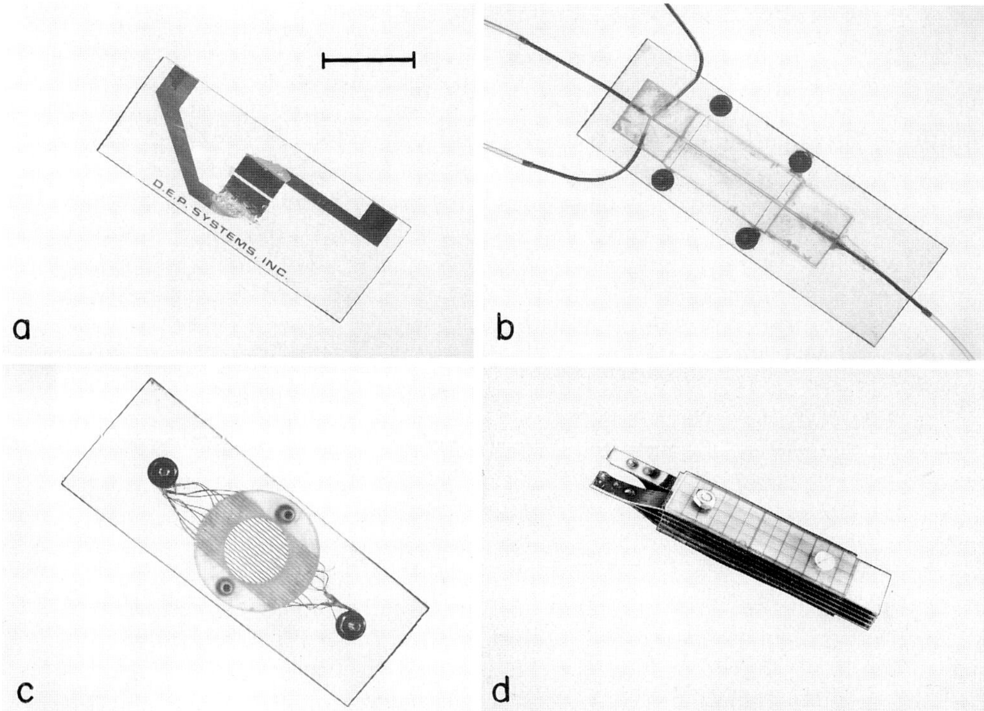


**Fig. 7a–d.** Photographs of different electrofusion chambers. All the chambers are shown at the same magnification (the bar in **a** represents 2 cm). **a** and **b** are micro-fusion chambers capable of holding only a few microliters of protoplast suspension. **c** and **d** are mass-fusion chambers capable of holding several hundred microliters of protoplasts. **a** D.E.P. Systems Inc. open fusion slide 2050S. Data obtained with this chamber are shown in Fig. 5 and Table 1. **b** A flow-through chamber with multiple inlet ports and a single pair of wire electrodes. This chamber is similar to that used by Bates et al. (1983). **c** The mass-fusion chamber of Bates (1985). Data obtained with this chamber are shown in Table 1. **d** A mass-fusion chamber having multiple flat-plate electrodes, after the design of Watts and King (1984)

with similar equipment, obtained fusion efficiencies ranging from 20% to 50% for a variety of different protoplast types.

Electrofusion can also be carried out without AC fields to align the cells. Chapel et al. (1986) used 2.5% PEG to aggregate *Nicotiana* protoplasts, which were then fused by application of a single DC pulse (1000 V/cm, 100 μs in duration). The fusion efficiency was 15%. Morikawa et al. (1986), also working with *Nicotiana* protoplasts, allowed cell-to-cell contacts to form before fusion simply by allowing the protoplasts to settle into a monolayer at the bottom of the fusion chamber. The protoplasts were then fused by means of a single, high-voltage DC pulse. Both groups used very simple fusion chambers consisting of two parallel metal plates, mounted in a plastic or glass box, giving chamber volumes up to 0.3 ml.

### 3.5 Culture of Electrically Fused Protoplasts and Recovery of Somatic Hybrids

Table 2 provides an overview of experiments in which the viability or culture of electrically fused protoplasts was assessed (see also de Vries and Tempelaar 1987).

In an early attempt at culture, Bates et al. (1983) reported that *Vigna* mesophyll protoplasts survived and divided after electrofusion. However, because no method was available to identify and follow the growth of the fusion products themselves, their viability remained in doubt. Similarly, Watts and King (1984), working with *Nicotiana*, and Zachrisson and Bornman (1984), working with *Brassica*, demonstrated that protoplasts subjected to electrofusion could be grown in culture but did not supply proof that the fusion products themselves would grow.

The first clear demonstrations that heterokaryons produced by electrical fusion could be grown came only when good selection systems were applied after fusion (Table 2). Kohn et al. (1985) fused protoplasts from two complementary nitrate-reductase-deficient lines of *N. tabacum*. Hybrid calli, recognized by their ability to grow in the absence of reduced nitrogen in the medium, were recovered with frequencies of 1%, a rate 10- to 100-fold higher than obtained after PEG-induced fusion of the same protoplasts. Although no other genetic markers were available to confirm the recovery of nuclear hybrids, these nitrate-reductase-deficient mutants are known to have very low rates of reversion.

Bates and Hasenkampf (1985) fused mesophyll protoplasts of *N. tabacum* with suspension-culture-derived protoplasts of *N. plumbaginifolia*. Fusion products were identified visually by the presence of green and nongreen sectors in the same cell, isolated by means of micropipettes, and cultured on conditioned media. The viability of these fusion products was underscored by their exhibiting plating efficiencies of 30%. Calli recovered from these cultured fusion products formed shoots but failed to root. However, isozyme profiles provided biochemical evidence of somatic hybridization. Failure of complete regeneration by these fusion products is probably attributable to the *N. plumbaginifolia* suspension culture used in these experiments, which was aneuploid and incapable of regeneration.

Morikawa et al. (1986) and Chapel et al. (1986) used electrofusion to produce somatic hybrids between *N. glauca* and *N. langsdorfii*. Hybrids were selected by their ability to grow on a hormone-free medium, which is a characteristic of this species combination. Hybrid calli were recovered with frequencies of 0.1%, which the authors state represents a tenfold increase over PEG fusion. In both these studies the hybrid nature of the fusion products was verified by electrophoretic analyses of fraction-I proteins. The selection system used here (hormone independence displayed by the hybrids) makes plant regeneration from hybrid calli difficult. However, Morikawa et al. (1986) did succeed in recovering some rooted plants.

A number of laboratories have now provided good evidence that somatic hybrid plants can be routinely regenerated from electrically fused protoplasts. Success has been based on the use of regenerable cell types and the application of a good biochemical or physical selection system. Puite et al. (1985, 1986) have

**Table 2.** Experiments in which the viability of culture of electrically fused protoplasts has been demonstrated

| Plant species | Protoplast origin | Fusion conditions | Selection system | Notes on culture or viability | Reference |
|---|---|---|---|---|---|
| *Brassica napus* + *B. napus* | Hypocotyl<br>Hypocotyl | AC: 65 – 80 V/cm, 1 MHz<br>DC: 0.9 – 1.8 kV, 50 µs | Fusion and culture of isolated pairs of karyoplasts and cytoplasts | Callus | Spangenberg and Schweiger (1986) |
| *N. glauca* + *N. langsdorfii* | Mesophyll<br>Mesophyll | Aggregation by 2.5% PEG<br>DC: 1 kV/cm, 100 µs | Hormone-independent growth | Callus | Chapel et al. (1986) |
| *N. glauca* + *N. langsdorfi* | Mesophyll<br>Mesophyll | Aggregation by high protoplast density<br>DC: 1 kV/cm, 100 – 200 µs | Hormone-independent growth | Plants | Morikawa et al. (1986) |
| *N. tabacum* + *N. tabacum* | Mesophyll<br>Mesophyll | AC: 150 V/cm, 1 MHz<br>DC: 1.5 V/cm, 55 µs | Vacuolate and evacuolate protoplasts were fused; heterokaryons recovered on isomotic density gradients | Fluorescein diacetate viability test | Naton et al. (1986) |
| *N. tabacum* + *N. tabacum* | Callus<br>Callus | AC: 200 V/cm, 0.9 MHz<br>DC: 1.2 – 1.5 kV/cm, 50 µs | Complementation of nitrate-reductase-deficient mutants | Shoots regenerated | Kohn et al. (1985) |
| *N. tabacum* + *N. tabacum* | Mesophyll<br>Mesophyll | AC: 66 V/cm, 1 MHz<br>DC: 0.9 kV/cm, 50 s | Fusion and culture of isolated pairs of protoplasts | Plants | Koop and Schweiger (1985) |
| *N. plumbaginifolia* + *N. tabacum* | Suspension-cell<br>Mesophyll | AC: 150 V/cm, 0.6 MHz<br>DC: 1 kV/cm, 50 µs | Heterokaryons were identified visually and isolated with micropipettes | Shoots | Bates and Hasenkampf (1985) |
| *N. plumbaginifolia* + *N. tabacum* | Mesophyll<br>Mesophyll | AC: 100 V/cm, 0.6 MHz<br>DC: 0.8 kV/cm, 25 µs | Kanamycin resistance | Plants | Bates et al. (1987c) |
| *N. plumbaginifolia* + *N. plumbaginifolia* | Callus<br>Mesophyll | AC: 60 V/cm, 0.5 MHZ<br>DC: 0.8 V/cm, 1 ms | Complementation of nitrate-reductase-defcient auxotroph mutants and amino acid | Callus | Negrutiu et al. (1986) |
| *N. plumbaginifolia* + *N. plumbaginifolia* | Mesophyll<br>Suspension-cell | AC: 150 V/cm, 1 MHz<br>DC: 1.2 – 2 kV/cm, 50 µs | Complementation of nitrate-reductase-deficient mutants | Callus | Puite et al. (1985) |
| *Solanum tuberosum* + *S. phureja* | Mesophyll<br>Mesophyll | AC: 150 V/cm, 1 MHz<br>DC: 1.2 – 2 kV/cm, 50 µs | Heterokaryons were identified by double fluorescent labeling and were isolated with micropipettes | Plants | Puite et al. (1986) |
| *S. brevidens* + *Nicotiana rustica* | Mesophyll<br>Suspension-cell | AC: 100 V/cm, 1 MHz<br>DC: 1.2 kV/cm, 50 µs | Heterokaryons embedded in agar were identified visually | Callus | Tempelaar and Jones (1985a) |

regenerated plants following fusion of complementary nitrate-reductase-deficient mutants of *N. plumbaginifolia*. In a second study Puite et al. (1986) reported the recovery of somatic hybrid plants from the electrofusion of *Solanum tuberosum* and *S. phureja* protoplasts. In this case, hybrid selection was achieved by visual identification of the fusion products (based on fluorescein and chlorophyll fluorescence in the two parental cell types) and their isolation by means of micropipettes and culture at low density. The hybrid nature of plants regenerated from these calli was confirmed by their intermediate morphology and by karyotype differences between the two parent species. Bates et al. (1987a, b) fused gamma-irradiated mesophyll protoplasts of a kanamycin-resistant *N. plumbaginifolia* seed line with unirradiated *N. tabacum* mesophyll protoplasts. Hybrids were recovered by their ability to grow on media containing kanamycin. In some experiments the frequency of recovery of resistant calli was as high as 0.1% (based on the number of *N. plumbaginifolia* protoplasts present before fusion). Resistant calli were regenerated, and the hybrid nature of the resulting plants was verified by isozyme analyses, nopaline detection, and Southern blotting to detect the kanamycin-resistance gene. Because of the irradiation that was applied to the *N. plumbaginifolia* protoplasts prior to fusion, many of these hybrids retained a reduced complement of *N. plumbaginifolia* chromosomes. The formation of such asymmetric hybrids was the principal goal of this study.

Finally, an interesting study by Koop and Schweiger (1985) demonstrates the unique opportunity afforded by electrofusion. They describe a system for the electrofusion of single pairs of protoplasts in microdrops. More than 90% of preselected pairs of tobacco protoplasts underwent fusion. The fusion products were cultured individually by simple addition of a microdrop of culture medium after fusion was complete. Fifty percent of the fusion products formed calli, and plants were regenerated from nearly all the calli. Spangenberg and Schweiger (1986) have now used this same system to fuse and culture specific pairs of *Brassica napus* protoplasts, cytoplasts, and karyoplasts.

## 4 Conclusions and Prospects

In the past few years electrofusion has developed from a biophysical novelty to being a practical protoplast-fusion technique. This transition required the development of fusion chambers capable of handling large numbers of protoplasts and careful attention to post-fusion selection systems for hybrid recovery.

Among the attributes of electrofusion are the speed, ease, and efficiency with which fusion can be accomplished. However, recent work suggests that electrofusion may also create entirely new experimental opportunities. These include the fusion of protoplasts that show toxic responses to PEG and the fusion of specific protoplast pairs, thus eliminating the need for a selection system. Exploration and amplification of these opportunities should give electrofusion a permanent place among the techniques used by plant somatic cell geneticists.

## References


Bates GW (1985) Electrical fusion for optimal formation of protoplast heterokaryons in *Nicotiana*. Planta 165:217–224

Bates GW, Gaynor JJ, Shekhawat NS (1983) Fusion of plant protoplasts by electric fields. Plant Physiol 72:1110–1113

Bates GW, Hasenkampf CA (1985) Culture of plant somatic hybrids following electrical fusion. Theor Appl Genet 70:227–233

Bates GW, Hasenkampf CA, Contolini CS, Piastuch WC (1987a) Limited gene transfer by somatic cell fusion in *Nicotiana*. Theor Appl Genet 74:718–726

Bates GW, Nea LJ, Hasenkampf CA (1987b) Electrofusion and plant somatic hybridization. In: Sowers AE (ed) Cell fusion. Plenum, New York, pp 479–496

Bates GW, Saunders JA, Sowers AE (1987c) Electrofusion, principles and applications. In: Sowers AE (ed) Cell fusion. Plenum, New York, pp 367–395

Chapel M, Teissie J, Alibert G (1984) Electrofusion of spermine-treated plant protoplasts. FEBS Lett 173:331–336

Chupel M, Montane M-H, Ranty B, Teissie J, Alibert G (1986) Viable somatic hybrids are obtained by direct current electrofusion of chemically aggregated plant protoplasts. FEBS Lett 196:79–86

Constabel F, Dudits D, Gamborg OL, Kao KN (1975) Nuclear fusion in intergeneric heterokaryons. A note. Can J Bot 53:2092–2095

de Vries SE, Tempelaar MJ (1987) Electrofusion and analysis of potato somatic hybrids. In: Bajaj YPS (ed) Biotechnology in agriculture and forestry 3. Potato. Springer, Berlin Heidelberg New York Tokyo, pp 211–222

Fromm ME, Taylor LP, Walbot V (1986) Stable transformation of maize after gene transfer by electroporation. Nature 319:791–793

Ghose TK, Golds TJ, Davey MR, Cocking EC (1986) An assessment of electrically and chemically induced protoplast fusion in forage legumes. Poster 434, VI Int Congr Plant tissue and cell culture, August 3–8, Univ Minnesota, Minneapolis, p 385

Kao KN (1975) A method for fusion of plant protoplasts with polyethylene glycol. In: Gamborg OL, Wetter LR (eds) Plant tissue culture methods. Prairie Research Laboratory, Saskatoon, Sask., Can, pp 22–27

Kinosita K, Tsong TY (1977) Formation and resealing of pores of controlled sizes in human erythrocyte membrane. Nature 268:438–441

Kinosita K, Tsong TY (1979) Voltage-induced conductance in human erythrocyte membranes. Biochim Biophys Acta 554:479–497

Kohn H, Schieder HR, Schieder O (1985) Somatic hybrids in tobacco mediated by electrofusion. Plant Sci 38:121–128

Koop H-U, Schweiger H-G (1985) Regeneration of plants after electrofusion of selected pairs of protoplasts. Eur J Cell Biol 39:46–49

Mehrle W, Zimmermann U, Hampp R (1985) Evidence for asymmetrical uptake of fluorescent dyes through electro-permeabilized membranes of *Avena* mesophyll protoplasts. FEBS Lett 185:89–94

Morikawa H, Sugino K, Hayashi Y, Takeda J, Senda M, Hirai A, Yamada Y (1986) Interspecific plant hybridization by electrofusion in *Nicotiana*. Biotechnology 4:57–60

Naton B, Mehrle W, Hampp R, Zimmermann U (1986) Mass electrofusion and mass selection of functional hybrids from vacuolate×evacuolate protoplasts. Plant Cell Rep 5:419–422

Nea LJ, Bates GW (1987) Protoplast electrofusion: factors affecting fusion efficiency and cell viability. Plant Cell Rep 6:337–340

Nea LJ, Bates GW, Gilmer PJ (1987) Facilitation of electrofusion of plant protoplasts by membrane active agents. Biochim Biophys Acta 897:293–301

Negrutiu I, De Brouwer D, Watts JW, Sidorov VI, Dirks R, Jacobs M (1986) Fusion of plant protoplasts: a study using auxotrophic mutants of *Nicotiana plumbaginifolia*, Viviani. Theor Appl Genet 72:279–286

Pohl HA (1978) Dielectrophoresis. Cambridge Univ Press, London

Puite KJ, van Wikselaar P, Verhoeven H (1985) Electrofusion, a simple and reproducible technique in somatic hybridization of *Nicotiana plumbaginifolia* mutants. Plant Cell Rep 4:274–276

Puite KJ, Roest S, Pijnacker LP (1986) Somatic hybrid potato plants after electrofusion of diploid *Solanum tuberosum* and *Solanum phureja*. Plant Cell Rep 5:262–265

Riggs CD, Bates GW (1986) Stable transformation of tobacco by electroporation: evidence for plasmid concatenation. Proc Natl Acad Sci USA 83:5602–5606

Ruzin SE, McCarthy SC (1986) The effect of chemical facilitators on the frequency of electrofusion of tobacco mesophyll protoplast. Plant Cell Rep 5:342–345

Scheurich P, Zimmermann U (1981) Electrically stimulated fusion of different plant cell protoplasts. Plant Physiol 67:849–853

Senda M, Takeda J, Abe S, Nakamura T (1979) Induction of cell fusion of plant protoplasts by electrical stimulation. Plant Cell Physiol 20:1441–1443

Spangenberg G, Schweiger H-G (1986) Controlled electrofusion of different types of protoplasts and subprotoplasts including cell reconstitution in *Brassica napus* L. Eur J Cell Biol 41:51–56

Tempelaar MJ, Jones MGK (1985a) Fusion characteristics of plant protoplasts in electric fields. Planta 165:205–216

Tempelaar MJ, Jones MGK (1985b) Directed electrofusion between protoplasts with different responses in a mass fusion system. Plant Cell Rep 4:92–95

Tempelaar MJ, Duyst A, De Vlas Sy, Krol G, Symmonds C, Jones MGK (1987) Modulation and direction of the electrofusion response in plant protoplasts. Plant Sci 48:99–105

Vienken J, Ganser R, Hampp R, Zimmermann U (1981) Electric field-induced fusion of isolated vacuoles and protoplasts of different developmental and metabolic provenience. Physiol Plant 53:64–70

Watts JW, King JM (1984) A simple method for large-scale electrofusion and culture of plant protoplasts. Biosci Rep 4:335–342

Zachrisson A, Bornman CH (1984) Application of electric field fusion in plant tissue culture. Physiol Plant 61:314–320

Zimmermann U (1982) Electric field-mediated fusion and related phenomena. Biochim Biophys Acta 694:227–277

Zimmermann U, Scheurich P (1981) High frequency fusion of plant protoplasts by electric fields. Planta 151:26–32

Zimmermann U, Buchner KH, Arnold WM (1984) Electrofusion of cells: recent developments and relevance for evolution. In: Allen MJ, Usherwoods PNR (eds) Charge and field effects in biosystems. Abacus Press, Normal, IL, pp 293–318

# XVI Versatile Apparatuses for Electrogene Mapping, Electrophoresis and Electrofusion

K. YOSHIDA[1] and T. KONDO[2]

## 1 Introduction

Electrophoresis is one of the most powerful and fundamental techniques for separation and analysis of proteins and nucleic acids which has been widely accepted for both preparative and analytical purposes (Andrews 1981; Hames and Rickwood 1981; Rickwood and Hames 1982; Maugh II 1983), especially in the fields of recombinant DNA and nucleic acid sequencing. There are only a few investigations that do not utilize electrophoretic techniques (Rickwood and Hames 1982; Weissbach and Weissbach 1986).

According to the progress of biological sciences and biotechnology, various techniques and modifications of electrophoresis have been introduced (Andrews 1981; Hames and Rickwood 1981; Rickwood and Hames 1982; Maugh II 1983; Weissbach and Weissbach 1986). The introduction and development of pulsed field gel electrophoresis (PFGE), which can separate large DNA's up to 10 Mbp (Schwartz and Cantor 1984; see review by Anand 1986 and by Barlow and Lehrach 1987), is the most recent and exciting advance in electrophoresis. This technique promises enormous applicabilities in the genetic and biotechnological study of both lower eukaryotes, whose intact chromosomal DNA's can be directly resolved as individual bands on a gel, and of higher eukaryotes, whose chromosomal DNA's can be cut to produce large fragments by rare cutting enzymes and subsequently analyzed by standard techniques for further study. The application of PFGE in plant biotechnology has just started.

In this chapter, we describe a versatile apparatus for conventional electrophoresis and electrically induced cell fusion, subsequently focused on PFGE, and describe the construction of a versatile apparatus for it.

## 2 Fundamental Theory and Physical Parameters in Electrophoresis

In spite of many efforts, the theory of electrophoresis is very complicated and still incomplete. Therefore, a simple description of electrophoresis is enough for the practical technique (van Holde 1971; Andrews 1981). Let us consider a particle

[1] Botanical Institute, Faculty of Science, Hiroshima University, Hiroshima 730, Japan
[2] National Institute for Basic Biology, Okazaki 444, Japan

Biotechnology in Agriculture and Forestry, Vol. 9
Plant Protoplasts and Genetic Engineering II (ed. by Y. P. S. Bajaj)

with a charge, q, which moves in an insulated medium under an electric field, E. The velocity of the particle, v, is determined by the balance between the electrical force, Eq, and the frictional resistance, fv, in which f is the frictional coefficient. Thus,

$$Eq = fv \ . \tag{1}$$

This can be converted to the following Eq. (2), in which V and L indicate voltage and electrode distance, respectively.

$$v = (q/f)\,E = (q/f)\,(V/L) \ . \tag{2}$$

Note that v is proportional to V and electrical current does not appear in Eq. (2). The electrophoretic mobility, U, which is very similar to a sedimentation coefficient in sedimentation, is defined as the velocity per unit field and expressed by Eq. (3),

$$U = v/E = q/f = (qD)/(kT) \ . \tag{3}$$

Here, D, k, and T indicate diffusion coefficient, constant, and absolute temperature, respectively. The last two equations in Eq. (3) can be derived by Eq. (2) and the equation $f = kT/D$.

Now let us consider the separability of two particles with mobility U1 and U2. In a field intensity E, the disengagement d after time t is expressed as follows:

$$d = (U1 - U2)\,Et = (U1 - U2)\,(V/L)\,t \ . \tag{4}$$

This equation indicates that a given linear separation in an electrophoretic system used always requires a constant amount of electricity. Because U depends on the frictional parameter as shown in Eq. (3), it should give information about the size and shape of the sample molecule. Unfortunately, the electrophoretic medium is normally not an insulator but an ionic electrolyte buffer which causes great complexity. The modification based on Debye-Hückel theory improves the complexity somewhat. However, it is still insufficient, so that the present theory of electrophoresis has not been successfully described as yet. Although electrophoresis does not provide much information about macromolecular structure, it has offered very powerful preparative and analytical tools because of its enormous separability of various molecules.

Careful users of electrophoresis have observed that the sample velocity is proportional to the current and not to the voltage. At first sight, this appears to be inconsistent with Eq. (2). This is explained by the fact that the velocity is in fact proportional to the voltage given on the molecules itself, but not to the applied voltage to the electrophoretic tray. In other words, the velocity depends on the value of the voltage across the gel, which is determined by the current.

The relationship among current, voltage, buffer concentration, and electrode distance in electrophoresis is given by the following Eq. (5), which we have introduced from Ohm's law and the theory of electrolyte conductance (Yoshida 1983c).

$$I = \Lambda_0 (\sqrt{1+4KC} - 1)(A/L) V \; . \tag{5}$$

Here, I, $\Lambda_0$, K, C, A, L, and V indicate the current (A), molar conductivity at infinite dilution (ohm/cm$^2$/mol), dissociation constant of buffer electrolyte, concentration of buffer (mol/dm$^3$), cross-sectional area (cm$^2$), length between electrodes (cm) and potential (V), respectively. Equation (5) also gives us information on the physical parameters that influence electrophoresis [see Fig. 6, which shows the computer plotting of Eq. (5) with experimental data]. At low C, Eq. (5) can be replaced by the following Eq. (6), in which k is constant.

$$I = k(C/L) V \; . \tag{6}$$

This equation can also be obtained by Taylor's expansion of Eq. (5) (Yoshida 1983c).

Another important physical parameter is heat generation during electrophoresis. This is given by Eq. (7):

$$Q = kRI^2t \; . \tag{7}$$

In the above equation, k is constant and Q is heat generated during electrophoretic run time, t.

By selecting the above physical parameters properly, one can obtain the maximum separability in electrophoresis.

## 3 Versatile Apparatus for Electrophoresis Including Isoelectrofocusing

During the past 20 years, a number of new techniques have been introduced in electrophoresis and isoelectrofocusing because of the response to the increasing demands for high resolution of various samples: disc, gradient, vertical, horizontal (flat-bed), and two-dimensional gel electrophoresis and isoelectrofocusing with or without gels, and so on. In order to accommodate these new techniques, various types of apparatuses have been designed and constructed (Andrews 1981; Hames and Rickwood 1981; Rickwood and Hames 1982; Weissbach and Weissbach 1986). However, each apparatus is usually constructed based on each one of the techniques.

In this section, we describe a versatile apparatus for both electrophoresis and isoelectrofocusing that accommodates nearly all the techniques used in molecular biology and biotechnology (Yoshida 1983a). The principle of design is to separate the apparatus into blocks or parts which usually have more than two roles in its assembly with removable bar electrodes, paper clips, and silicone rubber pads, In this way, the electrophoretic system can be used as an apparatus for the following: vertical slab, horizontal, and two-dimensional gel electrophoresis, and electrofocusing, preparative electrophoresis, and electrofocusing. Therefore, the apparatus is economical because it has in one unit the function of several devices.

Figure 1 shows a schematic diagram and the scale diagrams for the apparatus. Unless otherwise mentioned, each block of the apparatus was constructed from

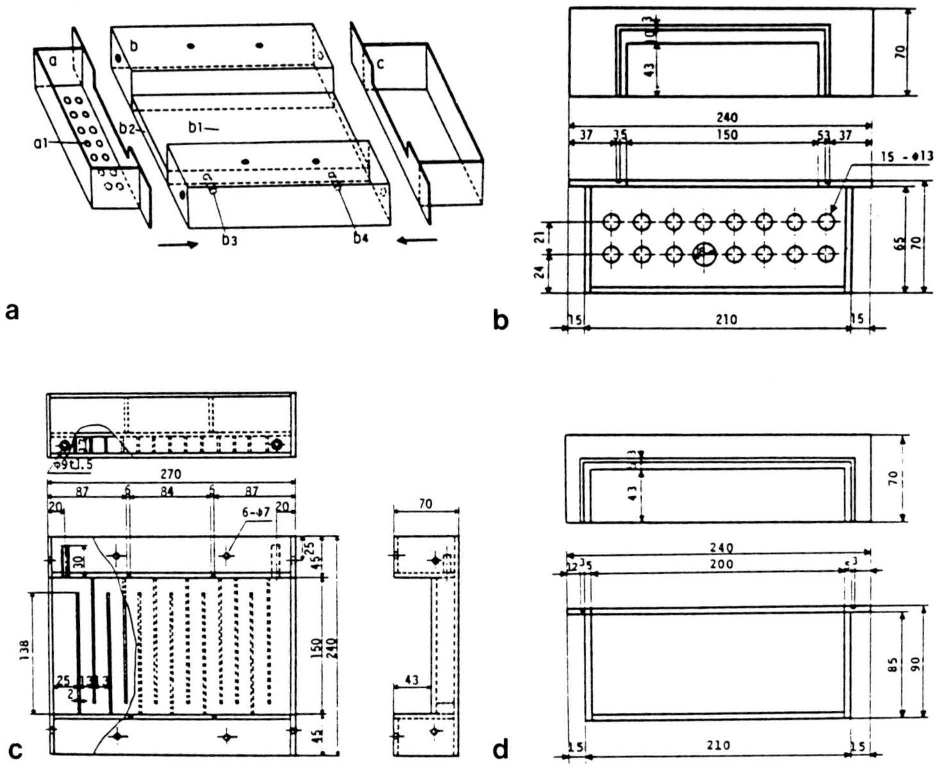


**Fig. 1a–d.** A versatile apparatus assembled for horizontal electrophoresis and electrofocusing. **a** Schematic diagram: *a* electrode buffer tray; *a1* hole for tube gel electrophoresis and electrofocusing; *b* flat-bed tray; *b1* electrophoresis tray; *b2* circulating cooling plate; *b3* inlet; *b4* outlet; *c* electrode buffer tray. The *arrows* indicate the direction of assembly. **b** Design drawing for block *a* in **a**. **c** Design drawing for block *b* in **a**. **d** Design drawing for block *c* in **a**. The figures are expressed in mm and each upper and lower drawing indicates the side and top view, respectively. The lower left in **c** is also the side view. (**a–d** Yoshida 1983a)

acrylic plates (5 mm thick). Figure 1a shows how the system is assembled for horizontal electrophoresis and flat-bed electrofocusing. The arrows indicate the direction of assembly of the apparatus. Figure 1b to c show the design drawing of the apparatus.

Figure 2 shows a removable bar electrode, which consists of a power supply joint and a holed acrylic bar plate to which a platinum wire was attached across the holes. The electrodes are usually used to be immersed in the electrode tray buffer except in the case of electrofocusing on a flat-bed gel.

The system assembly with additional blocks for vertical electrophoresis is shown in Figure 3a. In this case, the electrophoretic tray is used as a lower buffer tray together with side plate block, d, and bottom plate block, f. Figure 3b and c show the design drawing for the additional blocks. The bottom acrylic plates are 1 cm thick, while the other plates are 0.5 cm, as described above.

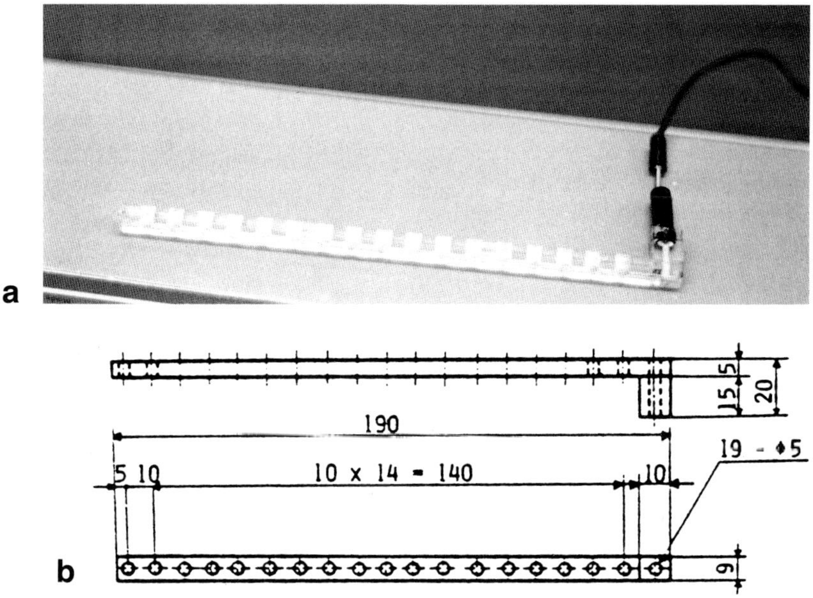


**Fig. 2a, b.** A removable bar electrode for the versatile apparatus. **a** Photograph of a removable electrode. **b** Design drawing of a removable electrode. The *upper* and *lower* drawing indicate the side and top view, respectively. A platinum wire (0.5 mm diameter) was attached across outlet holes for gas in the acrylic bar. (**a** Yoshida 1983a; **b** Yoshida 1986)

An assembly for preparative isoelectrofocusing and electrophoresis is shown in Fig. 4a and b. In order to save space, only one typical example of preparative isoelectrofocusing by the versatile apparatus is shown in Fig. 4c. The other assembly and performances including immunoelectrophoresis were described elsewhere (Yoshida 1983a). The detailed techniques in the application of conventional electrophoresis to plant molecular biology and biotechnology has been dealt with by Weissbach and Weissbach (1986).

## 4 Highly Simplified Horizontal Electrophoretic Apparatus Including a Simple Power Supply

As described above, electrophoresis is now one of the routine techniques in the fields of molecular biology and biotechnology. Therefore in these fields, at least one set of the electrophoretic apparatus with a power supply is needed for each scientist or laboratory bench. Moreover, the apparatus and power supply should be simple and compact with high cost performance.

For this purpose, we have introduced an extremely simplified apparatus including a power supply for flat-bed electrophoresis and isoelectrofocusing (Yoshida 1983b). As shown in Fig. 5, the apparatus consisted of a buffer tray with a plastic food container, in which a gel glass plate attached to two removable bar

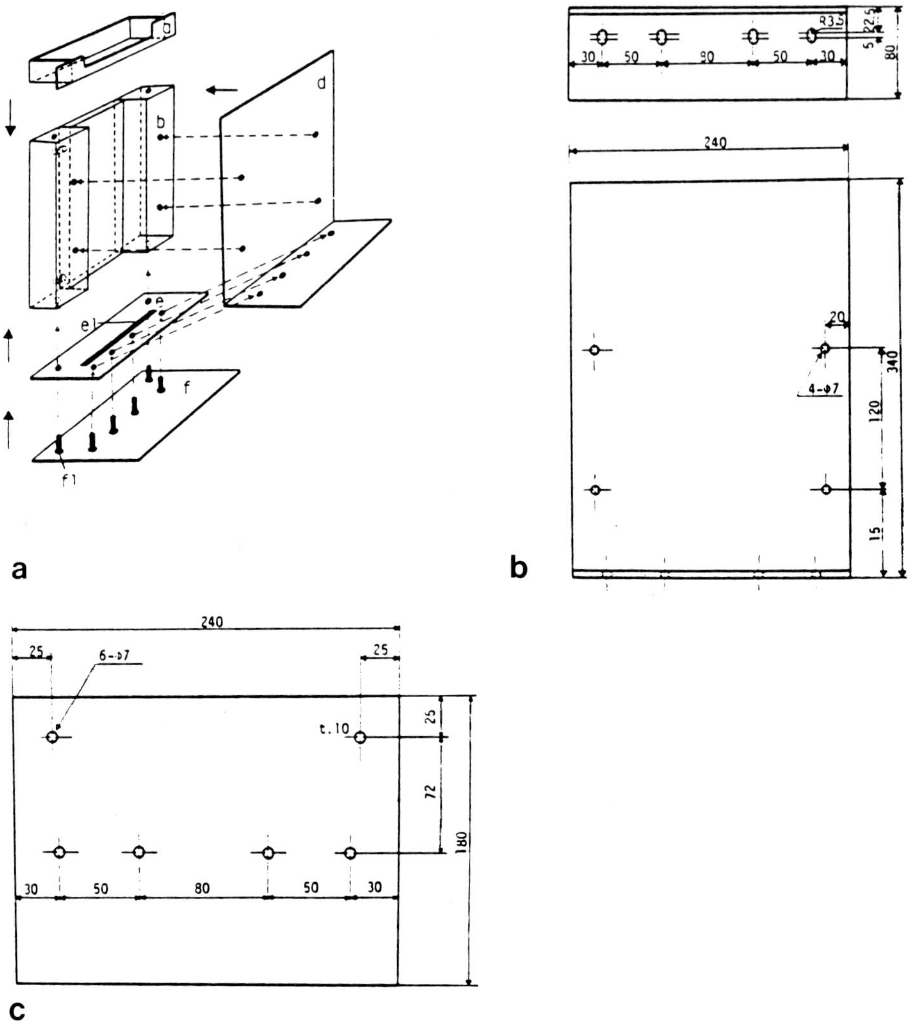


**Fig. 3a–c.** The versatile apparatus assembled for vertical slab gel electrophoresis. **a** Schematic diagram *a* upper electrode buffer tray; *b* lower electrode buffer tray; *d* side plate of the lower tray; *e* silicone rubber pad; *el* slot for lower electrode; *f* bottom plate; *fl* joint screw. The blocks *a* and *b* correspond to *a* and *b* of Fig. 1a. b Design drawing for block **d** in **a**. The *upper* and *lower* drawing indicate the top and side view, respectively. **c** Design drawing for block *f* in **a**. Top view. (**a–c** Yoshida 1983a)

electrodes with rubber bands, was immersed in a buffer in the case of submerged electrophoresis. The cooling plate in Fig. 5a is unnecessary for conventional agarose gel electrophoresis. The power supply essentially consists of a single bridge-diode which generates pulse direct current. Its output voltage can be more easily changed by the usage of a rehostat or thyristor without much loss of simplicity and cost performance. However, note that without the rehostat or thyristor, one can change electrophoretic current by buffer dilution or change of the elec-

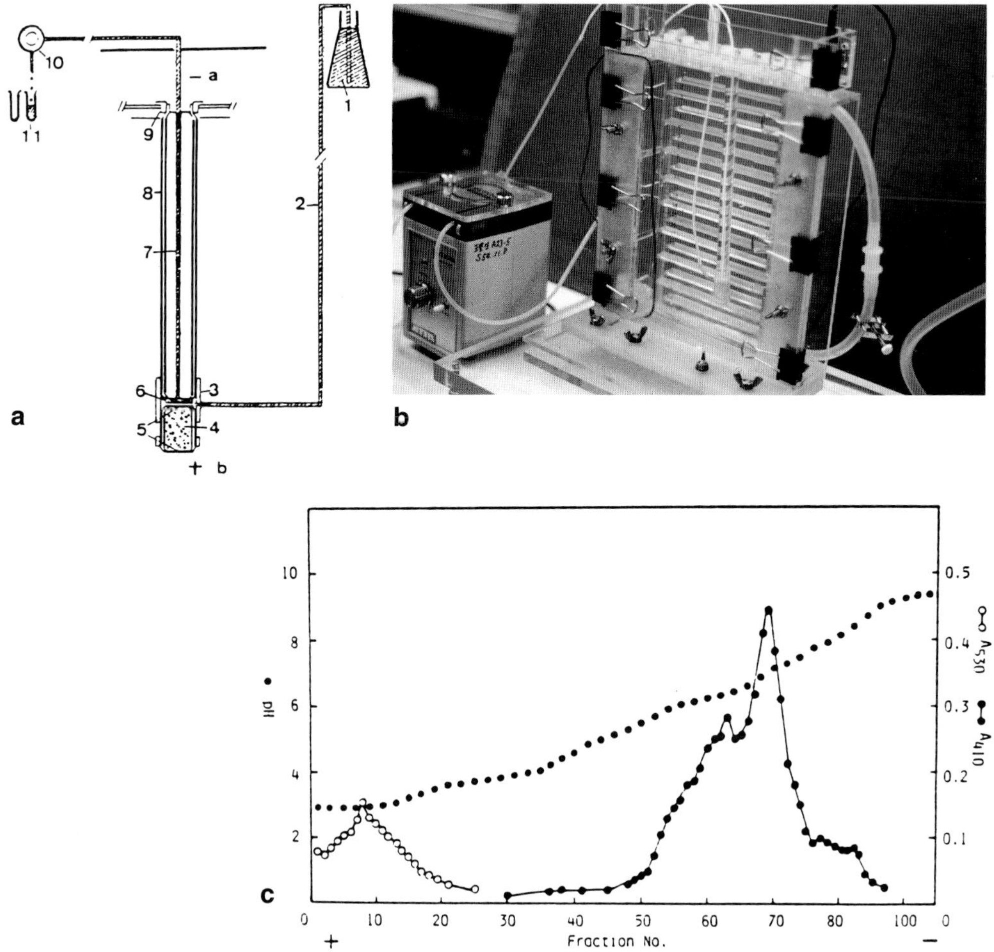


**Fig. 4a–c.** An assembly for preparative isoelectrofocusing and electrophoresis and its performance. **a** Schematic diagram of preparative column: *a* upper; *b* lower buffer tray; *1* reservoir of elution buffer; *2* connecting tubing; *3* silicone rubber tubing; *4* stopper gel; *5* dialyzing tubing membrane; *6* porous plastic plate; *7* polyethylene tubing; *8* glass preparative column; *9* silicone rubber connector of upper tray; *10* peristaltic pump; *11* test tube *1* to *9* were hand-made. The column *8* was submerged in the lower tray buffer **b** which was cooled by circulating chilled water. *b* Photograph of **a**. The *large* and *small arrows* indicate the separation column and magnetic stirring bar, respectively. **c** Preparative isoelectrofocusing in a versatile apparatus. Discontinuous density gradient in column (1×20 cm) made by 5%–40% sucrose in which carrier ampholyte (pH 3 to 10) (LKB) 1% was contained. The upper tray contained 0.25 N NaOH. The lower tray contained 0.3% phosphoric acid. After electrofocusing for 12 h at 350 V under the ice-cooled condition, each fraction (7 drops) was collected through a guiding tube *7* in **a**, by syphon action (in this case, the pump was not used). One milliliter of distilled water was added, then pH and absorbance of each fraction were measured. Electrofocusing samples were myoglobin (●——●) and methyl red (○ – – – ○). (**a–c** Yoshida 1983 a)

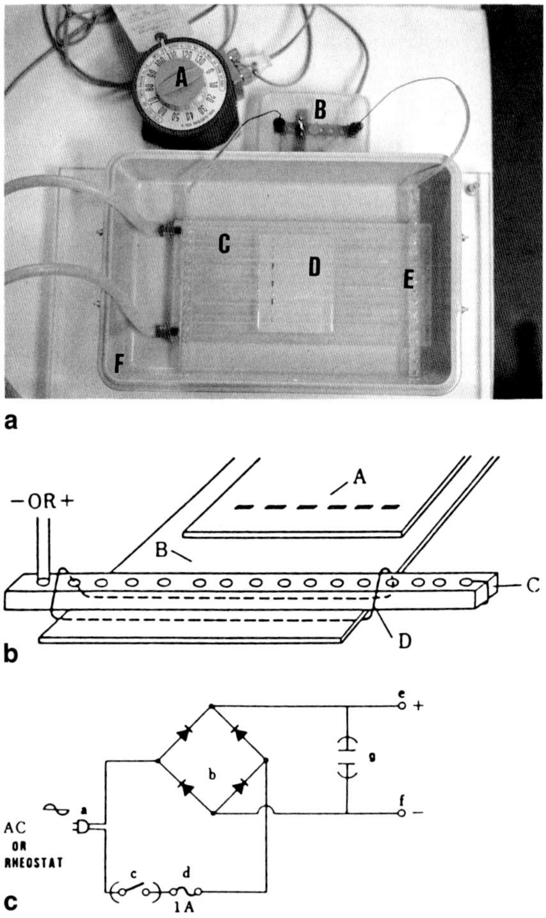


**Fig. 5a–c.** Highly simplified horizontal electrophoretic apparatus and power supply. **a** Photograph of a simplified apparatus with a power supply. *A* rehostat; *B* single bridge-diode-capacitor power supply; *C* circulating cooling plate; *D* gel; *E* removable bar electrode; *F* electrophoretic tray (a plastic food container 30×20×7 cm was used). **b** Assembly of the removable bar electrode and gel glass plate. *A* gel; *B* glass plate; *C* removable bar electrode; *D* rubber band. **c** Rectifying electrode circuit for the power supply in **a**: *a* plug to 100V or to a rheostat; *b* bridge-diode rectifier; *c* switch; *d* fuse; *f* DC-pulse dc output terminals; *g* electrolytic capacitor (removable) (**b, c** Yoshida 1983b)

trode distance according to the diagram in Fig. 6 or Eq. (5) (Yoshida 1983b, c). The above apparatus and power supply can be easily constructed and assembled by laboratory tools in a short time. The performance of the apparatus proved to be very satisfactory, in spite of its simplicity and inexpensiveness (Yoshida 1983a, b).

## 5 Protoplast Fusion by Electric Impulse Generated by the Simple Power Supply

Successful protoplast fusion in plants has been carried out by chemically mediated methods using high pH with calcium (Keller and Melchers 1973) or polyethylene glycol (Kao and Michayluk 1974). However, these chemical fusogens are not so reliable and are often toxic (Keller and Melchers 1973; Kao and Michayluk

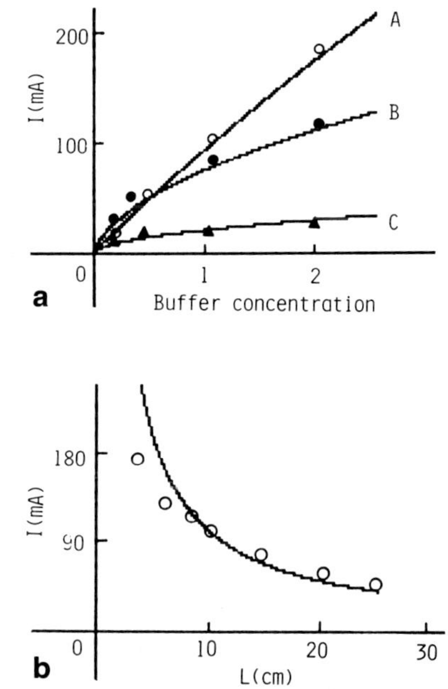


**Fig. 6a, b.** Relationship among electrophoretic current, electrode distance and buffer concentration. **a** Relationship between electrophoretic current I (mA) and buffer concentration C (dilution factor). *A, B,* and *C* indicate submerged, vertical slab, and disc gel electrophoresis, respectively. **b** Relationship between electrophoretic current I (mA) and electrode distance L (cm). In **a** and **b**, ○, ●, and △ indicate experimental data. (– – – – –) shows the computer plotting

1974; see review of Nagata 1984). Furthermore, it is difficult to obtain or select one-to-one fusants between different cell partners by the fusogen-mediated method because the method requires protoplast aggregates at high cell densities. As an alternative to the use of fusogenic agents, electrofusion, electrofield-induced cell fusion has been used (see reviews of Zimmermann 1982; Nagata 1984).

It is well known that the electric impulse causes great changes in the cell membranes. Zimmermann et al. (1974) introduced the following approximate expression for the potential difference Vm across a cell membrane,

$$Vm = fi \times ai \times E_0 \ , \tag{8}$$

where fi, ai, and $E_0$ indicate the shape factor, inner radius of a cell and external electric field strength, respectively. By adjusting Vm to 0.3 – 3V, reversible dielectric breakdown of cell membranes occurs and subsequently facilitates cell fusion or incorporation of exogenous DNA's.

Electrofusion consists of two main steps: cell contact and subsequent membrane fusion by electric pulse. In the first electrofusion by Senda et al. (1979), cell-to-cell contact was carried out by physically pushing two cells together with microelectrodes so that the yield of fusants was inevitably very small. Furthermore, it required special skill to handle microelectrodes under a microscope. Along this line, Neumann et al. (1980) and Morikawa et al. (1986) followed and reported the electrofusion of *Dictyostelium* cells and *Nicotiana* protoplasts, respectively. In their experiments, cell-to-cell contact was carried out by using a thick layer of cells or protoplasts and then cell fusion was induced by electric pulse. Their results indicate that the yield and efficiency are much improved.

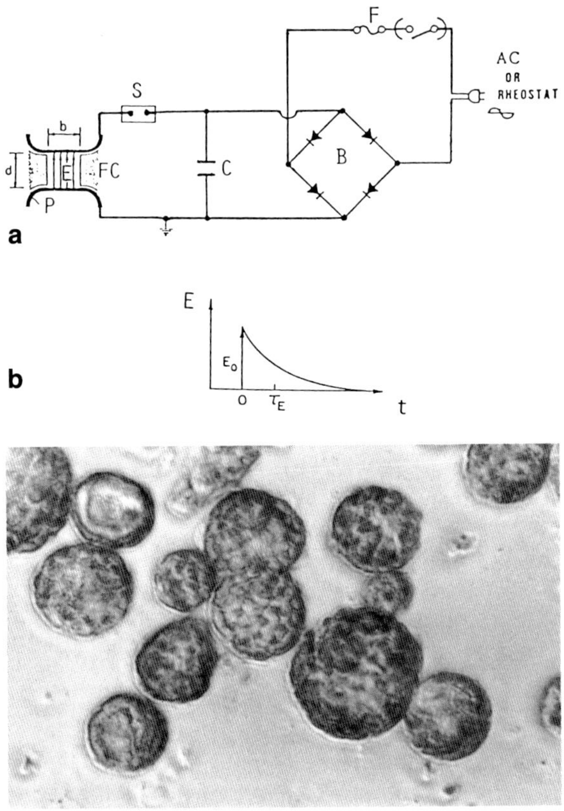


**Fig. 7a–c.** Cell fusion of tobacco protoplasts by an electric impulse. **a** Schematic diagram of the impulse device constructed from the fusion chamber (Neumann and Rosenheck 1972) and the simple power DC supply of Fig. 5c. *FC* fusion chamber; *P* platinum electrode; *d* distance; *b* length of the square cavity filled with cell suspension; *E* direction of homogeneous electric field; *S* spark gap (trigger); *C* capacitor; *B* bridge diode; *F* fuse. **b** Electric pulse generated by the device in **a**. $E_0$ initial field intensity; *t* time; $t_E$ relation time of field decay. **c** Photograph of *Nicotiana tabacum* protoplasts to which electric impulse was given by the device in **a**. In this case, $E_0 = 160$ V, C = 3.3 μF, b = 10 mm and d = 1 mm were used. Tobacco protoplasts were suspended in 2.5 mM $CaCl_2$ – 0.5 M sorbitol solution and layered in the fusion chamber (parallel platinum wire electrodes attached to a slide glass 2.5×7.5 cm) according to Morikawa et al. (1986). (**a** and **b** after Neuman and Rosenheck 1972)

Figure 7a shows a combined device of the Neumann's fusion chamber with our simple bridge diode-capacitor power supply for electrofusion. The decay of impulse with the initial intensity, $E_0$, in this device is determined by the following equation (Neumann et al. 1980):

$$E(t) = E_o \exp(-t/RC) \ . \tag{9}$$

Here, R and C indicate resistance of cell suspension and capacitance of discharge circuit, respectively. Equation (9) is plotted as shown in Fig. 7b.

The relationship among physical parameters in the fusion chamber is given as follows:

$$p = (V/I)(S/d) \ , \qquad (10)$$

in which p, V, I, S, and d indicate specific resistance of protoplast suspension, pulse intensity, current, area of electrodes, and distance between electrodes, respectively.

We examined the above device in the cell fusion of tobacco protoplasts according to Morikawa et al. (1986). Typical electrofusion of *Nicotiana tabacum* protoplasts is shown in Fig. 7c.

Instead of physical cell-to-cell contact, dielectrophoresis has been successfully introduced to electrofusion (see review of Zimmermann 1982). Electric field induces dipoles within cells, which result in the cell movement into the part of the highest field intensity. This process is known as dielectrophoresis (Pohl 1978). Dielectrophoresis finally results in the formation of cell chains (pearl chains) between the electrodes. Dielectrophoretic force, F, is expressed as follows:

$$F = k \times q \times \Delta E \ , \qquad (11)$$

in which k, q, and $\Delta E$ are constant, polarized charge of a cell and electric intensity across a cell, respectively. Under the nonhomogeneous AC field (in the case of point electrodes), polarized cells move toward the electrodes, since $\Delta E \neq 0$. On the other hand, polarized cells have essentially no movement toward electrodes under the homogeneous AC field (in the case of parallel bar electrodes) since $\Delta E \approx 0$. Moreover, the neighboring cells attract each other by Coulomb's force. This indicates that the homogeneous field is more effective for cell fusion than the nonhomogeneous (Zimmermann and Scheurich 1981). Improvement in yield and fusion efficiency including devices has since been carried out. Now a combined method of dielectrophoresis by high frequency AC and cell fusion induced by DC impulse has become an established technique for electrofusion (Watt and King 1984; Bates 1985; Bates and Hasenkampf 1985; Tempelaar and Jones 1985a, b; Watts et al. 1985).

Although the efficiency of electrofusion by the method in Fig. 7 is lower than that by the combined method of dielectrophoresis and cell fusion induced by DC pulse, the apparatus is very simple, inexpensive, and easy to make in a laboratory. Moreover, the above device is easily applied onto electric pulse-induced transformation, electroporation, with a slight modification because the device principle is essentially the same. It goes without mentioning that a safety device against electric shock should be included in the above apparatus.

## 6 Construction of a Versatile Apparatus for PFGE and Its Possible Application to "Electrogene Mapping"

DNA analysis by conventional gel electrophoresis is usually limited up to 20 kbp by using ordinary agarose gel and maximally up to 500 kbp by using very fragile diluted agarose gel (Rickwood and Hames 1982). In 1984, with a preparation technique of unbroken genomic DNA's inside agarose gels, Schwartz and Cantor introduced a new gel electrophoresis, pulsed field gel electrophoresis (PFGE). This can separate DNA molecules up to 2 Mbp (2000 kbp) by subjecting periodically to two approximately perpendicular electric fields. The development and modification to this technique have since progressed very rapidly and the size of DNA has now been resolved to more than 10 Mbp. In the yeasts *Saccharomyces cerevisiae* and *Schizosaccharomyces pombe*, of which the chromosomes have never been visibly condensed, almost all the chromosomal DNA's were separated as individual bands on gels and subsequently their electrophoretic karyotypes were established by identifying each band (Schwartz and Cantor 1984; Carle and Olson 1985b). At present, in the higher plants as well as in mammals, chromosomal DNA's are usually too large to be resolved as individual bands even by the most powerful PFGE technique (see Table 1). However, PFGE, used together with methods for cutting DNA's inside a gel into large fragments with restriction endonucleases, is highly useful in the construction of physical maps of the larger DNA, as recently proven in *Escherichia coli* bacteria (Smith et al. 1987).

The original apparatus by Schwartz and Cantor (1984) consisted of a square horizontal gel tray which was surrounded with vertical single point platinum electrodes (see Fig. 8a). Several efforts have since been made to improve especially the geometry of DNA tracks and the DNA size to be resolved (see Fig. 8). According to the electrode configuration and instrumentation, they are named as follows: "original" pulsed field gradient gel electrophoresis (Schwartz and Cantor 1984), orthogonal field alternation gel electrophoresis (Carle and Olson 1985a), double inhomogeneous field gradient gel electrophoresis (Smith and Cantor 1986), field inversion gel electrophoresis (Carle et al. 1986), vertical pulsed field gradient gel electrophoresis (Gardiner et al. 1986), contour clamped homogeneous electric field gel electrophoresis (Chu et al. 1986), homogeneous crossed field gel electrophoresis (see in Anand 1986), and so on. The original apparatus by Schwartz and Cantor (1984) gives diagonal DNA migration tracks, as expected in Fig. 8a. The subsequently developed apparatuses have been modified to give central straight migration tracks which make accurate DNA size estimation and comparison. Both vertical pulsed field gradient and homogeneous crossed field gel electrophoresis require more complicated special electrophoretic trays and devices (see Fig. 8e). On the other hand, field inversion gel electrophoresis yields very clear straight migration tracks in a conventional electrophoretic tray, but essentially requires a computer to gradually change the pulse time and thick platinum wire electrodes because the electrodes in this device corrode away so rapidly (see Fig. 8d). Although orthogonal field alternation gel electrophoresis requires rather simple devices and electrodes, it gives straight central tracks with some distortion on the side lanes (see Fig. 8b). The contour clamped homogeneous electric field gel electrophoresis yields fine straight tracks but requires a rather complicated switching unit and hexagonal electrode as shown in Fig. 8f. Some of the above devices are already commercially available. However, they are rather expensive and seem to leave much room for further improvement, since the PFGE technique itself is still under modification and DNA's longer than 20 Mbp are not resolved fully yet.

Although the theoretical principles for the PFGE are not yet established, some physical parameters are empirically known to have great effect on the separatabil-

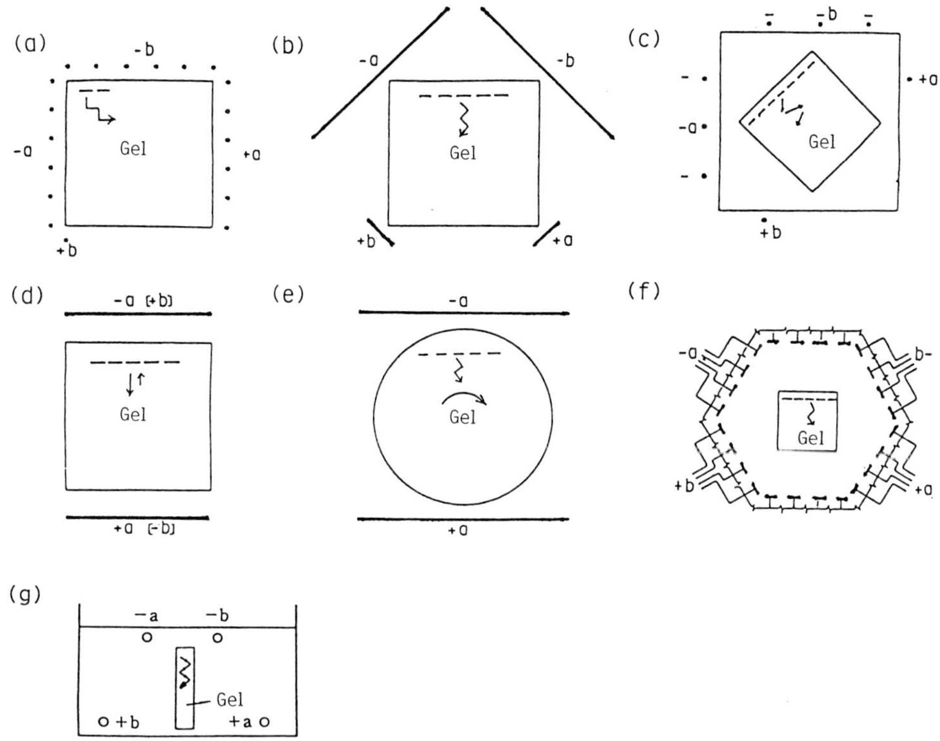


**Fig. 8a–g.** Various electrode configuration for PFGE. (●) and *bar* indicate diode-isolated point electrode and bar electrode, respectively. Electric polarities after switching are shown in brackets or ±a and ±b. **a** One uniform and one nonuniform field configuration of the original apparatus by Schwartz and Cantor (1984). **b** Both nonuniform fields configuration of orthogonal field alternation by Carle and Olson (1985a). **c** Modified double inhomogeneous field apparatus of **a** by Smith and Cantor (1986). **d** Field inversion field by Carle et al. (1986). **e** Rotatory gel under the uniform field by Southern et al. (see Anand 1986). **f** Configuration for contour-clamped homogeneous field by Chu et al. (1986). The *arrow* indicates the direction of DNA migration. **g** Vertical pulsed field gradient by Gardiner et al. (1986). Here, open circles indicate bar electrodes. In **a** to **f**, the top views of their electrode-gel configuration are shown, while the side view is shown in **g**

ity. DNA mixtures are separated according to their molecular sizes by repeating the electrophoretic run to another direction after the one direction, since the larger DNA's may take more time to turn the direction in the gel matrix. In other words, the reorientation time of the DNA is presumed to be proportional to their molecular size. Therefore, in addition to the physical parameters described above for conventional electrophoresis, the switching interval and the angle between two electrophoretic directions are very important to obtain an effective separation in PFGE. For this reason, the improvements in PFGE are mainly regarding the parameters (see Fig. 8 and review of Anand 1986).

The modification to the original preparation of intact DNA in agarose gel (solid-plug method) has lead to a micro-bead method (Cook 1984) that can more easily prepare many samples at one time.

## 6.1 Construction of Apparatus

By enlarging the size of our versatile apparatus for conventional electrophoresis (Figs. 1 to 4), it can be adapted to PFGE as well as DNA sequencing gel electrophoresis because of its versatile characteristics (Yoshida 1985; Yoshida and Kondo 1987). At present, it is not known which modification in Fig. 8 will survive as a standard PFGE. In consideration of this, together with the rapid progress in PFGE techniques, we have recently introduced an inexpensive versatile apparatus with removable electrodes. The device can be easily constructed by laboratory tools. All the modified PFGE of b, d and f in Fig. 8 are easily applicable to the device. It consists of a buffer tray, power switching unit, power supply, circulating cooling unit, microcomputer, and removable electrodes (see a block diagram of Fig. 9a). The power switching unit consists of a timer, microcomputer, driver, and relay. Besides the above switching unit, we have also constructed a power switching unit without a microcomputer. Therefore, we describe both switching devices.

### 6.1.1 Timing Control

The timing for providing electricity to a gel is controllable either by a wire-logic timer circuit or by a software controlled computer. Both timers are programable for the switching frequency of a charging voltage to the electrode pairs. The maximum limit of the frequency is between 0.2 Hz and 1 Hz for a magnetic relay which was used in this paper. For faster switching, a solid-state relay and/or a faster program described in PASCAL or ASSEMBLER should be used.

*1. Hardware Control Timer.* The timing circuit (Fig. 9b) consists of a programable oscillator, SEIKO 8651B, which generates a wide range of frequency with a 6-bit switch control. The output of the oscillator can be doubled by TTL 4-bit binary counter, 74LS93. A final frequency obtained through the controlling circuit is 10 to 1/3600 Hz. A duty ratio of each phase is 50% (even charging). Alternatively, recently available timer module (OMRON H3CA) is equivalent to this circuit. Although the device is not as flexible as a software-controlled one, it is comparatively inexpensive because of its not requiring a microcomputer.

*2. Software Control Timer.* The control by a microcomputer software has more flexibility but decreases cost performance somewhat. Without the use of microcomputer controlling, changing a switching frequency during electrophoresis and uneven charging of electricity to each electrode pair is essentially impossible for the above described timing circuit. This capability is very important and essential for field inversion gel electrophoresis (Carle et al. 1986), in which the current is reversed in an uneven fashion by changing periodicity as electrophoresis proceeds.

Because the switching frequency used in PFGE is relatively low, any computer which has an output port can be used in this device. In this paper, a lap-top 8-bit microcomputer HC-20 (EPSON, HX-20 outside Japan) was used. In this instance, a cassette recorder control port was used to switch the electricity output. The controlling program was written in a Microsoft BASIC (see listing in Appen-

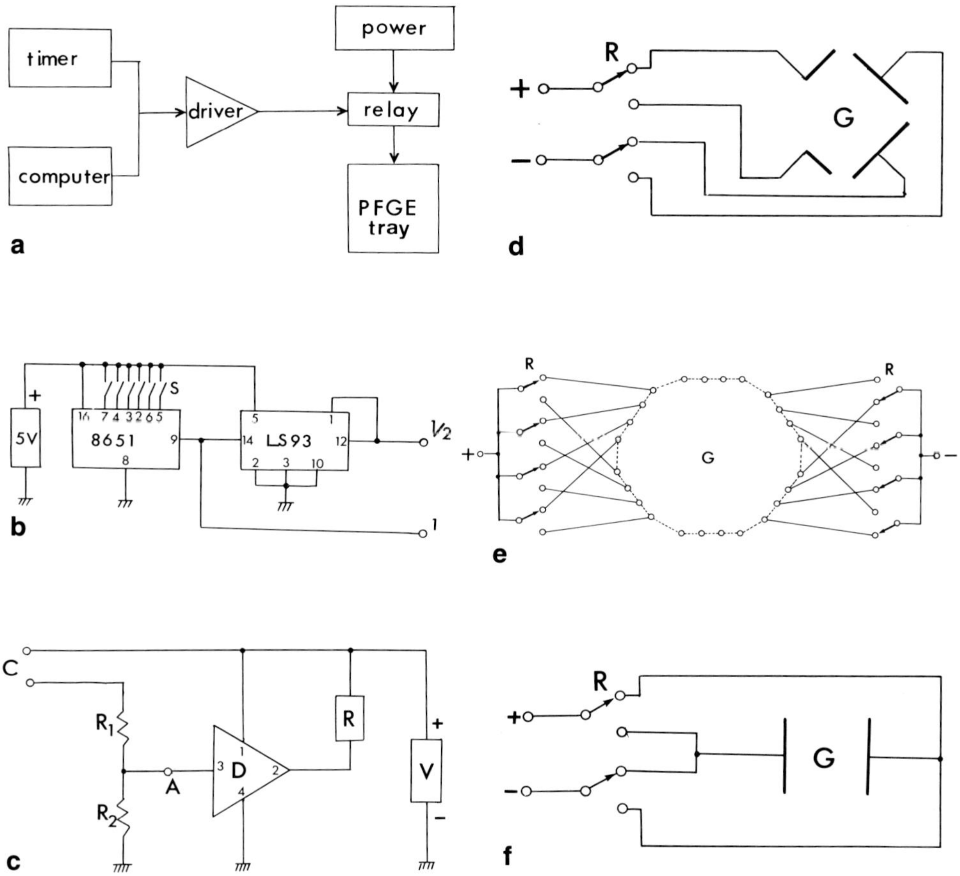


**Fig. 9a–f.** Block diagram of a versatile PFGE apparatus and its electric circuits. **a** Block diagram of a computer-controlled versatile PFGE apparatus. Output of timer circuit (see **b**) or software controlled computer (cassette recorder control or RTS/DTR of RS232C serial port) is amplified by driver (see **c**). Relay switches power supply to a PFGE tray. **b** Timer circuit. The circuit is not necessary for the computer-controlled system. By switching a 6-bit control switch S, 8651B can be changed. TTL of LS93 can optionally reduce the frequency to 1/2 of the frequency generated by 8651B. See data sheet of the IC. **c** Relay driving circuit. $R_1$ and $R_2$ is dependent on internal resistance of cassette recorder motor port (*C*) and the power voltage (*V*) for relay (*R*). If the relay operates at 12 V, 200 and 400 ohm registers should be for $R_1$ and $R_2$, respectively. **d** Switching by relay for orthogonal field alternation gel electrophoresis. **e** Switching by relay for contour-clamped homogeneous electric field gel electrophoresis. Two relays (*R*) are operated by the same driving circuit with opposite polarity of electrophoresis power. **f** Switching by the relay for the field inversion gel electrophoresis. In **d** to **f**, *G, R,* and *bar* indicate a gel, relay switch and electrode, respectively

dix). The program and interfacing circuit are applicable in any other computers whose cassette recorder motors can be controlled in BASIC. The following comments may not be necessary for the microcomputer-experienced scientists, but will be useful for those who are not familiar with this field. Because the computer executes subroutine line 1000 before starting each cycle, the modification of the functions in the lines 1010–1020 results in changing the frequency (variable

PERA in s) and the duty ratio (variable CUTA in s), as well as the following variables of charging cycle CN, PERS, PERE, CUTS, CUTE, DURATION and ELSP (in s) for field inversion gel electrophoresis (see Appendix). The program is designed to obtain a linearly increasing switching period from PERS to PERE and a charging phase against the electrode pairs from CUTS to CUTE during the programmed duration of electrophoretic run.

*3. Modification of Program for Other Computers.* The program listing in the Appendix is written for a HC-20 microcomputer. In order to use other computers, line 20 to 60 of the Appendix listing should be changed according to their Basic format if necessary. The TIMES function should adapt its format to obtain the correct time. Commands controlling the motor switch of the cassette recorder (MOTOR ON and MOTOR OFF equal to MOTOR0 and MOTOR1 in some BASIC) have to be changed to fit the computer used. Screen controlling commands below line 100 should also be changed according to the characteristics of the monitor display used.

In the case of no cassette recorder interface in the computer used, the game I/O port or two lines of RS232C serial port (RTS or DTR) interface can be used. MOTOR ON and MOTOR OFF commands in the listing should be changed to control these output ports to assert (5 V or higher) and to negate (0 V), respectively. OPEN command in some BASIC directly controls the status of two RS232C lines by selecting parameters (consult the manuals).

### 6.1.2 Relay Driver

The output from the timing circuit or the computer must be previously amplified to drive a relay (Fig. 9a). A TTL level output can be amplified by a Darlington transistor driver (ULN2064, Sprague) (Fig. 9c). The output of the cassette recorder interface can be converted to the TTL level signal (0 V or 5 V) by a register division of a power supply for the relay. If a driving relay is operated at 12 V, a simple register circuit (Fig. 9c) can generate TTL voltage. If an operating voltage of the relay is different, the register value should be adjusted to 5 V at point A of Fig. 9c in order to make the relay driver on. The output from the wire-logic timer is TTL level (Fig. 9 a). RTS or DTR of RS232C port can be connected to the open end of $R_1$ and ground after removing $R_2$ (see Fig. 9c). If the relay driver does not work, try to decrease the resistance value of $R_1$ until the driver works. In this device, a relay of MY-4 (OMRON) was used but another relay can be used as long as its capacity covers. A commercial AC adapter for a portable radio is better to use for the power supply to drive the relay if the voltage is matched. For the above PFGE, one relay is enough. However, for contour clamped homogeneous electric field gel electrophoresis, two relays should be operated in parallel. Note also that the electricity will continuously charge on the second electrode pair at the end of the programmed duration by software timer. Therefore, the automatic termination of electrophoresis is carried out by another independent timer to cut off the power supply.

### 6.1.3 Circuit for the Electrodes

For all PFGE except contour clamped homogeneous electric field gel electrophoresis, each side of the relay can be connected to the electrodes (see Fig. 9d and f). As shown in Fig. 8f, Chu et al. (1986) recommmend the use of the hexagonally arranged multi-electrodes, which are connected by a register network to make a uniform electric field for keeping side track distortion away (see Fig. 9e). For this reason, the value and power capacity of the register is very important. For example, registers network of 470 ohm (2 W) and 870 ohm (2 W) can be used for 150 V and 300 V power supply, respectively. It should be known that the configuration of the electrodes and conductance between electrodes (influenced by buffer concentration and electrode's shape) are also important. If there is an imbalance among the conductance between electrodes, the registers between electrode pairs with high resistance will be over-charged and this sometimes causes burning out. In order to avoid such a case, the adjustment of electrode configuration and/or the power dispersion capacity of registers are indispensable before the use of the system (see Fig. 9e). In Fig. 9f is shown the wiring between the relay to electrodes for field inversion gel electrophoresis. The moving of one electrode pair in Fig. 9d to the same position of the other pair will be equivalent to Fig. 9f.

## 6.2 Assembly of Apparatus and Its Performance

Figure 10a and b shows a photograph of the electrode configuration in this device for orthogonal field alternation gel electrophoresis and contour clamped homogeneous electric field gel electrophoresis, respectively. PFGE parameters including switching interval, time, and so on, are conversationally inputted into the microcomputer as shown in Fig. 10c.

As expected from the present capability of PFGE, genomic DNA's of liverwort moss were not clearly separated under the condition in which the yeast genomic DNA's were fully separated. This indicates the necessity for further improvement in separation techniques and instrumentation. As shown in Fig. 10c, however, PFGE profile of tobacco genomic DNA's has four distinct bands, of which the first three are considered to be satellite chromosomal DNA's by molecular hybridization (Yoshida and Kondo 1987).

As described above, the program is easily transportable to other microcomputers as long as their BASIC are Microsoft's. By changing the program and/or electrodes, our apparatus may easily follow the changes according to the rapid progress in PFGE techniques because of its software-controlled versatile characteristics and removable electrode configuration. Additionally, the apparatus can be used as a conventional electrophoretic device if necessary.

A safety lock should be attached to the above electrophoretic apparatus to avoid electric shock hazard because considerable high voltage and current have to be used in PFGE.

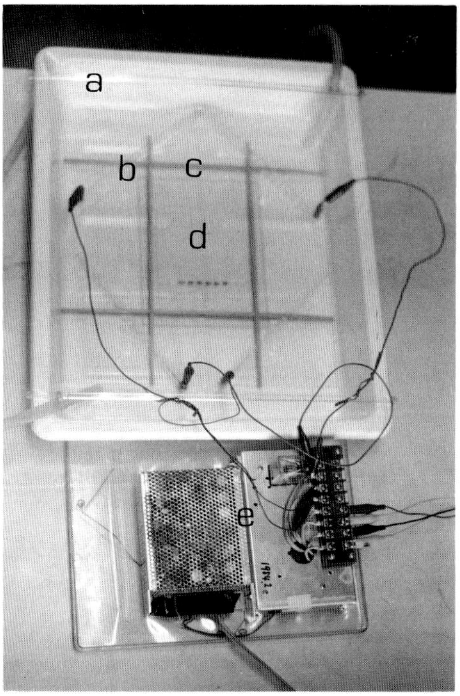


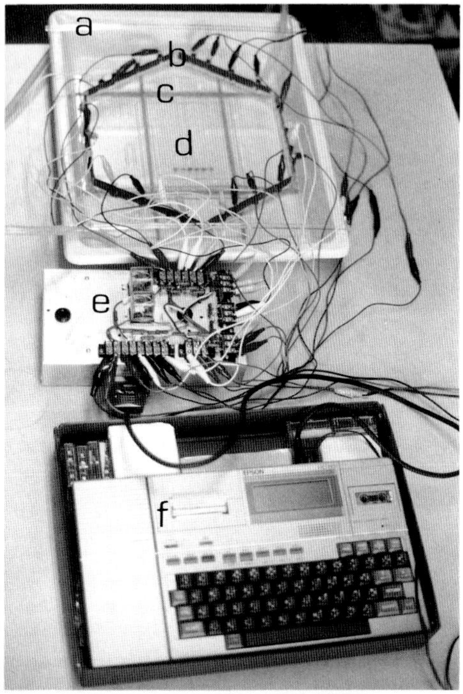


**a** **b**

```
PERIOD (STARTING, EN
DING in sec)=? 12,80

PHASE CHANGE (STARTI
NG,ENDING in sec)=?
9,60

DURATION in min=? 12
00
---------------
87/09/30 11:45:55
PERIOD:  12 - 80
PHASE CHANGE:  9 - 60
DURATION:  1200  min
PLANT DNA
K. YOSHIDA
300 V
PFGE
---------------
HIT G AND ACTIVATE P
OWER TO START
? G
---------------
 32            2
           A:ON
---------------
```

**c**

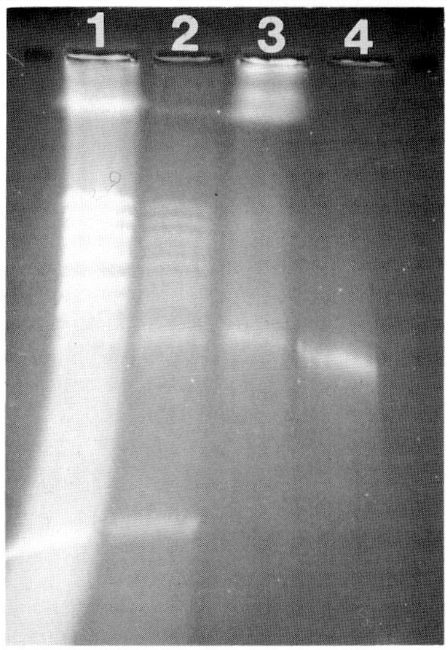


**d**

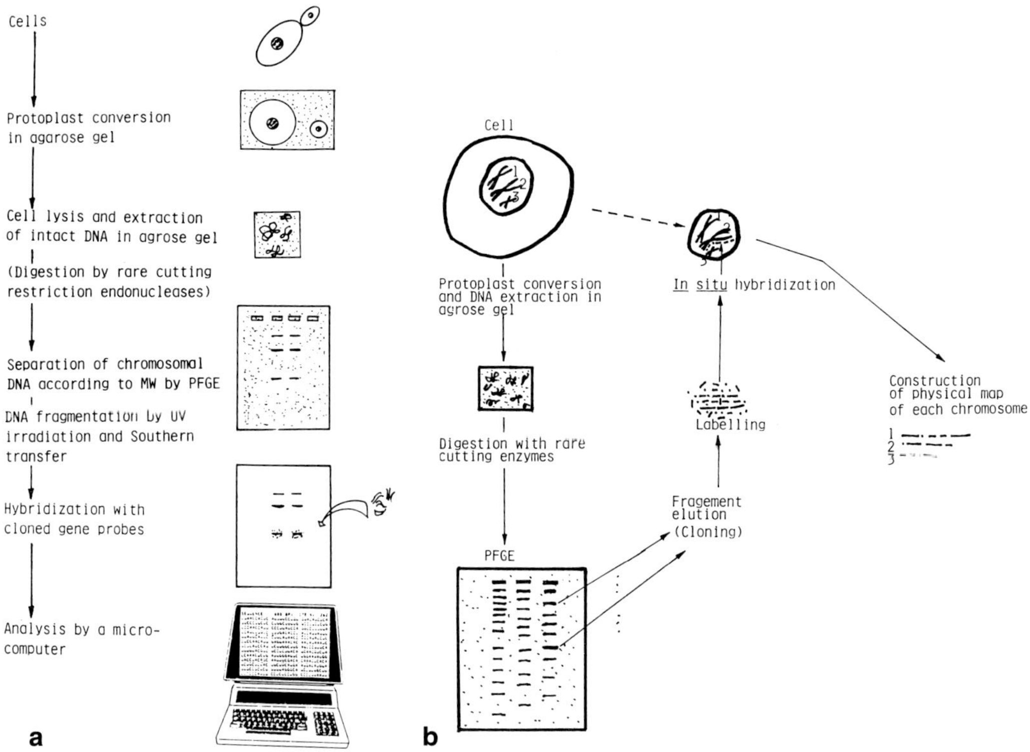


**Fig. 11.** Schematic diagram of electro-gene mapping. Electro-gene mapping (**a**) with and (**b**) without in situ chromosome hybridization

---

**Fig. 10a–d.** Photograph of the versatile PFGE apparatus and its performance. **a** Electrode configuration for orthogonal field alternation gel electrophoresis was adopted. In this, the field switching was carried out by a hardware. *a* buffer chamber (33×41×7 cm plastic tray for photograph used); *b* electrode (only one indicated) which was fixed with rubber bands as in Fig. 5b; *e*, rubber band to fix an agarose gel plate; *d* agarose gel; *e* switching unit. **b** Electrode configuration for contour-clamped homogeneous electric field gel electrophoresis was adopted. The switching was controlled by a microcomputer. *A* to *d* are same as in **b**. *e* switching unit which is controlled by a microcomputer (Epson HC-20 lap top computer), *f* through a RS232C interface. **c** Conversational inputting o PFGE parameters by the microcomputer. *a* to *d* are same as in **b**. *e* switching unit which is controlled by a microcomputer (Epson HC-20 lap top computer), *f* through a RS232C interface. **c** Conversational inputting of PFGE parameters by the microcomputer. **d** Photograph of plant DNA PFGE. Genomic DNA's of *Saccharomyces cerevisiae* were prepared by the solid plug method (Schwartz and Cantor 1984) or microbead method (Cook 1984). Plant DNA's were prepared by a modified plug method (Yoshida and Kondo 1987). Samples were applied onto 1% agarose gel and run at constant 240V and 140 s/cycle for 24 h at 15 °C. *1 S. cerevisiae*, strain T55 (solid plug); *2 S. cerevisiae*, strain T55 (microbead); *3 Nicotiana tabacum*; *4 Marchantia polymorpha* L., strain A18. The liverwort was cultured and converted to protoplasts according to Ono and Okamoto (1984)

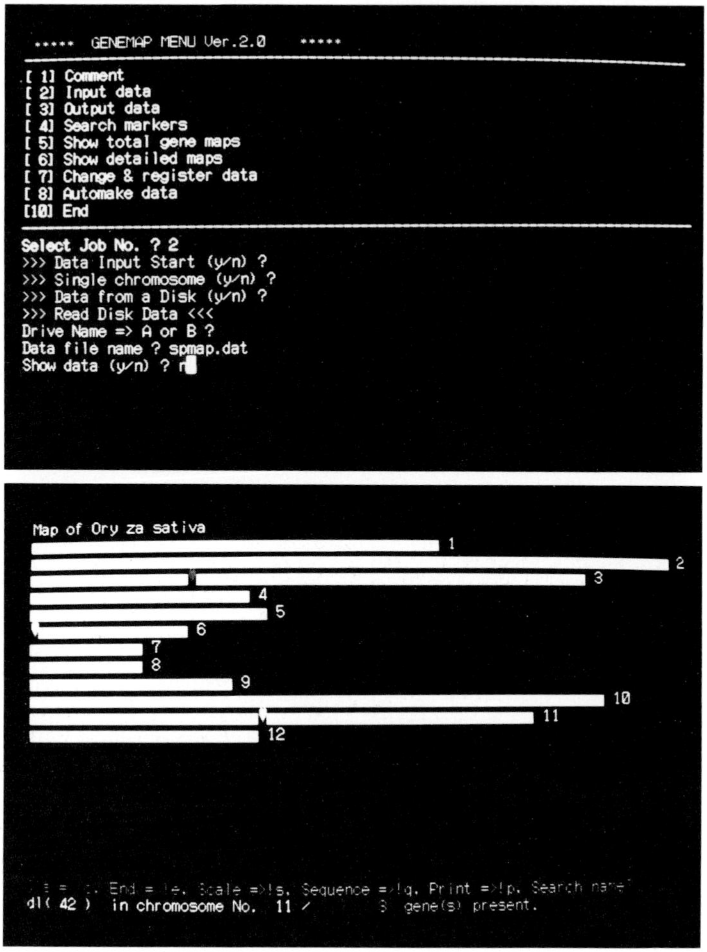


**Fig. 12a–d.** Databases for genes and its retrieval software (GENEMAP) for a microcomputer. **a** Initial menu; **b** chromosomal linkage map of *Oryza sativa* with gene data; **c** chloroplast physical and gene map of *Nicotiana tabacum* (data from Shinozaki et al. 1986); (d) calling DNA sequence DATA in GENEMAP. *Heart marks* in **b** and *bars* in **c** indicate genes' locations or cutting sites by restriction enzymes which were searched. A NEC PC9801 16 bit microcomputer with a color graphic monitor (640×400 dots) was used for GENEMAP execution. The GENEMAP was described in Basic compiler (Yoshida 1986)

## 6.3 Applications and Possibilities of PFGE in Plant Biotechnology

Dramatic separation of yeast chromosomal DNA's according to their molecular weights by PFGE makes possible "electrogene mapping" as schematically shown in Fig. 11. After the separation of chromosomal DNA's by PFGE, any cloned DNA probes can be assigned to chromosomes very easily by using specific hybridization techniques with or without in situ chromosome hybridization (Schwartz and Cantor 1984; Carle and Olson 1985b; Heiter et al. 1985; Johnson and Borst 1987; Yoshida et al. 1987).

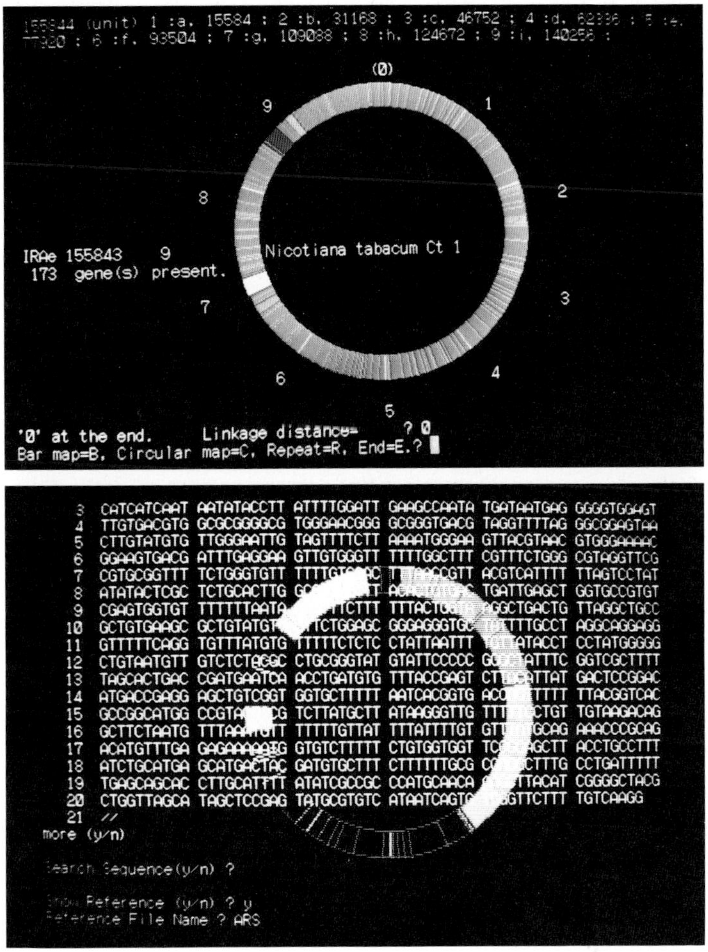


c

d

Fig. 12c, d

In connection with this, gene databases and their retrieval software for a microcomputer have been constructed and successfully applied to yeasts (Yoshida 1985, 1986). By simply changing the data, they are applicable to any organisms. The typical examples applied to the genetic data of *Oryza sativa* and *Nicotiana tabacum* are shown in Fig. 12. The analytical results are usually shown on a color graphic display.

On the other hand, the cytogenetic technique of in situ specific hybridization of cloned DNA probes to chromosomes is also powerful, and applicable to higher plants as well as to humans. However, the technique is much more limited and requires delicate skill in identifying each chromosome (Shen et al. 1987). Furthermore, high resolution mapping has not yet proven reliable. However, the PFGE and in situ hybridization techniques are complementary and therefore useful for gene mapping in higher plants. Other uses of PFGE include the identification or

**Table 1.** Haploid genome sizes and chromosome numbers in bacteria, yeasts and plants[a]

| Organisms | Chromosome numbers | DNA content per nucleus in megabase pairs | DNA content per chromosome in megabase pairs |
|---|---|---|---|
| Phage | | | |
| *lambda* | 1 | 0.05 | 0.05 |
| *T4* | 1 | 0.18 | 0.18 |
| Bacteria | | | |
| *Mycoplasma* | 1 | 0.75 | 0.75 |
| *Escherichia coli* | 1 | 4.5 | 4.5 |
| *Bacillus subtils* | 1 | 5.0 | 5.0 |
| fungi (yeasts) | | | |
| *Saccharomyces cerevisiae* | 17 | 14 | 0.82 (0.22 – 1.5) |
| *Schizosaccharomyces pombe* | 3 | 20 | 6.7 (3.0 – 12) |
| Plants | | | |
| *Arabidopsis thaliana* | 5 | 70 | 14 |
| *Phaseolus aureus* | 11 | 470 | 43 |
| *Gossypium indicum* | 13 | 780 | 60 |
| *Marchantia polymorpha* | 9 | 900 | 100 |
| *Marchantia polymorpha* chloroplast | | 0.12 | 0.12 |
| *Nicotiana tabacum* | 24 | 1600 | 67 |
| *Nicotiana tabacum* chloroplast | | 0.15 | 0.15 |
| *Glycine max* | 20 | 1800 | 90 |
| *Zea mays* | 10 | 4500 | 450 |
| *Pisum sativum* | 7 | 4500 | 643 |
| *Triticum aestivum* | 21 | 5900 | 281 |
| *Vicia faba* | 6 | 44000 | 7333 |
| *Disopyros kaki* | 45 | | |
| *Ipomoea batatas* | 45 | | |
| *Saccharum officinarum* | ca. 57 | | |
| *Abelomoschus esculentus* | 62 | | |
| *Piper nigrum* | ca. 64 | | |
| *Dioscorea batatas* | ca. 70 | | |
| *Ophioglossum reticulatum* | ca. 630 | | |
| Mammal | | | |
| *Homo sapiens* | 23 | 2900 | 126 (50 – 250) |

[a]Compiled from King (1974), Meyerowitz and Pruitt (1985), Scientific Tables (Rikagaku Nenpyou), Maruzen Co. (1987) (in Japanese), Ohyama et al. (1986) and Shinozaki et al. (1986). Figures in brackets indicate the maximum and minimum length of chromosomal DNA, respectively. The chromosome number indicates that for a haploid cell.

verification of large chromosomal aberrations and of satellite chromosomes, including translocations or deletions. Recent advance in PFGE facilitates megabase-scale direct DNA mapping in mammalian (Lawrence et al. 1987; see review by Barlow and Lehrach 1987) and bacterial chromosomes (Smith et al. 1987). Moreover, cloning vectors for DNA fragments of several hundred kilobase pairs have been recently constructed (Burke et al. 1987). PFGE has been inducing various new ideas and methods as well as new findings in the fields of molecular biology and biotechnology.

Considering the larger genomic sizes in higher plants as shown in Table 1, we must improve our techniques and instrumentation so we can accommodate the PFGE to higher plant DNA's, as in yeast DNA's. However, even at present, PFGE techniques developed in mammalian genomic and bacterial genomic DNA analyses, as described above, are very useful for molecular biological and genetic analyses of higher plants. In addition to this, PFGE may be applicable to higher plants which have enormously larger chromosome numbers (see Table 1). There is a possibility for DNA separation of those plants which are expected to have smaller genomic DNA's. Even if this is not the case, PFGE offers a great possibility to cover the cytological difficulties in those chromosome studies. Up to now, there has been no application of PFGE to higher plants, since it has just begun. However, PFGE is expected to have a great impact on plant sciences and biotechnology, as well as on other biological sciences.

*Acknowledgments.* We are grateful to Prof. Y.P.S. Bajaj for his encouragement and help. We also thank Prof. K. Ono of Kumamoto Univ. for his gift of the liverwort strain. Our study in this paper is partially supported by grants from the Tokai Gakujutu Shoreikai, the Ishida Foundation and the Ministry of Education, Science and Culture, Japan to K.Y.

## References

Anand R (1986) Pulsed field gel electrophoresis: a technique for fractionating large DNA molecules. Trends Genet 2:278–283

Andrews AT (1981) Electrophoresis: theory, techniques, and biochemical and clinical applications. Clarendon, New York Oxford

Barlow DP, Lehrach H (1987) Genetics by gel electrophoresis: the impact of pulsed field gel electrophoresis on mammalian genetics. Trends Genet 3:167–171

Bates GW (1985) Electrical fusion for optimal formation of protoplast heterokaryons in *Nicotiana*. Planta 165:217–224

Bates GW, Hasenkampf CA (1985) Culture of plant somatic hybrids following electric fusion. Theor Appl Genet 70:227–233

Burke DT, Carle GF, Olson CM (1987) Cloning of large segments of exogenous DNA into yeast by means of artificial chromosome vectors. Science 236:806–812

Carle GF, Olson MV (1985a) Separation of chromosomal DNA molecules from yeast by orthogonal-field-alternation gel electrophoresis. Nucleic Acids Res 12:5647–5664

Carle GF, Olson MV (1985b) An electrophoretic karyotype for yeast. Proc Natl Acad Sci USA 82:3756–3760

Carle GF, Frank M, Olson MV (1986) Electrophoretic separations of large DNA molecules by periodic inversion of the electric field. Science 232:65–68

Chu G, Vollrath D, Davis RW (1986) Separation of large DNA molecules by contour-clamped homogeneous electric fields. Science 234:1582–1585

Cook PR (1984) A general method for preparing intact nuclear DNA. EMBO J 3:1837–1842

Gardiner K, Laas W, Patterson D (1986) Fractionation of large mammalian DNA restriction fragments using vertical pulsed-field gradient gel electrophoresis. Somatic Cell Mol Genet 12:185–190

Hames BD, Rickwood D (eds) (1981) Gel electrophoresis of proteins. IRL, Oxford

Heiter P, Pridmore D, Hegemann JH, Thomas M, Davis RW, Philippsen P (1985) Functional selection and analysis of yeast centromeric DNA. Cell 42:913–921

Holde KE van (1971) Physical biochemistry. Prentics-Hall, Inglewood Cliffs

Johnson JJ, Borst P (1986) Mapping of VSG genes on large expression-site chromosomes of *Trypanosoma brucei* separated by pulsed-field gradient electrophoresis. Gene 43:213–220

Kao KN, Michayluk MR (1974) A method for high frequency intergenetic fusion of plant protoplast. Planta 115:355–367

Keller WA, Melchers G (1973) The effect of high pH and calcium on tobacco leaf protoplast fusion. Z Naturforsch 28C:737–741

King RC (ed) (1974) Handbook of genetics, vol 2. Plants, plant viruses, and protists. Plenum, New York London

Lawrance SK, Smith CL, Srivastava R, Cantor CR, Weissman SM (1987) Megabase-scale mapping of the HLA gene complex by pulsed field gel electrophoresis. Science 235:1387–1390

Maugh II TH (1983) A survey of separative techniques. Science 222:259–266

Meyerowitz EM, Pruitt RE (1985) *Arabidopsis thaliana* and plant molecular genetics. Science 229:1214–1218

Morikawa H, Sugino M, Yasuyuki H, Takeda J, Senda M, Hirai A, Yamada Y (1986) Interspecific plant hybridization by electrofusion in *Nicotiana*. Biotechnology 4:57–60

Nagata N (1984) Fusion of somatic cells. In: Linskens HF, Heslop-Harrison J (eds) Cellular interactions. Springer, Berlin Heidelberg New York, pp 491–507

Neumann E, Gerisch G, Opatz K (1980) Cell fusion induced by high electric impulses applied *Dictyostelium*. Naturwissenschaften 67:414–415

Neumann E, Rosenheck K (1972) Permeability changes induced by electric impulses in vesicular membranes. J Membr Biol 10:279–290

Ohyama K, Fukuzawa H, Kohchi T, Shirai H, Sano T, Sano S, Umesono K, Shiki Y, Takeuchi M, Chang Z, Aota S, Inokuchi H, Ozeki H (1986) Chloroplast gene organization deduced from complete sequence of liverwort *Marchantia polymorpha* chloroplast DNA. Nature (Lond) 322: 572–574

Ono K, Okamoto K (1984) Isolation and culture of protoplasts from the liverwort cell suspension cultures and the moss protonemata. J Hattori Bot Lab 56:201–207

Pohl HA (1978) Dielectrophoresis. Cambridge University Press, Cambridge

Rickwood D, Hames BD (eds) (1982) Gel electrophoresis of nucleic acid. IRL, Oxford

Schwartz DC, Cantor CR (1984) Separation of yeast chromosome-size DNA's by pulsed field gradient gel electrophoresis. Cell 37:67–75.

Senda M, Takeda J, Abe S, Nakamura T (1979) Induction of cell fusion of plant protoplasts by electrical stimulation. Plant Cell Physiol 20:1441–1443

Shen D, Wang Z, Wu M (1987) Gene mapping on maize pachytene chromosomes by in situ hybridization. Chromosoma 95:311–314

Shinozaki K, Ohme M, Tanaka M, Wakasugi T, Hayashida N, Matsubayashi T, Zaita N, Chunwongse J, Obokata J, Shinozaki K, Ohto C, Torazawa K, Meng BY, Sugita M, Deno H, Kamogashira T, Yamada K, Kusuda J, Takaiwa F, Kato A, Tohdoh N, Shimada H, Sugiura M (1986) The complete nucleotide sequence of the tobacco chloroplast genome: its gene organization and expression. EMBO J 5:2043–2049

Smith CL, Cantor CR (1986) Analysis of genome organization and rearrangements by pulsed field gradient gel electrophoresis. In: Setlow J, Hollaender A (eds) Genetic engineering, vol. 8. Plenum Press, New York London, pp 45–70

Smith CL, Econome JG, Schutt A, Kloc S, Cantor CR (1987) A physical map of *Escherichia coli* K12 genome. Science 236:1448–1453

Tempelaar MJ, Jones MGK (1985a) Fusion characteristics of plant protoplasts in electric fields. Planta 165:205–216

Tempelaar MJ, Jones MGK (1985b) Direct electrofusion between protoplasts with different responses in a mass fusion system. Plant Cell Rep 4:92–95

Watts JW, King JM (1984) A simple method for large scale electrofusion and culture of plant protoplast. Biosci Rep 4:335–342

Watts JW, Doonan JH, Cove DJ, King M (1985) Production of somatic hybrids of moss by electrofusion. Mol Gen Genet 199:349–351

Weissbach A, Weissbach H (eds) (1986) Methods in enzymology, vol 118. Plant molecular biology. Academic Press, New York London

Yoshida K (1983a) A versatile apparatus for electrophoresis including isoelectrofocusing and its performances. Ann Biochem 129:37–45

Yoshida K (1983b) A highly simplified horizontal electrophoretic apparatus including a handmade power supply and its application. Ann Biochem 130:246–256

Yoshida K (1983c) Appendix: Deviation of I-F relationship from Ohm's law and theory of electrolyte conductance and its computer simulation. Ann Biochem 130:256–259

Yoshida K (1985) Mitochondria transplantation and transformation in yeasts: Stepwise transformation as a general transformation method and electrogene mapping. IF Rep 13:143–150 (in Japanese)

Yoshida K (1986) Yeast genetic analysis programs (YGAP) for a microcomputer. Yeast 2:S438

Yoshida K, Kondo T (1987) Construction of a versatile micro-computer-controlled pulsed field electrophoretic apparatus and its application to DNA separation. Seikagaku 59:738 (in Japanese)

Yoshida K, Hisatomi T, Yanagishima N (1987) Sexual behavior and its pheromonal regulation in ascosporogenous yeasts (Plenary lecture). XIIth International Spec Symp Genetics of non-conventional yeasts, Weimar. Abstr PL1, p 1; Yoshida K, Hisatomi T, Yanagishima (1989) J Basic Microbiol 2:99–128

Zimmermann U (1982) Electric field-mediated fusion and related electrical phenomena. Biochem Biophys Acta 694:227–277

Zimmermann U, Scheurich P (1981) High frequency fusion of plant protoplast by electric fields. Planta 151:26–32

Zimmermann U, Pilwat G, Riemann F (1974) Dielectric breakdown of cell membranes. Biophys J 14:881–899

## Appendix. BASIC Program Listing for HC-20

```
  5 WIDTH 20, 100:CLS
 10 GOTO 400

 20 SEC = VAL (MID$ (TIME$, 1,2))*3600+VAL (MID$ (TIME$, 4,2))*60+
    VAL (MID$ (TIME$ 7,2))
 30 RETURN
 40 PRB = VAL (MID$ (TIME$, 7,2)): RETURN
 50 SOUND20, 10: MOTOR ON: LOCATE 10, 2, 0: PRINT "A:ON": RETURN
 60 SOUND10, 10: MOTOR OFF: LOCATE 10, 2, 0: PRINT "B:ON": RETURN

100 REM TIME COUNT ROUTINE
105 IF INKEY$ = "S" THEN GOSUB 60: STOP
110 GOSUB 20
120 IF PRB >50 AND SEC = 0 THEN SECO = SECO-24*3600
130 GOSUB 40
140 ELSP = SEC-SECO: LOCATE 0, 0, 0: PRINT ELSP
150 IF ELSP < > TMP THEN T = T+ELSP-TMP
160 TMP = ELSP: RETURN

400 REM-----START OF PROGRAM------
410 SEC = 0: GOSUB 60 : CLS

420 REM CONDI
430 INPUT "PERIOD (STARTING, ENDING in sec) = "; PERS, PERE
440 INPUT "PHASE CHANGE (STARTING, ENDING in sec) = "; CUTS, CUTE
450 INPUT "DURATION in min = "; DURATION :DURATION = DURATION*60
460 LPRINT "---------------"
464 LPRINT DATE$;" "; TIME$
468 LPRINT "PERIOD: "; PERS;"-"; PERE
470 LPRINT "PHASE CHANGE: "; CUTS;"-"; CUTE
472 LPRINT "DURATION: "; DURATION/60;" min"

480 INPUT "OK ?"; A$: IF A$ = "N" THEN 420
485 PERA = PERS: CUTA = CUTS
490 LINEINPUT "COMMENT (RET TO EXIT) = "; B$: IF B$< >"" THEN LPRINT B$:
    GOTO 490
```

```
500 PRINT "HIT G AND ACTIVATE POWER TO START": INPUT A$: IF A$< >"G"
    THEN GOTO 500
510 CLS: LOCATE 12, 0, 0: PRINT CN
530 GOSUB 20: SEC0 = SEC: GOSUB 50: CN = 0:T = 0

590 REM TIME COUNTING
600 GOSUB 100
610 IF T⇒CUTA THEN GOSUB 60
620 IF T⇒PERA THEN 660
630 GOTO 590
660 CN = CN+1: LOCATE 12, 0, 0: PRINT CN: IF ELSP>DURATION THEN GOSUB 60:
    END
670 GOSUB 1000
680 T = 0:GOSUB 50: GOTO 590

1000 REM SET NEXT CYCLE CONDITION
1010 PERA = PERS+(PERE-PERS)*ELSP/DURATION
1020 CUTA = CUTS+(CUTE-CUTS)*ELSP/DURATION
1090 RETURN
```

*Note Added in Proof*

Since the manuscript was completed early in 1987, several references especially in the field of PFGE should be added.

PFGE was applied to the physical mapping of large DNA fragments of tomato chromosome 9 (Ganal et al. 1989). Theoretical and quantitative investigations of PFGE were carried out by Southern et al. (1987), Cantor et al. (1988), Lande et al. (1988), Mathew et al. (1988a, b, c), Slater et al. (1988), and Deutsch and Madden (1989). More recently, the movement of individual DNA molecules during PFGE was directly observed by fluorescence microscopy (Schwartz and Koval 1989, Smith et al. 1989). Our versatile PFGE apparatus with its program has been extensively improved to be 1/5 in volume and 1/5 – 1/20 in cost by using a pocket calculating computer (Yoshida and Kondo 1988). Recent advanced techniques for genome analysis including PFGE are collected in the book edited by Davies (1988).

*References for Note Added in Proof*

Cantor CR, Gaal A, Smith CL (1988) High-resolution separation and accurate size determination in pulsed-field gel electrophoresis of DNA. 3. Effect of electrical field shape. Biochemistry 27:9216–9221

Davies KE (ed) (1988) Genome analysis – a practical approach. IRL, Oxford

Deutsch JM, Madden TL (1989) Theoretical studies during gel electrophoresis. J Chem Phys 90:2476–2485

Ganal MW, Young ND, Tanksley SD (1989) Pulsed field gel electrophoresis and physical mapping of large DNA fragments in the Tm-2a region of chromosome 9 in tomato. Mol gen genet 215:359–400

Lande M, Noolandi J, Turmel C, Brousseau R, Rousseau J, Slater GW (1988) Scrambling of bands in gel electrophoresis of DNA. Nucleic Acids Res 16:5427–5437

Mathew KM, Smith CL, Cantor CR (1988a) High-resolution separation and accurate size determination in pulsed-field gel electrophoresis of DNA. 1. DNA size standards and the effect of agarose and temperature. Biochemistry 27:9204–9210

Mathew KM, Smith CL, Cantor CR (1988b) High-resolution separation and accurate size determination in pulsed-field gel electrophoresis of DNA. 2. Effect of pulse time and electric field strength and implications for models of the separation process. Biochemistry 27:9210–9216

Mathew KM, Hui CF, Smith CL, Cantor CR (1988c) High-resolution separation and accurate size determination in pulsed-field gel electrophoresis of DNA. 4. Influence of DNA topology. Biochemistry 27:9222–9226

Schwartz DC, Koval M (1989) Conformational dynamics of individual DNA molecules during gel electrophoresis. Nature (London) 338:520–522

Slater GW, Rousseau J, Noolandi J (1988) On the stretching of DNA in the reptation theories of gel electrophoresis. Biopolymers 26:863–872

Smith SB, Aldridge PK, Callis JB (1989) Observation of individual DNA molecules undergoing gel electrophoresis. Science 243:203–206

Southern EM, Anand R, Fletcher DS (1987) A model for the separation of large DNA molecules crossed field gel electrophoresis. Nucleic Acids Res 15:5925–5943

Yoshida K, Kondo T (1988) Construction of a versatile apparatus for pulsed field gel electrophoresis. Protein Nucleic Acid Enzyme 34:2514–2526 (in Japanese)

# XVII Chloroplast Genomes as Genetic Markers

S. D. KUNG[1]

## 1 Introduction

All biological phenomena are the consequences of a coordinated expression of the genome. Many molecules have been identified as the products of certain genes. Thus, they can be used to study genomic structure, function, regulation, and evolution. Proteins of various sources are the major macromolecules which have for decades been used successfully as genetic markers (Kung 1976). The ultimate molecular marker for studying the evolution of genes is the gene itself. Therefore, DNA molecules are the ideal markers for phylogenetic investigations. In plants, there are three separate genomes: nuclear, chloroplast, and mitochondrial. In this chapter, the primary focus is placed on the properties and application of the chloroplast genome as genetic markers.

The existence of the chloroplast genome was first discovered in the 1950's (Chiba 1951) and subsequently confirmed in the 1960's (Kirk 1971). The advancement in biotechnology in the 1970's further stimulated the study of the chloroplast genome and ultimately facilitated the sequencing of two chloroplast genomes (Shinozaki et al. 1986; Ohyama et al. 1986). One of the many important contributions made through our study of chloroplast genome is the wide application of it as genetic marker. It is a useful marker and has tremendous potential for analyzing problems in plant genetics, breeding, and evolution. This creates a historical tunnel through which many past events may be unveiled. Examples of these applications include studying the sexual and somatic hybrids, tracing the inheritance pattern, facilitating gene transfer, measuring genetic variability, and estimating the rate of evolution. This reviewer will not attempt to cover all literature published in this field to date but rather to present some practical and representative examples as to how chloroplast genomes have been successfully used as genetic markers. For information on the general properties as well as on any specific aspects of chloroplast genomes, the reader is referred to many recent excellent reviews on this subject (Whitfield et al. 1983; Palmer 1985, 1987; Crouse et al. 1985; Gillham et al. 1985; Zurawski and Clegg 1987).

[1] Center for Agricultural Biotechnology, Maryland Institute of Biotechnology, and Department of Botany, The University of Maryland, College Park, Maryland 20742, USA

Biotechnology in Agriculture and Forestry, Vol. 9
Plant Protoplasts and Genetic Engineering II (ed. by Y. P. S. Bajaj)

## 2 Chloroplast Genome and Its Properties

A great number of chloroplast genomes have been analyzed and characterized. It is estimated that over 230 species belonging to 90 genera and 33 families of higher plants have been studied (Palmer 1985; Hirai et al. 1985). These chloroplast genomes have been analyzed at different levels of completeness. To date, two complete nucleotide sequences (*Nicotiana tabacum* and *Marchantia polymorpha*) and more than 40 physical maps and many restriction fragment patterns have been published (Crouse et al. 1985; Palmer 1985). In this article, only the salient features on the size and organization of the genomes, the structural components of some identified genes, and the sequence similarity and diversity, mode of inheritance, and mechanism of mutations of the chloroplast genomes of the chloroplast genomes related to its application as genetic markers will be described.

### 2.1 Genome Sizes

Most chloroplast genomes of higher plants have an average size of about 160 kb. This is represented by the tobacco chloroplast genome which consists of 155 844 bp (Shinozaki et al. 1986). However, significant variation exists. The genome size varies from 120 to 217 kb (Palmer 1985) depending on the presence

**Table 1.** The size of higher plant chloroplast genome and its inverted repeat

| Groups | Plants | Genome size | Inverted repeat | | Reference |
|---|---|---|---|---|---|
| | | | Size | % of total | |
| A | Geranium | 217 | 6 | | Palmer et al. (1987) |
| | Duckweed | 182 | 27 | 15 | Ee van et al. (1980) |
| | Tobacco[a] | 171 | 28 | 16 | Shen et al. (1982) |
| B | Evening primrose | 153 | 22.7 | 15 | Gordon et al. (1982) |
| | Spinach | 153.5 | 29.4 | 16 | Kolodner and Tewari (1979) |
| | Lettuce | 156.5 | 24.4 | 16 | Kolodner and Tewari (1979) |
| | Mustard | 158 | 23.7 | 15 | Link et al. (1981) |
| | Tobacco[b] | 155.8 | 26.4 | 17 | Shinozaki et al. (1986) |
| | Soybean | 151 | 23.9 | 16 | Chu and Tewari (1982) |
| | Nasturtium | 155 | 27 | 17 | Thompson et al. (1981) |
| | Mung bean | 150 | 23 | 15 | Palmer and Thompson (1981) |
| | Common bean | 150 | 22 | 15 | Mubumbila et al. (1983) |
| C | Wheat | 135 | 21 | 16 | Bowman et al. (1981) |
| | Corn | 136.1 | 22.5 | 16 | Kolodner and Tewari (1979) |
| | Barley | 133 | 20.9 | 16 | Poulsen (1983) |
| | Rice | 129.5 | 21.7 | 17 | Hirai et al. (1985) |
| D | Pea | 120 | | | Palmer and Thompson (1981) |
| | Broad bean | 121 | | | Koller and Delius (1980) |
| | Chick pea | 119 | | | Chu and Tewari (1982) |

[a] *N. acuminata*
[b] *N. tabacum*

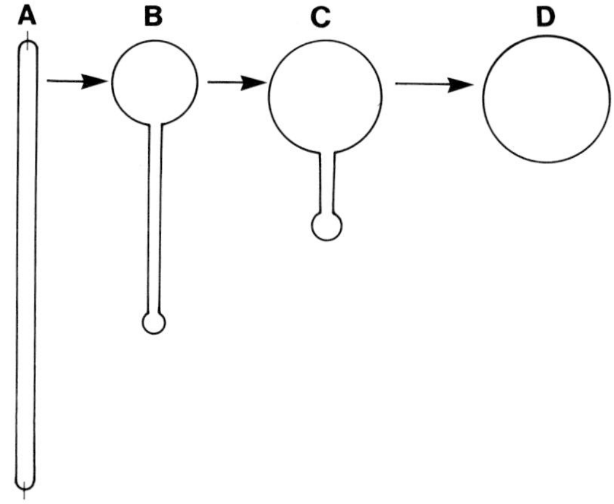


**Fig. 1A–D.** Diagram depicting a hypothetic sequence of chloroplast genome evolution as described in the text. **A** Total inverted repeat-two identical genomes jointed together in a head-to-head and tail-to-tail fashion. **B** The large and small single copy regions were generated by deletions. **C** Representing progressive deletions which resulted in the reduction of the size of the inverted repeat and consequently the entire genome. **D** Segments equivalent to one genome were progressively deleted. Thus a genome without the inverted repeat was created

or absence, as well as the size, of the inverted repeats. In most cases the size variation can be accounted for by changes in the inverted repeats (Palmer 1985; Tassopulu and Kung 1984). As listed in Table 1, the genome sizes can be divided into four groups. Group A is represented by the three largest sizes, 171, 180, and 217 kb. They all have very large inverted repeats. Group B includes the majority of the chloroplast genomes studied to date. They have an average size of 150–160 kb with the inverted repeats of 20–25 kb. Group C contains four smaller genomes of 135 kb associated also with a smaller inverted repeat about 20 kb. Group D is the smallest (120 kb) in genome size due to the lack of inverted repeats. It is quite evident that the expansion or reduction of the inverted repeats determines the overall size of the chloroplast genomes. Based on this consideration, it was proposed that the present chloroplast genomes may have evolved from a much larger one that consisted of two identical copies joined together in a head-to-head and tail-to-tail fashion (Tassopulu and Kung 1984). The entire genome was made up of two large inverted repeats. Every gene was duplicated. During the course of evolution, the size was reduced as deletions occurred repeatedly. As illustrated in Fig. 1, the present-day inverted repeats may represent a vestige of what once were two full copies of chloroplast genome. The deletion of various portions of the genome equal to one entire copy of the two genomes could give rise to a much smaller size without the inverted repeats. The chloroplast genome of geranium (Palmer et al. 1987), *N. acuminata* (Shen et al. 1982) and pea (Chu and Tewari 1982) are excellent examples representing the different levels or stages of such deletion (Fig. 1).

## 2.2 Genome Organization

All chloroplast genomes exist as a single circular molecule. The entire circular genome consists of four well-defined regions: the large and small single copy (LSC, SSC) separated by the inverted repeats ($IR_A$, $IR_B$). Figure 2 is the detailed physical map of the best-studied chloroplast genome, *N. tabacum*, for which the complete nucleotide sequence was determined (Shinozaki et al. 1986). Therefore, organization of the tobacco chloroplast genome will be referred to as a model. It has a contour length of 155 844 bp, with the LSC and SSC regions having 86 684 and 18 482 bp respectively. The $IR_A$ and $IR_B$ are identical, having 25 339 bp each. The genomic DNA is double-stranded. The DNA strand coding for the large subunit (LS) of RuBisCo was designated as A and the complementary strand as B (Deno et al. 1983). In *Nicotiana*, as in most plant genera, the chloroplast genome contains two copies of the 23S, 16S, 5S, and 4.5S rRNA genes in the inverted repeats, and 30 tRNA genes distributed throughout the genome. A chloroplast genome contains all the RNA genes for transcription and translation and thus it has all the machinery for protein synthesis. However, it does not contain all the genes for all the chloroplast proteins. For example, chloroplasts contain several hundred different proteins needed to perform the essential metabolic activities. Yet only about 50 such genes or reading frames on the chloroplast genome have been identified (Shinozaki et al. 1986). In tobacco, sequences for 20 genes or reading frames for ribosomal proteins, 21 genes or reading frames for thylakoid proteins and 6 genes for NADH dehydrogenase components have been

**Table 2.** Some salient features of the *Nicotiana* chloroplast genome and its genes. (Shinozaki et al. 1986)

| | | |
|---|---|---|
| The complete nucleotide sequence | | 155 844 bp |
| Identical inverted repeat (IR) | 25 339 bp | |
| Large single copy (LSC) | 86 684 bp | |
| Small single copy (SSC) | 18 482 bp | |
| The number of genes | | 146 |
| RNA's | | 34 |
| rRNA's | 4 | |
| tRNA's | 30 | |
| Proteins | | 88 |
| Ribosomal | 22 | |
| Stromal | 28 | |
| Thylakoid | 17 | |
| NADH | 7 | |
| Others | 14 | |
| General distribution of genes and open reading frames | | |
| IR | | 54 |
| SSC | | 15 |
| LSC | | 87 |
| A strand | | 62 |
| B strand | | 107 |
| General features | | |
| Eucaryotic-introns (15) | | |
| Procaryotic-promoters and terminators | | |

identified (Table 2). Most of the genes identified are associated with either ribosomal biogenesis or photosynthesis. For instance, among the chloroplast proteins are the LS of RuBisCo and polypeptides from the five functionally distinct thylakoid complexes. They are the photosystems I and II, light-harvesting chlorophyll-protein complex, the cytochrome b/f complex and the coupling factors.

As can be seen from Fig. 2, the overall organization of genes on chloroplast genome represents a mixed model of the prokaryotes and eukaryotes. The ribosomal RNA operon is identical in form for both chloroplast and eubacteria. So are the elements of the ATPase and certain ribosomal protein operons. On the other hand, the photosystem II proteins are encoded by widely scattered genes.

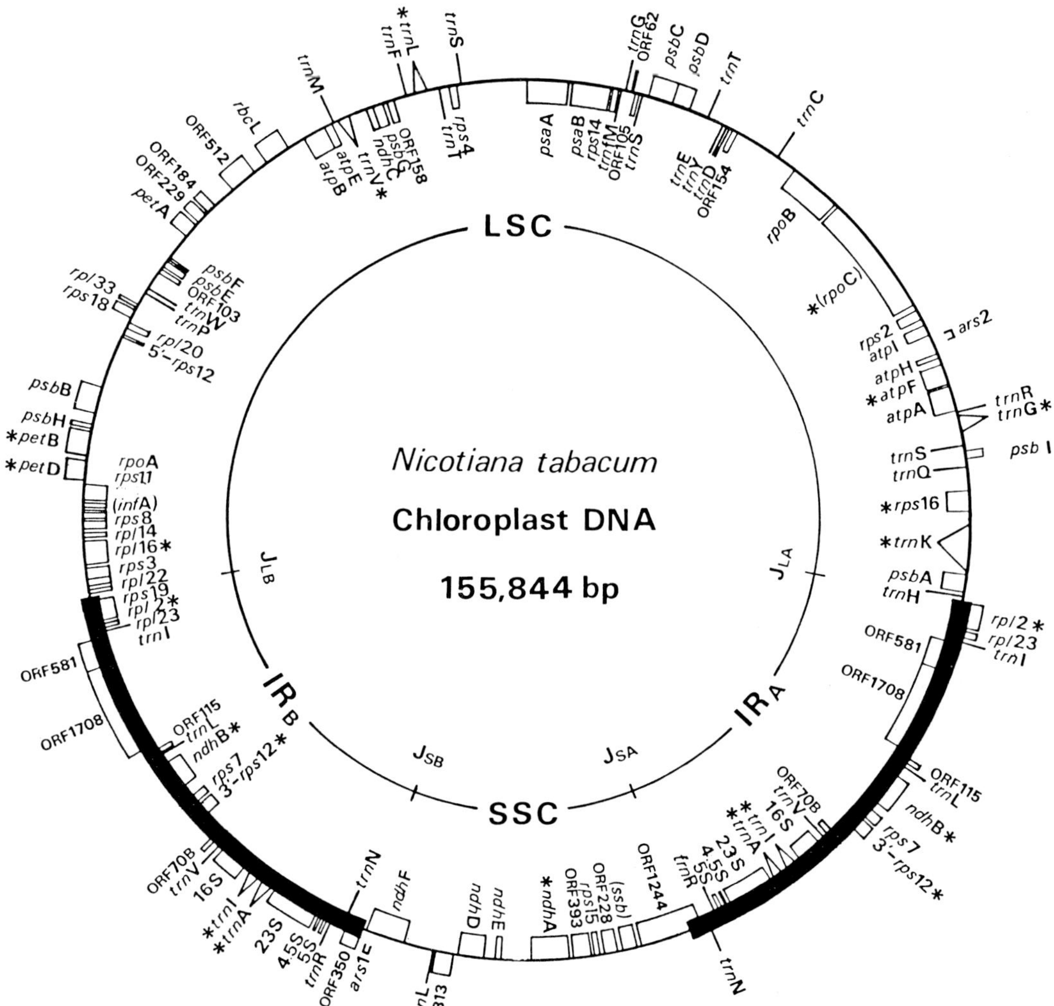


**Fig. 2.** The detailed physical map of *N. tabacum* chloroplast genome. (Shinozaki et al. 1986)

### 2.3 Gene Structures

The general structure of chloroplast genes is essentially prokaryotic in nature. All the important regulatory sequences such as promoters, terminators, and ribosome-binding sites are similar to those of prokaryotes. In fact, some of these functional components are exchangeable in vitro (Kong et al. 1984). Chloroplast promoters function efficiently in the prokaryotic system and vice versa. This is not surprising, since chloroplast promoters resemble those of prokaryotes in structure and function (Kung and Lin 1985). The promoters contain two conserved hexamers separated by a short stretch of about 17 bp. The first hexamer is TTGACA, resembling the -35 region of prokaryotic promoter sequences, in which the trimer TTG is highly conserved. In the second hexamer, the conserved sequence of TATAAT is identical to the -10 region (Rosenburg and Court 1979). Within this hexamer, three bases (TA---T) are highly conserved. The chloroplast promoters structurally identified are indeed functional promoters (Orozco et al. 1985; Kung and Lin 1985; Gruissem and Zurawski 1985). However, this is not entirely true for structurally identified terminators. In prokaryotes, the simple transcription terminators are broadly defined as a DNA region in which there is a small GC-rich segment of inverted repeats capable of forming a stem-loop structure followed by a series of A residues (Platt 1987). This permits the formation of intramolecular hairpin structure of the RNA transcript and causes the RNA polymerase to pause. The U residues of the RNA facilitate dissociation of the transcript from the template, since RNA-DNA hybrids with $_{r}U$-$_{d}A$ pairing are exceptionally unstable (Platt 1987). In chloroplasts, sequences capable of forming stem-loop structures at the 3′ flanking region of many chloroplast genes have been putatively identified as terminators (Shinozaki and Sugiura 1982; Zurawski and Clegg 1987). But such structures do not always function as effective transcription termination sites (Xu et al. 1989). They may well represent the site of RNA processing (Stern and Gruissem 1988).

### 2.4 Sequence Similarity and Diversity

Sequence similarity among the chloroplast genomes of higher plants has long been expected (Kung 1977). There is considerable evidence to support this conservative nature. Chloroplast genomes from various species of a given genus are very similar, if not identical, at the restriction fragment level. On the other hand, species diversity also exists. For instance, the restriction fragment patterns of many *Nicotiana* chloroplast genomes generated by a variety of enzymes exhibit a high degree of similarity and diversity within this genus. The similarity is usually associated with the genus, while the diversity is mainly related with the species. It is such a combination of genus-specific similarity and species-specific diversity that makes it the ideal genetic marker (Kung 1989a).

## 2.5 Mode of Inheritance

One major advantage of studying the chloroplast genome is that, in most cases, it is strictly inherited maternally. Maternal inheritance, also referred to as cytoplasmic or extra-chromosomal inheritance, describes the flow of genetic information in which only the organelle genes from the female parent are transmitted to the progeny. In the author's view, maternal inheritance is equivalent to or serves as the counterpart of meiosis in cell division. Otherwise, the number of types of organelles in any hybrid plant derived from many sexual crosses would be unimaginable. Although very little is understood about the molecular mechanism for such a device, several possible interpretations have been presented. A molecular mechanism of material inheritance is proposed by Sager's group (Royer and Sager 1979), who demonstrated that restriction-modification of the chloroplast genome is the key for maternal inheritance in *Chlamydomonas*. Dawid and Blacker (1972), on the other hand, observed that a simple exclusion or reduction of mitochondria from the male parent during fertilization may be the mechanism of maternal inheritance in frogs. In higher plants, a combination of exclusion and competition may form a cooperative operation dictating the maternal inheritance. In *Nicotiana*, the fertilized egg cell contained limited numbers of organelles from the male parent. The male organelles invariably undergo structural alterations and are eventually totally degenerated (Vaugh et al. 1980). The existence of only one, not both, chloroplast genome can be observed in progenies derived from sexual as well as from somatic hybridizations in *Nicotiana*. In the case of sexual hybridization the chloroplast genome is strictly maternal in type, while in the somatic hybrids there is an equal chance for the viability of chloroplasts of either type. The noncoexistence of organelles in a cell must bear important survival value selected throughout evolution (Kung 1989b).

## 2.6 Mechanisms of Mutation

Alteration of DNA is the main force in evolution. It is evident from the wide diversity of many chloroplast genomes that there have been considerable changes throughout the course of evolution. It has been demonstrated that the mechanism of chloroplast genome alteration involves point mutation, inversion, deletion, and duplication. Point mutations are primarily responsible for the observed gain and elimination of many restriction sites (Gordon et al. 1982). They occur rather frequently in relation to other mechanisms, and are clustered in one region which has been described as a "hot spot" (Kung et al. 1982; Tassopulu and Kung 1984). A large and a small deletion have also been discovered (Tassopulu and Kung 1984). The large deleted segment is about $7\times10^6$ Da. Whether this segment, which accounts for 7% of *Nicotiana* chloroplast genome, contains any structural genes is not known. The small deletion involves only about 500 bp (Salts et al. 1984). The example of chloroplast genome duplication is the existence of inverted and small tandem repeats (Whitfield and Bottomley 1983).

## 3 Chloroplast Genomes and Their Application as Genetic Markers

Among the three plant genomes, the chloroplast genome best serves as genetic marker, since its analysis is relatively simple, rapid, and inexpensive. The chloroplast genome represents a few percent of total plant DNA. It is relatively abundant in leaf cells and relatively easy to prepare in pure form. There are many isolation techniques tailored for almost every plant species (Kung and Williams 1969; Rhodes and Kung 1981; Zhao et al. 1989).

The size of the chloroplast genome is also the smallest among the three. It is suitable for restriction analysis and physical mapping. These and other physical properties make it an ideal genetic marker. In addition, the mutations so far found in chloroplast genomes are mostly point mutations which can be easily studied and compared at the restriction fragment level. This is another advantage to its use as genetic marker. Finally, the relatively slow rate of evolution and the maternal pattern of inheritance of the chloroplast genome all make the analysis for variations a much simpler task than otherwise.

### 3.1 Techniques Used for Genetic Analysis

It is the advancement of recombinant DNA technology which makes it possible to use chloroplast genome as genetic markers. There are three basic techniques used: restriction fragment analysis, physical map, and nucleotide sequence.

#### 3.1.1 Restriction Fragment Analysis

Recognition sequences by the restriction enzymes on DNA molecules provide ideal tools for analyzing and comparing chloroplast genomes. The size of all higher plant chloroplast genomes is well within the range amenable to direct restriction fragment analysis. Chloroplast genomes prepared by the standard method are pure enough to be suitable for direct restriction fragment analysis.

**Table 3.** Various fragments of *N. tabacum* chloroplast genome generated by various restriction enzymes

| Enzyme | Restriction site | # Sites |
|---|---|---|
| BamHI | GGATCC | 37 |
| BglI | GCCGGC | 10 |
| EcoRI | GAATTC | >40 |
| HindIII | AAGCTT | >20 |
| KpnI | GGTACC | >10 |
| PstI | CTGCAG | >14 |
| PvuII | CAGCTG | 14 |
| SalI | GTCGAC | 11 |
| SmaI | CCCGGG | 19 |
| XhoI | CTCGAG | 24 |

Table 3 lists the restriction enzymes used for and the number of restriction fragments generated on *Nicotiana* chloroplast genomes. For many *Nicotiana* species, EcoRI, BamHI, and SmaI are the enzymes of choice for restriction fragment analysis. All the restriction fragments generated by a given enzyme are separated on agarose gel, and direct comparison can be made between many samples (Herrmann and Whitfield 1982).

### 3.1.2 Physical Mapping

This method allows for the comparison of more-divergent chloroplast genomes than the restriction fragment analysis does. The construction of such physical maps can be readily accomplished by a hybridization approach of the cloned restriction fragments (Palmer and Thompson 1981) or sequential digestion (Gordon et al. 1981). Since different restriction enzymes cleave the DNA at different frequencies, the one which cuts the DNA infrequently provides the simplest tool for obtaining a map covering the entire genome. The most frequently used restriction enzymes for mapping chloroplast genomes are SalI, SmaI, PvuII. BglI, and XhoI.

### 3.1.3 Nucleotide Sequence Determination

The ultimate comparison of any genome is the nucleotide sequence determination. The complete nucleotide sequences of two chloroplast genomes and many chloroplast genes have been determined. The usefulness of such sequences in phylogenetic studies has been reviewed recently (Zurawski and Clegg 1987). Whenever the nucleotide sequence of a gene is determined, it provides invariably more valuable information on its evolution than on its function or regulation. For example, the nucleotide sequence of a gene tells us very little about its functional properties, since the secondary, tertiary, and quaternary conformations of both nucleic acids and proteins as well as their interactions are very important. However, it tells us a great deal about its evolution through the changes; such as the conserved regions versus the variable regions, the sequential alternations versus the randomized alterations and the mode and extent of mutations. They all contain valuable information in terms of evolution at the molecular level.

## 3.2 The Applications as Genetic Markers

Chloroplast genomes have been analyzed at all three levels: restriction fragment, physical map, and nucleotide sequence. The differences and similarities revealed at each level have been utilized as genetic markers for various studies. Some of the typical applications of chloroplast genome as genetic markers are summarized below:

### 3.2.1 At the Restriction Fragment Level

*1. Identifying the Cytoplasmic Germplasm of Somatic Hybrids.* Significant progress has been made in developing new somatic hybrids at both interspecific and intergeneric levels through protoplast fusion (Carlson et al. 1972). Fusion of plant and animal cells was also achieved but the fused cells failed to divide. The successful identification of the cytoplasmic germplasm of such fused products relies heavily on the availability of molecular markers. The first molecular marker used for such identification before the chloroplast genome could be conveniently analyzed at the restriction fragment level (Atchison et al. 1976) was the RuBisCo protein (Kung 1976). Chloroplast genomes have been used since to confirm the identity of the chloroplast genome donor in the interspecific somatic hybrid *N. glauca*×*N. langsdorfii* (Fig. 3) as well as in the intergeneric somatic hybrid between potato (*Solanum tuberosum*)×tomato (*Lycopersicon esculentum*) (Schiller et al. 1982). In both cases the chloroplast genome of each species is different from the other with respect to the restriction fragment pattern. Therefore, the chloroplast genome of each somatic hybrid can be compared and identified. Among the 14 mature *Nicotiana* interspecific somatic hybrids produced by protoplast fusion and examined by isoelectric focusing pattern of RuBisCo, eight had the *N. langsdorfii* type and six had the *N. glauca* type of chloroplast genome (Chen et al. 1977). Restriction fragment analysis of chloroplast genome from the *N. glauca* type confirmed its identity as determined by RuBisCo (Fig. 3). Similarly, of the 12 intergeneric somatic hybrids examined by restriction fragment analysis, eight contained potato and four contained tomato chloroplast genomes (Schiller et al. 1982) confirming the early identification through the use of RuBis-

**Fig. 3 A–E.** EcoRI restriction fragment patterns of chloroplast genome from five *Nicotiana* species. **A** and **B** are sexual and parasexual hybrids of *N. langsdorfii*×*N. glauca*, respectively. Both hybrids exhibited the *N. glauca* pattern of chloroplast genome which is different from that of *N. langsdorfii*. **C** *N. longiflora*. **D** *N. otophora*. **E** *N. langsdorfii*

Co (Metzlaff et al. 1981). There is only one or the other parental chloroplast genome in the somatic hybrids produced by protoplast fusion. Whether such fusion is interspecific or intergeneric, there seems to be an equal chance for either of these parental types to appear.

*2. Demonstrating the Mode of Maternal Inheritance.* Maternal inheritance is a widespread phenomenon in the plant kingdom in which the chloroplasts and mitochondria are transmitted only through the maternal parent to the progeny. The direct demonstration of this type of inheritance of organelles is to follow the mode of transmission of chloroplast or mitochondrial genomes from the parents to the hybrids. It has been well established that the chloroplast genome from *Nicotiana* species is strictly inherited via the maternal line. For example, the chloroplast genomes from *N. otophora* and *N. tomentosiformis* have different restriction fragment patterns (Fig. 4). In the reciprocal hybrids the SmaI restriction fragment patterns of their chloroplast genomes correspond to those of the female parents (Fig. 4). This is an unequivocal demonstration of maternal inheritance. Similar approaches have been employed to trace the mode of inheritance of organelles from many other plant species. EcoRI restriction fragment pattern of both chloroplast and mitochondrial genomes isolated from eight lines of *Triticale* differing in their cytoplasmic germplasm have been compared to those of the maternal parents. A strict mode of maternal inheritance was observed

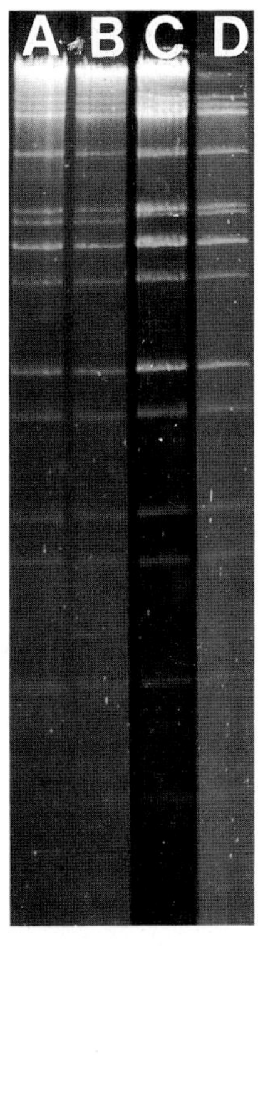


**Fig. 4A–D.** Chloroplast genome is inherited maternally as illustrated by their PstI restriction fragment patterns. **A** *N. tormentosiformis.* **B** *N. tormentosiformis* × *N. otophora.* **C** *N. otophora* × *N. tormentosiformis.* **D** *N. otophora*

(Vedel et al. 1981). In contrast to the uniparental inheritance of the cytoplasmic organelles described here, biparental inheritance of chloroplast genomes was also demonstrated. Using also the EcoRI restriction fragment patterns, biparental inheritance of chloroplast genomes exists in the genus *Pelargonium* (Metzlaff et al. 1981). Sexual crosses of *P. zonale* varieties with different chloroplast genome types provided direct evidence on the mode of biparental inheritance.

*3. Analyzing the Diversity of Chloroplast Genomes.* A general survey of the evidence accumulated to date indicates that there is a high degree of diversity at the restriction fragment level between the chloroplast genomes of many species within a genus. However, all the cultivars within a given species exhibit identical restriction patterns (Kung et al. 1981). In *Nicotiana*, for instance, such variability exists for every one of the 35 species examined except *N. sylvestris* and *N. tabacum* (Kung 1989a). They share the identical chloroplast genome as dictated by the mode of maternal inheritance. It has been well established that *N. sylvestris* was the maternal parent of *N. tabacum* in the original interspecific cross between *N. sylvestris*×*N. tomentosiformis* (Gray et al. 1974). Similar diversity exists in chloroplast genomes in the genera *Oenothera* (Gordon et al. 1982), *Glycine* (Shoemaker et al. 1986), *Linum* (Coates et al. 1982), *Daucus* (DeBonte et al. 1984), *Brassica* (Palmer et al. 1983a), *Triticum* (Vedel and Lebacq 1980; Bowman et al. 1983), *Coffea* (Berthon 1983), *Pennisetum* (Clegg et al. 1984), *Lycopersicon* (Palmer and Zamir 1982), *Solanum* (Hosaka et al. 1984) and *Heterogaura* (Sytsma and Gottlieb 1986), *Cucumis* (Perl-Treves and Galun 1985), and *Pisum* (Palmer 1985).

In general, this diversity is extremely useful as genetic marker in determining the proper taxonomic positions. In *Nicotiana*, there are five closely related species differing with each other by a sequential gain of a single SmaI site. This difference provides a clue as to how to arrange the five species in a proper evolutionary order. Thus, their evolutionary relationship can be readily established (Kung et al. 1982; Kung 1989a).

*4. Tracing the Origin of the Genus.* Restriction fragment analysis of the chloroplast genome has been used to examine the relationship of the monotypic genus *Heterogaura* to the related and speciose genus *Clarkia* (Sytsma and Gottlieb 1986). This analysis provided evidence that *Heterogaura* and *Clarkia dudleyana* shared a more recent common ancestor than either does with any other species. It is important to point out that these two species are more closely related than nearly all pairs of *Clarkia* species examined. In this instance it is obvious that the molecular marker is more reliable than the morphological data in tracing the origin of speciation. Using a similar approach, the origin of polyploid wheat cytoplasms was determined (Tsunewaki and Ogihara 1983).

### 3.2.2 At the Physical Map Level

The restriction fragment analysis measures mainly the point mutations which either give rise to or inflict the loss of a restriction site. Other alterations or rearrangements can be better detected by the physical maps. Here are some examples:

*1. Confirming the Lack of an Inverted Repeat in Some Chloroplast Genomes.* Most of the higher plant chloroplast genomes contain a large inverted repeat sequence (Whitfield and Bottomley 1983). The only exceptions to this pattern are pea (Chu and Tewari 1982), broad bean (Koller and Delius 1980) and chick pea (Palmer et al. 1983a). This was first detected by electron microscopy. However, whether they contain one or two copies of ribosomal RNA gene was not certain. The construction of a physical map of the pea chloroplast genome clearly demonstrated that it lacks the inverted repeat and contains only one copy of ribosomal RNA genes (Palmer et al. 1983a). A similar physical map constructed for broad bean (Ko et al. 1983; Shinozaki et al. 1984) supported this conclusion.

*2. Detecting Rearrangement Sequences in Chloroplast Genomes.* Restriction maps of pea and mung bean chloroplast genomes were constructed to study the extent of sequence homology. A hybridization approach revealed that sequences common to these two chloroplast genomes have become extensively rearranged (Palmer and Thompson 1981). The numbers of rearrangements are such that they totally scrambled the homologous sequences in the mung bean and pea chloroplast genomes. This may not be a common occurrence even in the legume family, since a complete colinearity exists between the mung bean and a common bean chloroplast genome (Palmer et al. 1983a). A similar linear order of sequence elements is also conserved between soybean and mung bean chloroplast genome with only some deletions/additions (Palmer et al. 1983a).

*3. Studying the Evolution and Phylogenetic Relationship.* Comparison of physical maps of chloroplast genomes from related species provides valuable information on their phylogenetic relationship. Figure 5 is the SmaI restriction map of chloroplast genomes from eight *Nicotiana* species (Tassopulu and Kung 1984). It illustrates the phylogenetic relationship of these *Nicotiana* species. For example, there is a single sequential addition of a SmaI site between species from *N. plumbaginifolia* to *N. langsdorfii* to *N. otophora* and to *N. tabacum* (Kung et al. 1982), indicating an evolutionary trend. This close relationship is supported by further analysis using several restriction enzymes (Table 4). Similar approaches have also been used to explore the evolution and the origin of amphidiploid *Brassica* species (Palmer et al. 1983a). From the large number of mutations analyzed it was possible to reach quantitative conclusions concerning both the relative time of hybridization and also the relative divergence of the diploid parents.

The results indicated that all the hybridization of the amphidiploids occurred very recently.

### 3.2.3 At the Nucleotide Sequence Level

The study of phylogeny of chloroplast genes at the nucleotide sequence level has recently been reviewed (Zurawski and Clegg 1987; Kung 1989a). Only the com-

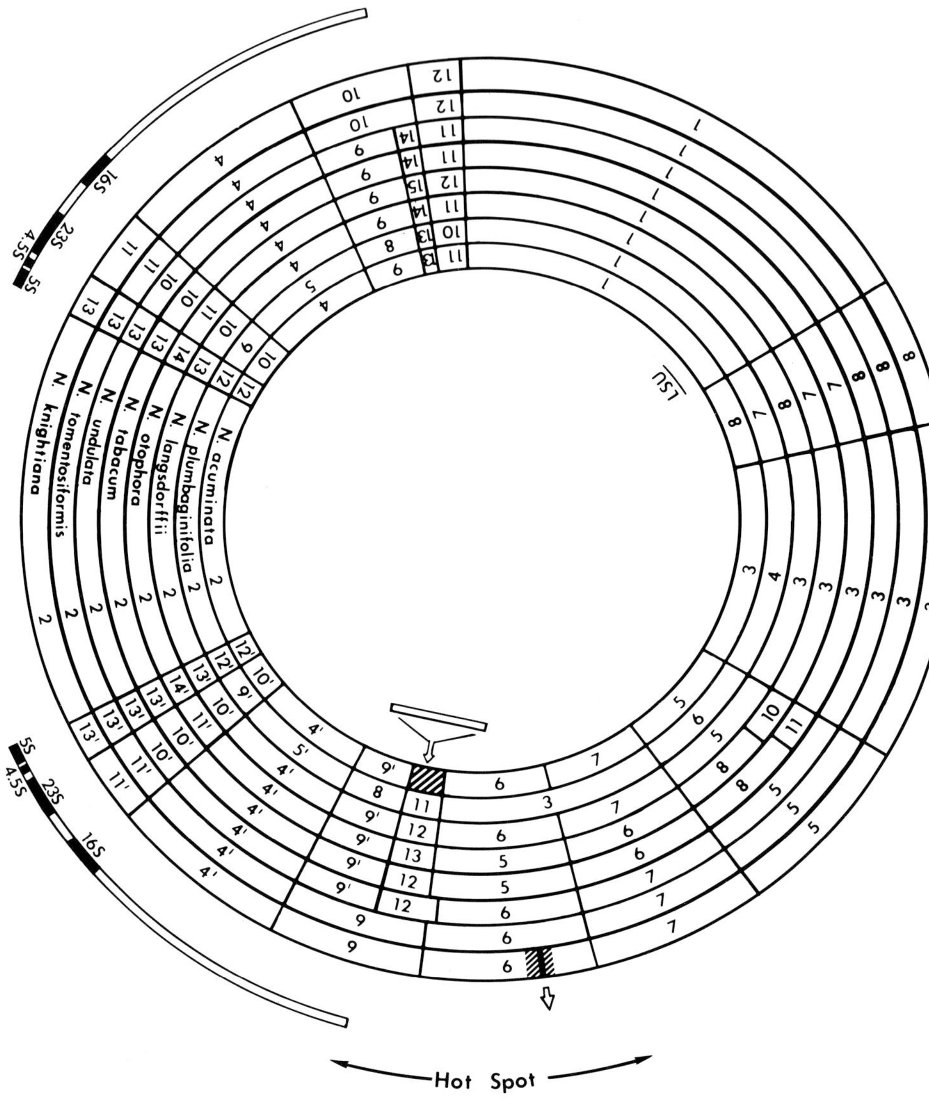


**Fig. 5.** The SmaI restriction map of chloroplast genome from eight *Nicotiana* species. The variations of SmaI sites are clustered in one area of the large single copy region. The location of the extra segment (11 kb) in *N. acuminata* (*shaded*) is identified. This region is designated as the "hot spot"

**Table 4.** The relationship of ct-DNA from five closely related species determined by restriction enzyme analysis. (Kung et al. 1982)

| Restriction enzymes | Distribution[a] | *N. tabacum* | N. sylvestris | N. longiflora | N. langsdorfii | N. plumbagi |
|---|---|---|---|---|---|---|
| SalI | T | 11 | 11 | 11 | 11 | 11 |
| | C | | 11 | 11 | 11 | 11 |
| | S | | 0 | 0 | 0 | 0 |
| SmaI | T | 19 | 19 | 19 | 19 | 18 |
| | C | | 19 | 19 | 18 | 16 |
| | S | | 0 | 0 | 1 | 2 |
| BamHI | T | 37 | 37 | 37 | 37 | 37 |
| | C | | 37 | 37 | 34 | 34 |
| | S | | 0 | 0 | 3 | 3 |
| EcoRI | T | 40 | 40 | 40 | 39 | 38 |
| | C | | 40 | 39 | 31 | 28 |
| | S | | 0 | 1[b] | 8 | 10 |

[a] The total (T), common (C) with *N. tabacum*, and species-specific (S) fragments
[b] This difference is so small that it can be considered identical

parative study at the nucleotide sequence level of the structural gene coding for the LS of RuBisCo and its essential regulatory sequences will be discussed here.

*1. The rbcL Genes.* rbcL genes from three species of *Nicotiana* (Shinozaki and Sugiura 1982; Lin et al. 1986), maize (McIntosh et al. 1980), spinach (Zurawski et al. 1981), and others (Zurawski and Clegg 1987) have been completely sequenced. The nucleotide sequences of maize and spinach have diverged by 16% and 10% from those of *Nicotiana*. The overall sequence is thus highly conserved. Because of the redundancy of the codons, the overall amino acid homology is 90% and 92%, respectively, for maize and spinach.

The degree of conservation is extremely high if the nucleotide sequences of the rbcL genes determined from different *Nicotiana* species are compared. The rbcL gene from *N. tabacum* (Shinozaki et al. 1986), *N. otophora*, and *N. acuminata* (Lin et al. 1986) contains 1431 bp coding for 477 amino acid residues. The overall nucleotide sequences of the rbcL gene from *N. otophora* and *N. acuminata* are very similar to that of *N. tabacum* (Shinozaki and Sugiura 1982). There are only eight base changes between *N. tabacum* and *N. otophora* and between *N. otophora* and *N. acuminata* resulting in merely two amino acid substitutions each. There are only four base changes between *N. acuminata* and *N. tabacum*, generating two amino acid replacements. This represents the first comparative sequence analysis of rbcL genes from three different species within a genus. There are only ten base alterations in the coding region comprising 1431 bp among three species. Six of these changes occurred at identical positions. The degree of sequence homology is in excess of 99.5%. This is much higher than the degree of sequence homology of the rbcL genes found among many plant species which ranges from 80% to 90%. For instance, there are 141 base changes between *Nicotiana* and spinach, and 224 between *Nicotiana* and maize.

*2. Promoters of rbcL Genes.* Recently, many chloroplast genes and their 5′-flanking regions have been sequenced and studied (Kung and Lin 1985). In

**Table 5.** The sequence homology of rbcL gene found among several plant species (after Zurawski and Clegg 1987)

| Plant species | Sequences<br>-35 -10 |
|---|---|
| *Zea mays* (Corn) | TTGGGTTGCGCtATAtcTATcAAAGAGTATAcAATAATgATG |
| *Hordeum vulgare* (Barley) | TTGGGTTGCGCtATAccTATcAAAGAGTATAcAATAATgATG |
| *Triticum aestivum* (Wheat) | TTGGGTTGCGCtATAtcTATcAAAGAGTATAcAATAATtATG |
| *Cenchrus setigerus* | TTGGGTTGCGCtATAtcTATcAAAGAGTATAcAATAATgATG |
| *Pisum sativum* (Pea) | TTGGGTTGCGCcATAcaTATgAAAGAGTATAgAATAATgATG |
| *Nicotiana tabacum* (Tobacco) | TTGGGTTGCGCtATAtaTATgAAAGAGTATAcAATAATgATG |
| *Spinacia oleracea* (Spinach) | TTGGGTTGCGCcATAtaTATgAAAGAGTATAcAATAATgATG |

**Table 6.** The psbA and rbcL promoters from different plant species (after Kung 1989a)

| Plant species | psbA promoter<br>-35 -10 |
|---|---|
| *S. nigrum* | TTGACAcgagtATATAAGtCATGTTATACT |
| Mustard | TTGACAtggctATATAAGtCATGTTATACT |
| Soybean | TTGACActggtATATAAGtCATGTTATACT |
| *N. debneyi* | TTGACAcgagtATATAAGtCATGTTATACT |
| *N. tabacum* | TTGACAcgagtATATAAGtCATGTTATACT |
| Spinach | TTGACAcgggcATATAAGgCATGTTATACT |

most cases, there are conserved promoter sequences in front of the chloroplast genes. The promoter sequences of rbcL genes from different *Nicotiana* species are identical (Table 5). Furthermore, the -35 and -10 regions of the rbcL promoters so far studied from several plant species are also identical (Table 5). This high degree of conservation of nucleotide sequences among the rbcL promoters may bear important biological implications and thus deserves further investigation. The other promoter, which resembles the prokaryotic type and is highly conserved, is that of psbA (Table 6). Available evidence suggests that both rbcL and psbA promoters are strong signals. Both contain highly conserved sequences: the TTG in the -35 region, and the first TA pair and the last T in the -10 region, with a near optimal distance of 18 bp. In fact, the psbA promoters are totally conserved in the region from -35 to -10 among plant species, being identical in all six plant species sequenced to date for rbcL and psbA. Both rbcL and psbA promoters are AT-rich (70%) and possess a TATA-like element between the -35 and -10 regions. This eukaryotic element is also functional in chloroplasts (Link 1984). In addition to the unique structural features, there are some obvious functional properties that support the notion that they are strong promoters. For example, rbcL and psbA mRNA's are the most abundant mRNA species in chloroplasts (Mattoo et al. 1984). Both proteins are synthesized in large amount in chloroplasts. It is conceivable that the structural features are responsible for their functional consequences.

*3. The rbcL Terminators.* In prokaryotic genes or operons there is a secondary stem-loop structure at the 3′ end to serve as the transcription termination signal (Rosenburg and Court 1979; Platt 1987). The region is relatively conserved and encodes in the mRNA the ability to form a stem-loop structure. Similar structural features were also found adjacent to the 3′ end coding regions of chloroplast transcriptional units (Shinozaki and Sugiura 1982; Kung et al. 1986). It is possible that such sequences serve as direct termination signals or it is equally possible that they serve as RNA processing signals (Platt 1987; Stern and Gruissem 1988). In contrast to rbcL promoters, the nucleotide sequences of the putative rbcL terminators are not highly conserved. There is less resemblance with the well-defined prokaryotic terminators. The similarity among the assumed rbcL terminators is in their ability to form stem-loop structures (Kung et al. 1986). The evolutionary implications of this property remains to be determined.

## 4 Conclusions

The recent advances of our knowledge of the chloroplast genome match the development of recombinant DNA technology. Both are traveling on the fast track. The outcome of every analysis of the chloroplast genome, whether it is at the restriction fragment, physical map, or nucleotide sequence level, invariably provides important information on its evolution. Consequently, a great amount of evidence on chloroplast genome evolution has been accumulated. It should also be pointed out that the structural analysis of chloroplast genomes yields more information on its phylogenetic traits than on its functional properties. Therefore, the use of chloroplast genome as a genetic marker represents a timely practical application of our knowledge of chloroplast genomes.

The chloroplast genome possesses many properties suitable to serve as a molecular marker for phylogenetic studies. It is relatively abundant in leaf cells; there are 20–50 copies of chloroplast genomes per chloroplast and 50–200 chloroplasts per cell. There are many devices tailored for isolation and purification of chloroplast genomes from different plant species which require special treatments (Kung and Williams 1969; Rhodes and Kung 1981; Zhao et al. 1989). The isolated chloroplast genomes in most cases are abundant enough and pure enough for direct analysis. More importantly, the size and complexity of the chloroplast genome are within the amenable range to handle with the current technology. However, there is still need to further simplify the techniques currently in use to suit researchers who are not trained specifically with the advanced technology. The ultimate goal is to make the analysis of the chloroplast genome as simple, rapid, and inexpensive as possible.

Currently, the restriction fragment analysis is most useful for the initial step. Although most restriction enzymes can recognize less than 0.1% of the entire sequence of any chloroplast genome, they can be used effectively to reflect the degree of relatedness. In all reported instances, the degree of sequence homology of many chloroplast genomes, based upon restriction fragment patterns, parallels that based upon hybridization data. The degree of sequence divergence for tobacco and spinach, and for tobacco and tomato agrees quite well (40% and 15% respectively) with that obtained from DNA-DNA hybridization studies (Bisaro and Siegal 1980). The estimates of sequence divergence between different *Nicotiana* species are also in line with their taxonomic positions. That *N. glauca* is less divergent from *N. plumbaginifolia* than from *N. excelsior* is expected, because both *N. glauca* and *N. plumbaginifolia* belong to American species, whereas *N. excelsior* is an Australian species (Goodspeed 1954).

The advantages in using chloroplast genomes as genetic markers are many. The practical applications are still in its infancy. There are more molecular biologists using this information for phylogenetic assessments as an additional contribution than evolutionists using this molecular technique as their primary approaches to seek answers. It should be emphasized that the molecular technology is here, is useful, is reliable, and we should make a good use of it in the study of various phases of plant evolution.

*Acknowledgments.* I thank Helen Phillips, Karen Teramura, Xin-Hua Feng, Mary Neville, Grace Kung and Neal Barnett for their help in preparing this manuscript.

## References

Atchison BA, Whitfield PR, Bottomley W (1976) Comparison of chloroplast DNAs by specific fragmentation with EcoRI endonuclease. Mol Gen Genet 148:263–269

Bisaro D, Siegal A (1980) Sequence homology between chloroplast DNAs of several higher plants. Plant Physiol 65:234–237

Bowman CM, Koller B, Delius H, Dyer TA (1981) A physical map of wheat chloroplast DNA showing the location of the structural genes for the ribosomal RNAs and the large subunit of ribulose-1,5-bisphosphate carboxylase. Mol Gen Genet 183:93–101

Bowman CM, Bonnard G, Dyer TA (1983) Chloroplast DNA variation between species of *Triticum* and *Aegilops*. Location of the variation on the chloroplast genome and its relevance to the inheritance and classification of the cytoplasm. Theor Appl Genet 65:247–262

Carlson PS, Smith HH, Dearing RD (1972) Parasexual interspecific plant hybridization. Proc Natl Acad Sci USA 69:2292–2294

Chen K, Wildman SG, Smith HH (1977) Chloroplast DNA distribution in parasexual hybrids as shown by polypeptide composition of fraction I protein. Proc Natl Acad Sci USA 74:5109–5112

Chiba Y (1951) Cytochemical studies on chloroplasts. I. Cytologic demonstration of nucleic acids in chloroplasts. Cytologia 16:259–264

Chu NM, Tewari KK (1982) Arrangement of the ribosomal RNA genes in chloroplast DNA of Leguminoseae. Mol Gen Genet 186:23–32

Clegg MT, Rawson JRY, Thomas K (1984) Chloroplast DNA variation in pearl millet and related species. Genetics 106:449–461

Coates D, Cullis CA (1987) Chloroplast DNA variability among *Linum* species. Am J Bot 74(2):260–268

Crouse EJ, Schmidt JM, Bohnert HJ (1985) Chloroplast and cyanobacterial genomes, genes and tRNAs: a compilation. Plant Mol Biol Rep 3:43–89

Dawid IB, Blacker AW (1972) Maternal and cytoplasmic inheritance of mitochondrial DNA in *Xenopus*. Dev Biol 29:152–161

DeBonte LR, Matthews BF, Wilson KG (1984) Variation of plastid and mitochondria DNAs in the genus *Daucus*. Am J Bot 71:932–940

Deno H, Shinozaki K, Sugiura M (1983) Nucleotide sequence of tobacco chloroplast gene for the alpha subunit of proton-translocating ATPase. Nucleic Acids Res 7:2185–2191

Ee JH van, Vos YJ, Planta RJ (1980) Physical map of chloroplast DNA of *Spirodela oligorhiza*: Analysis by the restriction endonucleases *PstI, XhoI* and *SacI*. Gene 12:191–200

Gillham NW, Boynton JE, Harris EH, Herrmann RG (1985) Evolution of plastid DNA. In: Cavalier-Smith T (ed) DNA evolution: natural selection and genome size. John Wiley & Sons, New York, pp 299–351

Goodspeed TH (1954) The genus *Nicotiana*. Chronica Botanica, Waltham, MA

Gordon KHJ, Crouse EJ, Bohnert HJ, Herrmann RG (1981) Restriction endonuclease cleavage site map of chloroplast DNA from *Oenothera parviflora* (*Euoenothera* plastome IV). Theor Appl Genet 59:281–296

Gordon KHJ, Crouse EJ, Bohnert HJ, Herrmann RG (1982) Physical mapping of differences in chloroplast DNA of the five wild-type plastomes in *Oenothera* subsection *Euoenothera*. Theor Appl Genet 61:373–384

Gray JC, Kung SD, Wildman SG, Sheen SJ (1974) Origin of *Nicotiana tabacum* L. detected by polypeptide composition of fraction I protein. Nature (Lond) 252:226–227

Gruissem W, Zurawski G (1985) Analysis of promoter regions for the spinach chloroplast *rbc*L, *atp*B, and *psb*A genes. EMBO J 4:3375–3383

Herrmann RG, Whitfield PR (1982) Restriction mapping of chloroplast DNA using low-melting temperature agarose in Methods in Chloroplast Molecular Biology. Edelman M, Hallick RB, Chua NH (eds) Elsevier Biomedical Press, Amsterdam, pp 451–468

Hirai A, Ishibashi T, Morikami A, Iwatsuki N, Shinozaki K, Sugiura M (1985) Rice chloroplast DNA: a physical map and the location of the genes for the large subunit of ribulose 1,5-bisphosphate carboxylase and the 32 photosystem II reaction center protein. Theor Appl Genet 70:117–122

Hosaka K, Ogihara Y, Matsubayashi M, Tsunewaki K (1984) Phylogenetic relationship between the tuberous *Solanum* species as revealed by restriction endonuclease and analysis of chloroplast DNA. Jpn J Genet 59:349–369

Kirk JTO (1971) Will the real chloroplast DNA please stand up. In: Autonomy and biogenesis of mitochondria and chloroplasts. Boardman NK, Linnane AW, Smillie RM (eds), Elsevier North Holland, Amsterdam, pp 267–276

Ko K, Strauss NA, Williams JP (1983) Mapping the chloroplast DNA of *Vicia faba*. Curr Genet 7:255–263

Koller B, Delius H (1980) *Vicia faba* chloroplast DNA has only one set of ribosomal RNA genes as shown by partial denaturation mapping and R-loop analysis. Mol Gen Genet 178:216–269

Kolodner RD, Tewari KK (1979) Inverted repeats in chloroplast DNA from higher plants. Proc Natl Acad Sci USA 76:41–45

Kong XF, Lovett PS, Kung SD (1984) *Nicotiana* chloroplast genome. IX. Identification of regions active as prokaryotic promoters in *E. coli*. Gene 31:23–30

Kung SD (1976) Tobacco fraction I protein: A unique genetic marker. Science 191:429–434

Kung SD (1977) Expression of chloroplast genomes in higher plants. Annu Rev Plant Physiol 28:401–437

Kung SD (1989a) In Molecular evolution of *Nicotiana* chloroplast genome. Kung SD, Artzen C (eds) Plant Biotech, Butterworths, Stoneham, MA, pp 373–391

Kung SD (1989b) Why does maternal inheritance exist? (manuscript)

Kung SD, Lin CM (1985) Chloroplast promoters from higher plants. Nucleic Acids Res 13:7543–7549

Kung SD, Williams JP (1969) Chloroplast DNA from broad bean. Biochim Biophys Acta 195:434–445

Kung SD, Zhu YS, Shen GF, Sisson VA (1981) *Nicotiana* chloroplast genome. II. Chloroplasts DNA alteration. Mol Gen Genet 183:20–24

Kung SD, Zhu YS, Shen GF (1982) *Nicotiana* chloroplast genome III. Chloroplast DNA evolution. Theor Appl Genet 61:73–79

Kung SD, Akada S, Mongkolsuk S, Kong XF, Lin CM, Lovett PS (1986) The promoters and terminators of chloroplast genes. The chondriome: Chloroplast and mitochondrial genomes. In: Mantell SH, Chapman GP, Street PFS (eds) John Wiley & Sons, New York, pp 142–163

Lin CM, Liu CC, Kung SD (1986) *Nicotiana* chloroplast Genome X. correlation between the DNA sequences of the gene and the isoelectric focusing pattern of the large subunits of RuBPCase. Mol Plant Biol 6:81–87

Link G (1984) DNA sequence requirements for the accurate transcription of a protein-coding plastid gene in a plastid in vitro system from mustard (*Sinapis alba* L.). EMBO J 3:1679–1704

Link G, Chambers SE, Thompson JA, Falk H (1981) Size and physical organization of chloroplast DNA from mustard (*Sinapis alba* L.). Mol Gen Genet 181:454–457

Mattoo A, Foffman-Falk H, Marder JB, Edelman M (1984) Regulation of protein metabolism: Coupling of photosynthetic electron transport to in vivo degradation of the rapidly metabolized 32-kilodalton protein of the chloroplast membranes. Proc Natl Acad Sci USA 81:1380–1384

McIntosh L, Poulsen C, Bogorad L (1980) Chloroplast gene sequence for the large subunit of ribulose bisphosphate carboxylase of maize. Nature (London) 288:556–560

Melchers G, Sacristan MD, Holder AA (1978) Somatic hybrid plants of potato and tomato regenerated from fused protoplasts. Carlsberg Res Commun 43:203–218

Metzlaff M, Borner T, Hagemann R (1981) Variations of chloroplast DNAs in the genus *Pelargonium* and their biparental inheritance. Theor Appl Genet 60:37–41

Mubumbila M, Gordon KHJ, Crouse EJ, Burkard G, Weil JH (1983) Construction of the physical map of the chloroplast DNA of *Phaseolus vulgaris* and localization of ribosomal and transfer RNA genes. Gene 21:257–266

Ohyama K, Fukuzawa H, Kohchi T, Shirai H, Sano T, Sano S, Umesono K, Shiki Y, Takeuchi M, Chang Z, Aota S, Inokuchi H, Ozeki H (1986) Chloroplast gene organization deduced from complete sequence of liverwort *Marchantia polymorpha* chloroplast DNA. Nature (Lond) 322:572–574

Orozco EM Jr, Mullet JE, Chua NH (1985) An in vitro system for accurate transcription initiation of chloroplast protein genes. Nucleic Acids Res 13:1283–1302

Palmer JD (1985) Comparative organization of chloroplast genomes. Annu Rev Genet 19:325–354

Palmer JD (1987) Chloroplast DNA evolution and biosystematic uses of chloroplasts DNA variation. Am Nat 130:S6–S29

Palmer JD, Thompson WF (1981) Rearrangements in the chloroplast genomes of mung bean and pea. Proc Natl Acad Sci USA 78:5533–5537

Palmer JD, Zamir D (1982) Chloroplast DNA evolution and phylogenetic relationships in *Lycopersicon*. Proc Natl Acad Sci USA 79:5006–5010

Palmer JD, Shields CR, Cohen DB, Orton TJ (1983a) Chloroplast DNA evolution and the origin of amphidiploid *Brassica* species. Theor Appl Genet 65:181–189

Palmer JD, Singh GP, Pillai DTN (1983b) Structure and sequence evolution of three legume chloroplast DNAs. Mol Gen Genet 190:13–19

Palmer JD, Nugent JM, Herbon LA (1987) Unusual structure of geranium chloroplast DNA: A triple-sized inverted repeat extensive gene duplications, multiple inversions, and two repeat families. Proc Natl Acad Sci USA 84:769–773

Perl-Treves R, Galun E (1985) The *Cucumis* plastome: physical map, intragenic variation and phylogenetic relationships. Theor Appl Genet 71:417–429

Platt T (1987) Transcription termination and the regulation of gene expression. Annu Rev Biochem 55:339–372

Poulsen C (1983) The barley chloroplast genome: Physical structure and transcriptional activity in vivo. Carlsberg Res Commun 48:57–80

Rhodes PR, Kung SD (1981) Chloroplast DNA isolation: purity achieved without nuclease digestion. Can J Biochem 59:911–915

Rosenburg M, Court D (1979) Regulatory sequences involved in the promotion and termination of RNA transcription. Annu Rev Genet 13:319–353

Royer HD, Sager R (1979) Methylation of chloroplast DNAs in the life cycle of *Chlamydomonas*. Proc Natl Acad Sci USA 76:5794–5798

Salts Y, Herrmann RG, Peleg N, Lavi U, Izhar S, Frankel R, Beckmann JS (1984) Physical mapping of plastid DNA variation among eleven *Nicotiana* species. Theor Appl Genet 69:1–14

Schiller B, Herrmann RG, Melchers G (1982) Restriction endonuclease analysis of plastid DNA from tomato, potato, and some of their somatic hybrids. Mol Gen Genet 186:453–459

Shen GF, Chen K, Wu M, Kung SD (1982) *Nicotiana* chloroplast genome. IV. *N. acuminata* has larger inverted repeats and genome size. Mol Gen Genet 187:12–18

Shinozaki K, Sugiura M (1982) The nucleotide sequence of the tobacco chloroplast gene for the large subunit of ribulose-1,5-bisphosphate carboxylase/oxygenase. Gene 20:92–102

Shinozaki K, Sun C, Sugiura M (1984) Gene organization of chloroplast DNA from the broad bean (*Vicia faba*). Mol Gen Genet 197:363–367

Shinozaki K, Ohme M, Tanaka M, Wakasugi T, Hayashida N, Matsubayashi T, Zaita N, Chunwongse J, Obokata J, Yamaguchi-Shinozaki K, Onto C, Torazawa K, Meng BY, Sugita M, Deno H, Damogashira T, Yamada K, Kusuda J, Takaiwa F, Kato A, Tohdoh N, Shimada H, Sugiura M (1986) The complete nucleotide sequence of the tobacco chloroplast genome: its gene organization and expression. EMBO J 5:2043–2049

Shoemaker RC, Hatfield PM, Palmer RG, Atherly AG (1986) Chloroplast DNA variation in the genus *Glycine* subgenus *soja*. J Hered 77:26–30

Stern DB, Gruissem W (1988) Control of Plastid gene expression: 3′ inverted repeats act as mRNA processing and stabilizing elements, but do not terminate transcription. Cell 51:1145–1157

Sytsma KJ, Gottlieb LD (1986) Chloroplast DNA evidence for the origin of the genus *Heterogaura* from a species of *Clarkia* (Onagraceae). Proc Natl Acad Sci USA 83:5554–5557

Tassopulu D, Kung SD (1984) *Nicotiana* chloroplast genome. 6. Deletion and hot spot – a proposed origin of the inverted repeat. Theor Appl Genet 67:185–193

Thompson JA, Hansmann P, Knoth R, Link G, Falk H (1981) Electron microscopical localization of the 23S and 16S rRNA genes within an inverted repeat for two chromoplast DNAs. Curr Genet 4:25–28

Tsunewaki K, Ogihara Y (1983) The molecular basis of genetic diversity among cytoplasms of *Triticum* and *Aegilops* species II. On the origin of polyploid wheat cytoplasms as suggested by chloroplast DNA restriction fragment patterns. Genetics 104:155–171

Vaugh KC, DeBonte LR, Wilson KG, Schaeffer GW (1980) Organelle alteration as a mechanism for maternal inheritance. Science 208:196–197

Vedel F, Lebacq P (1980) Cytoplasmic DNA variation and relationships in cereal genomes. Theor Appl Genet 58:219–224

Vedel F, Quetier F, Cauderon F, Dosba F, Doussinault G (1981) Studies on maternal inheritance in polyploid wheats with cytoplasmic DNAs as genetic markers. Theor Appl Genet 59:239–245

Whitfield PR, Bottomley W (1983) Organization and structure of chloroplast genes. Annu Rev Plant Physiol 34:279–310

Xu YQ, Akada S, Machii H, Kung SD (1989) (in preparation)

Zhao Y, Weng XH, Zou Q, Shen QF, Tang S, Chai JH, Wang XM, Kung SD (1989) Construction of cosmid library and detailed physical map of rice chloroplast DNA. (manuscript)

Zurawski G, Clegg MT (1987) Evolution of higher-plant chloroplast DNA-encoded genes: implications for structure-function and phylogenetic studies. Annu Rev Plant Physiol 38:391–418

Zurawski F, Perrot B, Bottomley W, Whitfield PR (1981) The structure of the gene for the large subunit of ribulose 1,5-bisphosphate carboxylase from spinach chloroplast DNA. Nucleic Acids Res 9:3251–3269

# XVIII Flow Cytometric Analysis and Sorting of Somatic Hybrid and Transformed Protoplasts

D. W. GALBRAITH[1]

## 1 Introduction

Considerable interest has developed in the potential of techniques of flow cytometry and cell sorting for providing insight into the processes of plant growth and development (Brown 1984; Galbraith 1984a, b; Petit et al. 1986; Glimelius 1987). The initial technical advance that culminated in the application of flow cytometry to plant cell systems was the development of procedures for the isolation of protoplasts from a wide variety of plant tissues. This is because a fundamental requirement for the operation of flow cytometers is the availability of suspensions of cells having isometric shapes. As detailed below, flow cytometry involves the analysis of the fluorescence and light-scattering properties of single cells during their passage within a narrow, precisely defined liquid stream through the focus of an intense light source. With the exception of pollen and microspores, protoplasts remain the sole example, within higher plant genera, of single-cell suspensions that are suitable for flow cytometric analysis.

A more proximal reason for the application of flow cytometry to plant protoplast systems involved the development of widespread interest in somatic hybridization through induced fusion of protoplasts. Interest was particularly stimulated by reports increasing the number of species for which protoplast regeneration into plants was possible. Furthermore, since the various fusion techniques appeared nonspecific with respect to the parental species or tissues used to provide the input protoplasts, the production of heterokaryons was envisaged as providing for analysis an almost-limitless set of combinations of different plant species. This would have broad implications for the development of useful systems of somatic cell genetics in the higher plants. In addition, the mingling of different cytoplasms during protoplast fusion was seen as leading towards the construction of desired assortments of cytoplasmic traits and the induction of new cytoplasmic traits through recombination, as well as allowing the transfer of linked genetic traits between different nuclear genomes. This initial enthusiasm has been tempered, particularly in view of the problems in achieving the compatible development of different nuclear genomes and the corresponding cytoplasms. Nonetheless, considerable optimism remains that practical benefits will accrue from the generation of novel plants derived from fused protoplasts.

[1] Department of Plant Sciences, University of Arizona, Tucson, Arizona 85721, USA

Biotechnology in Agriculture and Forestry, Vol. 9
Plant Protoplasts and Genetic Engineering II (ed. by Y. P. S. Bajaj)

From a technical viewpoint, it was quickly found that protoplast fusion treatments typically gave rise to low proportions of heterokaryons, and that not all of the heterokaryons subsequently develop into cell cultures and into complete plants. This made it impractical to simply regenerate all protoplasts contained within fusion mixtures, and then to screen for desired phenotypes. Considerable attention was therefore directed towards the development of procedures for the selective enrichment of heterokaryons during culture of the fused protoplasts (Evans et al. 1983; Davey and Kumar 1983; see Chap. XX, this Vol.). These included the use of mutants and genetic complementation, the use of metabolic complementation (involving the treatment of the parental protoplasts with different irreversible inhibitors), the identification of selective media requirements, and the use of visual markers for micromanipulation. Although most of the approaches described above have been used for the effective and successful selection of somatic hybrid plants, all of them suffered from certain limitations. In most cases, these limitations restrict the potential combinations of protoplasts that can be employed for heterokaryon production.

As an alternative and potentially more general approach, we wondered whether it might be possible to adapt for plant systems those concepts of fluorescence labeling employed in classic studies of animal cell hybrids (Frye and Edidin 1970). At that time, this approach appeared particularly attractive because of its possible combination with those methods that were being developed for the automated isolation of animal cell heterokaryons and hybrids through fluorescence-activated cell sorting (Melamed et al. 1979). The overall concept was simple in theory; the two populations of parental protoplasts would be identified through the presence of different fluorescent labels. Fusion would then give rise to heterokaryons that could be distinguished through the presence of both fluorescent labels. Flow sorting based on this criterion could then give rise to enrichment of the heterokaryons, which then would be regenerated into plants.

As reported both by my laboratory and elsewhere, it has indeed been possible to adapt techniques of flow cytometry and fluorescence-activated cell sorting for the analysis of plant protoplasts whilst maintaining their viability (Redenbaugh et al. 1982; Galbraith and Harkins 1982; Harkins and Galbraith 1984; Galbraith 1984a; Galbraith et al. 1984). It has also been possible to use these techniques for the selective enrichment of heterokaryons and heteroplasmons (Afonso et al. 1985; Alexander et al. 1985; Glimelius et al. 1986; Pauls and Chuong 1987). It has been possible to obtain hybrid callus from this sorted material (Afonso et al. 1985; Glimelius et al. 1986) and to regenerate hybrid plants (Afonso et al. 1985; Glimelius 1987). As a consequence, a reasonable understanding has been achieved at both the theoretical and practical level concerning the factors required for successful sorting of higher plant protoplasts (Harkins and Galbraith 1987; Galbraith et al. 1988).

In this chapter, the procedures that have been established for the isolation of somatic hybrid plants through fluorescence-activated sorting of fused protoplasts are outlined and recent developments in the area of the fluorescence characterization of protoplasts are described. These currently permit a sophisticated analysis of the process of gene expression within higher plant cells, and ultimately should lead to the direct isolation of transformed protoplasts. Other review articles chart the early progress in this area (Brown 1984; Galbraith 1984a, b), as well as providing a more recent perspective (Petit et al. 1986; Brown et al. 1987; Glimelius 1987).

## 2 Flow Analysis and Sorting of Somatic Hybrid Protoplasts

The unique design of flow cytometric instrumentation centers around the concept of rapid passage of single-cell suspensions through an area of intense, uniform illumination, to allow excitation of fluorochromes contained on or within the individual cells, as well as resulting in scattering of light. These signals are quantified analytically on a cell-by-cell basis, the information being accumulated in the form of population frequency distributions. The overall scheme is illustrated in Fig. 1. The cells travel within the center of a liquid jet, which emerges from a flow cell through a jeweled orifice of precisely defined diameter. The liquid jet, flowing at a rate of 5–10 m/s, intersects the focus of an intense light source, which is typically an argon ion laser. Fluorescence emission from the cells is detected using a system of lenses, with spectral discrimination between different fluorochromes being achieved through the use of conventional dichroic beam splitters and filters. The filtered signals are quantified through the use of photomultipliers, are digitized, and are stored in the memory of a microprocessor as histograms. Multiparametric histograms can be accumulated, in which the intensities of different fluorochrome emissions are plotted as frequency distributions. Electronic "windows" can then be placed around the different populations observed in these distributions in order to identify those cells specifically for sorting. The sorting process per se relies on a well-characterized physical property of liquid jets in air, the tendency of these to break into droplets. Linear theory indicates that if a periodic disturbance is applied to a liquid jet, this disturbance will exponentially propagate if its wavelength is greater than pi times the diameter of the jet (Harkins and Galbraith 1987). This synchronizes droplet formation at a precise distance below the point at which the disturbance is applied. In flow cytometric instrumentation, the disturbance is generated by a piezo-electric biomorphic crystal attached to the body of the flow cell, the relationship between the crystal drive frequency and the resultant undulation wavelength being described by the wave equation. Since the distance (hence the time interval) between the point of droplet formation and the point of laser excitation is fixed, it becomes possible to selectively charge the droplets during the process of droplet formation, by computer controlled switching of a potential difference that is applied to the jet. This switching process is very rapid, and is controlled by the positions of the windows placed on the frequency distribution histograms described above. Droplets with different charges are then physically separated by passage through an electrostatic field.

The conditions required for the successful analysis and sorting of hybrid protoplasts following induced fusion are as follows. Firstly, it is necessary to choose methods appropriate for the fluorescence labeling of the two different parental populations or protoplasts that will be involved in the formation of heterokaryons. These methods must allow the production of two protoplast populations

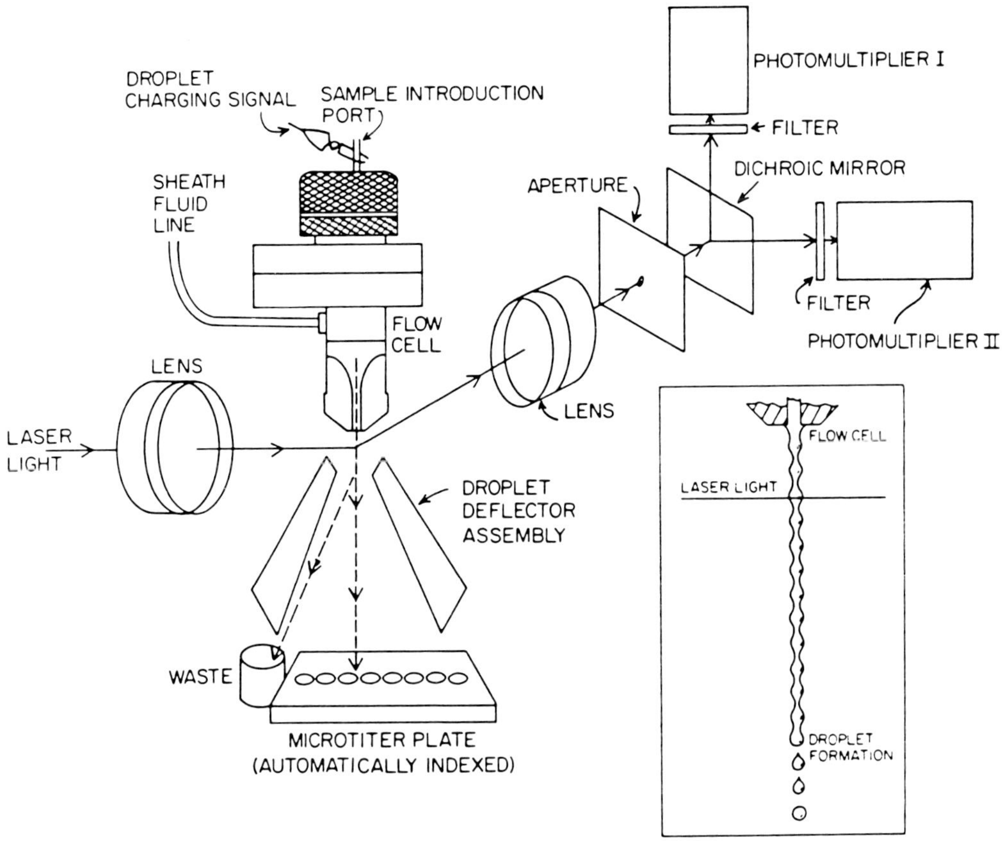


**Fig. 1.** Operation of the cell sorter in schematic form. The fluorescence emitted by individual protoplasts passing through the focus of the laser beam is analyzed by two photomultipliers, allowing two-color discrimination; for example, one for green (fluorescein) fluorescence and one for red (rhodamine, chlorophyll) fluorescence. A piezoelectric crystal attached to the body of the flow cell causes the fluid stream to break into droplets at a fixed distance below the laser intersection point (see *insert*). Pre-programmed sort criteria allow the automatic collection of those droplets containing heterokaryons, within the wells of a sterile microtiter plate. (Galbraith 1984a)

that can be unambiguously discriminated by the flow sorter. Secondly, methods for the induction of protoplast fusion must be selected. These should provide adequate levels of fusion without inducing significant leakage of the fluorescent labels. They must maintain protoplast viability and minimize clumping. Thirdly, conditions for the viable sorting and culture of heterokaryons must be achieved, typically at low or limiting cell densities. Finally, conditions for plant regeneration must be available or must be established.

## 2.1 Isolation of Labeled Protoplasts

Two methods can be employed for the isolation of protoplasts that contain fluorescent labels. Either the fluorescent labels can be added in the form of extrinsic probes, which attach to or permeate within the cells of protoplasts. Otherwise, use can be made of the naturally fluorescent compounds found within plant tissues, in particular chlorophyll. In both situations, a choice of pairs of labels is required; these pairs should have minimal spectral overlap, and should be conveniently excited by the available light sources of the flow cytometer. For instruments that employ argon ion lasers, available excitation occurs over a narrow region of the visible spectrum, from about 457 to 528 nm, with greatest intensities being found at 488 and 514 nm. UV excitation is not recommended for routine use, due to increased rate of plasma tube wear that is associated with UV operation.

### 2.1.1 Use of Exogenous Labeling Techniques

Two general approaches are available when exogenous labeling techniques are employed for the introduction of fluorescence into plant protoplasts, based on whether the labeling molecules react noncovalently or covalently with their targets. Noncovalent labeling molecules can be targeted to specific receptors, for example fluorochrome-conjugated antibodies or lectins, or to specific cellular locations such as the hydrophobic interiors of cellular membranes. In contrast, labeling molecules that react covalently will do so with those reactive species that are available within the cell; these in general are distributed rather uniformly throughout the cell.

*1. Noncovalent Labeling Procedures.* Plant tissues growing as cell suspension cultures are particularly amenable to labeling techniques that involve the addition of exogenous fluorescent dyes, since they typically contain low levels of natural fluorochromes. Growth in suspension culture also permits the use of extended periods of incubation that are required for the partition of the labels into their specific subcellular locations. Various intracellular sites can be targeted for labeling. In our first experiments, we employed the hydrophobic interior of cellular membrane as this target (Galbraith and Galbraith 1979). The membranes of soybean cell suspension cultures were readily labeled with fluorescent lipophilic compounds such as 5-(N-octadecanoyl)aminofluorescein ($F_{18}$) and octadecyl rhodamine B ($R_{18}$), supplied as dilute solutions in ethanol. In this case, direct access of the fluorescent dyes to the target for a relatively extended period of time is required for equilibration of the dye within the membranes. We found that the labeled protoplasts that could readily be prepared from the labeled cells were viable. After fusion was induced, heterologous fusion events were conspicuous. Importantly, a general leakage of probes between the two sets of parental protoplasts was not observed (Galbraith and Galbraith 1979).

Suspension culture protoplasts can also be labeled through the use of fluorogenic substrates, such as fluorescein diacetate (Alexander et al. 1985). In this system, penetration of the nonpolar, nonfluorescent fluorescein diacetate into the cellular cytoplasm is followed by enzymatic hydrolysis (Rotman and Papermaster 1966). This yields fluorescein, which is highly fluorescent and, due to its increased polarity (an ionic charge between $-1$ and $-2$, dependent on intracellular pH), is selectively retained by the semi-permeable plasma membrane. This retention is not complete, and the slow rate of leakage of fluorescein from labeled protoplasts may circumscribe the usefulness of this procedure in different protoplast systems. Use of carboxyfluorescein diacetate or carboxy-4′,5′-dimethylfluorescein diacetate (which upon hydrolysis yield derivatives with ionic charges between $-2$ and $-3$), perhaps in combination with carboxyeosin diacetate or carboxyerythrosin diacetate, may provide a higher degree of intracellular retention as well as spectral discrimination.

Other reports of procedures that may be suitable for labeling of suspension culture cells and protoplasts for use in flow analysis and sorting include the carboxyfluorescein and scopoletin (Kanchanapoom et al. 1985; Glimelius et al. 1986). These water-soluble fluorescent molecules appear to partition into the vacuoles of suspension culture cells of carrot during prolonged incubation. The labels are nontoxic and remain internalized during protoplast preparation and fusion. Carboxyfluorescein can also be used to label the vacuoles of leaf cells (Kanchanapoom et al. 1985). In this case, transpiration was used to imbibe the fluorescent species into the leaf tissues through the cut end of the petiole, prior to protoplast preparation. Carboxyfluorescein has also been employed for the labeling of *Brassica* hypocotyl protoplasts (Glimelius et al. 1986). Use of an overnight period of incubation during the process of protoplast production was required for adequate labeling by the fluorochrome.

Other types of noncovalent labeling techniques include the use of polyclonal antibodies raised against intact lead protoplasts of *Petunia* and *Lycopersicon* (Bergounioux et al. 1983) and of monoclonal antibodies directed against *Nicotiana tabacum* leaf protoplasts (Meyer et al. 1988). Other workers have reported accurate discrimination by monoclonal antibodies between suspension culture protoplasts of *N. glutinosa* and leaf protoplasts of *Phaseolus vulgaris* (Fitter et al. 1987).

*2. Covalent Labeling Procedures.* Alternative to the use of noncovalent fluorochromes is the direct labeling of plant cells and protoplasts with fluorescent dyes that are nucleophilic reagents. The success of this process relies on the extreme sensitivity with which fluorescent molecule can be detected in cells. The limit of detection of current flow cytometers approximates $1-2\times10^3$ molecules of fluorescein per cell (Muirhead et al. 1985); the corresponding limit for fluorescence microscopy is of a similar order of magnitude. At these levels of labeling, the covalent reaction of the fluorescent dyes with specific intracellular nucleophiles is unlikely to result in toxic effects. A variety of covalent labeling reagents have been used for labeling plant cells and protoplasts, including fluorescein isothiocyanate (FITC), rhodamine isothiocyanate (RITC), and various substituted derivatives of rhodamine isothiocyanate (TRITO, XRITC). As with all chemical reactions, optimal results will be obtained using fluorochromes that are chemically homogeneous; this depends both on the source and the storage conditions of the fluorochromes. A variety of novel reactive fluorochromes are now be-

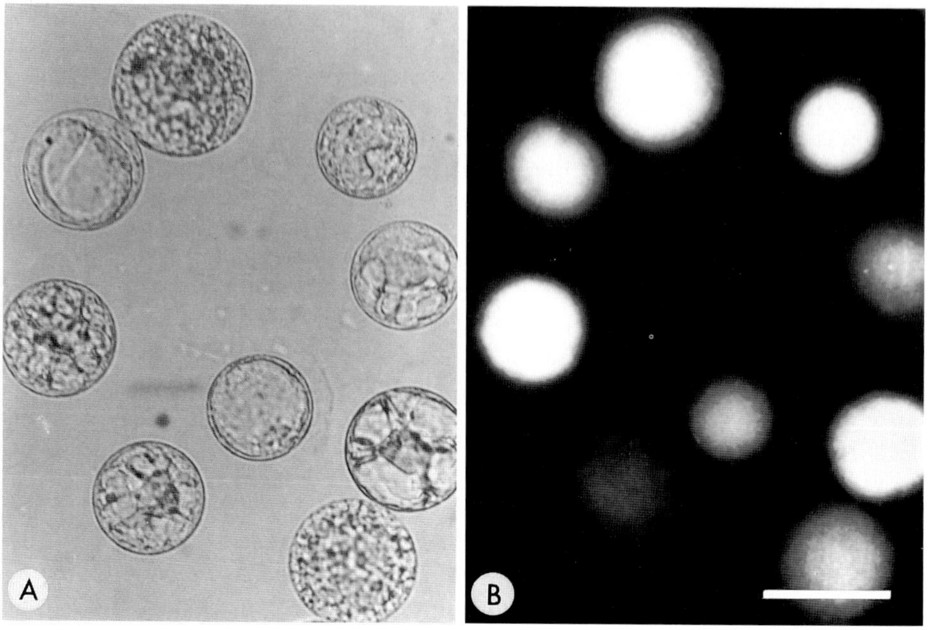


**Fig. 2 A, B.** Protoplasts prepared from FITC-labeled suspension cultures of *N. tabacum* (bar = 50 μm). **A** Bright-field illumination. **B** Epifluorescence illumination. (Galbraith et al. 1984)

coming available, and these may find applications in flow analysis (Haugland 1985).

We have employed nucleophilic fluorochromes such as FITC for the labeling of suspension culture cells and the protoplasts prepared from them (Galbraith et al. 1984; Afonso et al. 1985). This labeling procedure is nontoxic and, as illustrated in Fig. 2, viable protoplasts can be prepared from the labeled cells using conventional procedures (Galbraith et al. 1984). FITC can also be used to directly label protoplasts produced from suspension culture cells, in which case several cycles of centrifugation and resuspension are required for removal of nonbound FITC or its soluble reaction products. As compared to suspension culture cells, the cells within organized plant organs are essentially inaccessible to fluorochromes that are added to the external milieu, due to the complex architecture of the constituent tissues and the presence of a relatively impermeable cuticle. Two strategies for labeling have been developed. One is to label the protoplasts during the process of protoplast formation. This procedure, most commonly used with leaf tissues, involves the addition of the fluorochromes (FITC, RITC, TRITC or XRITC) to the polysaccharidase solutions used for protoplast preparation, typically during a long (15–18 h) period of incubation (Galbraith and Mauch 1980; Redenbaugh et al. 1982; Galbraith 1984a; Afonso et al. 1985). One advantage of this approach is that the protoplasts are subjected to minimal manipulation; unreacted dye can be removed using the standard procedures employed for protoplast purification (Galbraith 1984a). As discussed above, the relatively low level of the fluorochromes that are required for adequate labeling of the protoplasts minimizes toxicity to the protoplasts (Galbraith and Mauch 1980; Galbraith 1984a). The second strategy involves the direct labeling of the leaf protoplasts after purification. For fluorescent nucleophiles of low molecular weight, such as FITC, the approach used is similar to that described above for protoplasts derived from suspension cultures (Redenbaugh et al. 1982). Both of these experimental procedures have been successfully employed for the labeling of protoplasts prepared from etiolated seedlings of *Brassica napus*. The specific fluorescent labels have included FITC added during protoplast preparation (Pauls and Chuong 1987), and FITC or RITC added after protoplast purification (Glimelius et al. 1986). FITC and RITC have also been employed for the labeling of callus protoplasts of an albino mutant of *Petunia inflata* (Redenbaugh et al. 1982). In this case, an overnight incubation during the process of protoplast release gave higher levels of labeling than could be achieved with labeling of the protoplasts after purification.

#### 2.1.2 Use of Endogenous Fluorochromes as Labels

The major endogenous fluorochrome found within the cells of the green tissues of plants is chlorophyll. Mesophyll protoplasts derived from these tissues exhibit intense autofluorescence, and this can readily be detected using conventional flow cytometric instrumentation (Redenbaugh et al. 1982; Galbraith and Harkins 1982; Bergounioux et al. 1983; Harkins and Galbraith 1984, 1987; Galbraith 1984a; Afonso et al. 1985; Glimelius et al. 1986; Galbraith et al. 1987; Pauls and Chuong 1987). As discussed below, flow analysis of chlorophyll autofluorescence, in combination with a second fluorochrome (typically some derivative of fluorescein), has been used for the identification and sorting of heterokaryons formed between protoplasts of a variety of different plant species. Other natural fluorochromes can be detected through flow cytometric analysis of plant protoplasts. For example, flow analysis and sorting of protoplasts, prepared from suspension culture cells of *Catharanthus roseus*, has been achieved based on the fluorescence emission of serpentine (Brown et al. 1984), although this system has not as yet been used for the isolation of heterokaryons.

### 2.2 Induction of Protoplast Fusion

A variety of techniques have been developed for the induction of protoplast fusion. These include treatment with polyethylene glycol solutions, with high pH-high calcium solutions, and with electrical fields (Evans et al. 1983; Davey and Kumar 1983; Bates and Hasenkampf 1985; see Chap. XV, this Vol.). For compatibility with flow sorting techniques, it is important that the fusion technique provide adequate levels of fusion without damage to the protoplasts and without producing excessive amounts of cellular debris. It is also important to avoid clumping, since clumps can block the flow tip, although prefiltration can obviate this problem (Glimelius et al. 1986). Clumping also tends to increase the proportion of false-positive events identified as heterokaryons. With *Nicotiana*, we routinely employ a high pH-high calcium treatment for induction of fusion prior to sorting (Afonso et al. 1985). This typically provides heterokaryons at a level of 2% – 3%. Other workers have preferred the use of polyethylene glycol (Redenbaugh et al. 1982; Alexander et al. 1985; Glimelius et al. 1986; Pauls and Chuong 1987).

## 2.3 Flow Analysis and Sorting of Protoplasts

In order to successfully employ flow cytometry and fluorescence-activated cell sorting for the detection and automatic selection of fused protoplasts, several criteria must be satisfied. Firstly, the protoplasts must survive passage through the instrumentation, which must be capable of accurately and rapidly distinguishing between the different populations of labeled protoplasts. Secondly, the operation of the sorter must be achieved under sterile conditions. Finally, it must be possible to regenerate the protoplasts into plants for analysis.

For successful analysis and sorting of plant protoplasts and heterokaryons, the most important factor is the *quality* of the protoplast preparation. Ideally, the protoplasts should be fully viable (determined conveniently through FDA staining, or, where appropriate, through observation of cytoplasmic streaming) and there should be a minimum of subcellular debris. Use of density gradient centrifugation for protoplast purification is recommended. As illustrated in Fig. 3, for protoplasts that contain chloroplasts, flow cytometric monitoring of the proportions of free chloroplasts can rapidly provide an accurate determination of the numbers of broken protoplasts (Harkins and Galbraith 1984).

### 2.3.1 Instrument Modifications

The principle problems associated with the application of flow cytometry and cell sorting to plant protoplasts are a direct consequence of the large size of protoplasts as compared to animal cells. Plant protoplasts from tobacco leaf tissues typically range from 30 – 50 μm in diameter (Galbraith et al. 1988), although those from cultured cells can be much larger. In general, flow cytometric instrumentation was designed for optimal use with cells having sizes in the range of 5 – 10 μm. Some of these design features are essentially incompatible with the analysis and sorting of large plant protoplasts. For example, the use of conventional flow cells, equipped with 72-μm orifices, coupled to the high rate of flow of the fluid stream, introduces excessive shear forces which cannot be withstood by the relatively fragile protoplasts.

We found that the successful recovery of sorted protoplasts capable of regeneration in culture required use of flow cells having orifices of 100 – 200 μm in diameter (Fig. 4; Harkins and Galbraith 1984; Galbraith et al. 1984; Afonso et al. 1985). Optimization of this process has required a complete analysis of the interaction of the flow tip diameter with other hydrodynamic and optical factors governing the operation of flow cytometers (Harkins and Galbraith 1987), and these led to several modifications to the instrumentation.

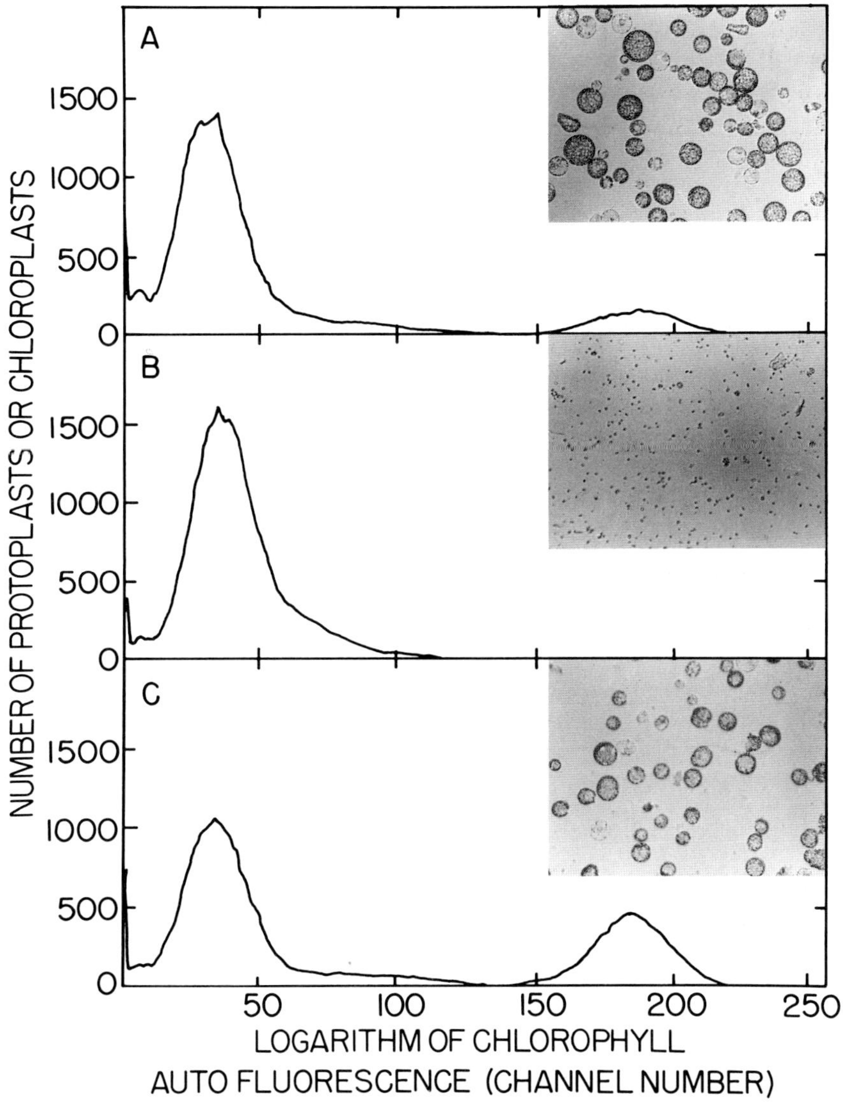


**Fig. 3A–C.** Histograms obtained by flow analysis of populations of protoplasts and chloroplasts based upon chlorophyll autofluorescence. **A** freshly prepared and purified protoplasts; **B** protoplasts lyzed by passage through a 15-μm filter; **C** freshly prepared and purified protoplasts subjected to chilling and hyperosmotic conditions. (Harkins and Galbraith 1984)

The most important instrument modifications derive from a required reduction in the system operating pressure. This reduction lowers the rate of sample flow, which consequently decreases the shear forces experienced by the protoplasts during passage through the instrumentation. However, the lowered velocity of the fluid jet affects the frequency with which the bimorphic crystal drive can be successfully used for droplet formation. As previously discussed, the wave-

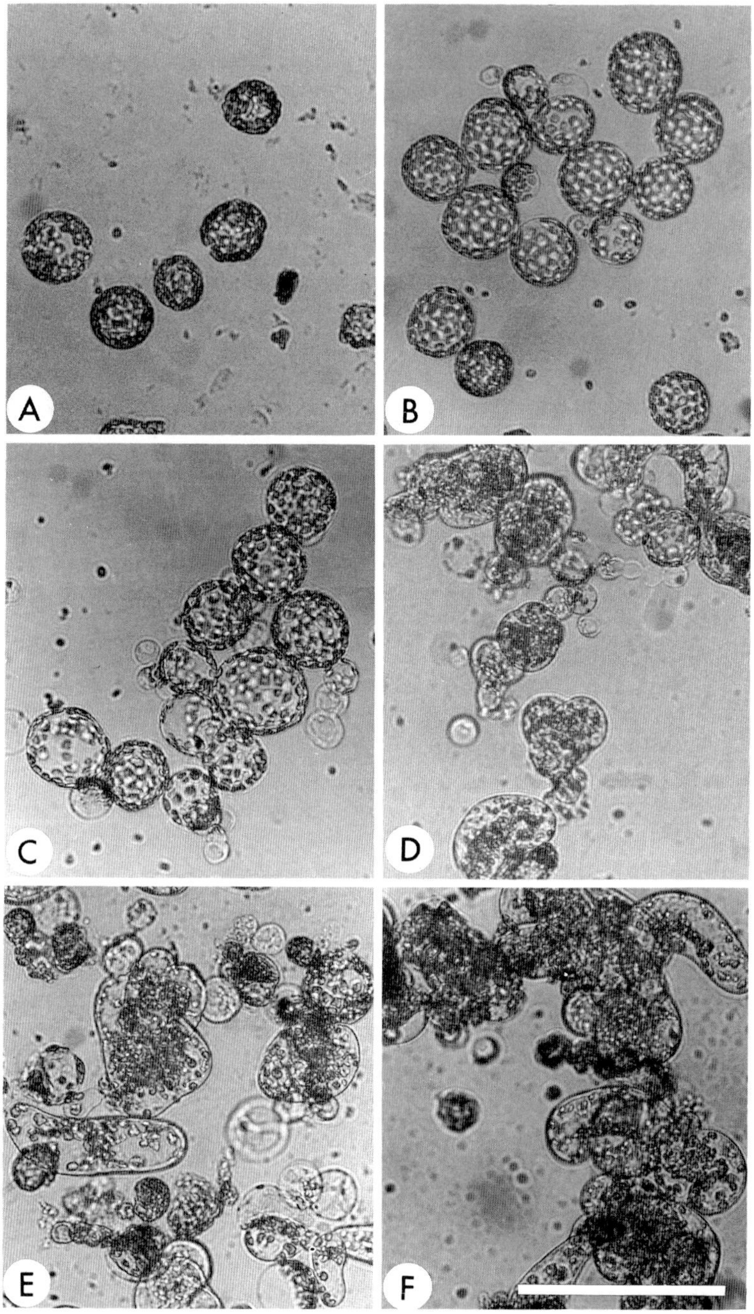


**Fig. 4 A–F.** Leaf protoplasts of *N. tabacum* cultured within microtiter plate wells after flow sorting: **A** immediately following sorting; **B** after 1 day; **C** after 2 days; **D** after 6 days; **E** after 9 days, and **F** after 16 days; *bar* = 100 μm. (Harkins and Galbraith 1984)

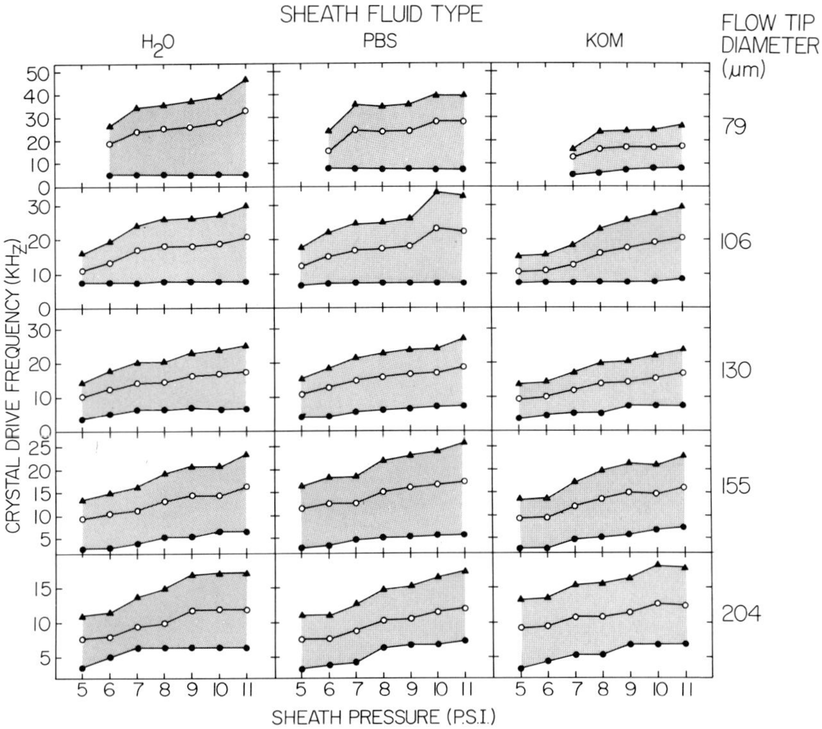


**Fig. 5.** Crystal drive frequency ranges for droplet formation as a function of sheath pressure and fluid type, and of flow cell tip diameter. (● – ● – ●), minimal frequencies; (○ – ○ – ○), theoretically optimal frequencies; (▲ – ▲ – ▲), theoretically maximal frequencies. (Harkins and Galbraith 1987)

length of the undulation induced on the fluid jet by the bimorphic crystal is related to the crystal drive frequency by the wave equation. For exponential growth of this undulation, leading to droplet break-off, this wavelength must be greater than a theoretical minimum (given by pi multiplied by the diameter of the fluid jet; Harkins and Galbraith 1987). At the other extreme, the longest wavelength that can be applied to the fluid jet is dictated by the observation that, as the undulation wavelength increases, the rate at which the undulation grows in amplitude (leading to break-off) decreases. Eventually, the point of droplet break-off can fall beneath the point at which droplet deflection occurs (the body of the deflection assembly; Fig. 1). A compilation of these useful ranges of crystal drive frequencies is given in Fig. 5 for three different sheath fluids and a variety of system pressures.

Interestingly, although droplet formation is possible over a wide range of frequencies for the large flow tips operated at low pressures, the recoveries after sorting (the sort efficiency) can be affected by the frequencies at which the crystal drives are operated. This effect was uncovered through the use of large indestructible biological particles (*Carya* and *Zea* pollen, diameters 51.1 and 95.3 μm) in a systematic standardization of conditions for the recovery of fragile plant pro-

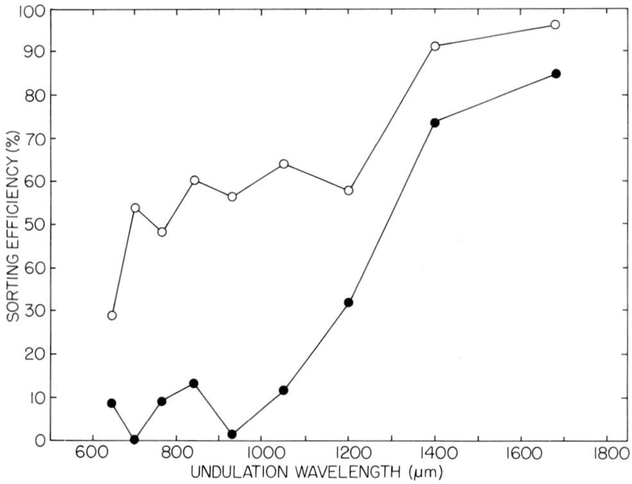


**Fig. 6.** Sorting efficiency of pecan and corn pollen as a function of undulation wavelength. (○ – ○ – ○), pecan pollen; (● – ● – ●), corn pollen. (Harkins and Galbraith 1987)

toplasts and heterokaryons (Harkins and Galbraith 1987). As the droplet undulation wavelengths decreased, although droplet break-off still occurred, the efficiencies of recovery of the largest particles dropped rapidly to zero (Fig. 6). The smaller pollen particles were recovered at higher overall efficiencies, but still exhibited this wavelength-dependent decrease. As particle sizes decreased further, this effect disappeared. These observations emphasize the importance of optimizing conditions for flow sorting with the large flow tips using standard particles that approximate the size of the protoplasts or heterokaryons. If small standard particles are used, it is unlikely that conditions optimal for large protoplasts or heterokaryons will be identified.

Other modifications required for flow sorting with the large flow tips include the use of enlarged sheath tanks equipped for constant pressure operation and, specifically for the Coulter EPICS V, the removal of a specific check valve and a lowering of the sort assembly. These minor modifications have been detailed elsewhere (Harkins and Galbraith 1987).

### 2.3.2 Sample Treatments

A reduction in protoplast size, through resuspension in hypertonic media (Galbraith 1984a; Harkins and Galbraith 1984), or through the preparation of mini-protoplasts (Glimelius 1987), can improve protoplast viability during sorting. Similarly, the use of cold treatments can be helpful (Harkins and Galbraith 1984; Glimelius et al. 1986). It should, however, be emphasized that protoplasts from different plant species and tissues can be expected to respond differently, and it is recommended that media conditions be systematically optimized for each protoplast system under study.

### 2.3.3 Flow Sorting Under Sterile Conditions

Operation of the flow sorter under sterile conditions can be achieved in a variety of ways. We routinely autoclave the sample handling vial, the sample introduction tubing, and the sheath filter and sheath tubing. We use sheath fluids that have been subjected to sterile filtration. The flow cell is cleaned by sonication in detergent solution and is rinsed in distilled water and 100% ethanol. The area around the flow cell body is swabbed thoroughly with 100% ethanol. After assembly of the system, sterile sheath fluid is passed through the flow tip during the period of instrumental alignment (a period of about 30 min). Problems regarding sterility are also avoided through use of single-cell deposition devices such as the Coulter Autoclone (operation of this instrument is described below), since contamination can be limited to single wells of the culture plates.

## 2.4 Heterokaryon Isolation and Culture

Several combinations of fluorescence labeling have been employed for the identification of the two parental protoplast populations in experiments involving protoplast fusion and the analysis and sorting of heterokaryons. Most popular has been the use of the natural chlorophyll autofluorescence of leaf mesophyll protoplasts for the identification of one set of parental protoplasts. The second set is then identified through use of a different fluorochrome, typically a derivative of fluorescein added exogenously. The main reason for the popularity of this approach is the good spectral separation between the emission maxima of fluorescein (ca. 520 nm) and chlorophyll (>610 nm), coupled to the ease of excitation of these fluorochromes using the available argon laser lines. However, this type of approach has one important limitation; it obviously excludes the use of mesophyll protoplasts in this second set, which are therefore restricted to protoplasts derived either from suspension cultures (Afonso et al. 1985; Alexander et al. 1985), or from etiolated plant organs (Glimelius et al. 1986; Pauls and Chuong 1987). It is possible that the developmental incompatibilities that can sometimes be observed in these heterokaryons may be due to differences in metabolic state, in cell cycle distribution and in cycle duration time and ploidy found between these populations of parental protoplasts (Afonso et al. 1985; Glimelius et al. 1986).

Flow analysis and sorting of heterokaryons formed between protoplasts from similar tissue sources requires the use of two different exogenous fluorochromes. Combinations based on the fluorescence emission of fluorescein and rhodamine have been successfully employed for the selection of heterokaryons produced between labeled leaf protoplasts (Afonso et al. 1985). The presence of chlorophyll autofluorescence clearly complicates the setting-up of the flow cytometric instrumentation, particularly when a single laser is used for fluorochrome excitation. This derives from the fact that it is impossible to achieve optimal excitation of both fluorochromes using a single laser excitation wavelength. The large number of variables that can be independently manipulated in this system include the choice of the laser excitation wavelength and of the specific emission barrier fil-

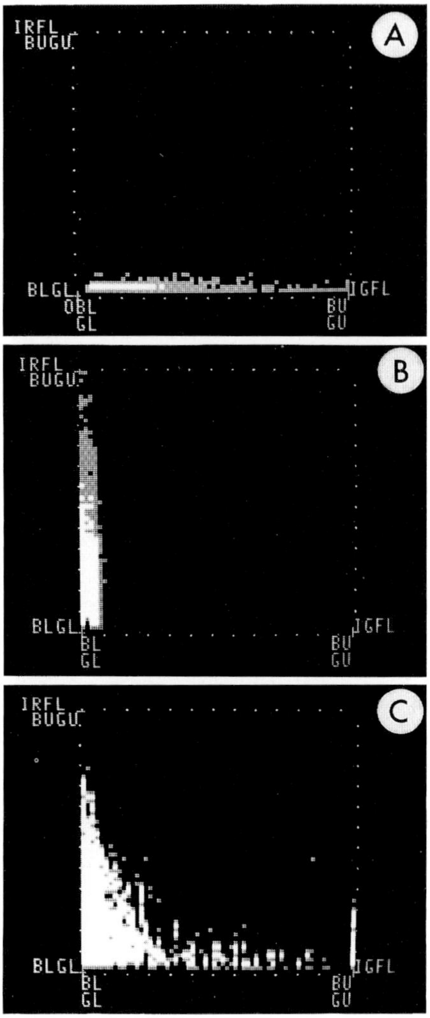


**Fig. 7 A–C.** Two-dimensional contour projections of the fluorescence of *Nicotiana* leaf protoplasts. **A** FITC-labeled protoplasts. **B** RITC-labeled protoplasts. **C** Protoplasts subjected to fusion with polyethylene glycol. (Galbraith 1984a)

ters and dichroic splitters, the amounts of the two different fluorochromes used for labeling the parental protoplasts, and the degree of electronic signal processing (subtraction) employed for eliminating cross-talk between the red and green channels (Galbraith 1984a). Nevertheless, conditions can be established in which the two parental protoplast populations fall on orthogonal axes of two-dimensional projections (Fig. 7). An alternative way to resolve this problem is the acquisition of a second laser (Melamed et al. 1979), in which case the emissions of fluorescein and rhodamine can be examined using independent light paths and signal processing.

A listing of the various combinations of fused protoplasts that have been subjected to flow analysis and flow sorting is presented in Table 1. Our work has employed flow sorting for the isolation of heterokaryons formed between unlabeled

**Table 1.** Heterokaryons, heteroplasmons, hybrids, and cybrids analyzed and isolated through flow cytometry and sorting

| Protoplast source | | Fusion protocol | Stage of regeneration | Reference |
|---|---|---|---|---|
| First | Second | | | |
| Analysis only | | | | |
| *E. lathyris* leaf (FITC) | *E. lathyris* leaf (RITC) | PEG | NR | Redenbaugh et al. (1982) |
| *B. napus* leaf | *B. napus* etiolated hypocotyls (FITC) | PEG | NR | Pauls and Chuong (1987) |
| Analysis and sorting | | | | |
| *N. tabacum* leaf (FITC) | *N. nesophila* leaf (RITC) | High pH/calcium | Plants, progeny | Afonso et al. (1985) |
| *N. tabacum* leaf (unlabeled) | *N. sylvestris* suspension culture (FITC) | High pH/calcium | Plants | Afonso et al. (1985) |
| *N. stocktonii* leaf (unlabeled) | *N. sylvestris* suspension culture (FITC) | High pH/calcium | Plants | Afonso et al. (1985) |
| *H. vulgare* leaf | *D. innoxia* suspension cultures (FDA) | PEG | NR | Alexander et al. (1985) |
| *B. campestris* leaf | *B. oleracea* etiolated hypocotyls (CF) | PEG | Shoots | Glimelius et al. (1986) |
| *N. tabacum* (cms *undulata*) leaf[a] | *N. tabacum* leaf | High pH/calcium | Plants | Ayres (1987) |
| *N. tabacum* (cms *glauca*) leaf[a] | *N. tabacum* leaf | High pH/calcium | Plants | Ayres (1987) |
| *N. tabacum* (cms *suaveolens*) leaf[a] | *N. tabacum* leaf | High pH/calcium | Plants | Ayres (1987) |

[a] Two sets of experiments were carried out, using FITC and RITC as labels in both possible combinations

*N. tabacum* or *N. stocktonii* leaf protoplasts with protoplasts isolated from FITC-labeled *N. sylvestris* suspension culture cells. We also isolated heterokaryons formed between FITC-labeled *N. tabacum* leaf protoplasts and RITC-labeled *N. nesophila* leaf protoplasts. In all cases, plants were regenerated from these heterokaryons and their hybridity was verified through conventional analyses (Afonso et al. 1985).

In all of our experiments, we have exclusively employed a single cell deposition device (the Coulter Autoclone) for the rapid and routine sorting of defined numbers of heterokaryons into 96-well culture plates. Technically, the Autoclone operates in a reverse-logic sort mode. Thus, all unwanted droplets carry charges and are deflected into a waste container. Droplets containing the desired heterokaryons are *un*charged and pass directly without deviation into the wells of the microtiter plates, which are then automatically indexed according to preprogrammed values. These wells are typically filled with approximately 50 µl of protoplast growth medium. In general, all of our work has employed a three-droplet sort mode; thus, for each positive event that is detected, three droplets are involved in the sort decision. Lower sort modes (two- and one-droplet modes) are available. Although these are more difficult to set up, since greater accuracy in aligning the charging process with that of droplet formation is required, they would have the advantage that greater numbers of protoplasts could be sorted before overflow from the well occurs.

It is well known that the process of cell wall deposition and cell division by protoplasts in culture often displays a dependence upon protoplast density. We

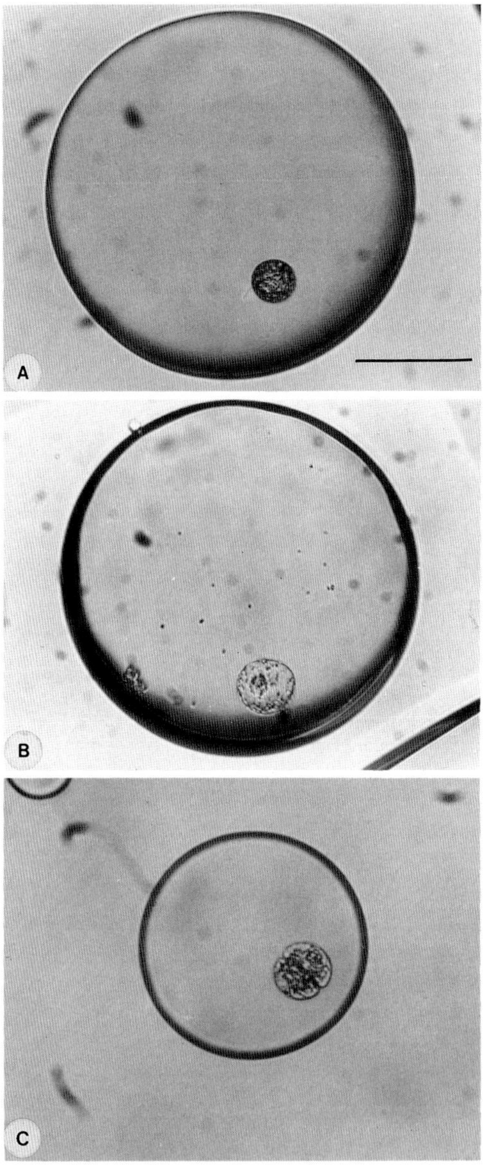


**Fig. 8A–C.** Microdroplet culture of *N. tabacum* leaf protoplasts. **A** Immediately after sorting into paraffin oil. **B** Identical field after 10 days in culture. **C** Nonsorted control, after 10 days in culture. (Courtesy C.L. Afonso and K.R. Harkins)

have found that the successful recovery of plants from sorted heterokaryons is also sensitive to protoplast density (Afonso et al. 1985). Several approaches can be employed to deal with this problem. The first involves the sorting of large numbers of heterokaryons (1000–10000) within the individual culture plate wells. After allowing the sorted heterokaryons to settle, the sample volume can then be reduced and if necessary the wells can be "sealed" by addition of small volumes of paraffin oil to prevent evaporation (Afonso et al. 1985). Under these circumstances, heterokaryon development parallels that observed with nonsorted samples (Afonso et al. 1985; Glimelius et al. 1986; Ayres 1987; Glimelius 1987), although callus formation and plantlet regeneration can be impeded, presumably due to incompatibilities between heterologous nuclei (Afonso et al. 1985; Glimelius et al. 1986). An alternative may be to sort individual protoplasts into wells containing paraffin oil (Fig. 8). Culture of protoplasts under oil in microdroplets apparently can lead to cell division if the microdroplet has a sufficiently small volume. This presumably reflects the successful "conditioning" of the growth medium contained within the microdroplet. It is therefore possible that this type of approach might allow the single-cell cloning of sorted heterokaryons. In other work, we have employed feeder cell layers (Ayres 1987), and this has allowed the successful culture of sorted heterokaryons and heteroplasmons at reasonably low initial densities.

## 2.5 Plant Regeneration and Progeny Analysis

The plants that we have regenerated from the sorted heterokaryons of *Nicotiana* illustrate several points. They demonstrate the feasibility of employing a variety of different labeling methods for identification of the heterokaryons. These permit the successful isolation of highly enriched populations of heterokaryons in a viable state using flow sorting, which can subsequently be regenerated into plants. In all of the combinations that were tested, the proportions of somatic hybrids that were obtained reflected the overall purity of the sorted heterokaryons, which argues that the development in culture of the heterokaryons is not significantly slower than that of the parental (nonfused) protoplasts. However, we consistently found somatic hybrid plants that contained more or less equal contributions of both parental nuclear genomes, as well as plants that contained little contribution from one of the parents (in these cases, the *tabacum* contribution always dominated). Obvious abnormalities (such as deformed and sterile flowers) were observed in the plants regenerated from heterokaryons in which one source of parental protoplasts was the *N. sylvestris* cell suspension culture; it is not clear whether this reflects cell cycle incompatibilities or the aneuploid nature of the cell culture. In contrast, the symmetrical somatic hybrid plants regenerated from the *N. tabacum/N. nesophila* leaf protoplast heterokaryons were self-fertile and through several seed generations displayed a stable phenotype that was intermediate between the two parents. These progeny plants could be successfully backcrossed to both *N. tabacum* and *N. nesophila* (Herman and Galbraith, unpubl.). This suggests that a novel amphidiploid species in the genus *Nicotiana* has been produced.

In recent experiments, we have regenerated plants from a series of sorted heteroplasmons, formed by fusion of leaf protoplasts using the FITC/RITC labeling protocol for identification. We employed three alloplasmic substitutions of *N. tabacum*, in which the cytoplasmic organelles were provided by *N. undulata, N. suaveolens*, and *N. glauca*, respectively. These alloplasmic substitutions display cytoplasmic male sterility (cms). Protoplasts from leaves of these plants were fused with protoplasts prepared from leaves of isogenic (male-fertile) *N. tabacum* plants. Most of the cybrid plants that were regenerated from the sorted heteroplasmons were diploid rather than tetraploid (Ayres 1987). This situation is similar to that observed by other workers employing micromanipulation (Gleba et al. 1984, 1985). We have also confirmed an apparent predominant survival of specific plastid types within these plants (Bonnett and Glimelius 1983); thus the plastids of *N. undulata* are preferentially retained in the regenerated plants at the expense of those of *N. tabacum*. This phenomenon is not affected by the choice of label, and since leaf tissues were employed as both sources of parental protoplasts, argues that cell cycle phase or phase duration times are not involved in the observed discrimination. Interestingly, we have been able to regenerate plants that are male-fertile from cybrids formed between different cms alloplasmic substitutions. We are currently attempting to establish the molecular basis for this restoration.

## 3 Flow Analysis and Sorting of Transformed Protoplasts

Another facet of our recent work has involved the analysis and sorting of protoplasts according to parameters relating to the expression of a transformed phenotype. The expression of a phenotype within the cell is a function of several parameters. These involve specific molecular processes such as the rate of transcription and translation of the particular gene and the rate of degradation of the gene products, as well as more global cellular descriptors, such as the size of the cell, its protein content and organelle complement, and its overall metabolic rate (i.e., the availability of ATP and reduced pyridine nucleotides, the availability of carbon sources, etc.). It is well known that plant tissues in general comprise a complex interspersion of different cell types (Esau 1965); the protoplasts derived from these tissues also are heterogeneous both in function and in size. It should further be appreciated that heterogeneities associated with cell size (i.e., diameter) are magnified when one considers parameters that must be functions of surface area or volume (such as gene expression).

We have addressed the question of size heterogeneity through the analysis of the time taken for protoplasts to pass through the focus of the laser beam. This flow cytometric parameter is termed the particle "time-of-flight" (TOF) and is typically measured from a defined point on the rise-side of the pulse of fluorescence emitted as the particle enters the beam, to a similarly defined point on the fall side (Fig. 9). The resultant TOF one-dimensional frequency distribution is illustrated in Fig. 10. By sorting tobacco leaf protoplasts according to a series of defined TOF parameter values and then measuring their sizes through microsco-

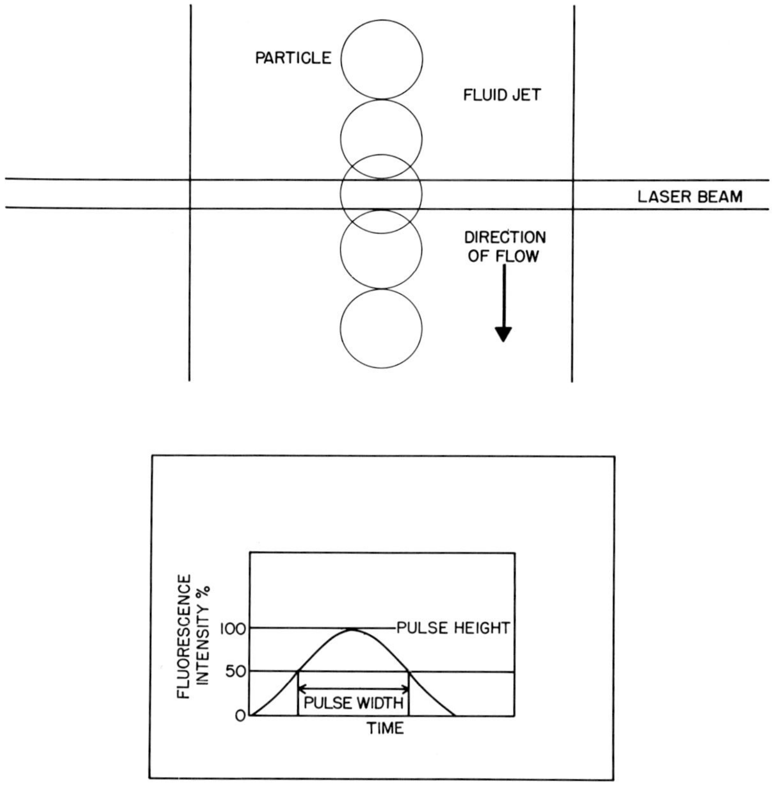


**Fig. 9.** Schematic scale description of the measurement of the pulse-width time-of-flight (PWTOF) parameter with the EPICS V system. (Galbraith et al. 1988)

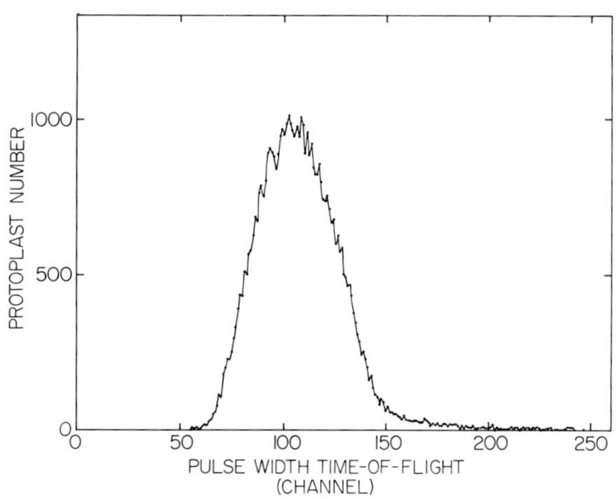


**Fig. 10.** Pulse-width time-of-flight analysis of leaf protoplasts, based on chlorophyll autofluorescence. (Galbraith et al. 1988)

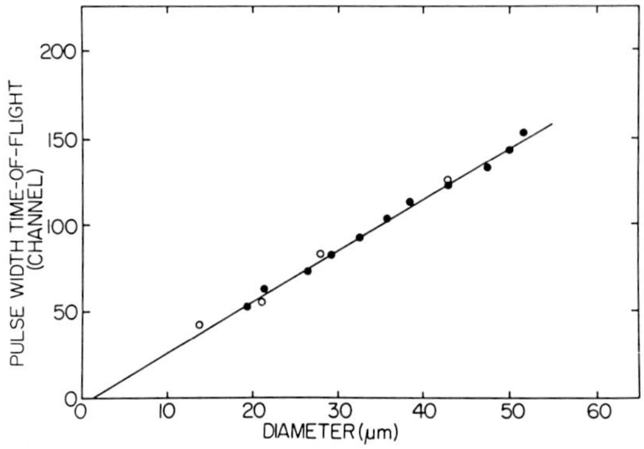


**Fig. 11.** Correlation between the pulse-width time-of-flight parameter and protoplast (● – ● – ●) or pollen/spore (○ – ○ – ○) diameters, based on chlorophyll or aniline blue-induced fluorescence. (Galbraith et al. 1988)

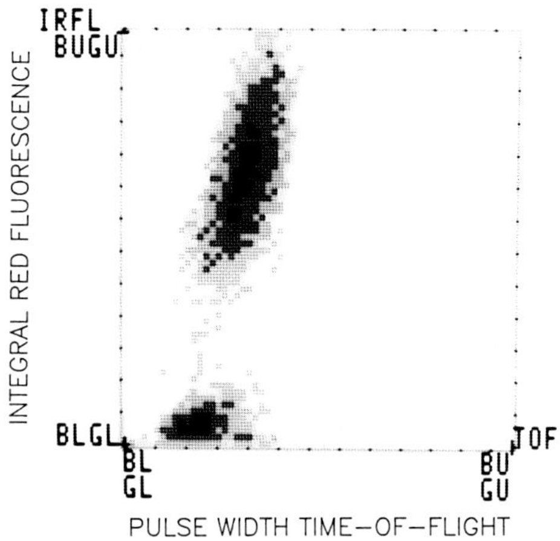


**Fig. 12.** Contour analysis of two-parameter measurement of the chlorophyll autofluorescence versus pulse-width time-of-flight characteristics of leaf protoplasts. FDA fluorochromasia was used as the sources of signals for the PWTOF analysis. The contour levels correspond to 5%, 15%, and 50% of the peak channel. (Galbraith et al. 1988)

py, we were able to show that the TOF parameter accurately and linearly reflects protoplast diameter (Fig. 11). This type of analysis can also by applied to protoplasts lacking chloroplasts using FDA fluorochromasia as the source of the signals (Galbraith et al. 1988). Using a similar approach, we have been also able to demonstrate that analysis of the emission of red autofluorescence (derived from chlorophyll contained in the chloroplasts of protoplasts from photosynthetic tissues) provides an accurate measure of the total chlorophyll content of the protoplast (Galbraith et al. 1988). Examination of two-dimensional histograms produced from tobacco leaf protoplasts, using TOF (based on FDA fluorochromasia) and chlorophyll autofluorescence as the flow parameters, revealed the presence of two major populations (Fig. 12). One population is derived from the nonchlorophyllous tissues of the leaf, which includes the epidermis and the perivascular parenchyma (Esau 1965). The other population is derived from the photosynthetic cells of the mesophyll, as confirmed by sorting.

Sorting based on size and chlorophyll content allows the isolation of populations of protoplasts that are homogeneous with respect to size, chlorophyll content, or both. This capability is important when one considers attempting a detailed analysis of cellular activities that are functions either of size or of chlorophyll content. Our results indicate that considerable heterogeneities exist in cellular chlorophyll content within a single leaf (Galbraith et al. 1988); increased heterogeneity exists between leaves and between plants (Galbraith et al. in prep.). Although TOF analysis indicates that cellular size heterogeneities are lower than those of chlorophyll content, it should be noted that cellular activities that are functions of surface area or volume will have arithmetically increased degrees of variation. The flow cytometric procedures described above allow one to control for these potential sources of variation. They also permit a direct analysis of the biological characteristics of different cell types within leaves. In collaboration with Drs. Jefferson, Kavanagh, and Bevan (Plant Breeding Institute, Cambridge, UK), we have examined the patterns of expression of heterologous chimeric genes in sorted protoplasts. We prepared protoplasts from leaf tissues of tobacco (*N. tabacum* L.) plants that had been transformed via *Agrobacterium tumefaciens* with a binary vector pBin19 (Bevan 1984), in which a chimeric gene had been inserted. This gene comprised the coding region of *E. coli* beta-glucuronidase (GUS) under the transcriptional control of the constitutive 35S promoter of cauliflower mosaic virus (CaMV), or of the light-regulated promoters from genes encoding ribulose bisphosphate carboxylase (small subunit) and the chlorophyll $a/\beta$-binding protein. The protoplasts were sorted in order to separate those containing chloroplasts from those lacking chloroplasts. We found that for the gene constructions involving light-regulated promoters, GUS expression was restricted to the protoplasts prepared from the mesophyll. In contrast, for the gene construction employing the "constitutive" CaMV 35S promoter, GUS expression was found in all types of protoplasts (Harkins et al. in prep.). Future goals of this research include the flow cytometric analysis and sorting of viable protoplasts using GUS expression as a further parameter. For this to be achieved, novel flow cytoenzymological substrates will most probably be required. The advantage of using flow sorting for the enrichment and isolation of transformed protoplasts relates to the observation that conventional selection (using drug resistance, typically against kanamycin) can adversely affect the process of plant regeneration.

## 4 Summary and Conclusions

The techniques of flow cytometry and cell sorting provide unique experimental methods for the manipulation of plant cells and protoplasts. The development of procedures for the fluorescence labeling of parental protoplasts has already led to the production of somatic hybrid and cybrid plants in several species. With the development of techniques for protoplast regeneration in additional species, one can be optimistic that flow sorting will provide a highly versatile and flexible means for the isolation of novel hybrid plants. Flow sorting has also the potential to allow the further development of techniques of somatic cell genetics in higher plants, even if regeneration remains recalcitrant. The use of flow cytometry and cell sorting for the characterization of the physical and physiological characteristics of plant protoplasts has broad general implications in plant cell and molecular biology. Since most processes in living cells are functions of the physical dimensions of the cell, and since most plant tissues comprise a variety of cell types of different sizes, an ability to measure and correct for size heterogeneity will remove an important source of variation in analyses of these processes.

It is clear that the techniques of flow cytometry and cell sorting will also continue to have a growing impact upon a wide variety of different problems in plant biology not included in the above discussion. For example, these techniques are particularly suited for the analysis of the cell division cycle, the measurement of nuclear genome sizes and the isolation of chromosomes (Galbraith et al. 1983; Galbraith 1984b; DeLaat and Blaas 1984; DeLaat and Shell 1986; Conia et al. 1987). Further applications can be envisaged in the area of mutant isolation, organelle and cell characterization, and the analysis of the protoplast cell surface.

*Acknowledgments.* I would like to thank K. Harkins and P. Herman for reading this manuscript and for providing helpful suggestions.

## References

Afonso CL, Harkins KR, Thomas-Compton M, Krejci AE, Galbraith DW (1985) Selection of somatic hybrid plants in *Nicotiana* through fluorescence-activated sorting of protoplasts. Biotechnology 3:811–816

Alexander RG, Cocking EC, Jackson PJ, Jett JH (1985) The characterization and isolation of plant heterokaryons by flow cytometry. Protoplasma 128:52–58

Ayres NM (1987) Inheritance of organelles in somatic cell hybrids and cybrids of *Nicotiana*. PhD Thesis, Univ Nebraska, Lincoln

Bates GW, Hasenkampf CA (1985) Culture of plant somatic hybrids following electrical fusion. Theor Appl Genet 70:227–233

Bergounioux C, Perennes C, Tabaeizadeh Z, Prevot P, Gadal P (1983) Characterization and separation of *Petunia hybrida* and *Lycopersicon esculentum* protoplasts based on natural fluorescence or FITC added to label antibodies in immunoreaction with the protoplast membrane. Experientia Suppl 45:262–263

Bevan M (1984) Binary *Agrobacterium* vectors for plant transformation. Nucleic Acids Res 12:8711–8721

Bonnett HT, Glimelius K (1983) Somatic hybridization in *Nicotiana*: behavior of organelles after fusion of protoplasts from male-fertile and male-sterile cultivars. Theor Appl Genet 65:213–217

Brown S (1984) Analysis and sorting of plant material by flow cytometry. Physiol Veg 22:341–349

Brown S, Renaudin J-P, Prevot C, Guern J (1984) Flow cytometry and sorting of plant protoplasts: technical problems and physiological results from a study of pH and alkaloids in *Catharanthus roseus*. Physiol Veg 22:542–554

Brown S, Jullien M, Coutos-Thevenot P, Muller P, Renaudin J-P (1987) Present developments of flow cytometry in plant biology. Biol Cell 58:173–178

Conia J, Bergounioux C, Perennes C, Muller P, Brown S, Gadal P (1987) Flow cytometric analysis and sorting of plant chromosomes from *Petunia hybrida* protoplasts. Cytometry 8:500–508

Davey MR, Kumar A (1983) Higher plant protoplasts retrospect and prospect. Int Rev Cytol Suppl 16:219–299

DeLaat AMM, Blaas J (1984) Flow cytometric characterization and sorting of plant chromosomes. Theor Appl Genet 67:463–467

DeLaat AMM, Shell JHN (1986) The integrity of metaphase chromosomes of *Haplopappus gracilis* (Nutt.) Gray isolated by flow cytometry. Plant Sci 47:145–151

Esau K (1965) Plant anatomy, 2nd edn. John Wiley & Sons, New York

Evans DA, Bravo JE, Gleba YY (1983) Somatic hybridization: fusion methods, recovery of hybrids and genetic analysis. Int Rev Cytol Suppl 16:143–159

Fitter MS, Norman PM, Hahn MG, Wingate VPM, Lamb CJ (1987) Identification of somatic hybrids in plant protoplast fusions with monoclonal antibodies to plasma-membrane antigens. Planta 170:49–54

Frye LD, Edidin M (1970) The rapid intermixing of cell surface antigens after formation of mouse-human heterokaryons. J Cell Sci 7:319–335

Galbraith DW (1981) Identification and sorting of plant heterokaryons. US Dep Commerce Patent 4,300,310

Galbraith DW (1984a) Selection of somatic hybrid cells by fluorescence-activated cell sorting. In: Vasil IK (ed) Cell culture and somatic cell genetics of plants. Academic Press, New York London Orlando, pp 433–447

Galbraith DW (1984b) Flow cytometric analysis of the cell cycle. In: Vasil IK (ed) Cell culture and somatic cell genetics of plants. Academic Press, New York London Orlando, pp 765–777

Galbraith DW, Galbraith JEC (1976) A method for identification of fusion of plant protoplasts derived from tissue cultures. Z Pflanzenphysiol 93:149–158

Galbraith DW, Harkins KR (1982) Cell sorting as a means for isolating somatic hybrids. In: Fujiwara A (ed) Plant tissue culture 1982. Maruzen, Tokyo, pp 617–618

Galbraith DW, Mauch TJ (1980) Identification of fusion of plant protoplasts. II. Conditions for the reproducible fluorescence labelling of protoplasts derived from mesophyll tissue. Z Pflanzenphysiol 98:129–140

Galbraith DW, Afonso CL, Harkins KR (1984) Flow sorting and culture of protoplasts: Conditions for high-frequency recovery, growth and morphogenesis from sorted protoplasts of suspension cultures of *Nicotiana.* Plant Cell Rep 3:151–155

Galbraith DW, Harkins KR, Jefferson RA (1988) Flow cytometric characterization of the chlorophyll contents and size distributions of plant protoplasts. Cytometry 9:75–83

Galbraith DW, Harkins KR, Maddox JM, Ayres NM, Sharma DP, Firoozabady E (1983) Rapid flow cytometric analysis of the cell cycle in intact plant tissues. Science 220:1049–1051

Gleba Y, Kolesnik NN, Meshkene IV, Cherep NN, Parokonny AS (1984) Transmission genetics of the somatic hybridization process in *Nicotiana.* I. Hybrids and cybrids among the regenerates from cloned protoplast fusion products. Theor Appl Genet 69:121–128

Gleba Y, Komarnitsky IK, Kolesnik NN, Meshkene I, Martyn GI (1985) Transmission genetics of the somatic hybridization process in *Nicotiana.* II. Plastome heterozygotes. Mol Gen Genet 198:476–481

Glimelius K (1987) Enrichment of somatic hybrids with flow cytometry. Int Assoc Plant Tissue Cult Newslett 52:2–10

Glimelius K, Djupsjobacka M, Fellner-Feldegg H (1986) Selection and enrichment of plant protoplasts heterokaryons of Brassicaceae by flow sorting. Plant Sci 45:133–141

Harkins KR, Galbraith DW (1984) Flow sorting and culture of plant protoplasts. Physiol Plant 60:43–52

Harkins KR, Galbraith DW (1987) Factors governing the flow cytometric analysis and sorting of large biological particles. Cytometry 8:60–70

Haugland R (1985) Handbook of fluorescent probes and research chemicals. Molecular Probes, Junction City, Oregon

Kanchanapoom K, Brightman AO, Grimes HD, Boss WF (1985) A novel method for monitoring protoplast fusion. Protoplasma 124:65–70

Melamed MR, Mullaney PF, Mendelsohn ML (1979) Flow cytometry and sorting. John Wiley & Sons, New York

Meyer DJ, Afonso CL, Harkins KR, Galbraith DW (1988) Characterization of plant plasma membrane antigens. In: Leaver CJ, Sze H (eds) Plant membranes: structures, function, biogenesis. UCLA Symp Molecular and cellular biology, New Ser, vol 63. Liss, New York, pp 123–140

Muirhead KA, Horan PK, Poste G (1985) Flow cytometry: present and future. Biotechnology 3:337–356

Pauls KP, Chuong PV (1987) Flow cytometric identification of *Brassica napus* protoplast fusion products. Can J Bot 65:834–838

Petit PX, Conia SC, Brown SC, Bergounioux C (1986) Cytométrie en flux et biotechnologies végétales. Un avenir prometteur. Biofutur 51:128–139

Redenbaugh K, Ruzin S, Bartholomew J, Bassham JA (1982) Characterization and separation of plant protoplasts via flow cytometry and cell sorting. Z Pflanzenphysiol 107:65–80

Rotman BR, Papermaster RO (1966) Membrane properties of living cells as studied by enzymatic hydrolysis of fluorogenic esters. Proc Natl Acad Sci USA 55:134–141

# XIX Isolation of Nuclei and Their Transplantation into Plant Protoplasts

P. K. SAXENA[1] and J. KING[2]

## 1 Introduction

During the last two decades, attempts have been made to use isolated plant nuclei as vehicles for gene transfer (see Potrykus and Lörz 1976; Giles 1978; Lörz 1985). The transfer of isolated nuclei provides an opportunity to study the interactions of a foreign nucleus with the organelles of a host in an unmodified cytoplasm or to combine two genotypes without encountering the barrier of cytoplasmic incompatibility, a distinct advantage over methods employing the fusion of two entire protoplasts. Questions such as those having to do with cytoplasmic sterility or the origin of transferred traits (nuclear vs. cytoplasmic) can also be addressed using nuclear transplantation.

The provision of adequate isolation techniques is crucial to any attempt to use organelles, e.g., nuclei, chromosomes, and chloroplasts in molecular studies. Large numbers of pure, intact organelles are required for most purposes. A major problem in this regard has been, and still is, the presence of the cell wall. Recent advances in the techniques of plant protoplast isolation have solved this problem to a major, but by no means complete, extent. In general, removing the cell wall before attempting to isolate the nucleus from a cell increases significantly the chances of retrieving the organelle with minimum damage. During the present decade several protocols have been reported for the isolation and transplantation of nuclei (Lörz and Potrykus 1978; Saxena et al. 1985a, 1986, 1987) and chromosomes (Szabados et al. 1981; Griesbach et al. 1982; Hadlaczky 1984; Griesbach 1987; DeLaat and Blaas 1987; Mii et al. 1987).

This chapter reviews the methods for the isolation of nuclei, primarily from protoplasts, and their use as vehicles for the transfer of genetic information by transplantation into other protoplasts.

## 2 Sources of Nuclei

The choice of starting material is extremely important in the isolation of nuclei, especially when sources other than protoplasts are used. Ideally, the tissue should be soft, pliable, and have a minimum of starch, chloroplasts, and inhibitors, such

[1] Department of Horticultural Science, University of Guelph, Guelph, Ontario, Canada NIG 2W1
[2] Department of Biology, University of Saskatchewan, Saskatoon, SK, Canada S7N OWO

Biotechnology in Agriculture and Forestry, Vol. 9
Plant Protoplasts and Genetic Engineering II (ed. by Y. P. S. Bajaj)

**Table 1.** Various protocols for the isolation of nuclei from protoplasts

| Plant | Method of disruption | Purification | Yield | Reference |
|---|---|---|---|---|
| *Glycine max* | Triton X-100 + homogenization in Dounce homogenizer | Centrifugation | 2.3% | Ohyama et al. (1977) |
| *Nicotiana tabacum* | Triton X-100 | Centrifugation | 90% | Hughes et al. (1977) |
| *Parthenocissus tricuspidata* | Deplasmolysis + Triton X-100 | Floatation over sucrose | 80% – 90% | Zuily-Fodil et al. (1978) |
| *Vigna unguiculata* | Deplasmolysis + passages through a needle of 26 gauge | Centrifugation | 30% – 40% | Tallman and Reeck (1980) |
| *Nicotiana tabacum* (partially digested cells) | Homogenization in Ultra-Turvax | Density gradient centrifugation | – | Willmitzer and Wagner (1981) |
| *Triticum monococcum* *Papaver somniferum* | Deplasmolysis + Triton X-100 + Transfers using a Pasteur pipette | Centrifugation | – | Hadlaczky et al. (1983) |
| *Brassica nigra* | Deplasmolysis + Triton X-100 + homogenization in Pyrex grinder | Filtration | 95% | Saxena et al. (1985a) |

as phenols and tannins. Meristematic cells, for example, offer these and some other advantages. To eliminate contamination by chloroplasts and starch, etiolated or embryonic cells have also been used (see Dunham and Bryant 1983).

Plant cell cultures are suitable sources for organelle isolation, as they consist of a relatively homogeneous population which can be easily grown in large quantities and which readily form protoplasts. A number of procedures employing protoplasts as the source of nuclei are listed in Table 1. There is no report of the direct isolation of nuclei from intact cultured cells, i.e., without partial or complete digestion of cell walls. A major problem associated with nuclear isolation from plant cell cultures is the low yield of nuclei as compared to animal cells, e.g., a nuclear yield from plant cell cultures equivalent to that obtained from 100 g of rat liver cells would require 4 kg of material (Willmitzer and Wagner 1981). It is therefore desirable to select appropriate source tissue capable of producing high yields of protoplasts.

## 3 Isolation Techniques

The two main steps in the process of isolation of nuclei are: (1) the rupture of cells to release the nuclei along with other cellular constituents and, (2) the separation and collection of the released nuclei. The process used to rupture cells is critical, as the shear forces needed to break cell walls may result in fragmentation of the organelles, nuclei in particular because of their relatively large size. Cells can be ruptured either enzymatically or nonenzymatically. The latter mechanical methods involve the use of such devices as the blender, the French pressure cell or sonicators, for cell breakage. Despite the overall low yield with these mechanical methods, they may be useful in certain experiments. Technical details of the shearing forces used in mechanical method and their efficacy have been discussed in earlier reviews (Price 1979; Dunham and Bryant 1983).

Enzymatic methods rely on the digestion of plant cell walls to convert cells into protoplasts. The rupture of protoplasts requires only low shearing force and their use eliminates the impurities of cell wall fragments in the nuclear preparations. The use of a Dounce homogenizer or a few passages through a needle of 18 to 26 gauge can cause disruption of protoplasts and the release of nuclei. Below is a discussion of the factors known to play an important role in the recovery of pure preparations of nuclei.

### 3.1 Detergent

As is apparent from Table 1, most methods employing protoplasts as the source of nuclei involve the use of a detergent to mediate cell lysis. Detergents such as Triton X-100 solubilize cell membranes and remove other cytoplasmic contamination. Triton X-100 was used originally for isolating rat liver nuclei (Blobel and Potter 1966). It facilitates the release of clean nuclei, but at the same time has been reported to damage the nuclear envelope (D'Alessio and Trim 1968), the extent of which is directly related to the concentration of the Triton X-100. In several studies where Triton X-100 at 0.1% – 1.0% was used, the nuclei isolated were seen to have severely damaged nuclear envelopes or none at all when examined under the electron microscope (Hughes et al. 1977; Hadlaczky et al. 1983). By contrast, using a combination of exposure to hypotonic buffer and low concentrations (0.01% – 0.02%) of Triton X-100, it is possible to isolate pure nuclei with largely intact nuclear envelopes (Saxena et al. 1985a, b).

It is important to note that the final yield, integrity, and cleanliness of nuclear preparations depends on the efficacy of protoplast disruption which, in turn, is greatly influenced by the pH of the isolation medium. For example, the yields of *Brassica* nuclei isolated at pH 5.3 were similar (65% – 68%) irrespective of the concentration of Triton X-100 used (0.01% – 0.1%) (Table 2) but only those nuclei

**Table 2.** Effect of the relationship between pH and various concentrations of Triton X-100 on the yield of nuclei. The nuclei were isolated by disrupting the protoplasts in the Nuclei Isolation Buffer (NIB) containing 10 mM MES, 0.2 M sucrose and different concentrations of Triton X-100 (0.01% – 0.1%). (Saxena et al. 1985a)

| | pH of NIB | | | | |
|---|---|---|---|---|---|
| | 3.3 | 4.3 | 5.3 | 6.3 | 7.3 |
| % Triton X-100 | Nuclear yields (%) | | | | |
| 0.01 | 6.0 | 9.5 | 65.5 | 8.5 | 1.0 |
| 0.05 | 16 | 18 | 62 | 7.0 | 0.0 |
| 0.1 | 31 | 40 | 68 | 5.0 | 0.5 |

isolated with 0.01% Triton X-100 had intact nuclear morphology (Saxena et al. 1985a). At low pH (3.3–4.3), plasma membrane fragments were not completely removed even with 0.1% Triton X-100 resulting in poor yields of nuclei (Table 2). The optimal pH range for isolation of nuclei from cell suspension cultures using a low detergent level (0.01%–0.02%) varies from 5.2–5.5 (Saxena et al. 1985b; Liu 1985). However, pH values between 6 and 8 have also been used in other protocols (Zuily-Fodil et al. 1978; Willmitzer and Wagner 1981).

### 3.2 Stabilizers and Protectants

Isolated nuclei, like protoplasts, require an osmotic stabilizer to maintain membrane integrity and prevent lysis. Sucrose has been the stabilizer of choice in most protocols, but because it is a highly viscous and strong dehydrating compound, its suitability in maintaining an ideal stabilizing environment is somewhat questionable (Price 1979). In our experience both sucrose and mannitol (or its isomer sorbitol) are equally effective (unpubl.). Some investigators have used high molecular weight polymers such as dextran, Ficoll, and gum arabic in combination with sucrose (Honda et al. 1966; Tautvydas 1971) as stabilizers.

A number of substances have been used as protective agents for nuclei. Divalent ions $Mg^{2+}$ and $Ca^{2+}$ are generally used to maintain the integrity of the chromatin. However, these ions cause aggregation of nuclei with plant cell constituents, and have been replaced by $Na^{+}$ and $K^{+}$ (Willmitzer and Wagner 1981; Saxena et al. 1985a). The use of the polyamines, spermine and spermidine, in nuclei isolation buffers was introduced by Hewish and Burgoyne (1973) for stabilizing chromatin in animal nuclei by inhibiting the effect of endogenous DNAse. Later, polyamines were reported to increase the nuclear yield by preventing the agglutination of the nuclei themselves or with other particulate material (Willmitzer and Wagner 1981; Saxena et al. 1985a).

Ethylene diamine tetraacetic acid (EDTA), a chelating agent, has also been used to inhibit the activity of DNAse as well as phenol oxidases, which require $Cu^{2+}$ for catalysis (Anderson 1968). Also, the inclusion of EDTA in nuclei isolation media tends to destabilize the plasmalemma, making the lysis of protoplasts easier (Kobza and Edwards 1984). Another problem related to the isolation of plant nuclei is the occurrence of phenolics that result from the rupturing of vacuoles containing secondary metabolites. Several compounds, known to remove phenolics or inhibit phenol oxidases, such as polyvinyl-polypyrrolidone, diethyldithiocarbamate, and mercaptobenzthiazol have been included in isolation media. Willmitzer and Wagner (1981) reported a satisfactory removal of browning with 10 mM mercaptoethanol, 5 mM EDTA, and 0.1% bovine serum albumin, together.

## 4 Isolation and Purification of Nuclei

Following the rupture of cells/protoplasts the lysate is passed through several layers of Miracloth or cheesecloth or nylon screens (Tautvydas 1971). The nuclei can be further purified by differential or density gradient centrifugation.

The large size of nuclei in comparison to other organelles can be used advantageously in their purification by low speed centrifugation to pellet nuclear fractions. Differential centrifugation is quick and saves the nuclei from mechanical damage because of the low speeds required. Quickness means that the nuclei have only a short exposure to degradative enzymes during isolation. However, the need for repeated centrifugation in this method may cause considerable lowering of the yields of the nuclei and the extraction of water-soluble components from nuclear matrices. The use of nonaqueous isolation medium can be useful in minimizing the leakage of water-soluble nuclear constituents.

Depending upon the use to which the nuclei are to be put, they can be purified further by density-gradient centrifugation. Ideally, the material used to prepare gradients should be dense, freely soluble, metabolically inert and nonviscous. Sucrose has been widely used in isopycnic gradients (Price 1979) because of its purity and low cost, but it has the disadvantage of giving rise to a high osmotic potential. Certain other compounds, less viscous than sucrose, such as silica solutions (e.g., Ludox), PVP-coated silica solutions (e.g., Percoll) and iodinated derivatives of benzoic acid (e.g., metrizamide), have been used successfully to obtain pure nuclei (Hendriks 1972; Pertoft and Laurent 1977; Willmitzer 1984).

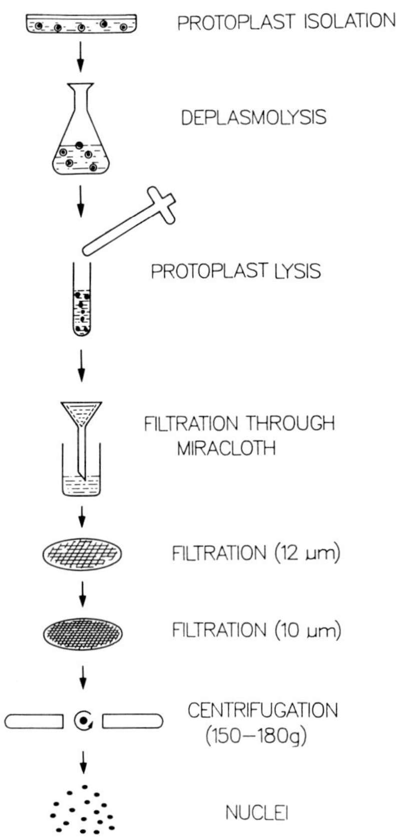


**Fig. 1.** Protocol for the isolation of nuclei from protoplasts. (Saxena et al. 1985a)

However, purification of these compounds before use may be necessary to obtain functional organelles (Price 1979; Kobza and Edwards 1984).

The quality of the nuclei, their purity, intactness, and biological activity, is as important as high yield. The first criterion by which to judge the quality of a nuclear preparation is to determine structural integrity which can be visualized by light (Nomarski optics) or electron microscopy (Hughes et al. 1977; Willmitzer and Wagner 1981; Saxena et al. 1985 a). The degree of contamination by other organelles can be measured either microscopically or by isolating total DNA. Separation of nucleic acids in a density gradient should give rise to a single band of DNA if the nuclei are pure (Ohyama et al. 1977). The functional criterion of

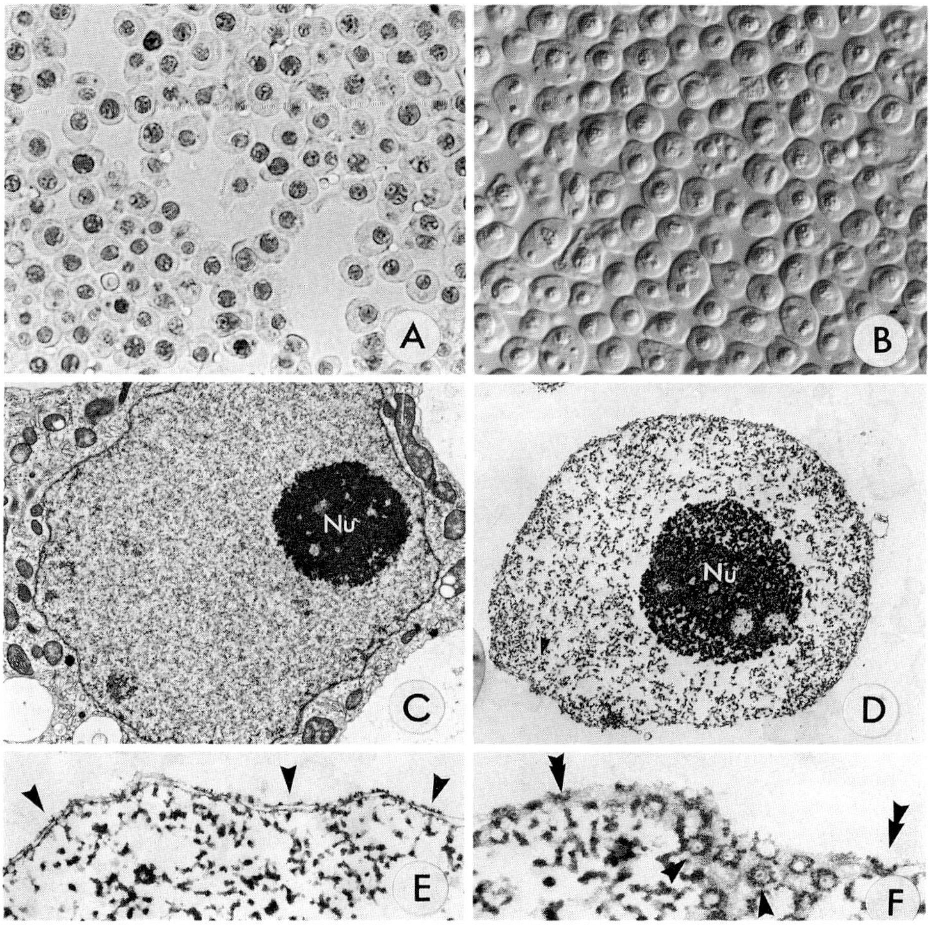


**Fig. 2 A–F.** Morphology of isolated nuclei: **A** light micrograph of isolated nuclei stained with toluidine blue. **B** Nomarski optics. **C** Electron micrograph of a nucleus of an intact protoplast (*Nu* nucleolus). **D** Electron micrograph of an isolated nucleus. **E** Electron micrograph of an isolated nucleus showing the nuclear envelope. **F** Electron micrograph of an isolated nucleus showing face view (*single arrows*) and cross-section (*double arrow*) of the nuclear envelope with nuclear pore complexes. (Saxena et al. 1985 a, b)

the quality of the nuclei is by the assay of the activity of enzymes associated with them, e.g., RNA and DNA polymerases, histone acetylase, poly-ADP-ribosylation, and protein kinase (see Willmitzer 1984).

Figure 1 shows the protocol for nuclear isolation developed in our laboratory. The salient features of this procedure are: (1) gentle rupturing of protoplasts by threefold destabilization, i.e., deplasmolysis in hypotonic buffer, exposure to detergent, and homogenization in a grinder rendering minimal damage to the nuclei, (2) purification via filtration thus avoiding repeated centrifugation which results in low yields and less intact nuclei, and (3) isolation at a pH which allows the lysis of protoplasts with 0.01% – 0.02% Triton X-100 and the use of a buffer supplemented with $Na^+$, $K^+$, EDTA and a polyamine to stabilize the nuclei.

The nuclei isolated with this procedure (Fig. 2A) were morphologically intact when viewed by Nomarski optics (Fig. 2B). The nuclear preparations were free of cytoplasmic contamination such as membrane fractions, live or damaged protoplasts, and other cell organelles. Contamination by starch grains was also insignificant. Ultrastructural examination of the isolated nuclei showed relatively well-preserved outer and inner membranes with intact nuclear pore complexes (Fig. 2D, E, F). The chromatin of the isolated (Fig. 2D) nuclei was more condensed than that of the intact cells (Fig. 2C). Besides condensation of chromatin, the extracted appearance of the nuclei may also be a result of the loss of water-soluble components of the nuclear matrix. In addition to the structural integrity, the isolated nuclei were biochemically active in the uptake of $^{14}C$ uridine and $^{14}C$ amino acids. Further, the DNA from the nuclei isolated with our procedure invariably formed only one band in a CsCl gradient (unpubl.).

## 5 Induction of Uptake of Nuclei by Protoplasts

The ability of protoplasts to take up foreign particles has stimulated attempts to use isolated cell organelles as vehicles for genetic transformation. Potrykus (1973) developed a method, referred to as the Sandwich Technique, to introduce nuclei into protoplasts. In this method, protoplasts and nuclei were pelleted alternately by centrifugation (i.e., protoplasts-nuclei-protoplasts) and the steps repeated several times. The factors responsible for inducing nuclear uptake were the difference in osmotic potential of the media in which protoplasts and nuclei were originally suspended, the presence of lysozyme (0.03%) in the nuclei isolation medium, and the formation of alternating layers. The frequency of uptake obtained by this technique varied from 0.1% to 1% (Potrykus 1973; Lörz and Potrykus 1978). The limitations of the Sandwich Technique were the requirement for different centrifugation speeds to pellet protoplasts and nuclei, ca. 600 and 1500 – 2000 rpm, respectively, changes in osmotic conditions, and severe damage to protoplasts by repeated centrifugation, particularly following the stressful process of deplasmolysis.

In any uptake experiment it is essential to differentiate the physical from the real biological uptake by assessing protoplast viability following uptake treatment. For example, in experiments to compare the efficiencies of different concentrations of calcium nitrate on nuclear uptake, Lörz and Potrykus (1978) found that the frequencies of uptake after 5- and 20-min treatments were 1.1% and 2.8%, but the viability maxima were 80% and 20%, respectively. However, of the several treatments, the use of polyethylene glycol (PEG) was found most suitable as it induced 3% – 5% uptake frequencies with 70% – 80% viability of the receptor protoplasts (Lörz and Potrykus 1978). Saxena et al. (1987) also optimized several factors and routinely obtained ca. 4% – 6% uptake frequencies with protoplasts and nuclei of different origin. The procedure involves incubating protoplasts and nuclei together for 15 – 20 min in plastic Petri dishes in an uptake inducing solution containing PEG (m.w. 4000), 0.1 M $CaCl_2 \cdot 2H_2O$, 0.05 M glycine (pH 6.8) and 5% dimethyl sulfoxide. The viability of protoplasts after nuclear uptake was 68% – 86%.

A common problem associated with the determination of nuclear uptake is to distinguish the nuclei merely attached to the surface of protoplasts from those actually taken up into the cytoplasm. A nucleus seen through a microscope will often appear to be inside, although it may be simply attached to the surface underneath or above the equatorial plane of the protoplast. Whether a nucleus is inside or outside can be determined by rolling the sample between the slide and coverslip so as to position the nucleus in the equatorial plane (Lörz and Potrykus 1978), a process which is time-consuming and not always accurate.

We have developed a relatively accurate method for quantitative estimation of uptake based on the differential staining of nuclei. Prior to uptake, nuclei are stained with a fluorescent dye, Hoechst 33258. Stained nuclei are then mixed with protoplasts, uptake induced and then the protoplast/nuclei mixture is stained with Evans Blue. The preparation is then examined microscopically in both ultraviolet and bright fields. Those nuclei that are attached to the protoplast surface appear blue in bright field (Fig. 3 A) and fluoresce in UV, whereas those inside the cytoplasm only fluoresce (Fig. 3 B) since they were not stained by Evans Blue, which does not penetrate into intact cells. The method thus can be used to evaluate a variety of factors affecting nuclear uptake (Saxena et al. 1987).

The mechanism by which nuclei are taken up by protoplasts is not clear. Plant protoplasts are capable of taking in a wide variety of foreign particles (Fowke and Gamborg 1980) ranging from ferritin molecules to algal protoplasts, a size range of 0.09 μm – 10 μm (Fowke 1986). Accordingly, three main pathways have been suggested for their uptake. Smaller particles such as ferritin have recently been shown to be endocytosed (Tanchak et al. 1984; Joachim and Robinson 1984) via coated vesicles (i.e., receptor-mediated) in a fashion similar to that observed in animal cells (Brown et al. 1983). Larger particles (0.1 – 3 μm diameter) are taken up by the invagination of certain uncoated regions of the plasma membrane followed by endocytosis. Polystyrene latex spheres (Suzuki et al. 1977) and bacteria (Hasezawa et al. 1983; Matsui et al. 1983) have been shown to be endocytosed in this manner. Still larger particles (3 – 16 μm diameter), e.g., algal protoplasts, are taken in by fusion (Fowke et al. 1981). Nuclei, on the basis of their size range and membrane-bound nature, are likely to be taken up either by fusion or through endocytosis following invagination of the plasma membrane. The ultrastructural studies to elucidate the mechanism(s) of nuclear uptake have not yet been carried out.

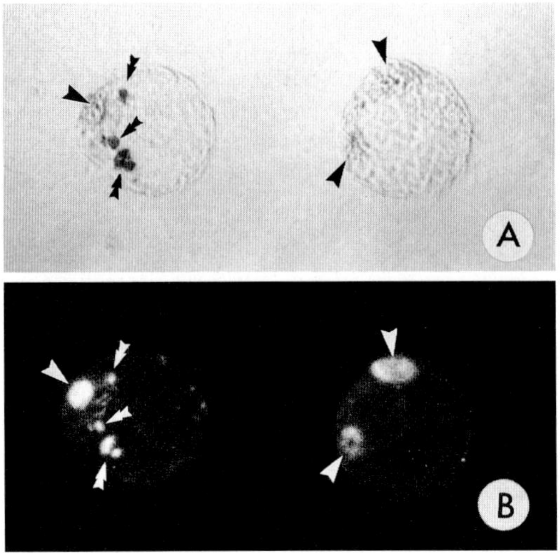

**Fig. 3A,B.** The technique of differential staining of nuclei to determine the uptake frequency. The uptake is carried out using the nuclei stained with Hoechst 33258 and the whole population is stained with Evans Blue. **A** The nuclei attached to the surface of protoplast are stained with Evans Blue (*left, double arrows*). Note that the nuclei taken up by protoplasts are not stained with Evans Blue (*arrow*). **B** The nuclei taken up fluoresce (*arrows*) in UV field. (Saxena et al. 1987)

The uptake of isolated metaphase chromosomes into plant protoplasts has also been achieved using PEG (Szabados et al. 1981). Recently, the method of microinjection (Steinbiss and Stabel 1983) has also been modified to achieve transfer of isolated chromosomes into protoplasts (Griesbach 1987; DeLaat and Blaas 1987). Microinjection appears to be a better technique, as it eliminates the harmful effects of PEG and other fusion-inducing chemicals. However, micromanipulations with microinjection set-up are relatively complicated, expensive to operate, and require accurate pre-determination of experimental conditions as compared to the PEG-mediated uptake of organelles.

## 6 Post-Transplantation Events

So far only two reports are available which describe post-transplantation events following nuclear uptake. In the first experiment, designed to seek proof of biological expression of the transferred nuclei, two recessive nuclear mutations of tobacco were used (Lörz and Potrykus 1978). These nuclear mutants, referred to as sublethal and virescent, were light-sensitive and were shown to complement by protoplast fusion (Melchers and Labib 1974). A total of $1.8 \times 10^7$ calli were recovered by incubating protoplasts of one mutant with the nuclei of the other and vice versa, in the presence of PEG. Of the $1.8 \times 10^7$ calli recovered from the combination of sublethal protoplasts and virescent nuclei, 40 calli were selected under high-light conditions. However, the plants regenerated from these calli were all sublethal type; thus biological proof of nuclear transfer was not demonstrated.

In the second approach, Saxena et al. (1986) used an auxotrophic mutant of *Datura innoxia* as receptor and isolated nuclei of *Vicia hajastana* as donor. The mutant (Pn1) had an absolute requirement for pantothenate (Savage et al. 1979)

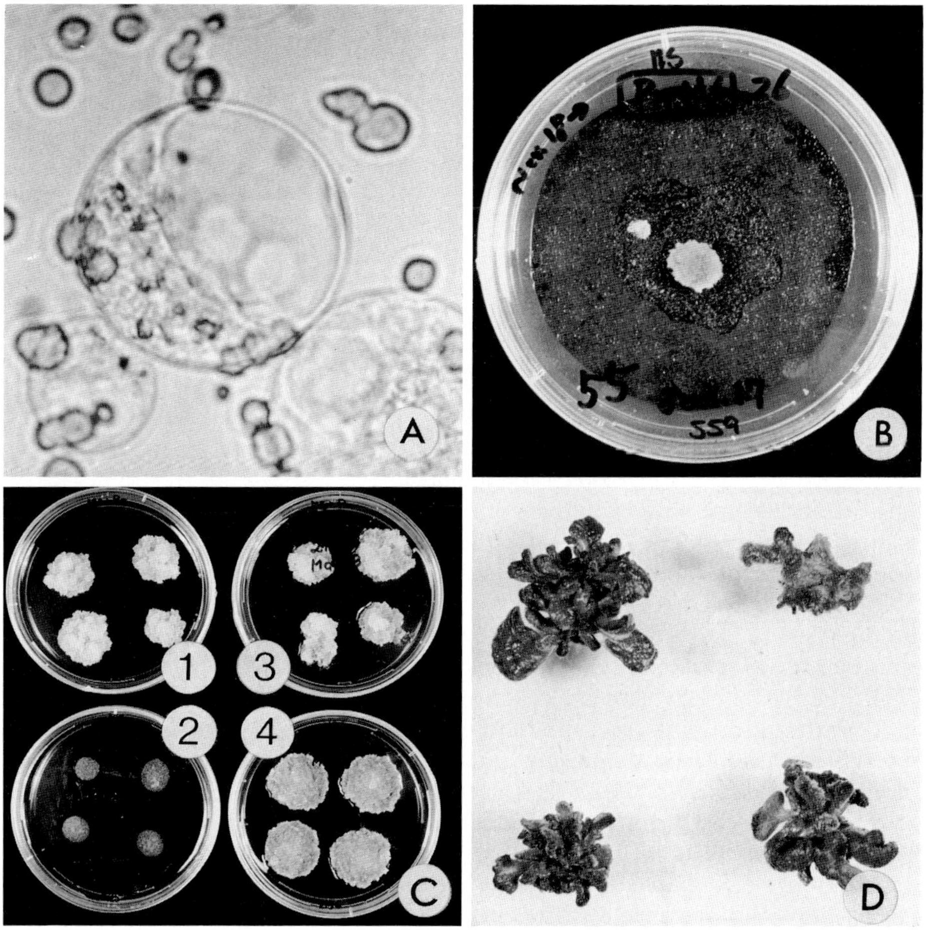


**Fig. 4A–D.** Recovery of prototrophic clones of *Datura innoxia* following transplantation of nuclei of *Vicia hajastana* into auxotrophic protoplasts. **A** Incubation of isolated protoplasts and nuclei in the presence of $Ca^{2+}$-PEG solution. **B** Selection of prototrophic clones on a feeder plate. Note that except for two survivors, the whole cell population is dead. **C** Comparison of the growth of prototrophic clones with that of the auxotroph. The auxotroph grew well on the pantothenate-supplemented medium (*1*) and died on minimal medium (*2*). The prototrophic clone grew equally well on the supplemented (*3*) and the minimal medium (*4*). **D** Regeneration of shoots from prototrophic clones. (Saxena et al. 1986)

for growth because of a block in the metabolic pathway leading to coenzyme A (King et al. 1980). In addition, Pn1 has never reverted to wild type since its isolation, thus making it a suitable selection system for nuclear transfer since only those cells transformed to a prototrophic state would grow on a medium lacking pantothenate. The uptake of nuclei into protoplasts (Fig. 4A) was induced by polyethylene glycol and $CaCl_2 \cdot 2H_2O$ at pH 6.8. Two key factors in the recovery of prototrophic clones following nuclear transplantation were the initial culture of protoplasts in pantothenate-supplemented medium and the use of feeder plates for selection to avoid the adverse effects of low cell density caused by the death of Pn1 cells on medium lacking pantothenate (Fig. 4B). Correction of auxotrophy (Fig. 4C), restoration of morphogenetic potential (Fig. 4D) of the otherwise nonregenerable Pn1 and detection of *Vicia* DNA in the Pn1 genome were the biological evidence for the expression of the transferred DNA. A great deal of variation was observed in the regeneration potential of the resulting prototrophic clones. Such variation was probably a result of random incorporation of unknown amounts of nuclear DNA.

## 7 Concluding Remarks

An overview of the information presently available on the uptake, integration, expression, and replication of nuclei leads to the conclusion that we are a long way from an adequate understanding of organelle transfer. Major constraints in the progress in this area have been the less than satisfactory methods of inducing uptake – compare the ca. 30%–40% frequency of protoplast fusions with the 5%–6% of nuclear uptake, and the lack of suitable selection systems and selectable nuclear markers. A general lack of enthusiasm for organelle uptake due to the rapid advances in *Agrobacterium*- and microprojectile-mediated and direct gene transfer (Krens et al. 1983; Herrera-Estrella et al. 1983; Paszkowski et al. 1984; Klein et al. 1988) has also resulted in slow progress.

The first priority at this point should be to develop conditions for efficient uptake with little or no damage to the biological activity of the nuclei and the receptor protoplasts. So far, there is no report of a comparison of the biological state of nuclei before, during, and after uptake. Generally, PEG causes damage to the nuclei, as seen with the light and electron microscopes (unpubl.), which may result in poor uptake. Considering the possibility of nuclear uptake by other means, the technique of electrofusion (Zimmermann and Scheurich 1981) may have merit. Microinjection of nuclei into protoplasts is also an interesting possibility. Steinbiss and Stabel (1983) developed, for the first time, the conditions to transfer a fluorescent dye, lucifer yellow, into plant cells by microinjection. The technique has been successfully applied to transfer whole chromosomes into protoplasts (Griesbach 1987; DeLaat and Blaas 1987). Microinjection of nuclei may be difficult because of their larger size, but may be suitable for smaller nuclei or for nuclear fragments. Similarly, it remains to be seen whether or not the recent advances in protoplast fusion (Kao 1986) and techniques of inducing uptake of DNA molecules, e.g., electroporation and heat shocks (Potrykus et al. 1985) can be, if at all, modified for nuclear uptake. A search for physical/chemical stimuli to counteract damage to plasma membrane due to the shocks involved in these approaches should be useful. A further alternative to protect the nuclei is to encapsulate them in liposomes before transfer, as has been successfully achieved in animal systems (Kondorosi and Duda 1980). Here again, the large size of the nuclei is a problem.

The choice of experimental systems is also very important, as shown previously (Saxena et al. 1986). The use of haploid nuclei and receptor protoplasts should be useful as a way to avoid the problems associated with polyploidy and the loss of regeneration potential in the resulting nuclear hybrids. In addition to the host selection system, the employment of nuclei with selectable markers, such as with resistance to an antibiotic (Horsch et al. 1984), is also desirable.

In conclusion, significant progress has been made during the last decade towards the isolation of pure nuclei from protoplasts and their transplantation into protoplasts. Given that it should be possible to work out optimal conditions of nuclear uptake and their incorporation into the host genome, it should be possible to study the interaction of isolated nuclei with foreign cells. It would also seem that nuclear uptake could become an additional method for genetic transformation, i.e., besides those involving direct, *Agrobacterium*- and microprojectible-mediated systems. The technique itself may be useful for creating genetic variability in highly inbred genomes, if not for transfer of specific genetic information, as it allows random and uncontrolled transfer of undefined amounts of DNA.

## 8 The Protocol for Isolation and Transplantation of Nuclei

Given below is the protocol being routinely used in our laboratory for the isolation of nuclei from a variety of cells and their transplantation into protoplasts. Some modifications may be needed to apply these procedures to different systems, depending upon their specific nature and requirements.

A) Isolate protoplasts using an appropriate enzyme solution of 0.5–0.6 M osmolarity and purify by floatation over a dense (20%) sucrose solution using centrifugation (120 g for 5 min). All subsequent steps of nuclear isolation should be carried out at 4 °C.

B) Suspend the protoplasts from which the nuclei are to be isolated in the Nuclei Isolation Buffer (NIB) containing 0.2 M sucrose, 10 mM MES, 10 mM each of NaCl and KCl, 0.1 mM spermine, 2.5 mM each of ethylenediaminetetraacetic acid and dithiothreitol and 0.01% Triton X-100 for 5–7 min.

C) Rupture the protoplasts by transferring the suspension to a glass homogenizer and applying 10–15 gentle strokes. Alternatively, pass the suspension through a needle of 18–26 gauge several times.

D) Filter the suspension through two layers of Miracloth followed by polycarbonate filters of 12 and 10-μm. Pellet the nuclei by centrifugation at 120–150 g for 10 min and resuspend in NIB lacking Triton X-100.

E) Mix the host protoplasts and isolated nuclei in a ratio of 1 : 50 and place 0.5 ml of this suspension in a 35-mm diameter Petri dish. Allow the protoplasts and the nuclei to settle for 5 min. Add five drops (100 μl each) of the Uptake Inducing Solution (UIS) at 30-s intervals. The UIS should contain 25% PEG (MW 4000), 0.1 M calcium chloride, 0.05 M glycine and 5% dimethylsulfoxide. Incubate for 15–20 min.

F) Dilute the UIS by adding 2 ml of 0.35–0.45 M mannitol solution containing 5 mM calcium chloride five times at intervals of 5 min each. Centrifuge the diluted suspension to pellet the protoplasts, purify over 20% sucrose solution and suspend in appropriate culture medium.

*Acknowledgments.* The research work included in this article was supported by an operating grant from the Natural Sciences and Engineering Research Council of Canada to JK. Photographic and typing assistance from Dennis Dyck and Evelyn Peters is gratefully acknowledged.

## References

Anderson JW (1968) Extraction of enzymes and subcellular organelles from plant tissues. Phytochemistry 7:1973–1988

Blobel G, Potter VR (1966) Nuclei from rat liver – isolation method that combines purity with high yield. Science 154:1662–1665

Brown MS, Anderson RGW, Goldstein JL (1983) Recycling receptors: The roundtrip itinerary of migrant membrane proteins. Cell 32:663–667

D'Alessio G, Trim AR (1968) A method for the isolation of nuclei from leaves. J Exp Bot 19:831–839

DeLaat AMM, Blaas J (1987) An improved method for protoplast micro-injection suitable for transfer of entire plant chromosomes. Plant Sci 50:161–169

Dunham VL, Bryant JA (1983) Nuclei. In: Hall JZ, Moore AL (eds) Isolation of membranes and organelles from plant cells. Academic Press, New York, pp 237–275

Fowke LC (1986) Ultrastructural cytology of cultured plant tissues, cells, and protoplasts. In: Vasil IK (ed) Cell culture and somatic cell genetics of plants. Academic Press, New York, pp 323–342

Fowke LC, Gamborg OL (1980) Applications of protoplasts to the study of plant cells. Int Rev Cytol 68:9–51

Fowke LC, Marchant HJ, Gresshoff PM (1981) Fusion of protoplasts from carrot cell cultures and green alga *Stigeoclonium*. Can J Bot 59:1021–1025

Giles KL (1978) The uptake of organelles and microorganisms by plant protoplasts: Old ideas but new horizons. In: Thorpe TA (ed) Frontiers of plant tissue culture 1978. University of Calgary Press, Calgary, Alberta, Canada, pp 67–74

Griesbach RJ (1987) Chromosome-mediated transformation via microinjection. Plant Sci 50:69–77

Griesbach RJ, Malmberg RL, Carlson PS (1982) An improved technique for the isolation of higher plant chromosomes. Plant Sci Lett 24:55–60

Hadlaczky G (1984) Isolation of organelles: Chromosomes. In: Vasil IK (ed) Cell culture and somatic cell genetics of plants, Vol 1. Academic Press, New York, pp 461–470

Hadlaczky G, Bisztray G, Praznovszky T, Dudits D (1983) Mass isolation of plant chromosomes and nuclei. Planta 157:278–285

Hasezawa S, Matsui C, Nagata T, Syono K (1983) Cytological study of the introduction of *Agrobacterium tumefaciens* spheroplasts into *Vinca rosea* protoplasts. Can J Bot 61:1052–1057

Hendriks AW (1972) Purification of plant nuclei using colloidal silica. FEBS Lett 24:101–105

Herrera-Estrella L, Depicker A, van Montagu M, Schell J (1983) Expression of chimaeric genes transferred into plant cells using a Ti-plasmid-vector. Nature 303:209–213

Hewish DR, Burgoyne LA (1973) Chromatin structure. The digestion of chromatin at regularly placed sites by a nuclear decarboxynuclease. Biochem Biophys Res Commun 52:504–510

Honda SI, Hongladurom J, Laties GG (1966) A new isolation medium for plant organelles. J Exp Bot 17:460–472

Horsch RB, Fraley RT, Rogers SG, Sanders PR, Lloyd A, Hoffmann N (1984) Inheritance of functional foreign genes in plants. Science 227:496–498

Hughes BG, Hess WM, Smith MA (1977) Ultrastructure of nuclei isolated from plant protoplasts. Protoplasma 93:267–274

Joachim SA, Robinson DG (1984) Endocytosis of cationic ferritin by bean leaf protoplasts. Eur J Cell Biol 34:212–216

Kao KN (1986) Fusion of plant protoplasts at the interface of a glucose and a sucrose-polyethylene glycol solution. J Plant Physiol 126:55–58

King J, Horsch RB, Savage AD (1980) Partial characterization of two stable auxotrophic cell strains of *Datura innoxia* (Mill.). Planta 149:480–484

Klein TM, Gradziel T, Fromm ME, Sanford JC (1988) Factors influencing gene delivery into *zea mays* cells by high-velocity microprojectiles. Bio/Technol 6:559–563

Kobza J, Edwards GE (1984) Isolation of organelles: Chloroplasts. In: Vasil IK (ed) Cell culture and somatic cell genetics of plants, Vol 1. Academic Press, New York, pp 471–482

Kondorosi E, Duda E (1980) Introduction of foreign genetic material into cultured mammalian cells by liposomes loaded with isolated nuclei. FEBS Lett 120:37–40

Krens FA, Wullems GJ, Schilperoot RA (1983) Transformation of plant protoplasts in vitro. In: Ciferri O, Dure L III (eds) Structure and function of plant genomes. Plenum, New York, pp 387–408

Liu Y (1985) The isolation of nuclei from cultured plant cells and their uptake into plant protoplasts. M Sc Thesis, Univ Saskatoon, Canada

Lörz H (1985) Isolated cell organelles and subprotoplasts – their role in somatic cell genetics. In: Dodds JH (ed) Plant genetic engineering. Cambridge University Press, London, pp 27–59

Lörz H, Potrykus I (1978) Investigations on the transfer of isolated nuclei into plant protoplasts. Theor Appl Genet 53:251–256

Matsui C, Hasezawa S, Tanaka N, Syono K (1983) Introduction of *Escherichia coli* cells and spheroplasts into *Vinca* protoplasts. Plant Cell Rep 2:20–32

Melchers G, Labib G (1974) Somatic hybridization of plants by fusion of protoplasts. I. Selection of light resistant hybrids of "haploid" light sensitive varieties of tobacco. Mol Gen Genet 135:277–294

Mii M, Saxena PK, Fowke LC, King J (1987) Isolation of chromosomes from cell suspension cultures of *Vicia hajastana* Grossh. Cytologia 52:523–528

Ohyama K, Pelcher PE, Horn D (1977) A rapid, simple method for nuclei isolation from plant protoplasts. Plant Physiol 60:179–181

Paszkowski J, Shillito RD, Saul MW, Mandak V, Hohn T, Hohn B, Potrykus I (1984) Direct gene transfer to plants. EMBO J 3:2717–2722

Pertoft H, Laurent TC (1977) Isopycnic separation of cells and cell organelles by centrifugation in modified colloidal silica gradients. In: Catsimpoolas N (ed) Methods of cell separation, Vol 1. Plenum, New York, pp 22–66

Potrykus I (1973) Transplantation of chloroplasts into protoplasts of *Petunia.* Z Pflanzenphysiol 70:364–366

Potrykus I, Lörz H (1976) Organelle transfer into isolated protoplasts. In: Dudits D, Farkas GL, Maliga P (eds) Cell genetics in higher plants. Academica Kiado, Budapest, pp 183–190

Potrykus I, Shillito RD, Saul MW, Paszkowski J (1985) Direct gene transfer: State of the art and future potential. Plant Mol Biol Rep 3:117–120

Price CA (1979) A note on isolation of plant nuclei. In: Reid E (ed) Plant organelles (Methodological survey in biochemistry Vol 9). John Wiley, New York, pp 200–206

Savage AD, King J, Gamborg OL (1979) Recovery of a pantothenate auxotroph from a cell suspension culture of *Datura innoxia* (Mill.). Plant Sci Lett 16:367–376

Saxena PK, Fowke LC, King J (1985a) An efficient procedure for isolation of nuclei from plant protoplasts. Protoplasma 128: 184–189

Saxena PK, Liu Y, Mii M, Fowke LC, King J (1985b) High nuclear yields from protoplasts of several plants. J Plant Physiol 121:193–197

Saxena PK, Mii M, Crosby WR, Fowke LC, King J (1986) Transplantation of isolated nuclei into plant protoplasts – A novel technique for introducing foreign DNA into plant cells. Planta 168:29–35

Saxena PK, Liu Y, King J (1987) Nuclear transplantation into protoplasts: Optimal conditions for induction and determination of nuclear uptake. J Plant Physiol 128:451–460

Steinbiss HH, Stabel P (1983) Protoplast-derived tobacco cells can survive capillary microinjection of the fluorescent dye lucifer yellow. Protoplasma 110:223–227

Suzuki M, Takebe I, Kajita S, Honda Y, Matsui C (1977) Endocytosis of polysterene spheres by tobacco leaf protoplasts. Exp Cell Res 105:127–135

Szabados L, Hadlaczky G, Dudits D (1981) Uptake of isolated plant chromosomes by plant protoplasts. Planta 151:141–145

Tallman G, Reeck GR (1980) Isolation of nuclei from plant protoplasts without the use of a detergent. Plant Sci Lett 18:271–275

Tanchak MA, Griffing LR, Mersey BG, Fowke LC (1984) Endocytosis of cationized ferritin by coated vesicles of soybean protoplasts. Planta 162:481–486

Tautvydas KJ (1971) Mass isolation of plant nuclei. Plant Physiol 47:499–503
Willmitzer L (1984) Isolation of organelles: Nuclei. In: Vasil IK (ed) Cell culture and somatic cell genetics of plants (Vol 1). Academic Press, New York, pp 454–460
Willmitzer L, Wagner KG (1981) The isolation of nuclei from tissue-cultured plant cells. Exp Cell Res 135:69–77
Zimmermann U, Scheurich P (1981) High frequency fusion of plant protoplasts by electric fields. Planta 151:26–32
Zuily-Fodil Y, Passaquet C, Fesnault R (1978) High yield isolation of nuclei from plant protoplasts. Physiol Plant 43:201–204

# XX Chromosome Transplantation and Applications of Flow Cytometry in Plants

A. M. M. DE LAAT[1], H. A. VERHOEVEN[1], and K. SREE RAMULU[1]

## 1 Introduction

Conventional plant breeding is gradually supplemented with novel techniques, e.g., in vitro propagation, selection at the cellular level and genetic manipulation. Genetic engineering has become realistic with the recent progress in molecular biology, cell biology, and tissue culture. By circumvention of sexual reproduction, the natural barriers of incompatibility can be bypassed. Whereas conventional breeding deals with combining whole genomes, differential transfer of genetic traits has become possible now. In the case where a monogenic trait has been characterized at the molecular level, the identified gene can be cloned and transferred into plant cells via DNA vectors (e.g., the *Agrobacterium*-derived Ti or Ri plasmids (Depicker et al. 1983) or viruses like CaMV (Brisson et al. 1984), or by direct DNA-uptake (Krens et al. 1982; Potrykus et al. 1985). The genetically modified cells are subsequently regenerated to transgenic plants.

The possibilities for identification of plant genes which code for economically important crop characters are limited. Moreover, many important traits like yield, stress tolerance, etc. are polygenically determined, and are therefore not amenable yet for handling by the DNA transformation techniques. Strategies for the identification of plant genes suffer from the inadequate knowledge of plant biochemistry, as well as from the size and complexity of the plant genome (the presence of noncoding sequences, silenced genes, etc.).

Protoplast fusion offers the possibility to transfer nonidentified or polygenically determined traits across sexual borders. However, as in conventional breeding, undesired characters have to be eliminated by repeated back-crossing of somatic hybrids. Irradiation of one of the parental lines before fusion enhances the loss of the foreign DNA (Famelaer et al. 1986). However, this occurs in a random manner, and might cause genetic damage.

Transplantation of individual chromosomes would partially solve this problem by the introduction of *a specific part* of a foreign genome. Moreover, chromosome transplantation might result in addition lines and provides a method for fine structural mapping of chromosomes. As proven in mammalian systems, the mass isolation and sorting of plant chromosomes by flow cytometry would allow the construction of chromosome-specific DNA libraries (Davies et al.

[1] Research Institute Ital, P.O. Box 48, 6700 AA Wageningen, The Netherlands

Biotechnology in Agriculture and Forestry, Vol. 9
Plant Protoplasts and Genetic Engineering II (ed. by Y. P. S. Bajaj)

1981). These will be very useful for the identification and molecular cloning of important plant genes.

In this chapter methods for isolation and transfer of plant chromosomes are presented, including the authors' recent results on the subject. Further, the application of flow cytometry for sorting of plant chromosomes and the prospects of chromosome transfer for plant breeding are discussed.

## 2 Flow Cytometry and Its Application

### 2.1 Instrumentation

Flow cytometry is derived from microdensitometry and microfluorimetry. Whereas in microscopy an object is positioned between a light source and an analysis system, in flow cytometry individual objects are introduced in an ordered liquid flow, which is intercepted by one or more light beams (e.g., lasers). Upon passage of an interception point, a particle may cause light scatter, absorption, and depolarization. Moreover, if excited at the appropriate wavelength, endogenous fluorescent compounds (e.g., aromatic secondary metabolites) or exogeneously applied fluorescent dyes with specificity for DNA, RNA, or specific proteins will give rise to fluorescence at a specific wavelength. These signals are detected and quantified by an optical system including photomultipliers (Fig. 1). The fluid stream might also contain a Coulter gate which allows measurements of cell volume by impedance changes. A connected computer allows analysis up to 5000 events per second and serves for data acquisition and statistical analysis. Multiple parameter analysis may yield information about several characteristics including size, shape, vitality, and endogenous content of DNA, RNA, proteins, secondary metabolites, etc.

Besides the analysis, current flow cytometers are also suitable for sorting. According to the object's signals and the setting of "sorting gates" for the relevant parameter(s), a sorting device is driven by the controlling computer.

Sorting may be based on hydrodynamic switching of the particle flow in a closed sorting flow chamber (Partec, Arlesheim, Switzerland). Alternatively, a vertical free flow stream is subjected to vibration to break into drops. Upon passage of a particle of interest, the fluid stream is charged temporarily at the moment that the object ends up in a free droplet. Only the charged droplets containing particles of interest are deflected in an electric field. For a more detailed

**Fig. 1.** Schematic presentation to the principle of flow cytometry and sorting. The sample is passed from the tube of the nozzle tip, where the hydrodynamic forces, caused by the passage of the sheath fluid, force the particles into a very narrow liquid stream. A vibrating piezo crystal forces the liquid stream to form small droplets. Before breaking into droplets, the particle containing liquid jet passes the laser beam, and the fluorescence of the particles along with the scattered light is detected by photomultiplier tubes and passed to an electronic processor. According to the machine settings, the selected sorting windows generate electric pulses for charging of the liquid droplets at the time of their formation. Thus, a negative charge will result in deflection to the right, whereas a positive charge will be deflected to the left. The uncharged droplets will not be deflected at all, and pass to the waste collection

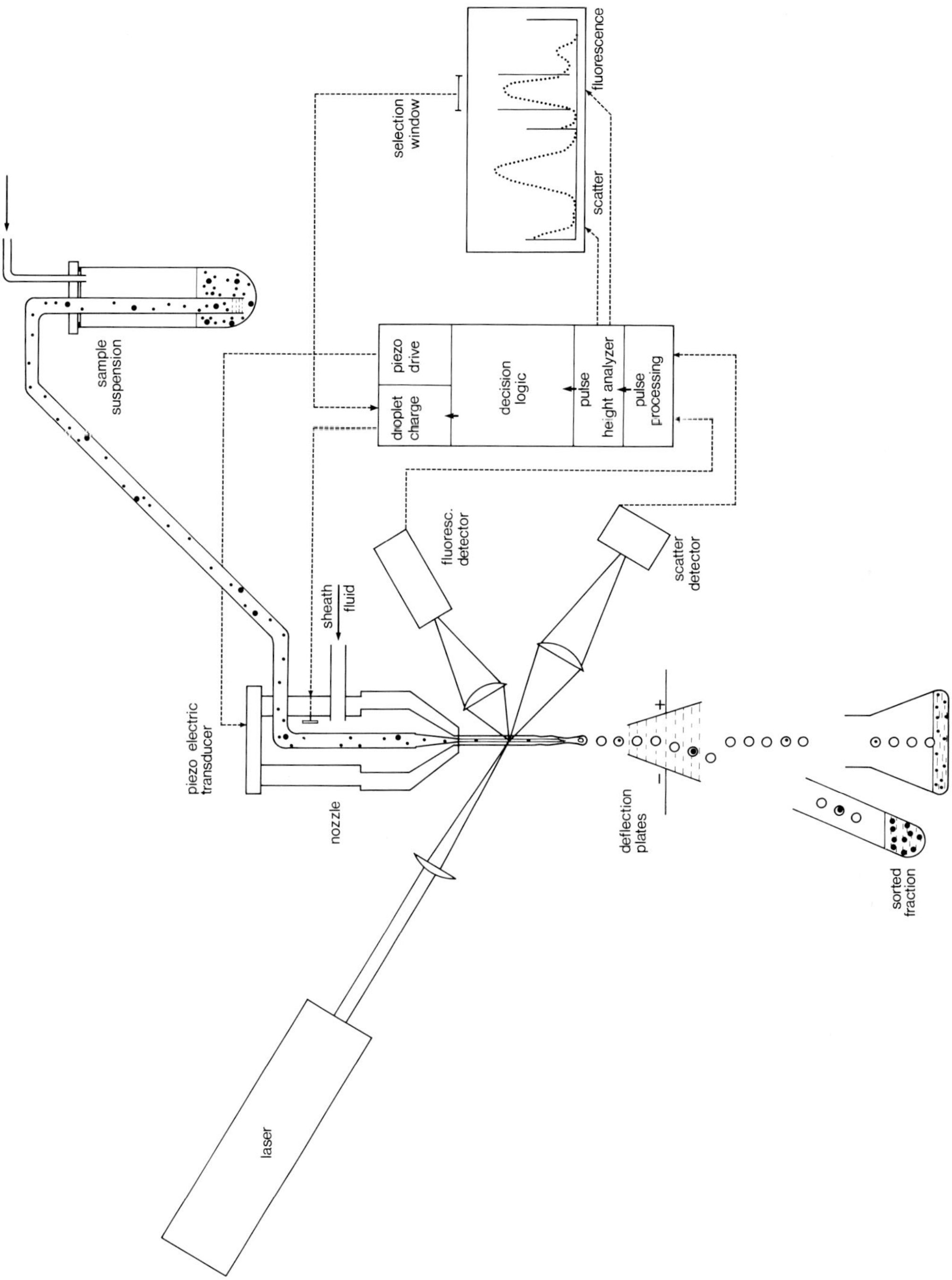

selection window
scatter
fluorescence
sample suspension
piezo drive
droplet charge
decision logic
pulse height analyzer
pulse processing
fluoresc. detector
scatter detector
sheath fluid
piezo electric transducer
nozzle
deflection plates
sorted fraction
laser

description of the principles of flow cytometry, and its potential in cell biological research, see Melamed et al. (1979) and Muirhead and Hovan (1984) respectively.

## 2.2 Application in Plant Biotechnology

A major limitation in flow cytometry is the maximum size for particles to be analyzed and sorted which does not allow the use of cell aggregates or tissues. With conventional cell sorters plant protoplasts are damaged severely by the high hydrodynamic forces due to their relatively large size and high fragility. Because of the prospects by cellular manipulations for genetic engineering of plants, special attention was given to modify the conventional instruments to be able to use plant protoplasts. Therefore, the flow tip orifices were changed from about 70 μm up to 200 μm (Jett and Alexander 1985; Harkins and Galbraith 1987). For this modification other technical changes had to be carried out (e.g., sheath pressure, vibration frequency, etc.).

Recently, an instrument using a closed flow chamber with a mechanical sorting system was developed. The company Partec (Arlesheim, Switzerland) claims that particles up to 500 μm size can be analyzed and sorted without damage due to shearing.

For some years now, flow cytometry has been applied successfully for plant materials, including pollen, protoplasts, nuclei, chromosomes, and mitochondria.

From the rapid progress that has been made in flow cytometry, it is clear that the technique is indeed highly useful in many different aspects of both fundamental plant sciences and applied plant biotechnology.

### 2.2.1 DNA Analysis

Flow cytometry can be used for rapid measurement of DNA content of nuclei or chromosomes in order to study synchronization of cell cultures, cell cycle patterns, ploidy levels, and karyotyping. In the last few years several authors have applied flow cytometry to various research areas (Table 1), documenting the unique potential of flow cytometry.

### 2.2.2 Identification and Sorting of Heterokaryons After Protoplast Fusion

Using differential fluorescent staining of both fusion partners, flow cytometry offers the opportunity for mass selection of heterokaryons from protoplast fusion mixtures, based on the coincidence of the two different fluorescent signals providing the desirable homogeneous populations of heterokaryons for quantitative development. Several factors, e.g., origin of protoplasts (mesophyll vs. cell suspension protoplasts), type of somatic hybridization (inter- or intraspecific, intergeneric) and the effects of irradiation or chemical treatment used in the case of asymmetric hybridization, might influence the interaction between the fusion partners during heterokaryon development.

**Table 1.** Recent achievements in the application of flow cytometry in plants

| Type of plant material | Achievements | Reference |
|---|---|---|
| Pollen | Detection and sorting of pollen mutants | Pinkel (1981) |
| Protoplasts | Adaptation of commercial flow cytometric instruments for viable sorting | Harkins and Galbraith (1984); Jett and Alexander (1985); Glimelius et al. (1986); Harkins and Galbraith (1987) |
| | Sorting of viable heterokaryons | Afonso et al. (1985); Alexander et al. (1985) |
| Cells, nuclei | Ploidy measurements | Galbraith and Shields (1982); Sharma et al. (1983); Carlberg et al. (1984); Sundberg and Glimelius (1986); Sree Ramulu and Dijkhuis (1986); Hänisch ten Cate and Sree Ramulu (1987) |
| | Cell cycle | Meadows (1982); Galbraith et al. (1983); Puite and Ten Broeke (1983); Galbraith (1984); Anderson et al. (1985); Karlsson and Vasil (1986); Bergounioux et al. (1986) |
| | Selection of cells containing specific secondary products | Brown et al. (1984, 1986) |
| Chromosomes | Karyotyping, sorting of individual chromosomes | de Laat and Blaas (1984); Petit et al. (1986); Conia et al. (1987) |

It has been shown that heterokaryons can be sorted with a purity of 80%–100% (Glimelius 1987). When producing somatic hybrids using polyethylene glycol and/or high pH, high $Ca^{2+}$ treatment, the protoplasts adhere to each other strongly (false hybrids). It is difficult to discriminate between the completely fused protoplasts and the false hybrids. Therefore, the fusion mixture has to be of high purity (review in Glimelius 1987). Heterokaryons can be directly sorted into culture vials.

## 3 Chromosome Isolation and Sorting

The work of Szabados et al. (1981) on the polyethylene glycol (PEG)-induced uptake of mass-isolated chromosomes has shown that at least the selectable marker on the donor chromosome can be transferred and maintained. Only after recovery of the transformed cells can the presence of a donor chromosome be identified cytologically. In the case where extensive chromosome rearrangements have taken place, karyotyping is difficult, making this approach less useful for chromosome mapping.

The application of flow cytometry can solve this problem by sorting the chromosomes of the donor parent in pure suspensions of single chromosomes. With such chromosome preparations, one can identify the transferred chromosomes belonging to a specific known population, allowing chromosome mapping.

Chromosome isolation and flow sorting could be successfully carried out in *Haplopappus gracilis* (de Laat and Blaas 1983; Puite et al. 1983; de Laat and Blaas 1984), *Nicotiana plumbaginifolia* (Verhoeven et al. 1987; van der Valk and

Verhoeven 1988) and *Petunia hybrida* (Conia et al. 1987). The basic requirements will be that the karyotype of the species has divergence in chromosome lengths and difference in DNA contents of the chromosomes, thus making the chromosomes amenable for sorting.

Procedures for mass isolation of chromosomes from species such as *Lycopersicon esculentum, Vicia faba, Zea mays, Pisum sativum, Hemerocallis* hybrids, *Allium cepa, Triticum monococcum, Papaver somniferum*, and *Lilium* have been standardized (Griesbach et al. 1982; Hadlaczky et al. 1983), but so far no sorting has been reported. In view of their importance, chromosome isolation and sorting procedures are given below in some detail for *H. gracilis*. With some modifications, the protocol can also be applied to *N. plumbaginifolia*.

### 3.1 Cell Cycle Synchronization

As is known, *H. gracilis* is a highly suitable model plant for studies on cell growth and cytology. It has a diploid number of four large chromosomes, the two pairs being distinguishable as meta- and acrocentrics (Jackson 1957).

Only metaphase cells are suitable for chromosome isolation. In plant cells, mitotic rate in culture is generally low, and therefore it is necessary to synchronize the cells, e.g., with inhibitors of DNA synthesis, such as hydroxyurea and aphidicolin. Therefore, log-phase cell cultures of *H. gracilis* were synchronized essentially according to Eriksson (1966) by incubation in 5 mM hydroxyurea to inhibit passage through S-phase. After 18 h of hydroxyurea treatment, the cells were thoroughly washed with culture medium, and resuspended in fresh culture medium. To arrest cells in metaphase, the spindle toxin colchicine was added immediately after washing (final concentration 0.05%).

Cells were incubated at 28 °C during 7 h, and afterwards, the temperature was reduced to 12 °C to prevent chromosome decondensation. After 8 h of incubation at the reduced temperature, mitotic indices of about 25% were routinely obtained (de Laat and Blaas 1984). To follow the process of cell synchronization, flow cytometric analysis of the cellular DNA content was carried out of burst protoplasts after various incubation periods. In log-phase suspensions, cells were mainly in G0/G1 phase (approx. 80%) whereas only a few cells were found in S (5%) and G2/M phase (15%). After incubation with hydroxyurea, all cells were accumulated at the G1/S border. After washing the cells free from hydroxyurea, most of the cells resumed DNA synthesis. Due to the action of colchicine, all dividing cells were arrested in metaphase.

Alternatively, Conia et al. (1987) used the spontaneously occurring synchrony of the first cell division of regenerating *Petunia* protoplasts, combined with a colchicine treatment to accumulate chromosomes at metaphase.

### 3.2 Isolation of Metaphase Chromosomes

Efforts to isolate chromosomes directly from the metaphase cells by mechanical disruption did not yield sufficient numbers of intact chromosomes. Moreover, high contamination of the chromosome isolates with nuclear fragments and cell debris occurred persistently. Since the cell wall is the main barrier for the isolation, protoplasts were prepared by treatment with cell wall digesting enzymes (3 h) (5% driselase, 0.5% pectolyase) in a 0.7 M mannitol solution (pH 5.6).

However, during protoplast preparation from mitotic cells at 28 °C, a rapid decondensation of metaphase chromosomes was observed, resulting in a marked decrease of the mitotic index. This could be prevented by changing the temperature from 28 °C to 12 °C and using a suitable buffer solution (3 mM MES, 5 mM $MgCl_2$, 2 mM $CaCl_2$, 370 mM mannitol and 250 mM glucose at pH 5.6) during a short, 2-h enzymatic treatment. After this short period, the cell wall was not removed completely, but it was weakened sufficiently to be disrupted easily. All subsequent steps were carried out on ice to prevent DNA degrading enzymes from digesting the metaphase chromosomes. Before disruption, the cells were thoroughly washed and swelled in 300 mM Mannitol during 10 min. The cells were pelleted and transferred into chromosome isolation buffer (15 mM HEPES, 1 mM EDTA, 15 mM DTT, 0.5 mM spermine, 80 mM KCl, 20 mM NaCl, 300 mM sucrose, 50 μg/ml EtBr (a DNA-specific intercalating fluorescent stain), and 1% Triton X-100, and disrupted by passage through a 26-G epidermic needle. Fluorescent microscopic observations revealed that these crude preparations contain large numbers of intact chromosomes, nuclei, and several low fluorescent debris particles of variable size.

### 3.3 Flow Cytometric Analysis and Sorting

Ethidiumbromide was used for flow cytometry, and during chromosome isolation, as DNA-stabilizing factor, which can be removed by washing the chromosomes, probably without affecting the biological activity of the chromosomes. The Triton X-100 (1%) was added to the chromosome isolation buffer to prevent clumping of the various constituents.

The flow cytometric analysis of crude protoplast extracts showed a DNA histogram with eight peaks (Fig. 2). The nature of these peaks was identified by sorting, and subsequent fluorescence microscope examination. Peak A consisted largely of debris, the combined peak BC contained the acrocentric chromosomes, whereas peaks D and E contained each one of the two homologous metacentric chromosomes. The measurements of chromosome lengths revealed that the double peaks resulted from differences between the homologous chromosomes already existing in $>$3-year-old cell suspension because of chromosome-structural alterations as a result of genetic instability (Sree Ramulu 1986). By precise length measurements of the chromosomes in a nonsynchronized cell suspension, a similar histogram for the chromosomes was obtained. These alterations thus did not occur during any stage of the cell cycle synchronization, or chromosome isolation and sorting procedure. The coefficient of variation of the chromosome peaks was routinely less than 5%. The remaining three peaks appeared to contain G1 nuclei (F), metaphase nuclei (aggregates of four chromosomes) (G), and G2 nuclei (H). The fluorescence intensity of metaphase nuclei (G) is significantly lower than the fluorescence of G2 interphase nuclei (H), although they all contain

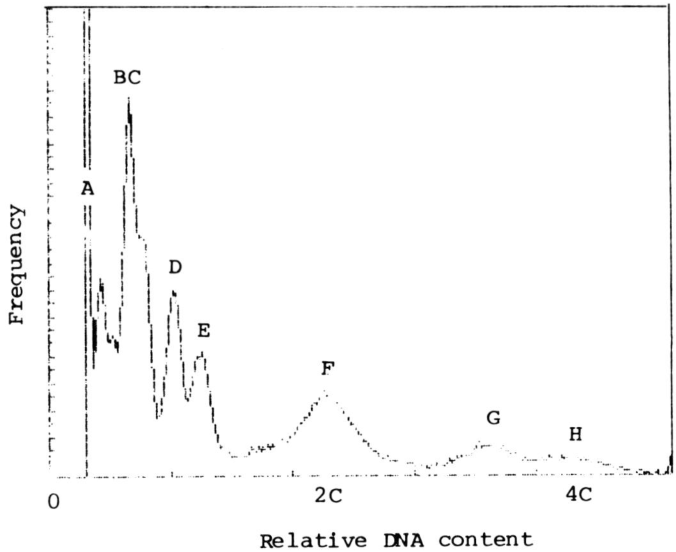


**Fig. 2.** Frequency distribution of the relative DNA content of ethidium bromide-stained chromosomes and nuclei from mechanically disrupted protoplasts derived from synchronized *H. gracilis* suspension cells. By flow sorting, peaks were identified as debris (*A*), acrocentric (*B/C*) and metacentric (*D/E*) metaphase chromosomes, G1 nuclei (*F*), mitotic nuclei (*G*) and G2 nuclei (*H*). (de Laat and Blaas 1984)

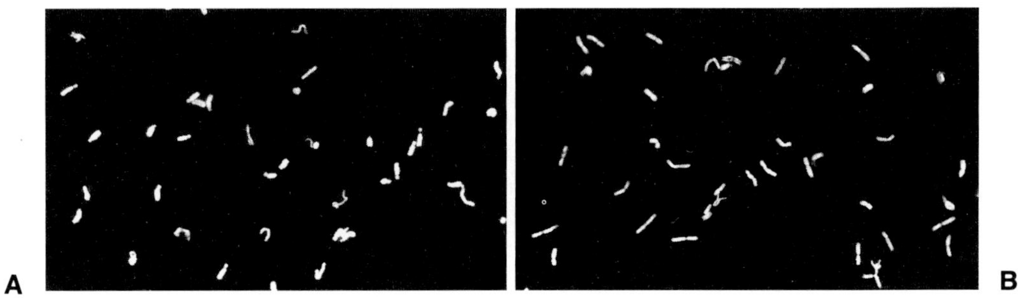


**Fig. 3.** Acrocentric (**A**) and metacentric (**B**) metaphase chromosomes of *H. gracilis*, sorted by flow cytometer. (de Laat and Blaas 1984)

the same amount of DNA. This might be a consequence of the strong condensation of the DNA in metaphase chromosomes, which interferes more strongly with the intercalation of EtBr than less condensed DNA. Also increased quenching of the excitation light by the higher spatial concentration of EtBr could be the cause of the reduced fluorescence.

Acrocentric chromosomes were sorted from peaks B and C (Fig. 3A), and metacentric chromosomes from peaks D and E (Fig. 3B), at a rate of 5000 chromosomes per min. Sorted chromosomes were collected in cooled tubes, restained with EtBr and collected on Teflon filters (Sartorius) with a pore size of

0.2 μm. The filters were embedded in glycerin, and examined by fluorescence microscopy, using an UG-2 exciter filter, a FL-500 dichroic mirror, and a LP-580 emission filter. The sorted fraction of the acrocentric chromosomes showed less than 5% contamination with the metacentric chromosomes, and vice versa.

In the case of *N. plumbaginifolia*, metaphase chromosomes were isolated from cell suspensions after hydroxyurea treatment (10 mM, 24 h) followed by colchicine (0.05%) or amiprophos-methyl (10 ppm) treatment. Colchicine-treated cells mostly resulted in chromosome clumping, whereas after amiprophos-methyl treatment chromosomes scattered throughout the cells, allowing the isolation of individual chromosomes (see Verhoeven et al. 1986; de Laat et al. 1987). The chromosome suspension was stained with ethidium bromide (stoichiometric for DNA content) and Hoechst 33258 (AT-specific) (10 μg/ml), and dual parameter flow cytometric analysis was performed with 360 nm excitation (Verhoeven et al. 1987). This resulted in a distinct three-dimensional pattern showing at least two populations of chromosomes with different AT/(AT+GC) ratios. Further advances in the techniques of dual staining for AT- and GC-rich regions of DNA would achieve progress in sorting of chromosomes of *N. plumbaginifolia* and of other species, which have relatively little variability in the size.

## 4 Characterization of Isolated, Flow-Sorted Chromosomes

### 4.1 Effect of Isolation and Sorting Procedures on Chromosome Integrity

The main objectives for plant chromosome isolation and sorting are the exploration of chromosome transplantation, the construction of chromosome-specific genome libraries, and fine structural mapping of chromosomes. The fate of a chromosome after transfer into a recipient plant cell will largely depend on the physical integrity of the isolated, flow-sorted chromosomes.

During the various steps in the isolation procedure, chromosome damage may occur. From studies with mammalian cells, for instance, it was suggested that hydroxyurea used for cell cycle synchronization causes DNA degradation (Coyle et al. 1970). Since cell walls are a major mechanical barrier for plant chromosome isolation, (partial) digestion of cell walls is needed.

The commercial preparations of cell wall-degrading enzymes (e.g., driselase and pectolyase, often used for protoplast and chromosome isolation) are impure, and substantially contaminated with DNase activity. Larkin and Scowcroft (1981) consider DNA damage in protoplasts due to these enzymes as a possible causative factor for genetic instability occurring during protoplast regeneration.

Several investigators have reported various buffers for plant chromosome isolation. Whereas some authors succeeded in isolating plant chromosomes in alkaline buffers (pH 8.4–8.6) (Matthews 1983), others used only neutral isolation (Griesbach et al. 1982; de Laat and Blaas 1984), which yielded high numbers of morphologically normal chromosomes. For mammalian chromosomes, pH effects were revealed for the intactness of the chromosomal DNA. Considerable DNA damage can occur at high pH (Blumenthal et al. 1979).

Mechanical shearing of the chromosomes occurs during passage of the hypodermic needle for chromosome release from mitotic protoplasts as well as during flow sorting as a consequence of the high hydrodynamic forces throughout the system.

To define the prospects for isolated chromosomes to be successfully used for chromosome transfer, the integrity of isolated, flow-sorted chromosomes (from synchronized cell suspensions of *H. gracilis*) was investigated.

## 4.2 Electron Microscopy and DNA Analysis of Sorted *H. gracilis* Chromosomes

To investigate the morphology of sorted *Haplopappus* chromosomes and the occurrence of double-stranded DNA breaks, EM studies were carried out. Therefore, chromosomes were spread on a hypophase of media, varying in pH and salt concentrations, according to Hadlaczky et al. (1981). After spreading,

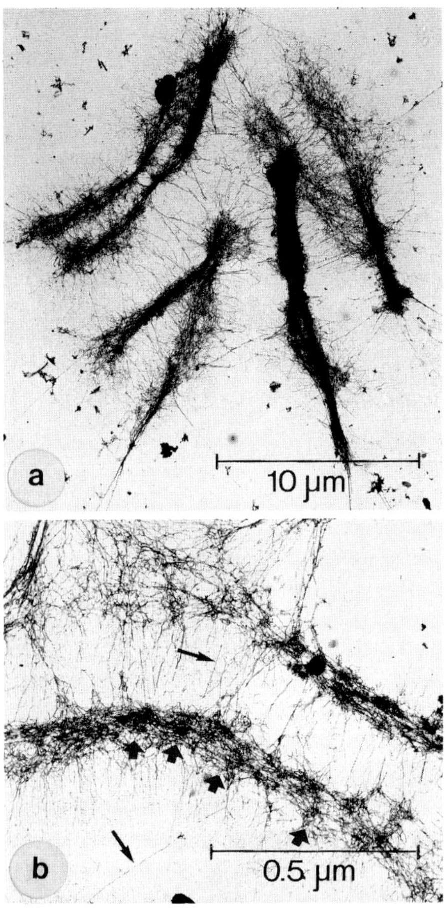


**Fig. 4.** Electron micrographs of whole mount spread of flow sorted mitotic chromosomes of *H. gracilis*. After incubation at pH 9.2 (1% sodium borate), centrifugation on a grid and critical point drying, thin chromatin fibers are spread all over the grid surface (*thin arrows*). On the remaining chromosome framework a banding pattern (*thick arrows*) is visible. (de Laat and Schell 1986)

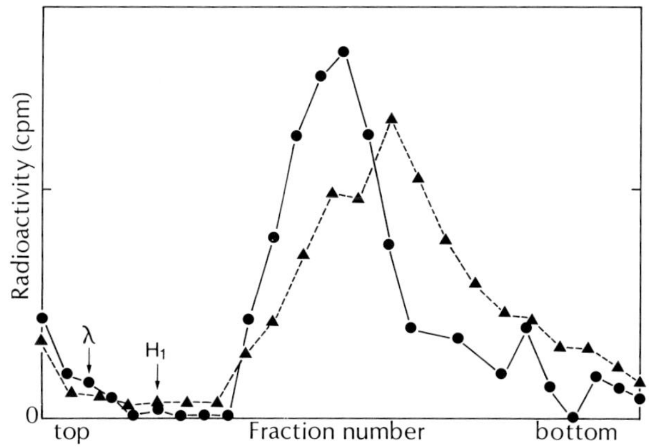


**Fig. 5.** Sedimentation pattern of DNA fragments from isolated nuclei (---) and chromosomes (—) isolated from nonsynchronized and synchronized cell suspensions of *H. gracilis* respectively (de Laat and Shell 1986). The preparations containing $5 \times 10^4$ nuclei and $10^5$ chromosomes respectively, were layered on top of a 5% – 20% alkaline sucrose gradient and centrifuged at 25000 g for 5 h at room temperature. The position of the calibration standards phage $\lambda$ (48 kb) and virus $H_1$ (110 kb) are indicated

chromosomes were recollected by centrifugation on a grid, and critical point dried. Figure 4 shows a spread chromosome (sodium borate treatment at pH 9.2) demonstrating a banding pattern of hetero- and euchromatin. The thin DNA fibers which leaked out of these chromosomes did not show significant damage; long DNA loops could be followed.

According to Blumenthal et al. (1979), who studied the average size of single-stranded DNA fragments isolated from isolated mammalian chromosomes, the DNA analyzed from isolated *Haplopappus* nuclei and chromosomes was studied (de Laat and Shell 1986). Therefore cell suspensions were labeled with tritiated thymidine and isolated, flow-sorted nuclei and chromosomes were lyzed in 0.5 M NaOH, and DNA fragments were separated on alkaline sucrose gradients (5% – 20% linear sucrose gradients in 0.1 M NaOH, 0.9 M NaCl and 0.1 M EDTA; 25000 g for 5 h). After liquid scintillation counting of the various fractions, the sedimentation pattern reveals the presence of very high MW single-strand DNA fragments for both, the nuclei isolated from nonsynchronized cell suspensions and the metaphase chromosomes (Fig. 5). The average fragment size was estimated as $0.4 - 2 \times 10^9$ Da.

Both the EM analysis and the DNA studies strongly indicate that neither the treatment of the suspension cells with the synchronizing agent hydroxyurea and colchicine, nor the treatment with cell wall-degrading enzymes, nor the procedures used for isolation and sorting caused a significant degradation of the chromosomal DNA. Therefore, such isolated chromosomes seem suitable for chromosome-transfer experiments.

## 5 Chromosome Transfer

### 5.1 Methodology

Cell walls form a physical barrier for high molecular weight molecules and particles to enter into cells. Except for the *Agrobacterium*-mediated DNA transfer system, at present only protoplasts are considered as a major target for the development of appropriate techniques for genetic manipulation. For the introduction of DNA, viral genomes, and cell organelles (mitochondria, plastids, nuclei, chromosomes) a variety of methods are currently under investigation. Some of these are outlined below, with the main emphasis on chromosome transfer.

### 5.2 Chemically Induced Uptake

PEG-induced uptake of isolated plant organelles e.g., nuclei (Lörz and Potrykus 1978; Saxena et al. 1986) and chromosomes (Szabados et al. 1981; Dudits and Praznovsky 1985)) has been described, but transfer frequencies were low. Efficient uptake of DNA and viral RNA was achieved by using chemical fusogens like polyethyleneglycol (PEG) (Krens et al. 1982) or by electroporation (Potrykus et al. 1985). Moreover, the PEG treatment seriously damaged the organelles and the transferred organelles are surrounded by a plasmalemma-derived membrane or are lacking the outer membrane due to endocytosis-like uptake or fusion with the plasmalemma respectively. For nuclei, stable transformation occurred at very low frequencies (Dudits and Praznovsky 1985; Saxena et al. 1986). The evidence from literature on mammalian chromosome transfer suggests that the transferred chromosomes are not integrated as a whole into the recipient genome, but only a fragment of the chromosome is integrated (McBride and Peterson 1980). The size of the transferred fragment (transgenome) ranged from less than 0.1% to 1% of the haploid genome. Dudits and Praznovsky (1985) described that in plants, the transplanted chromosome itself was soon lost. The genetic marker, however, localized on that chromosome, could be detected long afterwards.

### 5.3 Microinjection

Microinjection of plant protoplasts and cells is a promising technique (see Chap. XIV, this Vol.). The possibility of penetrating cell walls offers the opportunity to apply microinjection for transformation of crop species in which protoplast regeneration has not yet been achieved. Since the major monocot crops (e.g., cereals) are also out of the host range of *Agrobacterium*, several laboratories are involved now in microinjection of pollen, pollen tubes (Hepher et al. 1985), premature embryoids (Mathias et al. 1986), and embryonic suspension cells. A major drawback of microinjection is, however, that the number of injections per hour is very limited. In the optimal case, a few hundred cells can be treated in a day. In animals it has already been proven to be an efficient method for the introduction of foreign genes into the nucleus of egg cells leading to the production of transgenic animals (Hammer et al. 1985; Willadsen 1986). Recently, also in plants it has been shown that up to 25% of the injected protoplasts became stably transformed when injected directly into the nucleus. This supports the view, as with PEG-induced DNA uptake, that not only the cell membrane acts as a barrier, but also the nuclear membranes interfere with an optimal transformation.

By developing specific adaptations of the microinjection equipment, vital injection of plant chromosomes into protoplasts was achieved recently and the first results with plant cell transformation by the microinjection of plant chromosomes were published (Griesbach 1987). In this approach, the chromosomes are not selected or sorted, but injected at random. The combination of chromosome sorting and microinjection might offer more advanced gene-tagging possibilities. Our injection procedure developed recently is an improved microinjection set-up in which special attention has been put on the modifications prerequisite to the micromanipulation and injection of entire plant chromosomes (de Laat and Blaas 1987). Using this system, metaphase chromosomes isolated from a partially synchronized kanamycin-resistant suspension culture of *N. plumbaginifolia* were microinjected into recipient wild-type *N. plumbaginifolia* protoplasts. So far, only visual evidence was obtained. Currently, the fate of the transferred chromosomes is a major aspect of investigation.

## 6 Conclusions

The data presented in this chapter mainly deal with the methods for isolation of plant chromosomes, sorting by flow cytometry, characterization, and transfer. The technique of chromosome-mediated gene transfer in combination with flow cytometry has been extensively used in mammalian cell systems for studying gene expression, fine structural mapping of chromosomes, molecular characterization, and for the construction of chromosome-specific genome libraries. For molecular characterization of plant genome and for the benefit of plant breeding, the construction of chromosome-specific genome libraries in crop plants using flow sorting has been a major point of discussion at present (e.g., DOE/Los Alamos workshop on plant chromosome isolation, manipulation and genome sequencing, Los Alamos, 1986).

Unlike plant cells, the thick cell wall is absent in mammalian cells, which makes it easy to rupture and isolate the chromosomes for transfer. On the other hand, plant cells have the great advantage of being *totipotent*, thus allowing to generate a whole plant from a genetically manipulated cell. This makes it possible to analyze gene expression at plant level and heritable transmission to the progeny through meiosis.

During the last few years, optimum conditions for obtaining mass isolation of chromosomes from several plant species have been established by several investigators. Automated flow cytometry has been proven to be a powerful tool for plant cell biotechnology, e.g., for monitoring the synchronization of the cell cycle, the sorting of cellular components, the analysis of genetic instability (ploidy alterations, chromosome-structural modifications). Vital staining and the advanced dual parameter karyotyping through staining for AT- and GC-rich regions of DNA are now in progress, which should help to gain further knowledge in this technology.

The fate of the transplanted chromosomes has yet to be demonstrated in plants. Whether or not such chromosomes will remain mitotically stable and result in stable transformants will largely depend on their integrity. The duration of their residence in cytoplasm will be critical for integration. In this respect the stage of the cell cycle of the recipient is assumed to be of crucial importance. By partial synchronization, combined with vital DNA staining of the protoplasts, *specific cell cycle stages* (e.g., prophase) may be selected. This may be especially appropriate when using microinjection of chromosomes into the recipient protoplasts. This would probably limit the duration of time before entrapment in the protecting environment of one of the daughter nuclei occurs.

Alternatively, transfer of chromosomes in interphase, i.e., as restituted micronuclei induced by the spindle toxin amiprophos-methyl (de Laat et al. 1987; Sree Ramulu et al. 1988) would lead to a stable integration allowing the recovery of intact, single-donor chromosomes in the recipient genome. The combination of chromosome-mediated gene transfer technique with microcell fusion has already been proven to be valuable not only for the regional mapping of closely linked genes, but also for the transfer, isolation, and cloning of specific mammalian genes (Fournier 1982; Lugo and Fournier 1986). This approach, now under investigation in plants, can be exploited for genetic manipulation of specific chromosomes, chromosome segments, and genes.

## References

Afonso CL, Harkins KR, Thomas-Compton MA, Krejci AE, Galbraith DW (1985) Selection of somatic hybrid plants in *Nicotiana* through fluorescence-activated sorting of protoplasts. Biotechnology 3:811–816

Alexander RG, Cocking EC, Jackson PJ, Jett JH (1985) The characterization and isolation of plant heterokaryons by flow cytometry. Protoplasma 128:52–58

Anderson LK, Stack SM, Fox MH, Chuanshan Z (1985) The relationship between genome size and synaptonemal complex length in higher plants. Exp Cell Res 156:367–378

Bergounioux C, Perennes C, Miege C, Gadal P (1986) The effect of male sterility on protoplast division in *Petunia hybrida*. Cell cycle comparison by flow cytometry. Protoplasma 130:138–144

Blumenthal AB, Dieden JD, Kapp LN, Sedat JW (1979) Rapid isolation of metaphase chromosomes containing high molecular weight DNA. J Cell Biol 81:255–259

Brisson N, Paszkowski J, Penswick JR, Gronenborn B, Potrykus I, Mohn T (1984) Expression of a bacterial gene in plants by using a viral vector. Nature (Lond) 310:511–514

Brown S, Renaudin JP, Prevot C, Guern J (1984) Flow cytometry and sorting of plant protoplasts: technical problems and physiological results from a study of pH and alkaloids in *Catharanthus roseus*. Physiol Veg 22:541–554

Brown S, Jullien M, Coutos-Thevenat P, Muller Ph, Renaudin JP (1986) Present developments of flow cytometry in plant biology. Biol Cell 58:173–178

Carlberg I, Glimelius K, Eriksson T (1984) Nuclear DNA content during the initiation of callus formation from isolated protoplasts of *Solanum tuberosum* L. Plant Sci Lett 35:225–230

Conia J, Bergounioux C, Perennes C, Brown S, Muller Ph, Gadal P (1987) Chromosomes from protoplasts of *Petunia* hybrids, flow analysis and sorting. Biol Cell 58 Suppl: 5a

Coyle MB, Strauss B (1970) Cell killing and the accumulation of breaks in the DNA of HEp-2 cells incubated in the presence of hydroxyurea. Cancer Res 30:2314–2319

Davies KE, Young BD, Elles RG, Hill ME, Williamson R (1981) Cloning of a representative genomic library of the human X-chromosome after sorting by flow cytometry. Nature 293:374–378

de Laat AMM, Blaas J (1983) Isolation, purification and sorting by flow cytometry of metaphase chromosomes of *Haplopappus gracilis*. In: Potrykus I, Harms CT, Hinnen A, Hütter R, King PJ, Shillito A (eds) 6th Int Protoplast Symp, Poster Proc. Birkhäuser, Basel, pp 120–121

de Laat AMM, Blaas J (1984) Flow cytometric characterization and sorting of plant chromosomes. Theor Appl Genet 67:463–467

de Laat AMM, Blaas J (1987) An improved method for protoplast micro-injection suitable for transfer of entire plant chromosomes. Plant Sci 50:161–169

de Laat AMM, Shell JHN (1986) Integrity of metaphase chromosomes of *Haplopappus gracilis* (Nutt.) Gray isolated by flow cytometry. Plant Sci 47:145–151

de Laat AMM, Verhoeven HA, Sree Ramulu K, Dijkhuis P (1987) Efficient induction by amiprophos-methyl and flow cytometric sorting of micronuclei in *Nicotiana plumbaginifolia*. Planta 172:473–478

Depicker A, Montagu van M, Shell J (1983) Plant cell transformation by *Agrobacterium* plasmids. In: Kosuge T, Meredith CP, Hollaender A (eds) Genetic engineering of plants: an agricultural perspective. Plenum, New York, pp 143–176

Dudits D, Praznovzky T (1985) Intergeneric gene transfer by protoplast fusion and uptake of isolated chromosomes. In: Zaitlin M, Day P, Hollaender A (eds) Biotechnology in plant science: relevance to agriculture in the eighties. Academic Press, New York London Orlando, pp 115–128

Eriksson T (1966) Partial synchronization of cell division in suspension cells of *Haplopappus gracilis*. Physiol Plant 19:900–910

Famelaer Y, Cammaerts D, Karp A, Sidorov V, De Brouwer D, Negrutiu I, Jacobs M (1986) Cellular engineering by 'gamma fusion' and 'egg transformation': Recent experimental data and applications in plant breeding. In: Proc Symp Nuclear techniques and in vitro culture for plant improvement, IAEA, Vienna, pp 453–460

Fournier REK (1982) Microcell-mediated chromosome transfer. In: Shay J (ed) Techniques in somatic cell genetics. Plenum, New York, pp 309–327

Galbraith DW (1984) Flow cytometric analysis of the cell cycle. In: Vasil IK (ed) Cell culture and somatic cell genetics of plants, vol 1. Academic Press, New York London, pp 765–777

Galbraith DW, Shields BA (1982) The effect of inhibitors of cell wall synthesis on tobacco protoplast development. Physiol Plant 55:25–30

Galbraith DW, Harkins KR, Maddox JM, Ayres NM, Sharma DP, Firoozabady E (1983) Rapid flow cytometric analysis of the cell cycle in intact plant tissues. Science 220:1049–1051

Glimelius K (1987) Enrichment of somatic hybrids with flow cytometry. IAPTC Newslett 52:2–10

Glimelius K, Djupsjöbacka M, Fellner-Feldegg H (1986) Selection and enrichment of plant protoplasts heterokaryons of Brassicaceae by flow sorting. Plant Sci 45:133–141

Griesbach RJ (1987) Chromosome-mediated transformation via microinjection. Plant Sci 50:69–77

Griesbach RJ, Malmberg RL, Carlson PS (1982) An improved technique for the isolation of higher plant chromosomes. Plant Sci Lett 24:55–60

Hadlaczky G, Sumner AT, Ross A (1981) Protein-depleted chromosomes. I. Structure of isolated protein-depleted chromosomes. Chromosoma 81:537–555

Hadlaczky G, Bisztray G, Praznovszky T, Dudits D (1983) Mass isolation for plant chromosomes and nuclei. Planta 157:278–285

Hammer RE, Pursel VG, Rexroad CE, Wall RJ, Bolt DJ, Ebert KM, Palmiter RD, Brinster RL (1985) Production of transgenic rabbits, sheeps and pigs by microinjection. Nature (Lond) 315:680–683

Hänisch ten Cate CH, Sree Ramulu K (1987) Callus growth, tumour development and polyploidization in the tetraploid potato cultivar Bintje. Plant Sci 49:209–216

Harkins KR, Galbraith DW (1984) Flow sorting and culture of plant protoplasts. Physiol Plant 60:43–52

Harkins KR, Galbraith DW (1987) Factors governing the flow cytometric analysis and sorting of large biological particles. Cytometry 8:60–70

Hepher A, Gates P, Boulter D (1985) Microinjection of DNA into pollen, ovaries, and somatic cells. In: EC contractants Meet Biomolecular engineering programme, Copenhagen, Oct 9–10. Abstr, pp 30–31

Jackson RC (1957) New low chromosome number for plants. Sci 126:1115–1116

Jett JH, Alexander RG (1985) Droplet sorting of large particles. Cytometry 6:484–486

Karlsson SB, Vasil IK (1986) Growth, cytology and flow cytometry of embryogenic cell suspension cultures of *Panicum maximum* Jacq. and *Pennisetum purpureum* Schum. J Plant Physiol 123:211–227

Krens FA, Molendijk L, Wullems GJ, Schilperoort RA (1982) In vitro transformation of plant protoplasts with Ti-plasmid DNA. Nature (Lond) 296:72–74

Larkin PJ, Scowcroft WR (1981) Somaclonal variation – a novel source of variability from cell cultures for plant improvement. Theor Appl Genet 60:197–214

Lörz H, Potrykus I (1978) Investigations on the transfer of isolated nuclei into plant protoplasts. Theor Appl Genet 53:251–256

Lugo TG, Fournier REK (1986) Microcell fusion and mammalian gene transfer. In: Kucherlapati R (ed) Gene transfer. Plenum, New York, pp 79–93

Mathias RJ, Boyd LA, Gleddie SA, Flavell RB (1986) Transformation of wheat by microinjection. In: EC contractants Meet Biomolecular engineering programme, Wageningen, Jan 27–30. Abstr, pp 99–100

Matthews BF (1983) Isolation of mitotic chromosomes from partially synchronized carrot (*Daucus carota*) cell suspension cultures. Plant Sci Lett 31:165–172

Mc Bride OW, Peterson JL (1980) Chromosome-mediated gene transfer in mammalian cells. Annu Rev Genet 14:321–345

Meadows MG (1982) Characterization of cells and protoplasts of the B73 maize cell line. Plant Sci Lett 28:337–348

Melamed MR, Mullaney PF, Mendelsohn ML (1979) Flow cytometry and sorting. John Wiley & Sons, New York

Muirhead KA, Hovan PK (1984) Applications of flow cytometry to tissue culture systems. In: Maramorosch K (ed) Advances in cell culture, vol 3. Academic Press, New York London, pp 57–91

Petit P, Conia J, Brown SC, Bergounioux C (1986) Cytométrie en flux et biolotechnologies végétales. Un avenir prometteur. Biofutur 51:128–139

Pinkel D (1981) On the possibility of automated scoring of pollen mutants. Environ Health Perspectives 37:133–136

Potrykus I, Shillito RD, Saul MW, Paszkowski J (1985) Direct gene transfer. State of art and future potential. Plant Mol Biol Rep 3:117–128

Puite KJ, Ten Broeke WRR (1983) DNA staining of fixed and non-fixed plant protoplasts for flow cytometry with Hoechst 33342. Plant Sci Lett 32:79–88

Puite KJ, Gilissen LJW, de Laat AMM (1983) First steps to chromosome transplantation with *Hapopappus gracilis* using flow cytometry. In: Potrykus I, Harms CT, Hinnen A, Hütter R, King PJ, Shillito A (eds) 6th Int Protoplast Symp Poster Proc. Birkhäuser, Basel, pp 122–123

Saxena PK, Mii M, Crosby WL, Fowke LC, King J (1986) Transplantation of isolated nuclei into plant protoplasts. A novel technique for introducing foreign DNA into plant cells. Planta 168:29–35

Sharma DP, Firoozabady E, Ayres NM, Galbraith DW (1983) Improvement of anther culture in *Nicotiana*: Media, cultural conditions and flow cytometric determination of ploidy levels. Z Planzenphysiol 111:441–451

Sree Ramulu K (1986) Case histories of genetic variability in vitro: Potato. In: Vasil IK (ed) Cell culture and somatic cell genetics of plants, vol 3. Academic Press, New York London, pp 449–473

Sree Ramulu K, Dijkhuis P (1986) Flow cytometric analysis of polysomaty and in vitro genetic instability in potato. Plant Cell Rep 3:234–237

Sree Ramulu K, Verhoeven HA, Dijkhuis P (1988) Mitotic dynamics of micronuclei induced by amiprophos-methyl and prospects for chromosome-mediated gene transfer in plants. Theor Appl Genet 75:575–584

Sundberg E, Glimelius K (1986) A method for production of interspecific hybrids within Brassicaceae via somatic hybridization, using resynthesis of *Brassica napus* as a model. Plant Sci 43:155–162

Szabados L, Hadlaczky G, Dudits D (1981) Uptake of isolated plant chromosomes by plant protoplasts. Planta 151:141–145

Valk HCPM van der, Verhoeven HA (1988) Application of flow cytometry for the isolation of metaphase chromosomes from protoplasts of *Nicotiana plumbaginifolia*. In: 7th Int Protoplast Symp, Wageningen. Nijhoff, Dordrecht (in press)

Verhoeven HA, de Laat AMM, Blaas J (1986) Methodological developments for chromosome-mediated gene transfer in higher plants. In: Magnien E (ed) Biomolecular engineering in the European Community, achievements of the research programme (1982–1986). Final Rep, Nijhoff, Dordrecht, pp 977–981

Verhoeven HA, de Laat AMM, Sree Ramulu K, Dijkhuis P, Blaas J, Vloten-Doting L van (1987) Induction, isolation and flow cytometric sorting of metaphase chromosomes and micronuclei from cell suspension cultures of *Nicotiana plumbaginifolia*. EC contractant Meeting Biotechnology action programme, Louvain La Neuve, March. Abstr, pp 119–120

Willadsen SM (1986) Nuclear transplantation in sheep embryos. Nature (Lond) 320:63–65

# XXI Transfer of Mitochondria Through Protoplast Fusion

H. ICHIKAWA, L. TANNO-SUENAGA, and J. IMAMURA[1]

## 1 Introduction

The application of improved organelle or DNA isolation procedures and recombinant DNA technologies has led us to a more detailed understanding of higher plant mitochondrial genomes in the last decade. Mitochondrial genomes of higher plants are much larger in size than those of yeasts, fungi, and animals, and are variable, ranging from about 200 kb in *Brassica* and *Oenothera* to about 2400 kb in muskmelon (Ward et al. 1981; Leaver and Gray 1982; Levings 1983). Notwithstanding the larger size of plant mitochondrial DNA's (mtDNA's), the number of mitochondrial genes are thought to be similar compared with those of the other eukaryotes (Dawson et al. 1986; EcKenrode and Levings 1986). The mitochondrial genomes in some higher plants characterized so far, however, contain some direct repeat elements. Intramolecular recombinations through the direct repeats appear to produce the complex structure of plant mtDNA's (Palmer and Shields 1984; Pring and Lonsdale 1985; Quetier et al. 1985).

All mitochrondrial genomes encode tRNA's and rRNA's for their own translation systems and some proteins concerning electron transport. A few traits coded on mitochondrial genomes also have been reported, such as cytoplasmic male sterility (CMS), nonchromosomal stripe mutants of maize and T-toxin sensitivity of T (Texas) male-sterile cytoplasm in maize (Leaver and Gray 1982; Levings 1983; EcKenrode and Levings 1986; Newton and Coe 1986; Lonsdale 1987). CMS is a widespread phenomenon in many plant species and is agriculturally valuable for the production of F1-hybrid seed (Hanson and Conde 1985). However, in practice, an intensive and laborious back-crossing is needed to introduce CMS into a desired species. To bypass the back-crossing, direct transfer of isolated mitochondria or mtDNA carrying a CMS trait into fertile plants has been required; nevertheless the trials have not succeeded so far except for the experiments on successful transmission of CMS by somatic cell fusion.

Plastid and mitochondrial genomes or cytoplasmically encoded traits are usually inherited maternally by crossing in higher plants. In somatic cell fusion, however, not only nuclei but also cytoplasms from both parents are mixed in a fused cell. The fate of chloroplasts in fused cells has been studied (Maliga 1986). In many cases, parental chloroplasts were segregated randomly during the

[1] Plantech Research Institute, Kamoshida 1000, Midori-ku, Yokohama, 227 Japan

Biotechnology in Agriculture and Forestry, Vol. 9
Plant Protoplasts and Genetic Engineering II (ed. by Y. P. S. Bajaj)

developmental process of hybrid cells, and as a result of the segregation hybrid plants possessed only one type of the parental chloroplasts.

Belliard et al. (1979) first analyzed mitochondrial genomes from *Nicotiana tabacum* cybrid plants by restriction endonuclease analysis. MtDNA's from the cybrids showed novel restriction patterns which were different from those from each parental mtDNA and from the mixture of both parental patterns. They attributed the novel restriction patterns to intermolecular recombination between two mtDNA's. A CMS trait in a parental line was transferred to some of the tobacco cybrid plants. Rearrangements of mtDNA's have also been reported in other somatic hybrids or cybrids including Solanaceae, Umbelliferae, Cruciferae, Gramineae, Leguminosae, and Rutaceae. A list of these is summarized in Table 1. Rearrangements of mitochondrial genomes, however, do not always occur in fusion-derived plants. No mtDNA recombination has been indicated in *B. napus* cybrid plants (Yarrow et al. 1986; Barsby et al. 1987a, b; Chuong et al. 1988), somatic hybrids between potato and tomato (Shepard et al. 1983) and *Nicotiana* cybrid or somatic hybrid plants derived from homoplasmic fusions (Nagy et al. 1983) and carrot somatic hybrids (Kothari et al. 1986) as shown in Table 1.

**Table 1.** Somatic hybrid or cybrid plants possessing rearranged mitochondrial genomes or transferred mitochondrial genomes (or fragments)

| Fusion partners (Characteristics) | Mitochondrial type | Remarks (Regeneration ±) | Reference |
|---|---|---|---|
| Solanaceae | | | |
| 1. *Nicotiana tabacum* var. Techne (CMS) *N. tabacum* var. Xanthi | Recombinant | Cybrids(+) | Belliard et al. (1979) |
| 2. *N. tabacum* SR1, *N. knightiana* | Recombinant | Somatic hybrids(+) | Nagy et al. (1981) |
| 3. [D][a] *N. tabacum*, [R][a] Type A[b] (CMS) | Recombinant | Cybrids(+) | Galun et al. (1982) |
| 4a. [D] *N. tabacum* SR1, [R] *N. plumbaginifolia* | Recombinant | Cybrids(+) | Nagy et al. (1983) |
| 4b. *N. tabacum* var. Samsun (light-sensitive plastids) [R] *N. plumbaginifolia* | Recombinant | Cybrids(+) | Nagy et al. (1983) |
| 4c. [D] *N. plumbaginifolia* [R] *N. tabacum* | Recombinant | Cybrids(+) | Nagy et al. (1983) |
| 5. [D] *N. rustica* [R] *N. sylvestris* | Recombinant or [R] | Cybrids(+) | Aviv et al. (1984a) |
| 6a. [D] *N. undulata*, *N. bigelovii*, *N. alata* [R] *N. sylvestris* | Recombinant | Cybrids(+) | Aviv et al. (1984b) |
| 6b. [D] *N. undulata*, *N. bigelovii* [R] *N. tabacum* VBW (albino) | Recombinant | Cybrids(+) | Aviv et al. (1984b) |

[a] Donor[D]-recipient[R] protoplast fusion. Each donor parent was irradiated.
[b] Derived from fusion between *N. tabacum* possessing *N. suaveolens* cytoplasm (donor) and *N. sylvestris* (recipient).

**Table 1** (continued)

| Fusion partners (Characteristics) | Mitochondrial type | Remarks (Regeneration ±) | Reference |
|---|---|---|---|
| 7. [D] *N. gossei* [R] *N. plumbaginifolia* Np(LS1)41 (light-sensitive plastids) | Recombinant | Cybrids(+) | Medgyesy et al. (1985) |
| 8. [D] *N. debneyi* [R] *N. tabacum* cv. Consolation 402 | Recombinant | Cybrids(+) | Asahi et al. (1988) |
| 9. *Petunia hybrida* 3704, *P. parodii* 3688 (CMS) | Recombinant | Somatic hybrids(+) | Boeshore et al. (1983), Boeshore et al. (1985), Rothenberg et al. (1985) |
| 10. *P. hybrida* 3704, *P. inflata* (albino) | Recombinant | Somatic hybrids(+) | Clark et al. (1986) |
| 11. *Solanum tuberosum* 774[c] *Lycopersicon esculentum* cv. Rutgers or Nova | 774* | Somatic hybrids(+) | Shepard et al. (1983) |
| 12. *L. esculentum* cv. San Marzano, *S. rickii* LA1974 | Recombinant(?) | Somatic hybrids(+) | O'Connell and Hanson (1986) |
| 13. *S. tuberosum* 116 (albino · protoclone), *S. brevidens* | Recombinant or either | Somatic hybrids(+) | Kemble et al. (1986) |
| Cruciferae | | | |
| 14a. *Brassica napus* var. Brutor, *B. napus* C (Ogura CMS and chlorophyll-deficient) | Recombinant | Cybrids or somatic hybrids(+) | Chétrit et al. (1985) Vedel et al. (1986) Vedel et al. (1987) |
| 14b. *B. napus* var. Tower (triazine-resistant chloroplasts), *B. napus* C | Recombinant | Cybrids(+) | Chétrit et al. (1985) Vedel et al. (1987) |
| 15. *B. napus* var. Regent (Nap type CMS), *B. napus* var. Regent (triazine-resistant) | Nap CMS* | Cybrids(+) | Yarrow et al. (1986) |
| 16a. [D] *B. napus* var. Regent (triazine-resistant) [R] *B. napus* var. Regent (Polima type CMS) | Polima CMS* | A cybrid(+) | Barsby et al. (1987a) |
| 16b. *B. napus* var. Regent (Polima type CMS), *B. napus* var. Triton (triazine-resistant) | Polima CMS* | A cybrid(+) | Barsby et al. (1987a) |
| 17. [D] *B. napus* var. Regent (Polima type CMS) [R] *B. napus* AWR (*B. campestris* cytoplasm) | Either [D] or [R] | Cybrids(+) | Barsby et al. (1987b) |
| 18. [D] *B. napus* #1446 (Ogura type CMS) [R] *B. napus* cv. Olga | Recombinant | Cybrids(+) | Menczel et al. (1987) Morgan and Maliga (1987) |
| 19. *B. napus* (haploid and Polima CMS), *B. napus* (haploid and atrazine-resistant) | Polima CMS* | A cybrid(+) | Chuong et al. (1988) |

[c] A variegating protoclone of cv. Russet Burbank.
* No rearrangement of mtDNA's.

**Table 1** (continued)

| Fusion partners (Characteristics) | Mitochondrial type | Remarks (Regeneration ±) | Reference |
|---|---|---|---|
| 20. *B. oleracea* cv. Green Comet (Ogura CMS), *B. campestris* cv. Candle (atrazine-resistant) | Recombinant | A somatic hybrid(+) | Robertson et al. (1987) |
| Umbelliferae | | | |
| 21. *Daucus carota*, *D. capillifolius* | Recombinant | A somatic hybrid(−) | Matthews and Widholm (1985) |
| 22a. *D. carota* PR (glyphosate-resistant), *D. carota* HCl-3 (5-methyltryptophan-resistant) | Recombinant | Somatic hybrids(−) | Kothari et al. (1986) |
| 22b. *D. carota* C-1 (selenocysteine-resistant), *D. carota* HCl-3 | HCl-3* | Somatic hybrids(−) | Kothari et al. (1986) |
| 22c. *D. carota* PR, *D. carota* S232 (sodium selenate-resistant) | PR* | Somatic hybrids(−) | Kothari et al. (1986) |
| 23. [D] *D. capillifolius* [R] *D. carota* | Recombinant | Cybrids(+) | Ichikawa et al. (1987) |
| Gramineae | | | |
| 24. *Pennisetum americanum*, *Panicum maximum* | Recombinant | Somatic hybrids(−) | Ozias-Akins et al. (1987) |
| 25. *P. americanum*, *Saccharum officinarum* | Recombinant | Somatic hybrids(−) | Tabaeizadeh et al. (1987) |
| Leguminosae | | | |
| 26. *Medicago sativa* MSR12 (CMS), *M. falcata* 318 | Recombinant | Somatic hybrids(−) | D'Hont et al. (1987) |
| Rutaceae | | | |
| 27a. [D] PPT (Poorman × *Poncirus trifoliata*) [R] lemon (*Citrus limon* cv. Villafranca) | Similar to lemon (not identical) | Cybrids(+) | Vardi et al. (1987) |
| 27b. [D] PPT [R] Sour orange (*C. aurantium*) | Recombinant or [D](?) | Cybrids(+ ) | Vardi et al. (1987) |
| 27c. [D] Sour orange [R] Lemon | Recombinant or [D](?) | Cybrids(+) | Vardi et al. (1987) |

* No rearrangement of mtDNA's.

X- or gamma-irradiation of protoplasts is useful for the formation of cybrids (somatic hybrids possessing only one parental chromosome and both parental organelle genomes), since the irradiation treatment prevents colony formation from protoplasts and stimulates elimination of the irradiated nucleus from a fused cell (Zelcer et al. 1978; Menczel et al. 1982). Sidorov et al. (1981) reported that cybrids were selected effectively by the combination with X-irradiation to donor cells and iodoacetate treatment to recipient cells. Using the donor-recipient fusion technique, the whole or part of mitochondrial genome, or CMS traits have been introduced into recipient plants of tobacco (Zelcer et al. 1978; Aviv et al. 1984a, b; Fluhr et al. 1984; Medgyesy et al. 1985; Menczel et al. 1986;

Kumashiro and Kubo 1986), *Brassica* (Yarrow et al. 1986; Barsby et al. 1987 a, b; Menczel et al. 1987; Morgan and Maliga 1987) and *Daucus* (Ichikawa et al. 1987).

This chapter describes (1) a method of formation of *Daucus* cybrid plants with the selection on metabolic complementation, (2) characterization of these cybrid plants, and (3) analyses of cybrid mtDNA's.

## 2 Selection Method of Formation of *Daucus* Cybrid Plants by Donor-Recipient Protoplast Fusion

### 2.1 Plant Material and Protoplast Isolation

Suspension cultures of *D. carota* L. (line 35B, line 28K and cv. K5) were initiated from calli derived from seedling hypocotyls and maintained every 2 weeks in Murashige and Skoog (1962) medium (MS) containing 0.5 mg/l 2,4-D. Suspension cells of *D. capillifolius* Gilli. were kindly provided by Dr. Kameya of Tohoku University, and maintained in the same way as the *D. carota* cell lines. This *D. capillifolius* suspension cell line possessed no regeneration potential.

Three- to five-day-old subcultures were used for the source of protoplast preparations. The suspension cells were incubated in 1% driselase, 0.5% cellulase Onozuka RS and 0.01% pectolyase Y-23 dissolved in 0.5 M mannitol and 0.1% 2-(N-morpholino)-ethane sulfonic acid (MES) at pH 5.7. Incubation was carried out at 25 °C for 4 h with occasional shaking. Protoplasts were purified by passage through 37 and 20 μm stainless steel sieves, pelleted (100 g for 3 min), and washed twice with 10 ml of washing solution (0.5 M mannitol and 0.1% MES, pH 5.7). More than $10^7$ protoplasts were obtained from one g fr.wt. suspension cells.

### 2.2 Prefusion Treatment of Protoplasts

Protoplasts of *D. capillifolius* ($2 \times 10^6$/ml of washing solution) were X-irradiated with various dosages (1 krad/min) and cultured in MS plus 0.3 M sorbitol with

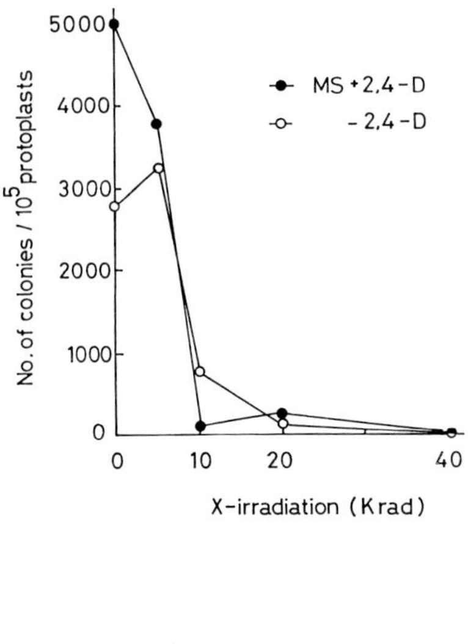


**Fig. 1.** Effect of X-irradiation dosage on colony formation of *D. capillifolius* protoplasts. Protoplasts were cultured for 1 month at a density of $10^5$/ml in a total volume of 4 ml MS with or without 0.1 mg/l 2,4-D. (After Ichikawa et al. 1987)

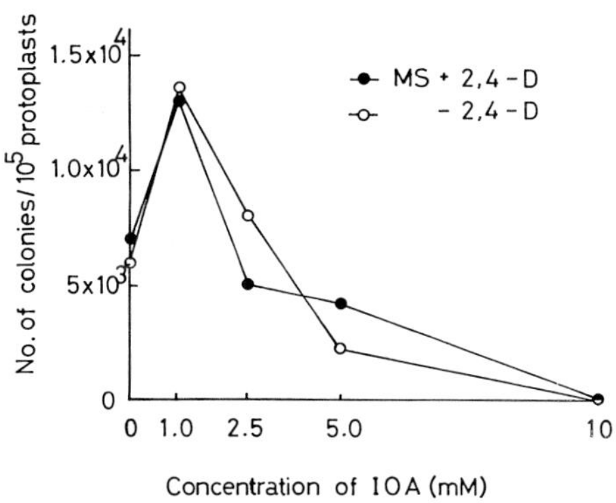


**Fig. 2.** Effect of iodoacetamide (IOA) concentration on colony formation of *D. carota* line 35B protoplasts. Protoplasts were cultured as in Fig. 1. (After Ichikawa et al. 1987)

or without 0.1 mg/ml 2,4-D for about a month. Number of colonies per plate were counted. As shown in Fig. 1, usually more than 40 krad exposure to X-ray inhibited colony formation in our experiments. Pretreatment of donor protoplasts was performed with 60 krad X-irradiation.

*D. carota* line 35B protoplasts ($10^6$/ml of washing solution) were treated with various concentrations of iodoacetamide (IOA) for 10 min at room temperature and cultured for a month. No protoplast division was observed for the 10 mM IOA treatment (Fig. 2). Protoplasts of *D. carota* (recipient) were then inactivated before fusion with 15 mM IOA for 10 min.

### 2.3 Protoplast Fusion and Culture

IOA-treated recipient protoplasts were mixed with X-irradiated donor protoplasts in a 1:1 ratio and fused in the presence of polyethylene glycol (PEG) according to the method of Kao and Michayluk (1974). The fused protoplasts were cultured in MS supplemented with 0.3 M sorbitol and 0.1 mg/l 2,4-D at a density of $1.25\times10^6$/ml at 25 °C.

As control experiments the X-irradiated protoplasts or IOA-treated protoplasts were cultured separately with or without PEG treatment. Furthermore the mixtures of both kinds of protoplasts were cultured without PEG treatment. Neither formed colonies. Although the X-irradiated protoplasts with PEG treatment divided and formed a few colonies occasionally, the colonies were never regenerated into plants.

After about 1 month of culture, hundreds of colonies were obtained from mixed and PEG-treated protoplasts. The plates containing fusion-derived colonies were gradually substituted with MS plus 0.1 M sorbitol without 2,4-D. The resultant somatic embryos were transferred to twofold diluted MS containing 0.2% Gelrite for further growth. Subsequently the plants grown enough were transplanted into soil.

# 3 Characterization of Putative Cybrid Plants

## 3.1 Leaf Morphology

*D. capillifolius* can be easily distinguished from *D. carota* by its leaf characteristics, as shown in Fig. 3A and C. Dudits et al. (1977) produced somatic hybrid plants between albino *D. carota* and green *D. capillifolius* by protoplast fusion without irradiation pretreatment. The interspecific somatic hybrids showed intermediate leaf shape. On the other hand, the regenerated plants derived from our fusion experiments between *D. capillifolius* (donor) and *D. carota* line 35B (recipient) had the same leaf morphology as the parental *D. carota* plants, while the characteristics of leaves from those regenerated plants were clearly different from those of the other parental *D. capillifolius* (Fig. 3). Since the leaf morphology could not be descriminated between *D. carota* line 28K and cv. K5, it was impossible to characterize leaf morphological traits of the regenerated plants from donor-recipient fusion between them.

**Fig. 3A–C.** Leaf morphology of *D. capillifolius*, *D. carota* line 35B and their cybrid plant. **A** *D. capillifolius*; **B** cybrid; **C** line 35B. (After Ichikawa et al. 1987)

### 3.2 Chromosome Observation

Parental cell suspensions of *D. capillifolius* (no regeneration potential) or *D. carota*, and root tips of regenerated plants were incubated with 2 mM 8-hydroxy-quinoline, then fixed in ethanol:acetic acid (3:1), and treated with an enzyme solution containing 4% cellulase Onozuka RS, 1% pectolyase Y-23, 7.5 mM KCl and 7.5 mM EDTA, pH 4.0 (Nishibayashi and Kaeriyama 1986). They were gently spread on slidge glass and stained with aceto-orcein for microscopic observation.

The results of chromosome counting are listed in Table 2. The chromosome number of *D. carota* line 35B suspension cells used in the experiments was 17 (2n−1). One chromosome might be deleted from the genome in the cells during culture (Bayliss 1975). The *D. capillifolius* cell line possessed more than 60 chromosomes, presumably because of the prolonged culture of several years. The chromosome numbers of the five regenerated plants were counted. Four plants: A-22, A-27, A-31, and A-39 possessed 17 chromosomes, like the *D. carota* line 35B, and the other one (A-59) had 34 (4n−2), twice as many. The chromosomal and leaf morphological observations strongly suggested that the regenerated plants possessed chromosomes of only one of the fusion parents, *D. carota* and the chromosomes derived from *D. capillifolius* cells were eliminated from the fused cells during protoplast culture.

Chromosome numbers of fusion products between *D. carota* line 28K and *D. carota* cv. K5 were also analyzed. Both parental lines possessed chromosome numbers of 18 (2n). Chromosome numbers of six fusion-derived plants were in the range of 17−46 (Table 2). Three of the six regenerated plants had 18 chromosomes. The frequency of cybrid formation in this combination was lower than that in the *D. capillifolius* and *D. carota* 35B fusion.

**Table 2.** Chromosome numbers of fusion-derived *Daucus* plants and their parental lines

| Plant no. or Parental line | | Chromosome no. |
|---|---|---|
| (i) | 35B (recipient) | 17 |
| (ii) | *D. capillifolius* (donor) | >60 |
| Fusion (i) and (ii) | A-22 | 17 |
| | A-27 | 17 |
| | A-31 | 17 |
| | A-39 | 17 |
| | A-59 | 34 |
| (iii) | cv. K5 (recipient) | 18 |
| (iv) | 28K (donor) | 18 |
| Fusion (iii) and (iv) | D-1-27 | 17 |
| | E-2-8 | 18 |
| | E-2-9 | 18 |
| | E-2-14 | 18 |
| | E-2-26 | 46 |
| | A-8-1 | 24 |

# 4 Restriction and Southern Analyses of mtDNA's from Carrot Cybrids

## 4.1 Isolation and Restriction Endonuclease Analysis of mtDNA from Suspension Cells

The cell suspensions of fusion-derived plants were prepared from the shoot segments. Five to ten g of cell suspension were incubated with enzyme solution as mentioned above. Protoplasts were collected by centrifugation, washed once with 0.5 M mannitol plus 0.1% MES (pH 5.7), and resuspended in CP lysis buffer (DeBonte and Matthews 1984) using 2 ml ice-cold buffer per g tissue. Cells were ruptured by a blender at 15000 rpm for 5 s three times. Nuclei and cell debris were sedimented from the cell extract by centrifugation at 3000 g for 5 min. The supernatant was centrifuged at 6000 g for 5 min, and this step was repeated once more. Mitochondrial pellet was collected from the supernatant by centrifugation at 15000 g for 5 min. The mitochondrial fraction was resuspended in DNase buffer (0.3 M sucrose, 0.05 M Tris-HCl, pH 7.5) with 10 µg/g fr.wt. DNase I and 0.01 M $MgCl_2$, and incubated at 4 °C for 30 min. The mitochondrial suspension was layered on 5 ml of shelf buffer (0.6 M sucrose, 0.02 M EDTA, 0.01 M Tris-HCl, pH 7.2) and centrifuged at 15000 g for 10 min. This step was repeated once to remove DNase I completely. The pellet was resuspended in Sarkosyl buffer [1% N-lauryl sarcosine in 0.05 M Tris-HCl, 0.02 M EDTA (pH 8)] with 50 µg/ml proteinase K and incubated at 37 °C for 1 h. MtDNA was isolated by phenol-chloroform extraction and precipitated by ethanol in the presence of 0.3 M sodium acetate. The average yield of mtDNA was about 50 µg from 10 g fr.wt. of suspension cells. MtDNA (1–2 µg) was digested to completion with restriction enzymes according to the supplier's instructions for at least 4 h at 37 °C. The restriction fragments were separated on 0.5%–0.7% agarose gels and stained with ethidium bromide.

## 4.2 Stability of mtDNA's from Carrot Cell Suspensions Derived from Pretreated Protoplasts

Matthews and DeBonte (1985) pointed out the stability of carrot organelle genomes in cell suspension cultures maintained for 10 years and in regenerated carrot plants from a cell culture. In our experiments, the mtDNA restriction patterns of *D. carota* line 28K suspension cell lines also did not change over a subculture of one and a half years. The results suggest that carrot mtDNA's isolated from suspension cells are stable with prolonged periods of culture.

In order to study the effects of pretreatments on mtDNA, protoplasts of *D. carota* line 35B were isolated from a cell suspension and treated with IOA under lethal dose (5 mM). The protoplasts were cultured and regenerated to plants through somatic embryogenesis. Suspension cells were prepared from shoot segments of the regenerated plants. MtDNA preparations isolated from the cell suspensions were compared with a mtDNA of 35B suspension cells originated from seeds by restriction enzymes. The restriction patterns of IOA-treated and regenerated plants were homologous to that from the original suspension except for an imperceptible difference in fragment stoichiometry (unpubl.). MtDNA restriction patterns of "escaped" cells from X-irradiation were also analyzed and showed indistinguishable changes from those of cell lines initiated from seeds (data not shown). These results showed that the pretreatments to carrot protoplasts and following culture hardly affected the stability of mtDNA's at least qualitatively. Carrot cultured cells were concluded to be a suitable material for the isolation of mtDNA.

### 4.3 Restriction Endonuclease Analysis of mtDNA's from Putative Cybrid Plants

Prior to the mtDNA analysis of regenerated plants derived from protoplast fusion, the restriction fragment patterns of mtDNA's extracted from both parental cell lines were compared (*D. capillifolius* and *D. carota* 35B in Fig. 4; *D. carota* 28K and cv. K5 in Fig. 5). Different restriction patterns were detected between the mtDNA's from both parents.

The restriction patterns of mtDNA's from fusion-derived plants (A-22, A-27, A-31, and A-39) were identical to each other, and that of mtDNA from A-59 was slightly different from those of the former mtDNA's (Fig. 4). MtDNA's from those cybrid plants possessed unique restriction patterns compared to those of their parental cell lines. For example, as shown in Fig. 4A, seven unique Sal I fragments not present in both parental mtDNA's, five fragments present only in *D. carota*, and one of only *D. capillifolius* were observed in mtDNA's from the fusion-derived A-22, A-27, A-31, and A-39 plants. A total of 25 cybrid plants were

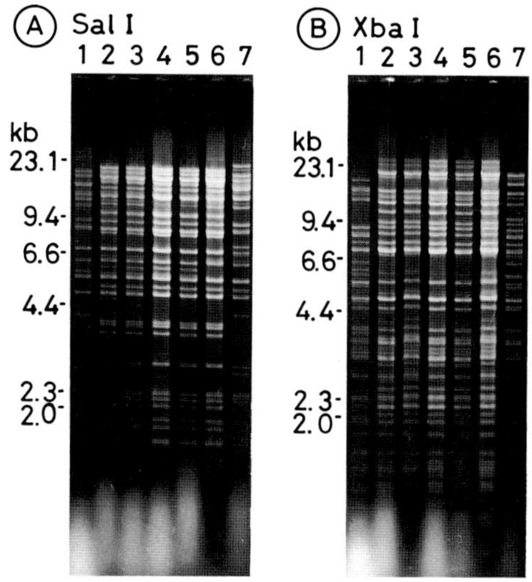


**Fig. 4.** Sal I (**A**) and Xba I (**B**) restriction endonuclease analysis of mtDNA's from *D. capillifolius*, *D. carota* line 35B and their cybrid plants. *D. capillifolius* (*lane 1*), cybrid plants: A-22 (*2*), A-27 (*3*), A-31 (*4*), A-39 (*5*) and A-59 (*6*), and *D. carota* line 35B (*7*). (After Ichikawa et al. 1987)

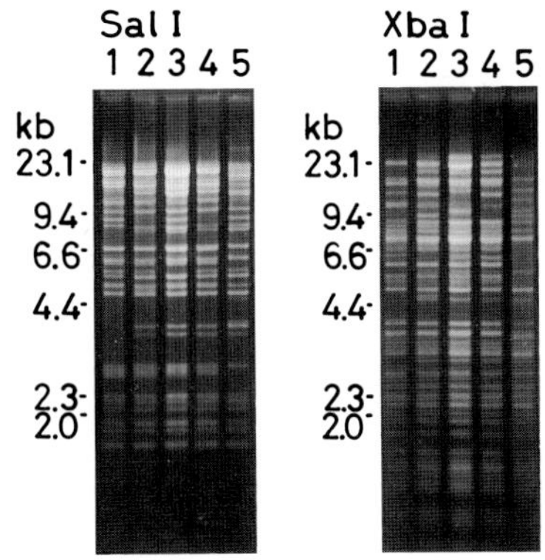


**Fig. 5.** Sal I and Xba I restriction endonuclease analysis of mtDNA's from *D. carota* line 28K, *D. carota* cv. K5 and their cybrid plants. *D. carota* 28K (*lane 1*), cybrid plants: E-1-6 (*2*), E-2-9 (*3*) and A-1-20 (*4*), and cv. K5 (*5*)

analyzed for their mtDNA's. The mtDNA restriction fragment patterns of those plants were classified into two types, namely I (ex. A-22, A-27, A-31, A-39) and II (ex. A-59) as shown in Fig. 4, although the two patterns were nearly identical. The novel mtDNA restriction patterns in *Daucus* have been observed only in heteroplasmic fusion combinations. These results indicated that the novel mtDNA band patterns in *Daucus* fusion products were the results of cytoplasmic mixing and interspecific mtDNA recombination. It was concluded accordingly that A-22, A-27, A-31, A-39, and A-59 plants possessing leaf morphology and chromosomes from only their recipient parent were cybrids between *D. capillifolius* and *D. carota* 35B, and that mtDNA sequences characteristic for *D. capillifolius* were introduced into *D. carota* plants by cell fusion after X-irradiation of *D. capillifolius* protoplasts.

Interestingly, the novel restriction patterns in those cybrid plants between *D. capillifolius* and *D. carota* line 35B showed correspondence to the mtDNA pattern of another Japanese cultivar *D. carota* cv. K5. The relationship between them was unknown.

MtDNA's were also isolated from suspension cells obtained from fusion-derived plants between X-irradiated *D. carota* 28K protoplasts and IOA-treated cv. K5 protoplasts. The mtDNA's showed unique and rearranged restriction patterns as shown in Fig. 5.

### 4.4 Southern Hybridization Analysis of mtDNA's from Cybrid Plants

To examine further the fate of mitochondrial genomes, namely, segregation or recombination of them in carrot cybrid plants, Southern hybridization experiments were performed by using mitochondrial gene probes.

The mitochondrial restriction fragments separated on agarose gels (Fig. 5) were transferred to Zeta-Probe blotting membrane (Bio-Rad). Three cloned genes from pea mtDNA containing 26S rRNA gene (pMTS44), cytochrome oxidase subunit II gene (COXII, pMTB62) and ATPase $\alpha$-subunit gene (ATP$\alpha$, pMTB17) were labeled with $^{32}P$ by multiprime DNA labeling systems (Amersham) according to the supplier's instructions and were used as probes. The mitochondrial clones were kindly provided by Drs. K. Nakamura and A. Morikami of Nagoya University. Southern blot hybridization of three mitochondrial probes to mtDNA's from three of the cybrid plants between *D. carota* line 28K and cv. K5 was performed in the presence of formamide (Maniatis et al. 1982).

A COXII gene probe hybridized to identically sized fragments in both parents and in the cybrids, E-1-6, E-2-9 and A-1-20 (Fig. 6A, Table 3). Moreover, mtDNA of a cybrid plant, A-1-20, contained specific fragments (2.8 kb: Sal I and 5.2 kb: Xba I). An ATP-$\alpha$ probe hybridized to a 14.5 kb Sal I fragment in three cybrids (Fig. 6B, Table 3). According to close observation, this fragment differed in size from either of the parental-specific fragments. Interestingly, the ATP-$\alpha$ probe, as well as COXII, hybridized to the specific band of A-1-20 mtDNA, a 2.8 kb SalI fragment and a 5.2 kb Xba I fragment. This result suggested that these fragments should contain a crossing-over point for recombination between ATP$\alpha$ and COX-II genes. Each parental specific band was detected for 26S rDNA probe (Fig. 6C, Table 3). A cybrid, E-2-9, contained both of these bands. The stoichiometry of

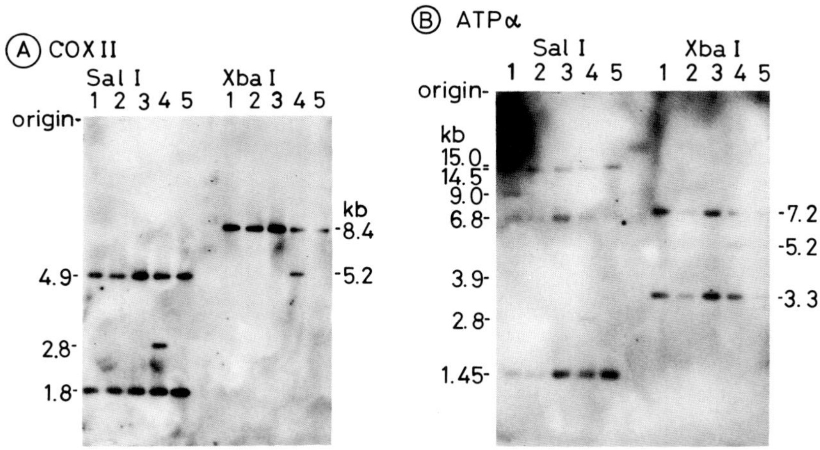


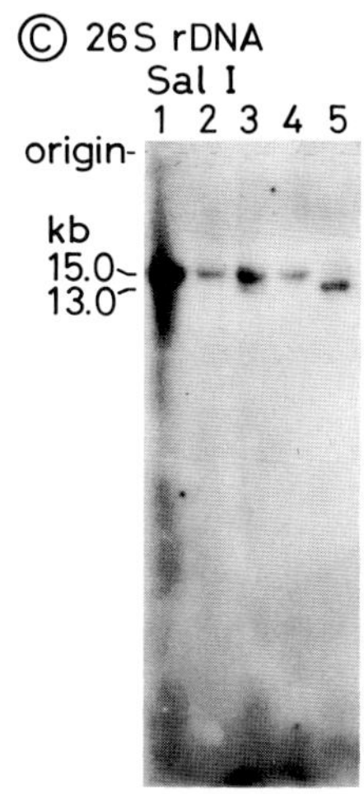


**Fig. 6A–C.** Southern blot hybridization of mitochondrial gene probes to Sal I and Xba I fragments of mtDNA's from *D. carota* line 28K, cv. K5 and their cybrids. **A** Cytochrome oxidase subunit II gene probe (COXII); **B** ATPase $\alpha$-subunit gene probe (ATP$\alpha$); **C** 26S rRNA gene probe (26S rDNA). Lanes are shown as in Fig. 5

**Table 3.** Carrot mtDNA fragments (kb) hybridized to mitochondrial gene probes

| ATP*α* | Sal I | | | | | | | Xba I | | |
|---|---|---|---|---|---|---|---|---|---|---|
| K5 | 15.0 | | | 6.8 | 3.9 | | 1.45 | 7.2 | | 3.3 |
| A-1-20 | | 14.5 | | 6.8 | | 2.8 | 1.45 | 7.2 | 5.2 | 3.3 |
| E-2-9 | | 14.5 | | 6.8 | | | 1.45 | 7.2 | | 3.3 |
| E-1-6 | | 14.5 | | 6.8 | | | 1.45 | 7.2 | | 3.3 |
| 28K | | | 9.0 | 6.8 | | | 1.45 | 7.2 | | 3.3 |
| COXII | Sal I | | | XbaI | | | | 26S | Sal I | |
| K5 | 4.9 | | 1.8 | 8.4 | | | | K5 | | 13.0 |
| A-1-20 | 4.9 | 2.8 | 1.8 | 8.4 | 5.2 | | | A-1-20 | 15.0 | |
| E-2-9 | 4.9 | | 1.8 | 8.4 | | | | E-2-9 | 15.0 | 13.0 |
| E-1-6 | 4.9 | | 1.8 | 8.4 | | | | E-1-6 | 15.0 | |
| 28K | 4.9 | | 1.8 | 8.4 | | | | 28K | 15.0 | |

a donor-specific band (15 kb: Sal I) was much higher than a recipient-specific band (13 kb) in the cybrid. The other two cybrid mtDNA's, however, possessed only a donor-specific fragment. The recipient-specific 26S rDNA would be eliminated from their mtDNA's in the course of recombination. From the Southern analysis, the mtDNA's from carrot cybrid plants were also found to possess complex combination of both parental mtDNA's as well as a *Brassica* somatic hybrid (Robertson et al. 1987), *Brassica* cybrids (Vedel et al. 1986; Morgan and Maliga 1987), tobacco cybrids (Nagy et al. 1981; Galun et al. 1982), *Petunia* somatic hybrids (Boeshore et al. 1985; Rothenberg et al. 1985), and *Panicum* + *Pennisetum* somatic hybrids (Ozias-Akins et al. 1987).

## 5 Conclusion

Cybrids are successfully formed in *Daucus* by donor-recipient protoplast fusion, and mitochondrial traits detected by Southern hybridization can be transferred from donor to recipient cells. It is indicated that the selection systems based on metabolic complementation could be effective in the introduction of cytoplasmic traits which exhibit maternal inheritance in *Daucus*.

MtDNA recombinations were detected in the *Daucus* cybrid plants. Nagy et al. (1983) suggested that mtDNA rearrangements could be detected only in heteroplasmic fusion combinations. The factors which affect mtDNA recombinations in fusion-derived plants, however, have not been elucidated. Direct repeat elements are thought to be involved in rearrangements in mtDNA of *Zea* (Lonsdale et al. 1984) and *Brassica* (Palmer and Shields 1984). There might be some relationships between the degree of mtDNA recombinations in the cybrids and the number or structure of repeat elements within those mtDNA's. By examining intra- or intermolecular mtDNA rearrangements further, the mechanisms of mtDNA recombinations would be elucidated in more detail. CMS, mtDNA restriction patterns and nonchromosomal stripe mutants of maize have been known as mitochondrial markers until now (Leaver and Gray 1982; Levings 1983; Newton and Coe 1986; Lonsdale 1987). In order to regulate desirable combinations of organelles in fused cells, isolation of mitochondrial mutants with the selectable markers should be required.

*Acknowledgments.* Gratitude is expressed to all the members of Plantech Research Institute for their advice and discussions.

## References

Asahi T, Kumashiro T, Kubo T (1988) Constitution of mitochondrial and chloroplast genomes in male sterile tobacco obtained by protoplast fusion of *Nicotiana tabacum* and *N. debneyi*. Plant Cell Physiol 29:43–49

Aviv D, Bleichman S, Arzee-Gonen P, Galun E (1984a) Interspecific cytoplasmic hybrids in *Nicotiana*. Identification of plastomes and chondriomes in *N. sylvestris*+*N. rustica* cybrids having *N. sylvestris* nuclear genomes. Theor Appl Genet 67:499–504

Aviv D, Arzee-Gonen P, Bleichman S, Galun E (1984b) Novel alloplasmic *Nicotiana* plants by "donor-recipient" protoplast fusion: cybrids having *N. tabacum* or *N. sylvestris* nuclear genomes and either or both plastomes and chondriomes from alien species. Mol Gen Genet 196:244–253

Barsby TL, Chuong PV, Yarrow SA, Wu S-C, Coumans M, Kemble RJ, Powell AD, Beversdorf WD, Pauls KP (1987a) The combination of Polima cms and cytoplasmic triazine resistance in *Brassica napus*. Theor Appl Genet 73:809–814

Barsby TL, Yarrow SA, Kemble RJ, Grant I (1987b) The transfer of cytoplasmic male sterility to winter-type oilseed rape (*Brassica napus* L.) by protoplast fusion. Plant Sci 53:243–248

Bayliss MW (1975) The effect of growth in vitro on the chromosome complement of *Daucus carota* (L.) suspension cultures. Chromosoma 51:401–411

Belliard G, Vedel F, Pelletier G (1979) Mitochondrial recombination in cytoplasmic hybrids of *Nicotiana tabacum* by protoplast fusion. Nature (Lond) 281:401–403

Boeshore ML, Lifshitz I, Hanson MR, Izhar S (1983) Novel composition of mitochondrial genomes in *Petunia* somatic hybrids derived from cytoplasmic male sterile and fertile plants. Mol Gen Genet 190:459–467

Boeshore ML, Hanson MR, Izhar S (1985) A variant mitochondrial DNA arrangement specific to *Petunia* stable sterile somatic hybrids. Plant Mol Biol 4:125–132

Chétrit P, Mathieu C, Vedel F, Pelletier G, Primard C (1985) Mitochondrial DNA polymorphism induced by protoplast fusion in Cruciferae. Theor Appl Genet 69:361–366

Chuong PV, Beversdorf WD, Powell AD, Pauls KP (1988) Somatic transfer of cytoplasmic traits in *Brassica napus* L. by haploid protoplast fusion. Mol Gen Genet 211:197–201

Clark E, Schnabelrauch L, Hanson MR, Sink KC (1986) Differential fate of plastid and mitochondrial genomes in *Petunia* somatic hybrids. Theor Appl Genet 72:748–755

Dawson AJ, Jones VP, Leaver CJ (1986) Strategies for the identification and analysis of higher plant mitochondrial genes. Meth Enzymol 118:470–488

DeBonte LR, Matthews BF (1984) Rapid isolation and purification of plastid and mitochondrial DNA from carrot cell suspensions. Plant Mol Biol Rep 2:32–36

D'Hont A, Quetier F, Teoule E, Dattee Y (1987) Mitochondrial and chloroplast DNA analysis of interspecific somatic hybrids of a Leguminosae: *Medicago* (alfalfa). Plant Sci 53:237–242

Dudits D, Hadlaczky Gy, Levi E, Fejer O, Haydu Zs, Lazar G (1977) Somatic hybridization of *Daucus carota* and *D. capillifolius* by protoplast fusion. Theor Appl Genet 51:127–132

EcKenrode VK, Levings CS III (1986) Maize mitochondrial genes. In Vitro Cell Dev Biol 22:169–176

Fluhr R, Aviv D, Galun E, Edelman M (1984) Generation of heteroplastidic *Nicotiana* cybrids by protoplast fusion: analysis for plastid recombinant types. Theor Appl Genet 67:491–497

Galun E, Arzee-Gonen P, Fluhr R, Edelman M, Aviv D (1982) Cytoplasmic hybridization in *Nicotiana*: mitochondrial DNA analysis in progenies resulting from fusion between protoplasts having different organelle constitutions. Mol Gen Genet 186:50–56

Hanson MR, Conde MF (1985) Functioning and variation of cytoplasmic genomes: lessons from cytoplasmic-nuclear interactions affecting male fertility in plants. Int Rev Cytol 94:213–267

Ichikawa H, Tanno-Suenaga L, Imamura J (1987) Selection of *Daucus* cybrids based on metabolic complementation between X-irradiated *D. capillifolius* and iodoacetamide-treated *D. carota* by somatic cell fusion. Theor Appl Genet 74:746–752

Kao KN, Michayluk MR (1974) A method for high frequency intergeneric fusion of plant protoplasts. Planta 115:355–367

Kemble RJ, Barsby TL, Wong RSC, Shepard JF (1986) Mitochondrial DNA rearrangements in somatic hybrids of *Solanum tuberosum* and *Solanum brevidens*. Theor Appl Genet 72:787–793

Kothari SL, Monte DC, Widholm JM (1986) Selection of *Daucus carota* somatic hybrids using drug resistance markers and characterization of their mitochondrial genomes. Theor Appl Genet 72:494–502

Kumashiro T, Kubo T (1986) Cytoplasm transfer of *Nicotiana debneyi* to *N. tabacum* by protoplast fusion. Jpn J Breed 36:39–48

Leaver CJ, Gray MW (1982) Mitochondrial genome organization and expression in higher plants. Annu Rev Plant Physiol 33:373–402

Levings CS III (1983) The plant mitochondrial genome and its mutants. Cell 32:659–661

Lonsdale DM (1987) Cytoplasmic male sterility: a molecular perspective. Plant Physiol Biochem 25:265–271

Lonsdale DM, Hodge TP, Fauron CM (1984) The physical map and organization of the mitochondrial genome from the fertile cytoplasm of maize. Nucleic Acids Res 12:9249–9261

Maliga P (1986) Cell fusion to introduce genetic information coded by chloroplasts and mitochondria in flowering plants. In: Bogorad L (ed) Molecular developmental biology. Liss, New York, pp 45–53

Maniatis T, Fritsch EF, Sambrook J (1982) Molecular cloning: a laboratory manual. Cold Spring Harbor Lab, New York

Matthews BF, DeBonte LR (1985) Chloroplast and mitochondrial DNAs of the carrot and its wild relatives. Plant Mol Biol Rep 3:12–16

Matthews BF, Widholm JM (1985) Organelle DNA composition and isoenzyme expression in an interspecific somatic hybrid of *Daucus*. Mol Gen Genet 198:371–376

Medgyesy P, Golling R, Nagy F (1985) A light sensitive recipient for the effective transfer of chloroplast and mitochondrial traits by protoplast fusion in *Nicotiana*. Theor Appl Genet 70:590–594

Menczel L, Galiba G, Nagy F, Maliga P (1982) Effect of radiation dosage on efficiency of chloroplast transfer by protoplast fusion in *Nicotiana*. Genetics 100:487–495

Menczel L, Polsby LS, Steinback KE, Maliga P (1986) Fusion-mediated transfer of triazine-resistant chloroplasts: Characterization of *Nicotiana tabacum* cybrid plants. Mol Gen Genet 205:201–205

Menczel L, Morgan A, Brown S, Maliga P (1987) Fusion-mediated combination of Ogura-type cytoplasmic male sterility with *Brassica napus* plastids using X-irradiated CMS protoplasts. Plant Cell Rep 6:98–101

Morgan A, Maliga P (1987) Rapid chloroplast segregation and recombination of mitochondrial DNA in *Brassica* cybrids. Mol Gen Genet 209:240–246

Murashige T, Skoog F (1962) A revised medium for rapid growth and bioassay with tobacco tissue cultures. Physiol Plant 15:473–497

Nagy F, Torok I, Maliga P (1981) Extensive rearrangements in the mitochondrial DNA in somatic hybrids of *Nicotiana tabacum* and *Nicotiana knightiana*. Mol Gen Genet 183:437–439

Nagy F, Lazar G, Menczel L, Maliga P (1983) A heteroplasmic state induced by protoplast fusion is a necessary condition for detecting rearrangements in *Nicotiana* mitochondrial DNA. Theor Appl Genet 66:203–207

Newton KJ, Coe EH Jr (1986) Mitochondrial DNA changes in abnormal growth (nonchromosomal stripe) mutants of maize. Proc Natl Acad Sci USA 83:7363–7366

Nishibayashi S, Kaeriyama J (1986) Structural stability of chromosomes in rice (*Oryza sativa* L.) plants regenerated from somatic tissue culture. Plant Tissue Cult Lett 3:31–34

O'Connell MA, Hanson MR (1986) Regeneration of somatic hybrid plants formed between *Lycopersicon esculentum* and *Solanum rickii*. Theor Appl Genet 72:59–65

Ozias-Akins P, Pring DR, Vasil IK (1987) Rearrangements in the mitochondrial genome of somatic hybrid cell lines of *Pennisetum americanum* (L.) K. Schum+*Panicum maximum* Jacq. Theor Appl Genet 74:15–20

Palmer JD, Shields CR (1984) Tripartite structure of the *Brassica campestris* mitochondrial genome. Nature (Lond) 307:437–440

Pring DR, Lonsdale DM (1985) Molecular biology of higher plant mitochondrial DNA. Int Rev Cytol 97:1–46

Quetier F, Lejeune B, Delorme S, Falconet D (1985) Molecular organization and expression of the mitochondrial genome of higher plants. In: Douce R, Day DA (eds) Encyclopedia of plant physiology, New Ser, vol 18. Springer, Berlin Heidelberg New York Tokyo, pp 25–36

Robertson D, Palmer JD, Earle ED, Mutschler MA (1987) Analysis of organelle genomes in a somatic hybrid derived from cytoplasmic male-sterile *Brassica oleracea* and atrazine-resistant *B. campestris*. Theor Appl Genet 74:303–309

Rothenberg M, Boeshore ML, Hanson MR, Izhar S (1985) Intergenomic recombination of mitochondrial genomes in a somatic hybrid plant. Curr Genet 9:615–618

Shepard JF, Bidney D, Barsby T, Kemble R (1983) Genetic transfer in plants through interspecific protoplast fusion. Science 219:683–688

Sidorov VA, Menczel L, Nagy F, Maliga P (1981) Chloroplast transfer in *Nicotiana* based on metabolic complementation between irradiated and iodoacetate treated protoplasts. Planta 152:341–345

Tabaeizadeh Z, Pring DR, Vasil IK (1987) Analysis of mitochondrial DNA from somatic hybrid cell lines of *Saccharum officinarum* (sugarcane) and *Pennisetum americanum* (pearl millet). Plant Mol Biol 8:509–513

Vardi A, Breiman A, Galun E (1987) *Citrus* cybrids: production by donor-recipient protoplast-fusion and verification by mitochondrial-DNA restriction profiles. Theor Appl Genet 75:51–58

Vedel F, Chétrit P, Mathieu C, Pelletier G, Primard C (1986) Several different mitochondrial DNA regions are involved in intergenomic recombination in *Brassica napus* cybrid plants. Curr Genet 11:17–24

Vedel F, Mathieu C, Chétrit P, Pelletier G, Primard C (1987) Mitochondrial DNA variation in cytoplasmic male sterile somatic hybrids of *Brassica napus*. Plant Physiol Biochem 25:249–257

Ward BL, Anderson RS, Bendich AJ (1981) The mitochondrial genome is large and variable in a family of plants (Cucurbitaceae). Cell 25:793–803

Yarrow SA, Wu SC, Barsby TL, Kemble RJ, Shepard JF (1986) The introduction of CMS mitochondria to triazine tolerant *Brassica napus* L., var. "Regent", by micromanipulation of individual heterokaryons. Plant Cell Rep 5:415–418

Zelcer A, Aviv D, Galun E (1978) Interspecific transfer of cytoplasmic male sterility by fusion between protoplasts of normal *Nicotiana sylvestris* and X-ray irradiated protoplasts of male-sterile *N. tabacum*. Z Pflanzenphysiol 90:397–407

# XXII Incorporation of the Firefly Luciferase Gene into Plant Cells

D. W. Ow[1] and S. H. Howell[2]

## 1 Introduction

Reporter, or indicator genes such as *lac*Z (beta-galactosidase) and *CAT* (chloramphenicol acetyltransferase), encode for enzymes that allow convenient and sensitive biochemical detection of gene activity. For this reason, they have been used extensively for analysis of gene regulation. Recently, luciferases, enzymes that catalyze light emission, have been successfully used to measure gene expression in bacteria, mammalian cells and plant cells (Engebrecht et al. 1985; Ow et al. 1986, 1987; de Wet et al. 1987). In principle, a luciferase gene (or gene operon) is the biological equivalent of an indicator light. By placing it under the control of a biological switch, such as when fused behind a gene promoter, light emission becomes a reflection of promoter activity. Not only does it indicate when and where the biological switch is turned on, but the intensity of light emission measures transcriptional strength. Aside from monitoring gene promoter activity, a biological indicator light can also be put to many other uses. For example, it can serve as a screen-able marker in the transfer of DNA into eukaryotic cells, or as a convenient tag for studies of transmission and population genetics. This chapter reviews our recent work on the development of the firefly luciferase gene as a reporter of gene expression in plants.

## 2 The Firefly Luciferase

The use of light to measure gene expression was first accomplished using the *Vibrofischeri lux* operon fused to bacterial promoters (Engebrecht et al. 1985). The bacterial luciferase is composed of two nonidentical subunits, encoded by genes *luxA* and *luxB*. For its use as a reporter enzyme in a eukaryotic system, fusions to both *lux* genes would be necessary. In contrast, the firefly (*Photinus pyralis*) luciferase is a single polypeptide and, hence, only a single promoter-gene construct is necessary for expression in eukaryotic cells.

The firefly luciferase light-emitting reaction requires the substrates $O_2$, Mg-ATP, and a small heterocyclic carboxylic acid known as luciferin (reviewed by

[1] Plant Gene Expression Center, U.S. Dept. of Agriculture, and University of California at Berkeley, 800 Buchanan Street, Albany, CA 94710, USA
[2] Boyce Thompson Institute, Cornell University, Tower Road, Ithaca, NY 14853, USA

Biotechnology in Agriculture and Forestry, Vol. 9
Plant Protoplasts and Genetic Engineering II (ed. by Y. P. S. Bajaj)

McElroy and DeLuca 1983). As shown below, this series of reactions begins with the adenylation of the enzyme-bound luciferin. Next, luciferin is oxidatively decarboxylated, leaving the enzyme-bound oxyluciferin in an excited state. Return to ground state releases light energy with a high quantum yield (>0.8).

luciferase + Mg-ATP + luciferin → luciferase-luciferyl-AMP + PPi
luciferase-luciferyl-AMP + $O_2$ → luciferase-oxyluciferin* + $CO_2$ + AMP
luciferase-oxyluciferin* → luciferase + oxyluciferin + light (560 nm).

Because oxyluciferin has high affinity for the enzyme, the kinetics of the reaction is typical of product inhibition; and the intensity of the burst of light is directly proportional to enzyme concentration over four orders of magnitude.

Assaying for luciferase activity in vitro is simple and rapid. The components of the reaction are commercially available and comparatively inexpensive. A luminometer, which is a photomultiplier tube connected to a signal amplifier, can be used to measure light output. One of the most sensitive luminometers on the market today (Monolight 2001) can detect the light emission from about $5 \times 10^5$ molecules of active enzyme. In comparison, the minimal detectable signal for a standard chloramphenicol acetyltransferase assay requires about $3 \times 10^8$ molecules of active enzyme. Depending on the instrumentation used, assaying for luciferase activity has been estimated to be anywhere from 10- to 1000-fold more sensitive than is a $^{14}C$ assay for CAT (de Wet et al. 1987). Moreover, the measurement for luciferase is linear over a four log range.

In 1985, the firefly luciferase cDNA was cloned and expressed in *E. coli* (de Wet et al. 1985). Recently, the entire cDNA sequence has been determined (de Wet et al. 1987). Our laboratory was interested in using this cDNA as a reporter gene in plant cells, particularly for use in transient expression assays. Because gene transfer into plant cells has not been as efficient as that into mammalian cells, there is a need to compensate for this deficiency by using a more sensitive reporter gene. In addition to increasing sensitivity, we were also fascinated with the possibility that a transgenic plant would emit light if substrates were provided. If this were possible, it would allow the visualization of the spatial and temporal patterns of gene expression in a growing plant.

## 3 Expression of the CaMV 35S-Luciferase Chimeric Gene

To obtain expression in plant cells, the luciferase (*luc*) cDNA was fused behind the Cauliflower Mosaic Virus (CaMV) 35S RNA promoter (Ow et al. 1986). Although this ~ 1800 bp cDNA fragment contains a consensus polyadenylation (polyA) signal, it was difficult to predict how well a firefly polyA signal would function in plant cells. To insure expression, a 3′ end fragment from the nopaline synthase (*nos*) gene was fused behind the luciferase cDNA. This *nos* 3′ end fragment contains a polyA signal known to function in plant cells. The entire CaMV-*luc-nos* chimeric gene was inserted into the multicopy plasmid vector pUC19.

The prototype construct (pDO432) was functionally tested in an electroporation transient-expression assay (Fromm et al. 1985). This technique uses an elec-

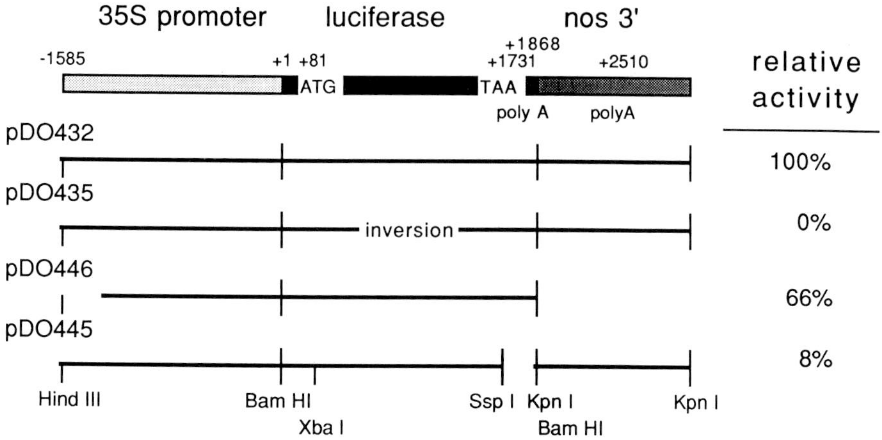


**Fig. 1.** Relative activity of CaMV 35S-luciferase constructs in transient expression assays. (Ow et al. 1986)

tric pulse, released from the discharge of a capacitor, to allow penetration of DNA through cell membranes. In a typical experiment, 10 µg of plasmid DNA were mixed with ca. 5 million protoplasts from a carrot suspension cell line (W001c). Up to 25000 light units per million protoplasts have been detected in many recent experiments. This is greater than 2500-fold above the background noise level. Since each light unit in our luminometer (LKB model 1250) is equivalent to the activity of about $3 \times 10^5$ molecules of active enzyme, the amount of light output is about 7500 molecules per cell, on the average, or about 50 pg of enzyme per µg protein in the extract.

Activity is not detected when the *luc* cDNA fragment is cloned in the reverse orientation, indicating that luciferase expression is driven by the CaMV 35S promoter (Fig. 1). When the *nos* 3′ end fragment is omitted, activity drops by about a third. When the polyA signal present on the 3′ end of the luciferase cDNA is omitted, activity is reduced by about tenfold. Although the reason for this is not clear, one explanation may be that the more proximal firefly polyA signal is preferred over the more distal signal within the *nos* gene 3′ end fragment. The combination of two polyA signals gives the highest level of expression.

The construct producing the most luciferase activity, pDO432, was used for stable gene transfer experiments. pDO432 was linearized with a restriction endonuclease and inserted between the right and left T-DNA borders of pBIN19, a binary vector derived from the tumor-inducing (Ti) plasmid of *Agrobacterium tumefaciens* (Bevan 1984). A *nos-npt*II chimeric gene, which serves as a selective marker (kanamycin resistance) for plant cells, is also contained within pBIN19 between the T-DNA borders. This pBIN19::pDO432 cointegrate was conjugated into an *Agrobacterium* host, that also harbors a disarmed Ti plasmid to provide DNA transfer functions in trans. The *Agrobacterium* exconjugant was used to inoculate tobacco leaf discs according to established procedures (Horsch et al. 1985). In brief, this involves soaking leaf explants in plant medium containing the desired *Agrobacterium* strain. After sufficient time is allowed for DNA transfer, infecting bacteria are eliminated with antibiotics, and the leaf explants transferred to shoot-inducing medium. Only transformed cells, having acquired the closely linked kanamycin resistance gene, would regenerate into shoots in the presence of kanamycin. Shoots are then excised and moved to hormone-free medium to induce root formation.

In this fashion, transgenic plants were selected for kanamycin resistance and screened for luciferase activity. Initially, plants were screened for luciferase activity by assaying leaf extracts. As it turned out, an easier method was to excise a small piece of leaf explant (ca. 3 $mm^2$), drop it into buffer containing luciferin, and monitor for light production in a luminometer. Apparently, luciferin is able to diffuse into plant cells, where there are sufficient amounts of ATP and $O_2$ to allow light production. The ease of this procedure allows the screening of hundreds of plants in a few hours. In species where the use of antibiotic selection is

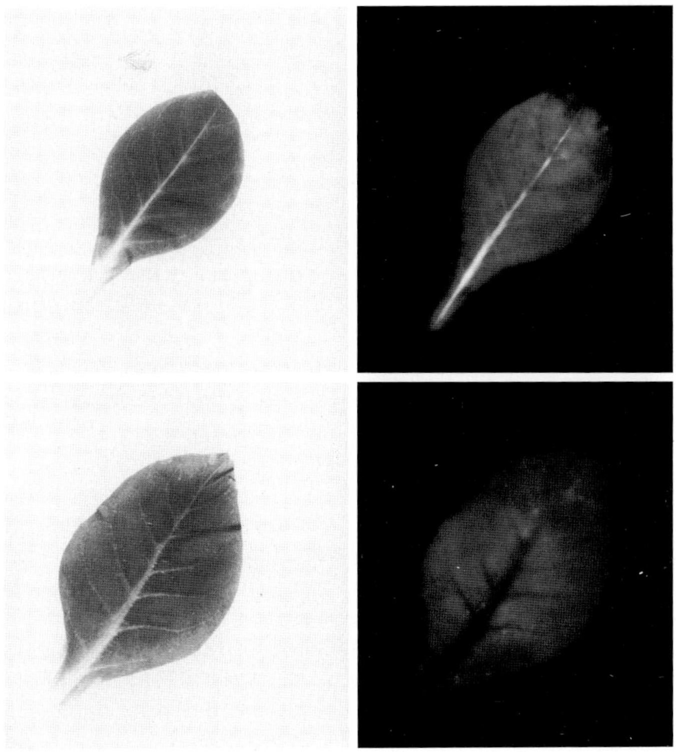

a

b

**Fig. 2a, b.** Spatial pattern of luminescence in transgenic plant leaves. **a** (*top*) Leaf from a tobacco (*Nicotiana tabacum*) primary transformant, harboring the CaMV 35S promoter-luciferase gene fusion, photographed with ordinary light (*left photo*) and by its own illumination (*right photo*). To detect light emission, the leaf petiole was dipped into a 1 mM luciferin solution for an hour and exposed by contact with X-ray film for 3 min. A positive image was made from the X-ray film. **b** (*bottom*) As in **a**, but the leaf was obtained from a primary transformant harboring the *rbc*S promoter-luciferase gene fusion. Due to relatively low luciferase activity, exposure time on the X-ray film was extended to 3 h

**Fig. 3.** Light production from a transgenic tobacco plant harboring the CaMV 35S promoter-luciferase gene fusion. The roots of the plant were dipped into a 1 mM solution of luciferin for 4 h and exposed by contact to Kodak Ektachrome 200 film for 24 h. (Ow et al. 1986)

difficult, screening a large number of putative transformants for the luciferase marker may be a worthwhile alternative.

Since luciferin can diffuse into plant cells, we tried dipping the petioles of whole leaves (ca. 2.5×4.5 cm) into luciferin solution (1 mM) and photographing the light emission by contact exposure to film. On short time exposures, however, saturation of the X-ray film can make it difficult to differentiate the relative light intensity from different tissues within the leaf. Figure 2 a shows a positive print made from a typical short time exposure on X-ray film. Illuminated areas reveal the light-emitting tissues of the leaf, with the veins showing relatively greater luciferase activity.

An entire plant was also photographed in a similar fashion. Figure 3 shows a young transgenic tobacco plant (ca. 12 cm in height) photographed by its own light. In this experiment, the roots of the plant were soaked in a 1 mM luciferin solution for 6 h, covered with a plastic wrap, placed on a piece of foam, and contact exposed overnight to a sheet of Kodak Ektachrome ASA 200 film. As shown, the relative intensity of light is strongest in the vascular system. It also appears that the younger uppermost leaves emitted more light than the older, lower ones.

## 4 Spatial Pattern of Gene Expression

The whole plant and leaf shown in Fig. 2a and Fig. 3, as well as in other similar experiments, show that light is emitted in a specific pattern. This pattern of luciferase activity suggested several possible explanations. Among these are (1)

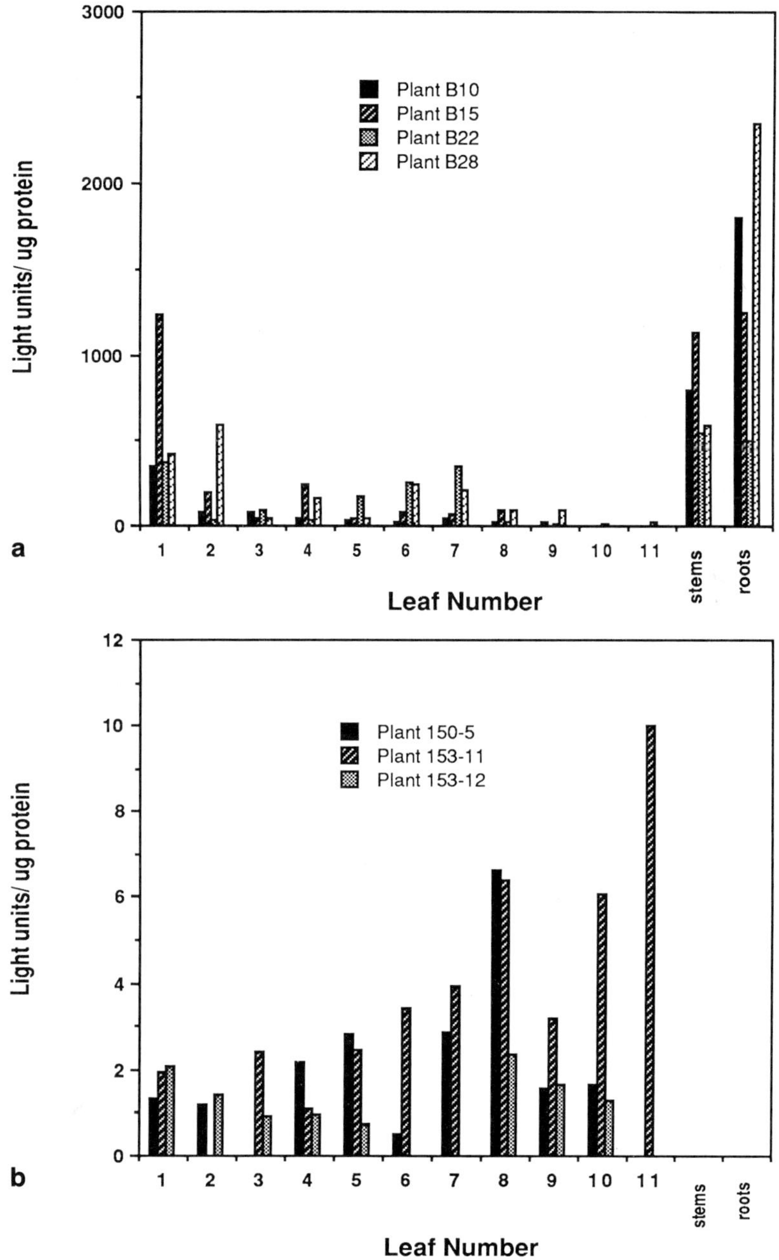


**Fig. 4a, b.** Distribution of luciferase activity in the organs of transgenic plants. Plants were aseptically grown in glass jars until a height of 12 to 15 cm. Extracts were prepared from plant organs by grinding a small amount of tissue in liquid $N_2$. Luciferase activity was determined as described in Fig. 1. Leaf number is the position of the leaf on the plant, starting from the younger leaves at the top of the plant progressing to the older leaves on the bottom. Not all plants have the same number of leaves. **a** Four primary transformants (B10, B15, B22, B28) harboring the CaMV 35S promoter-luciferase gene fusion. **b** Three primary transformants (150-5, 153-11, 153-12) harboring the *rbc*S promoter-luciferase gene fusion. (Ow et al. 1986)

differential gene expression, (2) relative accessibility of the substrates luciferin, ATP, and/or $O_2$, and (3) localization of luciferase to vascular tissues. Since there were no previous data on the spatial pattern of activity of the CaMV 35S RNA promoter, we were unable at the time to differentiate among these possibilities.

Recently, we have obtained two lines of evidence that suggest that the pattern of light emission detected by photographic means could be a reflection of the tissue-specific pattern of gene activity. First, luciferase activities found in extracts made from different tissues correlate with the pattern of light emission seen in Fig. 3. Figure 4a shows the tissue distribution of luciferase activity found in extracts of four primary transformants. As was seen using photographic techniques, greater activity was found in extracts from roots, stems, and younger leaves. This argues against the possibility that substrate accessibility was the predominant factor in determining the pattern of light emission seen in Fig. 3.

The second line of evidence was derived from recent studies on a well-defined leaf-specific promoter (Ow and Howell in prep.). We exchanged the CaMV 35S RNA promoter in pDO432 with the ribulose-bisphosphate carboxylase small subunit (*rbs*S) promoter from tobacco (Mazur and Chui 1985) and transferred this construct (pDO607) into whole plants. The tissue distribution of luciferase activity of three transgenic plants shows that transcription from the *rbc*S promoter is leaf-specific (Fig. 4b). Although expression levels were relatively low, light emission from individual leaves could be detected by contact exposure to X-ray film. Figure 2 b shows a positive print made from one such film. Contrary to the pattern seen with the CaMV 35S promoter, light emission was undetectable from the veins of the leaf. This absence of luciferase activity in vascular tissues is strong evidence that the pattern of light emission found with the CaMV 35S promoter cannot be attributed predominantly to substrate accessibility or localization of enzyme to vascular tissues. We conclude that the CaMV 35S promoter, aside from being a strong promoter, also has properties which allow it to be more active transcriptionally in vascular tissues.

## 5 Expression of Luciferase in Cultured Cells

Incorporating a sensitive reporter gene into a plant cell line can open many research opportunities. For example, protoplasts derived from such a line can be used for protoplast fusion experiments. Alternatively, a light-emitting cell line can be used to study the mutation rate of cultured cells by monitoring the variation in luciferase activity. The rate and molecular basis of the variation may shed light on the molecular nature of somaclonal variation, a phenomenon of frequent occurrence in cultured plant cells.

From a transgenic plant, a cell line was initiated from leaf protoplasts (Ow et al. 1986). Extracts from cells grown in suspension showed moderate activity (210 light units per microgram protein). To monitor for light production in vivo, cells were distributed into the wells of a microtiter dish using approximately 375 to 3000 cells per well. Luciferin, sterilized by filtration, was added to a final molarity of 16 to 400 μM. The dish of cells was then placed on X-ray film for an hour to

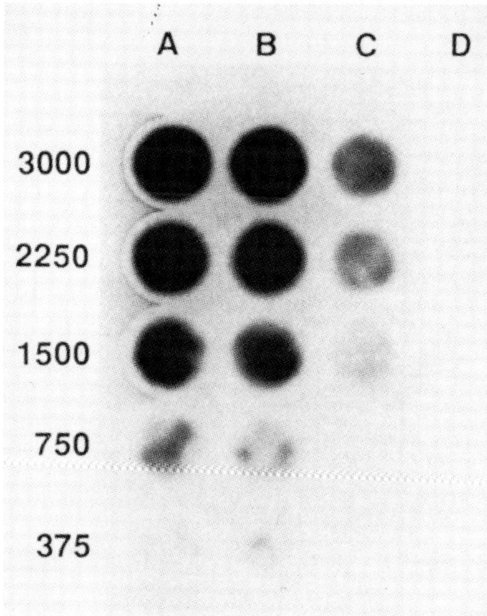


**Fig. 5.** Luminescence from cultured tobacco cells harboring the CaMV 35S promoter-luciferase fusion. Suspension culture lines were derived from a transgenic plant (columns *A, B,* and *C*) and an untransformed tobacco (column *D*). The approximate number of cells, indicated by the *numbers to the left* of each row, were distributed into the wells of a microtiter dish, supplemented with luciferin (columns *A* and *D* 400 μM, *B* 80 μM, and *C* 16 μM), and exposed to X-ray film for 1 h. The edges of the microtiter dish were milled so that the bottom of the wells were in direct contact with the film. (Ow et al. 1986)

detect light production. Figure 5 shows that 80 μM of luciferin was sufficient for optimal detection of activity. The cells survived this concentration of luciferin in long-term incubation (up to several weeks). Hence, at a luciferin concentration of 80 μM, this assay constitutes a simple, nondisruptive procedure for detecting luciferase activity from a large number of independent cell lines.

## 6 Transient Expression Studies

Transient expression assays allow rapid analysis of gene activity from a large collection of DNA constructs, and can be useful for comparing the intrinsic strength determined by functional elements of a gene promoter. For example, the functions of the TATA box, the CCAAT box, and the enhancer elements of many eukaryotic promoters have been examined through transient expression studies (e.g., see Banerji et al. 1981; Graves et al. 1986; Myers et al. 1986).

We have explored the use of luciferase gene fusions in the analysis of functional regions of the CaMV 35S RNA promoter (Ow et al. 1987). A set of 5′ end deletions of the CaMV 35S promoter was generated and introduced into carrot protoplasts by electroporation. Figure 6 shows that this promoter can be functionally defined by three regions. The proximal region consists of the minimal promoter that extends to −73 bp relative to the cap site. This region contains a TATA box centered at −28. The medial region is defined as the segment between −73 and −89. Inclusion of this segment gives an additional 30-fold increase in activity over the minimal promoter. Full activity is obtained with the inclusion of both the medial and distal regions (to −148).

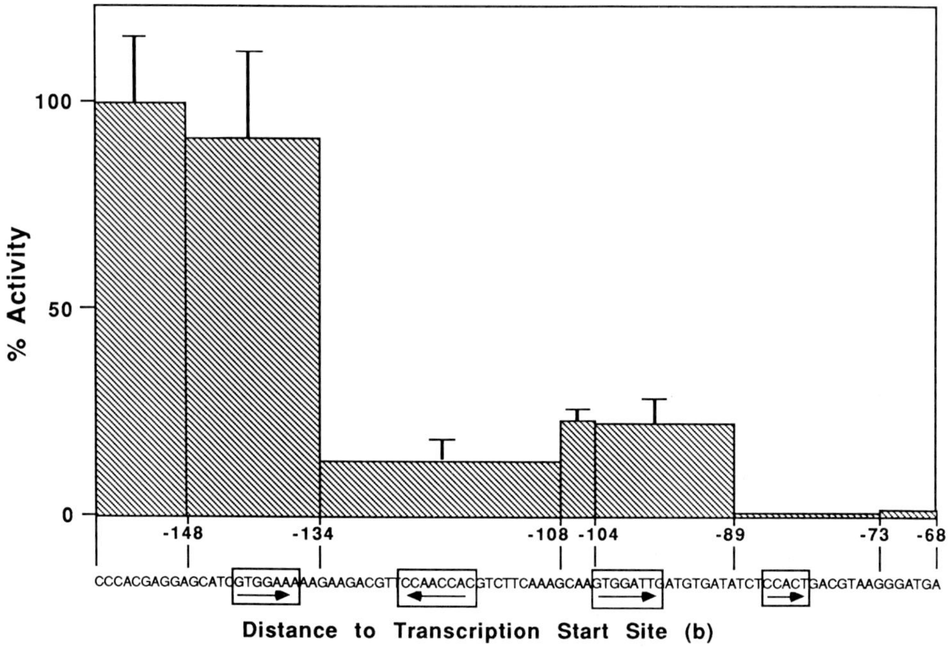


**Fig. 6.** Relative activity of 5′ end deletions through the distal and medial regions of the CaMV 35S RNA promoter. Relative mean values and population standard deviations, obtained from six or more transient expression samples, are normalized to a construct with a deletion end point at −148 bp. Deletion end points, and homologs of the consensus core enhancer and the CCAAT box are indicated in the DNA sequence of the promoter. (Ow et al. 1987)

The distal region contains three homologs to the consensus enhancer core sequence found in many animal and viral promoters (Fig. 6). This region has some of the properties of a eukaryotic enhancer element. For example, the segment from −148 to −89 can be inverted with respect to the core promoter (medial plus proximal regions) without abolishing its enhancing activity (Fig. 7). In addition, multiple copies of the distal region can further increase the transcriptional strength of its own core promoter, as well as a heterologous promoter, the 19S RNA promoter from CaMV (Fig. 8). Its transcriptional enhancement effect, however, is dependent on position. It loses effectiveness when fused farther upstream of the core promoter, or downstream from the transcription unit (Figs. 7, 8). In this respect, the distal region, at least in a single copy, more closely resembles a typical upstream activation sequence found in eukaryotic promoters.

## 7 Conclusions and Prospects

The luciferase assay is a powerful tool for assessing gene expression in plants. The assay is highly sensitive, permitting measurement of low levels of luciferase production. There is no detectable background in extracts from nontransformed

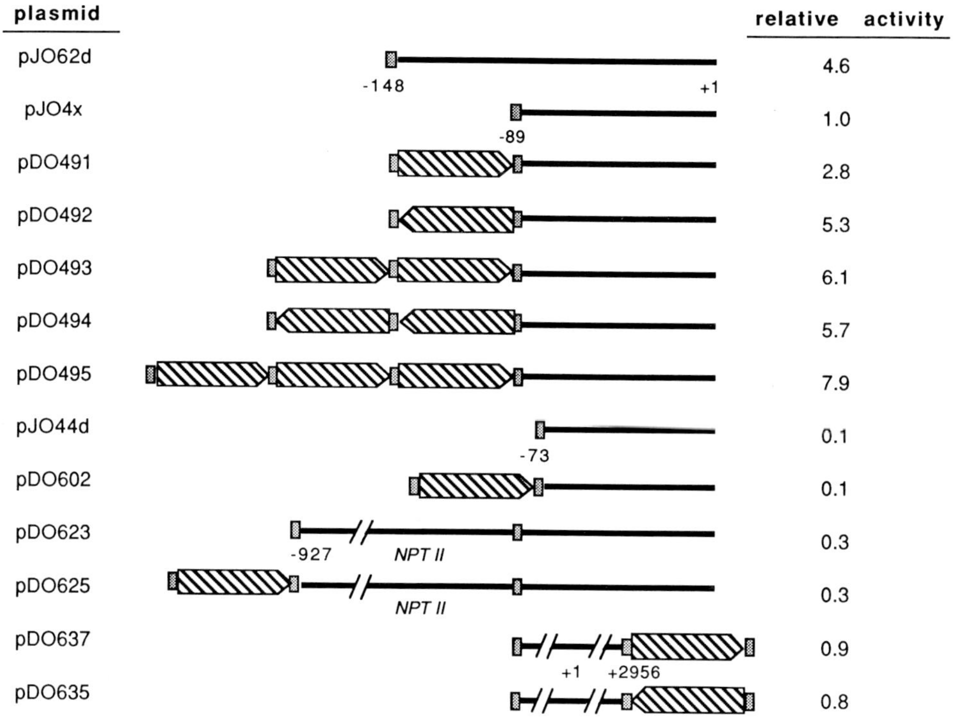


**Fig. 7.** Inversion and duplication of the CaMV 35S promoter distal region segment (−148 to −89, shown as *hatched arrow*) flanked by *Xho*I and *Sal*I linkers (shown as *rectangular boxes*). Relative activity, normalized to a construct with a deletion end point at −89 bp, represents the mean value from six or more transient expression samples. (Ow et al. 1987)

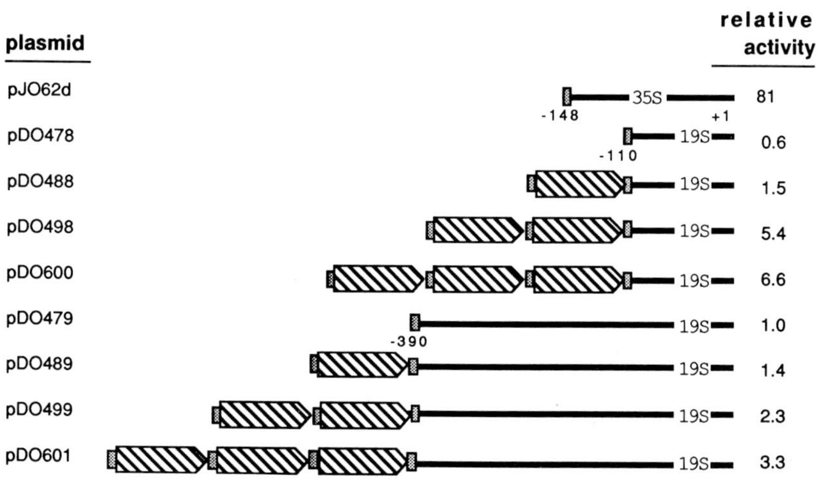


**Fig. 8.** Fusion of the CaMV 35S RNA promoter distal region segment (indicated as in Fig. 7) to the CaMV 19S RNA promoter. Relative activity, normalized to the 19S RNA promoter with a 5′ end point at −390, represents the mean value of six or more transient expression samples. (Ow et al. 1987)

plant cells. The enzyme is stable in extracts from the plant cells that we have examined (tobacco, carrot, and maize), since extracts left at 4 °C for a week retained nearly full activity. Most importantly, since the luciferase assay is rapid and inexpensive, multiple samples can be tested, as in our transient expression studies, to obtain more quantitative data.

The luciferase gene can be considered a potentially ideal reporter gene for studying the spatial and temporal regulation of gene expression. By using simple X-ray film exposure or by more elaborate devices such as image-intensifying video equipment, the spatial pattern of luciferase activity in plant tissues or whole plants can be visualized. Because detection of activity can be nondestructive and noninvasive, luciferase activity can be assessed periodically over time, such as through the course of plant development.

Finally, the use of this gene can be extended beyond studies of gene expression. For plant species where the use of an antibiotic selection has met with little success, screening for a luciferase-positive phenotype could be a worthwhile alternative. Plant tissues from a large number of putative transformants could be screened by quantitative enzyme assays or simply by an exposure to X-ray film. In another example, one can imagine incorporating the luciferase gene into the genome of *Drosophila* or plants, hence allowing population biologists or transmission geneticists to trace the spread of a conveniently assayable trait.

## 8 Protocol

To place the luciferase cDNA under the transcriptional control of the CaMV 35S promoter, an expression vector was first constructed placing a *Hind*III-*Bam*HI fragment containing the CaMV 35S promoter and a *Bam*HI-*Eco*RI fragment containing the 3′ end of the nopaline synthase gene (to provide a plant polyadenylation signal) into pUC19. The luciferase cDNA was then inserted between these two fragments at the unique *Bam*HI site. This prototype plasmid (pDO432) was used for both transient and stable gene transfer experiments. For transient assays, plasmid DNA were introduced into carrot protoplasts (from W001c suspension line) by electroporation (275 V, 100 ms discharge from a 320 μF capacitor). Luciferase activity in protoplast extracts were assayed in 14 mM glycylglycine buffer (pH 7.8), 14 mM $MgCl_2$, 6 mM ATP, and 0.2 mM luciferin. Light intensity was measured in a luminometer (LKB model 1250). For stable incorporation of the luciferase gene into the tobacco genome, pDO432 was ligated to the *Hind*III site of pBIN19, between the *Agrobacterium tumefaciens* T-DNA borders (Bevan 1984). The cointegrate pBIN19::pDO432 plasmid was mobilized with helper plasmid pRK2013 (Ditta et al. 1980) into *A. tumefaciens* strain GV3111 bearing the helper plasmid pTiB63SE (Horsch et al. 1985). Transconjugants were selected with antibiotics [kanamycin (50 μg/ml), chloramphenicol (10 μg/ml), carbenicillin (30 μg/ml)], and used to inoculate *Nicotiana tabacum* (Wisconsin-38 cultivar). Small leaf explants (about 5 $mm^2$) from aseptically grown plants were soaked in 10 ml of MS media containing bacteria at a cell density of approximately $1 \times 10^5$ per ml. After 2 days, the leaf explants were transferred into fresh MS agar plates containing hormones for shoot induction (0.1 μg/ml naphthyleneacetic acid, 1 μg/ml benzyladenine), and antibiotics for counter-selection of *Agrobacteria* (500 μg/ml cefotaxime). When microsprouts begin to appear in about a week, the leaf explants were transferred to fresh MS shoot-inducing medium further supplemented with kanamycin (100 μg/ml) to select for transformed sprouts. Shoots of 1 to 2 cm in height were excised from the explant and transferred into MS agar medium to allow for root formation. Tissues from putative transformed plants were assayed for luciferase activity.

*Acknowledgments.* We are grateful to our colleagues K. Wood, M. DeLuca, J. de Wet, D. Helinski, J. Jacobs, and S. Lee, who have contributed to the various aspects of this work. This project was supported by the USDA-SEA competitive grants program (85-CRCR-1-1540), by the University of California Biotechnology Research and Education Program, and by an NSF postdoctoral fellowship in plant biology.

## References

Banerji J, Rusconi S, Schaffner W (1981) Expression of a beta-globin gene is enhanced by remote SV40 DNA sequences. Cell 27:299–308

Bevan M (1984) Binary *Agrobacterium* vectors for plant transformation. Nucleic Acids Res 12:8711–8720

de Wet JR, Wood KV, Helinski DR, DeLuca M (1985) Cloning of firefly luciferase cDNA and the expression of active luciferase in *Escherichia coli*. Proc Natl Acad Sci USA 82:7870–7873

de Wet JR, Wood KV, DeLuca M, Helinski DR, Subramani S (1987) Firefly luciferase gene: structure and expression in mammalian cells. Mol Cell Biol 7:725–737

Ditta G, Stanfield S, Corbin D, Helinski DR (1980) Broad host range DNA cloning system for gram-negative bacteria: construction of a gene bank of *Rhizobium mellioti*. Proc Natl Acad Sci USA 77:7347–7351

Fromm M, Taylor LP, Walbot V (1985) Expression of genes transferred into monocot and dicot plant cells by electroporation. Proc Natl Acad Sci USA 82:5824–5828

Engebrecht J, Simon M, Silverman M (1985) Measuring gene expression with light. Science 227:1345–1347

Graves BJ, Johnson PF, McKnight SL (1986) Homologous recognition of a promoter domain common to the MSV LTR and the HSV tk gene. Cell 44:565–576

Horsch RB, Fry JE, Hoffmann NL, Eichholtz D, Rogers SG, Fraley RT (1985) A simple and general method for transferring genes into plants. Science 227:1229–1231

Mazur BJ, Chui CF (1985) Sequence of a genomic DNA clone for the small subunit of ribulose bisphosphate carboxylase-oxygenase from tobacco. Nucleic Acids Res 7:2373–2386

McElroy W, DeLuca M (1983) Firefly and bacterial luminescence: basic science and applications. J Appl Biochem 5:197–209

Myers RM, Tilly K, Maniatus T (1986) Fine structure genetic analysis of a beta-globin promoter. Science 232:613–618

Ow DW, Wood KV, DeLuca M, de Wet J, Helinski DR, Howell SH (1986) Transient and stable expression of the firefly luciferase gene in plant cells and transgenic plants. Science 234:856–859

Ow DW, Jacobs JD, Howell SH (1987) Functional regions of the CaMV 35S RNA promoter determined by use of the firefly luciferase gene as a reporter of promoter activity. Proc Natl Acad Sci USA 84:4870–4874

# XXIII Uptake of Viruses by Plant Protoplasts and Their Use as Transforming Agents

A. C. CASSELLS[1]

## 1 Introduction

Genetically engineered plant viruses have potential, as nonintegrating transforming agents, to alter the plant phenotype, cross-protect against virulent virus strains or exploit the plant as an expression vehicle for the production of novel compounds (Cassells 1986; Schilperoort 1986). The development and exploitation of nonintegrating transforming agents involves consideration of the strategy outlined in Fig. 1 as well as health and safety, ecological and commercial risk evaluation (Cassells 1986). The choice of the progenitor of the transforming agent, having considered the above, will depend on its amenability to genetic engineering, and also on the development of a system in which its expression can be studied. It is in the latter regard that studies on virus uptake and expression in protoplasts relate to the subject of this volume.

In so far as animal virologists led the way in elucidating the events in viral replication in eukaryotes (e.g., Burke and Russell 1975), it is logical that plant virologists would seek to confirm the existence of parallel events in plant virus replication. Thus plant virologists speculated about receptor sites on the plasmalemma, uncoating sites, etc. (Fig. 2). Shaw (1985) has listed four routes whereby it is suggested that viruses enter the plant cell (Table 1). As is pointed out, direct evidence for any of these pathways is lacking. A priori, since most plant viruses are vector-transmitted, it may be reasonably hypothesized that the encapsulated genome is introduced directly into the cytoplasm and that no virus-specific receptor sites exist on the exterior of the plasmalemma (Fig. 2). This assumption was confirmed in initial studies on virus replication in isolated protoplasts; most plant

**Table 1.** Postulated channels for the entry of plant viruses into cells

1. Direct penetration of the cell wall and plasmalemma following wound damage (Rawlins and Tompkins 1936).
2. Deposition of virus on the plasmalemma following disruption of the cell wall, uptake of the virus particles by pinocytosis (Mundry 1963).
3. Entry via subcuticular electron-lucid areas in the wall caused by abrasion and forming a transient fluid bridge with the plasmalemma (Favali et al. 1977).
4. Entry via ectodesmata (Merkens et al. 1972).

[1] Department of Plant Science, University College, Cork, Ireland

Biotechnology in Agriculture and Forestry, Vol. 9
Plant Protoplasts and Genetic Engineering II (ed. by Y. P. S. Bajaj)

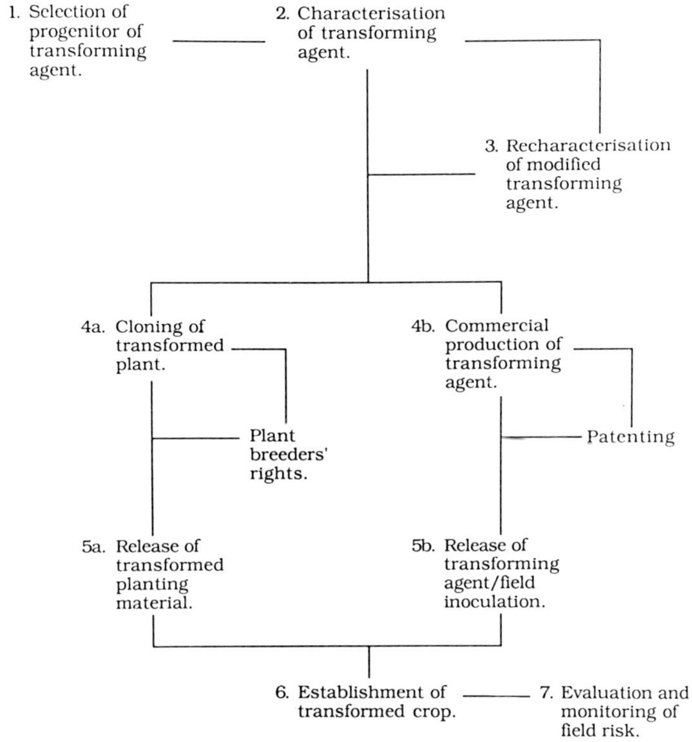


**Fig. 1.** Stages in the development of nonintegrating transforming agents. 1: Selection of the progenitor of the transforming agent involves consideration of the taxonomy, genetics, pathology and ecology of the plant viruses. 2: Characterization continues the broad studies of 1 to the virus strain level. 3: Following genetic engineering of the progenitor, the transforming agent is recharacterized, may include studies on genome expression in protoplasts. 4, 5: Commercialization may be by (a) the cloning and release of transformed plants or (b) by patenting of the agent for spray application. 6, 7: Involve establishment of the crop and subsequent monitoring etc. (for further aspects of the model see Cassells 1986)

viruses did not bind actively to the plasmalemma, that is initiate infection (Takebe 1977). Brome mosaic virus (BMV) and pea enation mosaic virus (PEMV) (Motoyoshi and Hull 1974; Motoyoshi et al. 1974a), which had a net positive charge on the particle, did bind and initiate infection, but at low frequency. Takebe and Otsuki (1969) first reported the successful chemical stimulation of inoculation of tobacco protoplasts which overcame the charge repulsion between TMV and the plasmalemma. Their method involved the clumping of tobacco mosaic virus particles with poly-L-ornithine (PLO), a polycation; clumped virus particles were then mixed with tobacco protoplasts and infection resulted. This procedure also increased the efficiency of inoculation with BMV (Motoyoshi et al. 1974a). An inherent problem with the PLO method is that the protoplast is exposed to large numbers of particles of the order of thousands, one or several of which may be required to initiate infection. These high background particle numbers mask the early events in replication (see below).

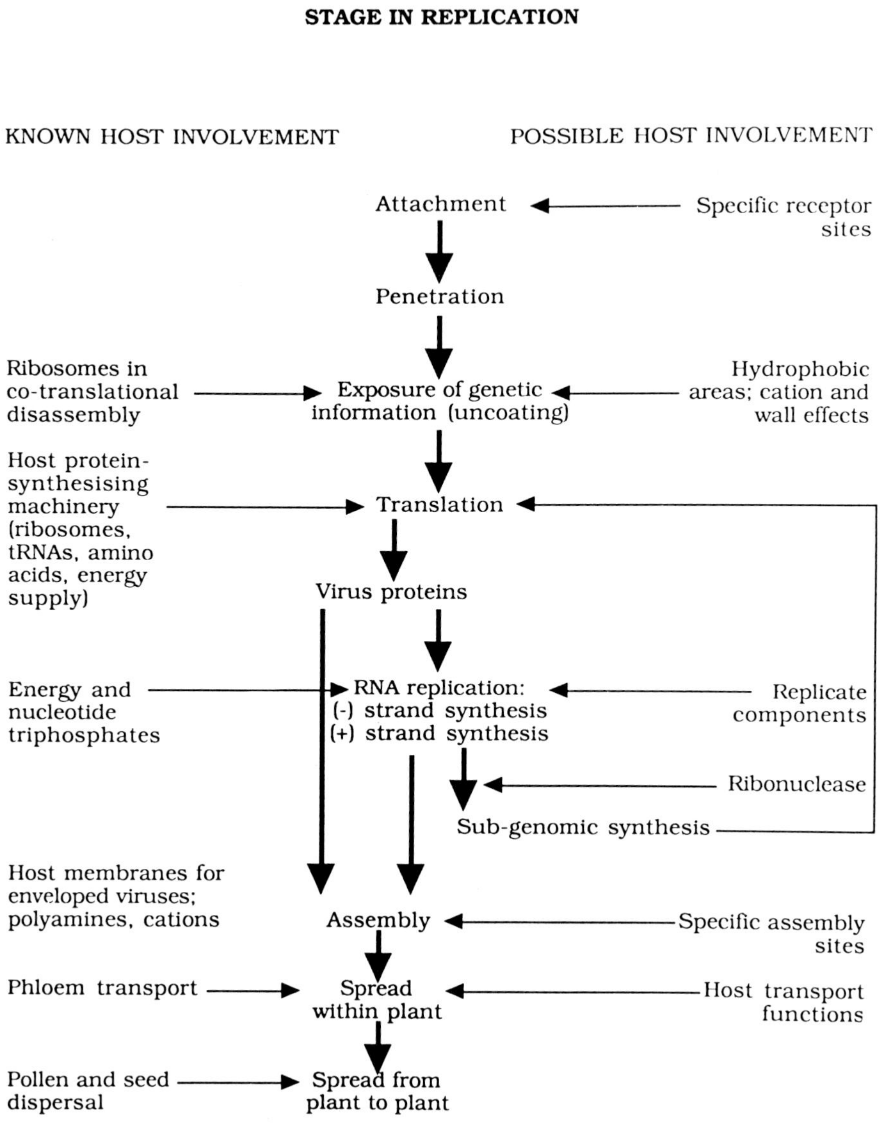


**Fig. 2.** Known and possible involvement of host-specified functions in the multiplication of a representative virus with a positive-sense RNA genome. Virus-controlled functions are shown by *heavy arrows,* and host functions by *light arrows.* (Fraser 1987)

Here, chemical, physical, and physicochemical methods of virus inoculation of protoplasts with viruses including virus-based agents are discussed below in terms of the following criteria.

1. Applicability – the method should be widely applicable to viruses of all groups.
2. Input multiplicity – the method should be efficient, i.e., the input multiplicity should be known and inoculation should be achieved with, or approaching, the

minimum theoretical number of virus particles – one particle in the case of unipartite genomes and so on (see Matthews 1981).
3. Percentage infection – the method should be capable of achieving high percentage infection of the protoplasts population under routine laboratory conditions.
4. Simulation of in vivo replication – the inoculation process should mimic the natural infection process and the events therein.
5. Nontoxic – the process should not adversely effect the ability of the inoculated protoplasts to support virus replication.

## 2 Chemical Methods for the Inoculation of Plant Protoplasts

### 2.1 The Poly-L-Ornithine Procedure

This was developed by Takebe and Otsuki (1969) and subsequently refined (Otsuki et al. 1972). The method has been used widely and successfully for the inoculation of protoplasts from a wide range of plant species, with representative viruses from the majority of the virus groups (see Muhlbach 1983; Table 2). Divalent cations act synergistically with PLO in the inoculation of protoplasts from soybean cell suspension cultures (Lesney and Murakishi 1981a). A protocol for the method is given in Table 3.

The mechanism of PLO-stimulated inoculation has been the subject of EM and biochemical investigation. Takebe and co-workers are of the opinion that adsorbed virus particles are taken up by pinocytosis and released into the cytoplasm where replication occurs (e.g., Honda et al. 1974; Otsuki et al. 1972; Takebe 1977). This is in agreement with reports that PLO enhances pinocytotic uptake by animal cells (Ryser 1968). Burgess et al. (1973a, b) and Kassanis et al. (1977) have suggested that virus may enter the cytoplasm through lesions induced by PLO treatment. Further support for a passive mechanism of uptake comes from the metabolic inhibitor studies of Motoyoshi et al. (1974a, b).

Notwithstanding the usefulness of the PLO inoculation procedure in elucidating the time sequence of viral nucleic acid and protein synthesis (see, e.g., Dougherty and Hiebert 1985), it has a major limitation in so far as thousands of virus particles are adsorbed to each protoplast; for example, Motoyoshi et al. (1974b) estimated that $7 \times 10^4$ TMV particles and $3-5 \times 10^5$ cowpea chlorotic mottle virus particles were adsorbed per protoplast. Thus the method has revealed little about the early events in plant virus replication (Shaw 1985).

### 2.2 The Polyethylene Glycol (PEG) Method

Kao and Michayluk (1974) developed this procedure for protoplast fusion studies. It was first applied to virus inoculation studies by Cassells and Barlass (1978a, b). PEG has also been used extensively for the stimulation of viral RNA uptake

**Table 2.** Virus groups representative members of which have been shown to replicate in protoplasts (see also Table 5 and Takebe 1983)

| Group | Representative virus(es) | Reference |
|---|---|---|
| Tobravirus | Tobacco rattle | Kubo et al. (1974) |
| Tobamovirus | Tobacco mosaic | Takebe and Otsuki (1969) |
| | Cucumber green mottle mosaic | Sugimura and Ushiyama (1975) |
| Potexvirus | Potato virus X | Otsuki et al. (1974) |
| | Cymbidium mosaic | Burgyan et al. (1986) |
| | Clover yellow mosaic | Rao and Hiruki (1978) |
| Potyvirus | Potato virus Y | Barker and Harrison (1984) |
| Luteovirus | Tobacco necrotic dwarf | Kubo and Takanami (1979) |
| | Barley yellow dwarf | Oshima et al. (1984) |
| Tymovirus | Turnip yellow mosaic | Renaudin et al. (1975) |
| Sobemovirus | Southern bean mosaic | Jarvis and Murakishi (1980) |
| | Turnip rosette | Morris-Krsinich and Hull (1983) |
| Tombuvirus | Cymbidium ringspot | Russo and Gallitelli (1985) |
| | Carnation mottle | Kluge et al. (1983) |
| Comovirus | Cowpea mosaic | Hibi et al. (1975) |
| | Cowpea severe mosaic | Beier et al. (1981) |
| | Bean pod mottle | Lesney and Murakishi (1981 a, b) |
| Nepovirus | Raspberry ringspot | Kubo et al. (1974) |
| | Tomato blackring | Fritsch et al. (1978) |
| Pea enation | Pea enation mosaic | Motoyoshi and Hull (1974) |
| Cucomovirus | Cucumber mosaic | Otsuki and Takebe (1973) |
| | Peanut stunt | Linthrost and Kaper (1984) |
| Bromovirus | Brome mosaic | Motoyoshi et al. (1974a) |
| | Cowpea chlorotic mottle | Motoyoshi et al. (1973) |
| Alfalfa mosaic virus | Alfalfa mosaic | Motoyoshi et al. (1975) |
| Rhabdovirus | Sonchus yellow net | Van Beek et al. (1985) |
| Geminivirus | Bean golden mosaic | Furusawa et al. (1980) |
| Caulimovirus | Cauliflower mosaic | Howell and Hull (1978) |
| Unassigned | Broadbean wilt | Kage et al. (1975) |
| | Sweet clover necrotic mosaic | Pappu and Hiruki (1986) |
| | Blackeye cowpea mosaic | Dijkstra et al. (1987) |
| | Velvet tobacco mottle | Wu et al. (1985) |

**Table 3.** Protocol for the inoculation of protoplasts using the poly-L-ornithine procedure. (Based on Takebe 1984)

1. *Preparation of inoculum:* mix together the following in a 50-ml centrifuge tube, all in 0.6 M D-mannitol: 2 ml of freshly prepared 0.1 μM PLO (MW 100000), 2 ml of 10 μg/ml TMV, 1 ml of 0.2 M citrate buffer pH 5.2 to a total volume of 10 ml. Incubate at 25 °C for 10 min.
2. *Inoculation:* pour a freshly prepared protoplast suspension ($4 \times 10^6$ cells in 10 ml 0.6 M mannitol) on to the inoculum solution. Incubate at 25 °C for 10 min with occasional agitation.
3. *Washing:* collect the protoplasts by centrifugation (ca. 100 g 2 min) and carefully resuspend in 10 ml fresh 0.6 M mannitol. Transfer to a $17 \times 105$ mm centrifuge tube. Wash twice by centrifugation, resuspend in 10 ml 0.6 M mannitol.
4. *Culture:* resuspend the washed inoculated protoplasts in 10 ml culture medium (Takebe 1984). Transfer to a 100-ml conical flask, incubate at 28 °C, 18-h photoperiod ca. 3000 lx.

**Table 4.** Protocol for the inoculation of protoplasts using the polyethylene glycol procedure. (Based on Cassells and Cocker 1980)

1. *Virus precipitation onto protoplast surface:* to 0.4 ml 0.66 M mannitol pH 5.8 containing $10^6$ protoplasts add 10 µl TMV (12 mg/ml virus protein in 0.5 ml 0.66 M mannitol buffered with 5 mM phosphate at pH 6.5) and 0.1 ml 83.3 mM PEG 6000 in 0.66 M mannitol. Incubate at 4 °C for 20 min.
2. *Protoplast aggregation:* to 0.3 ml 83.3 mM PEG 6000 in 0.66 M mannitol add 0.7 ml of 0.66 M mannitol in 3.5 mM phosphate buffer pH 6.5 containing 11 mM $CaCl_2$). Incubate at 25 °C for 1 h.
3. *Protoplast washing:* Allow the protoplasts to precipitate under gravity. Resuspend the protoplasts in 50 mM $CaCl_2$ in 0.66 M mannitol pH 10.5. Precipitate the protoplasts under gravity (approx. 10 min). Resuspend the pellet in liquid culture medium (Nagata and Takebe 1970) pH 6.5. Allow to settle under gravity.
4. *Post-inoculation incubation:* Resuspend the protoplasts (approx. $10^6$) in 5 ml liquid culture medium (as above). Incubate in growth room (23 °C, 16-h photoperiod, 8.6 W/$m^2$).

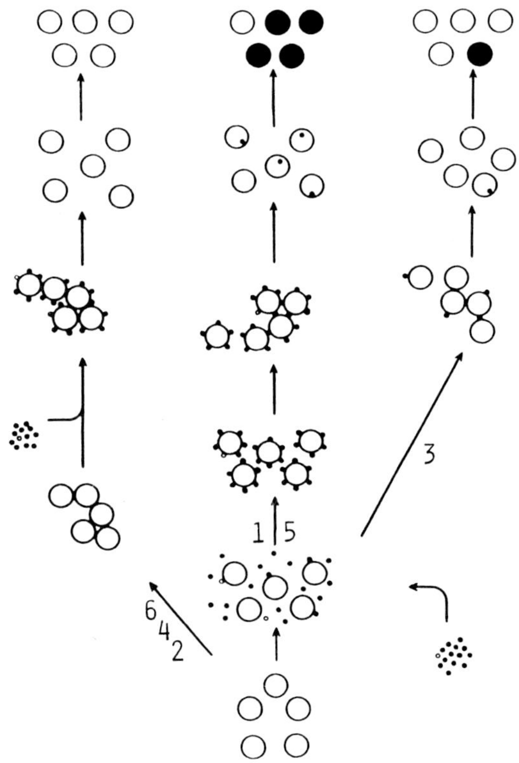


**Fig. 3.** The relation between virus precipitation, protoplast clumping and virus infection of tobacco mesophyll protoplasts. In procedures *1, 3,* and *5,* virus precipitation or entrapment between clumping protoplasts is possible. In procedures *2, 4,* and *6,* protoplast aggregation occurs before virus addition and/or precipitation. *Small black circles* represent infective virus, *small open circles* represent defective virus; *large open circles* represent non-infected protoplasts, *large black circles* represent infected protoplasts. (Cassells and Cocker 1980)

by protoplasts (see, e.g., Muhlbach 1983) and in liposome studies (Rollo 1982) (see below). A protocol for its use is given in Table 4.

The mechanism of PEG stimulation of virus uptake by protoplasts has been investigated by Cassells and Cocker (1980). By altering the sequence of addition of PEG, virus, and protoplasts, it was suggested that for infection to occur, virus

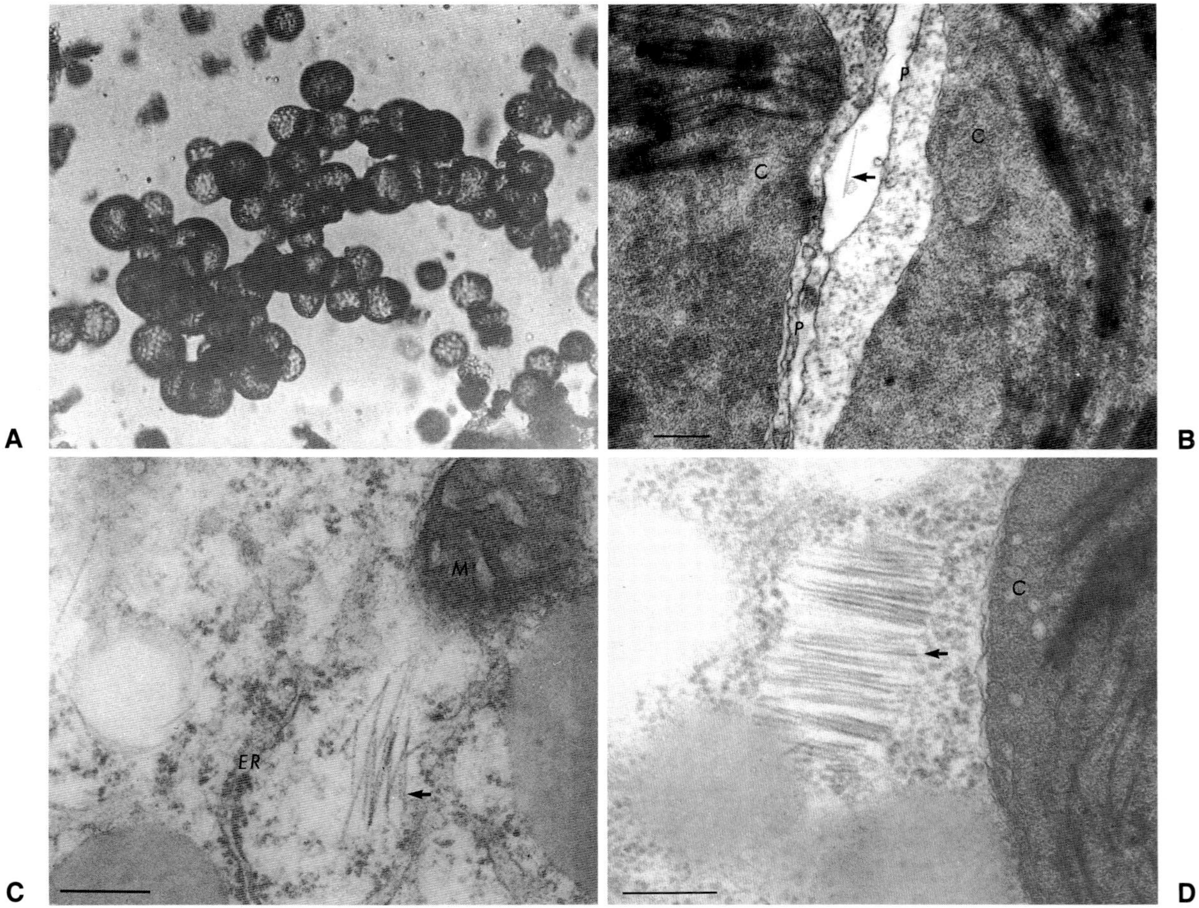


**Fig. 4 A–D.** Inoculation and replication of TMV in tobacco protoplasts. **A** Transient clumping of protoplasts in the PEG inoculation procedure; **B** a TMV particle trapped in the void between adjacent areas of localised membrane fusion; **C** early progeny virus production 10 h post inoculation; **D** accumulation of progeny particles

must first be precipitated/adsorbed on the protoplast surface followed by aggregation of the protoplasts. On washing with high calcium at high pH, transient membrane fusion is postulated to occur in the zones of protoplast-to-protoplast adhesion and virus enters through the perturbed membranes to initiate infection (Figs. 3, 4). In contrast with the PLO method, virus entry may be localized in the PEG procedure, however, virus inoculum is adsorbed over the total protoplast surface and thus infective particles are indistinguishable from virus trapped in the relatively large fusion zones which remains latent.

Other polymers have been used successfully also to stimulate virus uptake by plant protoplasts, but no studies have been made of their modes of action (see Muhlbach 1983).

## 3 Electrical and Physicochemical Methods for the Inoculation of Plant Protoplasts

Interest in genetic manipulation and genetic engineering of plants has provided impetus for the development of refined methods for protoplast fusion and for the introduction of nucleic acids into protoplasts. Electrofusion and electroporation are examples of methods for the latter (see review by Zimmerman et al. 1981; Hofmann and Evans 1986). In electrofusion, charge repulsion between protoplasts is overcome by polarizing the protoplasts in an alternating current (AC) field, following alignment, fusion is achieved by local destabilization of the protoplast membranes by exposure to a direct current (DC) pulse. Electroporation involves the creation of pores in the plasmalemma by exposure to a DC pulse. Nucleic acids may enter the protoplast through these transient artificial channels (Fromm et al. 1985).

### 3.1 The Electroporation Method

Electroporation has been used successfully to inoculate viral RNAs into plant protoplasts (Okada et al. 1986). Watts et al. (1987) have recently reported studies on the inoculation of protoplasts with brome mosaic virus (BMV) and cowpea chlorotic mottle virus (CCMV) by electroporation. These viruses have equivalent diameters, the former has a net positive charge and the latter a net negative charge. Under the experimental conditions described, BMV but not CCMV inoculation was achieved, the former resulted in low percentage infection. The authors suggested that charge in the case of CCMV, and paucity of pores of adequate size, may explain the poor result.

In this laboratory, following PEG precipitation of TMV on the tobacco protoplast surface, infection comparable to that reported for the PEG protocol (Table 4) was obtained by applying two DC pulses of 1.5 kv/cm for 40 μs to the virus-coated protoplasts (substituting for the high calcium, high pH wash) (Prendergast and Cassells unpubl.). This result suggests that a hybrid PEG-electroporation procedure may increase the efficiency of inoculation in electroporation. The general criticism, however, still applies, namely, that there is a high background particle count in the procedure.

### 3.2 Microprojectiles

Another physical method of introducing nucleic acids into protoplasts which overcomes the practical problems associated with microinjection (Crossway et al. 1986), is the use of microprojectiles (Klein et al. 1987). This method has been successfully used to introduce foreign DNA into onion cells. The DNA was expressed and an efficiency of 90% cell transformation was reported for onion epidermal cells.

## 4 Liposome Carriers for the Inoculation of Protoplasts

Liposomes are artificial membranes surrounding an aqueous phase (vesicles) (Bangham et al. 1965; Fig. 5). They can be constructed as multilamellar (MLU)

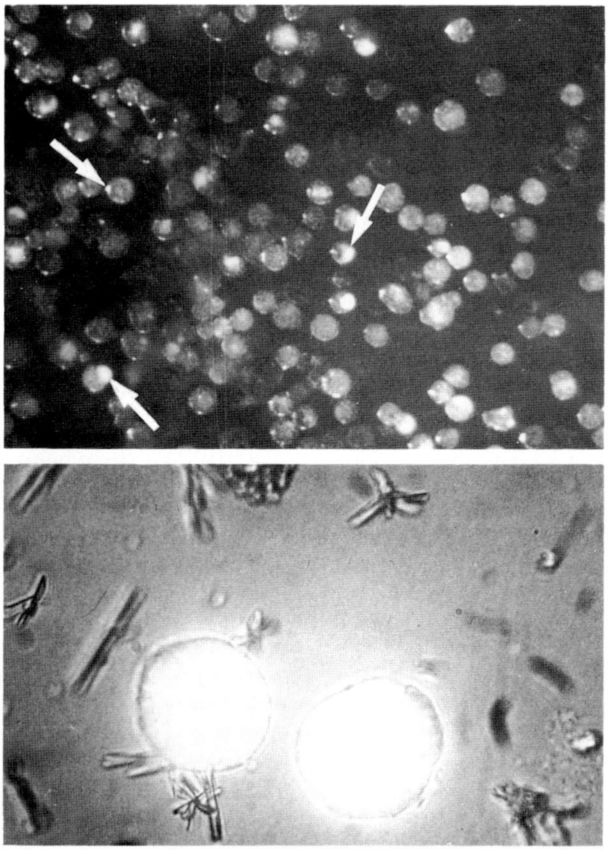

A

B

**Fig. 5A, B.** Uptake of liposomes by plant protoplasts. **A** Adhesion of fluorescent labeled liposomes to the plasmalemma; **B** discharge of fluorescent marker dye into protoplasts or unilamellar structures with varying surface charge characteristics (for reviews see Makins 1983; Matthews 1983; Ostro 1983; Straubinger and Papahadjopoulos 1983). Unilamellar vesicles are characterized on size as large (LUV's) or small (SUV's).

LUV's are most commonly used in carrier studies as they have a relatively higher capture volume (volume of aqueous phase encapsulated) than SUV's and MLV's and are less stable than the latter. This is important where the objective is to introduce the carrier phase of the liposome into the cytoplasm of the target cell. MLU's may be taken up by pinocytosis and merged with the lysosomal apparatus, resulting in hydrolysis of labile components (Ostro and Giacomoni 1985).

Early studies on liposome-protoplast interactions sought to exploit the potential to produce positively charged vesicles which would spontaneously bind, and fuse with the negatively charged protoplast (Cassells 1979). Now, however, PLO, PEG, or polyvinyl alcohol (PVA) are used as fusogens to enhance uptake (Makins 1983).

Rollo and Hull (1982) used MLV delivery stimulated by PEG to inoculate turnip rosette virus. In recent studies LUV's have been used to inoculate protoplasts with TMV-RNA (Nagata et al. 1981; Fukunaga et al. 1981; Fraley et al. 1982; Watanabe et al. 1982).

Liposome-mediated delivery has been shown to be an important method for the introduction of nucleic acid into protoplasts (Lurquin 1984; Nagata 1984) offering protection against nucleases. It has an advantage over the commonly used methods discussed above in that the process can be accurately quantified (Cutler et al. 1984).

A concern in using liposomes for virus inoculation is that the procedures used during liposome capture may involve sonication leading to fracture of the virus particles and exposure to chemical environments which can result in virus uncoating (see liposome preparation in Deamer and Uster 1984; Makins 1983; Straubinger and Papahadjopoulus 1983) (see below).

## 5 Efficiency of Inoculation Procedure

The efficiency of the procedure used to introduce virus or virus-derived agents into protoplasts can be monitored by two parameters viz. (1) the number of sites of initiation of virus replication per protoplasts expressed relative to the number of infective (complete) genome copies entering, and (2) the percentage of competent protoplast infected. For reasons given above, except for liposome-mediated delivery, the former of these parameters is difficult to monitor. The latter can be determined by staining with fluorescent antibody (Otsuki and Takebe 1969) or following suppression of background synthesis by autoradiography (Cocker and Cassells 1983, Fig. 6).

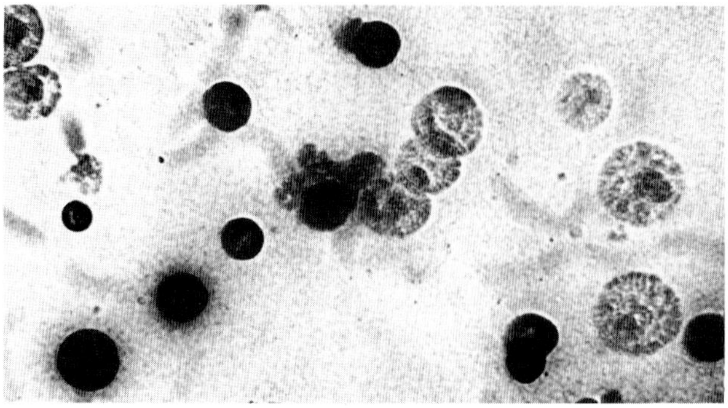

**Fig. 6.** Use of $^{32}P$ autoradiography to detect the replication of viral RNA as a parameter of protoplast inoculation (darkened protoplasts)

## 6 Virus Uptake and Transformation

Plant viruses are currently classified into artificial groups based on type of nucleic acid in the genome, morphology, and presence or absence of an envelope on the particle (Table 5). Numerically, and also in breadth of host range, (+) ssRNA ge-

**Table 5.** Groupings of plant viruses based on type of genome and particle morphology. (See Francki 1983)

| Properties | Virus groups |
|---|---|
| ss-RNA, helical, no envelope | Tobravirus |
| | Hordeivirus |
| | Tobamovirus |
| | Potexvirus |
| | Carlavirus |
| | Potyvirus |
| | Closterovirus |
| ss-RNA, icosahedral, no envelope | Luteovirus |
| | Maize chlorotic dwarf virus |
| | Tymovirus |
| | Sobemovirus |
| | Tobacco necrosis virus |
| | Tombusvirus |
| | Comovirus |
| | Nepovirus |
| | Pea enation mosaic virus |
| | Dianthovirus |
| | Cucumovirus |
| | Bromovirus |
| | Ilarvirus |
| | Alfalfa mosaic virus |
| ss-RNA, helical, envelope | Rhabdoviridae |
| | Tomato spotted wilt virus |
| ds-RNA, icosahedral, no envelope | Reoviridae |
| | (Phytoreovirus and Fijivirus) |
| ss-DNA, icosahedral, no envelope | Geminivirus |
| ds-DNS, icosahedral, no envelope | Caulimovirus |

**Table 6.** Preliminary grouping of plant RNA viruses based on genome replication strategy. (Based on Dougherty and Hiebert 1985)

1. Viral genome functions as polycistronic messenger (possible members):
   Tobacco necrosis virus
   Potexvirus group
2. Viral genome functions as monocistronic messenger RNA for polyproteins
   Comovirus group
   Nepovirus group
3. Viral genome expression is limited to five prime regions; subgenomic RNA's function as messengers for other viral proteins:
   Tobamovirus group
   Tobravirus group
   Bromovirus group
   Ilarvirus group
   Cucumovirus group
   Alfalfa mosaic virus
4. Viral genome is expressed by polyprotein synthesis processed by proteolysis; subgenomic messengers also function
   Tymovirus group
   Turnip mottle virus (unassigned)

nome viruses dominate (Matthews 1981). Within the (+) ssRNA viruses, four genome replication strategies have been postulated (Dougherty and Hiebert 1985; Table 6).

In choosing the progenitor for a nonintegrating transforming agent, in addition to aspects discussed above (see Sect. 1, also Cassells 1986), two factors are generally considered viz. amenability to genetic engineering and packaging constraint in the particle. Considering amenability to genetic engineering techniques, Cauliflower Mosaic Virus (CaMV) has been the widely chosen progenitor. CaMV has a dsDNA genome and appropriate restriction sites and is consequently compatible with methodology developed for the *Agrobacterium* Ti plasmid (see Yadav 1986; Maniatis et al. 1982). CaMV has been shown to infect turnip protoplasts (Howell and Hull 1978) and more recently a chimeral CaMV carrying methotrexate resistance has been shown to transform turnip plants (Brisson et al. 1984).

A strategy for the engineering of (+) ssRNA viruses has been described by Ahlquist and coworkers. The technique involved cloning cDNA copies of the 3 BMV genome RNA's, insertion of bacterial chloramphenicol acetyltransferase (CAT) into the RNA3. Infectious RNA1, RNA2 and chimeric RNA3 containing CAT, were generated using the ppMI vector (Ahlquist and Janda 1984). Barley protoplasts were infected with the ppMI transcripts and CAT activity was detected (French et al. 1986).

## 7 Summary and Conclusions

Virologists, and those developing nonintegrating transforming agents, seek to unravel the events in replication from entry to genome expression, and ideally, require a synchronous cell virus replication system which satisfies the criteria listed in Section 2 above.

Isolated mesophyll protoplasts satisfy the requirements of the cell system being relatively uniform genetically and physiologically (Table 2). Callus-derived intact cells have also attracted attention as a model system (see Murakishi et al. 1971), but a caveat must be raised against the latter, or protoplasts derived from same, which may be genetically unstable and physiologically/developmentally variable (Street 1977). Cell monolayers, for example as used by Klein et al. (1987), may also have potential, provided cell-to-cell signals or gradients do not result in heterogeneity of response (Pratt and Matthews 1971).

As stressed in the introduction, no virus receptor sites are known to occur on the plasmalemma. Consequently, artificial methods must be employed to introduce virus into plant protoplasts, albeit this may parallel the situation in natural transmission of plant viruses. Chemical methods, whether based on wound damage or pinocytosis, can be used to inoculate protoplasts and virus-protoplast studies have yielded valuable information about virus replication strategies (see, e.g., Dougherty and Hiebert 1985) but the early events are confused by the high multiplicity of viral inoculum. Of the newer methods, liposome-mediated delivery or use of microprojectiles have the potential to achieve infection with low and controlled inoculum inputs. Liposome preparation, however, may result in virus particle uncoating, due to the lipophilic environment and thus the early events of in vivo uncoating may be bypassed. As yet it is not clear if similar limitations apply to the use of microprojectiles.

Several studies have shown that foreign DNA including chimeric gene constructs may be incorporated into the eukaryote nuclear genome (Paszkowski et al. 1984; Schilperoort 1986) or, appropriately modified, targeted on the organellar genomes (Boutry et al. 1987). Such DNA may be introduced by a variety of strategies (as outlined in Sect. 4). In contrast, viral-based transforming agents may be designed to be nonintegrating. In the development of the latter, protoplasts have an important role in facilitating studies on genome expression as in virus studies (see Sect. 1). With wide host ranges RNA viruses have potential as transforming agents. Transacting divided genome RNA viruses (see Table 6) may be engineered using, e.g., the BMV study discussed above as a model. In vitro adventitious regenerants from transformed protoplasts (inoculated with naked RNA as in Sect. 4) may be micropropagated for clonal multiplication and release or the transforming agent may be spray inoculated onto seedlings for growing on (see Fig. 1 and Cassells 1986). In the latter case, the engineered viral genome may require encapsidation or pseudovirions may be made (Lurquin and Kado 1977) and the agents introduced by mechanical (spray) inoculation. Encapsidation in liposomes or use of microprojectiles may circumvent limitations imposed by the carrier capacity of the native virus particles.

In conclusion, efficient quantifiable delivery systems for the delivery of viruses and engineered viruses into protoplasts should satisfy the criteria listed in Section 2. In this regard, liposome delivery systems potentially satisfy all the criteria. For less rigorous studies, electroporation or the precipitation electrofusion procedures described in Section 4 may suffice. Functional uptake (expressed as % transformation) can be monitored by fluorescent antibody staining where adequate

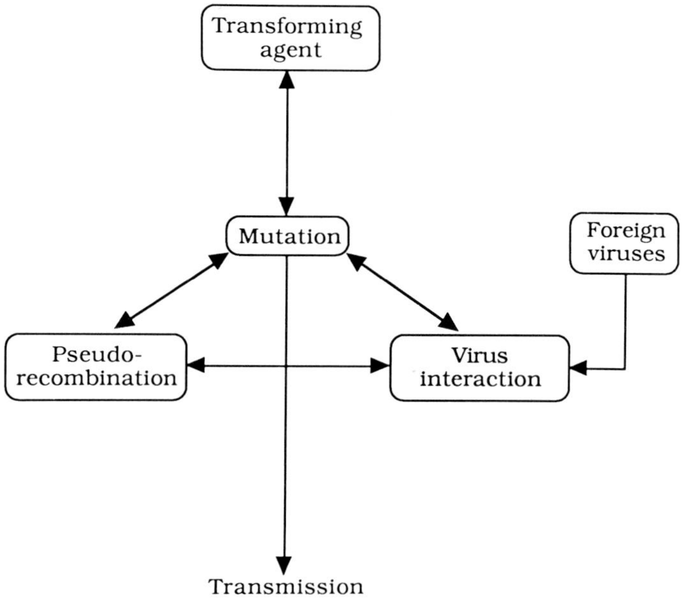


**Fig. 7.** Factors which may alter the characteristics of a genetically engineered virus-based transforming agent in the field. These factors are related to change in transforming agent and interaction with other viruses infecting the crop

levels of viral genome or chimeric proteins are expressed or alternatively by autoradiography following suppression of background incorporation.

However, studies on transforming agents in protoplasts, while giving valuable information about basic aspects of the agents performance, cannot substitute for information gained from field trials. Figure 7 shows some of the events which may occur under field conditions and which may result in alteration in the performance and characteristics of the agent (for a fuller discussion of this aspect see Cassells and Herrick 1977; Cassells 1986).

## References

Ahlquist P, Janda M (1984) cDNA cloning and in vitro transcription of the complete brome mosaic virus genome. Mol Cell Biol 4:2876–2882

Bangham AD, Standish MM, Watkins JC (1965) Diffusion of univalent ion across the lamellae of swollen phospholipids. J Mol Biol 13:238–252

Barker H, Harrison BD (1984) Expression of genes for resistance to potato virus Y in potato plants and protoplasts. Ann Appl Biol 105:539–546

Beier H, Issinger OG, Denschle M, Mundry KW (1981) Translation of the RNA of cowpea severe mosaic virus in vitro and in cowpea protoplasts. J Gen Virol 54:379–390

Boutry M, Nagy F, Poulsen C, Aoyagi K, Chua NH (1987) Targeting of bacterial chloramphenicol acetyltransferase to mitochondria in transgenic plants. Nature 328:340–342

Brisson N, Paszkowski J, Penswick JR, Gronenberg B, Potrykus I, Hohn T (1984) Expression of a bacterial gene in plants by using a viral vector. Nature 310:511–514

Burgess J, Motoyoshi F, Fleming EN (1973a) Effect of poly-L-ornithine on isolated tobacco mesophyll protoplasts: evidence against stimulated pinocytosis. Planta 111:199–208

Burgess J, Motoyoshi F, Fleming EN (1973b) The mechanism of infection of plant protoplasts by viruses. Planta 112:323–332

Burgyan J, Russo M, Gallitelli D (1986) Translation of cymbidium ringspot virus RNA in cowpea protoplasts and rabbit reticulocyte lysates. J Gen Virol 67:1149–1160

Burke DC, Russell WC (eds) (1975) Control Processes in Virus Multiplication. Cambridge Univ Press, Cambridge, pp 449

Cassells AC (1979) Uptake of lipid vesicles by isolated tomato protoplasts. Nature 275:760

Cassells AC (1986) *Pelargonium peltatum* (ivy-leaf pelargonium) Harlequin type: the use of beneficial infective agents and implications for genetic engineering of plants by non-integrating vectors. Acta Hortic 182:229–236

Cassells AC, Barlass M (1978a) A method for the isolation of stable mesophyll protoplasts from tomato leaves throughout the year under standard conditions. Physiol Plant 42:236–242

Cassells AC, Barlass M (1978b) The initiation of TMV infection in isolated protoplasts by polyethylene glycol. Virology 87:459–462

Cassells AC, Cocker FM (1980) TMV inoculation of tobacco protoplasts in the presence of protoplast fusion agents. Z Naturforsch 35c:1057–1061

Cassells AC, Herrick CC (1977) Cross protection between mild and severe strains of tobacco mosaic virus in doubly inoculated tomato plants. Virology 78:253–260

Cocker FM, Cassells AC (1983) Autoradiography of $^{32}P$ and $^{14}C$ incorporated into protoplasts as a means of determining the percentage of virus infected protoplasts. J Virol Methods 6:311–318

Crossway A, Oakes JV, Irvine JM, Ward B, Knauf VC, Shewmaker CK (1986) Integration of foreign DNA following microinjection of tobacco mesophyll protoplasts. Mol Gen Genet 202:179–185

Cutler AJ, Constabel F, Kurz WGW, Shargool PD (1984) Quantitation of the delivery of liposome contents into plant protoplasts. Anal Biochem 139:482–486

Deamer DW, Uster PS (1984) Liposome preparation: methods and mechanisms. In: Ostro MJ (ed) Liposomes. Marcel Dekker, New York, pp 27–52

Dougherty WG, Hiebert E (1985) Genome structure and gene expression of plant RNA viruses. In: Davies JW (ed) Molecular plant virology, Vol II. CRC Press, Boca Raton, Florida, pp 23–82

Dijkstra J, van Beek NAM, Lohuis D, van Helden M, Meyer R (1987) Is a helper factor necessary for infection of cowpea protoplasts with blackeye cowpea mosaic virus? Neth J Plant Pathol 93:43–47

Favali MA, Conti GG, Bassi M (1977) Some observations on virus-induced local lesions by transmission and scanning electron microscopy. Acta Phytopathol Acad Sci Hung 12:141–150

Fraley RT, Dellaporta SC, Papahadjopoulos D (1982) Liposome-mediated delivery of tobacco mosaic virus RNA into tobacco protoplasts: A sensitive assay for monitoring liposome-protoplast interactions. Proc Natl Acad Sci USA 79:1859–1863

Francki RIB (1983) Current problems in plant virus taxonomy. In: Matthews REF (ed) A critical appraisal of viral taxonomy. CRC Press, Florida, pp 63–104

Francki RIB (1985) The viruses and their taxonomy. In: Francki RIB (ed) The plant viruses, Vol I. Plenum Press, New York, pp 1–18

Fraser RSS, Letchworth WK (1987) Biochemistry of virus-infected plants. Research Studies Press, p 259

French R, Janda M, Ahlquist P (1986) Bacterial gene inserted in an engineered RNA virus: efficient expression in monocotyledonous plant cells. Science 231:1294–1297

Fritsch C, Mayo MA, Murant AF (1978) Translation of the satellite RNA of tomato blackring virus in vitro and in tobacco protoplasts. J Gen Virol 40:587–594

Fromm M, Taylor LP, Walbot V (1985) Expression of genes transferred into monocot and dicot plant cells by electroporation. Proc Natl Acad Sci USA 82:5824-5828

Fukunaga Y, Nagata T, Takebe I (1981) Liposome-mediated infection of plant protoplasts with tobacco mosaic virus RNA. Virology 113:752–760

Furusawa I, Yamaoka N, Okuno T, Yamamoto M, Kuhno M, Kunoh H (1980) Infection of turnip (*Brassica rapa* cultivar Perviribis) protoplasts with cauliflower mosaic virus. J Gen Virol 48:431–436

Hibi T, Rezelman G, van Kammen A (1975) Infection of cowpea mesophyll protoplasts with cowpea mosaic virus. Virology 64:308–318

Hofmann GA, Evans GA (1986) Electronic, genetic, physical and biological aspects of cellular electromanipulation. IEEE Eng Med Biol Mag 5:6–24

Honda Y, Matsui C, Otsuki Y, Takebe I (1974) Ultrastructure of tobacco mesophyll protoplasts inoculated with cucumber mosaic virus. Phytopathology 64:30–34

Howell SH, Hull R (1978) Replication of cauliflower mosaic virus and transcription of its genome in turnip leaf protoplasts. Virology 86:468–481

Jarvis NP, Murakishi HH (1980) Infection of protoplasts from soybean cell culture with southern bean mosaic and cowpea mosaic viruses. J Gen Virol 48:365–376

Kage T, Ozaki T, Inoue T (1975) Horse bean leaf protoplast production and broad bean wilt virus infection. Ann Phytopathol Soc Jpn 41:107–108

Kao KN, Michayluk MR (1974) A method for high-frequency intergeneric fusion of plant protoplasts. Planta 115:355–367

Kassanis B, White RF, Turner RH, Wood RD (1977) The mechanism of virus entry during infection of tobacco protoplasts with TMV. Phytopathol Z 88:215–228

Klein TM, Wolf ED, Wu R, Sanford JC (1987) High velocity microprojectiles for delivering nucleic acids into living cells. Nature 327:70–73

Kluge S, Kristen U, Qertel C (1983) Infection of *Dianthus caryophyllus* cultivar Williams protoplasts with carnation mottle virus. J Gen Virol 64:2485–2488

Knight DE, Scrutton MC (1986) Gaining access to the cytosol: the technique and some applications of electropermeabilization. Biochem J 234:497–506

Kubo S, Takanami Y (1979) Infection of tobacco mesophyll protoplasts with tobacco necrotic dwarf virus, a phloem limited virus. J Gen Virol 42:387–398

Kubo S, Harrison BD, Robinson DJ (1974) Effects of phosphate on the infection of tobacco protoplasts by tobacco mottle virus. Intervirology 3:382–387

Lesney MS, Murakishi HH (1981a) Infection of soyabean protoplasts from cell suspension culture with bean pod mottle virus. J Gen Virol 47:387–395

Lesney MS, Murakishi HH (1981b) Effects of buffer and amendments on infection of soybean callus protoplasts with bean pod mottle virus. Phytopathology 71:236

Linthrost HJM, Kaper JM (1984) Replication of peanut stunt virus and its associated RNAs in cowpea protoplasts. Virology 139:317–329

Luria SE, Darnell JE, Baltimore D, Campbell A (1978) General Virology, 3rd edn. Wiley, New York, p 578

Lurquin PF (1984) Liposomes as carriers for the transfer of nucleic acids into higher plant protoplasts. In: Beers RF Jr, Bassett EG (eds) Cell fusion, gene transfer and transformation. Raven, New York, pp 227–236

Lurquin PF, Kado CI (1977) Genetic engineering of plants: Uptake of plasmid and pseudo-virions by plant protoplasts. Arch Int Physiol Biochim 85:999–1000

Makins JF (1983) The theory and practical applications of liposome-protoplast interactions. In: Potrykus I, Harms CT, Hinnen A, Hütter R, King PJ, Shillito RD (eds) Protoplasts 1983 – Lecture Proceedings. Birkhäuser, Basel, pp 197–207

Maniatis T, Fritsch EF, Sambrook J (1982) Molecular cloning: A laboratory manual. Cold Spring Harbor Laboratory, New York

Matthews B (1983) Liposome-mediated delivery of DNA to plant protoplasts. In: Evans D, Sharp WR, Ammirato PV, Yamada Y (eds) Handbook of plant cell culture, Vol 1. Macmillan, New York, pp 520–540

Matthews REF (1981) Plant virology, 2nd edn. Academic Press, New York, p 897

Merkens WSW, de Zoeten GA, Gaard G (1972) Some observations on ectodesmata and the virus infection process. J Ultrastruct Res 41:397–406

Morris-Krsinich B, Hull R (1983) Replication of turnip rosette virus RNA in inoculated turnip *Brassica rapa* cultivar Just Right protoplasts. J Gen Virol 64:2661–2668

Motoyoshi F, Hull R (1974) The infection of tobacco protoplasts with pea enation mosaic virus. J Gen Virol 24:89–100

Motoyoshi F, Bancroft JB, Watts JW, Burgess J (1973) The infection of tobacco protoplasts with cowpea chlorotic mottle virus and its RNA. J Gen Virol 20:177–193

Motoyoshi F, Bancroft JB, Watts JW (1974a) The infection of tobacco protoplasts with a variant of brome mosaic virus. J Gen Virol 25:31–36

Motoyoshi F, Watts JW, Bancroft JB (1974b) Factors influencing the infection of tobacco protoplasts by cowpea chlorotic mottle virus. J Gen Virol 25:245–256

Motoyoshi F, Hull R, Flack IH (1975) Infection of tobacco mesophyll protoplasts by alfalfa mosaic virus. J Gen Virol 27:263–266

Muhlbach HP (1983) The use of protoplasts in plant virus research. In: Potrykus I, Harms CT, Hinnen A, Hütter R, King PJ, Shillito RD (eds) Protoplasts 1983 – Lecture Proceedings. Birkhäuser, Basel, pp 111–122

Mundry KW (1963) Plant virus-host cell interactions. Annu Rev Phytopathol 1:173–196

Murakishi HH, Hartmann JX, Beachy RN, Pelcher LE (1971) Growth curve and yield of tobacco mosaic virus in tobacco callus cells. Virology 43:62–68

Nagata T (1984) Interaction of liposomes and protoplasts as a model system of protoplast fusion. In: Beers RF Jr, Bassett EG (eds) Cell fusion, gene transfer and transformation. Raven, New York, pp 217–226

Nagata T, Takebe I (1970) Cell wall regeneration and cell division in isolated tobacco mesophyll protoplasts. Planta 93:301–308

Nagata T, Okada K, Takebe I, Matsui C (1987) Delivery of tobacco mosaic virus RNA into plant protoplasts mediated by reverse-phase evaporation vesicles (liposomes). Mol Gen Genet 184:161–165

Ohsima K, Matsubara A, Uyeda I, Shikata E (1984) Infection of barley mesophyll protoplasts with barley yellow dwarf virus. Mem Fac Agric Hokkaido Univ Jpn 14:98–102

Okada K, Nagata T, Takebe I (1986) Introduction of functional RNA into plant protoplasts by electroporation. Plant Cell Physiol 27:619–626

Ostro MJ (ed) (1983) Liposomes. Marcel Dekker, New York, p 397

Ostro MJ, Giacomoni D (1985) Liposomes as a tool in molecular biology: A comparison to other methodologies. In: Ostro MJ (ed) Liposomes. Marcel Dekker, New York, pp 145–208

Otsuki Y, Takebe I (1969) Fluorescent antibody staining of tobacco mosaic virus antigen in tobacco mesophyll protoplasts. Virology 38:497–499

Otsuki Y, Takebe I (1973) Infection of tobacco mesophyll protoplasts by cucumber mosaic virus. Virology 52:433–438

Otsuki Y, Takebe I, Honda Y, Matsui C (1972) Ultrastructure of infection of tobacco mesophyll protoplasts by tobacco mosaic virus. Virology 49:188–194

Otsuki Y, Takebe I, Honda Y, Kajita S, Matsui C (1974) Infection of tobacco protoplasts by potato virus X. J Gen Virol 22:375–385

Pappu HR, Hiruki C (1986) Infection of mesophyll protoplasts of *Nicotiana clevelandii* with sweet clover necrotic mosaic virus. Phytopathology 76:1121

Paszkowski J, Shillito RD, Saul M, Mandak V, Hohn T, Hohn B, Potrykus I (1984) Direct gene transfer to plants. EMBO J 3:2717–2722

Potrykus I, Saul MW, Shillito RD, Paszkowski J (1986) Techniques for transferring genes into plants. BCPC Monogr 34:21–26

Pratt MJ, Matthews REF (1971) Non-uniformities in the metabolism of excised leaves and leaf disks. Planta 99:21–36

Rao DV, Hiruki C (1978) Infection of cowpea mesophyll protoplasts with clover yellow mosaic virus. J Gen Virol 38:303–312

Rawlins TE, Tompkins CM (1936) Studies on the effect of Carborundum as an abrasive in plant virus inoculations. Phytopathology 26:578

Renaudin J, Bove JM, Otsuki Y, Takebe I (1975) Infection of brassica leaf protoplasts by turnip yellow mosaic virus. Mol Gen Genet 141:59–68

Rollo F (1982) Liposomes as a tool for introducting biologically active viral nucleic acids into plant protoplasts. In: Lurquin PF, Kleinhofs A (eds) Genetic engineering in eukaryotes. Plenum, New York, pp 179–185

Rollo F, Hull R (1982) Liposome-mediated infection of turnip protoplasts with turnip rosette virus and RNA. J Gen Virol 60:359–363

Russo M, Gallitelli D (1985) Infection of cowpea protoplasts with cymbidium ringspot virus. J Gen Virol 66:2033–2038

Ryser HJP (1968) Uptake of protein by mammalian cells: an underdeveloped area. Science 159:390–396

Schilperoort RA (1986) Integration, expression and stable transmission through seed of foreign genes in plants. In: Horn W, Jensen CJ, Odenbach W, Schieder O (eds) Genetic manipulation in plant breeding. de Gruyter, Berlin, pp 837–858

Shaw JG (1985) Early events in plant virus infections. In: Davies JW (ed) Molecular plant virology, Vol II. CRC Press, Boca Raton, Florida, pp 1–22

Straubinger RM, Papahadjopoulos D (1983) Liposomes as carriers for intracellular delivery of nucleic acids. Methods Enzymol 101:512–527

Street HE (1977) Applications of cell suspension cultures. In: Reinert J, Bajaj YPS (eds) Applied and Fundamental Aspects of Plant Cell, Tissue, and Organ Culture. Springer, Berlin Heidelberg New York, pp 649–667

Sugimura Y, Ushiyama R (1975) Cucumber green mottle mosaic virus infection and its bearing on cytological alterations in tobacco mesophyll protoplasts. J Gen Virol 29:93–98

Szoka F, Papahadjopoulos D (1978) Procedure for preparation of liposomes with large internal aqueous space and high capture by reverse phase evaporation. Proc Natl Acad Sci USA 75:4194–4198

Takebe I (1977) Protoplasts in the study of plant virus replication. In: Fraenkel-Conrat H, Wagner RW (eds) Comprehensive virology, Vol II. Plenum, New York, pp 237–274

Takebe I (1983) Protoplasts in plant virus research. Int Rev Cytol Suppl 16:89–110

Takebe I (1984) Inoculation of protoplasts with plant viruses. In: Vasil IK (ed) Cell culture and somatic cell genetics of plants, Vol I. Academic Press, New York, pp 492–502

Takebe I, Otsuki Y (1969) Infection of tobacco mesophyll protoplasts by tobacco mosaic virus. Proc Natl Acad Sci USA 64:843–848

Van Beek NAM, Dijkstra J, Peters D (1985) The development of sonchus yellow net virus and festuca leaf streak virus in *Vigna unguiculata* protoplasts. Virus Res Suppl 1:13

Watanabe Y, Ohno T, Okada Y (1982) Virus multiplication in tobacco protoplasts inoculated with tobacco mosaic virus encapsulated in large uni-lamellar vesicle liposomes. Virology 120:478–480

Watts JM, King JM, Stacey NJ (1987) Inoculation of protoplasts with viruses by electroporation. Virology 157:40–46

Wilson T, Papahadjopoulos D, Taher R (1979) The introduction of polio virus into cells via lipid vesicles (liposomes). Cell 17:77–84

Wu J, Lu W, Titan B, Qiu B (1985) Multiplication of velvet tobacco mottle virus encapsidated virusoid RNA in *Nicotiana clevelandii* protoplasts. Acta Microbiol Sin 25:221–226

Yadav NS (1986) Molecular biology of plant cell transformation. In: Hennig W, Reinert J (eds) Results and Problems in Cell Differentiation, Vol 12. Springer, Berlin Heidelberg New York Tokyo, pp 109–144

Zaitlin M, Hull R (1987) Plant virus-host interactions. Annu Rev Plant Physiol 38:291–315

Zimmerman U, Scheurich P, Pilwat G, Benz R (1981) Cells with manipulated functions: new perspectives for cell biology, medicine and technology. Angew Chem Int Ed Engl 20:325–344

# XXIV Uptake of Fungal Protoplasts by Plant Protoplasts

P.T. Lynch[1], S. Isaac[2], and H.A. Collin[2]

## 1 Introduction

A range of relationships can be formed between fungi and higher plants. Fungi are heterotrophic and grow by absorbing easily assimilable substrates directly or by first breaking down compounds by means of extracellular enzymes. Nutrients may be obtained from nonliving material by saprotrophy; from living tissues symbiotically, conferring mutual nutritional benefit or alternatively by parasitism through biotrophy or necrotrophy, causing disease and often resulting in plant death.

Mutualistic relationships between fungi and plant roots involve a range of physiological associations with different forms of nutritional interdependence. The ectomycorrhizal fungi form a mycelial sheath around plant roots, penetrating the outer layers of the root tissue intercellularly. Endomycorrhizal fungi form intracellular nutrient absorbing structures and therefore have a physically close relationship with the plant cells which are invaded. Although the fungus is a drain on the carbon resources of the plant, nutrient flow is a two-way process (Harley and Smith 1983) and these associations significantly increase plant growth, indeed the presence of vesicular-arbuscular mycorrhizae is of great economic significance with regard to crop plant species.

Fungi are very successful plant parasites and penetrate host tissues either intercellularly or intracellularly. Host plants subsequently act as the nutrient substrate for continued fungal growth and development (Hancock and Huisman 1981). Necrotrophs obtain nutrients from dead cells, killed on hyphal invasion or in advance by fungal toxins or enzymes. Biotrophs have a closer and more specialized relationship with plant tissues and obtain nutrients from living cells by means of modified feeding hyphae or haustoria. Biotrophs are a drain on plant resources (Hancock and Huisman 1981).

The responses of host plants to such fungal invasions are complex and have been described at three levels by Ralton et al. (1985). Tissue responses result in pathogen isolation (Beckman and Talboys 1981). Cellular responses include changes in cell wall structure, increasing mechanical strength and resistance to enzymic attack (Ride 1980; Kolaltukudy 1984). Cytoplasmic responses include the

[1] Plant Genetic Manipulation Group, Department of Botany, University of Nottingham, Nottingham, NG7 2RD, UK
[2] Department of Botany, University of Liverpool, P.O. Box 147, Liverpool L69 3BX, UK

Biotechnology in Agriculture and Forestry, Vol. 9
Plant Protoplasts and Genetic Engineering II (ed. by Y.P.S. Bajaj)

hypersensitive reaction, which results in host cell death on contact with the invading pathogen. Changes in host cell membrane permeability occur (Kiraly 1980) and antimicrobial phytoalexins are produced in adjacent healthy tissue (Keen 1982).

Plant tissue culture and cell suspension culture techniques provide useful experimental systems for the study of those physiological interactions which occur between higher plants and fungi (Helgeson 1983). Isolated plant protoplasts are easily manipulated and can be maintained in a homogeneous and controlled environment, allowing the influence of fungal activity on the plant plasmamembrane and cytoplasm to be investigated in biochemical and morphological terms. Two particular characteristics of protoplasts can be exploited for the examination of plant-fungal interactions. Firstly, protoplasts are capable of taking up material, such as microbes (Harding and Cocking 1986), and secondly, protoplasts can fuse with each other. Additionally, the use of plant and fungal protoplasts together would enable close contact to be set up between the plant plasmalemma and the fungal plasmamembrane with the possibility of membrane fusion and cytoplasmic mixing.

Plant protoplasts have been shown to take up a wide range of particles, e.g., latex particles (Grout et al. 1973), polystyrene spheres (Suzuki et al. 1977), naked DNA (Steinbiss and Broughton 1983), viruses (Takebe 1983), organelles (Lörz 1985), and intact microbes (Cocking 1984). Davey and Cocking (1972) suggested that the uptake of microbes into plant protoplasts might result in the establishment of new symbiotic relationships, and attempts have been made to establish nitrogen fixing relationships by this route (Davey and Cocking 1972; Giles and Vasil 1980). Uptake of cyanobacteria has also been achieved (Davey and Power 1975; Burgoon and Bottino 1976) although extended metabolically active associations have not always resulted (Meeks et al. 1978).

Potentially fungal protoplasts could act as eukaryotic gene vectors for plant cell transformation. A range of DNA vectors have been tested; DNA incorporated into plasmids (Lörz et al. 1985), viruses (Howell et al. 1980; Lebeurier et al. 1980), organelles (Galun and Aviv 1986) and microbes (Wullems et al. 1984; Wullems and Schilperoort 1985). Hasezawa et al. (1981), and have reported the transformation of plant protoplasts by the introduction of *Agrobacterium tumefaciens* spheroplasts.

Protoplast fusion provides a system for the transfer of genetic characters (Smith and Mastrangelo-Hough 1979). Constabel et al. (1982) have attempted to transfer the ability for secondary metabolite synthesis in culture from fungi, using fungal protoplasts, into protoplasts from cultured *Catharanthus roseus* cells. Adhesion between viable protoplasts was observed although no gene transfer was detected.

Detailed morphological and biochemical investigations of those interactions which occur between plant and fungal protoplasts require a reliable experimental system. Celery (*Apium graveolens*) is an important commercial crop plant both as a fresh vegetable and in convenience foods (Orton 1984), which has been grown very successfully in culture (Williams and Collin 1976a, b) and methods for the liberation of protoplasts have been developed (Orton 1984; Lynch 1987). Similarly, reliable techniques and extensive information concerning protoplast physiology and regeneration have been described for the saprophytic fungus *Aspergillus nidulans* (Isaac 1985) and for the potential pathogen of celery *Fusarium oxysporum* (Lynch et al. 1985).

This chapter describes the fusion of protoplasts from these species, using several techniques and considers the factors affecting such experimental systems and their potential application.

## 2 Protoplast Isolation

### 2.1 Celery Protoplasts

The physiological condition of the source material is perhaps the most important factor for long-term experiments and affects both the yield and viability of the protoplasts released (Davey 1983). Cultured cells have been considered as potentially ideal source material for protoplast isolation because the growth environment can be easily controlled and also because such cultures are inherently axenic. However, with repeated subculture such cells do gradually lose totipotency; aneuploidy and polyploidy increase (Bayliss 1980), and structural changes occur in chromosomes (Muarata and Orton 1983).

For the investigation of plant-fungal interactions, several considerations influenced the choice of source material for the plant protoplast component. Viable protoplasts were required routinely for experimental purposes and although methods were developed for protoplast isolation from axenic celery seedling mesophyll tissue, cell suspension cultures were a more dependable source (Lynch 1987). Protoplasts were successfully obtained from late lag phase suspension cultures, but not from cells in exponential phase. Increased culture age is known to contribute to wall thickening, which reduces susceptibility to enzyme lysis and older cultures also contain a higher proportion of aged, senescing cells (Evans and Bravo 1983). Protoplasts were isolated after lytic incubation overnight, using the protocol given in Fig. 1. It has been suggested that such relatively long periods of lytic incubation, required for the isolation of protoplasts from cultured cells, may aid eventual protoplast fusion (Kao 1977).

### 2.2 Fungal Protoplasts

There is a continuous and characteristic variation in the morphological organization and in the biochemical properties of intact fungal hyphae from apical to progressively sub-apical regions. The disintegration of mycelium by lytic digestion leads to the division of cytoplasm into discrete protoplast units and must also result in the disruption of metabolic systems which are normally fully integrated. Many protoplasts may be released from one hyphal compartment and, in coenocytic species, may contain several nuclei. During lytic incubation protoplasts are released first from apical regions and subsequently from regions progressively further from the hyphal apex (Isaac et al. 1979; Isaac 1985). Isolated

**CELLERY SEED**

*Apium graveolens* Linn. var. French Dimant

surface sterilize: 1) immerse 8 min in 5% sodium hypochlorite
2) wash (x3) in sterile distilled water

germinate : MS medium* + sucrose ($30 g.l^{-1}$)+ Bacto Agar ($10 g.l^{-1}$)
1) vials under continuous light 4°C for 48h
2) vials under continuous light 22°C for 3-4 weeks.

seedlings : 6-8 weeks old; 4-6 leaves.

↓

**CALLUS CULTURE**

inoculate : petiole section (1cm long) excised aseptically
placed on MS supplemented medium**+ agar.

incubate : 20°C with 12h light/dark cycle

callus formation : after 37 days

subculture : monthly

↓

**CELL SUSPENSION CULTURE**

inoculate : friable, undifferentiated callus into liquid
MS supplemented medium**

incubate : 20°C with shaking (100 r.p.m)
12h light/dark cycle

subculture : 3 week intervals. Select for fast growing dispersed callus.

↓

**PROTOPLAST ISOLATION**

source material : 3 day old cell suspension cultures (late lag phase)

lytic mixture : tissue (1g fresh weight) resuspended in
stabilizer (0.6M mannitol) and
lytic enzyme (0.1% Pectolyase Y23♦ and 2% Cellulysin♦♦)

lytic incubation : 30°C with gentle agitation for 16h.

↓

**PROTOPLAST HARVEST**

harvest protoplasts : by filtration through pre-wetted 100 μm nylon net
to remove cell debris.

wash : by centrifugation (200g for 1 min) in stabilizer solution to remove
lytic enzyme mixture.

---

* MS medium : MURASHIGE & SKOOG (1962)

** MS supplemented medium : MS medium plus 2, 4-D ($0.25 mg.l^{-1}$)
and Kinetin ($0.6 mg.l^{-1}$)

♦ Petolyase Y23 : R.W. Unwin & Co. Ltd.

♦♦ Cellulysin : Cambridge Bioscience

**Fig. 1.** Protocol for tissue culture and protoplast isolation from celery (*Apium graveolens*)

**FUNGAL SPORES**

| *Aspergillus nidulans* | *Fusarium oxysporum* |
|---|---|
| inoculation : conidia 1 x $10^6$.$cm^{-3}$ final concentration | inoculation : conidia 1 x $10^3$.$cm^{-3}$ final concentration |
| medium : Vogels* liquid minimal medium | medium : liquid salts medium** |
| incubation : 30°C with shaking (200 rpm) for 18h | incubation : 25°C with shaking (100rpm) for 17h |

↓

**MYCELIUM**

| | |
|---|---|
| harvest : exponential phase by vacuum filtration | harvest : exponential phase by vacuum filtration |
| plasmolysis : resuspend in 0.2M phosphate buffer pH 5.8 plus 1.2M KCl | |

↓

**LYTIC INCUBATION**

| | |
|---|---|
| source material : mycelium 0.02g fresh weight.$cm^{-3}$ | source material : mycelium 0.02g fresh weight.$cm^3$. |
| lytic mixture : 0.2M phosphate buffer pH 5.8 plus 0.6M KCl stabilizer and Novozym 234♦ (1mg.$cm^{-3}$) | lytic mixture : 0.2M phosphate buffer pH 5.8 plus 0.6M mannitol stabilizer and Novozym 234♦ (7.5mg.$cm^{-3}$) |
| incubation : 30°C with gentle shaking for 3h | incubation : 30°C with gentle shaking for 3h |

↓

**PROTOPLAST HARVEST**

harvest : by filtration through porosity 1 sintered glass filter; wash (by centrifugation at 800g for 5 mins) and resuspend in buffer-stabilizer

* Vogels medium : VOGEL, 1964

** *Fusarium* salts medium : (g.$l^{-1}$) glucose 0.7; $MgSO_4$ $7H_2O$ 0.5; $KH_2PO_4$ 0.2; $NH_4NO_3$ 0.1.

♦ Novozym 234 : Novo Industri, Denmark

**Fig. 2.** Protocol for protoplast isolation from fungal mycelium

fungal protoplasts are metabolically active (Isaac 1985) and a proportion of those which contain nuclei have the capacity to regenerate a new cell wall and give rise to normal mycelium in culture.

As for plant cell systems the release of protoplasts from any fungal species is influenced by a range of factors, e.g., temperature, pH, stabilizer type and concentration, the mixture of lytic enzymes used, and the physiological state of the source mycelium (Davis 1985). Investigation of these parameters and the maximization of yields has been the subject of much research to date and the isolation of fungal protoplasts can be achieved easily and routinely. Protocols used for protoplast isolation from *A. nidulans* and *F. oxysporum* are described in Fig. 2.

## 3 Initiation of Fusion Between Plant and Fungal Protoplasts

Although plant protoplasts have been shown to fuse spontaneously during incubation (Withers and Cocking 1972), chemical and electrical methods have been used to increase the frequency of fusion events and the range of materials taken up by protoplasts. The use of polyethylene glycol (PEG) was developed by Kao and Michayluk (1974) for plant protoplast fusions and is now the most commonly used chemical fusogen for plant material (Lazar 1983). It is also a very successful treatment for fungal protoplast fusion (Ferenczy 1985) and has been used to stimulate the uptake of microbes into plant protoplasts (Davey and Power 1975). Electrofusion methods have also led to successful plant protoplast fusion (Bates 1985; Morikawa et al. 1986) and also fusion of microbial protoplasts, especially those derived from yeast cells (Halfmann et al. 1982). Electrofusion procedures can result in high fusion frequencies (Zimmermann and Scheurich 1981; Bates and Hasenkampf 1985) and are potentially more controllable and efficient than chemical treatments.

An investigation of the biochemical interactions between plant and fungal protoplasts requires relatively large amounts of material. We have investigated both PEG-stimulated fusion and electrofusion of plant and fungal protoplasts in order to maximise the frequency of fusion events. Several methods were used to detect fusion so as to quantify the levels of fusion achieved.

### 3.1 The Use of Polyethylene Glycol as a Fusogenic Chemical Agent

The stimulation of protoplast fusion by PEG occurs as a two-stage process. The initial mixing of protoplasts with PEG in solution caused an enhanced aggregation and adhesion of fungal protoplasts to celery protoplasts. Subsequent dilution of PEG concentration led to the uptake of fungal protoplasts into plant protoplast cytoplasm. The protocol given in Figure 3 yielded maximum levels of fusion with this material.

Direct observation of PEG-treated protoplasts were made using light microscopy in the first instance (Fig. 4). Celery protoplasts were spherical with diameters in the range 10 – 40 μm. The largest protoplasts were most vacuolate. Almost all contained large lobed nuclei which fluoresced brightly under ultraviolet illumination after staining with 4, 6-diamidino -2-phenylindole (DAPI) (Linn et al. 1977). Fungal protoplasts were also spherical, but were much smaller in diameter (0.5 – 7 μm). It was only possible to detect the presence of fungal protoplasts within plant protoplast cytoplasm after DAPI staining, when nuclear fluorescence was observed (Fig. 4). *A. nidulans* protoplasts were observed adhering to a maximum of 54.0% ± 1.4% celery protoplasts and, using nuclear fluorescence as the method of detection, a maximum of 14.0% ± 1.6% celery protoplasts contained fungal nuclei. Similar data were obtained with *F. oxysporum*. Brightly fluorescing plant nuclei may have obscured the much smaller fungal nuclei; this method may therefore have underestimated the fusion frequency.

Ultrastructural observations were made in order to detect the presence of fungal protoplasts within celery protoplast cytoplasm and to assess any long-term

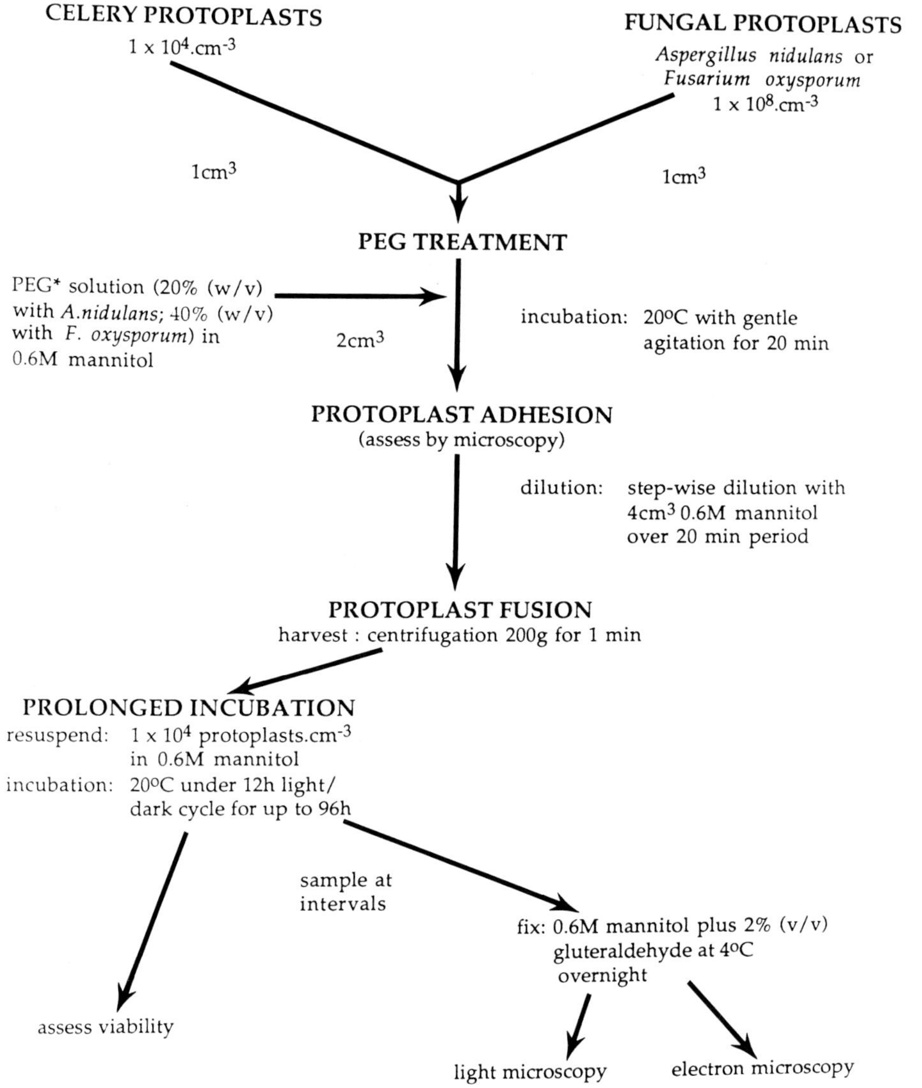


*PEG : M. Wt 8,000; Sigma Chemical Co. Ltd.

**Fig. 3.** Protocol for polyethylene glycol (PEG)-stimulated fusion between plant and fungal protoplasts

morphological changes on prolonged incubation. Quantitative assessment of fusion using electron microscopy was made over a sample of 800 sections; of these, a maximum of 1.5% celery protoplast sections contained *A. nidulans* protoplasts and in a similar experiment 0.6% contained *F. oxysporum* protoplasts. These are lower values than those obtained by fluorescent microscopy, and may have been due to difficulty in identification of fungal protoplasts in thin sections. The main

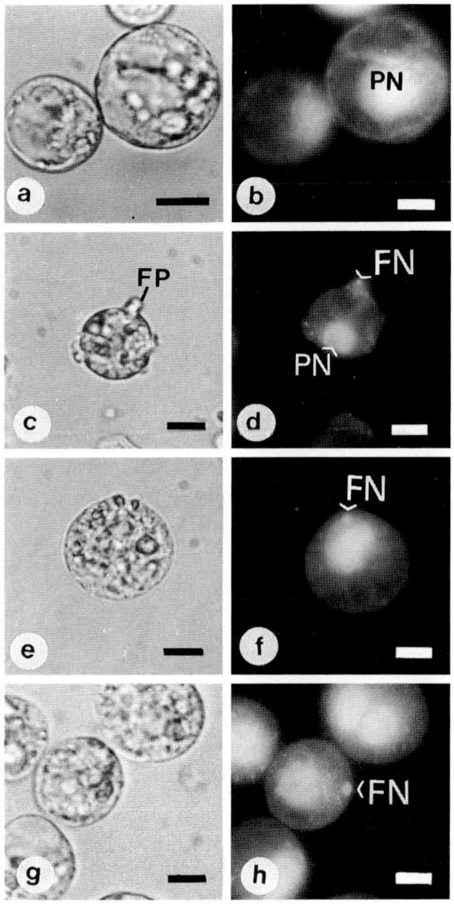


**Fig. 4a–h.** Light micrographs to show the effect of polyethylene glycol (PEG) treatment on celery and fungal protoplasts incubated together. **a, b** Celery protoplasts isolated in 0.6 M mannitol. **c, d** Celery protoplasts with *A. nidulans* protoplasts adhering. **e, f** Celery protoplasts containing *A. nidulans* protoplasts immediately after PEG dilution. **g, h** Celery protoplasts containing *F. oxysporum* protoplasts immediately after PEG dilution. **a, c, e, g** Light micrographs; **b, d, f, h** corresponding micrographs viewed under ultraviolet light after DAPI staining. *PN* plant nucleus; *FN* fungal nucleus; *bar* = 10 μm

criteria used to positively identify structures within celery protoplasts which corresponded to fungal protoplasts which had been taken up, were size and subcellular organization, i.e., the size ranges and the presence of organelles were related to those of isolated fungal protoplasts. Some structural differences were also observed between fungal cytoplasm and plant cytoplasm.

Ultrastructural observations confirmed that protoplasts were bounded by membrane and lacked cell wall material. Some celery protoplasts appeared to lack vacuoles (Fig. 5a), some contained small vacuoles throughout the cytoplasm, whereas others had several large vacuoles (Fig. 5b). Otherwise celery protoplasts were morphologically homogeneous; large nuclei were visible and a range of proplastids were seen. *A. nidulans* protoplasts were much smaller and showed greater morphological heterogeneity (Fig. 5). Some were small and cytoplasmically dense (Fig. 5c), others were larger and more vacuolate with granular cytoplasm (Fig. 5d). Such heterogeneity has been described previously for *A. nidulans* protoplast populations (Isaac et al. 1979). Ultrastructural observations of mixed protoplast material were made during adhesion (Fig. 6a) and after PEG-stimulated fusion

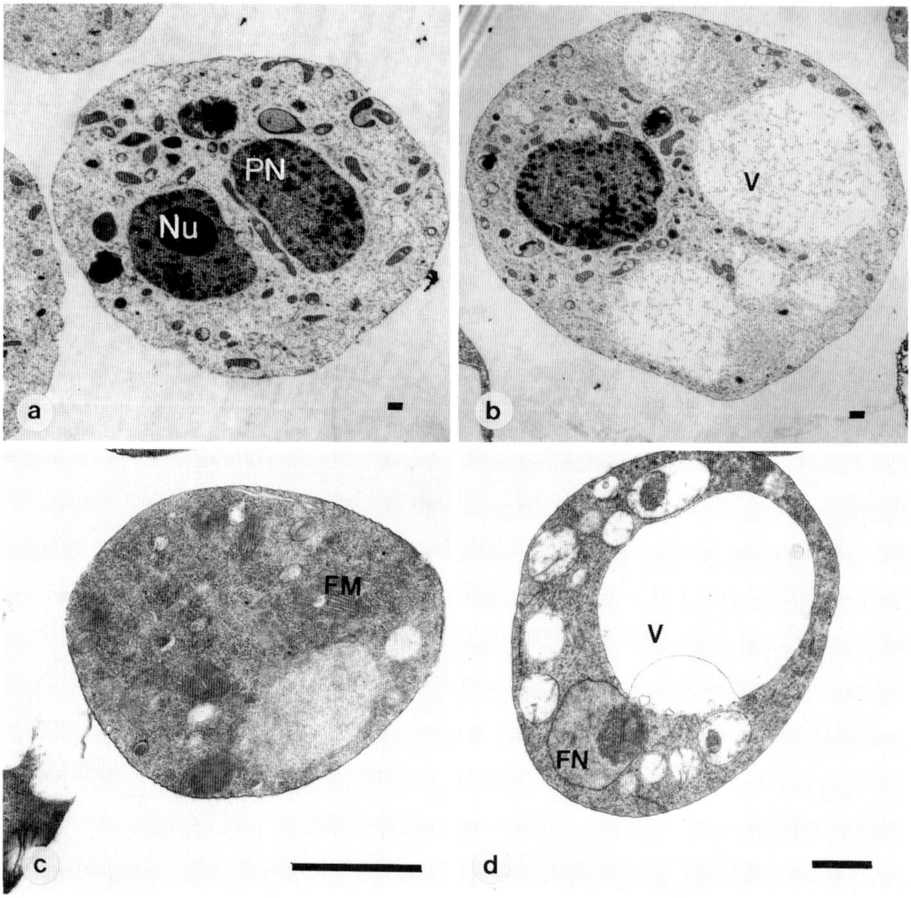


**Fig. 5a–d.** Electron micrographs to show protoplast ultrastructure. **a** Protoplast isolated from celery in 0.6 M mannitol. **b** Extensively vacuolate celery protoplast. **c** Protoplast isolated from *A. nidulans*; small protoplast with dense cytoplasm. **d** Larger, vacuolate *A. nidulans* protoplast. *PN* plant nucleus; *FN* fungal nucleus; *Nu* nucleolus; *FM* fungal mitochondrion; *V* vacuole; *bar* = 1 μm

(Fig. 6b). It was interesting that those fungal protoplasts adhering to and taken up by celery protoplasts were predominantly the small dense *A. nidulans* protoplasts which probably arise from apical regions of hyphae. It is not clear why this apparently preferential uptake occurred. In general, these fungal protoplasts were small and may therefore have been taken up by endocytosis more easily or may have been more stable during the uptake process. Both *A. nidulans* and *F. oxysporum* protoplasts identified within celery protoplasts were bounded by a plasmamembrane and in some cases a second membrane was also present.

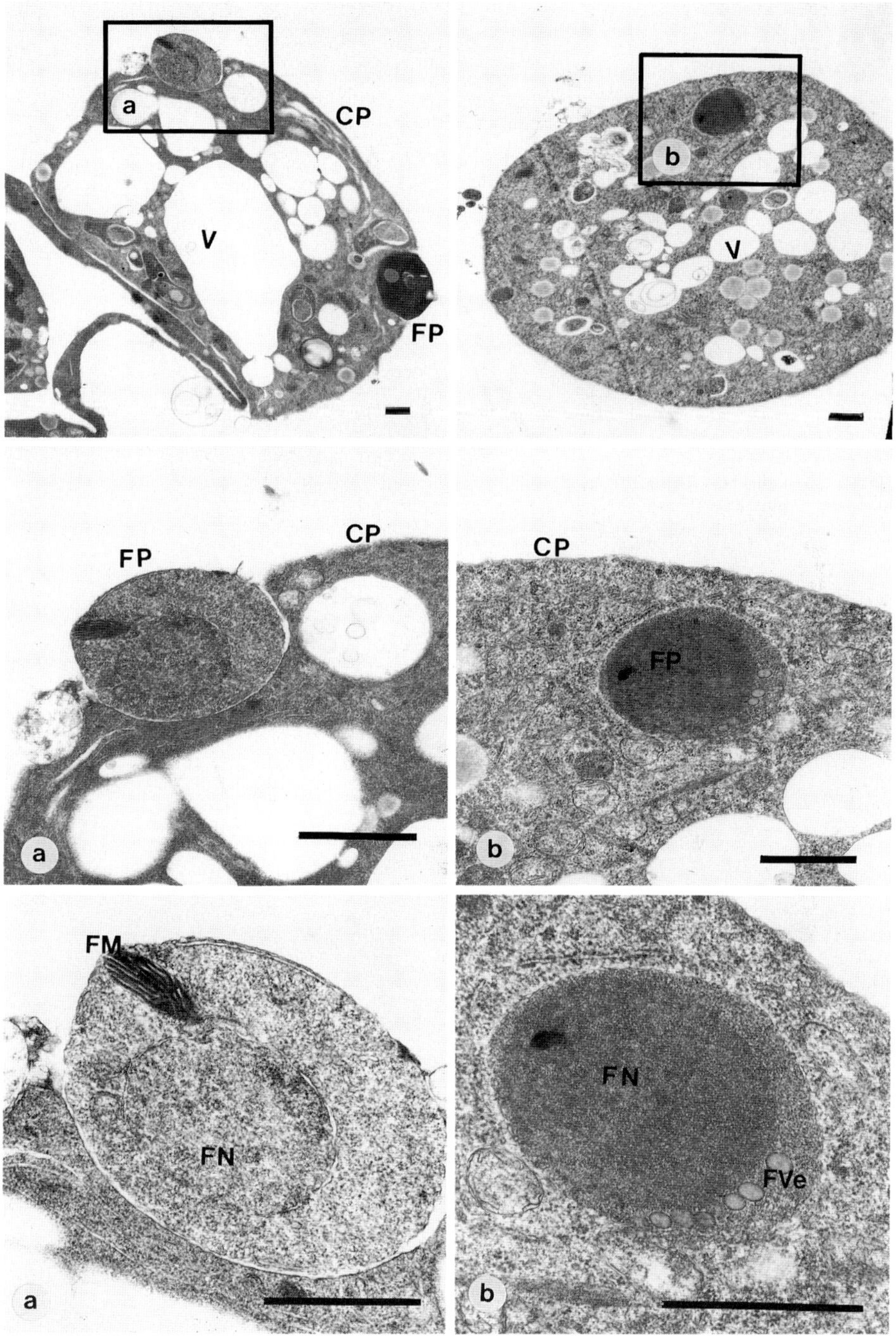


**Fig. 6a, b.** Electron micrographs showing **a** adhesion and **b** fusion between celery protoplasts and small, cytoplasmically dense *A. nidulans* protoplasts. *CP* celery protoplast; *FP* fungal protoplast; *FN* fungal nucleus; *FM* fungal mitochondrion; *FVe* fungal vesicles; *V* vacuole; *bar* = 1 μm

## 3.2 Electrofusion

Electrofusion is also a two-stage process. Initially protoplasts placed in the alternating current (AC) alignment field protoplasts formed pearl chains (Fig. 8). In general, celery protoplasts adhered to the electrodes and *A. nidulans* protoplasts associated with the end of celery protoplast pearl chains. Subsequent DC pulse treatment resulted in fusion between the two protoplast types and after fusion protoplasts gradually rounded up (Fig. 9). By direct microscopic observation in

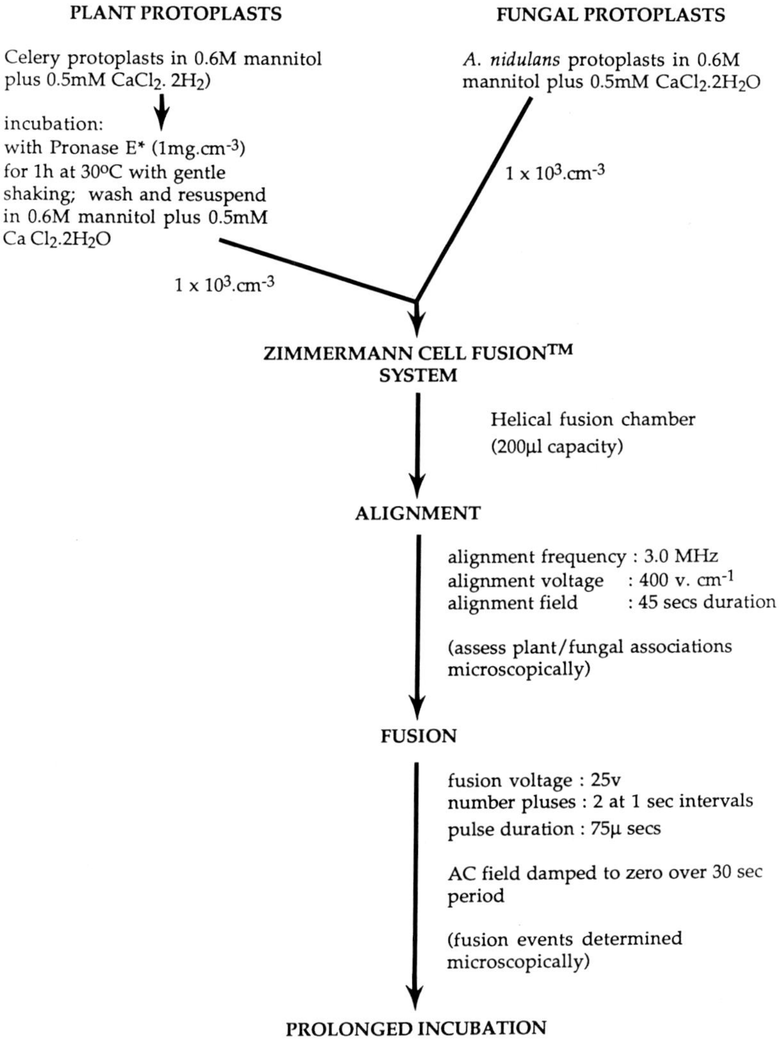


*Pronase E : protease, Type XXV; Sigma Chemical Co. Ltd.

**Fig. 7.** Protocol for electrofusion of plant and fungal protoplasts

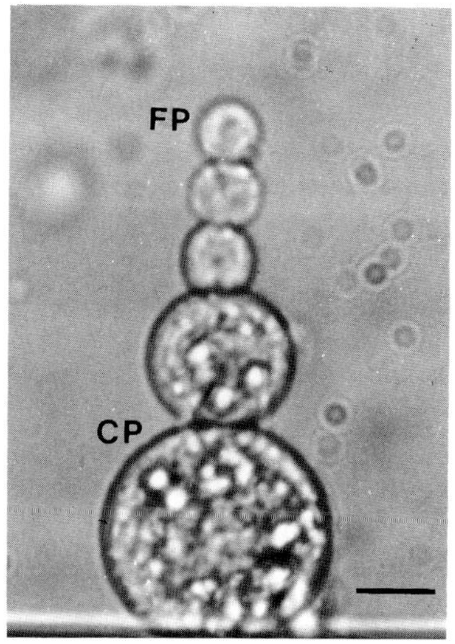


**Fig. 8.** Light micrograph to show pearl chaining of celery protoplasts with *A. nidulans* protoplasts adhering in AC alignment field (3 MHz, 400 V/cm). *CP* celery protoplast; *FP* fungal protoplast; *bar* = 10 μm

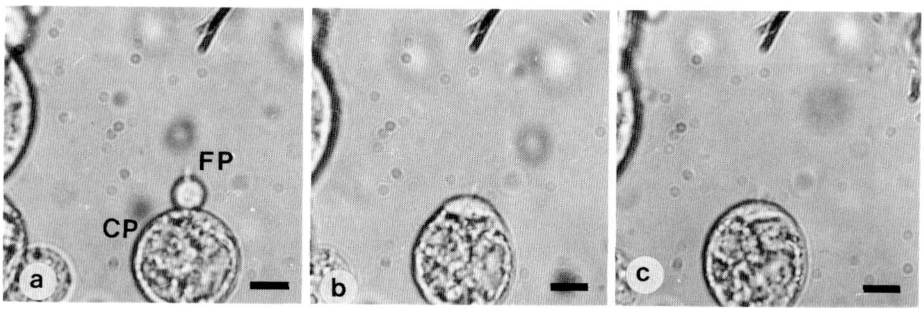


**Fig. 9 a – c.** Electrofusion of celery and *A. nidulans* protoplasts. **a** Adhesion of celery and *A. nidulans* protoplast in AC alignment field. **b** Immediately after fusion pulses; celery and *A. nidulans* protoplasts fused, alignment field present. **c** 5 min after fusion pulses; fused protoplast round up. *CP* celery protoplast; *FP* fungal protoplast; *bar* = 10 μm

excess of 40% celery protoplasts were seen to fuse with *A. nidulans* protoplasts. In general, small cytoplasmically dense, *A. nidulans* protoplasts were involved in fusion. Adjustment of electrical parameters was made in order to maximize the frequency of one-to-one plant-fungal protoplast fusions (Lynch 1987). The protocol used for successful plant-fungal protoplast electrofusion is given in Fig. 7.

Stronger electrical fields were required for fusion of fungal protoplasts, probably as a result of the small size (Zimmermann 1982). The treatment of celery protoplasts with pronase E and the presence of $Ca^{2+}$ during the electrical treatment procedures were important for successful fusion resulting in increased stability at the higher voltages required and greater capacity for fusion. Fusion of cells of different sizes after such treatment have been reported elsewhere (Salhani et al. 1985). The effects have been related to the extension of lipid domains in the membranes, by proteolytic removal of membrane protein, which aids membrane resealing (Zimmermann and Vienken 1982) and the binding of $Ca^{2+}$ to the plasmalemma which enhances membrane stability (Abe and Takeda 1986).

## 4 Prolonged Incubation of Plant-Fungal Protoplasts

After fusogenic treatment protoplast samples were incubated over a prolonged period in stabilizer and in stabilized complete medium. During incubation ultrastructure, viability and respiration rates were investigated to ascertain the extent to which any changes could be attributed to the presence of fungal protoplasts within celery protoplast cytoplasm.

### 4.1 Ultrastructural Changes

For 24 h after PEG treatment, the numbers of celery protoplasts containing fungal protoplasts, detected by electron microscopy, did not change. Subsequently, fewer were observed, although *A. nidulans* protoplasts were detected within celery protoplasts 72 h after fusion. Little morphological change was seen in protoplasts from either *A. nidulans* or *F. oxysporum* which persisted in celery protoplast cytoplasm (Fig. 10). Celery protoplasts showed some loss of integrity which was not attributable to fungal activity, since degeneration also occurred in control treatments lacking fungal protoplasts; and was not related to the nutrient status of the incubation medium. Treatment with PEG has been shown to affect the structural integrity of spinach mitochondria (Benbadis and Davy de Virville 1982); such specific affects were not observed in celery material. Wall regeneration was not observed. This may be attributed to inappropriate culture conditions and not necessarily to the quality of the protoplasts involved (Fowke and Gamborg 1980). Regeneration may be achieved more easily on solid medium (Takeuchi and Komamine 1982).

### 4.2 Metabolic Activities

#### 4.2.1 Viability after PEG Treatment

Viability of celery protoplasts was indicated by the ability to exclude Evans Blue vital dye and was assessed during prolonged incubation (Fig. 11). Most celery protoplasts were viable on isolation but viability decreased during the first 12 h of incubation. After this initial loss, viability subsequently declined, but more slowly. Incubation in mannitol has been shown to have adverse affects on plant protoplast viability (Thompson et al. 1986). A larger proportion of celery protoplasts were viable for a longer time when incubated in complete medium. The viability of celery protoplasts, after treatment in PEG, showed a similar pattern on prolonged incubation, but the viability of all samples was consistently lower than for untreated protoplasts. This data supports reports concerning the toxicity of PEG,

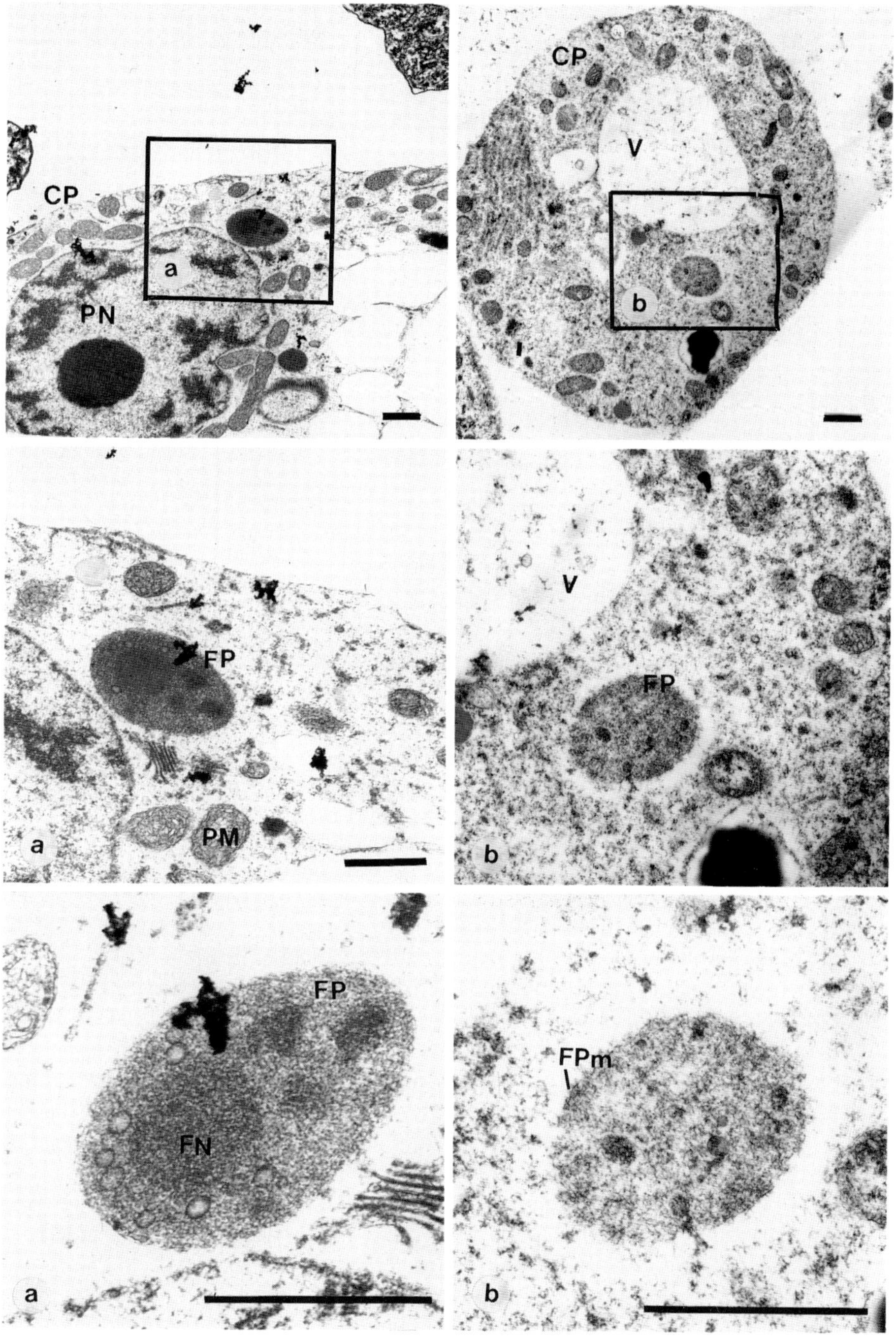


**Fig. 10a,b.** Electron micrographs to show fungal protoplasts present within celery protoplast cytoplasm during prolonged incubation. **a** Dense, nucleate *A. nidulans* protoplast within celery protoplast after 24-h incubation in complete medium. **b** *F. oxysporum* protoplast within celery protoplast after 12-h incubation in stabilizer. *CP* celery protoplast; *FP* fungal protoplast; *PN* plant nucleus; *FN* fungal nucleus; *PM* plant mitochondrion; *FPm* fungal plasmamembrane; *V* vacuole; *bar* = 1 μm

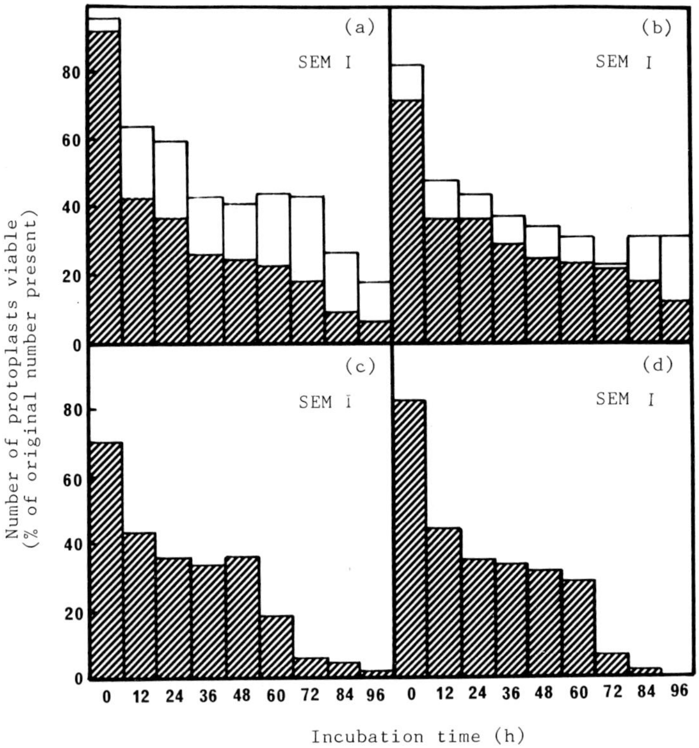


**Fig. 11 a–d.** Viability of celery protoplasts during prolonged incubation in: **a** 0.6 M mannitol stabilizer solution. **b** Complete medium. **c** 0.6 M mannitol stabilizer after uptake of *A. nidulans* protoplasts. **d** Complete medium after uptake of *A. nidulans* protoplasts. □ untreated material; ▨ PEG-treated protoplasts. Values are means of triplicate determinations. No standard error (SEM) exceeded the maximum value indicated

which causes leakage of materials from treated cells (Hahn-Hagerdal et al. 1986) and affects viability. A higher percentage of celery protoplasts treated with PEG in the presence of *A. nidulans* protoplasts (Fig. 11 c, d) were viable after 48 h incubation than after PEG-control treatment. After 60 h viability was low in both stabilizer and complete medium. After fusion with *F. oxysporum* protoplasts, celery protoplast viability was significantly depressed after 12 and 24 h (data not shown); incubation was not continued longer. It is interesting to speculate on the reasons for this; it is not clear if the loss of viability is a form of hypersensitive reaction in the plant protoplasts, or if it is the direct result of activity of the fungal pathogen. Fungal penetration does give rise to cell disorganization, electrolyte leakage, and necrosis (Kiraly 1980) in plant cells; this is an area which clearly warrants further investigation.

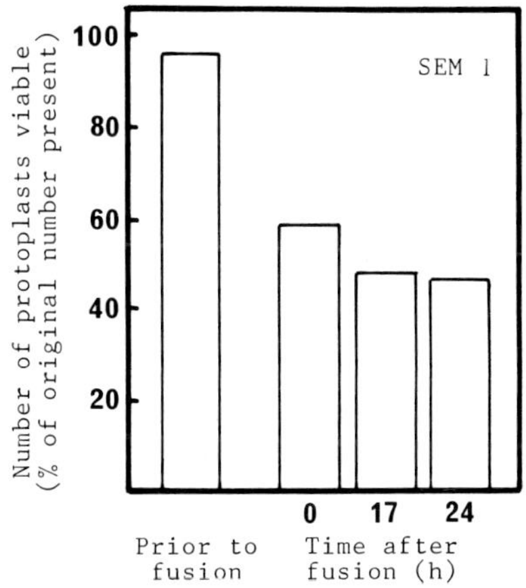


**Fig. 12.** Celery protoplast viability after electrofusion with *A. nidulans* protoplasts. Values are means of triplicate determinations. No standard error (SEM) exceeded the maximum value indicated

### 4.2.2 Viability After Electrofusion

The electrofusion procedure reduced celery protoplast viability very quickly but during 24 h incubation after fusion viability decreased by only a further 10% (Fig. 12). Subjection to the higher voltages required for fusion between plant and fungal protoplasts than for plant-plant fusions may have resulted in some damage to celery protoplasts. Zimmermann et al. (1980) reported that substantial current flow through protoplasts resulted in damage to cellular organelles and eventually to deterioration in cells. Membrane breakdown may also result in ion loss, although it has been suggested that electrofusion procedures cause less leakage of amino acids, protein, and DNA from protoplasts than PEG-fusion treatments (Hahn-Hagerdal et al. 1986). Shearing effects during the use of the helical fusion chamber (200 μl capacity), used to provide larger volumes of treated protoplasts than the flat chamber system (10 μl capacity), caused mechanical rupture of some protoplasts, particularly those with larger diameters. Celery protoplasts were more stable in strong electrical fields following pronase E treatment.

### 4.2.3 Respiration After Electrofusion

The normal respiration of isolated plant protoplasts (Taylor and Hall 1976) and respiration of fungal protoplasts over prolonged periods (Isaac and Gokhale 1983) have been reported previously. The respiration rates of untreated and electrofused celery protoplasts during a 24 h incubation period are shown in Fig. 13. The respiration rate of isolated celery protoplasts maintained in mannitol stabilizer, showed a slight decline during incubation. Pronase E slightly depressed respiration immediately following treatment. Electrofusion procedures did not have a great effect on respiration rates where celery protoplasts were treated alone. However, following electrofusion with *A. nidulans* protoplasts respiration rates were very significantly elevated during the first 3 h incubation. Subsequently,

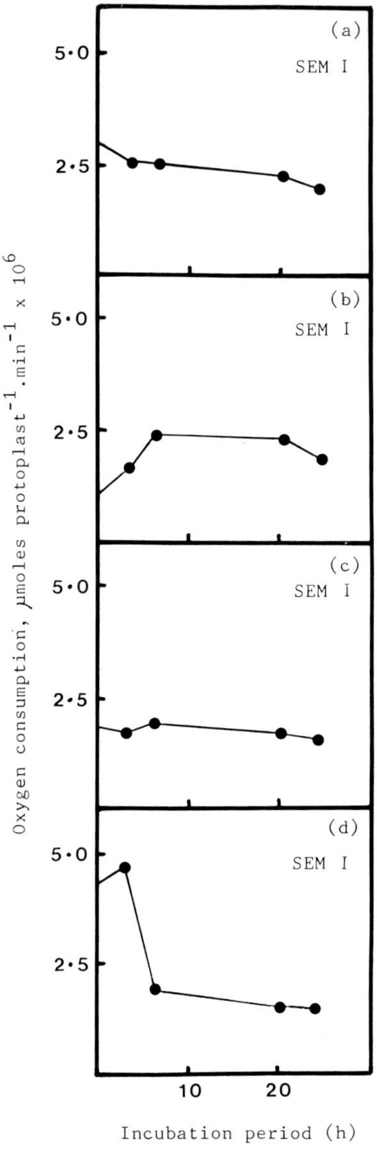


**Fig. 13a–d.** Oxygen consumption of celery protoplasts during prolonged incubation. **a** In 0.6 M mannitol stabilizer. **b** After Pronase E treatment. **c** After electrofusion. **d** After electrofusion with *A. nidulans* protoplasts. Values are means of duplicate determinations. No standard error (SEM) exceeded the maximum value indicated

rates were similar to those of unfused protoplasts. It is not clear to what extent the marked stimulation was due to the mixing of plant and fungal cytoplasm or possibly to the additive effect of fungal mitochondria.

## 5 The Fate of Plant-Fungal Protoplasts

It was possible to induce celery protoplasts to take up intact fungal protoplasts by stimulation with PEG. In this case it was probable that the smaller fungal protoplasts were taken up into plant protoplast cytoplasm by endocytosis with vesicles, since two limiting membranes were seen in electron micrographs. In the long term, these vesicles may interact with lysosomes, resulting in breakdown of the fungal protoplasts. During prolonged incubation some plant-fungal protoplasts persisted, although there was an overall decline in the numbers observed. It was not clear if this decline was due to fungal protoplast breakdown after uptake or to disintegration of the combined plant-fungal protoplasts. With similar material Meeks et al. (1978) observed the uptake of cyanobacteria into tobacco protoplasts; subsequently cyanobacteria were expelled and the tobacco protoplasts disintegrated.

Where protoplasts were held in vesicles it is unlikely that cytoplasmic mixing could occur immediately. Harding (1984) was unable to demonstrate transfer of antibiotic resistance to tobacco protoplasts via *E. coli* spheroplasts and suggested that since spheroplasts were taken up into vesicles no genetic exchange could take place. However, Hasezawa et al. (1981) showed that *Vinca* callus, derived from protoplasts which had taken up *Agrobacterium* spheroplasts, was able to synthesize octopine, indicating the occurrence of genetic exchange.

After electrofusion and cytoplasmic mixing, with fungal protoplasts, celery protoplasts retained viability and showed elevated levels of respiration shortly after fusion. Following PEG-stimulated uptake of *A. nidulans* protoplasts, celery protoplasts showed improved levels of viability. These results may indicate positive interaction between the two. Uptake of *F. oxysporum* protoplasts, however, resulted in depression of celery protoplast viability. It is not clear to what extent this depression was due to the interaction between a potential host and pathogen. This reaction may indicate some form of hypersensitive response by the plant or the effect of fungal activity. The synthesis of phytotoxic compounds by fungal protoplasts is very unlikely in such a short period of time and in a comparatively hostile environment. It has been shown that fungal infections increase the secondary metabolite synthesis of host plant cells (Mansfield et al. 1974) although, in the intact plant, such stimulation can also occur in cells which are not penetrated directly by the fungus. It is unlikely that such activity could account for the changes measured here.

## 6 Conclusions

Biotrophic fungi develop very close associations with plant tissues, invading by means of specialized hyphae (haustoria) which grow into living host cells without directly penetrating the cytoplasm. The host cell plasmalemma becomes invaginated around growing haustoria and a large area of surface contact is formed where nutrient exchange can occur. Ultrastructural examinations of the membrane formed between the invading fungus and plant cytoplasm have shown that it differs in character from normal plant plasmalemma. This extrahaustorial membrane is thickened (Gill and Gay 1977) and lacks both ATPase and $\beta$-glycerophosphate activity (Spencer-Phillips and Gay 1981). Very close membrane contact between plant and fungal protoplasts was required for fusion to occur after stimulation by either PEG or electrical treatments. This may have resulted in some mixing of membrane components, particularly after electrofusion. It is not clear whether changes occurred in the biochemical activities associated with this altered membrane structure and how this relates to the natural situation. After PEG treatment fungal protoplasts were bounded by two membranes within the celery protoplast cytoplasm. It is interesting to consider the extent to which the uptake of fungal protoplasts may represent the formation of artificial biotrophic structures.

The study of biochemical aspects of cellular interactions between plants and fungi is difficult in the intact plant. The use of protoplast material to investigate such relationships provides a system in which the initiation of interactions can be precisely defined. Additionally, the potential for maintenance of such material in controlled conditions for prolonged periods has been demonstrated.

## 7 Summary Protocol

*1. Plant Protoplast Isolation.* Protoplasts were isolated from late lag phase cell suspension cultures of celery suspended in a lytic mixture containing stabilizer (0.6 M mannitol) and lytic enzyme (0.1% Pectolyase Y23 and 2% Cellulysin). Lytic incubation was at 30 °C with gentle agitation for 16 h (for details see Fig. 1).

*2. Fungal Protoplast Isolation.* Protoplasts were isolated from exponential phase mycelium of *A. nidulans* grown in liquid culture suspended in a lytic mixture containing buffer stabilizer (0.2 M phosphate buffer pH 5.8 plus 0.6 M KCl) and lytic enzyme (1 mg $cm^3$ Novozym 234). Lytic incubation was at 30 °C with gentle shaking for 3 h (for details see Fig. 2).

*3. Fusion Treatment.* Plant and fungal protoplast suspensions were mixed and treated at 20 °C with gentle agitation for 20 min with PEG solution to aggregate protoplasts. Stepwise dilution of this mixture, over a 20-min period, resulted in protoplast fusion (for details see Fig. 3). Electrofusion was also successful for this material (see Fig. 7).

## References

Abe S, Takeda J (1986) Possible involvement of calmodulin and the cytoskeleton in electrofusion of plant protoplasts. Plant Physiol 81:1151–1155

Bates GW (1985) Electrical fusion for optimal formation of protoplast heterokaryons in *Nicotiana*. Planta 165:217–224

Bates GW, Hasenkampf CA (1985) Culture of plant somatic hybrids following electrical fusion. Theor Appl Genet 70:227–233

Bayliss MW (1980) Chromosomal variation in plant tissues in culture. Int Rev Cytol Suppl II A:113–144

Beckman CH, Talboys PW (1981) Anatomy of resistance. In: Mace ME, Bell AA, Beckman CH (eds) Fungal wilt diseases of plants. Academic Press, New York London, pp 487–518

Benbadis A, Davy de Virville J (1982) Effects of PEG treatment, used for protoplast fusion and organelle transplantation, on the functional and structural integrity of mitochondria isolated from spinach leaves. Plant Sci Lett 26:257–264

Burgoon AC, Bottino PJ (1976) Uptake of the nitrogen fixing blue green alga *Gleocapsa* into protoplasts of tobacco and maize. J Hered 67:223–226

Cocking EC (1984) Plant cell fusion: transformations using plant and bacterial protoplasts. In: Beers RF, Bassett B (eds) Cell fusion: gene transfer and transformation. Raven, New York, pp 139–144

Constabel F, Kurz WGW, Kutney JP (1982) Variation in cell cultures of periwinkle *Catharanthus roseus*. In: Fujiwara A (ed) Plant tissue culture, 1982. Maruzen, Tokyo, pp 301–304

Davey MR (1983) Recent developments in the culture and regeneration of plant protoplasts. In: Potrykus I (ed) Protoplasts 1983. Proc 6th Int Protoplast Symp, Basel. Birkhäuser, Basel, pp 19–30

Davey MR, Cocking EC (1972) Uptake of bacteria by isolated higher plant protoplasts. Nature (Lond) 239:455–456

Davey MR, Power JB (1975) PEG-induced uptake of microorganisms into higher plant protoplasts: an ultrastructural study. Plant Sci Lett 5:269–274

Davis B (1985) Factors influencing protoplast isolation. In: Peberdy JF, Ferenczy L (eds) Fungal protoplasts – applications in biochemistry and genetics. Dekker, New York, pp 45–71

Evans DA, Bravo JE (1983) Plant protoplast isolation and culture. In: Giles KL (ed) Plant protoplasts. Academic Press, New York London, pp 33–53

Ferenczy L (1985) Protoplast fusion in yeasts. In: Peberdy JF, Ferenczy L (eds) Fungal protoplasts – applications in biochemistry and genetics. Dekker, New York London, pp 279–306

Fowke LC, Gamborg OL (1980) Application of protoplasts to the study of plant cells. Int Rev Cytol 68:9–51

Galun E, Aviv D (1986) Organelle transfer. Methods Enzymol 118:595–611

Giles KL, Vasil IK (1980) Nitrogen fixation and plant tissue culture. Int Rev Cytol Suppl 11B:81–99

Gill F, Gay JL (1977) Ultrastructural and physiological properties of the host interfacial components of haustoria of *Erysiphe pisi* in vivo and in vitro. Physiol Plant Pathol 10:1–12

Grout BWW, Willison JHM, Cocking EC (1973) Interactions at the surface of plant cell protoplasts; an electrophoretic and freeze-etch study. J Bioenerget 4:311–328

Hahn-Hagerdal B, Ho Ono K, Zachrisson A, Bornman CH (1986) PEG and electric field treatment of plant protoplasts: characterization of some membrane properties. Physiol Plant 67:359–364

Halfmann HJ, Rocken W, Emeis CC, Zimmermann U (1982) Transfer of mitochondrial function into a cytoplasmic respiratory-deficient mutant of *Saccharomyces* yeast by electrofusion. Curr Genet 6:25–28

Hancock JG, Huisman OC (1981) Nutrient movement in host-pathogen systems. Annu Rev Phytopathol 19:309–331

Harding K (1984) Studies on plant genetic transformation using tobacco protoplasts. Ph D Thesis, Univ Nottingham

Harding K, Cocking EC (1986) The interaction between *Escherichia coli* spheroplasts and plant protoplasts: A proposed procedure to deliver foreign genes into plant cells. Protoplasma 130:153–161

Harley JL, Smith SE (1983) Mycorrhizal symbiosis. Academic Press, New York London

Hasezawa S, Nagata T, Syōno K (1981) Transformation of *Vinca* protoplasts mediated by *Agrobacterium* spheroplasts. Mol Gen Genet 182:206–210

Hasezawa S, Matsui C, Nagata T, Syōno K (1983) Cytological study of the introduction of *Agrobacterium tumefaciens* spheroplasts into *Vinca rosea* protoplasts. Can J Bot 61:1052–1057

Helgeson JP (1983) Studies of host-pathogen interactions in vitro. In: Helgeson JP, Deverall BJ (eds) Use of tissue culture and protoplasts in plant pathology. Academic Press, New York London, pp 9–38

Howell SH, Walker LL, Dudley RK (1980) Cloned cauliflower mosaic virus DNA infects turnips (*Brassica rapa*). Science 208:1265–1267

Isaac S (1985) Metabolic properties of protoplasts. In: Peberdy JF, Ferenczy L (eds) Fungal protoplasts, applications in biochemistry and genetics. Dekker, New York, pp 171–188

Isaac S, Gokhale AV (1983) Respiration of protoplasts from *Aspergillus nidulans*. Exp Suppl 45:334–335

Isaac S, Briarty LG, Peberdy JF (1979) The stereology of protoplasts from *Aspergillus nidulans*. In: Ferenczy L, Farkas GL (eds) Advances in protoplast research. Pergamon, Oxford, pp 213–219

Kao KN (1977) Chromosomal behaviour in somatic hybrids of soybean – *Nicotiana glauca*. Mol Gen Genet 150:225–230

Kao KN, Michayluk MR (1974) A method for high frequency intergeneric fusion of plant protoplasts. Planta 115:355–367

Keen NT (1982) Specific recognition in gene-for-gene host-parasite systems. In: Ingrams DS, Williams PH (eds) Advances in plant pathology, vol 1. Academic Press, New York London, pp 35–82

Kiraly Z (1980) Defenses triggered by the invader: hypersensitivity. In: Horsfall JG, Cowling EG (eds) Plant pathology. An advanced treatise, vol 5. Academic Press, New York London, pp 201–224

Kolaltukudy PE (1984) Fungal penetration of defensive barriers of plants. In: Dugger WM, Bartnicki-Garcia S (eds) Structure, function and biosynthesis of plant cell walls. Waverley, Baltimore, pp 302–343

Lazar GB (1983) Recent developments in plant protoplast fusion and selection technology. In: Potrykus I (ed) Protoplasts 1983. Proc 6th Int Protoplast Symp, Basel. Birkhäuser, Basel, pp 61–68

Lebeurier G, Hirth L, Hohn T, Hohn B (1980) Infectivities of native and cloned DNA of cauliflower mosaic virus. Gene 12:139–146

Linn MS, Comings DE, Alfi OS (1977) Optical studies of the interaction of 4,6-diamidino-2-phenylindole with DNA and metaphase chromosomes. Chromosoma 60:15–20

Lörz H (1985) Isolated cell organelles and subprotoplasts: their roles in somatic cell genetics. In: Dodds JH (ed) Plant genetic engineering. Univ Press, Cambridge

Lörz H, Baker B, Shell J (1985) Gene transfer to cereal cells mediated by protoplast transformation. Mol Gen Genet 199:178–182

Lynch PT (1987) The uptake of fungal protoplasts by plant protoplasts. Ph D Thesis, Univ Liverpool

Lynch PT, Collin HA, Isaac S (1985) Isolation and regeneration of protoplasts from *Fusarium tricinctum* and *Fusarium oxysporum*. Trans Brit Mycol Soc 85:135–140

Mansfield JW, Hargreaves JA, Boyle FC (1974) Phytoalexin production by live cells in broad bean leaves infected with *Botrytis cinerea*. Nature (Lond) 252:316–317

Meeks JC, Malmberg RL, Wolk CP (1978) Uptake of autotrophic cells of a heterocyst-forming cyanobacterium by tobacco protoplasts, and the fate of their association. Planta 139:55–60

Morikawa H, Sugino K, Hayashi Y, Takeda T, Senda M, Hirai A, Yamada Y (1986) Interspecific plant hybridization by electrofusion in *Nicotiana*. Biotechnology 4:57–60

Muarata M, Orton TJ (1983) Chromosome structural changes in cultured celery cells. In Vitro 19:83–89

Murashige T, Skoog F (1962) A revised medium for rapid growth and bioassays with tobacco tissue culture. Physiol Plant 15:473–497

Orton TJ (1984) Celery. In: Sharp WR, Evans DA, Ammirato PV, Yamada Y (eds) Handbook of plant cell culture, vol 2. Collier Macmillan, London, pp 243–267

Ralton JE, Howlett BJ, Clarke AE (1985) Receptors in host-pathogen interactions. In: Chadwick CM, Garrod DT (eds) Hormone receptors and cellular interactions in plants – intercellular and intracellular communication, vol 1. Univ Press, Cambridge, pp 281–318

Ride JP (1980) The effect of induced lignification on the resistance of wheat cell walls to fungal degradation. Physiol Plant Pathol 16:187–196

Salhani N, Vienken J, Zimmermann U, Ward M, Davey MR, Clothier RH, Balls M, Cocking EC, Lucy JA (1985) Haemoglobin synthesis and cell wall regeneration by electric field-induced interkingdom heterokaryons. Protoplasma 126:30–35

Smith HH, Mastrangelo-Hough IA (1979) Genetic variability available through cell fusion. In: Sharp WR, Larsen PO, Paddock EF, Raghavan V (eds) Plant cell and tissue culture – principles and applications. Ohio State Univ Press, Columbus, pp 265–286

Spencer-Phillips PTN, Gay JL (1981) Domains of ATPase in plasmamembranes and transport through infected plant cells. New Phytol 89:393–400

Steinbiss HH, Broughton WJ (1983) Methods and mechanisms of gene uptake in protoplasts. In: Giles KL (ed) Plant protoplasts. Academic Press, New York London, pp 191–208

Suzuki M, Takebe I, Kajita S, Honda Y, Matsui C (1977) Endocytosis of polystyrene spheres by tobacco leaf protoplasts. Exp Cell Res 105:127–135

Takebe I (1983) Protoplasts in plant virus research. In: Giles KL (ed) Plant protoplasts. Academic Press, New York London, pp 89–110

Takeuchi Y, Komamine A (1982) Effects of culture conditions on cell division and composition of regenerated cell walls in *Vinca rosea* protoplasts. Plant Cell Physiol 23:249–255

Taylor ARD, Hall JL (1976) Some physiological properties of protoplasts isolated from maize and tobacco tissues. J Exp Bot 27:383–391

Thompson MR, Douglas TJ, Obata-Sasamoto H, Thorpe TA (1986) Mannitol metabolism in cultured plant cells. Physiol Plant 67:365–369

Vogel HJ (1964) Distribution of lysine pathways among fungi; evolutionary implications. Am Nat 98:435–446

Williams L, Collin HA (1976a) Embryogenesis and plantlet formation in tissue cultures of celery. Ann Bot (London) 40:325–332

Williams L, Collin HA (1976b) Growth and cytology of celery plants derived from tissue cultures. Ann Bot (London) 40:333–338

Withers LA, Cocking EC (1972) Fine structural studies on spontaneous and induced fusion of higher plant protoplasts. J Cell Sci 11:59–75

Wullems GJ, Schilperoort RA (1985) Plant protoplast transformation by *Agrobacterium* in relation to plant biotechnology. In: Fowke LC, Constabel F (eds) Plant protoplasts. CRC, Boca Raton, Florida, pp 205–230

Wullems GJ, Krens FA, Peerbolte R, Schilperoort RA (1984) T-DNA in transformants obtained in vitro and in intraspecific somatic hybrids. In: Beers RF, Bassett EG (eds) Cell fusion: gene transfer and transformation. Raven, New York, pp 237–258

Zimmermann U (1982) Electrical field induced fusion of biological cells. Biochim Biophys Acta 696:227–277

Zimmermann U, Scheurich P (1981) High frequency fusion of plant protoplasts by electrical fields. Planta 151:26–32

Zimmermann U, Vienken J (1982) Electric field-induced cell-to-cell fusion. J Membr Biol 67:165–182

Zimmermann U, Vienken J, Pilwat G (1980) Development of drug carrier systems: electrical field induced effects in cell membranes. Bioelectrochem Bioenerget 7:553–574

# XXV Chloroplast Uptake and Genetic Complementation

K. L. Giles[1]

## 1 Chloroplast Uptake

Rewriting the chapter more than 10 years after its original publication (Giles 1977) is not made easy by the exciting changes that have occurred in closely associated fields over that time. To a large degree the uptake of foreign DNA, using *Agrobacterium tumefaciens* as a vector has eclipsed organelle uptake studies in the past few years. This trend is understandable because of the increased flexibility and range of genetic manipulations possible using the *Agrobacterium* system, since it effectively allows the manipulation of the whole genome rather than the plastoms alone. Chloroplast uptake and exchanges, however, continue to have a useful role in studies involving nuclear-chloroplast interaction, the physiology of isolated plastids and experiments investigating the transfer of herbicide resistance, particularly to the triazines, between varieties and species.

It is interesting to speculate that although the processes of evolution impact the nuclear genome where there are only a limited number of genes regulating each trait, the chloroplast genome is arranged and inherited very differently. Chloroplasts have a large gene copy number for each trait and are inherited in relatively large numbers during sexual reproduction.

Analysis of chloroplast DNA, using restriction enzymes, of *Nicotiana* has revealed that although there was a high degree of conservation within the genus, there were numerous species-specific differences (Salts et al. 1984). After a review of the chloroplast DNA structure it can be concluded that only a slow rate of genetic change occurs in the chloroplast genome.

Because of the differential rates of genetic change between nucleus and chloroplast, and since the nuclear plastid harmony must be maintained to allow for efficient function, accordingly it might be expected that the range of viable interactions between different chloroplast populations in foreign cytoplasm might be quite broad.

Such a hypothesis is discussed by Glimelius and Bonnett (1986), who describe the production and partial characterization of cybrids between *Nicotiana tabacum* and *Petunia hybrida*. The success of their experiments demonstrated that protoplast fusion can be used to obtain vigorously growing intergeneric cybrids, but they depended on the development of techniques to achieve efficient interspecific organelle transfer. Zelcer et al. (1978) used gamma-irradiation to in-

[1] Vipont Research Laboratories (Canada) Inc., c/o Plant Biotechnology Institute, National Research Council of Canada, 110, Gymnasium Road, Saskatoon, SK, Canada, S7NOW9

Biotechnology in Agriculture and Forestry, Vol. 9
Plant Protoplasts and Genetic Engineering II (ed. by Y. P. S. Bajaj)

activate the nucleus of the protoplast chosen as the organelle donor species. Menczel et al. (1982) used gamma-irradiation – the method used by Glimelius and Bonnett (1986). Glimelius et al. (1981) showed that albino protoplasts could be restored to photoautotrophy by protoplast fusions and Medgyesy et al. (1985) used such a chloroplast mutant together with gamma-irradiation to select for interspecific chloroplast transfer by production of cybrids of *Nicotiana* resulting from somatic cell fusion. These two papers confirm earlier observations made by the author (Giles 1973).

Glimelius and Bonnett concluded that sorting of chloroplast types after fusion was complete, there being no variegated plants regenerated and no segregation with respect to tentoxin sensitivity. They suggest that, together with transformation, organelle uptake or transfer has a role to play in the full utilization of genetic diversity in plant breeding.

Factors affecting uptake of particles, microorganisms and plastids into protoplasts seem to be relatively uniform across a wide variety of structures. Yamada and Sakaguchi (1981) concluded that the factors most favorable for bacteria uptake into yeast protoplasts were very similar to those effective in fusion of fungal protoplasts. This led them to suggest that uptake involved a membrane fusion mechanism as part of the uptake event. They found maximum uptake, 2.3 bacterial cells per protoplast, was obtained when yeast protoplasts were mixed with about 100-fold bacterial cells, treated with 30% PEG 4000 solution containing 50 to 60 mM $CaCl_2$ at 30 °C for 30 min, and then washed and incubated in 50 mM Tris-HCl (pH 8.0) containing 50 to 60 mM $CaCl_2$.

When looking at the uptake of blue-green algae into yeast protoplasts, Yamada and Sakaguchi (1981) found that there was a rapid formation of nonspecific aggregates. There followed one of two different uptake events. In the first, the cyanobacterial cell found itself in a slight depression in the plasma membrane of the protoplast. Plasmolysis of the protoplast by the bathing PEG solution caused the depression to deepen and finally the membrane sealed over the top of the blue-green algae cell, thus leaving it in a membrane-bound vesicle. In the second form of uptake, the cyanobacteria was surrounded by yeast protoplasts on all sides and due to fusion of the protoplasts was left again in membrane-bound vesicules within the cytoplasm of the fused protoplast. Such membrane associations may be important in endosymbiotic systems, and may help in the establishment of stable relationships.

The growing understanding of uptake physiology and long-term stable relationships of plastids in foreign cytoplasm will increasingly allow the complete exploitation of genetic diversity both at the cell and whole plant level. What is needed is a development to simplify the selection of foreign chloroplasts, which is easier to operate than present methods, and uptake that involves less risk of membrane fusion and subsequent chloroplast disintegration.

## 1.1 The Uptake of Chloroplasts by Animal Cells

The occurrence of plant chloroplasts in animal cells was first demonstrated by Kawaguti and Yamasu (1965) in the marine gastropod *Elysia atroviridis*. The evidence that the structures observed were chloroplasts and not endosymbiotic algae was based on electron microscope examination and on comparison with the plastid structure in the siphonaceous alga *Codium fragile* on which the animals feed. No evidence was presented in this report to show that the chloroplasts were functional. The ultrastructure of chloroplasts in the digestive gland of *Elysia viridis* was studied by Taylor (1968), who showed that in this species, too, there was considerable similarity between the chloroplasts in the slug and those of *Codium tomentosum*, the alga upon which it feeds.

Yonge and Nicholas (1940) reported that the sacoglossan opisthobranch gastropod, *Tridachia crispata* (Bergh), had symbiotic, unicellular zooxanthellae within the tissues of the gut. Upon re-examination by Trench (1969), these zooxanthellae were found to be absent, but the slugs did possess free chloroplasts, not whole algal cells, as a functional endosymbiont.

The chloroplasts were 2–3 microns in diameter and were in the cells lining the tubules of the digestive diverticulum. Electron microscopy revealed that the plastids were bound by a double membrane and possessed thylakoids. Osmiophilic granules, and what were interpreted as starch grains, were also seen in the chloroplasts. That these plastids were photosynthetically active was demonstrated by $^{14}CO_2$ pulsing of the slug in the light and dark, and autoradiography of histological preparations. Experimental animals in the light-fixed appreciable amounts of $^{14}C$, whilst dark control specimens fixed negligible amounts.

The life span of these plastids in the cell was estimated to be at least 6 weeks and possibly longer (Trench et al. 1969). The chloroplasts appear to originate from siphonaceous algae and are not inherited but are acquired anew by each generation. That they are not passed on in the eggs as proplastids cannot be completely ruled out, since electron microscope investigations have not been made.

Endosymbiotic chloroplasts have not been demonstrated in vertebrate tissues, but Nass (1969) showed that mouse fibroblasts in culture would incorporate isolated chloroplasts of spinach and African violet. These organelles were isolated and taken up by phagocytosis into the mouse cells, but despite this they were not contained in vacuoles or digestive vesicles. They could be seen segregating during mouse cell mitosis but were not seen to divide themselves, and were thus markedly diluted relative to the number of mouse L cells present within a period of 5–6 days. It was claimed that the plastids retained their structural integrity, though the electron microscope pictures included do not support this belief, the outer membrane having disappeared in all cases. Hill activity and the ability to fix carbon dioxide were reportedly maintained for 24 h.

Although very tight endosymbiosis appears to exist between some invertebrates and chloroplasts from the algae they eat, there is no suggestion that this relationship is dependent upon any form of genetic complementation. Giles and Sarafis (1971, 1972) demonstrated that chloroplasts from *Caulerpa sedoides* (Brown in Turner) C. Agardh could be maintained in chicken egg whites for prolonged periods during which repeated chloroplast division occurred. Structural integrity was strictly maintained, and carbon fixation was demonstrated 96 h after isolation. These findings suggest that chloroplasts from at least some of the siphonaceous algae are autonomous for some of their metabolic functions, and this could partly account for their association with gastropod molluscs, whose gut diverticulae must provide an environment suitable for the maintenance of these activities, to the ultimate advantage of the gastropod. The fact that these chloroplasts are not readily digested by the slug may be due to the rigescent integument (Giles and Sarafis 1974) which surrounds this type of plastid. It is interesting to note that when taken up by phagocytosis into mouse L cells, plastids of *C. sedoides* are maintained for several days and there is some suggestion that plastid division may occur within the animal cells. However, when the plastids are taken up by primary cultures of fetal lamb or bovine cells, the animal cells react by forming autosomal vesicles and die, releasing the chloroplasts into the medium. There does, therefore, seem to be some specificity of reaction to chloroplast uptake, at least in cultured mammalian cells.

The reverse experiment, introducing higher plant plastids into gastropod mollusc cells in vivo or in vitro, has not been attempted. It has proved difficult to date to find molluscs "uninfected" with chloroplasts in the wild. Maintaining animals in the dark for long periods does not rid them of plastids but frequently leads to death of the slug. Although gastropod mollusc larvae are free from chloroplasts as far as can be determined by optical microscopy, they die if they are not infected with plastids as they grow older. It would be of interest to determine the range of plastids capable of becoming the endosymbiont within these larvae.

### 1.2 The Uptake of Chloroplasts by Plant Cells and Protoplasts

Reports of the uptake of chloroplasts or other organelles by plant cells have necessarily been few. The rigid cell wall surrounding the plant cell is an effective barrier to all but the most minute particles, viruses normally entering the cell through plasmodesmata after damage to a neighboring cell has allowed entry into the plant symplast. However, the slime molds, members of the *Acrasiales*, have no such rigid cell walls and are capable of pinocytosing small particles. It is perhaps surprising that organisms such as the slime molds have not been exploited more for this kind of study, since they provide a useful environment for the investigation of chloroplast survival in foreign plant cytoplasm. The ability of genetically dissimilar chloroplasts to survive in the same cell has been demonstrated in *Spirogyra* (van Wisselingh 1920), where in so-called mixed cells there exist chloroplasts both with and without pyrenoids. Such differences in chloroplast structure are maintained for many generations showing such plastids to be genetically different. In the iojap mutation of maize green and mutant white plastids are found in some cells (Rhoades 1943) (see Sect. 3.3 for a fuller account). In *Oenothera*, because of the biparental transmission of plastids during fertilization, Stubbe (1957) also could report mixed cells in the cotyledons and early leaves, though rarely in later leaves, of hybrid variegated plants. These findings add credence to the idea that genetically different chloroplasts may survive in the same cytoplasm.

An interesting communication from Carlson (1973a) reported the introduction of green tobacco chloroplasts into protoplasts from leaf mesophyll cells of a maternally inherited variegating albino mutant of *Nicotiana tabacum* (Burk and Grosso 1963). These protoplasts are albino and contain no functional chloroplasts. Carlson (1973a) reported that they were capable of spontaneously pinocytosing normal green chloroplasts, and under suitable culture would regenerate whole plants, with a normal complement of green functional plastids. This, it was claimed, demonstrated that the chloroplasts were capable of utilizing information encoded by the protoplast nucleus. It is unfortunate that no methods or techniques used in this quite remarkable experiment were given in the publication, since to be acceptable as science it must be repeatable. The isolation of intact functional chloroplasts from tobacco is difficult, and no details of how this problem was overcome were given. The uptake of chloroplasts does not seem to occur spontaneously without the addition of some external compounds to the medium, but no details of methods used to induce plastid uptake were given. No information about the percentage of protoplasts taking up chloroplasts was given either.

Potrykus (1973) put forward some valid criticisms of Carlson's interpretation of this experiment. First, from experience with extrachromosomal, maternally inherited, stable, mutant plastids one should expect to get variegated plants when growing protoplasts via callus to a plant. One would expect from Carlson's (1973a, c) experiments to produce at least some variegated plants along with white and green ones. Secondly, since many extrachromosomally inherited, variegating, albino plants are periclinal chimeras of the "green/green/white" type (Kirk and Tilney-Bassett 1967) the plastids of the epidermis are white because of their position but are genetically green. Such a cell would yield a protoplast capable of regenerating a green plant without taking up any chloroplasts. Thirdly, the pictures presented by Carlson (1973b) show protoplasts that are apparently damaged and probably dead. The chloroplasts are ill-defined and seem to be adhering to the tonoplast or cytoplasm. This latter comment is concurred with by Bonnett and Eriksson (1974).

Potrykus and Hoffmann (1973) succeeded in transplanting isolated nuclei of *Petunia hybrida* into protoplasts of *P. hybrida*, *Nicotiana glauca* and *Zea mays*. About 0.5% of the protoplasts appeared to take up stained nuclei after treatment with 0.03% lysozyme in slightly hypotonic mannitol solution. That the nuclei were truly inside the protoplasts was demonstrated by following the position of the nucleus under the microscope while rolling the protoplast under the coverslip. Furthermore, when the protoplasts were burst under slight pressure it could be seen that the incorporated nucleus extruded together with the original organelles. This is a convincing demonstration of uptake and incorporation by the protoplast, but it is still important that uptake be proven without doubt. It is always difficult to be sure that particles, be they chloroplasts or nuclei, are truly within the plasmalemma when using only optical microscopy. The picture shown by Potrykus and Hoffmann (1973), taken using oblique light and fluorescence, is not completely satisfactory. The ultimate demonstration of uptake of plastids must be by examination of serial thin sections by electron microscopy. Serial sections are necessary because particles could be taken in by a pinocytotic vesicle open to the exterior of the cell by a narrow neck, which might not be seen in a single section. Such a demonstration is still wanting.

Potrykus (1973) reported the transplantion of *P. hybrida* (cv. Cyatype) chloroplasts into the albino protoplasts from an extrachromosomally inherited, variegated *P. hybrida*. The chloroplasts were isolated by osmotic shock of isolated protoplasts, collected by centrifugation and resuspended in either 0.35 M mannitol or equiosmolar 0.21 M sodium nitrate. The osmolarity at which the protoplasts burst, and hence the osmolarity to which the chloroplasts were first subjected, was not given. The chloroplasts were taken up by the protoplasts using either lysozyme or sodium nitrate treatment (see Sect. 2.3).

Isolation of chloroplasts by protoplast rupture, as reported by Potrykus (1973), is in our experience not likely to yield only single isolated plastids, but also groups of plastids. The groups tend to be held together, often with mitochondria, surrounded by membrane of either plasmalemma or tonoplast origin. This means that the sodium nitrate treatment, reported to be the more effective of the two methods tried, may well have had its effect by causing fusion of these subprotoplasts with the albino protoplasts. This conclusion is borne out by the pictures presented, showing the "uptake" of eight chloroplasts into a very localized area of the protoplast, a condition that indicates that fusion may have been involved. It is also mentioned that up to 20 plastids were taken in by some protoplasts, an unlikely event if they were taken in singly.

The most convincing report, and certainly the best pictorial record, of chloroplast uptake to date is that of Bonnett and Eriksson (1974). Using polyethylene glycol treatment, they succeeded in transferring chloroplasts of the alga *Vaucheria dichotoma* with high frequency into the protoplasts of *Daucus carota. Vaucheria* plastids were chosen for several reasons, the most important being their small size and their ability to withstand the necessary manipulations. It is significant that the chloroplasts of *V. dichotoma* possess a rigescent integument as described by Giles and Sarafis (1974) and it is this structure that gives the plastid its structural stability at varying osmolarity. A great deal of trouble is taken to convince the reader that the chloroplasts are indeed taken into the cell cytoplasm; none was seen in the vacuole. Although further proof in the form of serial EM sections would be desirable, their evidence is convincing, especially their photograph (Fig. 1 a), kindly lent for reproduction here, showing a plastid in a cytoplasmic bulge into the vacuole. Figure 1 b shows another protoplast with a chloroplast near to the vacuole and well within the protoplast cytoplasm. Unfortunately, there is no report of any attempt to ascertain whether the chloroplasts were still active in any way before or after uptake.

Application of cell culture techniques in attempts to produce crop plant species resistant to the triazine herbicides (photosystem II herbicides) by mutation of the chloroplast gene *psb-A* has been the focus of work in Maliga's laboratory at Advanced Genetic Sciences Inc. (Maliga 1985). Mutant selection in populations of protoplast-derived colonies has yielded triazine-resistant *Nicotiana* plants. The mutants are similar to resistant weeds found under field conditions. Chloroplast transfer via protoplast fusion has been shown to be a practical way of transferring herbicide resistance between related plant species. Genetic recombination between the chloroplast genomes has been described and the potential of this method to transfer chloroplast genes from weeds to sexually incompatible crop plants appears an exciting possibility for the future.

Cseplo et al. (1986) used protoplast cultures of *Nicotiana plumbaginifolia* to isolate mutants resistant to antibiotics and to photosynthesis-inhibiting herbi-

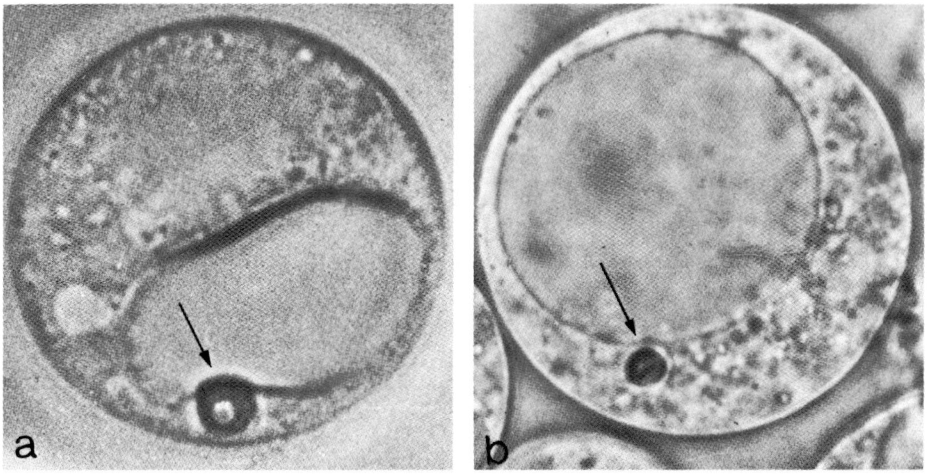


**Fig. 1a, b.** The induced uptake of chlorplasts of *Vaucheria dichotoma* (L.) Ag. by protoplasts of *Daucus carota* (L.) **a** A chloroplast (*arrowed*) in the cytoplasm of the protoplast causing a bulge into the vacuole; a clear indication that the plastid has been taken up. **b** Another chloroplast (*arrowed*) near to the vacuole and well within the protoplast cytoplasm. (**a** × 2500; **b** × 1600). (Bonnett and Eriksson 1974)

cides. Inhibitors of bacterial protein synthesis were used in the selection of many chloroplast mutants under standard culture conditions, while special systems had to be established to select mutants to the photosynthetic herbicides. Under the appropriate conditions bleaching could be observed and the resistant mutants identified. These authors also demonstrated that protoplast fusion was suitable to rescue the mutant chloroplasts from undesirable nuclear backgrounds, or to transfer them into different species.

The surprising difference between workers using animal cells and those using plant protoplasts is that the latter have ignored the obvious importance of the biological activity of the incorporated chloroplasts. Apart from mentioning that the plastids are functional Carlson (1973a, b) gave no measurements of carbon fixation. Potrykus and Hoffmann (1973) did not attempt a demonstration of activity of the nuclei. As will be discussed in a later section, it is extremely unlikely that any of the plastids used so far in uptake studies with higher plant protoplasts are functional. It is quite probable that it is the type of chloroplast eaten by gastropods, those symbiotic with marine siphonaceous algae, that offer the best hope of continued activity in both animal and foreign plant cytoplasm.

## 2 The Preparation of Material

### 2.1 The Preparation of Protoplasts

Three criteria must be upheld during the preparation of protoplasts for use in uptake experiments. The protoplasts must be (1) without any trace of cell wall, (2) highly cytoplasmic, and (3) alive. Traces of cell wall material attached to the protoplast interfere with uptake and can damage, and often rupture, intact isolated plastids. Relatively, cytoplasmic protoplasts seem more likely to take up particles than highly vacuolate ones, possibly because there is more cytoplasm to accommodate the added volume. Dead protoplasts are incapable of pinocytosis and as such incapable of meaningful uptake. Incorporation of particles into the cytoplasm of broken protoplasts is obviously not significant in the present context.

Any of the published methods, both in this Volume and elsewhere, will give high yields of viable protoplasts, though slightly different osmotic conditions and enzymes might have to be used with different species. Highly cytoplasmic protoplasts can be found from meristematic tissues of both whole plant and callus origin. Both sources have their drawbacks however. Meristems from whole plants are time-consuming to dissect, and because of their size yield only a small number of protoplasts. The cells of the meristem are closely packed and thus slow the penetration of the cell wall-degrading enzymes, increasing the time necessary for exposure to enzymes. Rapidly growing meristematic callus gives a mixed population of protoplasts – some highly cytoplasmic, others highly vacuolate –; the proportions of each depending on the precise growth conditions.

## 2.2 The Isolation of Chloroplasts

The methods used for the isolation of active chloroplasts from siphonaceous green algae and from higher plants differ only in detail, related to the different structures of the whole plant and plastids. The main consideration in both cases is the speed of isolation of the chloroplast from the cytoplasm and vacuolar contents. The latter is particularly important since in many plants heavy metal ions and phenolic compounds contained in the vacuole will cause almost immediate inhibition of the chloroplast enzymes.

Giles and Sarafis (1971) described a method for the extraction and isolation of chloroplasts from the marine siphonaceous alga *C. sedoides*. In an attempt to simulate cell cytoplasm during the isolation, egg white was added to their medium, but subsequent experiments have shown that for the maintenance of activity the mineral salts medium described below is quite satisfactory.

The method favored for the isolation of spinach chloroplasts is modified from Bucke et al. (1966). Spinach is grown hydroponically in Hoagland's nutrient medium to minimize the build-up of heavy metal ions. Plants 2–3 weeks old are routinely used and only the youngest leaves chosen. These are immediately chilled in iced water and exposed for 10 min to bright light 250 W $m^2$. They are then quickly torn into thumb-nail sized pieces and put in a Virtis blender flask containing the solution detailed below. This solution is stored frozen and used in a half-thawed state. Just before the leaf material is put in the blender, 2 mM ascorbate is added to the mixture (Table 1).

As soon as the ascorbate and leaf material are added, the whole is blended with the half-thawed medium for 5 s and immediately the slurry is filtered through three layers of Miracloth into a centrifuge tube free from all traces of detergent, which might lyze or break the chloroplasts, and centrifuged at 2000 g for

**Table 1.** Isolation medium

| | | | |
|---|---|---|---|
| $MgCl_2$ | 1 mM | EDTA | 2 mM |
| $MnCl_2$ | 1 mM | Sorbitol | 0.33 M |
| NaCl | 10 mM | MES 50 mM | pH 6.5 |
| $NaHCO_3$ | 20 mM | | |

20 s. The pellet is resuspended in the above medium and its activity measured with an oxygen electrode.

Activities of 50–80 μ—M oxygen/mg chlorophyll/h are routinely recorded with this method even at pH 6.5, a low pH for photosynthetic activity. Levels as high as 300–350 μM/mg/h can be recorded for spinach chloroplasts prepared in the same way but maintained at pH 8.2 (Lyttelton pers. commun.) This method works well for spinach, pea and *Chenopodium*, but does not yield intact active chloroplasts from all species. Treated with the same degree of loving care and attention, tobacco chloroplasts are normally stripped of their outer membrane and are rapidly inactivated, although the stromal lamellae and grana hang together. There are no methods available that will give intact, active plastids of all plant species and, as yet, those species mentioned above are the only ones from which it can be said that good preparations can be made.

The preparation of chloroplasts from protoplasts has been suggested as a useful method of isolation. Most protoplasts after isolation are actively photosynthetic, though this activity decreases over a period of hours. Preparation of chloroplasts from protoplasts is suggested as being by simple lysis of the protoplast in hypotonic solution and subsequent release of the chloroplasts. In this laboratory such a method leads to the formation of groups of plastids, often accompanied by mitochondria, embedded in small pieces of cytoplasm, the whole being surrounded by a membrane, presumably of either tonoplast or plasmalemma in origin. When one speaks of uptake in this case one is really referring to fusion events, and these will be discussed as such later.

## 2.3 The Induction of Chloroplast Uptake into Protoplasts

Several methods for the induction of chloroplast uptake into plant protoplasts have been published, all of them depending either on some form of osmotic effect drastically plasmolysing the protoplast or on hypothesized changes in the charge pattern on the plasmalemma. Carlson (1973 a, b) found it unnecessary to use any agent to promote chloroplast uptake, but this is not the general finding. Davey and Cocking (1972) when introducting bacteria into pea protoplasts stressed the importance of having the bacteria present during the plasmolysis of the cells. This does not seem to be necessary in the case of chloroplasts.

Potrykus (1973) described two methods for the transplantation of chloroplasts of *Petunia*. In the first method protoplasts and chloroplasts were centrifuged in alternating layers together with 0.03% lysozyme, and then subjected to a hypotonic shock. The second method consisted in incubating the isolated single cells of *Petunia*, prepared by pectinase treatment, with 2% cellulase in 0.35 M mannitol, pH 5.4 at 20 °C for 60 min. When the first protplasts appeared, the single cells were transferred into a suspension of chloroplasts in 2% cellulase in 0.21 M sodium nitrate (equiosmolal with 0.35 M mannitol). The mixed suspension was incubated at 20 °C for 15–30 min in a swing-out rotor at low speed at about 10 g. The pellet was resuspended every 5 min. When all the cells had formed protoplasts, the mixed suspension was washed clean of cellulase and free chloroplasts and the protoplasts examined. The possibility that the so-called uptake under these circumstances may actually be fusion as a result of the preparation of chloroplasts from the breakage of protoplasts in the presence of sodium nitrate has already been mentioned. The second of these two methods gave the best uptake but the condition of the chloroplasts after 15–30 min in 2% cellulase at 20 °C is questionable. Isolated chloroplasts are normally maintained at, or near, 0 °C to retain their activity, and an incubation of this length at 20 °C is not calculated to maintain their integrity or metabolic activity, especially since cellulase preparations frequently contain appreciable amounts to protease.

Bonnett and Eriksson (1974) used a method involving plasmolysis by polyethylene glycol (PEG) to induce uptake of *Vaucheria* chloroplasts by carrot protoplasts. A dense mixture of protoplasts and chloroplasts was made up to a volume of 0.5 ml with protoplast culture medium (Wallin et al. 1974). A volume of 1.5 ml of a solution containing 3 ml of protoplast culture medium and 7 ml of a 56% aqueous solution of PEG was added to the protoplast-chloroplast mixture. The final concentration of PEG was about 29%, a concentration known to promote cell aggregation and cell fusion without reducing cell viability (Eriksson et al. 1974). After 10 min the mixture was diluted to 10 ml with a solution of 0.1 M $CaCl_2$ in 0.3 M sorbitol, and centrifuged for 3 min at 150 g. The pellet was resuspended in 10 ml of the same solution, recentrifuged and resuspended in 2 ml of the protoplast culture medium.

As already mentioned (Sect 1.2) this method seems to give good results, with up to 16% of the protoplasts taking up chloroplasts. Kao and Michayluk (1974) observed some of the effects of PEG on protoplasts and mention a reduction in protoplast volume with a temporary wrinkling of the plasmalemma, aggregation of the protoplasts into groups, and eventual protoplast fusion. It is easy to see how any of these events might bring about chloroplast uptake (Yamada and Sakaguchi 1981).

New methods causing higher percentages of protoplasts to take in chloroplasts, and more importantly taking in larger numbers of chloroplasts must be sought. This last point is important because it may be that the number of plastids taken up may be critical in the further culture of the protoplast.

## 3 Genetic Complementation

### 3.1 How Autonomous Is the Chloroplast?

The genetic and developmental autonomy of the chloroplast has for long been in dispute. In the light of evidence of the autonomy of plastids (Ridley and Leech 1970; Giles and Sarafis 1971, 1972) it seems probable that chloroplasts synthesize their own fatty acids. In the intact leaf most of the fatty acid synthesis takes place in chloroplasts (Brooks and Stumpf 1965). The cofactor requirements of synthesis, and the fact that malonate is incorporated, suggest that the pathway of fatty acid synthesis is similar to that worked out for animals (Alberts et al. 1965).

Since protein is the major chloroplast constituent it is interesting to note that as early as 1959, Tolbert showed that the products of photosynthesis formed from $^{14}CO_2$ by isolated chloroplasts included alanine and aspartate. Isolated spinach chloroplasts will incorporate radioactivity from $^{14}C$-acetate into an amino acid fraction which includes glutamate (Mudd and McManus 1965). Chloroplasts seem therefore to be able to synthesize some of their amino acids. Synthesis of proteins in isolated chloroplasts of *Euglena gracilis* has been described (Eisenstadt and Brawerman 1963, 1964). It is of particular relevance to the present subject that incorporation of $^{14}C$-leucine into protein took place rapidly during the first 15 min after isolation, but had virtually ceased after 30 min, indicating that the activity of isolated chloroplasts is short-lived.

The protein present in the largest quantity in the chloroplast is Fraction I protein. This name was first used by Wildman and Bonner (1947) to describe a high molecular weight soluble protein purified from spinach leaves. This protein can account for up to 50% of the total soluble protein in leaf extracts and is homogeneous by electrophoretic and centrifugal criteria (Kawashima and Wildman 1970). It is located in the stroma of the chloroplasts (Lyttleton and Ts'o 1958), where it catalyzes the $CO_2$-fixing step in photosynthesis (Weissbach et al. 1956; Park and Pon 1961).

The reported molecular weight of Fraction I protein purified from eukaryotic cells or blue green algae varies between $4.7 \times 10^5$ and $6.6 \times 10^5$ Da (McFadden 1973). The most likely molecular weight is about $5.25 \times 10^5$ Da. This value is calculated from Fraction I crystallized from *Nicotiana* which could not be distinguished from Fraction I isolated from *Spinacia* (Kawashima and Wildman 1971). The intriguing point is that treatment of Fraction I protein with either urea (Moon and Thompson 1969) or sodium dodecyl sulfate (Rutner and Lave 1967) dissociates the molecule into large subunits (MW $5.2-6.0 \times 10^4$) and small subunits (MW $1.2-1.8 \times 10^4$). It seems from the work of Rutner (1970), that the most likely composition of the undissociated Fraction I protein molecule is 8 large subunits with 8–10 small subunits.

Comparative studies carried out both between large and small subunits of Fraction I from a single species and between subunits from different species, have shown by amino acid compositions (Kawashima 1969; Moon and Thompson 1969; Sugiyama et al. 1971), tryptic peptide analysis (Kawashima and Wildman 1971; Kawashima et al. 1971), and immunological studies (Kawashima et al. 1971) that although the large subunit shares no similarities with the small subunit in the same species and small subunits from different species are highly dissimilar, the large subunits from different species are closely related. These findings may be of profound importance in studies concerned with chloroplast uptake into foreign cytoplasm.

Cycloheximide, an inhibitor of 80 S ribosomal protein synthesis, has been used in studies on the synthesis of Fraction I protein in *Chlamydomonas reinhardi* (Armstrong et al. 1971; Kawashima and Wildman 1972) and higher plants (Criddle et al. 1970; Zucker 1972) and the results have suggested the involvement of cytoplasmic ribosomes. Criddle et al. (1970) have shown that the incorporation of labeled amino acids into the large subunit of Fraction I protein in greening barley leaves is preferentially inhibited by chloramphenicol, while that into the small subunit is inhibited preferentially by cycloheximide. These results indicate that the large subunit is synthesized by chloroplast ribosomes, and the small subunit by cytoplasmic ribosomes. These findings are supported by studies using isolated chloroplasts (Blair and Ellis 1972, 1973) and by Kawashima's (1970) finding that $^{14}CO_2$ is incorporated preferentially into the large subunit during short-term photosynthesis by *Nicotiana* leaves.

The nature of the mechanism regulating the synthesis of the two subunits in the different cellular compartments is still not clear. The large subunit synthesized in isolated chloroplasts does not interchange with that in preformed Fraction I protein, suggesting that the chloroplasts lack a pool of the small subunits (Blair and Ellis 1972, 1973).

A particularly useful review by Schell (1987) on transgenic plants as tools for the study of molecular organization of plant genes discusses the role of the genes controlling the regulation of the genes involved in ribulose diphosphate carboxylase fractions. Some of the genes react to environment conditions, such as light quality and flux. Secondly, these genes are developmentally controlled and it has been shown that the nuclear gene is expressed in response to changes in chloroplast structure.

To test whether light-regulated enhancer elements are responsible for the light- and chloroplast-regulated expression of other nuclear genes as well, such a gene of unknown function was isolated from a potato genomic library. The regulatory element was sufficient to confer all qualitative regulatory traits. Destruction of chloroplasts by photooxidation, after treatment with the herbicide Norflurazon, resulted in a drastic reduction in all light-induced nuclear genes. It was concluded that a signal monitoring the developmental and functional integrity of the chloroplasts must therefore be involved in the regulation of the expression of these genes.

The central role of Fraction I in the functioning of the chloroplast presents a very real problem to the successful integration of the plastid in foreign cytoplasm. The reported long-lived activity of the chloroplasts of siphonaceous algae in animal cells presumably represents a long-lived form of the protein or possibly a form of it different from that found in higher plants. A ray of hope may be gleaned from the fact that the larger subunit seems very constant between species, suggesting that the production of hybrid forms of the enzyme might not be impossible.

### 3.2 Genetic Complementation in Chloroplasts by Protoplast Fusion

Because of the easy identification of chlorophyll under the optical microscope, the chloroplast has been used in several demonstrations of genetic complementation by protoplast fusion.

The first demonstration of complementation was reported by Giles (1973) and involved the use of mutant lines of *Zea mays*. Fusion of protoplasts containing green chloroplasts from the striped mutant iojap, with protoplasts of the white deficient mutant containing only white undeveloped plastids, caused the greening of the white deficient chloroplasts over a period of 72 h. Fusion of white and green protoplasts from the iojap plants caused no such greening of the mutant iojap plastids.

In order to explain the type of genetic complementation occurring in the iojap-white deficient heterokaryons, it is necessary to understand something of the genetics involved. Maize plants homozygous for the recessive gene iojap (ijij) exhibit a chlorophyll striping or variegation (Jenkins 1924). The ij gene is situated on chromosome 7; its position in the linkage group has been determined with some precision through 3-point linkage tests. Cytological examination of the mesophyll cells of the white regions of ijij plants show that not only do the chloroplasts lack chlorophyll, but they are also much smaller than those of the normal green areas (Rhoades 1943). The observation that $F_1$ plants of ijij constitution may be white or striped suggests that the modified plastids have a genetic continuity unaffected by the genetic constitution of the cell, and can thus be considered due to plastid mutation. This was shown to be correct when it was demonstrated that the aberrant plastids could remain and cause striping in IjIj plants (Rhoades 1943). Thus it was shown that the iojap gene causes a high frequency of irreversible plastid mutations. The white deficient plants (wdwd) are due to a homozygous deletion of the terminal portion of a rearranged chromosome 9, caused by X-ray treatment (McClintock 1944). The deletion is caused by the "breakage-fusion-bridge cycle" (McClintock 1941) which occurs throughout the gametophyte mitoses, so that a resulting egg or pollen nucleus possesses a chromosome 9 with a freshly broken end, thus being deficient for minute terminal portions of the short arm of chromosome 9. After fertilization the cycle ceases to operate and the chromosome behaves normally. Plants homozygous for the chromosome 9 lacking the most terminal portions are pale yellow (pyd) rather than green, whereas those homozygous for a slightly larger deficiency lack both chlorophyll and carotenoids (wd).

The complementation that occurs in the iojap-white deficient heterokaryons is presumably due to the normal chromosome 9 donated by the iojap nucleus complementing the deficienency in the white deficient chromosome. Thus a lack of genetic information in one nucleus of the heterokaryon has been complemented by its presence in the other, allowing at least partial development of the chloroplasts. The failure of the green iojap to effect the formation of chlorophyll in the white iojap protoplast is because the condition is due to a plastid mutation, unaffected by the nucleus, and is not complemented by the presence of normal plastids. This latter point is shown by the existence in the plant of cells with mixed plastid populations. This explanation of the complementation is slightly complicated by the existence of the R locus of the allele $R^r$. This allele controls, in conjunction with other factors, the pigmentation of both plant and seed (Stadler 1946). It is possible that the greening that takes place in the white deficient protoplasts is at least in part due to interaction and complementation at the R locus between the two genomes in the heterokaryon permitting greater pigmentation. However, this does not alter the fact that complementation of at least one, and possibly more, alleles occurs in these heterokaryons.

Because of the present impossibility of culturing maize protoplasts, no whole plants could be produced after complementation. Using mutants of tobacco, however, Melchers and Labib (1974) have reported complementation between certain chlorophyll-deficient and light-sensitive varieties. They used the following varieties $sl_1$, $sl_2$ (sublethal) (Povilaitis and Cameron 1963), vi-A (virescent) (Smith 1975), yg (yellow green) (Nolla 1934), and yc (yellow Crittenden) (Valleau 1957). No fusion hybrids were found using yg and yc. Using sublethal and virescent, haploid plants were produced by anther culture, and the plants maintained in a low light intensity (800 ×), at 28 °C with 80% – 90% humidity. After protoplast production and fusion, the protoplasts were plated and regenerated to whole plants as described by Takebe et al. (1971). For 48 h they were cultivated at approximately 300 lx, and after that at 3000 lx (Enzmann-Becker 1973). After 2 – 3 weeks, cultures forming calli were picked out and grown at a light intensity of 8000 – 10000 lx. The hybrid plants expressed complementation for normal chlorophyll content and were not light-sensitive. That these plants were not simply revertants of the sublethal or virescent was demonstrated by the segregation of the progeny of the fusion hybrid into green, white and yellow seedlings.

This experiment is a clear indication of the usefulness of the genetic complementation of chloroplasts as a method for the selection of fusion products. However, Melchers and Labib (1974) do point out that in the case of two calli chosen for regeneration by their color the plants were not dark green, showing that the method is not entirely effective.

Similar experiments using chlorophyll mutants of tobacco were described by Shvidkaya and Gleba (1974). Here, too, it was shown to be possible to complement chlorophyll deficiency by protoplast fusion. While the protoplasts were still at the single heterokaryon stage no complementation was observed (Butenko pers. commun.) and the reason for this is unclear. It is possible that the dedifferentiation of the tobacco protoplasts prior to the formation of callus precludes the expression of chlorophyll synthesis, whereas in maize, where the protoplasts do not divide to form callus, dedifferentiation does not proceed to this point and chlorophyll synthesis may occur. However this may be, whole green plants could be regenerated after fusion of the protoplasts from mutant lines.

## 4 A Summary – The Agronomic Potentials of Chloroplast Uptake

From the foregoing sections it is abundantly clear that it is too early to assess the potential importance that chloroplast introduction might have for the improvement of agronomic species.

There are several reasons for this reductance to evaluate the potential. Firstly, there are very few convincing reports of chloroplast uptake by protoplasts. In no case has biological activity been shown to exist in the chloroplasts, before or after uptake. In only one case has the protoplast as yet been shown to be capable of division after uptake (Carlson 1973), but the fact that chloroplasts were ever taken up in this experiment is open to some doubt.

Once active plastids have been introduced into foreign cytoplasm the question of genetic complementation arises. That isolated chloroplasts are capable of certain autonomous biological activity, when kept in ideal conditions at low temperatures over periods of a few hours at most, again raises the problem as to whether after introduction they would be capable of complementation or any activity. Prolonged periods at room temperature, or higher, are calculated only to reduce chloroplast activity perhaps to such a low level that genetic complementation becomes impossible. Nobody has suggested that apparently dead plastids, capable of no activity, can undergo a "miracle cure" in foreign cytoplasm and become active again. They may, however, serve as useful carriers of genetic information which, by a process akin to transformation (Ledoux and Huart 1968; Hess 1969a, b), might be expressed in the host cell. There has been no demonstration as yet of any expression of genetic material carried by isolated chloroplasts after their uptake by protoplasts.

In the cases of complementation by protoplast fusion, chloroplasts are obviously capable of expressing the genetic information donated by their new environment. In fusion experiments, however, they are transferred in their own cytoplasm to a hybrid form and are never as vulnerable to damage as the isolated plastid. At this stage of development it is very difficult to evaluate the potential of this type of corrective complementation, especially since its application is limited by the inability to culture the protoplasts of many species of both agronomic importance and otherwise.

An immediate use for chloroplast uptake studies is in the examination of the autonomy and genetics of the chloroplast. The hybrid nature of the Fraction I protein that must be built up in the cases of interspecific chloroplast uptake is an example of the kind of topic that can be studied now. Although not of direct agronomic application, this sort of study would certainly be of interest and quite possibly have far reaching implications for the whole field of somatic hybridization. It was pointed out a few years ago that one of the immediate applications of somatic hybridization might be as a genetic tool for the dissection of just this sort of problem (Giles 1973). Although it would be as ill advised to say that chloroplast uptake has no agronomic potential as it would be to say that it has great potential, it can be said that such a potential has not yet been demonstrated.

## 5 The Potential for the Uptake of Blue Green Algae

Since the earliest speculation concerning the origin of chloroplasts the blue green algae, because their organization is so similar to that of the chloroplast, have been put forward as the free-living ancestors of the chloroplast. They offer the advantage for uptake experiments that being free-living they are autonomous, indeed some of them, being capable of nitrogen fixation, are even more autonomous than most higher plants. They also possess a rigid cell wall which makes them much more amenable to manipulation than a plastid. Encouragement for their future use in uptake studies can be gained from the fact that there do exist natural symbioses within higher plants involving intracellular blue green algae. *Gunnera* has intracellular blue green algae actively fixing nitrogen, localised in mucilage glands at each node and at the growing tip. There is a tendency for the vegetative cells of the alga to change to heterocysts, a form lacking photosynthetic ability but seemingly having a higher concentration of the nitrogen-fixing enzymes. The algae are enclosed in an interconnecting meshwork of membrane, possibly derived from the plasmalemma. The mode of uptake of these algae into the cells of *Gunnera* is still obscure, but the study of their uptake by protoplasts instead of the uptake of isolated chloroplasts has yielded some interesting results.

Yamada and Sakaguchi (1981) failed in forming a stable nitrogen-fixing association between *Saccharomyces* and *Azotobacter*. They identify, however, three conditions in addition to not being digested initially, required for the establishment of a stable relationship: (1) a supply of nutrient and growth factors from the host, (2) cell division in harmony with that of the host, and (3) harmlessness to the host. It is interesting to note that *Anacystis nidulans* were not expelled from *Saccharomyces* protoplasts as had been noted by Meeks et al. (1978). It appeared that the cyanobacterial cells were digested in a heterophage-like process involving several enzymes.

## 6 Conclusions and Prospects

The examination of the autonomy and genetics of chloroplast function remains the most attractive prospect for further work. The recent description of the interaction between nuclear and chloroplast genomes in the control of light-mediated proteins is a case in point. An understanding of the genetic recombination occurring between plastids as uncovered by herbicide-resistant mutant investigations provides a new insight into the future exploitation of plastid genetic resources.

Although the physical uptake of isolated chloroplasts and blue green algae continues to reveal new aspects of cellular activity, it is protoplast fusion which remains the central and most promising method for chloroplast interchange and exchange between varieties and species. Such investigations involving the production and isolation of mutant plastid lines will lead to both a better understanding of the chloroplast genome and its interaction with the nucleus.

## References

Alberts AA, Majerus PK, Vagelos PR (1965) Acyl carrier protein VI. Purification and properties of -ketoacyl acyl carrier protein synthetase. Biochemistry 4:2265–2274

Armstrong JJ, Surzyki SJ, Moll B, Levine RP (1971) Genetic transcription and translation specifying chloroplast components in *Chlamydomonas reinhardi*. Biochemistry 10:692–701

Blair GE, Ellis RJ (1972) Light driven synthesis of the large subunit of Fraction I proteins by isolated chloroplasts. Biochem J 127:42 pp

Blair GE, Ellis RJ (1973) Protein synthesis in chloroplasts. I. Light driven synthesis of the large subunit of Fraction I protein by isolated pea chloroplasts. Biochim Biophys Acta 319:223–234

Bonnett HT, Eriksson T (1974) Transfer of algal chloroplasts into protoplasts of higher plants. Planta 120:71–79

Brooks JL, Stumpf PK (1965) A soluble fatty acid synthesizing system from lettuce chloroplasts. Biochim Biophys Acta 98:210–213

Bucke C, Walker DA, Baldry CW (1966) Some effects of sugars and sugar phosphates on carbon dioxide fixation by isolated chloroplasts. Biochem J 101:636–641

Burk JG, Grosso JJ (1963) Plasmagenes in variegated tobacco. J Hered 54:23–25

Carlson PS (1973a) The use of protoplasts for genetic research. Proc Natl Acad Sci USA 70:598–602

Carlson PS (1973b) Towards a parasexual cycle in higher plants. In: Protoplastes et fusion de cellules somatiques végétales. Int Coll Centre Natl Rech Sci, Paris, pp 497–501

Criddle RS, Dau B, Kleinkopf GE, Huffaker RC (1970) Differential synthesis of ribulose diphosphate carboxylase subunits. Biochem Biophys Res Commun 41:621–627

Cseplo A, Medgyesy P, Marton L (1986) In vitro induction, isolation and transfer of chloroplast mutations in *Nicotiana*. Proc Nuclear techniques and in vitro culture for plant improvement. IAEA, Vienna, pp 137–146

Davey MR, Cocking EC (1972) Uptake of bacteria by isolated higher plant protoplasts. Nature (Lond) 239:455–456

Eisenstadt JM, Brawerman G (1963) The incorporation of amino acids into the protein of chloroplasts and chloroplast ribosomes of *Euglena gracilis*. Biochim Biophys Acta 76:319–321

Eisenstadt JM, Brawerman G (1964) The protein synthesizing systems from the cytoplasm of *Euglena gracilis*. J Mol Biol 10:392–402

Enzmann-Becker G (1973) Plating efficiency of protoplasts of tobacco in different light conditions. Z Naturforsch 28:470–471

Giles KL (1972) An interspecific aggregate cell capable of cell wall regeneration. Plant Cell Physiol 13:207–210

Giles KL (1973) Attempts to demonstrate genetic complementation by the technique of protoplast fusion. In: Protoplastes et fusion de cellules somatiques végétales. Coll Int Centre Natl Rech Sci, Paris, pp 485–495

Giles KL (1974) Complementation by protoplast fusion using mutant strain of maize. Plant Cell Physiol 15:281–285

Giles KL (1977) Chloroplast uptake and genetic complementation. In: Reinert J, Bajaj YPS (eds) Applied and fundamental aspects of plant cell, tissue and organ culture. Springer, Berlin Heidelberg New York, pp 536–550

Giles KL, Sarafis V (1971) On the survival and reproduction of chloroplasts outside the cell. Cytobiosynthesis 4:61–74

Giles KL, Sarafis V (1972) Chloroplast survival and division in vitro. Nature New Biol 236:56–58

Giles KL, Sarafis V (1974) Implications of rigescent integuments as a new structural feature of some algal chloroplasts. Nature (Lond) 248:512

Giles RE, Ruddle FH (1973) Production and characterisation of proliferating somatic cell hybrids. In: Kruse PF, Patterson MK (eds) Tissue culture – methods and applications. Academic Press, New York London, pp 475–500

Glimelius K, Bonnett HT (1986) *Nicotiana* cybrids with *Petunia* chloroplasts. Theor Appl Genet 72:794–798

Glimelius K, Chen K, Bonnett HT (1981) Somatic hybridization in *Nicotiana*: Segregation of organellar traits among hybrid and cybrid progeny. Planta 153:504–510

Hess D (1969a) Versuche zur Transformation an höheren Pflanzen: Induktion und konstante Weitergabe der Anthocyansynthese bei *Petunia hybrida*. Z Pflanzenphysiol 60:348–353

Hess D (1969b) Versuche zur Transformation an höheren Pflanzen: Wiederholung der Anthocyan-Induktion bei *Petunia* und erste Charakterisierung des transformierenden Prinzips. Z Pflanzenphysiol 61:286–298

Jenkins MT (1924) Heritable characteristics of maize XX. Iojap striping a chlorophyll defect. J Hered 15:472–497

Kao KN, Michayluk MR (1974) A method for high-frequency intergeneric fusion of plant protoplasts. Planta 115:355–367

Kawaguti S, Yamasu T (1965) Electron microscopy on the symbiosis between an elypsioid gastropod and chloroplasts of a green alga. Biol J Okayama Univ II:57–62

Kawashima N (1969) Comparative studies from Fraction I protein from spinach and tobacco leaves. Plant Cell Physiol 10:31–40
Kawashima N (1970) Non-synchronous incorporation of $^{14}CO_2$ into amino acids of the two subunits of Fraction I protein. Biochem Biophys Res Commun 38:119–124
Kawashima N, Wildman SG (1970) Fraction I protein. Annu Rev Plant Physiol 21:325–358
Kawashima N, Wildman SG (1971) Studies on Fraction I protein. II. Comparison of physical, chemical, immunological and enzymatic properties between spinach and tobacco. Fraction I protein. Biochim Biophys Acta 229:749–760
Kawashima N, Wildman SG (1972) Studies on Fraction I protein. IV. Mode of inheritance of primary structure in relation to whether chloroplast or nuclear DNA contains code for a chloroplast protein. Biochim Biophys Acta 262:42–49
Kawashima N, Kwok S, Wildman SG (1971) Studies on Fraction I protein. III. Comparison of the primary structure of the large and small subunits obtained from five species of *Nicotiana.* Biochim Biophys Acta 236:578–586
Kirk JTO, Tilney-Bassett RAE (1967) The Plastids. Freeman, London San Francisco
Ledoux L, Huart R (1968) Integration and replication of DNA of *Micrococcus lysodeikticus* in DNA of germinating barley. Nature (Lond) 218:1256–1259
Lyttleton JW, Ts'o POP (1958) The localisation of Fraction I protein of green leaves in the chloroplasts. Arch Biochem Biophys 73:120–126
Maliga P (1985) Cell culture isolation and characterization of agronomically useful mutants of higher plants. Biotechnology in international agricultural research; inter-center seminar on international agricultural research centers and biotechnology, Manila, Philippines, 1984, pp 111–120
McClintock B (1941) The stability of broken ends of chromosome in *Zea mays.* Genetics 26:234–282
McClintock B (1944) The relation of homozygous deficiencies to mutations and allelic series in maize. Genetics 29:478–502
McFadden BA (1973) Autotrophic $CO_2$ assimilation and the evolution of ribulose diphosphate carboxylase. Bacteriol Rev 37:289–319
Medgyesy P, Golling R, Nagy F (1985) A light sensitive recipient for the effective transfer of chloroplast and mitochondrial traits by protoplast fusion in *Nicotiana.* Theor Appl Genet 70:590–594
Meeks JC, Malmberg RL, Wolk PC (1978) Uptake of auxotrophic cells of a heterocyst-forming Cyanobacterium by tobacco protoplasts and the fate of their associations. Planta 139:55
Melchers G, Labib G (1974) Somatic hybridisation of plants by fusion of protoplasts. I. Selection of light resistant hybrids of "haploid" light sensitive varieties of tobacco. Mol Gen Genet 135:277–294
Menczel L, Galiba G, Nagy F, Maliga P (1982) Effect of radiation dosage on efficiency of chloroplast transfer by protoplast fusion in *Nicotiana.* Genetics 100:43–52
Moon KE, Thompson EOP (1969) Subunits from reduced and S-carboxymethylated ribulose diphosphate carboxylase (Fraction I protein). Austr J Biol Sci 22:463–470
Mudd JB, McManus TT (1965) Relationship of the synthesis of lipid and water soluble acids by chloroplast preparations. Plant Physiol 40:340–344
Nass MMK (1969) Uptake of isolated chloroplasts by mammalian cells. Science 165:1128–1131
Nolla JAB (1934) A study of glaucous and yellow characters in *Nicotiana tabacum* L. J Agric Univ Puerto Rico 18:443–562
Park RB, Pon NG (1961) Correlation of structure with function in *Spinacea oleracea* chloroplasts. J Mol Biol 3:1–10
Potrykus I (1973) Transplantation of chloroplasts into protoplasts of *Petunia.* Z Pflanzenphysiol 70:364–366
Potrykus I, Hoffmann F (1973) Transplantation of nuclei into protoplasts of higher plants. Z Pflanzenphysiol 69:287–289
Povilaitis B, Cameron DR (1963) A mutation causing chlorophyll deficiency in *Nicotiana tabacum.* Can J Genet Cytol 5:233–238
Rhoades HH (1943) Genic induction of an inherited cytoplasmic difference. Proc Natl Acad Sci USA 29:327–329
Ridley SD, Leech RM (1970) Division of chloroplasts in an artificial environment. Nature (Lond) 227: 463–465
Rutner AC, Lave MD (1967) Non-identical subunits of ribulose diphosphate carboxylase Biochim Biophys Res Commun 28:531–537

Rutner AC (1970) Spinach-D 5-phosphoribose isomerase purification and properties of the enzyme. Biochemistry 9(1):178–184

Salts Y, Hermann RG, Peleg N, Lavi U, Izhar F, Frankel R, Beckmann JS (1984) Physical mapping of plastid DNA varieties among eleven *Nicotiana* species. Theor Appl Genet 69:1–14

Shell JS (1987) Transgenic plants as tools to study the molecular organization of plant genes. Science 237:1176–1182

Shvidkaya LG, Gleba YY (1974) Plating of mesophyll protoplasts, isolated from different photosynthetic mutants of *Nicotiana tabacum*. L. 3rd Int Congr Plant cell and tissue culture, Abstr 195, Univ Leicester

Smith HH (1972) Plant genetic tumors. Progr Exp Tumor Res 15:138–164

Smith HH (1974) Model systems for somatic cell plant genetics. BioScience 24:269–276

Smith HH (1975) *Nicotiana*. In: King RC (ed) Handbook of genetics. Plenum, London

Stadler LJ (1946) Spontaneous mutation of the R locus of maize. I. The aleurone colour and plant colour effects. Genetics 31:377–394

Stubbe W (1957) Dreifarbenpanaschierung bei Oenothera. I. Entmischung von 3 in der Zygote vereinigten Plastidomen. Ber Dtsch Bot Ges 70:221–226

Sugiyama T, Tomoki I, Azakawa T (1971) Subunit structure of ribulose 1,5-diphosphate carboxylase from *Chlorella ellipsoidea*. Biochemistry 10:3406–3411

Takebe I, Otsuki Y (1969) Infection of tobacco mesophyll protoplasts by tobacco mosaic virus. Proc Natl Acad Sci USA 64:843–848

Takebe I, Otsuki Y (1973) Fine structure of isolated mesophyll protoplasts of tobacco. Planta 113:21–27

Takebe I, Otsuki Y (1974) Multiple infection of tobacco mesoplasts by plant viruses. 3rd Int Congr Plant cell and tissue culture, Abstr 36, Univ Leicester

Takebe I, Otsuki Y, Aoki S (1968) Isolation of tobacco mesophyll cells in intact and active state. Plant Cell Physiol 9:115–124

Takebe I, Labib G, Melchers G (1971) Regeneration of whole plants from isolated mesophyll protoplasts of tobacco. Naturwissenschaften 58:318–320

Taylor DL (1968) Chloroplasts as symbiotic organelles in the digestive gland of *Elysia viridis* (Gastropoda: Opisthobrancha). J Mar Biol 48:1–15

Tolbere NE (1958) Secretion of glycolic acid by chloroplasts. In: The photochemical apparatus. Its structure and function. Brookhaven Symp Biol 11:271–275

Trench RK (1969) Chloroplasts as functional endosymbionts in the mollusc *Tridachia crispata* (Bergh). Nature (Lond) 222:1071–1072

Trench RK, Green RW, Bystrom BG (1969) Chloroplasts as functional organelles in animal tissues. J Cell Biol 42:404–407

Valleau WD (1957) Yellow Crittenden – a mutant in dark tobacco. Tab Sci 1:91–92

Wallin A, Glimelius K (1974) The induction of aggregation and fusion of *Daucus carota* protoplasts by polyethylene glycol. Z Pflanzenphys 74(1):64–80

Weissbach A, Horecker BL, Hurwitz J (1956) The enzyme formation of phosphoglyceric acid from ribulose diphosphate and carbon dioxide. J Biol Chem 218:795–810

Wildman SG, Bonner J (1974) Proteins of green leaves. I. Isolation, enzymic properties and auxin content of spinach cytoplasmic proteins. Arch Biochem Biophys 14:381–413

Wisselingh C van (1920) Über Variabilität und Erblichkeit. Z Indukt Abst Vererbungsl 22:65

Yamada T, Sakaguchi K (1981) Polyethylene glycol induced uptake of bacteria into yeast protoplasts. Agric Biol Chem 45:2301–2309

Yonge CM, Nicholas HM (1940) Structure and function of the gut and symbiosis with zooxanthellae in *Tridachia crispata*. Pap Tortugas Lab 32:289–303

Zelcer A, Aviv D, Galun E (1978) Interspecific transfer of cytoplasmic male sterility by fusion between protoplasts of normal *Nicotiana sylvestris* and x-ray irradiated protoplasts of male sterile *N. tabacum*. Z Pflanzenphysiol 90:397–407

Zucker (1972) Light and enzymes. Annu Rev Plant Physiol 23:133–156

# XXVI Cytoplasts: Isolation and Uses

C. R. LANDGREN

## 1 Introduction

During the production and handling of protoplasts, a variety of plasmalemma-bound fragments may be produced. These fragments contain a portion of the cytoplasm and may or may not contain the nucleus (see Figs. 1–4). Enucleate plasmalemma-bound protoplast fragments have been observed in protoplasts preparations from explants of *Pisum sativum* leaves (Wallin et al. 1978), *Allium cepa* epidermis (Bradley 1978; Bracha and Sher 1981), *Brassica oleracea* cotyledons (Vatsya and Bhaskaran 1981), *Lycopersicon esculentum* fruits (De and Swain 1983), and *Brassica napus* hypocotyls (Spangenberg 1985; Spangenberg and Schweiger 1986; Spangenberg et al. 1986). Enucleate membrane-bound protoplast fragments form when protoplasts are derived from elongated unicellular filaments including *Funaria hygrometrica* protonema (Gwozdz and Waliszewska 1979), *Nicotiana* and *Hyoscyamus muticus* pollen tube cells (Lörz and Potrykus 1980a) and *Gossypium hirsutum* fibers (Gould et al. 1986). Enucleate membrane-bound protoplast fragments have also been produced from protoplasts derived from suspension cultured cells of *Nicotiana tabacum*, *N. sylvestris*, *Daucus carota* (Wallin et al. 1978), *Zea mays*, *H. muticus* (Lörz et al. 1981), *Nicotiana plumbaginifolia* (Hoffmann 1981), *Solanum nigrum* (Lesney et al. 1983), *Parthenocissus tricuspidata*, *Psophocarpus tetragonolobus* (Fakhrai and Evans 1986), *Lycopersicon peruvianum*, *L. penelli*, *Petunia hybrida*, and *Solanum lycopersicoides* (Tan 1987).

Enucleate plasmalemma-bound protoplast fragments have been called nuclear-free protoplasts (Binding and Kollmann 1976), subprotoplasts (Wallin et al. 1978; Vatsya and Bhaskaran 1981), enucleated protoplasts (Bradley 1978; Izhar and Power 1979), cytoplasts (Lörz et al. 1981; Archer et al. 1982), and anucleate protoplasts (Gould et al. 1986). The term cytoplast seems to be gaining favor among those who work with protoplast fragments, perhaps because of its brevity and the emphasis it places on the cytoplasmic origin of the structure.

Protoplast fragments containing the nucleus, plastids, and mitochondria are genetically identical to the parental cells, since the genetic information in a plant cell is located in these three structures. Protoplast fragments containing a nucleus generally have sufficient cytoplasm so that the plastome (plastid genetic complement) and the chondriome (mitochondrial genetic complement) are well repre-

Department of Biology, Middleburgy College, Middlebury, VT 05753, USA

Biotechnology in Agriculture and Forestry, Vol. 9
Plant Protoplasts and Genetic Engineering II (ed. by Y. P. S. Bajaj)

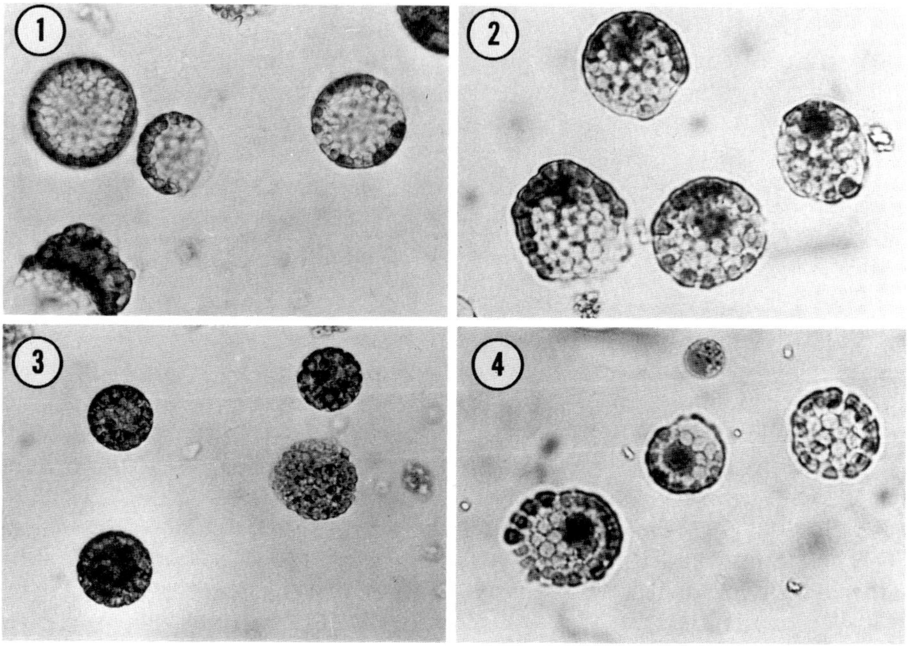


**Figs. 1–4.** Protoplasts and cytoplasts of *Nicotiana (deb)tbc* after centrifugation at 200 g for 5 min on a two-step Percoll gradient. **1, 2** Material floating above 28% Percoll. **1** Presumed protoplasts directly from the gradient. **2** Protoplasts with prominent nuclei after staining with carbol fuchsin. **3, 4** Cytoplasts and protoplasts from the interface of the 28% and the 50% Percoll layers. **3** Denser, less-vacuolate structures than the typical protoplasts. **4** Two protoplasts, a large vacuolate cytoplast, and a smaller less-vacuolate cytoplast. The bar in Fig. **10** represents 10 μm in **1–4**. (**2** and **4** from Archer et al. 1982; **1** and **3** courtesy of K. Archer and H.T. Bonnett)

sented. Nucleate protoplast fragments are therefore essentially miniature protoplasts. This identity with the parental protoplast population decreases the utility of nucleate protoplast fragments in experiments involving genetic analysis or the engineering of plant cell lines.

Nucleate protoplast fragments provide a cytoplasm-enriched, vacuole-depleted, nucleate body for cell fusions (Lörz and Potrykus 1980b). Densely protoplasmic structures are more suited to fusions (Power et al. 1977), although advances in fusion technology may have decreased the particular advantages associated with the small vacuolar volume.

## 2 Cytoplasts

### 2.1 Spontaneous Cytoplast Formation

Cytoplasts are formed spontaneously during protoplast isolation (see Table 1 and Figs. 5–11). During plasmolysis, the change in volume apparently precedes the

**Table 1.** Plant species from which cytoplasts have formed spontaneously during protoplast isolation

| Species | Origin of cells | Plasmolyticum | Enzyme treatment | Isolated structures | Purification procedures | Reference |
|---|---|---|---|---|---|---|
| *Allium cepa* | Bulb epidermis | 0.7 M mannitol | 0.5% cellulysin<br>0.1% pectinase | C, N, P | None | Bradley (1978) |
| *Brassica oleracea* | Cotyledons | 0.33 – 0.77 M mannitol | 2% cellulase Onozuka R10<br>1% macerozyme | C, P | None | Vatsya and Bhaskaran (1981) |
| *Funaria hygrometrica* | Protonema | 0.3 M mannitol | 4% cellulase Onozuka R10<br>3% mMacerozyme R10<br>3% Rohzyme HP150 | C, P | 100 μm filter | Gwozdz and Waliszewska (1979) |
| *Gossypium hirsutum* | Ovule fibers | 0.6 M mannitol equivalents | 1.4% cellulase "CEL" | C | Float cytoplasts on 4.5% Ficoll | Gould et al. (1986) |
| *Lycopersicon esculatum* | 6 – 7 week fruits | 0.4 M mannitol | None | C, N, P, V | None | De and Swain (1983) |
| *Nicotiana glutinosa* | Susp. culture cells | 0.5 M sorbitol | 1% Cellulysin | C, P | Sediment in two-phase Percoll | Archer et al. (1982) |
| *Nicotiana tabacum* | Pollen tube cells | – | – | C, N | None | Lörz and Potrykus (1980a) |

Abbreviations used for isolated structures: C = cytoplasts, P = protoplasts, N = nucleate membrane-bound subprotoplast, V = vacuoplast

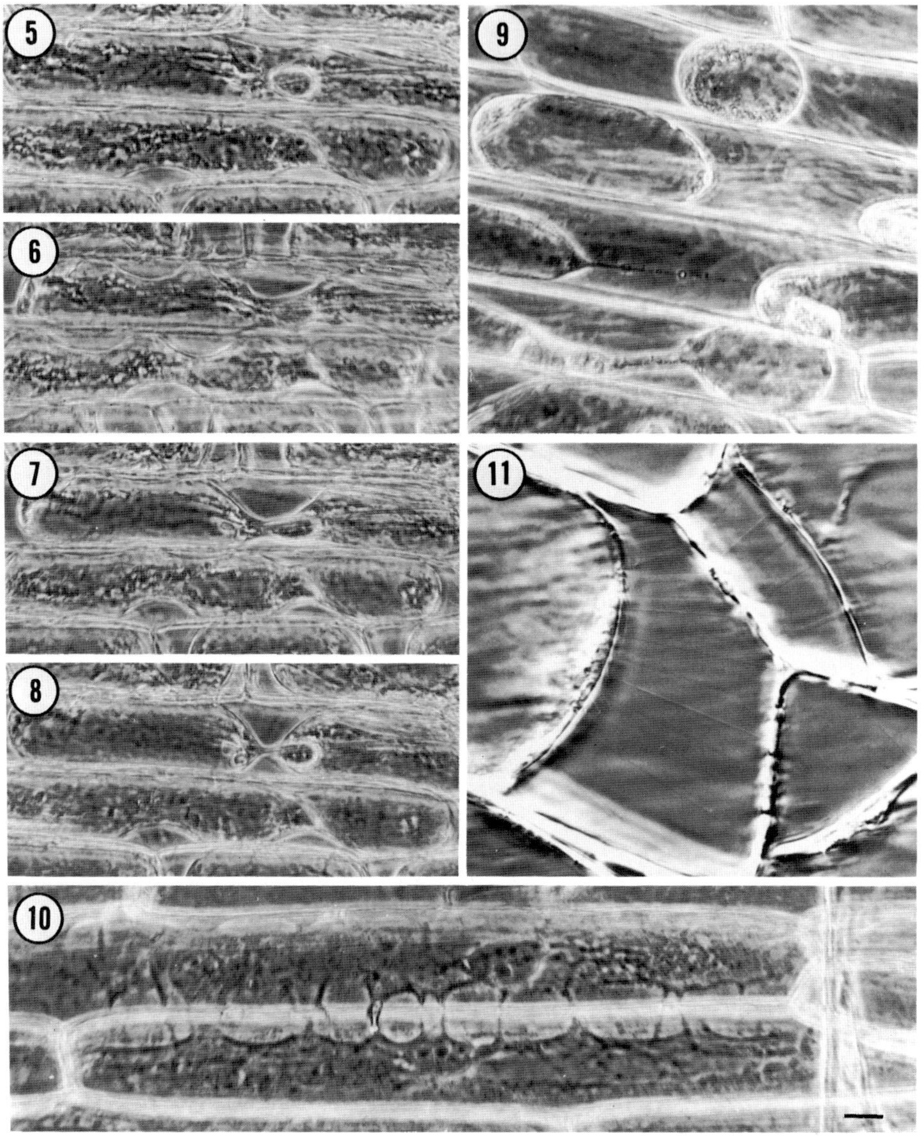


A

**Figs. 5–11. A** Onion bulb epidermis tissues plasmolyzed in 1 M sorbitol. **5–10** Phase contrast optic. **11** Hoffmann interference contrast optics. **5–8** illustrate the time course of early plasmolysis in onion epidermis, representing respectively 30 s, 2 min, 12 min, and 15 min after exposure to 1 M sorbitol in CPW salts. The formation of a bridge between a nascent cytoplast and miniprotoplast is evident in one of the cells illustrated. The removal of water from the cytoplasm seems to precede its removal from the vacuole. **9** shows cells from a similar preparation after 30 min in the plasmolyzing solution. **10–11** The role of plasmodesmatal connections in cytoplast formation. **10** Heavy plasmodesmata locally restrict the shrinking of the protoplast. **11** Thinner plasmodesmata do not restrict cytoplasmic shrinking. **B** In the line drawings, the cell walls are indicated by *heavy lines*, the protoplasmic mass by *light stipple*, and the nucleus by *heavier stipple*. *Bar* = 10 μm for **5–8**, 6 μm for **9**, **10**, and 2.5 μm for **11**. (**5–7**; **9–11** Archer et al. 1982; **8** courtesy of C. R. Landgren and H. T. Bonnett)

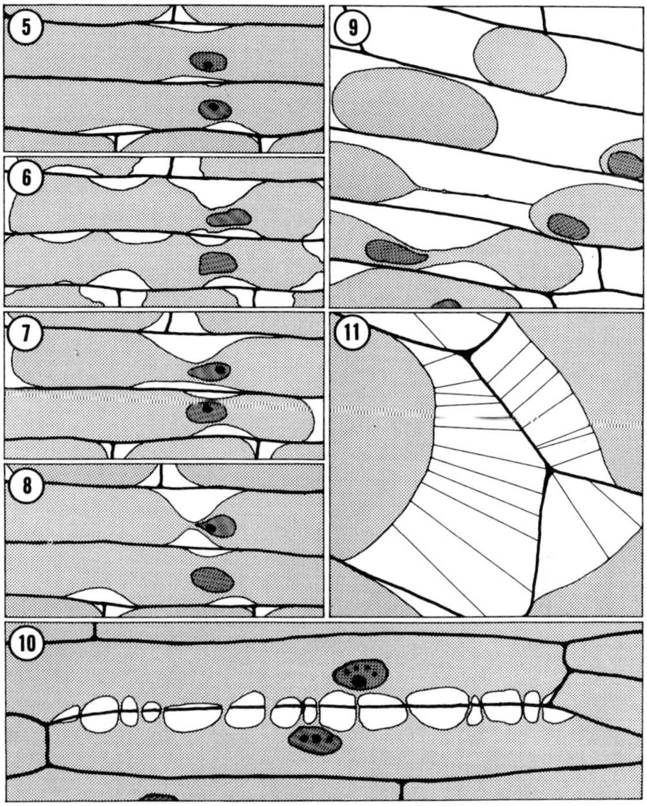


**Figs. 5–11 B**

change in surface area, resulting in an excess of membrane. A rapidly shrinking protoplast may divide into two spheres to maintain a surface area approaching that of the original cell. Cytoplasts form more abundantly in solutions of higher osmotic concentrations (Vatsya and Bhaskaran 1981). In *Brassica* cotyledonary protoplasts there is an almost linear increase in the fraction of cytoplasts (from 5%–35%) in protoplasts isolated in mannitol concentrations from 0.44 M to 0.77 M.

Spontaneous cytoplast formation in situ occurs in elongate cells even without the degradation of cell walls when *Elodea* leaves are immersed in hypertonic solutions (Yoshida 1961). Large numbers of cytoplasts form in cotton fibers, whose length may be 60 or more times greater than their 30-μm diameter. The cytoplasts isolated from cotton fibers continue the metabolism associated with their growth for several days following isolation (Gould et al. 1986). *N. glutinosa* suspension culture cells are elongate and form cytoplasts freely, while more isodiametric *N. debneyi* (*tabacum*) suspension culture cells do not (Archer et al. 1982). Moss protonema are also more likely to form cytoplasts when the cells are elongated (Gwozdz and Waliszewska 1979).

The process of spontaneous cytoplast formation (Figs. 5–11) is evident in onion epidermis undergoing rapid plasmolysis (Archer et al. 1982). Plasmolysis events in cells which are protected from physical disruption (shaking and stirring) are initially highly localized. The location of plasmolysis seems to be related to the absence of plasmodesmata, to the vacuolar anatomy, and to the thickness of the cytoplasm (Figs. 5–11). Protoplasts undergoing plasmolysis round up when stirred or shaken. The largest number of cytoplasts are formed when the cells, isolated from physical stress, are plasmolyzed in 0.7 to 1 M mannitol solutions. Eight percent of the "protoplasts" formed from onion epidermis plasmolyzed in 0.7 M mannitol and B5 salts lack nuclei (Bradley 1978).

Spontaneous formation of cytoplasts in culture occurs when buds are expelled and detach from the parent *N. tabacum* pollen tetrad protoplasts (Bajaj 1974). This has also occurred in cultures of protoplasts of *N. plumbaginifolia* (Hoffmann 1981). Substantial budding is observed in pollen tetrad and pollen mother cell protoplasts of *Atropa belladonna*, *N. tabacum* and *Triticum aestivum* (Bajaj 1974). Cytoplasts forming from buds appear to be the result of growth in an unhealthy environment. No wall formation is evident in detached cytoplasts of *N. plumbaginifolia* during culture over a period of several months.

## 2.2 Centrifugal Cytoplast Production

Centrifugal forces have been used to separate cytoplasts and miniprotoplasts in plant (see Table 2 in this article and Figs. 1–4 in Chap. XXVIII, this Vol.) and animal systems. Centrifuged sea urchin (*Arbacia*) eggs break into cytoplasts and nucleate protoplast fragments. The eggs elongate as the dense nucleus drags half the cytoplasm down while the enucleate portion remains buoyant. The cell breaks into two fragments when the length is greater than $\pi$ times the diameter measured perpendicular to the long axis. In this respect the cell behaves as a spherical droplet not readily miscible with the medium (Harvey 1931).

The medium in which the protoplasts are centrifuged has a significant effect on the success of experimental production of cytoplasts. Discontinuous density gradients of sucrose and Percoll have been used for cytoplast production. Nucleated protoplast fragments sediment to the interface at the top of the most dense step. This is usually saturated sucrose (Wallin et al. 1978; Lesney et al. 1983) or Percoll in concentrations as high as 50% (Lörz et al. 1981). The cytoplasts remain in the less dense steps. Cytochalasin B (30–50 μg/ml) and dimethylsulfoxide (0.5%) with (Wallin et al. 1978; Spangenberg et al. 1985) and without cytochalasin (Lesney et al. 1986) have been added to the centrifuging medium to increase cytoplast yields.

Cytoplasts separate from nucleate protoplast fragments at 20000 g to 40000 g after 15 min (Wallin et al. 1978) to 90 min (Lörz et al. 1981) centrifugation at 12 °C (Lörz et al. 1981; Lesney et al. 1986) to 37 °C (Wallin et al. 1978). The cytoplasts produced are viable for at least 1 week (Fakhrai and Evans 1986). Contamination of cytoplasts produced under these conditions ranged from 2% (Maliga et al. 1982) to 25% nucleate structures (Fakhrai and Evans 1986). Sixty-two to 72% of the *Solanum nigrum* protoplasts layered onto a 1 M mannitol with 0.5% dimethylsulfoxide, 33% sucrose, saturated sucrose, step gradient are enucleated following a 23000 g centrifugation for 60 min (Lesney et al. 1983).

**Table 2.** Plant species from which cytoplasts have been formed by high speed centrifugation

| Species | Origin of cells | Plasmolyticum | Isolated structures | Isolation procedures | | Reference |
|---|---|---|---|---|---|---|
| | | | | Centrifuge speed | Gradient | |
| *Allium cepa* | Bulb epidermis | 0.35 M mannitol | C, N | 17000 rev/min | CB, D, S, M | Bracha and Sher (1981) |
| *Brassica napus* | Hypocotyl | 0.3 M mannitol | C, N | 20000 g | CB, D, S, M | Spangenberg et al. (1986) |
| *Daucus carota* | Susp. culture cells | 0.4 M sorbitol | C, N | 40000 g | CB, D, S | Wallin et al. (1978) |
| *Hyoscyamus muticus* | Cultured cells | 0.25 M mannitol, 0.1 M $CaCl_2$ | C, N | 20000 – 40000 g | PL, M, CA | Lörz et al. (1981) |
| *Lycopersicon peruvianum* | Susp. culture cells | 0.5 M mannitol | C, N | 26000 g | PL, M, CA | Tan (1987) |
| *Lycopersicon penelli* | Susp. culture cells | 0.5 M mannitol | C, N | 26000 g | PL, M, CA | Tan (1987) |
| *Nicotiana sylvestris* | Susp. culture cells | 0.4 M sorbitol | C, N | 20000 g | CB, D, S | Wallin et al. (1978) |
| *Nicotiana tabacum* | Susp. culture | 0.3 M sorbitol | C, N | 40000 g | CB, D, S | Wallin et al. (1978) |
| | | 0.25 M mannitol, 0.1 M $CaCl_2$ | C, N | 20000 – 40000 g | PL, M, CA | Lörz et al. (1981) |
| *Parthenocissus tricuspidata* | Susp. culture cells | – | C, P* | – | PL, M, CA | Fakhrai and Evans (1986) |
| *Petunia hybrida* | Susp. culture cells | 0.5 M mannitol | C, N | 26000 g | PL, M, CA | Tan (1987) |
| *Pisum sativum* | Leaves | 0.4 M sorbitol | C, N | 20000 g | CB, D, S | Wallin et al. (1987) |
| *Psophocarpus tetragonolobus* | Susp. culture cells | – | C, P* | – | PL, M, CA | Fakhrai and Evans (1986) |
| *Solanum lycopersicoides* | Susp. culture cells | 0.5 M mannitol | C, N | 26000 g | PL, M, CA | Tan (1987) |
| *Solanum nigrum* | Susp. culture cells | 0.5 M mannitol | C, N | 23000 g | S, M, D | Lesney et al. (1983) |
| *Zea mays* | Susp. culture cells | 0.25 M mannitol 0.1 M $CaCl_2$ | C, N | 20000 – 40000 g | PL, M, CA | Lörz et al. (1981) |

Abbreviations used for isolated structures: C = cytoplasts, P = protoplasts, P* = nucleate subprotoplasts identified as protoplasts, N = nucleate membrane-bound subprotoplast, V = vacuoplast.
Abbreviations used for centrifugation gradient composition: S = sucrose, CA = $CaCl_2$, CB = cytochalasin B, D = dimethylsulfoxide, PL = Percoll, M = mannitol.

The subcultured cell lines have provided better yields of cytoplasts than the older cultures (Lesney et al. 1986).

The centrifugation used to produce cytoplasts also separates the cytoplasts from protoplasts and nucleate protoplast fragments. The cytoplasts produced by high speed centrifugation are generally less dense than the protoplasts and nucleate protoplast fragments. This contrasts with some spontaneously produced cytoplasts which are more dense than the protoplasts with which they were isolated (Archer et al. 1982).

### 2.3 Identification of Cytoplasts

The definitive techniques used to distinguish cytoplasts from protoplasts and miniprotoplasts are destructive. Most workers favor the relatively simple carbol fuchsin procedure developed by Miller et al. (1971). Acridine orange/fluorescence microscopy (Lörz et al. 1981; Vatsya and Bhaskaran 1981), 6-diamidino-2-phenylindole (DAPI) (Fakhrai and Evans 1986; Gould et al. 1986) and Hoechst stain (Tan 1987) have also been used to determine the proportion of cytoplasts in an isolate. Cytoplasts have a relatively normal electron microscopic anatomy (Lörz et al. 1981), but can only be demonstrated to be enucleate by serial section, a procedure too tedious to be generally practical.

The sizes of protoplast-derived structures can suggest the number of cell fragments in an isolate (Vatsya and Bhaskaran 1981), but cannot differentiate between nucleate protoplast fragments and cytoplasts. The effects of plasmolysis on normal cell size must be taken into consideration when using the size of protoplast-derived structures to indicate the fragmentation of protoplasts.

## 3 Uses of Cytoplasts

Cytoplasts provide a means for the transfer of organelles and, more specifically, the transfer of the plastome and chondriome to a foreign genomic environment. This allows the study of the interactions of genomes with particular cytoplasmic genetic elements and the study of the interactions of populations of chloroplasts or mitochondria with different genetic characters. Streptomycin resistance in *Nicotiana* has been transferred by interspecific fusions of cytoplasts and protoplasts. Cytoplast transfer of plastome and chondriome eliminates the difficulties associated with organelle transplantation by direct fusion of the organelle with the protoplast (Bonnett and Eriksson 1974; Bonnett 1976; Davey et al. 1976).

The transfer of chondriome or plastome by cytoplast-protoplast fusions normally involves the transfer of both genetic complements and also increases the possibility of nuclear transfer over that observed in direct organelle fusion procedures. Intergeneric cybridization between *Petunia* and *Lycopersicon* produced one plant which contained a mixture of plastomes, while only the *Lycopersicon* chondriome was present. The mixture of plastomes was unstable, and one was lost over time (Tan 1987). This instability of mixed plastomes was also observed in

H. Bonnett's laboratory (unpubl.) when plastomes of *Nicotiana* were mixed by protoplast fusions. The only plant with two plastomes was chimeral, with each sector characterized by a single plastome. The remainder of the cybrid plants each had only one plastome, though in some cases it was the foreign rather than the host plastome.

Cytoplasts are useful for the engineering by fusions of protoplasts with particular genetic complements (Maliga 1982; Spangenberg and Schweiger 1986; Spangenberg et al. 1986; see also Chap. XXVIII, this Vol.). Cytoplast-protoplast fusions and cytoplast fusions with nucleate protoplast fragments produce cells which regenerate calli with about one half the frequency of callus regeneration from protoplast-protoplast fusions (Spangenberg and Schweiger 1986). The production of protoplast fragments seems to have limited the recovery of the fusion products, since fusion of any protoplast fragment with an intact protoplast reduced the callus formation in the fusion products. Nonetheless, the fusion products involving *Brassica* cytoplasts are able to produce minicalli approximately one third (30% – 34%) of the time.

Reconstructed cells, produced by the fusion of cytoplasts and nucleate protoplast fragments, are less stable than protoplasts. Twenty-two percent of the reconstructed cells survived DNA microinjection, while 82% of the protoplasts survived the microinjection (Spangenberg et al. 1986). Here it appears that the fusion and not the protoplast fragmentation are important to the unique character of the fusion products. Isolated nucleate protoplast fragments survive microinjection just as often (83% of the time) as protoplasts. On the other hand, karyoplasts are particularly cytoplasm-rich, while cytoplasts may have a large vacuole which will contribute to instability.

Cytoplasts are functionally similar to irradiated protoplasts used for transfer of cytoplasmic heritable factors (Aviv and Galun 1980; Kumashiro and Kubo 1986). They have the advantage of carrying none of the nuclear genetic material, whether it is active or not.

Cytoplasts have been used to condition the medium used for the culture of single protoplasts. The fraction of microcalli forming in media conditioned by cytoplasts (29%) was comparable to the fraction which grew in media conditioned by karyoplasts (nucleate protoplast fragments) (32%), protoplasts (26%), and culture cells (25%). Vacuoles were not effective conditioning agents. Fewer microcalli formed in medium conditioned by vacuoles than in medium which was not conditioned (Spangenberg et al. 1986). This suggests that the diffusible conditioning factors are present in cytoplasts in concentrations approaching those of intact cells, but are absent from isolated vacuoles.

## 4 Conclusions and Prospects

Nucleate and enucleate, membrane-bound protoplast fragments form spontaneously during protoplast isolation (see Table 1). Their formation can be promoted by using cell lines which are not isodiametric, by overplasmolysis, or by using cell lines with substantial plasmodesmatal connections. Agitation during protoplast production decreases the chances of cytoplast formation by breaking connections between the cell wall and the protoplast. Shearing forces, applied to severely plasmolyzed cells after incubation in an environment without physical disruption, break the tenuous connections between the nucleate and enucleate fragments.

Protoplast fragments can be enriched by taking advantage of their small size or their differences in density from intact protoplasts. Cytoplasts can be filtered through small mesh nylon, although this will not effectively eliminate nucleate protoplast fragments from the fraction, nor will it eliminate smaller protoplasts. Cytoplasts can also be isolated by density gradients. Cytoplasts produced by centrifugation are generally less dense than protoplasts and nucleate protoplast fragments, while those produced spontaneously may be more or less dense than the protoplasts and nucleate protoplast fragments.

Cytoplasts and nucleate membrane-bound protoplast fragments can also be produced by high-speed centrifugation (see Table 2). The production procedures also separate the nucleate and enucleate protoplast fragments. The descent of the cytoplasmically rich portion of the protoplast is slower than that of the nucleate portion. Generally, suspension culture cells are suitable for centrifugal production of cytoplasts, but explants of intact plants which produce protoplasts readily upon enzymic digestion may also be used. The particular enzyme regime used to produce protoplasts has no apparent effect on the later production of cytoplasts. Washed or unwashed protoplast preparations are layered onto a discontinuous gradient with saturated sucrose (or up to 50% Percoll) as a cushioning layer at the bottom of the gradient. The gradient is centrifugated at 20000 – 40000 g. Cytoplasts are separated from the nucleate portion of the protoplast after less than 2 h of centrifugation. Cooling or cooler centrifugal temperatures (4 – 12 °C) seem to improve the quantity or survival of the cytoplasts. Cytochalasin B speeds up the release of the cytoplasts, but may decrease their viability.

Cytoplasts are removed from the interface between the least dense bands or from the air-liquid interface either by pipetting or by the collection of fractions from the bottom of the tube. Nucleate, membrane-bound protoplast fragments are isolated from the top of the cushioning (most dense) layer. Centrifugal production of cytoplasts yields larger numbers of generally less contaminated cytoplasts than can be isolated from preparations where cytoplasts form spontaneously.

There appears to be no limit to the species from which cytoplasts can be produced. Plant material which can be used as a source of isolated protoplasts can be used as a source of protoplasts for centrifugal production of cytoplasts. Extremely large vacuoles may make the cytoplasts fragile. The cytoplasts produced by centrifugation maintain the metabolic activities associated with their cytoplasm and remain metabolically active for some time.

The fusion of cytoplasts with protoplasts or with nucleate, membrane-bound cytoplasmic fragments provides a tool for the engineering of cells with unique genome/chondriome or genome/plastome combinations. Electrofusions seem particularly useful for the careful engineering of new or reconstructed cells. Cell reconstruction by electrofusion may provide a useful tool for study of interorganelle genetic interactions. Cell reconstruction is less likely to be useful in breeding programs built on embryogenesis in cultures derived from reconstructed cells. The limits imposed on engineering by gross recombination of organelles are greater than those imposed on the more subtle molecular genetic engineering techniques involving viral or bacterial vectors. Vector-mediated modification of organelles in cytoplasts used for cell reconstructions may ultimately lead to more productive uses of cytoplasts in breeding programs.

## References

Archer EK, Landgren CR, Bonnett HT (1982) Cytoplast formation and enrichment from mesophyll tissues of *Nicotiana* spp. Plant Sci Lett 25:175–185

Aviv D, Galun E (1980) Restoration of fertility in cytoplasmic male sterile *Nicotiana sylvestris* by fusion with x-irradiated *Nicotiana tabacum* protoplasts. Theor Appl Genet 58:121–128

Bajaj YPS (1974) Isolation and culture studies on pollen tetrad and pollen mother cell protoplasts. Plant Sci Lett 3:93–99

Binding H, Kollmann R (1976) The use of subprotoplasts for organelle transplantation. In: Dudits D, Farkas GL, Maliga P (eds) Cell genetics in higher plants. Académiai Kiado', Budapest, pp 191ff

Bonnett HT (1976) On the mechanism of the uptake of *Vaucheria* chloroplasts by carrot protoplasts treated with polyethylene glycol. Planta 131:229–233

Bonnett HT, Eriksson T (1974) Transfer of algal chloroplasts into protoplasts of higher plants. Planta 120:71–79

Bracha M, Sher N (1981) Fusion of enucleated plant protoplasts of *Allium cepa* L. Plant Sci Lett 23:95–101

Bradley PM (1978) Production of enucleated plant protoplasts of *Allium cepa*. Plant Sci Lett 13:287–290

Davey MR, Frearson EM, Power JB (1976) Polyethylene glycol-induced transplantation of chloroplasts into protoplasts: an ultrastructural assessment. Plant Sci Lett 7:7–16

De DN, Swain D (1983) Protoplasts, cytoplasts and sub-protoplasts from ripening tomato fruits: their nature and fusion properties. In: Sen SK, Giles KL (eds) Plant cell culture in crop improvement. Plenum, New York, pp 201–208

Fakhrai H, Evans P (1986) A productive method for bulk preparation of viable cytoplasts from suspension culture-derived protoplasts of *Parthenocissus tricuspidata* and *Psophocarpus tetragonolobus*. Plant Physiol 80 (suppl):38

Gould JH, Palmer RL, Dugger WM (1986) Isolation and culture of anucleate protoplasts from cotton (*Gossypium hirsutum*) fiber: assessment of viability and analysis of regenerated wall polymers. Plant Cell Tissue Organ Cult 6:61–72

Gwozdz EA, Waliszewska B (1979) Regeneration of enzymatically isolated protoplasts of the moss *Funaria hygrometrica* L. Sibth. Plant Sci Lett 15:41–45

Harvey EN (1931) The tension at the surface of marine eggs, especially those of the sea urchin, *Arbacia*. Biol Bull 61:273–279

Hoffmann F (1981) Formation of cytoplasts from giant protoplasts in culture. Protoplasma 107:387–391

Izhar S, Powers JB (1979) Somatic hybridization in *Petunia*: a male sterile cytoplasmic hybrid. Plant Sci Lett 14:49–55

Kumashiro T, Kubo T (1986) Cytoplasm transfer of *Nicotiana debneyi* to *Nicotiana tabacum* by protoplast fusion. Jpn J Breed 36:39–48

Lesney MS, Callow PW, Sink KC (1983) A simplified method for bulk production of cytoplasts from suspension-culture derived protoplasts of *Solanum nigrum* L. In: Potrykus I, Harms CT, Hinnen A, Hütter R, King PJ, Shillito RD (eds) Protoplasts. Birkhäuser, Basel, pp 116–117

Lesney MS, Callow PW, Sink KC (1986) A technique for bulk production of cytoplasts and miniprotoplasts from suspension culture-derived protoplasts. Plant Cell Rep 5:115–118

Lörz H, Potrykus I (1980a) Isolation of subprotoplasts for genetic manipulation studies. In: Ferenczy L, Farkas GL, Lazar G (eds) Advances in protoplast research (Proc 5th Int Protoplast Symp) 7.1979 Szeged, Hungary. Pergamon, Oxford, pp 377–382

Lörz H, Potrykus I (1980b) Cytoplasts and mini-protoplasts: Isolation, characterization, culture and fusion of plant cell-derived sub-protoplasts. Eur J Cell Biol (abstr) 22:502

Lörz H, Paszkowski J, Dierks-Ventling C, Potrykus I (1981) Isolation and characterization of cytoplasts and miniprotoplasts derived from protoplasts of cultured cells. Physiol Plant 53:385–391

Maliga P (1982) Protoplasts of flowering plants and their use in cell genetic manipulation. Acta Bot Neerl 31:249

Maliga P, Lörz H, Lazar G, Nagy F (1982) Cytoplast-protoplast fusion for interspecific chloroplast transfer in *Nicotiana*. Mol Gen Genet 185:211–215

Miller RA, Gamborg OL, Keller WA, Kao KN (1971) Fusion and division of nuclei in multinucleated soybean protoplasts. Can J Genet Cytol 13:347–353

Powers JB, Evans PK, Cocking EC (1977) Fusion of plant protoplasts. Cell Surf Rev 8:370–385

Spangenberg G (1985) Different types of protoplasts from *Brassica napus* L.: analysis of conditioning effects at the single cell level. Eur J Cell Biol 39:41–45

Spangenberg G, Schweiger HG (1986) Controlled electrofusion of different types of protoplasts and subprotoplasts including cell reconstitution in *Brassica napus* L. Eur J Cell Biol 41:51–56

Spangenberg G, Neuhaus G, Schweiger HG (1986) Expression of foreign genes in cell reconstitution of a microinjected karyoplast and a cytoplast. Eur J Cell Biol 42:236–238

Tan MLMC (1987) Somatic hybridization and cybridization in Solanaceous species. Academisch Proefschrift, Vrije Universiteit te Amsterdam

Vatsya B, Bhaskaran S (1981) Production of sub-protoplasts in *Brassica oleracea* var. Capitata – a function of osmolarity of the media. Plant Sci Lett 23:277–282

Wallin A, Glimelius K, Eriksson T (1978) Enucleation of plant protoplasts by cytochalasin B. Int J Plant Physiol 87:333–340

Yoshida Y (1961) Role of nucleus in cytoplasmic activities with special reference to the formation of surface membranes in *Elodea* leaf cells. Plant Cell Physiol 2:139–156

# XXVII Vacuoles: Isolation, Purification, and Uses

A. MARETZKI and M. THOM[1]

## 1 Introduction

Since plant vacuoles were first successfully isolated by Cocking in 1960, interest in them as a subject of research has steadily increased. Over the past few years the availability of vacuoles from an increasing number of plant species has narrowed gaps in our knowledge about intracellular compartmental distribution, electrochemical properties of the tonoplast, and membrane transport of primary and secondary metabolites associated with these properties. Vacuoles are being used extensively to investigate both enzymes functioning exclusively or primarily within the vacuole and enzymes associated with the tonoplast. Direct access to this organelle has at times merely confirmed prior inferences or conclusions reached from studies with cells or tissues; but, more frequently, new perspectives have been gained. Either way, access to isolated vacuoles has become an important asset for the plant physiologist.

Techniques for the isolation of vacuoles have undergone some evolution, but basically remain similar to those initially developed. Updates on new modifications in procedures have been published at relatively frequent intervals (Matile 1979; Leigh 1981; Alibert and Boudet 1982; Wagner 1983; Boller and Wiemken 1986; Wagner in press). We will concentrate here less on isolation per se and uses of vacuoles and rather on the purification of vacuoles once they have been released from tissue or protoplasts. As more definitive information is sought about this organelle, its degree of purity becomes an important factor for many investigations. Recent developments in purification techniques and how they relate to the final objectives of any particular study are also covered. This information is presented so that methods in use at this time can be comprehensively compared to provide investigators with a useful guide for their particular problem.

## 2 General Methods of Isolation

### 2.1 Mechanical Slicing

This procedure, first introduced by Leigh and Branton (1976), is still a preferred way of releasing vacuoles from fresh tissue. Unfortunately, relatively few tissues,

[1] Hawaiian Sugar Planters' Association, P.O. Box 1057, Aiea, Hawaii 96701, USA
Supported in part by NSF grant # DMB 8613973.

Biotechnology in Agriculture and Forestry, Vol. 9
Plant Protoplasts and Genetic Engineering II (ed. by Y.P.S. Bajaj)

such as beet and horseradish roots (Leigh and Branton 1976; Willenbrink and Doll 1979; Grob and Matile 1979), lend themselves to this approach. Preplasmolysis of horseradish cells in 1.5 M sucrose (Grob and Matile 1979) permitted subsequent slicing in 0.55 M NaCl. In more succulent tissues, vacuoles do not appear to be released intact in acceptable yields without some enzymatic digestion. Slicing is a tedious procedure if carried out manually with a razor blade Using the slicing apparatus developed by Leigh and Branton (1976), yields, while still low, become high enough to carry out experiments requiring substantial quantities of intact vacuoles. In terms of purity, vacuole release by slicing may be the most desirable method, but direct comparison of this and any of the methods depending on preferential protoplast rupture in the same plant tissue are not available.

### 2.2 Direct Isolation

This method is similar to the above mechanical slicing technique. In *Hevea brasiliensis* latex (D'Auzac 1975) it is possible to collect intact vacuoles directly after cutting the tissue. However, this represents a rather special case with limited significance in terms of isolation of vacuoles from other plants.

### 2.3 Protoplast-Based Methods

Preferential rupturing of the plasma membrane, leaving the vacuole intact, has been the method of choice for most tissues as well as for cell-cultured material. Other methods falling under this heading can be adapted for most, though not necessarily all, plant species. Of course, overall success depends on whether the tissue of choice or cultured material lends itself to making protoplasts. It is in conjunction with these methods that most progress has been made toward achieving higher purity of the vacuoles. The methods fall into six categories.

#### 2.3.1 Osmotic Shock

A procedure developed by Wagner and Siegelman (1975) provided the first method by which quantities of protoplasts could be isolated from leaves, petals, pedicels, fruits, and stems of a variety of plants. Basically, the method relies upon altering the osmotic conditions of the protoplasts by rapid addition to the protoplast preparation of a medium of lower osmotic concentration. With a correct solute concentration difference between cells and added medium, differential breakage between plasma membrane and tonoplast can be achieved. The method works most successfully when the protoplasts in the preparation are quite uniform in terms of osmotic potential. Trapping of cytoplasm and other organelles between fragments of plasma membrane and the vacuole (a problem not necessarily confined to this method) may result.

### 2.3.2 Polybase Lysis

This method relies on altering the charge on the external surface of the plasma membrane by the addition of DEAE dextran, followed by rapid removal of excess DEAE with dextran sulfate. This procedure was first introduced by Buser and Matile (1977) for protoplasts from *Bryophyllum* leaves. Modifications of this method have, since then, become a preferred way to isolate vacuoles, although not all protoplasts from all sources lend themselves to this approach. An estimated requirement of 1 ng DEAE dextran addition per protoplast (Hopp et al. 1985) seems to be a reasonable starting point for applying this method to a new species. A number of the purification procedures we will describe use polybase lysis in a combination approach.

### 2.3.3 Chelating of Plasma Membrane

The addition of EDTA or EGTA chelates the $Ca^{2+}$ on the plasma membrane, thereby making this membrane highly susceptible to breakage, but leaving the tonoplast intact. Like polybase lysis, this method also alters the charge on the plasma membrane.

### 2.3.4 Shearing

First used with higher plants by Lörz et al. (1976), shearing refers to the selective rupture of plasma membrane when protoplasts are subjected to a high centrifugal force. Thus, the protoplasts are layered over a solution containing a high concentration of Ficoll and, as breakage occurs, released vacuoles float either to the top of the gradient or remain at the Ficoll/buffered osmoticum interface. This approach has also recently been used in direct combination with purification. Shearing can be done in other ways (e.g., Kaiser et al. 1982), but it is not a procedure applicable to all protoplasts. Shearing has the advantage over DEAE dextran in that less damage may result from contact with a relatively inert substance such as Ficoll, although, depending on the ultimate use of the vacuoles for experimentation, Ficoll also may interfere. However, it can be readily removed by washing.

### 2.3.5 Differential Enzyme Digestion

For some types of tissues, e.g., *Nicotiana* cell suspensions (Schloss et al. 1987) this method seems to work well. The tissue is preplasmolyzed and then subjected to enzymatic cell wall digestion. Large numbers of vacuoles are sometimes released without any further intervention, other than some agitation of the digested material.

### 2.3.6 Digitonin Lysis of Plasma Membrane

A recent isolation method, the preferential destabilization of plasma membrane by digitonin (Le-Quoc et al. 1987) might provide a method for wider application. For *Acer pseudoplatanus* cell cultures, the optimum concentration of digitonin after protoplasting of the cells was 5.25 mg per ml and treatment was for 5 min. Up to 85% of the protoplasts were lyzed and the remainder were removed by centrifugation.

Most of the above more generally applicable procedures for isolation have been described in previous reviews (Matile 1979; Boller and Wiemken 1986). Table 1 includes references listing the use of each of the above methods since they first became available to investigators, as well as some offshoots of these methods. We have not attempted to include every report in the table. Rather, we have endeavored to include representative examples from each of the groups that are presently engaged in vacuole isolation on the theory that new investigators then have a convenient starting point to search the literature further.

**Table 1.** Procedures used for vacuole isolation

| General approach[a] | Objective(s) | Plant species and tissue | Reference |
|---|---|---|---|
| A | Vacuole composition changes | *Acer pseudoplatanus* cells | Alibert et al. (1982) |
| A | Membrane potential | *Acer pseudoplatanus* cells | Gibrat et al. (1985) |
| H | Isolation of vacuoles | *Acer pseudoplatanus* cells | Le-Quoc et al. (1987) |
| A, D | ATPase activity comparison | *Acer pseudoplatanus* cells | Pugin et al. (1986) |
| G | Electrical characterization of tonoplast | *Actinidia chinensis* fruit | Chedhomme and Rona (1986) |
| E | Protein analysis | *Ananas comosus* leaf | Kenyon and Black (1986) |
| A, C | Hydrolytic enzymes in vacuoles | *Ananas comosus/Tulipa* petals and leaves; *Nicotiana tabacum* cell cultures | Boller and Kende (1979) |
| B | Membrane potential of vacuole | *Beta vulgaris* roots | Barbier and Guern (1982) |
| B | Sucrose uptake | *Beta vulgaris* roots | Doll et al. (1979) |
| B | Isolation of vacuoles | *Beta vulgaris* roots | Leigh and Branton (1976) |
| B | Membrane potential | *Beta vulgaris* roots | Miller et al. (1984) |
| B | Vacuolar constituents | *Beta vulgaris* roots | Salyaev et al. (1983) |
| A | Determination of malic acid in vacuole | *Bryophyllum* leaves | Buser and Matile (1977) |
| E | Alkaloid uptake | *Catharanthus roseus* cells | Deus-Neumann and Zenk (1984) |
| A | Efflux of malate and influx of nitrate | *Catharanthus roseus* cells | Marigo et al. (1985) |
| A | Leakage from vacuole | *Catharanthus roseus* cells | Renaudin et al. (1987) |
| A | ATP-dependent membrane potential | *Chenopodium rubrum* cells | Bentrup et al. (1986) |
| A | Uptake of acylated anthocyanin | *Daucus carota* cells | Hopp and Seitz (1987) |
| E | Fructan localization | *Helianthus tuberosis* | Frehner et al. (1984) |
| G | ATPase characterization | *Hevea brasiliensis* | D'Auzac (1975) |

**Table 1** (continued)

| General approach[a] | Objective(s) | Plant species and tissue | Reference |
|---|---|---|---|
| D | Compartmentation of photosynthetic products | *Hordeum vulgare* leaves | Kaiser et al. (1982) |
| D | Nitrate determination in vacuoles | *Hordeum vulgare* leaves | Martinoia et al. (1981) |
| D | Malate uptake | *Hordeum vulgare* leaves | Martinoia et al. (1985) |
| C | Oxalic acid deposition | *Hordeum vulgare* | Wagner (1981) |
| A | Accumulation of coumarylglucosides | *Hordeum vulgare* leaves | Werner and Matile (1985) |
| A | GA uptake and compartmentation | *Hordeum vulgare* and *Vigna sinensis* leaves | Ohlrogge et al. (1980) |
| B | Vacuolar localization of glucosinolates | Horseradish root | Grob and Matile (1979) |
| C | Vacuolar and tonoplast ATPase | *Kalanchoë daigremontiana* | Aoki and Nishida (1984) |
| A | Vacuolar ATPase activity | *Kalanchoë daigremontiana* leaves | Smith et al. (1984) |
| C | RNase activity | *Lycopersicon esculentum* cells | Abel and Glund (1986) |
| C | Compartmentation analysis | *Malus pumila* fruit | Yamaki (1984) |
| C | Isolation of vacuoles | Many sources of petals, leaves, fruit, and stems | Wagner and Siegelman (1975) |
| D | Isolation of vacuoles | Many sources | Lörz et al. (1976) |
| A | Isolation of vacuoles | *Melilotus alba* mesophylls | Boudet et al. (1981) |
| C | Isolation of tonoplast | *Nicotiana glutinosa* cells | Briskin and Leonard (1980) |
| C | Compartmental analysis | *Nicotiana rustica* | Saunders (1979) |
| F | Determination of peroxidases | *Nicotiana tabacum* leaves and cells | Schloss et al. (1987) |
| D | Alkaloid localization | *Papaver bracteatum* | Kutchan et al. (1986) |
| A, C | Uptake of flavonoids | *Petroselinum hortense* cells | Matern et al. (1986) |
| A | Vacuole localization of ethylene-induced chitinase | *Phaseolus vulgaris* leaves | Boller and Vögeli (1984) |
| D | Sugar transport | *Pisum sativum* leaves | Guy et al. (1979) |
| C | Synthesis of malic acid esters | *Raphanus sativum* leaves | Strack and Sharma (1985) |
| E | Measurement of pH in vacuole | *Ricinus communis* endosperm | Nishimura (1982) |
| D | Isolation of vacuoles | *Saccharum* sp. cells | Thom et al. (1982) |
| D | Isolation of vacuoles | *Sedum telephium* leaves | Kringstad et al. (1980) |
| C | Isolation of vacuoles | *Suaeda maritina* | Dracup and Greenway (1985) |
| C | Vacuolar proton pumping | *Tulipa* petals | Wagner and Lin (1982) |
| C | Membrane potential | *Valerianella locusta* | Weigel and Weiss (1984) |
| D | Proton transport | *Zea mays* coleoptiles | Mandala and Taiz (1985) |
| D | Glucose transport in tonoplast | *Zea mays* coleoptiles | Rausch et al. (1987) |
| C | Isolation of vacuoles | *Daucus carota* roots | Keller (1988) |

[a] A, polybase lysis; B, mechanical slicing; C, osmotic shock; D, shearing; E, EDTA or EGTA treatment; F, preplasmolysis followed by differential enzyme digestion; G, direct isolation; H, digitonin lysis of plasma membrane.

## 3 Purification Procedures

Vacuoles from those tissues which lend themselves to isolation by one or another general method listed above must, for almost all purposes, be subjected to a clean-up procedure in order to remove cytoplasmic components and plasma membrane fragments. Some of the clean-up methods that have been developed over the past few years are distinct advances, in terms of both simplification of the overall procedures and effectiveness. We will narrow our focus to detailed descriptions of procedures which have been used in recent years, since these most nearly represent the state of the art and are most likely to be useful to new investigators. Again, distinct differences among the species so far used for vacuolar studies are likely to exist; differences also extend to the type of tissues being used. To some extent the effectiveness of a clean-up procedure is dependent upon the choice of initial isolation method, and certainly, specific gravity of the vacuoles may dictate a particular procedure.

### 3.1 Evaluating Contamination

For nearly all analyses carried out with vacuoles or tonoplast preparations, it is not only essential to reduce the contaminating factors to a minimum but also to know what contaminants are present in any given preparation. Not only must reliable markers for vacuoles/tonoplast be available, but also similarly reliable markers for the organ, membraneous, or cytosolic contaminants suspected to be associated with the preparation, in order to avoid possible misinterpretation of the data.

Broken or damaged vacuoles can influence results, as, for instance, when a compartmental analysis is being sought. The addition of neutral red in place of a natural colored marker, such as betacyanin in red beet has been used by a number of workers (e.g., Guy and Kende 1984; Kaiser and Heber 1984). The location of the vacuolar band (or bands) on a gradient can thereby be visually identified, and it can also indicate leakage of vacuolar content due to loss of intactness. Alibert et al. (1982) used a concentration of 50 to 100 μg neutral red per ml in their protoplast preparations. Fluorescein diacetate solutions, because of their low pH, do not penetrate into the vacuole and can, therefore, be very useful in showing cytoplasm trapped between resealed plasma membrane fragments and the vacuole (Admon et al. 1980). This phenomenon is a distinct problem, both for osmotic lysis and for shearing approaches. Admon et al. (1980) have done a direct comparison of osmotic lysis and mechanical slicing and were able to demonstrate that when beet tissue is sliced immediately after exposure to a hypotonic solution the vacuoles are relatively more pure because less plasma membrane adheres to the tonoplast.

A clearcut universal marker for the tonoplast is still lacking. Where present, a hydrolase associated only with the vacuole, such as $\alpha$-mannosidase, is useful as a distinct marker (Boller and Kende 1979). In parsley (Matern et al. 1986) and in sugarcane (unpubl.) $\alpha$-mannosidase is not present in the vacuole. On the other hand, both these species have all their acid phosphatase associated with the vacuole. An $\alpha$-galactosidase distinctly associated with the vacuole has been reported in *Petunia hybrida* petals (Aerts and Schram 1985) but, so far, nowhere else. These hydrolases are all in the soluble pool of the vacuoles and therefore cannot be utilized as markers for tonoplast preparations, at least when this is being accomplished by direct isolation via microsomal fractions.

Until an unequivocally precise tonoplast marker becomes readily available, researchers must rely largely on the specificity of markers for the potentially contaminating cytosol and membranes. Theoretically, a tonoplast-bound ATPase should be an ideal direct marker, provided it has characteristics that distinguish it clearly from other ATPases. Some investigators (Briskin et al. 1985; Poole et al. 1984) use ATPase and capitalize on the fact that tonoplast ATPase can be specifically inhibited by $NO_3$. Unfortunately, the $NO_3$ inhibition does not appear to be as effective as one would like to have it for a marker in all species. A recent report of antibodies specific to subfractions of the tonoplast ATPase (Mandala and Taiz 1985) is promising.

Table 2 summarizes the approaches being used to achieve purification of vacuoles. Most of the presently popular procedures have evolved as improvements or adaptations of methods used earlier, often by the same group. Relying on differential density properties, a gradient, either continuous or discontinuous, is usually used and vacuoles are either sedimented or floated through the gradient. Free flow electrophoresis is only just beginning to be tapped as a means of purifying plant vacuoles. Two-phase (dextran/polyethylene glycol) systems have found some favor but are not widely used. One problem that becomes apparent from the percent contamination ranges listed in Table 2 is the wide discrepancy often found among replicate experiments even in the hands of the same experimenters. This point is demonstrated specifically by Smith et al. (1984). Yields can also vary widely. This can be a problem, particularly in cell-cultured material. Alibert et al. (1982) reported vacuole yields of 60% in cells of *Acer pseudoplatanus* 4 days after transfer and only 25% yields in 15-days-old cultures. When horseradish root was mechanically sliced to 0.2 mm thickness, vacuole recovery yields were greater than when thinner slices were cut (Grob and Matile 1979). Presumably, thinner cuts resulted in a larger number of vacuoles sliced open.

Choice of both isolation and purification procedures profoundly influence the success with which desired determinations can be made on a vacuole preparation. In protoplast-based methods, problems can arise due to contamination introduced by the enzyme preparations used for digestion of the cell wall. Boller and Kende (1979) showed that Cellulysin, commonly used for release of protoplasts, contains large amounts of many hydrolases. These must be effectively removed from the protoplasts to which they adhere (but do not penetrate) by washing the protoplast preparation thoroughly prior to release of vacuoles. Only then can a reliable determination of hydrolytic enzyme distribution in vacuole and cytoplasm be made. A similar assessment of protease distribution may be interfered with even more by contaminants in Cellulysin because these enzymes do penetrate the plasma membrane (Boller and Kende 1979). Proteases can be removed by treatment of the Cellulysin with phenylmethylsulfonyl fluoride prior to its use. ATPase activity of vacuoles can be reduced severely by metrizamide (Doll et al. 1979), although the vacuoles can be effectively rinsed free of the metrizamide. Pugin et al. (1986) have demonstrated that vacuoles of *Acer pseudoplatanus* cells prepared

**Table 2.** General purification procedures for vacuole preparations

| Isolation procedure[a] | Purification procedure | Final yield of vacuoles (%) | Contamination[b] (%) | Objective | Reference |
|---|---|---|---|---|---|
| A | Ficoll step gradient | n.g. | Mit (9); Per (2); Protos (0) | Efflux of malate and influx of nitrate | Marigo et al. (1985) |
| A | Ficoll step gradient | 40 | Protos (4 – 9) | Photosynthetic products in vacuole | Boller and Alibert (1983) |
| B | Urografin step gradient | 20 | n.g. | Localization of glucosinolates | Grob and Matile (1979) |
| A | Ficoll step gradient | 44 | Per (6.3); Mit (2.5); Cyt (6.5) | Isolation of vacuoles | Kringstad et al. (1980) |
| A | Sucrose step gradient | 20 – 25 | Cyt (0); Chl (0); Protos (3) | Chitinase localization | Boller and Vögeli (1984) |
| A | Ficoll step gradient | 25 – 40 | Total contaminants (5) | Isolation, enzyme distribution | Boudet et al. (1981) |
| C | Ficoll step + sucrose linear gradients | 11 | ER (0.1); Mit (0.9) | Isolation of vacuoles | Mettler and Leonard (1979) |
| D | Ficoll step gradient | n.g. | Cyt (19); Mit (11); ER (11); PM (18.5) | Isolation of vacuoles | Thom et al. (1982) |
| C | Ficoll step + sucrose linear gradients | 5 | Protos (3); other contaminants negligible | Isolation of tonoplast | Briskin and Leonard (1980) |
| E | Sucrose cushion | n.g. | All contaminants (<1) | Compartmentation of hydrolases | Nishimura and Beevers (1978), Nishimura (1982) |
| C | Centrifugation | n.g. | Chl (0.4); none other | Metabolite distribution | Wagner (1979) |
| C | Ficoll step gradient | 35 | Mit (5); Cyt (5 – 7); ER (5 – 7) | Isolation of vacuoles | Yamaki (1984) |
| C | Ficoll step gradient | 35 | Chl (1.6); Mit (9); ER (11.4) | Nicotine distribution | Saunders (1979) |
| D | Ficoll step gradient | 26 | Chl (<1); Cyt (5); Mit (3) | Nitrate distribution | Martinoia et al. (1981) |
| D | Percoll step gradient | 10 – 15 | Chl (0.4); Cyt (2 – 5) | Photosynthetic products in vacuole | Kaiser et al. (1982) |

**Table 2** (continued)

| Isolation procedure[a] | Purification procedure | Final yield of vacuoles (%) | Contamination[b] (%) | Objective | Reference |
|---|---|---|---|---|---|
| A | Ficoll step gradient | 25 – 60[3] | Mit (4.6); microbodies (3.8); nuclei (5.6); ribosomes (2.3) | Biochemical composition | Alibert et al. (1982) |
| A | Urografin gradient | 3 – 15 | ER (15); Cyt (6) | Uptake of anthocyanin | Hopp and Seitz (1987) |
| D | Dextran linear gradient | n.g. | Mit (4.5 – 5.5); Golgi (7); ER (16.5) | Proton transport | Mandala and Taiz (1985) |
| A | Single step on Ficoll gradient | 12 | Chl (5); Mit (20); Per (25); ER (0); Cyt (37) | ATPase activity | Smith et al. (1984) |
| C | Ficoll step gradient | n.g. | ER (7.1); Mit (3.9); Cyt (3.3); Chl (2.3) | ATPase activity | Aoki and Nishida (1984) |
| D | Percoll step gradient | n.g. | Overall <2% | Transport of anions | Martinoia et al. (1986) |
| E | Glycine-betaine gradient | n.g. | Mit (2); microbodies (12); Cyt (16) | Fructan localization | Frehner et al. (1984) |
| C | Sucrose step gradient | 9.4 | Chl and Mit (traces); Cyt (0.75) | Oxalic acid deposition | Wagner (1981) |
| H | Ficoll step gradient | 20 | Cyt, Mit, Per (all below level of detection) | Isolation of vacuoles | Le-Quoc et al. (1987) |
| D | Ficoll step gradient | n.g. | Mit (5 – 15); Chl (5 – 9); Cyt (14 – 27); PM (none detected) | Ethylene formation | Guy and Kende (1984) |
| C | Ficoll cushion | 49 | Chl (0 – 2); Cyt (inconclusive); PM (slight); Protos (2) | Nitrate storage | Graustedt and Huffaker (1982) |
| D | Ficoll step gradient | n.g. | Chl (<1); Mit (<1); Cyt (5.4 – 12.3); Per (24) | Ethylene formation | Mayne and Kende (1986) |
| C | Low speed wash | Approx. 50 | Mit (0.9); ER (1.1); Cyt (4.4) | Isolation of vacuoles | Keller (1988) |

[a] A, polybase lysis; B, mechanical slicing; C, osmotic shock; D, shearing; E, EDTA or EGTA treatment; F, preplasmolysis followed by differential enzyme digestion; G, direct isolation; H, digitonin lysis of plasma membrane.

[b] Abbreviations for contaminants: Mit – mitochondria; Chl – chloroplasts; Per – peroxisomes; ER – endoplasmic reticulum; Protos – protoplasts; PM – plasma membrane.

[c] Depending on age of cells in culture.

by shearing in Ficoll retained ATPase activity that was much greater than in vacuoles isolated by the DEAE dextran method. Isolation and/or purification of vacuoles under high salt conditions can change the pH and membrane potential of the vacuoles.

## 3.2 Gradient Separations

For consistency and ease in following details of procedures that we describe as examples of a particular approach, we will designate the lowermost phase as A and move upward through B, C, etc. We will spell out details such as buffer, centrifugation, and osmotic conditons for the methods we will use as examples. However, it should be realized that these conditions may have been developed by a research group for a particular tissue and, while the general approach may be the same in another case, the details may differ markedly.

### 3.2.1 Sucrose

*1. Step Gradient.* Using vacuoles isolated directly from preplasmolyzed, enzymatically digested leaf tissue of *Nicotiana tabacum*, Schloss et al. (1987) loaded crude vacuole preparations over a step gradient of sucrose (w/v) in isoosmotic concentrations of KCl (0.35 M), $MgCl_2$ (0.245 M), $CaCl_2$ (0.245 M), pH 6.0 as follows: (A) 20%, (B) 10%, (C) 5%, (D) 2.5%. The gradient was centrifuged at 80000 g for 2 h and purified vacuoles were recovered from the A/B interface. A very similar method was used to clean up vacuoles from *Nicotiana tabacum* cell suspensions. In that case the purified vacuoles were recovered from the C/D interface. Markers for chloroplasts, mitochondria, and cytoplasm gave negative reactions, but between 10% and 25% contamination by endoplasmic reticulum was found.

*2. Floatation from Single Layer.* Wagner (1981) has reported a method whereby vacuoles from osmotically lyzed *Hordeum* protoplasts in a suspension were made 17% with respect to sucrose by addition of a 60% sucrose solution. The resulting suspension was overlayered with 8% (w/v) sucrose, 20 mM Hepes-NaOH, pH 8.0. After centrifugation at 5000 g for 10 min at 20 °C, vacuoles floated to the top of the 8% sucrose layer. Wagner (pers. commun.) has recently utilized a similar procedure for *Datura* vacuoles, in this case overlayering with 2 ml of 8% (w/v) sucrose in the same buffer but using, in addition, 0.76 M mannitol in the solution. This preparation was centrifuged at 2000 g for 10 min, using a slow acceleration rate. Vacuoles floated to the top and contained less than 5% lysate contamination.

### 3.2.2 Ficoll

Ficoll is a synthetic, highly branched, spherical, highly water-soluble polymer (mw 400000) of sucrose and epichlorohydrin, and has low intrinsic viscosity at high concentrations. Gradients of Ficoll are widely employed as a means of removing contaminant from vacuole preparations. In a method described by Yamaki (1984) for vacuoles released by osmotic lysis from apple pulp, a Ficoll gradient was layered over (A) 2 ml of lysate as follows: (B) 3 ml of 10% and (C) 4 ml of 5% Ficoll, each in 0.7 M mannitol and 2 mM Tris-HCl. (D) consisted of 1 ml of 1.4% Ficoll only. The D layer was discarded and replaced repeatedly after centrifugation at 100 g at 0 °C. Final centrifugation at 2000 g for 20 min brought the purified vacuoles to the C/D interface.

While vacuoles in the above example were floated up, in many other procedures they are spun down through one or more layers of Ficoll. In a procedure reported by Aoki and Nishida (1984), vacuoles prepared by osmotic rupturing from protoplasts of *Kalanchoë* were purified on a Ficoll gradient consisting of (A) 15%, (B) 10%, (C) 7%, (D) 2.5%, each in 0.2 M mannitol. When this preparation was subjected to centrifugation at 6000 g for 35 min, purified vacuoles could be recovered at the C/D interface. Similar methods are being used by others, but it should be stressed again that exact conditions must be adjusted, depending on the origin of the vacuoles and their intrinsic properties. Thus, for instance, for vacuoles also obtained by osmotic lysis, Weigel and Weiss (1984) used a gradient of (A) 10%, (B) 5% and (C) 0.5% Ficoll, each in 0.5 M mannitol. The vacuoles loaded on top of this gradient banded at the A/B interface when spun at a much higher speed than in the prior example, i.e., at 28000 g for 30 min.

In recent years, a Ficoll gradient has been most commonly combined into a one-step procedure for vacuoles released protoplasts. Some of these methods will be discussed separately.

### 3.2.3 Percoll

This substance is polyvinylpyrrolidone (PVP) coated with colloidal silica. Kaiser and Heber (1984) and Martinoia et al. (1985) have switched to the use of Percoll instead of Ficoll for their unique method of vacuole isolation from barley mesophyll protoplasts after lysis of plasma membrane of barley mesophyll protoplasts by forcing the protoplast through a fine, long, hypodermic needle. (A) was the protoplast lysate in 0.45 M sorbitol and 18% (v/v) Percoll dissolved in a 30 mM Hepes-imidazole buffer also containing 0.1% (w/v) BSA, 0.1% (w/v) PVP (10 kd), pH 7.4, (B) was 2% Percoll in the above solution, while (C) was the solution without Percoll and with sorbitol replaced by 0.45 M glycine-betaine. The centrifugation conditions were 100 g for 2 min, followed by 1100 g for 3 min. Vacuoles floated to the B/C interface. These vacuoles were further purified by resuspension in (A) 8% Percoll which was overlayered by B and C, consitituted as above. The second round of centrifugation was at 650 g for 4 min and vacuoles were recovered from C. In corresponding experiments (Kaiser and Heber 1984) lysate was prepared in 12.5% instead of 18% Percoll and some salt additions were made in the protoplast lysate, i.e., $MgCl_2$ and $MnCl_2$ each at 1 mM; NaCl, 10 mM; and EDTA, 2 mM. Although no marker enzyme analysis was reported, these vacuoles were successfully used to demonstrate electrogenically driven uptake of malate (Martinoia et al. 1985) and non-energy-dependent sucrose uptake (Kaiser and Heber 1984).

The work of the above group provides an excellent example, both of evolution of methodologies for isolation and purification and for modifications of methods adapted within a single tissue (barley mesophyll) to accomplish a number of different objectives with the vacuoles obtained. Thus, for compartmental analysis of vacuoles in conjunction with ion uptake and efflux measurements, Martinoia et al. (1986) modified the above Percoll approaches further by using a Percoll gradient [(A) 30% Percoll (v/v) and 0.5 M sorbitol in an Mes (5 mM) buffer also containing 1 mM $CaCl_2$, pH 6.0; (B) 20%, and (C) 5% Percoll in the above solution] to select a protoplast population of more homogeneous specific density. The released vacuoles could, in turn, be subjected to purification by spinning them into water (for instant breakage) above a silicone ill layer (AR 200) in a microfuge (9600 g). The process required only 15 s and hence prevented metabolic changes in the internal pool. Contamination, measured by marker enzymes, of the intact vacuoles isolated by these means was reported to be much less than 2%.

### 3.2.4 Nycodenz

A neutral triiodinated derivative of benzoic acid, Nycodenz is another useful substance for vacuole purification because of its low viscosity, which permits rapid migration of vacuoles during centrifugation. To the best of our knowledge, its use has been restricted to one group so far (Gibrat et al. 1985). Vacuoles released from *Catharanthus roseus* protoplasts by Na-EDTA induced lysis were purified by floatation on a two-step gradient. (A) consisted of the protoplast lysate, 0.31 M sorbitol, and 7.7% Nycodenz in a Hepes buffer, 10 mM, pH 7.1. (B) contained 0.42 M sorbitol and 2% Nycodenz in the Hepes buffer. Vacuoles floated to the top of B, either when centrifuged at 1 g for 30 min or at 160 g for 3 min, 4 °C. Purity was assessed by marker enzymes and found to be greater than 95%.

### 3.2.5 Urografin

Urografin (N-methylglucamine) has been found to be superior to sucrose or Ficoll by some investigators (Grob and Matile 1979; Hopp and Seitz 1987). Grob and Matile (1979), working with sliced horseradish root from which vacuoles were directly released, layered 0.7 M NaCl and 25% Urografin (w/v) under the vacuole extract and centrifuged for 7 min at about 500 g. This procedure sedimented the vacuoles to the interface and contaminating starch to the bottom of the tube. The vacuoles were removed and diluted with 25% Urografin. Upon a second centrifugation under similar conditions, the vacuoles floated into a 0.7 M NaCl layer at the top. Although there was a substantial loss of vacuoles during this purification procedure, a number sufficient for the desired compartmental analysis were recovered.

### 3.2.6 Metrizamide

For the purification of mechanically isolated beet vacuoles (Boller and Kende 1979), a gradient of Metrizamide [2-(3-acetamido-5-N-methylacetamido-2,4,6-tri-iodobenzamido)deoxy-D-glucose] has been used. In (A) the vacuoles, suspended in the appropriate buffer, are mixed in 15% Metrizamide, (B) consists of 10% Metrizamide in the same buffer. (C) is 2.5% Metrizamide in buffer and (D) buffer only. The majority of the vacuoles were located at the B/C interface. In a later modification (Leigh 1981) the 2.5% layer was eliminated.

## 3.3 Isolation-Purification via a Single Step

Combination techniques suitable for release and purification of vacuoles have become more popular in recent years, although only one has basically been available for some time (Boudet et al. 1981). Polybase lysis of protoplasts seems to lend itself particularly well for this purpose, since DEAE dextran and dextran sulfate can be incorporated into adjacent layers of a gradient through which the protoplasts are sedimented and vacuoles recovered at an interface below the DEAE dextran and dextran sulfate layers (Boudet et al. 1981). Recently, a modification of this basic procedure, which could become an important advance for systems other than *Chenopodium rubrum*, for which it was used, was introduced by Bentrup et al. (1986). It is based on a six-layer gradient, with the intact protoplasts introduced at the bottom and floated up into the top layer. The gradient is as follows: (A) 20% Ficoll and 0.3 M mannitol in a 20 mM Mes-KOH buffer, pH 6.3, containing also 25 mM KCl, 1 mM DTT, 0.5 mM $MgCl_2$ and 0.1% BSA; (B) 17% Ficoll and 0.3 M mannitol in the same buffer as A but at pH 6.0 and containing all the additions of A but replacing BSA with 1 mM NaEDTA; (C) 15% Ficoll, 0.1 M mannitol and 0.8% (w/v) DEAE-dextran (500 kDa) in 20 mM Hepes-KOH, pH 6.0, and containing all the other additions as in B; (D) 5% Ficoll, 0.3 M mannitol and 0.4% dextran sulfate 5000 in 20 mM Hepes-KOH, pH 7.0, containing also 1 mM DTT and 0.5 mM $MgCl_2$; (E) 2.5% Ficoll and 0.3 M mannitol in 50 mM Hepes-KOH, pH 7, and containing also 1 mM DTT, 0.5 mM $MgCl_2$ and 25 mM KCl; (F) 0.3 M mannitol in buffer and with all the constituents of E but without Ficoll. The gradient was centrifuged at 22500 g for 60 min and vacuoles were recovered from layer F. No marker enzymes were determined, but it must be assumed that the vacuoles, which have been used successfully for patch clamp experiments (Bentrup et al. 1985) are relatively clean.

## 3.4 Two-Phase Procedure

An aqueous two-phase partition procedure which separates vacuoles according to their surface properties, rather than their size and density, has been investigated recently by Murphy et al. (1986). Rather surprisingly, vacuoles partitioned predominantly into the lower, PEG layer, rather than into the upper phase consisting of dextran. Most promising was a system containing 6.3% of each dextran and

PEG in an 8 mM $SO_4$-Tris buffer, pH 7.8, containing also 0.08% BSA and 0.4 M sorbitol. The vacuole suspension was added after vigorous mixing of the two phases and the phases were allowed to separate by standing for 30 min or centrifugation at 850 g for 5 min. Several factors contributed to optimization of vacuole recoveries of about 75% in the PEG phase. Experiments carried out at temperatures of 18–20 °C were better than when separations were conducted at 4 °C and a pH of 7.5 was better than pH 5.8. Unfortunately, as the authors point out, the method suffers from a couple of serious problems at this stage of development: plasma membrane is partitioned into the PEG phase as well and it is well known that in most vacuole isolation procedures, excepting possibly a mechanical release, pieces of plasma membrane tend to associate with the vacuoles. These workers also found that upon dilution of the PEG phase, a step necessary to recover the vacuoles, they tend to lyze extensively. Furthermore, since separation of the phases is slow, alteration in vacuolar content may take place.

## 3.5 Tonoplast Isolation and Purification

### 3.5.1 Sucrose Step Gradient

A sucrose step gradient has been used by Rausch et al. (1987) for the purification of tonoplast from a microsomal fraction. In this case the starting preparation was loaded over (A) 45%, (B) 36%, (C) 31%, (D) 26%, and (E) 18% (w/v) sucrose in the extraction buffer and a tonoplast-enriched fraction was recovered from the D/E interface after centrifugation at 100000 g for 120 min. Enrichment of the tonoplast in this fraction was verified by enzyme marker analysis but the data were not shown.

### 3.5.2 Linear Sucrose Gradient

Yoshida et al. (1983) have described a linear gradient of sucrose, 50%–15% (w/w) in a total volume of 34 ml of 10 mM Tris-maleate containing 1 mM EDTA, pH 7.3, which cleans up plasma membrane, but also purifies tonoplast vesicles at a different interface. The vesicle preparation in 4 ml of 10% sucrose, 10 mM Tris-maleate and 3 mM EDTA, pH 7.3, was layered over the sucrose gradient and centrifuged at 96000 g for 13 h at 2 °C (Beckman SW27) and the tonoplasts were collected in a 1.2-ml aliquot from the top. Similar methods have been used by Sze (1980), Briskin et al. (1985), Du Pont et al. (1982), and Poole et al. (1984).

### 3.5.3 Free-Flow Electrophoresis

This method has recently been explored by Sandelius et al. (1986) for the separation of tonoplast and other plant membranes. It has so far been primarily employed for membrane separations in animal tissue. While not all membrane types separate as cleanly as one might like, tonoplast and plasma membrane separate as well-defined fractions from Golgi dictyosomes, endoplasmic reticulum, mitochondria, plastids, and nuclear envelope on either side of the main peak containing these other cytoplasmic membranes. The identification of tonoplast was on the basis of membrane thickness, tonoplast having a thickness of between 7 and 9 nm. Now that antibodies to subunits of the tonoplast ATPase are available (Mandala and Taiz 1986), the identification of a tonoplast fraction, on the basis other than thickness or differential $NO_3$ sensitivity (Briskin et al. 1985; Poole et al. 1984), will presumably be used. The chief drawback to the appealing prospect of clean tonoplast separations from microsomal fractions by free-flow electrophoresis is high cost of the necessary equipment.

## 4 Isolation and Purification Scheme

A procedure that lends itself to recovery of a vacuole preparation is illustrated in Fig. 1. It shows the mechanical isolation and clean-up of vacuoles from red beet roots. Modifications of this method have been used by Leigh and Branton (1976), Willenbrink and Doll (1979), and Thom et al. (1986).

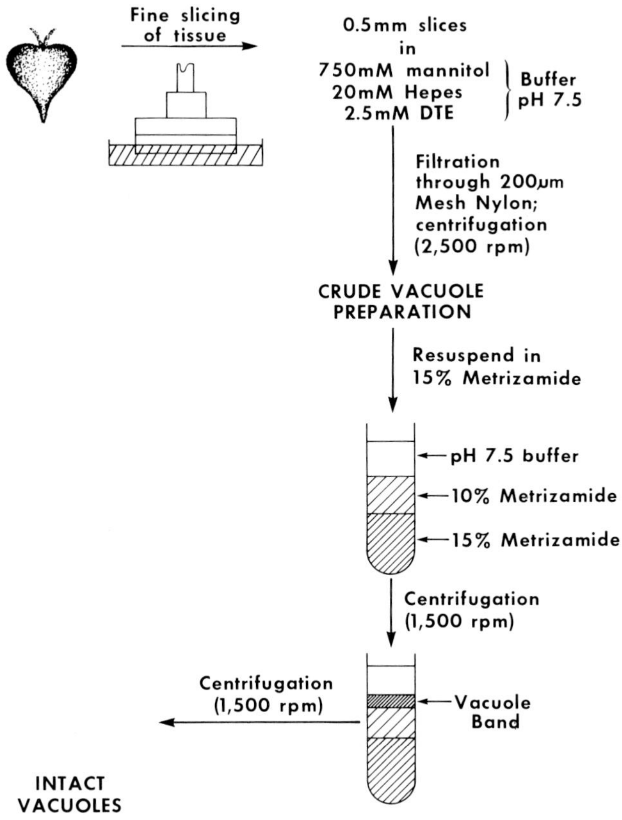


**Fig. 1.** Isolation and purification of red beet root vacuoles

# 5 Uses of Vacuoles

A glance through Tables 1 and 2 shows the range of objectives that have been sought, and for the most part, achieved, through the use of isolated vacuoles.

## 5.1 Compartmentation

Intracellular compartmentation analyses were incomplete until it became possible to isolate this organelle, which can occupy a major part of a plant cell's total volume. Ability to determine the distribution of solutes between cytoplasm and vacuole has greatly facilitated our understanding of intermediary metabolism. For instance, although it has long been realized that nitrate must accumulate in some cellular compartment, its localization within the vacuole is a finding that had to await the isolation of this organelle (Martinoia et al. 1981). The localization of basic peroxidases in the vacuole, rather than in the cell wall, has only recently been confirmed (Schloss et al. 1987). Similarly, the intracellular location of enzymes catalyzing steps in pathways of products which accumulate in vacuoles was previously uncertain. Examples are enzymes in fructan (Frehner et al. 1984) and in ethylene formation (Boller and Kende 1979). To determine reliably the percent distribution of intracellular solutes, ratios of total cellular protein to $\alpha$-mannosidase when present in the vacuole, and an appropriate cytoplasmically localized enzyme can be calculated; from these ratios the distribution of a given solute can be assessed (Boller and Alibert 1983).

## 5.2 Transport

Electrochemical properties of vacuoles provide the key to many of its dynamic functions in cellular distribution of charged and neutral metabolites. Electrogenic properties are a characteristic of both the tonoplast and the soluble vacuolar pool. Consequently, both tonoplast vesicles and intact vacuoles are used to investigate such properties. Measurements of membrane potential, pH values, and ATPase-proton-mediated solute transfers each provide facets that can be approached directly through the availability of isolated vacuoles. Recent progress in patch clamping techniques with vacuoles of higher plants (Hedrich et al. 1986) is beginning to give us more specific insights into membrane channeling of ions. Investigating the electrochemical properties of vacuoles used to be the almost exclusive domain of researchers working with giant algal cell (e.g., Findley and Hope 1976) because microelectrodes could be inserted into vacuoles or cytoplasm of *Chara* or *Nitella* species through the intact cell. Direct access to vacuolar properties via intact cells, isolated vacuoles, tonoplast vesicles, or patch clamping of vacuoles has provided a much broader perspective, presently ranging over at least 25 higher plant species, to evaluate the role of the vacuole in accumulation and distribution of metabolites. A new appreciation has evolved for vacuole-associated regulatory functions in metabolite adjustments of cells undergoing aging changes, stress conditions, or diurnal fluctuations.

### 5.3 Biotechnological Applications

The stage is, therefore, being set for biotechnological applications in the broadest sense. This could, for instance, include improved recovery of a commercially desirable vacuolar hydrolytic enzyme by a continuous or immobilized cell culture process, the optimization of a pharmacologically important secondary product enzymatically produced in the vacuole, or the better recovery of such products from the vacuole. The tonoplast is the site of many transport-related proteins and, increasingly, there is interest in identification of genetic sites coding for these proteins (Mandala and Taiz 1986). Perhaps, one day, it may be desirable to clone transport-related genes for ATPase (Mandala and Taiz 1986), PPases (Wang et al. 1986), or a specific carrier (e.g., Deus-Neumann and Zenk 1984) to increase storage of sought-after secondary products in the vacuole. The UDPglucose-dependent group translocator for vacuolar accumulation of sucrose (Thom and Maretzki 1985) is a potential future candidate for cloning to increase sucrose deposition. Fructans, which are partially synthesized and then accumulated in the vacuole, may play a role in cold protection of plants (Chatterton et al. 1986) and could come under consideration for gene modifications through cloning techniques. One of the functions of ethylene, which is synthesized in the vacuole (Boller and Kende 1979) is to increase chitinase induction and chitinase can protect against invading pathogenic fungi that contain chitin (Broglie et al. 1986). Seen only from this point of view, positive regulation of ethylene formation might become desirable.

## 6 Conclusions

Undoubtedly, as we learn more about involvement of the vacuole in maintenance or promotion of physiological balances, speculations such as we have indulged in the above section will incease. In the meantime, a primary focus should continue on how to obtain the best possible material with which progress in our basic knowledge about vacuole composition and functions can be made. How to improve and extend the options for isolation and purification of vacuoles from a greater number of plant species remains a challenge.

As described, there are many highways and byways that lead to functionally intact vacuoles. Because it is very difficult to make value judgements on the methodologies that are being utilized by present investigators, these have been presented in a rather indiscriminate manner. To date there have been relatively few comparative studies across several species and tissues of some of the more popular approaches used, both for isolation of vacuoles, as well as for their purification. However, a few such comparisons have been made. Wagner and Siegelman (1975) compared their osmotic lysis method in several species and in leaves, petals, pedicels, stems, and styles within a species. Lörz et al. (1976) reported the Ficoll shearing method to release vacuoles from protoplasts of a number of both dicotyledenous and monocotyledenous plant species and several types of tissues. Boller and Kende (1979) made a direct comparison of vacuole preparations from tobacco protoplasts via polybase and osmotic lysis, the latter giving the better yield. On the other hand, in *Bryophillum* (Buser and Matile 1977), the opposite was found to be the case. More recently, Matern et al. (1986) demonstrated that *Petroselinum* vacuoles isolated by the Boudet et al. (1981) modified DEAE dextran procedure were recovered in much higher yields than via osmotic lysis. Similarly, Aerts and Schram (1985) reported that *Petunia* petal vacuoles isolated via osmotic shock were leaky, whereas those isolated by the DEAE dextran method were not. Unfortunately, the emphasis on positive results in most of the reported literature deemphasizes description and discussion of negative data from approaches that undoubtedly have been tried by most investigators before embarking on a method that gave satisfactory results. After all, in most studies carried out more recently, the isolation and purification of vacuoles merely represent a means to an end. From our experience with sugarcane we know for a fact that methods for isolation by osmotic lysis, or polybase lysis do not work nearly as well as shearing through Ficoll, at least not without seeking extensive modifications of reported procedures. Similarly, many of the purification procedures that are so successfully used for other plants and tissues do not lend themselves directly to sugarcane. Even in the isolation of vacuoles from sugarcane cell cultures (Thom et al. 1982) and vacuole isolation from the fresh stalk tissue (Maretzki and Thom 1987), considerable modifications of methodologies were necessary. While we do not wish to imply that there are no universally applicable approaches across tissues and plant species, every investigator confronted with a need to isolate and purify vacuoles and/or tonoplast in a plant so far not subjected to any of the known procedures may plan to explore several of the presently available methods.

## References


Abel S, Glund K (1986) Localization of RNA-degrading enzyme activity within vacuoles of cultured tomato *Lycopersicon esculentum* cultivar lucullus cells. Physiol Plant 66:79–86

Admon A, Jacoby B (1980) Assessment of cytoplasmic contaminations in isolated vacuole preparations. Plant Physiol 65:85–87

Aerts JMFG, Schram AW (1985) Isolation of vacuoles from the upper epidermis of *Petunia*-hybrid petals. 1. A comparison of isolation procedures. Z Naturforsch 40c:189–195

Alibert G, Boudet AM (1982) Progrès, problèmes et perspectives dans l'obtention et l'utilization de vacuoles isolée. Physiol Veg 20:289–302

Alibert G, Carrasco A, Boudet AM (1982) Changes in biochemical composition of vacuoles isolated from *Acer pseudoplatanus* L. during cell culture. Biochim Biophys Acta 721:22–29

Aoki K, Nishida K (1984) ATPase activity associated with vacuoles and tonoplast vesicles isolated from the CAM plant, *Kalanchoë daigremontiana*. Physiol Plant 60:21–25

Barbier H, Guern J (1982) Transmembrane potential of isolated vacuoles and sucrose accumulation in *Beta vulgaris* roots. In: Marmé D, Marré E, Hertel R (eds) Plasmalemma and tonoplast: their function in the plant cell. Elsevier Biomedical Press, Amsterdam, pp 233–239

Bentrup FW, Hoffmann B, Gogarten-Boekels M, Gogarten JP (1985) A patch clamp study of tonoplast electrical properties in vacuoles isolated from *Chenopodium rubrum* suspension cells. Z Naturforsch 40c:886–890

Bentrup FW, Gogarten-Boekels M, Hoffmann B, Gogarten JP, Baumann C (1986) ATP-dependent acidification and tonoplast hyperpolarization in isolated vacuoles from green suspension cells of *Chenopodium rubrum* L. Proc Natl Acad Sci USA 83:2431–2433

Boller T, Alibert G (1983) Photosynthesis in protoplasts from *Melilotus alba*: Distribution of products between vacuole and cytosol. Z Pflanzenphysiol 110:231–238

Boller T, Kende H (1979) Hydrolytic enzymes in the central vacuole of plant cells. Plant Physiol 63:1123–1132

Boller T, Vögeli U (1984) Vacuolar localization of ethylene-induced chitinase in bean leaves. Plant Physiol 74:442–444

Boller T, Wiemken A (1986) Dynamics of vacuolar compartmentation. Annu Rev Plant Physiol 37:137–164

Boudet AM, Canut H, Alibert G (1981) Isolation and characterization of vacuoles from *Melilotus alba* mesophyll. Plant Physiol 68:1354–1358

Briskin DP, Leonard RT (1980) Isolation of tonoplast vesicles from tobacco protoplasts. Plant Physiol 66:684–687

Briskin DP, Thornley WR, Wyse R (1985) Membrane transport in isolated vesicles from sugar beet taproot. I. Isolation and characterization of energy dependent, $H^+$-transporting vesicles. Plant Physiol 78:865–870

Broglie KE, Gaynor JJ, Broglie RM (1986) Ethylene regulated gene expression: molecular cloning of the genes encoding an endochitisase from *Phaseolus vulgaris*. Proc Natl Acad Sci USA 83:6820

Buser C, Matile P (1977) Malic acid in vacuoles isolated from *Bryophyllum* leaf cells. Z Pflanzenphysiol 82:462–466

Chatterton NJ, Harrison PA, Bennett JH (1986) Environmental effects on sucrose and fructan concentrations in leaves of *Agropyron* spp. In: Cronshaw J, Lucas WL, Giaquinta RT (eds) Phloem Transport. Plant Biology Vol I, Allan Liss, NY, pp 471–476

Chedhomme F, Rona JP (1986) Isolation and electrical characterization of tonoplast vesicles from the Kiwi fruit (*Actinidia chinensis*). Physiol Plant 67:29–36

Cocking EC (1960) A method for the isolation of plant protoplasts and vacuoles. Nature 187(4741):962–963

D'Auzac J (1975) Caracterisation d'une ATPase membranaire en presence d'une phosphatase acide dans les lutoides du latex d'*Hevea brasiliensis*. Phytochemistry 14:671–675

Deus-Neumann B, Zenk MH (1984) A highly selective alkaloid uptake system in vacuoles of higher plants. Planta 162:250–260

Doll S, Rodier F, Willenbrink J (1979) Accumulation of sucrose in vacuoles isolated from red beet tissue. Planta 144:407–411

Dracup MNH, Greenway H (1985) A procedure for isolating vacuoles from leaves of the halophyte *Suaeda maritima*. Plant Cell Environ 8:149–154

DuPont FM, Bennett AB, Spanswick RM (1982) Localization of a proton-translocating ATPase on sucrose gradients. Plant Physiol 70:1115–1119

Findley GP, Hope AB (1976) Electrical properties of plant cells: methods and findings. In: Lüttge U, Pitman MG (eds) Transport in Plants II. Encyclopedia of Plant Physiology, New Series 2 (A). Springer, Berlin Heidelberg New York, pp 53–92

Frehner M, Keller F, Wiemken A (1984) Localization of fructan metabolism in the vacuoles isolated from protoplasts of Jerusalem Artichoke tubers (*Helianthus tuberosis* L.). J Plant Physiol 116:197–208

Gibrat R, Barbier-Brygoo H, Guern J, Grignon C (1985) Membrane potential difference of isolated plant vacuoles: Positive or negative I. Evidence for membrane binding of cationic probes. Biochim Biophys Acta 819:206–214

Graustedt RC, Huffaker RC (1982) Identification of the leaf vacuole as a major nitrate storage pool. Plant Physiol 70:410–413

Grob K, Matile P (1979) Vacuolar location of glucosinolates in horseradish root cells. Plant Sci Lett 14:327–335

Guy M, Kende H (1984) Conversion of 1-amino cyclo propane-1-carboxylic acid to ethylene by isolated vacuoles of *Pisum sativum*. Planta 160:281–287

Guy M, Reinhold L, Michaeli D (1979) Direct evidence for a sugar transport mechanism in isolated vacuoles. Plant Physiol 64:61–64

Hedrich R, Flügge UI, Fernandez JM (1986) Patchclamp studies of ion transport in isolated plant vacuoles. FEBS Lett 204:228–232

Hopp W, Seitz HU (1987) The uptake of acylated anthocyanin into isolated vacuoles from a cell suspension culture of *Daucus carota*. Planta 170:74–85

Hopp W, Hinderer W, Petersen M, Seitz HU (1985) Anthocyanin-containing vacuoles isolated from protoplasts of *Daucus carota* cell cultures. In: Pilet PE (ed) The Physiological Properties of Plant Protoplasts. Springer, Berlin Heidelberg New York, pp 122–132

Kaiser G, Heber U (1984) Sucrose transport into vacuoles isolated from barley mesophyll protoplasts. Planta 161:562–568

Kaiser G, Martinoia E, Wiemken A (1982) Rapid appearance of photosynthetic products in the vacuoles isolated from barley mesophyll protoplasts by a new, fast method. Z Pflanzenphysiol 107:103–113

Keller F (1988) A large-scale isolation of vacuoles from protoplasts of mature carrot taproots. J Plant Physiol 132:199–203

Kenyon WH, Black CC (1986) Electrophoretic analysis of protoplast, vacuole, and tonoplast vesicle proteins in crassulacean acid metabolism plants. Plant Physiol 82:916–924

Kringstad R, Kenyon WH, Black CC Jr (1980) The rapid isolation of vacuoles from leaves of crassulacean acid metabolism plants. Plant Physiol 66:379–382

Kutchan TM, Rush M, Coscia CJ (1986) Subcellular localization of alkaloids and dopamine in different vacuolar compartment of *Papaver bracteatum*. Plant Physiol 81:161–166

Leigh RA (1981) Methods, progress and potential for the use of isolated vacuoles in studies of solute transport in higher plant cells. Physiol Plant 57:390–396

Leigh RA, Branton D (1976) Isolation of vacuoles from root storage tissue of *Beta vulgaris* L. Plant Physiol 58:656–662

Le-Quoc K, Le-Quoc D, Pugin A (1987) An efficient method for vacuole isolation using digitonin for plasmalemma lysis. J Plant Physiol 126:329–335

Lörz J, Harms CT, Potrykus I (1976) Isolation of "vacuoplasts" from protoplasts of higher plants. Biochem Physiol Pflanz 169:617–620

Mandala S, Taiz L (1985) Proton transport in isolated vacuoles from corn coleoptile. Plant Physiol 78:104–109

Mandala S, Taiz L (1986) Characterization of the subunit structure of the maize tonoplast ATPase. J Biol Chem 261:12850–12855

Maretzki A, Thom M (1987) UDP-glucose-dependent sucrose translocation in tonoplast vesicles from stalk tissue of sugarcane. Plant Physiol 83:235–237

Marigo G, Bouyssou H, Belkoura M (1985) Vacuolar efflux of malate and its influence on nitrate accumulation in *Catharanthus roseus* cells. Plant Sci 39:97–103

Martinoia E, Heck U, Wiemken A (1981) Vacuoles as a storage compartment for nitrate in barley leaves. Nature 289:292–294

Martinoia E, Flügge UI, Kaiser G, Heber U, Heldt HW (1985) Energy-dependent uptake of malate into vacuoles isolated from barley mesophyll protoplasts. Biochem Biophys Acta 806:311–319

Martinoia E, Schramm MJ, Kaiser G, Kaiser WM (1986) Transport of anions in isolated barley vacuoles. I. Permeability to anions and evidence for a $Cl^-$uptake system. Plant Physiol 80:895–901

Matern V, Reichenbach C, Heller W (1986) Efficient uptake of flavonoids into parsley *Petroselinum hortense* vacuoles requires acylated glycosides. Planta 167:183–189

Matile P (1979) Biochemistry and function of vacuoles. Annu Rev Plant Physiol 29:193–213

Mayne RG, Kende H (1986) Ethylene biosynthesis in isolated vacuoles of *Vicia faba*. Requirement for membrane integrity. Planta 167:159–165

Mettler IJ, Leonard RT (1979) Isolation and partial characterization of vacuoles from tobacco protoplasts. Plant Physiol 64:1114–1120

Miller AJ, Brimelow JJ, John P (1984) Membrane-potential changes in vacuoles isolated from storage roots of red beet (*Beta vulgaris* L.). Planta 160:59–65

Murphy TM, Widell S, Hellergren J (1986) On the use of phase partition to isolate vacuoles from cultured cells. Plant Cell Tissue Organ Cult 7:205–216

Nishimura M (1982) pH in vacuoles isolated from castor bean endosperm. Plant Physiol 70:742–744

Nishimura M, Beevers H (1978) Hydrolases in vacuoles from castor bean endosperm. Plant Physiol 62:44–48

Ohlrogge JB, Garcia-Martinez JL, Adams D, Rappaport L (1980) Uptake and subcellular compartmentation of gibberelin $A_1$ applied to leaves of barley and cowpea. Plant Physiol 66:422–427

Poole RJ, Briskin DP, Kratky Z, Johnstone RM (1984) Density gradient localization of plasma membrane and tonoplast from storage tissue of growing and dormant red beet. Characterization of proton-transport and ATPase in tonoplast vesicles. Plant Physiol 74:549–556

Pugin A, Montrichard F, Le-Quoc K, Le-Quoc D (1986) Influence of the method for the preparation of vacuoles on the vacuolar ATPase activity of isolated *Acer pseudoplatanus* cells. Plant Sci 47:165–177

Rausch I, Butcher DN, Taiz L (1987) Active glucose transport and proton pumping in tonoplast membrane of *Zea mays* L. coleoptiles are inhibited by anti $H^+$-ATPase antibodies. Plant Physiol 85:996–999

Renaudin JP, Brown SC, Barbier-Brygoo H, Guern J (1987) Quantitative characterization of protoplasts and vacuoles from suspension-cultured cells of *Catharanthus roseus*. Physiol Plant 68:695–703

Salyaev RK, Kuzevanov VY, Ozolina NV (1983) Content of lipids, proteins, and carbohydrates in the membrane of isolated vacuoles of red beet. Sov Plant Physiol 29:718–724 (English translation)

Sandelius AS, Penel C, Auderset G, Brightman A, Millard M, Morré DJ (1986) Isolation of highly purified fractions of plasma membrane and tonoplast from the same homogenate of soybean hypocotyls by free flow electrophoresis. Plant Physiol 81:177–185

Saunders JA (1979) Investigation of vacuoles isolated from tobacco. I. Quantitation of nicotine. Plant Physiol 64:74–78

Schloss P, Walter C, Maeder M (1987) Basic peroxidases in isolated vacuoles of *Nicotiana tabacum* L. Planta 170:225–229

Smith JAC, Uribe EG, Ball E, Lüttge U (1984) ATPase activity associated with isolated vacuoles of the crassulacean acid metabolism plant *Kalanchoë daigremontiana*. Planta 162:299–304

Strack D, Sharma V (1985) Vacuolar localization of the enzymatic synthesis of hydroxycinnamic acid esters of malic acid in protoplasts from *Raphanus sativus* var sativus leaves. Physiol Plant 65:45–50

Sze H (1980) Nigericin-stimulated ATPase activity in microsomal vesicles of tobacco callus. Proc Natl Acad Sci USA 78:5578–5582

Thom M, Maretzki A (1985) Group translocation as a mechanism for sucrose transfer into vacuoles from sugarcane cells. Proc Natl Acad Sci USA 82:4697–4701

Thom M, Maretzki A, Komor E (1982) Vacuoles from sugarcane suspension cultures. I. Isolation and partial characterization. Plant Physiol 69:1315–1319

Thom M, Leigh RA, Maretzki A (1986) Evidence for the involvement of a UDP-glucose-dependent group translocator in sucrose uptake into vacuoles of storage roots of red beet. Planta 167:410–413

Wagner GJ (1979) Content and vacuole/extravacuolar distribution of neutral sugars, free amino acids, and anthocyanin in protoplasts. Plant Physiol 64:88–93

Wagner GJ (1981) Vacuolar deposition of ascorbate-derived oxalic acid in barley. Plant Physiol 67:591–593

Wagner GJ (1983) Higher plant vacuole and tonoplasts. In: Hall JL, Moore AL (eds) Isolation of Membranes and Organelles from Plant Cells. Academic Press, New York, pp 83–118

Wagner GJ (1988) Isolation of higher plant mature vacuoles: General principles, criteria for purity and integrity. In: Methods in Enzymology, Academic Press, New York (in press)

Wagner GJ, Lin W (1982) An active proton pump of intact vacuoles isolated from *Tulipa* petals. Biochim Biophys Acta 689:261–266

Wagner GJ, Siegelman HW (1975) Large scale isolation of intact vacuoles and isolation of chloroplasts from protoplasts of mature plant tissue. Science 192:1298–1299

Wang Y, Leigh RA, Kaestner KM, Sze H (1986) Electrogenic $H^+$ pumping pyrophosphatase in tonoplast vesicles of oat roots. Plant Physiol 81:497–502

Weigel HJ, Weiss E (1984) Determination of the proton concentration difference across the tonoplast membrane of isolated vacuoles by means of 9-aminoacridine fluorescence. Plant Sci Lett 33:163–175

Werner C, Matile P (1985) Accumulation of coumarylglucosides in vacuoles of barley mesophyll protoplasts. J Plant Physiol 118:237–249

Willenbrink J, Doll S (1979) Characteristics of the sucrose uptake system of vacuoles isolated from red beet tissue. Kinetics and specificity of the sucrose uptake system. Planta 147:159–162

Yamaki S (1984) Isolation of vacuoles from immature apple fruit flesh and compartmentation of sugars, organic acids, phenolic compounds and amino acids. Plant and Cell Physiol 25:151–166

Yoshida S, Uemura M, Niki T, Sakai A, Gusta LV (1983) Partition of membrane particles in aqueous two-polymer phase system and its practical use for purification of plasma membrane from plants. Plant Physiol 72:105–114

# XXVIII Miniprotoplasts and Their Use in Genetic Engineering

A. WALLIN, S. WAARA, and T. ERIKSSON[1]

## 1 Introduction

Protoplasts are an excellent system for the induction of genetic variation in plant cells: through somaclonal variation, somatic hybridization, cybrid formation, transformation and incorporation of isolated chromosomes or isolated DNA. By fragmentation of protoplasts, small-sized, nucleus-containing, nearly vacuole-free subprotoplasts with high stability (miniprotoplasts) can be formed. Due to their small size, the absence of major vacuoles, and their low cytoplasm content, miniprotoplasts have clear advantages over ordinary protoplasts in certain types of cell manipulation.

The preparation of miniprotoplasts and cytoplasts has been reviewed by Lörz (1984) and the subprotoplast terminology by Bradley (1983). In this chapter, terminology and characterization of miniprotoplasts, as well as practical procedures for subprotoplast preparation, are presented. Moreover, the applications of miniprotoplasts and techniques in which they are of special interest are discussed.

## 2 Types of Subprotoplasts

At high speed centrifugation of protoplasts through a density gradient, two main types of subprotoplasts are formed (Fig. 1). Fragmentation of protoplasts is based on internal density differences. When exposed to high centrifugal force, the high-density organelles of the protoplast, surrounded by plasma membrane, are drawn out from the rest of the protoplast until only a thin cytoplasmic strand remains

**Fig. 1A, B.** Formation of miniprotoplasts and cytoplasts by high speed centrifugation. **A** Schematic outline of protoplast fragmentation. **B** Fragmentation of protoplasts isolated from cell suspension cells of *Nicotiana sylvestris*. Enucleation was carried out as described in Table 2 in a cytochalasin B-containing medium at 20000 g, 37 °C. **a** Freshly isolated protoplasts. **b** Drawn-out protoplast. **c** Cytoplasts in protoplast growth medium. **d** Miniprotoplasts in 1.5 M sorbitol. **e** Miniprotoplasts in protoplast growth medium. (Wallin et al. 1978)

[1] Department of Physiological Botany, University of Uppsala, Box 540, S-75121 Uppsala, Sweden

Biotechnology in Agriculture and Forestry, Vol. 9
Plant Protoplasts and Genetic Engineering II (ed. by Y. P. S. Bajaj)

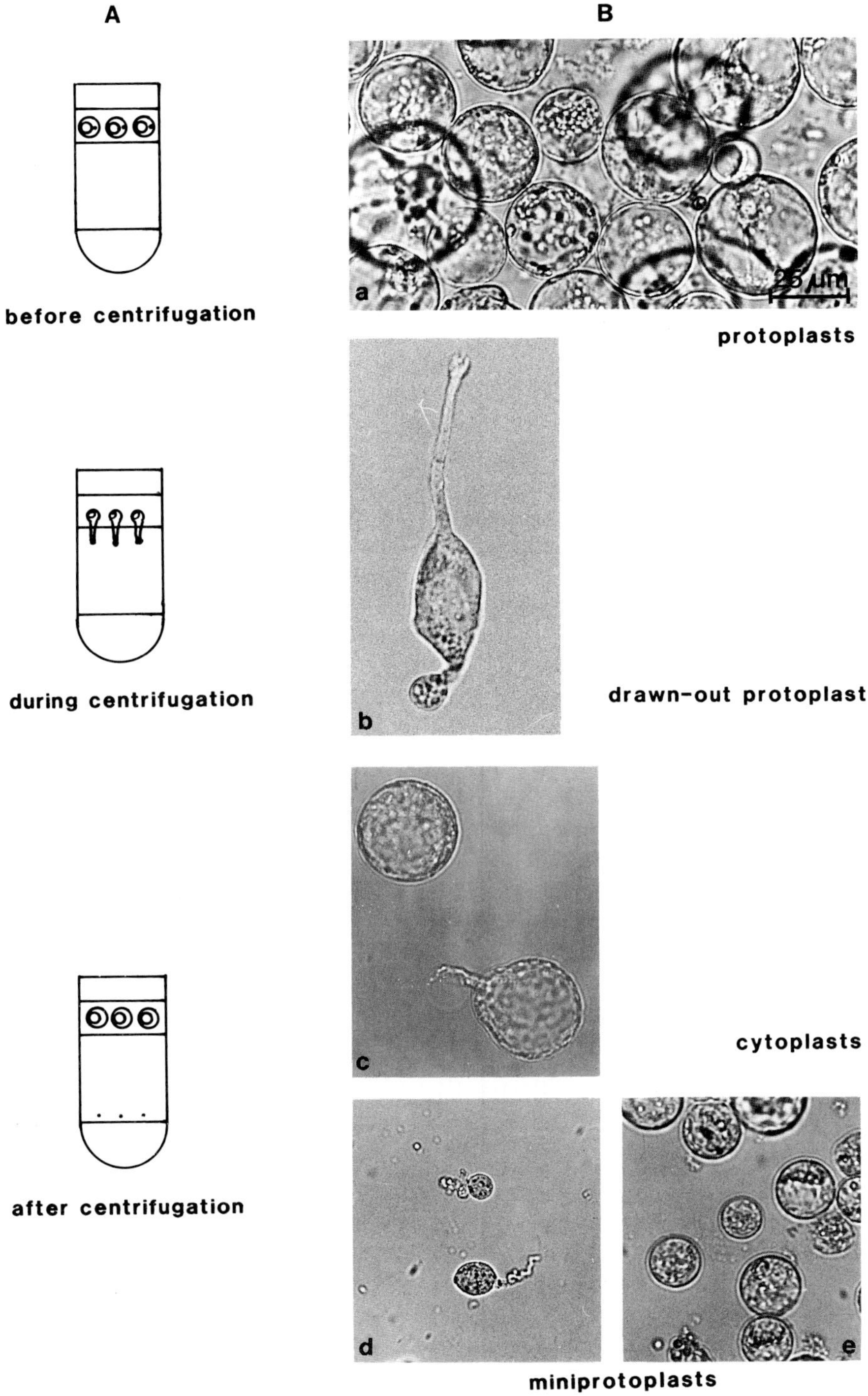
A
B
before centrifugation
during centrifugation
after centrifugation
a
25 µm
protoplasts
b
drawn-out protoplast
c
cytoplasts
d
e
miniprotoplasts

Fig. 1

**Table 1.** Subprotoplast terminology

| A. Subprotoplast with nucleus normally without vacuole(s) | Reference |
|---|---|
| Karyoplast | Lörz and Potrykus (1980) |
| Miniprotoplast | Wallin et al. (1978) |
| | Lörz et al. (1981) |
| Subprotoplast | Binding and Kollman (1976) |
| Evacuolated protoplast | Griesbach and Sink (1983) |
| Nucleoplast | Gleba and Sytnik (1984) |
| Nucleate cytoplast | Bradley (1983) |
| Nucleate microplast | Bilkey et al. (1982) |
| Nucleated miniprotoplast | Bracha and Sher (1981) |
| Nucleated, lowcytoplasm Protoplast | Lesney et al. (1983) |

| B. Subprotoplast without nucleus with vacuole(s) | Reference |
|---|---|
| Cytoplast | Lörz and Potrykus (1980), Hoffman (1981), Archer et al. (1982) |
| Nuclear-free subprotoplast | Binding and Kollman (1976) |
| Enucleated protoplast | Bradley (1978), Wallin et al. (1978), Izhar and Power (1979) |
| Anucleate protoplast | Vatsya and Bhaskaran (1981) |
| | Bilkey et al. (1982) |
| Enucleate microplast | Bradley (1983) |

and will finally be pinched off (Fig. 1). The subprotoplasts will then accumulate in bands at their corresponding density. The subprotoplast containing the nucleus will move towards the bottom of the gradient and the vacuole-containing subprotoplast towards the surface of the gradient. The heavy subprotoplast, the miniprotoplast, is characterized by a nucleus surrounded by some cytoplasm and a plasma membrane. The light subprotoplast contains the major vacuoles and the rest of the cytoplasm enclosed in a plasma membrane. Vacuoles have not been found in miniprotoplasts directly after centrifugation.

The term miniprotoplast was chosen by Wallin et al. (1978) as equivalent to the term minicells used by Ege et al. (1974) for animal cells. Other designations are based on the name of the most characteristic organelle in the subprotoplast, e.g., karyoplast (Lörz and Potrykus 1980) and nucleoplast (Gleba and Sytnik 1984), or on the way in which it has been formed, e.g., evacuolated protoplasts (Griesbach and Sink 1983). The most common terms for the different types of subprotoplasts are presented in Table 1.

At high speed centrifugation, the protoplasts are highly drawn out. If centrifugation is interrupted before separation has taken place or if other conditions such as speed, temperature, or gradient composition are not optimal for fractionation of the protoplast, drawn-out protoplasts, as shown in Fig. 1 b, are formed.

## 3 Methods for Fragmentation of Protoplasts

In all published methods in which protoplasts are fractionated into viable subprotoplasts, high-speed centrifugation through some type of density gradient has been used (Table 2). Whichever method is used, about two-thirds of the healthy protoplasts are transformed to miniprotoplasts (Wallin et al. 1978; Lörz et al. 1981; Griesbach and Sink 1983). By using a discontinuous, iso-osmotic density gradient produced by mixing $CaCl_2 \cdot H_2O$, mannitol, and Percoll, Lörz et al. (1978) could avoid shrinkage of miniprotoplasts, which otherwise occurs when osmolality increases. This is actually the case in the methods where sucrose gradients or a continuous silica gradient is used. In the system developed by Griesbach and Sink (1983), differential sedimentation of *Petunia* mesophyll protoplasts took place through a continuous Percoll gradient. Their method is the easiest and most rapid of the procedures but with the same efficiency. In our laboratory it has worked very well also on mesophyll protoplasts of potato (Sylvia Waara unpubl.). The evacuolation, which occurs in three stages, is described in detail by Griesbach and Sink (1983). Firstly, the protoplasts are enriched in a layer at their corresponding density. This takes about 5 min. Secondly, the chloroplasts accumulate within the protoplast, but the protoplast remains in its position in the gradient. Thirdly, the vacuole pinches off from the plasmalemma and floats to the surface of the Percoll gradient.

Cytochalasin B (CB), a fungal metabolite used in the first published work on miniprotoplast formation (Wallin et al. 1978, 1979), is not necessary for good protoplast fragmentation at high-speed centrifugation (Lesney et al. 1983) and neither is dimethylsulfoxide (DMSO), the solvent for CB (Lesney et al. 1986). However, CB and DMSO influence the plasma membrane and/or cytoplasm fluidity, in some cases perhaps speeding up separation of miniprotoplast from cytoplast. In a CB medium, 60% of the *Nicotiana sylvestris* protoplasts produced miniprotoplasts after 15 min centrifugation, but in the controls with only DMSO or without any supplement, just a few percent of the protoplasts were pinched off to real miniprotoplasts and the remaining 40% or 60% respectively of the living protoplasts had only reached the drawn-out stage (Wallin et al. 1978). By prolonging the centrifugation time to 60–65 min in a DMSO-free gradient and keeping a low temperature, Lesney et al. (1986) obtained 70%–90% enucleation. In any case, if using CB, the protoplasts should not be treated for more than a few minutes before centrifugation. When high dose (50 µg/ml) CB was added to *Nicotiana sylvestris* protoplasts, cytoplasmic streaming was inhibited in a few minutes (Wallin 1977). DMSO is known to have a drastic immediate effect on plasma membrane (McIntyre et al. 1974). DMSO down to 0.01% instantly changed protoplast conformation. The spherical shape disappeared, but in a few minutes the protoplasts recovered (Wallin 1977). A prolonged treatment in CB resulted in inhibition of cell wall synthesis and blocking of cytokinesis (Wallin 1977; Meyer and Herth 1978). After 1-h treatment, the time needed for fragmentation, the inhibition obtained by CB was entirely reversed. Now, we routinely use Percoll gradients for miniprotoplast preparation.

The fragmentation method reported by Lesney et al. (1986) resulted, in their description, in bands of high quality cytoplasts. These cytoplasts with a large

**Table 2.** Fractionation and distribution of protoplast fragments by high speed centrifugation in density gradients

| Gradients for enucleation of protoplasts (Media from top to bottom) | Distribution profile after centrifugation | Origin of protoplasts | Reference |
|---|---|---|---|
| a. Protoplasts in protoplast culture medium + 50 μg/ml (0.5% DMSO)<br>b. 1.5 M sorbitol + 50 μg/ml CB (0.5% DMSO)<br>c. Saturated sucrose 20000 – 40000 g, 37 °C, 15 min | 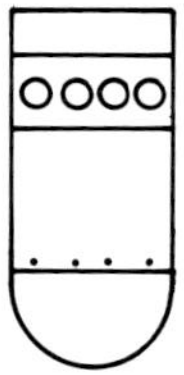 | *Nicotiana tabacum* (c.s.c.)<br>*Nicotiana sylvestris* (c.s.c.)<br>*Pisum sativum* (m.c.)<br>*Daucus carota* (c.s.c.) | Wallin et al. (1978) |
| a. Protoplasts in CPW 0.35 M mannitol + 0.5% DMSO<br>b. 1.0 M mannitol 50 μg/ml CB (0.5% DMSO)<br>c. 33% sucrose<br>d. 45% sucrose<br>e. Saturated sucrose 17000 rpm, 37 °C, 20 min | 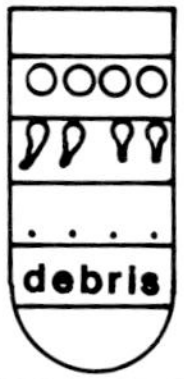 | *Allicum cepa* (m.c. and bulb scale cells) | Bracha and Sher (1981) |
| a. Protoplasts in 0.33 mannitol + 0.5% DMSO<br>b. 1 M mannitol + 0.5% DMSO<br>c. 33% sucrose<br>d. Near-saturated sucrose 23000 g, 23 °C→10 °C, 60 min | 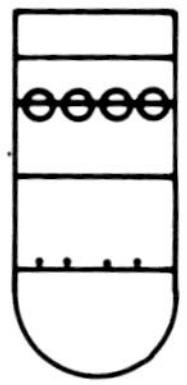 | *Solanum nigrum* (c.s.c.) | Lesney et al. (1983) |
| a. Protoplasts in 0.35 M, mannitol<br>b. 1 M mannitol<br>c. 33% sucrose<br>d. Near-saturated sucrose 23000 g, 0 – 4 °C, 65 min | 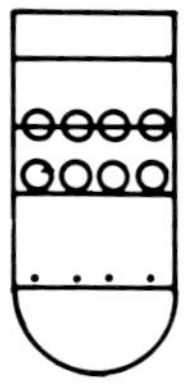 | | Lesney et al. (1986) |

**Table 2** (continued)

| Gradients for enucleation of protoplasts (Media from top to bottom) | Distribution profile after centrifugation | Origin of protoplasts | Reference |
|---|---|---|---|
| a. Protoplasts in 0.22 M $CaCl_2$<br>b. 0.5 M mannitol 5% Percoll<br>c. 0.48 M mannitol 20% Percoll<br>d. 0.45 M mannitol 50% Percoll 40000 g, 12 °C, 45 – 90 min |  | *Zea mays* (c.s.c.)<br>*Hyoscyamus muticus* (c.s.c.)<br>*Nicotiana tabacum* (c.s.c.) | Lörz et al. (1981) (Details of the distribution profile see the authors) |
| a. 0.5 ml protoplasts in Percoll (100%), 100 mM $CaCl_2$, 5 mM Hepes, 0.49 M mannitol, 40000 rpm, 23 °C, 60 min | 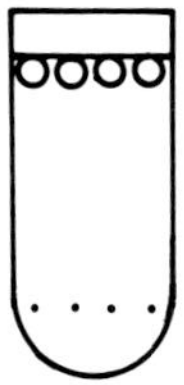 | *Petunia parodii* (m.c.) | Griesbach and Sink (1983) |
| a. Protoplasts in W5<br>b. 0.5 ml $CaCl_2$ + 0.5 ml 0.5 M mannitol<br>c. 0.15 ml Percoll + 2.85 ml 0.52 M mannitol<br>d. 0.6 ml Percoll + 2.4 ml 0.61 M mannitol<br>e. 1.5 ml Percoll + 1.5 ml 0.92 M mannitol 26000 rpm, 23 °C, 30 min | 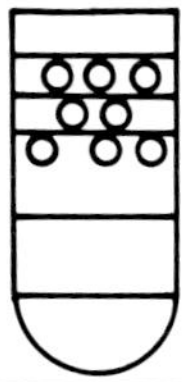 | *Nicotiana tabacum* (c.s.c.)<br>*Petunia hybrida* (c.s.c.)<br>*Lycopersicon peruvianum* (c.s.c.)<br>*L. penelli* (c.s.c.)<br>*Solanum lycopersicoides* (c.s.c.) | Tan (1987) |

c.s.c. = cell suspension cells, m.c. = mesophyll cells. Cytoplast O, Drawn-out protoplast ♀, Miniprotoplast •.

amount of cytoplasm accumulated in a specific band in the gradient, separated from strongly vacuolated cytoplasts. The high quality cytoplasts are suggested to be superior in fusion experiments (Lesney et al. 1986). High yield of cytoplasts is also reported by Tan (1987) in a Percoll gradient.

All fractionation methods presented in Table 2 result in large quantities of miniprotoplasts and/or cytoplasts (90% – 100%). However, it is difficult to compare the yields of the various methods because different ways of calculation and different plant materials have been used. In discontinuous gradients, it seems important that the gradient has enough steps to enable each type of subprotoplasts, and preferably the debris, to accumulate separately. Saturated or near-saturated sucrose at the bottom of the centrifuge tube serves as a muffle. A swing-out rotor is preferable to an angle rotor (own experience).

Besides gradient composition, temperature is an important factor for efficient fragmentation. Interspecific differences in temperature preference for optimal formation of cytoplasts have been reported by Tan (1987). For cytoplast formation from cell suspension protoplasts, 12 °C was optimal for *Solanum lycopersicoides*, 37 °C for *Lycopersicon peruvianum* and *L. pennellii*, and no temperature preference was found in *Petunia hybrida* or *Nicotiana tabacum*.

The viability and yield of subprotoplasts is largely dependent on the type and quality of the original protoplasts. Protoplasts derived from suspension cells, mesophyll cells, onion bulb scale cells (Table 2) and hypocotyl cells (Hurlbut et al. 1985) have been successfully fragmentated into viable subprotoplasts. However, protoplasts isolated from suspensions or hypocotyls seem to be superior to the fragile mesophyll protoplasts, when using the methods we have tested, especially for cytoplast preparation. Lesney et al. (1986) reported the highest yield of potato cytoplasts from protoplasts isolated from cell suspension cells 3 days after subculture. No difference in viability of cytoplasts prepared 2, 3, or 4 days after subculture was found. Other factors essential for subprotoplast fragmentation and for the distribution profile are the size homogeneity and the concentration of the original protoplasts (Lörz et al. 1981; Tan 1987).

## 4 Characterization of Miniprotoplasts

Just after fragmentation, the miniprotoplast is drop-shaped, with remnants of cytoplasmic strands (Fig. 2). In density gradients with high osmolality in the bottom layers of the centrifuge tube, the miniprotoplasts are highly shrunk during sedimentation (Fig. 2). The original osmolality of the protoplasts can be restored by slowly diluting the medium surrounding the miniprotoplasts with protoplast growth medium or salt solution. The deformed miniprotoplasts round up and obtain a water potential close to that of the original protoplasts (Fig. 3). Miniprotoplasts prepared in an isoosmotic gradient contained only 15% of the original protoplast volume (Lörz et al. 1981). The volume of drop-shaped miniprotoplasts (Fig. 2) was almost one-twentieth of the original protoplast volume (Wallin et al. 1979). (Compare size in Fig. 1!) Detailed analyses of the proportions of basic cytoplasm and cytoplasmic organelles in miniprotoplasts and

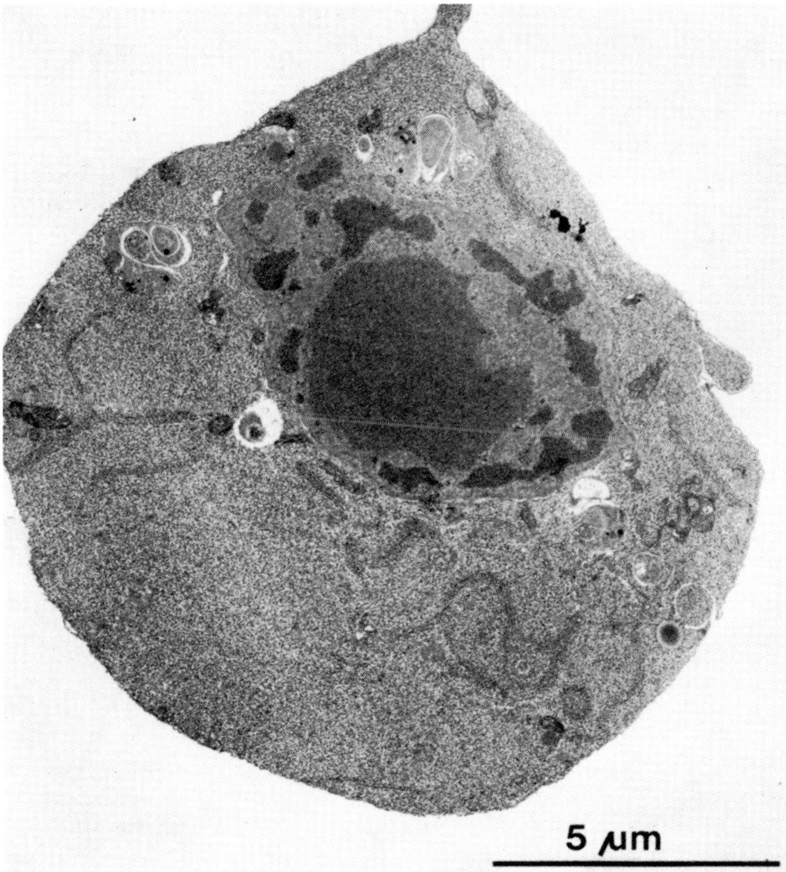


**Fig. 2.** A miniprotoplast isolated from a tobacco protoplast of a cell suspension cell directly after centrifugation at 40000 g in 1.5 M sorbitol containing 50 μg cytochalasin B/ml. (Wallin et al. 1979)

cytoplasts are still lacking. In mesophyll protoplasts from mature leaves, cytoplasts seem to have a density close to that of the nucleus and accumulate together with miniprotoplasts. Cytoplasts from such protoplasts consist of a large vacuole, some residual cytoplasm, and single chloroplasts. These cytoplasts are normally fragile.

High metabolic activity has been measured in miniprotoplast (Lörz et al. 1981). Protein synthesis, studied by uptake and incorporation of $^3$H-leucine, was significantly higher in miniprotoplasts than in protoplasts.

The viability of miniprotoplasts has also been confirmed by the ability of miniprotoplasts to enlarge, form a cell wall, divide, and regenerate into plants. Viable miniprotoplasts have an exceptional ability to increase in size during cultivation. Miniprotoplasts are able to synthesize a cell wall and to divide. In corn, a division frequency of 2% – 3% was obtained (Lörz et al. 1981). Plants have been regenerated from miniprotoplasts derived from suspension cultures of *Solanum*

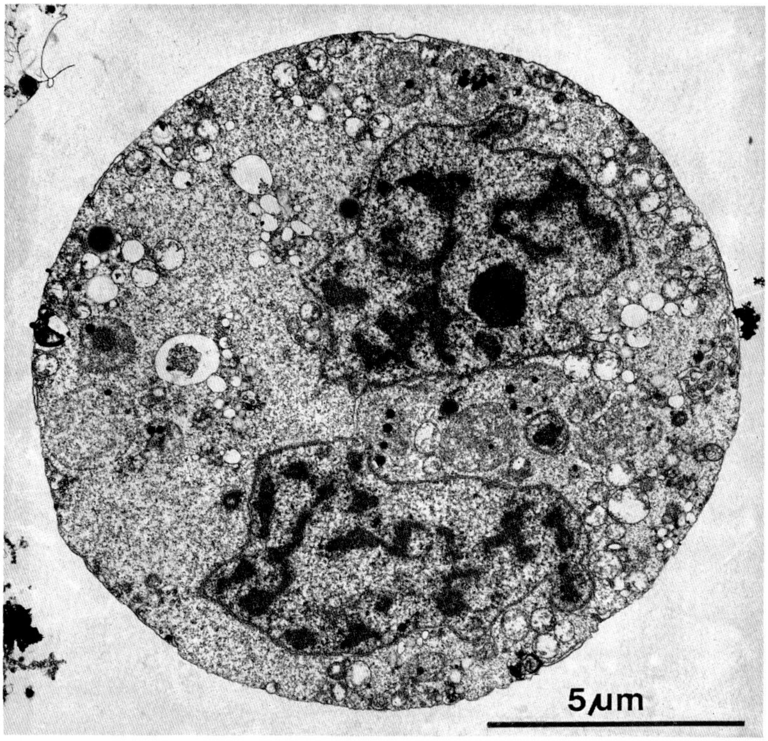


**Fig. 3.** The same type of miniprotoplast as in Fig. 2 in protoplast growth medium. (Wallin et al. 1979)

*nigrum* (L.) (Lesney et al. 1986) and from fusion products of miniprotoplasts (Wallin unpubl.). In order to distinguish nucleated from enucleated protoplast fragments, carbol fuchsin (Lesney et al. 1986), acridine orange (Lörz and Potrykus 1978), acetocarmine, or Hoechst 3332 (Tan 1987) have been utilized.

## 5 Use and Significance of Miniprotoplasts in Cell Manipulation and Genetic Engineering

### 5.1 Somatic Hybridization

Evacuolated protoplasts can be used either as one or both of the fusion partners in somatic hybridization. Hybrid cells have been formed by transfer of nuclei via fusion of miniprotoplasts from cell lines of nitrate reductase deficient tobacco (Wallin et al. 1979). Two different mutant lines, both lacking nitrate reductase and auxotrophic for reduced nitrogen, were used. Hybrids were selected by the ability to grow in media containing nitrate. Miniprotoplasts from one of these mutant cell lines were fused with miniprotoplasts or complete protoplasts from the other line and fusion products were formed. Some of the hybrid cells later regenerated into plants (Wallin unpubl.). Thus, nucleus transfer resulted in complementation and repair of nitrate reductase deficiency. It is impossible to analyze the fate of organelles in this system since both lines are tobacco. Unfortunately, no analysis of organelles has been carried out in any system involving miniprotoplasts. The number of organelles in a mini-protoplast is insufficiently known (and may differ between various types of protoplast and between various isolation procedures). However, the organelle content is significantly reduced in most miniprotoplasts. According to Lörz et al. (1981), only 15% of the cytoplasm remained in the miniprotoplasts.

Miniprotoplasts might also serve as recipients of cytoplasmic organelles. If a miniprotoplast is fused with a complete protoplast, a hybrid may be formed which has nuclei from both parents but presumably plastids and mitochondria from the complete protoplast only. However, it is not impossible that the miniprotoplast contains some organelles, in which case a mixture of the chondriomes will result. On the other hand, in somatic hybrids between complete protoplasts, it is known that plastids from one of the parental lines normally prevail after repeated cell divisions and are conserved in their original form in the hybrid cell (Clark et al. 1985). Thus the numerical superiority of the protoplast plastids in a hybrid cell between a protoplast and a miniprotoplast may give the protoplast plastids a good chance to take over. In cytoplasmic hybrids, recombination of chloroplast DNA has not been observed but in mitochondria, DNA recombines (Boeshore et al. 1983; Rothenberg et al. 1985) and the genome can also be rearranged (Nagy et al. 1981). Statistically, there is a relatively low chance for mitochondria of a miniprotoplast in a fusion product between a protoplast and a miniprotoplast to survive and to be involved in recombination, but there is a chance. The reliability in such a system could be tested by using mitochondria coding for CMS or by DNA markers.

For selection of somatic hybrids, flow cytometry is an excellent method that does not require mutants or specific culture conditions of the parental cells (Glimelius 1987). An alternative method that does not require complicated equipment is the use of miniprotoplasts. By gradient centrifugation, bands of protoplasts, miniprotoplasts, and fusion products (protoplasts × miniprotoplasts) can be separated due to different density. By fusing an evacuolated mesophyll protoplast of *Petunia parodii* with a complete suspension protoplast of *Petunia inflata*, Griesbach and Sink (1983) created a system where, after centrifugation, they obtained a band with over 50% fusion products. Controlled electrofusion has successfully been used by Spangenberg and Schweiger (1986) for fusion of subprotoplasts with protoplasts of rapeseed. Defined pairs of protoplasts and subprotoplasts pretreated in proteinase were fused by single or multiple negative dc-pulses. The fusion yield between karyoplast and protoplast, cytoplast and protoplast or karyoplast and cytoplast in microdroplets was 30% – 34%. Routinely, 10% – 20% of the fusion products of a karyoplast and a cytoplast formed microcalli.

Bracha and Sher (1981) reported fusion between miniprotoplasts and cytoplasts. They fused purified miniprotoplasts originating from onion mesophyll protoplasts with cytoplasts from bulb epidermis protoplasts. No further development of these fusion products has been reported.

Miniprotoplasts have properties which correspond well with properties postulated for high fusion frequency (Cocking 1973), i.e., dense cytoplasm, lack of vacuole, and small size. However, no obvious difference in reactions to fusion agents was observed by Lörz et al. (1981) in fusion experiments carried out with miniprotoplasts, cytoplasts, and protoplasts of corn and tobacco. In fusion with polyethylene glycol, it can be difficult to obtain the appropriate proportions of the two types (Glimelius pers. commun.) and in electrofield fusion, difference in density of the fusion partners can make them form two separate layers, which obstructs an efficient fusion.

### 5.2 Flow Cytometry

Plant cells cannot be manipulated as easily as animal cells. Often it is the vacuole which restricts the manipulation of plant cells. In flow cytometry, cells, nuclei or chromosomes in a liquid jet have to pass through a small orifice intersect. In the first attempts to sort plant protoplasts by flow cytometry, instruments designed for animal cells were used. When plant protoplasts were forced through the narrow channel, they were damaged. To overcome this problem, Hurlbut et al. (1985) used miniprotoplasts from hypocotyl protoplasts of *Brassica napus*. By eliminating the vacuole, not only substantial reduction in size but also an increased stability was obtained, and both these conditions improved the survival after flow sorting. Today, modifications of the instruments by using large flow tip orifices have enhanced sorting of plant protoplasts (Glimelius 1987).

### 5.3 Microinjection

For transfer of isolated organelles, chromosomes, DNA, etc. from one plant to another, micropipette injection into plant protoplasts is a possible method. Such a technique has been used in animal cell manipulation for many years (Ansorge 1982). However, with plant cells, several practical problems are associated with microinjection. Firstly, the protoplasts have to be attached in a fixed position to make it possible to penetrate the plasma membrane and secondly, the needle has to be aimed correctly. Not only does the large vacuole make it difficult to hit the cytoplasm, the tonoplast can also be damaged so that toxic compounds from the vacuole leak into the cytoplasm (Griesbach and Sink 1983). The acid pH of the vacuole can also destroy cytoplasmic enzymes.

The use of miniprotoplasts to improve microinjection has been proposed by several authors (e.g., Steinbiss and Stabel 1983; Griesbach and Sink 1983; Burgess 1984). Recently, Spangenberg et al. (1986) have demonstrated the expression of foreign genes in karyoplasts after microinjection. Two gene constructions (pLGV23neo or pSV2neo) coding for the neomycin phosphotransferase II (NPTII) were injected into karyoplasts prepared from *Brassica napus* hypocotyl protoplasts. Microinjected karyoplasts as well as reconstituted cells (electrofusion-mediated products obtained by fusion of a microinjected karyoplast and a cytoplast) survived injection and showed expression of microinjected DNA. The survival rate in protoplasts and karyoplasts was similar (83% and 82%, respectively) but that of reconstituted cells essentially lower (22%). The expression of injected DNA was analyzed by immunofluorescence. A positive immunoreaction was obtained by attachment of NPTII specific antibodies to NPTII in protoplasts (70%), karyoplasts (61%) and reconstituted cells (61%). However, the number of cells with expression of microinjected DNA that formed microcalli was somewhat higher for karyoplasts (31%) than for protoplasts (27%) and reconstituted cells (18%).

### 5.4 Other Fields of Interest

Miniprotoplasts may be a tool for improved selection of amino acid auxotrophs. If intact cells are used for mutant selection it takes a long time for the cellular pool of amino acids to become depleted and for true amino acid auxotrophs to be isolated (Matile 1978). Griesbach and Sink (1983) suggest the use of evacuolated protoplasts to keep mutagens within the cytoplasm or nucleus for a longer period since the vacuoles may accumulate the mutagen. This hypothesis is interesting, but has to be proved. New vacuoles are formed very quickly when miniprotoplasts grow in an appropriate medium. In any case, the evacuolated protoplasts may be an appropriate system for studying vacuole formation biochemically and histologically.

For genetic engineering of animal cells, not only minicells with the whole nucleus but also microcells, with just a single or a few chromosomes surrounded by a very small amount of cytoplasm enclosed in a membrane, have been formed and used for chromosome transfer (Ege and Ringertz 1974; Ege 1980). Microcells were produced by inducing micronucleation in cells by colchicine treatment and then drawing out the micronuclei by centrifugation in the presence of cytochlasin B. Similarly, microplasts of plant cells might be formed and utilized for transfer of single chromosomes. This system does not admit transfer of specific chromosomes, however, but it can still be of interest for species with well-analyzed and a low number of chromosomes. Recently, Verhoeven et al. (1987) reported micronucleation in plant cells. Suspension cells of *Nicotiana plumbaginifolia* were synchronized in hydroxyurea and were than treated in amiprophosphomethyl, which affects microtubuli and induces formation of micronuclei. High speed centrifugation (100000 g) in a Percoll gradient resulted in bands of protoplasts with nuclei with different numbers of chromosomes. One fraction contained protoplasts with 1–2 chromosomes, another contained protoplasts with 3–4 chromosomes. This procedure opens prospects for application of micronuclei for transfer of specific, intact chromosomes in plants.

## 6 Summary and Conclusions

For optimal fragmentation of protoplasts, the preparation methods have to be adjusted for each specific plant material. High yields of viable subprotoplasts are necessary if they are to be useful in somatic hybridization, microinjection, flow cytometry etc. Miniprotoplasts may be unique materials for certain types of cell manipulation. However, the advantages of subprotoplasts over protoplasts have to be considered in each specific case. The use of miniprotoplasts may not be the best alternative for every purpose, since each step in the handling of the plant material will reduce viability. The potential of miniprotoplasts in genetic engineering is still uncertain. New hybrids and cybrids, as well as manipulated cells, have to be formed and analyzed in order to estimate their value for the future.

*Acknowledgments.* We thank Prof. S. Izhar for fruitful discussions concerning miniprotoplasts as recipients of cytoplasmic organelles, and Fil. mag. L. Grönroos for her critical reading of this manuscript.

## References

Ansorge W (1982) Improved system for capillary microinjection into living cells. Exp Cell Res 140:31–37

Archer EK, Landgren CR, Bonnett HT (1982) Cytoplast formation and enrichment from mesophyll tissues of *Nicotiana* spp. Plant Sci Lett 25:175–185

Aviv D, Galun E (1980) Restoration of fertility in cytoplasmic male sterile *Nicotiana sylvestris* by fusion with X-irradiated *N. tabacum* protoplasts. Theor Appl Genet 58:121–128

Bilkey PC, Davey MR, Cocking EC (1982) Isolation, origin and properties of enucleated plant microplasts. Protoplasma 110:147–151

Binding H, Kollman R (1976) The use of subprotoplasts for organelle transplantation. In: Dudits D, Farkas GL, Maliga P (eds) Cell genetics in higher plants. Académiai Kiadó, Budapest, pp 191–206

Boeshore ML, Lifshitz I, Hanson MR, Izhar S (1983) Novel composition of mitochondrial genomes in *Petunia* somatic hybrids derived from cytoplasmic male sterile and fertile plants. Mol Gen Genet 190:459–467

Bracha M, Sher N (1981) Fusion of enucleated protoplasts with nucleated miniprotoplasts in onion (*Allium cepa* L.). Plant Sci Lett 23:95–101

Bradley PM (1978) Production of enucleated plant protoplasts of *Allium cepa*. Plant Sci Lett 13:287–290

Bradley PM (1983) The production of higher plant subprotoplasts. Plant Mol Biol Rep 1(3):117–123

Burgess J (1984) The revolution that failed. New Sci 1429:26–29

Clark E, Izhar S, Hanson MR (1985) Independent segregation of chloroplast DNA and cytoplasmic male sterility in somatic hybrids of *Petunia*. Mol Gen Genet 199:440–445

Cocking EC (1973) Plant cell modification: Problems and perspectives. Coll Int Cent Natl Rech Sci (Paris) 212:327–337

Ege T (1980) Fusion of cell fragments as a method in cell genetics. In: Celis JE, Graessmann A, Loyter A (eds) Transfer of cell constituents into eukaryotic cells. Plenum Press, New York London, pp 201–233

Ege T, Ringertz NR (1974) Preparation of microcells by enucleation of micronucleate cells. Exp Cell Res 87:378–382

Ege T, Hamberg H, Krondahl U, Eriksson J, Ringertz NR (1974) Characterization of minicells (nuclei) obtained by cytochalasin enucleation. Exp Cell Res 87:365–377

Gleba YY, Sytnik KM (1984) Protoplast fusion, genetic engineering in higher plants. Springer, Berlin Heidelberg New York, pp 27–28

Glimelius K (1987) Enrichment of somatic hybrids with flow cytometry. Newslett Int Assoc Plant Tissue Cult 52:2–10

Grisbach RJ, Sink KC (1983) Evacuolation of mesophyll protoplasts. Plant Sci Lett 30:297–301

Hoffman F (1981) Formation of cytoplasts from giant protoplasts in culture. Protoplasma 107:387–391

Hurlbut P, Fellner-Feldegg H, Glimelius K (1985) Sorting of miniprotoplasts by flow cytometry. Heredit Suppl 3:142

Izhar S, Power JB (1979) Somatic hybridization in *Petunia*: A male sterile cytoplasmic hybrid. Plant Sci Lett 14:49–55

Lesney MS, Callow PC, Sink KC (1983) A simplified method for bulk production of cytoplasts from suspension-culture derived protoplasts of *Solanum nigrum* L. In: Potrykus I, Harms CT, Hinnen A, Hütter R, King PJ, Shillito RD (eds) Protoplasts 1983. Birkhäuser, Basel, pp 116–117

Lesney MS, Callow PW, Sink KC (1986) A technique for bulk production of cytoplasts and miniprotoplasts from suspension culture-derived protoplasts. Plant Cell Rep 5:115–118

Lörz H (1984) Enucleation of protoplasts: Preparation of cytoplasts and miniprotoplasts. In: Vasil IK (ed) Cell culture and somatic cell genetics of plants, vol 1. Academic Press, New York London, pp 448–453

Lörz H, Potrykus I (1978) Investigation on the transfer of isolated nuclei into plant protoplasts. Theor Appl Genet 53:251–256

Lörz H, Potrykus I (1980) Isolation of subprotoplasts for genetic manipulation studies. In: Ferenczy L, Farkas GL, Lazar G (eds) Advances in protoplast research. Pergamon, Oxford, pp 377–382

Lörz H, Paszkowski J, Dierks-Ventling C, Potrykus I (1981) Isolation and characterization of cytoplasts and miniprotoplasts derived from protoplasts of cultured cells. Physiol Plant 53:385–391

Maliga P, Lörz H, Lazar G, Nagy F (1982) Cytoplast-protoplast fusion for interspecific chloroplast transfer in *Nicotiana*. Mol Gen Genet 185:211–215

Matile P (1978) Biochemistry and function of vacuoles. Annu Rev Plant Physiol 29:193–213

McIntyre JA, Gilula NB, Karnovsky MJ (1974) Cryoprotectant-induced redistribution of intramembranous particles in mouse lymphocytes. J Cell Biol 60:192–203

Medgyesy P, Menzel L, Maliga P (1980) The use of cytoplasmic streptomycin resistance. Chloroplast transfer from *Nicotiana tabacum* into *Nicotiana sylvestris* and isolation of their somatic hybrids. Mol Gen Genet 179:693–698

Meyer Y, Herth W (1978) Chemical inhibition of cell-wall formation and cytokinesis, but not of nuclear division in protoplasts of *Nicotiana tabacum* L. cultivated in vitro. Planta 142:253–262

Nagy F, Török I, Maliga P (1981) Extensive rearrangements in the mitochondrial DNA in somatic hybrids of *Nicotiana tabacum* and *Nicotiana knightiana*. Mol Gen Genet 183:437–439

Rothenberg M, Boeshore ML, Izhar S, Hanson MR (1985) Intergenomic recombination of mitochondrial genomes in a somatic hybrid plant. Curr Genet 9:615–618

Siderov VA, Menczel L, Nagy F, Maliga P (1981) Chloroplast transfer in *Nicotiana* based on metabolic complementation between irradiated and iodoacetate treated protoplasts. Planta 152:341–345

Spangenberg G, Schweiger H-G (1986) Controlled electrofusion of different types of protoplasts and subprotoplasts including cell reconstitution in *Brassica napus* L. Eur J Cell Biology 41:51–56

Spangenberg G, Neuhaus G, Schweiger H-G (1986) Expression of foreign genes in a higher plant cell after electrofusion-mediated cell reconstitution of a microinjected karyoplast and a cytoplast. Eur J Cell Biology 42:236–238

Steinbiss HH, Stabel P (1983) Protoplast derived tobacco cells can survive capillary microinjection of the fluorescent dye lucifer yellow. Protoplasma 116:223–227

Tan M-L MC (1987) Somatic hybridization and cybridization in Solanaceous species. Acad Proefschr, Vrije Univ Amsterdam. Free Univ Press, Amsterdam

Vatsya B, Bhaskaran S (1981) Production of subprotoplasts in *Brassica oleracea* var. Capitata – a function of osmolarity in the media. Plant Sci Lett 23:277–282

Verhoeven HA, Sree Ramulu K, van der Valk HPCM, Dijkhuis P, Blaas J (1987) Factors involved in micronucleation in plant cells. In: Abstr. 7th International Protoplast Symposium, Wageningen, The Netherlands

Wallin A (1977) Isolation, culture and fusion of protoplasts from cell suspensions of higher plants. Acta Univ Upsaliensis 405

Wallin A, Glimelius K, Eriksson T (1978) Enucleation of plant protoplasts by Cytochalasin B. Z Pflanzenphysiol 87:330–340

Wallin A, Glimelius K, Eriksson T (1979) Formation of hybrid cells by transfer of nuclei via fusion of miniprotoplasts from cell lines of nitrate reductase deficient tobacco. Z Pflanzenphysiol 91:89–94

Zelcher A, Aviv D, Galun E (1978) Interspecific transfer of cytoplasmic male sterility by fusion between protoplasts of normal *Nicotiana sylvestris* and X-ray-irradiated protoplasts of male sterile *N. tabacum*. Z Pflanzenphysiol 90:397–407

# Subject Index